Mechanisms of Homogeneous Catalysis from Protons to Proteins

Mechanisms of Homogeneous Catalysis from Protons to Proteins

MYRON L. BENDER
Department of Chemistry
Northwestern University

WILEY-INTERSCIENCE

A Division of John Wiley & Sons, Inc.
New York · London · Sydney · Toronto

Copyright © 1971, by John Wiley & Sons, Inc.

All rights reserved. Published simultaneously in Canada.

No part of this book may be reproduced by any means, nor transmitted, nor translated into a machine language without the written permission of the publisher.

Library of Congress Catalogue Card Number: 73-153080

ISBN 0-471-06500-5

Printed in the United States of America.

10 9 8 7 6 5 4 3 2 1

To my father and my mother, Dr. Averam B. Bender and Fannie L. Bender, who first kindled the spark of scholarship, and to three great teachers, Drs. Henry B. Hass, Paul D. Bartlett, and Frank H. Westheimer, who furthered it.

Preface

During my years in graduate school, I observed that teachers of organic chemistry never mentioned the existence of physical chemistry and vice versa.

Therefore, Hammett's book, *Physical Organic Chemistry*, came as a great revelation. I read it from cover to cover.

Biochemistry has now come of age, so this book can perhaps help to create still another field—physical-organic-biochemistry, if you will. In other words, what I am endeavoring to do in this book is to achieve a synthesis of all these areas, and specifically to show that a bridge exists between organic and enzymic reactions.

Others have been of the same opinion, and such books as Bruice and Benkovic's *Bio-Organic Mechanisms*, Kosower's *Molecular Biochemistry*, and Jencks' *Catalysis in Chemistry and Enzymology* have already appeared, as well as Waley's *Mechanisms of Organic and Enzymatic Reactions*, Ingraham's *Biochemical Mechanisms*, and Westley's *Enzymatic Catalysis*.

This effort of mine represents an attempt to update the classical work of Bell, *The Proton in Chemistry*.

The first chapters of the book were written in 1965 to 1966, and the remainder from 1968 to 1970, with a hospital stay and convalescence in-between.

I wish to acknowledge the hospitality of Sir Ewart Jones and Dr. Jeremy Knowles of the Dyson Perrins Laboratory, University of Oxford, England, where some of this book was written. My most profound thanks go also to my former student, now an esteemed colleague in the field, Professor Ferenc J. Kézdy, without whose help and advice the manuscript could not have been completed, and to John V. Killheffer, whose editorial talents proved to be invaluable.

My sincere appreciation is extended also to Mrs. Helen M. Brandt, my secretary, and others for their diligent labors in typing and retyping the manuscript, and to my wife, Muriel, whose understanding was essential.

Of course, many other people helped, too, and I thank all of them, but, as usual, the final criticisms must be borne by me.

MYRON L. BENDER

Evanston, Illinois
January 1971

Contents

1. Introduction   1

   Part One   Acid-Base Catalysis

2. Proton Transfers   19
3. Oxonium and Hydroxide Ions in Catalysis   37
4. General Acid-Base Catalysis: Theory   72
5. General Acid-Base Catalysis: Mechanism   95

   Part Two   Organic and Inorganic Catalysis

6. Nucleophilic and Electrophilic Catalysis   147
7. Catalysis by Fields   194
8. Metal Ion Catalysis   211

   Part Three   Bridging Nonenzymic and Enzymic Catalysis

9. Intramolecular Catalysis   281
10. Multiple Catalysis   321
11. Catalysis by Complexation   351

    Part Four   Enzymic Catalysis

12. Enzymes: Classification and Determination   397

13.	Structure of Enzymes	456
14.	The Concept of the Active Site	476
15.	Enzyme Kinetics	487
16.	Enzyme Mechanisms: Proteins	503
17.	Enzyme Mechanism: Coenzymes	539
18.	Enzyme Catalysis: Binding and Specificity	622
19.	Theories of Enzyme Catalysis	642
	Postscript	660
	Index	663

Mechanisms of Homogeneous Catalysis from Protons to Proteins

Chapter 1

INTRODUCTION

1.1 Definitions   2
1.2 The Basis of Catalytic Action   9
1.3 Catalysis and the Equilibrium Constant 11
1.4 Kinetics of Catalysis 12
1.5 Classification of Catalysis 15

Understanding homogeneous catalysis is one of the most intriguing challenges of modern chemistry. Its manifestations range from the microscopic scale of vital processes occurring within single plant and animal cells to industrial operations involving many thousands of pounds of material—from reactions between the simplest molecules to those between the most complex. Catalysis has been recognized almost since chemistry itself has been studied; it is presently undergoing a most intensive investigation, and it promises not only to be better understood in times to come, but also to become more and more valuable in understanding other facets of chemistry. In this book the history of homogeneous catalysis is traced, a contemporary exposition of it is presented, and what might be expected of this field in the future is projected.

Berzelius, over a century ago, alluded to a "catalytic force,"[1] but it was not until 1900 that the concept of catalysis (by oxonium and hydroxide ions) was placed on a firm experimental and theoretical basis. During the twenties and thirties, with the advent of new ideas about acids and bases, the concept of general acid-base catalysis was developed. Within the last decade, enzymes as chemical catalysts have received much attention. At the same time, new ideas about catalysis by polyfunctional organic molecules and by metal ion complexes have arisen. Thus the variety and sophistication of catalysis for both organic and inorganic reactions have undergone the exponential growth characteristic of other freshly stimulated scientific fields.

Millions of chemical compounds are now known. Therefore, there must be innumerable combinations among them; that is, innumerable conceivable reaction pathways. As one method of coping with such complexity, a comprehension of catalysis is essential: it is the guide for recognizing the few operative pathways among the countless possibilities.

Catalysis is a large field that can be subdivided in many ways. The book concentrates on homogeneous catalysis, processes that occur in solution, and organic reactions, but mentions heterogeneous catalysis, inorganic reactions, and processes that occur in the gas or solid phase.

The study of catalysis can be approached from several directions: from a physical chemical description; from the kinetics of the reactions; or from an organic chemical classification of the kinds of reactions that are catalyzed. This book focuses on the mechanisms of catalysis. Questions are asked as to how and why catalytic reactions proceed. Of course, one must know what the catalysts are, but the most provocative questions inquire into how catalysts operate mechanistically and how one goes about building better ones.

1.1 DEFINITIONS

Berzelius first used the word "catalyst" to denote "substances [which] are able to awaken affinities which are asleep at [this] temperature by their mere presence and not by their own affinity."[1] He was concerned with a "catalytic force" that would account for the fact that a series of repetitive reactions could be brought about by an external agent. Berzelius could not specify the nature of this force, but contented himself with saying that it was a special manifestation of the normal electrochemical properties of substances. Identification of the nature of catalysts in modern chemical terms was made when investigations of reaction velocity required that certain catalysts be chemical reagents. Ostwald, in 1895,[2] defined catalysts as "substances which change the velocity of a given reaction *without* modification of the energy factors of the reaction." Later he gave a definition more pointedly concerned with the kinetics of the reaction: "a catalyst is a substance which alters the velocity of the chemical reaction without appearing in the end product of the reaction."[3]

Definitions

In each of these *qualitative* definitions, catalysis is intimately tied to reaction velocity and to stoichiometry, both of which are amenable to *quantitative* experimentation. These two very important characteristics of a chemical transformation can indeed be expressed in the form of equations. For example, a simple uncatalyzed reaction might be described by the stoichiometric equation

$$A \longrightarrow \text{products} \tag{1.1}$$

and by the rate of the uncatalyzed reaction, V_u,

$$V_u = \frac{-d[A]}{dt} = k[A] \tag{1.2}$$

where A is the reacting substance, [A] its concentration, t the time, and k the rate constant characteristic of the reaction. The definitions of catalysis quoted in the preceding paragraph state that when the same transformation occurs under the influence of a catalyst, C, then the catalyst itself is not chemically modified and appears in the stoichiometric equation as both reactant and product:

$$A + C \longrightarrow \text{products} + C \tag{1.3}$$

Furthermore, since the catalyst increases the rate of the reaction, V_c must contain the concentration of the catalyst raised to some positive power, for example,

$$V_c = \frac{-d[A]}{dt} = k_{\text{cat}}[A][C] \tag{1.4}$$

where k_{cat} is the rate constant of the catalyzed reaction.† Since, by definition, the catalyst is not used up ([C] = constant), its concentration may be combined with k_{cat} into a single constant characterizing the catalytic process. On the basis of these more quantitative expressions of the influence of catalysis on chemical change, a more quantitative definition of catalysis is possible: "a substance is said to be a catalyst for a reaction in a homogeneous system when its concentration occurs in the velocity expression to a higher power than it does in a stoichiometric reaction."[4]

Alternatively, a catalyst may be said to increase the rate of approach of a system to any state that is chemically and thermodynamically possible in its absence.[5] The question then arises as to whether the free energy change of the overall process from reactants to products can be changed at all by

---

† The superficiality of these kinetics will be seen later in the detailed treatment of the kinetics of catalyzed reactions.

a catalyst. Rigorous interpretation of this definition of catalysis would exclude any modification of the extent of reaction; the use of a finite amount of catalyst, however, will in general modify it by a finite amount. A more realistic definition would then require the catalyst not to alter the standard free energy change of the reaction by more than a small fraction of its original value.[5]

A corollary of eq. 1.3 is that a catalyst may be effective when present in small proportion relative to the reactants; this property follows from the fact that the catalyst is not used up in the reaction, and so is available to act again and again. From a mechanistic point of view, however, a more important property—also implicit in eq. 1.3—is that the catalyst functions generally by reacting chemically with one or more of the reactants.

If the catalyst indeed participates in the reaction, and effects a change in its rate, then a modification of the reaction pathway must occur. Thus, expressing catalysis in mechanistically meaningful and useful terms, the essential role of a catalyst that acts according to the scheme of eq. 1.3 is to make available to the system an alternate reaction pathway. The requirements of this new pathway are that (1) it circumvents the rate-limiting step of the original pathway; (2) it leads to products that differ from the original set only by the presence of the catalyst; and (3) the catalyst is easily removed from the products by exchange with reactants.

The first of these conditions assures that the catalyzed reaction is faster than the uncatalyzed one, thereby satisfying the definition of a catalyst as an accelerator. The second condition assures that the effect of the catalyst is limited to the velocity: it does not affect the course of the reaction (i.e., the stoichiometry). The third condition assures that the catalytic substance is regenerated, and therefore is effective when present in small amount relative to the reactants (and is, at least in principle, recoverable after the reaction is complete).

The effectiveness of a catalyst is determined by the relative velocities of the original and the alternate pathways. Each pathway usually has some particular step that controls the overall rate: the rate of this step is determined by its free energy of activation, which is the highest free energy barrier along the route from reactants to products. The most important factor in catalytic efficiency is the amount by which these critical free energies of activation differ for the catalyzed and uncatalyzed reactions.

The frequent references to the function of a catalyst as a substance that lowers the activation energy of a reaction are somewhat misleading (Fig. 1.1). The correct statement is that the catalyst changes the pathway of the reaction to a pathway having a lower standard free energy of activation (Fig. 1.2). Hinshelwood expressed this idea concisely, defining catalysis as "a process providing an alternative and more speedy reaction route."[6] A

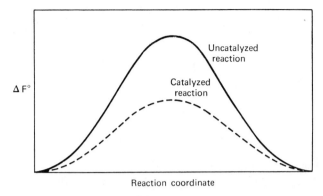

Fig. 1.1 Hypothetical catalysis without changing the pathway of a reaction.

corollary of this definition is that catalysis can be explained in terms of "normal" chemical reactions involving the catalyst and the substrate. A straightforward example of a catalyzed reaction that parallels a normal reaction in every respect is the iodide ion-catalyzed racemization of 2-iodooctane,[7] which proceeds by a nucleophilic substitution (S_N2) process.

Catalyzed reactions ordinarily proceed through the formation of intermediates that undergo further transformations yielding products and regenerating the catalyst. A catalytic reaction may be described as a chain reaction in which the catalyst is used up in one step but is regenerated in a succeeding one. Sometimes, this chain process can be dissected into its

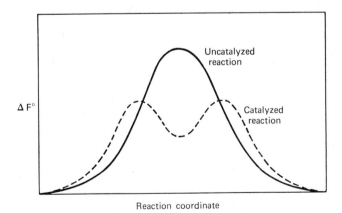

Fig. 1.2 Hypothetical catalysis with a change in the pathway of reaction.

individual component reactions, as has the set of coupled reactions shown in scheme 1.5,

$$\text{substrate—H}_2 + \text{nicotinamide adenine dinucleotide} \xrightleftharpoons{\text{dark}}$$
$$\text{substrate} + \text{reduced nicotinamide adenine dinucleotide}$$
$$\text{reduced nicotinamide adenine dinucleotide} + \text{methylene blue} \xrightleftharpoons{h\nu}$$
$$\text{nicotinamide adenine dinucleotide} + \text{reduced methylene blue} \quad (1.5)$$
$$\text{reduced methylene blue} + O_2 \xrightleftharpoons{\text{(aeration)}} \text{methylene blue} + H_2O$$

each step of which is influenced by an external variable.[8]

The chain character of catalysis is further shown in the effect of iodine upon many decompositions in the gas phase. For example, the thermal decomposition of acetaldehyde gives methane and carbon monoxide. The uncatalyzed reaction is itself probably a chain process; its order is between $3/2$ and 2 and its activation energy is between 46 and 50 kcal/mole, depending on the pressure of the system. However, when a small amount of iodine is added, the rate is increased several thousandfold and the reaction becomes precisely first-order in acetaldehyde and first-order in iodine, with an activation energy of 32 kcal/mole. At the start of the catalyzed reaction, almost all of the iodine disappears. Its concentration remains low until near the end of the reaction, when it all reappears. The catalyzed reaction takes place in two stages: (1) fast formation of methyl iodide, hydrogen iodide, and carbon monoxide; and (2) slow reaction of the methyl iodide and hydrogen iodide to form methane and regenerate the iodine.[9] A free-radical mechanism for the process may be written:

$$\begin{aligned} I_2 &\longrightarrow 2I\cdot \\ I\cdot + CH_3CHO &\longrightarrow HI + CH_3CO\cdot \\ CH_3CO\cdot &\longrightarrow CH_3\cdot + CO \\ CH_3\cdot + I_2 &\longrightarrow CH_3I + I\cdot \end{aligned} \quad (1.6)$$

Equation 1.6 accounts for the fast production of methyl iodide, hydrogen iodide, and carbon monoxide. This set of reactions is followed by a slower reaction, probably not of the free-radical type, which gives the final products, methane, and iodine:

$$CH_3I + HI \longrightarrow CH_4 + I_2 \quad (1.7)$$

In this overall process, the intermediate species are relatively stable and build up to concentrations where they may be observed. In many catalyzed reactions, however, the intermediates are reactive and never attain appreciable (observable) concentrations.

A complication in the discussion of catalytic reactions concerns the definition of substances that accelerate chemical reactions but are not regenerated. The use of biacetyl instead of iodine to accelerate the decomposition of acetaldehyde exemplifies this phenomenon. Biacetyl is not regenerated in the last step of a multistep scheme in contrast to iodine which is regenerated. The mechanism of the biacetyl acceleration, however, parallels that of the iodine catalysis: biacetyl forms two acetyl radicals in an initiation step; each acetyl radical decomposes to carbon monoxide and a methyl radical. How is one to treat those reactions exhibiting both an acceleration and a mechanism comparable to those of catalyzed reactions? The usual answer to this question is to call such processes *sensitized* reactions, and to call the chemical substances sensitizers (or promoters or initiators). The attitude taken here is that these substances should be considered together with true catalysts when they bring about processes that fit a common mechanistic pattern.

Some limitation, though, must be placed upon the space devoted to substances that are not true catalysts, so that they do not divert us too far from our chosen topic. Many reactions are accelerated by acids or bases that are not regenerated because of a subsidiary prototropic equilibrium. For example, in "base-catalyzed" ester hydrolysis, the carboxylic acid produced consumes the base in a reaction incidental to the hydrolysis. In other prototropic reactions, such as the "base-catalyzed" bromination of acetone, the base is converted to its conjugate acid. In the "aluminum chloride-catalyzed" acetylation of benzene, aluminum chloride forms a complex with the acetophenone formed and so is not available for further catalytic action. We have adopted the following viewpoint regarding these situations. For some purposes, it is useful to regard any substance that changes the rate of a desired reaction as a catalyst, regardless of its fate. However, if the substance is not regenerated, it is not a catalyst but a reactant; it is then more appropriate to recognize two distinct—though similar—reactions than to speak of one reaction, catalyzed or uncatalyzed. The reaction involving the new reactant, although the same as the old one in many respects, may differ profoundly in its standard free energy change. If this quantity is negative, whereas for the original reaction it was positive, the effect of the added reactant has been no less than replacing a thermodynamically unfavored process with a favored one—a change of fundamental consequence. On the other hand, if a reaction can be somehow only *slightly* altered, then its standard free energy change will likewise be little affected, and the sensitizer, promoter, or initiator will be included as part of a coherent mechanistic description.

In the preceding paragraph, we considered some substances that accelerate reactions but are not catalysts. It is worth mentioning some reactions accelerated by substances that actually are catalysts but are sometimes not recognized as such. For example, the solvolysis of benzyl chloride in aqueous solution is accelerated by mercuric ion. This acceleration is *not* caused by displacement of equilibrium by removal of the product chloride ion in a mercuric chloride complex. It has already been stated that catalysis does not appreciably alter the position of an equilibrium: here we see the converse, also true, that displacing an equilibrium by some means does not constitute catalysis of the processes involved in that equilibrium. What then is the mechanism by which small concentrations of mercuric ion increase the rate several hundredfold? This reaction proceeds through a carbonium ion intermediate; mercuric ion (or complexes thereof) accelerates the formation of the benzyl cation in the rate-determining step by interaction with the halogen atom in the transition state. Thus stabilization of the transition state rather than the product ground state leads to the acceleration by mercuric ion.

Along the same lines, information has been widely disseminated on catalysis by concentrated sulfuric acid. Chemical folklore has repeatedly suggested that sulfuric acid speeds a reaction by removing water from the system, thereby displacing its equilibrium. In fact, sulfuric acid accelerates many reactions because of its great ability to donate protons in a medium of high dielectric constant. This catalysis is thus again one of stabilization of a transition state rather than of a product ground state.

Some circumstances leading to acceleration of reaction will be excluded from this discussion. For example, light and heat are not substances and therefore cannot catalyze reactions; they have profound effects on reaction rates but will not be considered here. A more difficult problem is concerned with the solvent: changes in the medium can also have profound effects on the rates of reactions. Whether the solvent participates in the transition state, however, is often a moot question. A rigorous attitude would exclude solvents from a discussion of catalysis, but it seems fitting to extend the scope of this treatment to solvent and salt effects, on the grounds that these are important accelerations by substances that do not participate in the reaction.

On the basis of the foregoing survey of ideas concerning catalysis, it is possible and desirable at this point to state a definition of it upon which extensive further development of the subject can be rationally based. A definition that will serve as such a reference point for the remainder of this book is the following: catalysis of a chemical reaction is an acceleration brought about by a substance that is not consumed in the overall reaction; this substance—the catalyst—usually functions by interacting with a starting material to yield an alternate set of species that can react by a pathway involving a lower free energy of activation to give the products and to regenerate the catalyst.

1.2 THE BASIS OF CATALYTIC ACTION

Although no general correlation exists between the *tendency* of reactions to occur (thermodynamics) and the *rates* at which they occur (kinetics), the standard free energy of a reaction is directly related to its standard free energy of activation often enough for exceptions to attract attention. Catalytic reactions constitute such exceptions, either because (1) the catalyzed reaction is fast compared to expectation from equilibrium, or (2) the uncatalyzed reaction is abnormally slow compared to expectation from equilibrium.[10] One such abnormally slow reaction is that of ceric ion with thallous ion:

$$2Ce^{+4} + Tl^{+1} \longrightarrow Tl^{+3} + 2Ce^{+3} \qquad (1.8)$$

Since Tl^{+2} does not exist, this process requires a three-body collision and is therefore extremely slow. If manganous ion is added to this system, however, a series of two-body reactions can occur, each with a normal velocity; the effect of catalysis is to *convert an abnormally slow reaction to one of normal speed*. In this case, it is interesting to note that the alternate pathway afforded by the catalyst apparently derives its efficiency from the higher frequency factors of the individual steps compared to the lower factor of the uncatalyzed reaction.

$$\begin{aligned} Ce^{+4} + Mn^{+2} &\longrightarrow Ce^{+3} + Mn^{+3} \\ Mn^{+3} + Ce^{+4} &\longrightarrow Ce^{+3} + Mn^{+4} \\ Mn^{+4} + Tl^{+1} &\longrightarrow Mn^{+2} + Tl^{+3} \end{aligned} \qquad (1.9)$$

Catalysis that is fast compared to expectation from equilibrium is exemplified by numerous reactions catalyzed by acids or bases. Acid-base catalysis is effective because it depends upon proton transfers, which are fast compared to the making and breaking of most other chemical bonds. In fact, proton transfers are quite rapid compared to reactions of comparable free energy change. Consider the two reactions

$$CH_3OCH_3 + H_2O \rightleftarrows CH_3O^- + CH_3OH_2^+ \quad \text{slow} \qquad (1.10)$$

$$CH_3OH + H_2O \rightleftarrows CH_3O^- + H_3O^+ \quad \text{fast} \qquad (1.11)$$

There is only 3 kcal/mole difference in the standard free energy of these two reactions, the equilibrium constants differing by only a hundredfold. The *rates*, however, differ by a tremendous amount: the former is too slow to measure, while the latter has a finite rate constant. The difference between these two processes is one of steric hindrance. That is, the reaction of water with dimethyl ether is sterically hindered with respect to the reaction of water with methanol because of the repulsion of nonbonded atoms (the hydrogen atoms of the methyl group).

Although many proton transfers to or from oxygen are fast, not all such processes occur at the same speed (Chapter 2). Consider the reaction involved in the autoprotolysis of water

$$H_2O + H_2O \rightleftharpoons H_3O^\oplus + OH^\ominus \tag{1.12}$$

This reaction is endothermic as written. The heat of the reverse reaction is 13.8 kcal/mole. Therefore the activation energy of the forward reaction must be at least this amount. Thus an unfavorable equilibrium constant necessarily makes this reaction slow, even though on other grounds it should proceed rapidly.

Easily polarizable nucleophiles react at greater rates than expected from thermodynamics. Nucleophiles such as iodide ion, thiosulfate ion, thiourea, and carbanions react with alkyl iodides at rates higher than their basicities predict. The explanation of this phenomenon has been given many times; namely, that the outer-shell electrons of these species are easily polarized. Their distribution can be distorted in the field of an electrophilic reagent so that a new bond can be formed without bringing the rest of the system into close contact. Hence steric repulsion is reduced and reactivity is enhanced. An easily polarizable nucleophile of high reactivity together with the possibility that the product formed from its reaction may be more unstable than the reactants leads to the phenomenon of nucleophilic catalysis. Iodide ion is such a nucleophile; moreover, it is an easily displaced leaving group. This conjunction of properties leads to the utility of iodide as a catalyst for such reactions as hydrolysis of alkyl chlorides (eq. 1.13) and formation of Grignard reagents from alkyl chlorides.

$$\begin{aligned} I^\ominus + RCl &\longrightarrow RI + Cl^\ominus \\ RI + H_2O &\longrightarrow ROH + I^\ominus + H^\oplus \end{aligned} \tag{1.13}$$

Iodide ion is a more active nucleophile than water because of polarizability, even though its basicity is much lower. Further, the alkyl iodide intermediate is more susceptible to hydrolysis than is the alkyl chloride; together these factors produce catalysis.

Metal ion-donor atom reactions are often fast. This fact leads to the possibility that a metal ion can be introduced into an organic molecule through coordination to an oxygen or a nitrogen atom, leading to an overall polarization of the system and possible catalysis.

These observations give an inkling of the basis of catalytic action. A more complete elucidation will, however, take a considerably more detailed discussion: it is in fact the subject of this book.

1.3 CATALYSIS AND THE EQUILIBRIUM CONSTANT

We have said before that a catalyst speeds up the attainment of equilibrium but does not change the position of equilibrium. A catalyst cannot change the yield since it cannot combine with the product although it can combine with the reactant. These statements follow from thermodynamic arguments since the standard free energy of reaction, and thus the equilibrium constant, depends only on the initial and final stages and is independent of the path of reaction. A proof of the independence of equilibrium from reaction mechanism comes from a consideration of the principle of microscopic reversibility. Microscopic reversibility stems from statistical considerations that state that if a system is at equilibrium and there are a number of molecular transitions occurring between the various states within the system, each transition must separately be at equilibrium.[12] Thus for both an uncatalyzed reaction and the corresponding catalyzed reactions, eqs. 1.14 and 1.15, one can write the equilibrium constant 1.16. Thus the equilibrium constant has not been changed in going from the uncatalyzed reaction to the catalyzed reaction,

$$A \underset{k_{-1}}{\overset{k_1}{\rightleftarrows}} B \qquad (1.14)$$

$$A + C \underset{k_{-2}}{\overset{k_2}{\rightleftarrows}} X \underset{k_{-3}}{\overset{k_3}{\rightleftarrows}} B + C \qquad (1.15)$$

$$K_{eq} = \frac{a_B}{a_A} \qquad (1.16)$$

where a is activity. From eq. 1.16 it follows that any change the catalyst brings about in the rate constant of the forward reaction is accompanied by a corresponding change in the rate constant of the reverse reaction.

Further consideration of microscopic reversibility indicates that if the mechanism of the forward reaction is known, the mechanism of the reverse reaction must also be known. For example, if a reaction goes through an intermediate X in the forward direction, the reaction proceeds via the same intermediate X in the reverse direction. In other words, if the energy barriers are known in the forward direction then they are known (in reverse) for the reverse direction.[11]

The principle of microscopic reversibility, however, does not rule out multiple pathways.[12] Consider an isotopic exchange reaction in which the reactant and product differ only by isotopic substitution. The system must proceed through a single symmetrical pathway but this symmetrical pathway can consist of two individual unsymmetrical pathways (assuming they do not cross) as well as one symmetrical pathway. Microscopic reversibility will be satisfied for each system, as shown in Fig. 1.3.

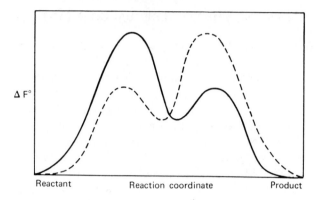

Fig. 1.3 A two-path isotopic exchange reaction in which two mirror image pathways each comprise 50% of the total reaction in each direction. Reprinted from R. L. Burwell, Jr. and R. G. Pearson, *J. Phys. Chem.*, **70**, 300 (1966). © (1966) by the American Chemical Society. Reprinted by permission of the copyright owner.

1.4 KINETICS OF CATALYSIS

We have seen earlier that a catalyst accelerates a reaction by providing an alternate mechanism. Let us assume that in the catalytic pathway an intermediate compound between the catalyst and the substrate forms, then undergoes a reaction to yield the product and regenerate the catalyst. Using this description of the catalytic process, it is of interest to develop some general and simple kinetic equations for it.

Let us assume a stoichiometry

$$R + R' \xrightarrow{C} P \tag{1.17}$$

(where R and R' are reactants, C is the catalyst, and P is the product), and a mechanism

$$R + C \underset{k_{-1}}{\overset{k_1}{\rightleftharpoons}} X$$

$$X + R' \xrightarrow{k_2} P + C \tag{1.18}$$

where X is the intermediate. For most systems represented by eq. 1.18, useful kinetic expressions can be derived by employing the steady-state approximation; namely, that the rate of change in [X] is negligible compared to those in [R] or [P]:

$$\frac{d[X]}{dt} \ll \frac{d[R]}{dt} \quad \text{or} \quad \frac{d[P]}{dt} \tag{1.19}$$

Kinetics of Catalysis

For our system, this approximation takes the form

$$\frac{d[X]}{dt} = k_1[R][C] - k_{-1}[X] - k_2[X][R'] = 0 \tag{1.20}$$

hence

$$[X] = \frac{k_1[R][C]}{k_{-1} + k_2[R']} \tag{1.21}$$

If k_{-1} is much larger than either k_1 or $k_2[R']$, X will exist in a low concentration and will be virtually at equilibrium with R and C; such an intermediate is called an Arrhenius intermediate.

If k_{-1} and $k_2[R']$ are comparable and much larger than k_1, the concentration of X will be low but not simply related to [R] and [C]; in this case, X is called a van't Hoff intermediate.

Now, from eq. 1.21,

$$\frac{d[P]}{dt} = k_2[X][R'] = \frac{k_1 k_2[R][R'][C]}{k_{-1} + k_2[R']} \tag{1.22}$$

Often it is more convenient to consider only the beginning of the reaction, when the approximations[13]

$$[C] = [C]_0 - [X] \quad \text{and} \quad [R] = [R]_0 - [X]$$

are valid. Substituting these values into eq. 1.20 and dropping terms in $[X]^2$ gives

$$v_0 = \left(\frac{d[P]}{dt}\right)_{t=0} = \frac{k_1 k_2 [R]_0 [R']_0 [C]_0}{k_{-1} + k_1([C]_0 + [R]_0) + k_2[R']_0} \tag{1.23}$$

For a catalyzed reaction, the catalyst concentration is usually much lower than the reactant concentration; that is, $[C]_0 \ll [R]_0 \sim [R']_0$, and thus $k_1[C]_0 \ll k_1[R]_0$. Then

$$v_0 = \frac{k_1 k_2 [R]_0 [R']_0 [C]_0}{k_{-1} + k_1[R]_0 + k_2[R']_0} \tag{1.24}$$

If $k_2[R']_0 \ll k_{-1} \sim k_1[R]_0$, a condition which may hold sometimes but need not always obtain, the intermediate X is an Arrhenius intermediate and

$$v_0 = \frac{k_1 k_2 [R]_0 [R']_0 [C]_0}{k_{-1} + k_1[R]_0} \tag{1.25}$$

Defining $K = k_{-1}/k_1$,

$$v_0 = \frac{k_2 [R]_0 [R']_0 [C]_0}{K + [R]_0} \tag{1.26}$$

Equation 1.26 has the form of the Michaelis-Menten equation[14] for enzyme-catalyzed reactions, in which the initial formation of an adsorptive complex between enzyme and substrate is usually assumed, or the Langmuir isotherm in heterogeneous catalysis.[15] As does the Michaelis-Menten equation, (1.26) indicates that the initial rate is directly proportional to $[C]_0$ and $[R']_0$ but of variable order with respect to $[R]_0$. This can be most easily seen in Fig. 1.4, a plot of eq. 1.26 in the form of v_0 versus $[R]_0$.

At very low $[R]_0$ ($\ll K$), the reaction is first-order in $[R]_0$; at very high $[R]_0$ ($\gg K$), the reaction is zero-order in $[R]_0$.

This phenomenon is usually described as a "saturation" of the catalyst by the substrate. Alternatively, we could have assumed that $[C]_0 \gg [R]_0 \sim [R']_0$ and that $k_2[R']_0 \ll k_{-1}$. Then eq. 1.27 results.

$$v_0 = \frac{k_2[R]_0[R']_0[C]_0}{K + [C]_0} \quad (1.27)$$

Equation 1.27 has the same form as eq. 1.26. Thus, at very low $[C]_0$ ($\ll K$) the reaction is first-order in $[C]_0$; at very high $[C]_0$ ($\gg K$) the reaction is zero-order in $[C]_0$. This time a "saturation" of the substrate by the catalyst is seen (Fig. 1.5). Thus the phenomenon of "saturation" is not specific to the catalyst; it can occur with either catalyst or reactant in any reaction forming an unstable intermediate in equilibrium with reactant.

The treatment given above implies that the catalytic process can proceed no faster than the first step. Alternatively the catalyst must react with the reactant more rapidly than the reactant normally decomposes.

For a chain mechanism involving initiation, propagation, and termination

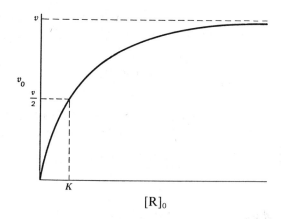

Fig. 1.4 A plot of eq. 1.26.

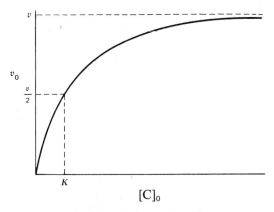

Fig. 1.5 A plot of eq. 1.27.

reactions, the assumption of the stationary state can again be used to define the kinetics. For the system,

$$\text{catalyst} \underset{}{\overset{k_i}{\rightleftarrows}} X \qquad \text{initiation}$$
$$X + R + R' \xrightarrow{k_p} P + X \qquad \text{propagation} \qquad (1.28)$$
$$X + I(X) \xrightarrow{k_t} \text{Reaction of } X \qquad \text{termination}$$

the assumption of the stationary state will lead to the rate law

$$\text{rate} = k_p k_i \frac{[C][R][R']}{k_t[I]} \qquad (1.29)$$

Equation 1.29 shows that the rate of the overall reaction (due essentially to the propagation step) can be many times greater than the rate of initiation (first step). This conclusion is contrary to the conclusion derived above for a nonchain reaction. Thus a chain process should be the best kind of catalytic reaction from the point of view of accelerating the reaction. However, a propagation step is a phenomenon that occurs only under special mechanistic circumstances.

1.5 CLASSIFICATION OF CATALYSIS

With this brief survey of some of the basic questions involved in homogeneous catalysis, let us discuss catalysis by looking at some of its manifestations.

We shall begin with a consideration of what is perhaps the oldest, and superficially at least, the simplest of homogeneous catalyses—acid-base catalysis. We shall then advance through more complicated catalyses to a discussion of enzymic catalysis, the ultimate in catalytic efficiency and specificity.

REFERENCES

1. J. Berzelius, *Jahresber*, **15**, 237 (1835); *Ann. Chim. Phys.*, **61**, 146 (1836); quoted in J. E. Jorpes, *Jac. Berzelius, His Life and Work*, Almqvist and Wiksell, Stockholm, 1966, p. 112.
2. W. Ostwald, *Chem. Betrach., Die. Aula.*, no. 1 (1895); quoted in R. P. Bell, *Acid Base Catalysis*, Oxford University Press, 1941, p. 2.
3. W. Ostwald, *Phys. Z.*, **3**, 313 (1902).
4. R. P. Bell, *Acid Base Catalysis*, Oxford University Press, Oxford, 1941, p. 3.
5. P. G. Ashmore, *Catalysis and Inhibition of Chemical Reactions*, Butterworth and Co., Ltd., 1963.
6. C. N. Hinshelwood, *The Structure of Physical Chemistry*, Oxford University Press, Oxford, 1951, p. 369.
7. E. D. Hughes, F. Juliusberger, S. Masterman, B. Topley, and J. Weiss, *J. Chem. Soc.*, 1525 (1935).
8. F. Wagner, J. Convit, E. Bernt, and M. Nelbock, *Angew. Chem., Int. Ed.*, **3**, 587 (1964).
9. R. F. Kaull and G. K. Rollefson, *J. Amer. Chem. Soc.*, **58**, 1755 (1936).
10. A. A. Frost and R. G. Pearson, *Kinetics and Mechanism*, John Wiley and Sons, New York, 2nd ed., 1961; K. B. Yatsimirski and A. Filipov, *Kinet. Catal.*, **6**, 599 (1965).
11. R. C. Tolman, *The Principles of Statistical Mechanics*, Oxford University Press, Oxford, 1938, pp. 163, 165.
12. R. L. Burwell, Jr., and R. G. Pearson, *J. Phys. Chem.*, **70**, 300 (1966).
13. K. J. Laidler and I. M. Soquet, *J. Phys. Chem.*, **54**, 519 (1950).
14. L. Michaelis and M. L. Menten, *Biochem. Z.* **49**, 333 (1913).
15. I. Langmuir, *J. Amer. Chem. Soc.*, **40**, 1361 (1918).

Part One

Acid-Base Catalysis

Chapter 2

PROTON TRANSFER

2.1	Hydroxide and oxonium ions	19
2.2	Relationship between Rate Constants and Equilibria	22
2.3	Generalized Acid-Base Systems	25
2.4	Solvent Participation in Proton Transfer	28
2.5	Chain Proton Transfer	34
2.6	Proton Transfers in Nonaqueous Media	34

Proton transfer is the essence of acid-base catalysis. In recent years, the experimental description and theoretical interpretation of proton transfer (and therefore acid-base catalysis) has flowered with the advent of modern relaxation and spectroscopic methods for the measurement of the extremely fast reactions in which the proton participates.[1-6] While a consideration of the elegant methods and theoretical framework used in the determination of the kinetics of these reactions, whose half-lives range from 10^{-5} to 10^{-11} sec, is beyond the scope of this work, the results are essential to a proper consideration of acid-base catalysis.

2.1 HYDROXIDE AND OXONIUM IONS

Table 2.1 lists some representative rate constants for the reactions of acids with hydroxide ion (k_{OH}) and of bases with oxonium ion (k_H). Many of these rate constants are of the order of magnitude calculated for diffusion control (ca. $10^{10}\ M^{-1}\ \text{sec}^{-1}$). Bases including hydroxide ion, imidazole, fluoride ion, and water, representing a range of 16.7 pK units, react with oxonium ion with essentially identical rate constants. Likewise, water, imidazolium ion, hydrogen fluoride, and oxonium ion react with hydroxide ion with essentially identical rate constants. The rate of reaction of an oxonium ion with hydroxide ion ($1.4 \times 10^{11}\ M^{-1}\ \text{sec}^{-1}$) is even faster than that of oxonium ion with a solvated electron ($2.3 \times 10^{10}\ M^{-1}\ \text{sec}^{-1}$). The simplest explanation of this rate difference is that proton transfers are not slowed by solvent reorganization whereas electron transfers are impeded by this process. The high rate constants of proton transfer are explained by eq. 2.1, which describes proton transfer as: (1) the diffusion-controlled formation of a hydrogen bond between the acid, H—Y, and the base, X:; (2) the transfer of a proton to make a new hydrogen-bonded

Table 2.1

Rate Constants for the Reaction of Some Acids with Hydroxide Ion and Of Their Conjugate Bases with Oxonium Ion[7,a]

Acid	Conjugate Base	pK_a	k_H ($M^{-1}\ \text{sec}^{-1}$)	k_{OH} ($M^{-1}\ \text{sec}^{-1}$)
H_2O	$HO^\ominus$	15.75	1.4×10^{11}	
H_2O(ice)	$HO^\ominus$	21.4	8.6×10^{12} [b]	
D_2O	$DO^\ominus$	16.5	8.4×10^{10}	
$ImH^\oplus$	Imidazole	6.95	1.5×10^{10}	2.5×10^{10}
H_2S	$HS^\ominus$	7.24	7.5×10^{10}	
HF	$F^\ominus$	3.15	1.0×10^{11}	
(salicylate structure)	Salicylate dianion	11.05		1.4×10^{7} [c]
(azo structure)	(Dianion)	11.90		4.75×10^{5}
Enol of acetylacetone	Enolate ion	8.24	3.1×10^{10}	1.9×10^{7}
Acetylacetone	Enolate ion	9.0	1.2×10^{7} [c]	4×10^{4}
Acetone	$CH_3COCH_2^\ominus$	~20	~5×10^{10}	2.7×10^{-1}
H_2CO_3	$HCO_3^\ominus$	6.35	5.6×10^{4}	1×10^{4}
Tropylium ion ($+H_2O$)	Tropanol	4.75	6.6×10^{4}	

[a] Temperature = 298° K unless otherwise specified.
[b] Temperature = 263°.
[c] Temperature = 285°.

complex; and (3) the diffusion-controlled dissociation of this hydrogen-bonded product.

$$X: + H-Y \rightleftharpoons \underset{1}{X\cdots H-Y} \rightleftharpoons \underset{2}{X-H\cdots Y} \rightleftharpoons \underset{3}{X-H + :Y} \quad (2.1)$$

Equation 2.1 will explain the preceding rate difference since the proton transfer is guided by a hydrogen bond whereas the electron transfer is not. In addition, the distance over which the proton is transferred is small compared with the electron jump distance, and the field of the surrounding solvent is relatively much less altered as a result of the transfer. As seen in Table 2.1 the reaction of an oxonium ion and a hydroxide ion in ice is faster than the corresponding reaction in water by a factor of approximately 60. This difference can be readily explained using the hypothesis of eq. 2.1, since ice contains a perfect hydrogen-bonded lattice. Further evidence pertaining to this point will be noted later.

In discussing rate constants of the oxonium and hydroxide ion reactions of Table 2.1, the following factors need to be considered: (1) spatial (symmetry and steric) factors; (2) electrostatic interactions; (3) the hydrogen-bonded structure of the substrate; and (4) electronic redistribution during the reaction. The reaction of oxonium ion with fluoride ion is slightly faster than the reaction of oxonium ion and hydrosulfide ion. This small difference can be explained on a statistical basis since the fluoride ion has four electron pairs that can accept the proton whereas the hydrosulfide ion has only three. Electrostatic interactions cause only small perturbations of the rate constant, presumably because of the high dielectric constant of water, the solvent for these proton transfers. Roughly speaking, the rate constant of proton transfer from oxonium ion decreases by a factor of 2 for each positive charge introduced into systems of equal size. For example, the rate constants for the reaction of oxonium ion with three metal ion complexes of varying charge are $HOCu(H_2O)_5^{\oplus}$, 10^{10} M^{-1} sec^{-1}; $HOCo(NH_3)_5^{2\oplus}$, 5×10^9 M^{-1} sec^{-1}; $HNRPt(en)_2^{3\oplus}$, 1.9×10^9 M^{-1} sec^{-1}.

In contrast to these small effects, the presence of a hydrogen bridge in the substrate leads to a large diminution in the rate constant of proton transfer to the solvent or another molecule. Two examples of this phenomenon may be seen in Table 2.1: the reactions of hydroxide ion with salicylate ion and with the *ortho*-hydroxyazobenzene derivative. The effect of the internal hydrogen bond on these reactions amounts to a 10^3 to 10^5 decrease in the rate constant compared to "normal" proton transfers. Mechanism 2.1 implies that proton transfer between these two substrates and hydroxide ion can occur only after the internal hydrogen bond (bridge) is broken. The rate constant of proton transfer will reflect the difference in stability constants of the internal and external hydrogen bridges, as compared to a rate constant of a diffusion-controlled reaction without this complication.

Another large effect on the rate constant of proton transfer reactions is seen with compounds in which the acid differs from its conjugate base in electronic distribution and molecular structure. That is, if the reaction of the acid with hydroxide ion or the reaction of the conjugate base with oxonium ion is associated with a large redistribution of electronic charge through a resonance interaction, the reaction rate may be much lower than that of a diffusion-controlled reaction. Consider the reactions of the carbon acids acetylacetone and acetone and the reactions of carbon dioxide and tropylium ion of Table 2.1. The carbon acids in contrast to the "normal" oxygen and nitrogen acids show rate constants much lower than those of diffusion-controlled reactions. The reaction of the enolate ion of acetylacetone with an oxonium ion, to give the enol, which does not involve appreciable redistribution of electronic charge, proceeds with a diffusion-controlled rate constant of 10^{10} M^{-1} sec^{-1}. On the other hand, the reaction of the enolate ion with oxonium ion to give the ketone, which involves considerable redistribution of electronic charge, has a rate constant of only 10^7 M^{-1} sec^{-1}. Furthermore, the reaction of hydroxide ion with the enol of acetylacetone, which contains an internally hydrogen-bonded proton, has a rate constant of 10^7 M^{-1} sec^{-1}, whereas the reaction of the ketone, which does not contain an internally hydrogen bonded proton, with hydroxide ion has a rate constant of only 4×10^4 M^{-1} sec^{-1}. This low figure is a consequence of the large redistribution of electronic charge that must accompany the transformation of the ketone to the corresponding enolate ion. The reactions of carbon dioxide and tropylium ion with hydroxide ion similarly involve large redistributions of electronic charge; this fact is reflected in their small prototropic rate constants.

The low rate constants of proton transfer involving carbon acids are of special significance to organic chemistry. Two interrelated factors apparently account for these low rate constants: (1) a carbon acid does not form hydrogen bridges readily; and (2) considerable electronic redistribution is ordinarily associated with transfer of a proton from a carbon acid.

2.2 RELATIONSHIP BETWEEN RATE CONSTANTS AND EQUILIBRIA

Let us now consider the relationship between the ionization constants of acids and bases and their abilities to accept or donate protons. As a measure of this ability we will use the rate constant for the donation of a proton from an acid to water or the rate constant for the abstraction of a proton by a base from water. Tables 2.2 and 2.3 give the rate constants

Table 2.2
Rate Constants for the Reaction of Some Bases with Water[5,7,a]

B	k_f (M^{-1} sec^{-1})	k_r (M^{-1} sec^{-1})	pK_a of BH$^\oplus$	log k_f
OH$^\ominus$	5×10^9	5×10^9	15.74	9.69
NH$_3$[b]	6×10^5	3.4×10^{10}	9.25	5.78
Imidazole	2.2×10^3	2.5×10^{10}	6.95	3.34
H$_2$O	2.5×10^{-5}	1.4×10^{11}	−1.74	−4.21

[a] 25° C unless otherwise noted.
[b] 20° C.

Table 2.3
Rate Constants for Reactions of Some Acids with Water[5,7,a]

HA	k'_f (M^{-1} sec^{-1})	k'_r (M^{-1} sec^{-1})	pK_a of HA	log k'_f
H$_3$O$^\oplus$	1×10^{10}	1×10^{10}	−1.7	−10.0
H$_2$SO$_4$	$\sim 10^9$	$\sim 1 \times 10^{11}$	2	~ 9
HF	7.1×10^7	1×10^{11}	3.15	7.85
CH$_3$COOH	8.2×10^5	4.5×10^{10}	4.74	5.91
H$_2$S	4.3×10^3	7.5×10^{10}	7.24	3.63
ImH$^\oplus$	1.8×10^3	1.5×10^{10}	6.95	3.26
(CH$_3$)$_3$NH$^\oplus$	1×10^1		9.8	1.00
Glucose	$< 5 \times 10^{-3}$	$> 10^{10}$	12.3	< -2.7
H$_2$O	2.5×10^{-5}	1.4×10^{11}	15.74	−4.61
H$_2$SO$_3$	3.2×10^6	2×10^3	1.8	6.51
H$_2$CO$_3$	0.025	5.6×10^4	3.60	−1.20

[a] 25° C.

of proton transfer between several bases and water, according to eq. 2.2, and between several acids and water, according to eq. 2.3.

$$B + H_2O \underset{k_r}{\overset{k_f}{\rightleftharpoons}} BH^\oplus + OH^\ominus \qquad (2.2)$$

$$HA + H_2O \underset{k'_r}{\overset{k'_f}{\rightleftharpoons}} H_3O^\oplus + A^\ominus \qquad (2.3)$$

The data in Table 2.2 show that a direct relationship exists between the pK_a of the conjugate acid of the base and the logarithm of the rate constant for proton abstraction from water by the base. Likewise, Table 2.3 and Fig. 2.1 show a direct relationship between the pK_a of the acid and the logarithm of the rate constant for proton donation to water by the acid. The reason for both these relationships is that the reverse reactions are essentially diffusion-controlled proton transfers having either a zero or a negative free

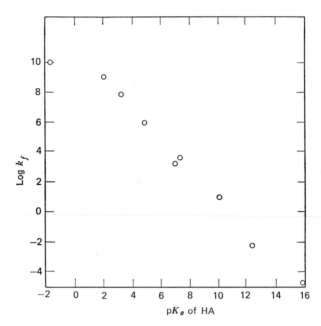

Fig. 2.1 The relationship between the rate constant of proton transfer from an acid to water and pK_a of the acid.

energy change. The direct relationship between the logarithm of the rate constants of proton transfer and the pK_a's of the bases or acids may be expressed by eqs. 2.4 and 2.5, respectively.[7] In fact, these equations were used in calculating a few of the quantities shown in Tables 2.2 and 2.3.

$$k_f = k_r \times 10^{-(15.5 - pK_a)} \qquad (2.4)$$
$$k_f' = k_r' \times 10^{-pK_a} \qquad (2.5)$$

Table 2.3 includes two slow proton transfer reactions, involving sulfurous acid and carbonic acid. The low values of these rate constants probably reflect the large electronic redistribution accompanying the ionization process.

The ionizations of carbon acids must be probed in more detail. The rates of prototropic reactions of a family of ketones are represented by Fig. 2.2, which shows a monotonic nonlinear variation of k_r and k_r' as a function of the pK of the ketone. There is, however, no simple relation between k_r or k_r' and the pK when the tabulation includes carbon acids other than ketones (Table 2.4). As mentioned earlier, all of these rate constants are much smaller than the diffusion-controlled limit. On the other hand, the reaction

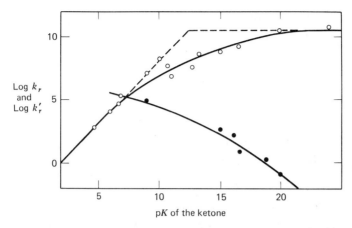

Fig. 2.2 Dependence of the recombination rate constants k'_r (○ – ○) and k_r (● – ●) on pK in the protolysis of ketonic compounds. Experimental data are circles. The broken line denotes the linear extrapolation. From M. Eigen, *Angew. Chem. Int. Ed.*, **3**, 12 (1964).

Table 2.4
Rate Constants of Reactions of Some Carbon Acids with Water[5,7,a]

HA	pK	log k_f[b]	log k_r[b]
CH_3COCH_3	20	−9.3	10.7
$CH_2(CO_2C_2H_5)_2$	13.3	−4.6	8.7
$CH_2(CN)_2$	11.2	−1.8	9.4
$CH_3COCH_2COCH_3$	9.0	−1.8	7.2
$C_2H_5NO_2$	8.6	−7.4	1.2
$CH_3COCH_2NO_2$	5.1	−1.4	3.7
$CH_2(NO_2)_2$	3.6	−0.1	3.5

[a] k_f determined by bromination.
[b] Rate constants M^{-1} sec^{-1} at 18 to 25°.

rates of enolate ions with oxonium ion do approach this limit (Table 2.1); these processes, of course, do not strictly involve carbon acids, since they produce the enols and not the ketones.

2.3 GENERALIZED ACID-BASE SYSTEMS

Proton transfers in generalized acid-base systems may be described by the equation

$$XH + Y^\ominus \rightleftharpoons X^\ominus + YH \qquad (2.6)$$

For the reactions of hydroxide ion with HX or HY, or for the reactions of oxonium ion with $X^\ominus$ or $Y^\ominus$, one would expect, on the basis of the previous discussion, diffusion-controlled rate constants, assuming that the various species are "normal" acids and bases. For reactions proceeding according to eq. 2.6, one would expect that the rates would likewise be diffusion-controlled when $pK_{HY} > pK_{HX}$, that is, when $Y^\ominus$ is a stronger base than $X^\ominus$. This statement may be alternatively expressed by saying that the reaction should be diffusion-controlled when the reaction as written is exergonic. Certainly the faster proton transfer must occur in the direction of the exergonic reaction. Further, a proton transfer can be very fast even in a symmetrical system when the difference in pK between the two reactants is zero, that is, when the free energy of the reaction is zero. Tables 2.2 and 2.3, for example, show that the rate constants of the symmetrical reactions between oxonium ion and water and between hydroxide ion and water are essentially diffusion-controlled. Those prototropic reactions in which the free energy of the reaction is negative would certainly be expected to proceed with an even higher rate, which can only be diffusion-controlled.

When a whole series of prototropic reactions of noncarbon acids is exergonic, all the rates should be diffusion-controlled and therefore independent of the pK's of the bases involved. When this is true, $\log k$ of the reverse reactions must be a linear function of the difference between the pK's of the two reactants. This conclusion may be easily seen since the equilibrium constant for eq. 2.6 can be expressed as either of two quantities:

$$K_{eq} = \frac{K_{HX}}{K_{HY}} = \frac{k_f}{k_r} \qquad (2.7)$$

Equation 2.7 can be transformed into its logarithmic counterpart

$$\log k_f - \log k_r = pK_{HY} - pK_{HX} = \Delta pK \qquad (2.8)$$

When $\log k_f$ is constant, as it must be in these exergonic reactions, eq. 2.8 becomes

$$\log k_r = \text{constant} - \Delta pK \qquad (2.9)$$

Thus a proton transfer reaction conforming to eq. 2.6 can be represented by Fig. 2.3: when the pK of the acceptor is greater than the pK of the donor (when ΔpK is positive), the rate of the proton transfer is independent of ΔpK and diffusion-controlled; when ΔpK is approximately zero, a transition from independence of ΔpK to a linear dependence on ΔpK occurs, the forward reaction getting progressively slower and the reverse reaction becoming diffusion-controlled. Thus, two limiting slopes are reached for each reaction of zero and ± 1.

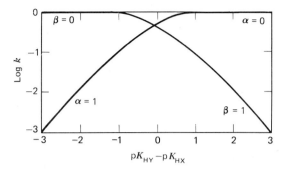

Fig. 2.3 Idealized log k versus $\Delta pK = pK_{HY} - pK_{HX}$ dependence for proton transfer for a symmetrical charge type. From M. Eigen, *Angew. Chem. Int. Ed.*, **3**, 14 (1964).

For a proton transfer reaction involving a separation of charge, such as,

$$XH + Y \rightleftharpoons X^{\ominus} + YH^{\oplus} \qquad (2.10)$$

the relationship between log k and ΔpK is asymmetric, because the limiting value for the diffusion-controlled reaction will not be attained in the reaction direction involving separation of charge. A hypothetical example of the relationship between rate constants and equilibrium constants of reactions related to eq. 2.10 is shown in Fig. 2.4, which depicts a reaction diffusion-controlled in one direction but not in the other. Many real examples of this relationship have been collected by Eigen.[7] One set of reactions, involving imidazole as acceptor with several donors, is shown in Fig. 2.5, which corresponds closely to the hypothetical situation shown in Fig. 2.4 for the reaction of a neutral base with neutral acids. The curvature in the relationship

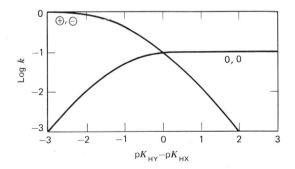

Fig. 2.4 Idealized log k versus $\Delta pK = pK_{HY} - pK_{HX}$ dependence for proton transfer in "normal" acid-base systems: for charge neutralization. From M. Eigen, *Angew. Chem. Int. Ed.*, **3**, 14 (1964).

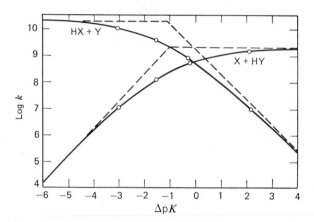

Fig. 2.5 Log k versus ΔpK-dependence for some proton transfer reactions. Experimental points refer to order from left (negative ΔpK) to right (positive ΔpK). Acceptor (X): imidazole; donor (HY). From M. Eigen, *Angew. Chem. Int. Ed.*, **3**, 15 (1964).

between log k and ΔpK in both Figs. 2.3 and 2.4 can be expressed by a McLaurin series as shown in eq. 2.11.

$$\log k = f(\Delta pK) \simeq 0 + \alpha\,\Delta pK + \alpha'(\Delta pK)^2 + \alpha''(\Delta pK)^3 + \cdots \tag{2.11}$$

In proton transfer reactions following eqs. 2.6 or 2.10, all the factors pertaining to proton transfers involving oxonium or hydroxide ions come into play. A simple relationship between log k and pK would be expected when the donor and acceptor atoms consist of classical hydrogen bond-forming atoms such as oxygen and nitrogen, when no internal hydrogen bond exists, and when electronic and spatial configurations are such that the acids and their conjugate bases are comparable. Conversely, reactions of carbon acids and bases cannot be described by such a simple relationship. A typical example is the reaction of acetylacetone with various bases as shown in Fig. 2.6. Qualitatively, proton transfers to or from carbon have the characteristics of the reactions of "normal" (oxygen and nitrogen) acids and bases. However, there are quantitative differences; most importantly, the transition from a slope of zero to one occurs over a very much wider range of ΔpK than with "normal" acid and base reactions.

2.4 SOLVENT PARTICIPATION IN PROTON TRANSFER

By means of proton magnetic resonance it is possible in favorable cases to measure not only the number of solvent molecules that participate in a given process, but also the rate of proton transfer between solute and solvent.

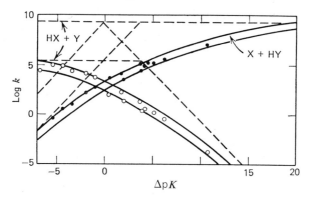

Fig. 2.6 Log k versus ΔpK dependence for proton transfer between acetylacetone (HX) or its enolate (X) and a series of bases (Y) or acids (HY). Acetylacetone reacts in its keto form. The acceptors (Y) are (from left to right): HO^-; glycerol (anion); mannose (anion); glucose (anion); trimethylphenoxide, phenoxide, pyrophosphate, chlorophenoxide, p-nitrophenoxide, HS^-; dimedone (enolate); pyridine; acetate; benzoate; hydroxyacetate; chloroacetate; H_2O. From M. Eigen, *Angew. Chem. Int. Ed.*, **3**, 17 (1964).

Consider a reaction in which an acid molecule HA reacts with n solvent molecules SH in such a way that $n + 1$ protons are transferred. The corresponding rates of proton exchange associated with these reactions will be $d[HA]/dt$ in the acid and $d[SH]/dt$ in the solvent. But $d[SH]/dt$ equals n times $d[HA]/dt$, since n solvent molecules exchange a proton every time one HA molecule exchanges a proton. If we can measure both $d[SH]/dt$ and $d[HA]/dt$, the ratio is equal to n, which is thereby evaluated.

The first use of the p.m.r. method to study the kinetics of proton exchange involved aqueous solutions of methylammonium ion.[9] Since then, these studies have been extended to a number of amines and a number of hydroxylic solvents, covering a wide range of dielectric constant.[10] In solvents of high dielectric constant, for example, water and methanol, the major processes in the low pH range include: (1) transfer of a proton from $BH^\oplus$ to the solvent, (2) direct proton transfer from $BH^\oplus$ to B, and (3) proton transfer from $BH^\oplus$ to B involving one solvent molecule as shown in eqs. 2.12 to 2.14.

$$BH^\oplus + ROH \underset{k_{-1}}{\overset{k_1}{\rightleftarrows}} B + ROH_2^\oplus \qquad (2.12)$$

$$BH^\oplus + B \xrightarrow{k_2} B + HB^\oplus \qquad (2.13)$$

$$\underset{R}{BH^\oplus + OH} + B \xrightarrow{k_3} B + \underset{R}{HO} + HB^\oplus \qquad (2.14)$$

In solvents of low dielectric constant, proton transfer processes are affected by ionic association. In *t*-butyl alcohol ($D = 12.47$ at 25°) the reactive species is the solvated ion pair rather than the solvated free ion.

$$BH^{\oplus}X^{\ominus} + \underset{t\text{-Bu}}{OH} + B \xrightarrow{k} B + \underset{t\text{-Bu}}{HO} + X^{\ominus}HB^{\oplus} \qquad (2.15)$$

In this reaction, a proton hydrogen-bonded to the anion in the ion pair is transferred much less readily than a proton hydrogen-bonded to a solvent molecule. In glacial acetic acid ($D = 6.22$ at 25°) methylamines are converted almost completely to their acetate salts, which exist in this solution largely in the form of ion pairs, $BH^{\oplus}OAc^{\ominus}$. Proton exchange is first-order in this ion pair and probably involves a two-step mechanism with initial formation of undissociated amine.[11]

Rate constants of proton exchange of methyl-substituted amines in water according to eqs. 2.12 to 2.14 are shown in Table 2.5. The values of k_{-1} ($= k_1/K_a$) for these substrates are of such large magnitude as to suggest diffusion-controlled processes. Values of k_1 (in contrast to those of k_{-1}) are quite sensitive to the presence of inert salts, decreasing when salt is added. Furthermore, there is a striking reduction in k_1 in the presence of strong acids; k_1 probably goes to zero at very high acidity. On the basis of this evidence, step 1 should be written as the two-step mechanism

$$\begin{aligned} BH^{\oplus}\text{---}(OH_2)_n &\rightleftharpoons B\text{---}(HOH)_m + H(OH_2)^{\oplus}_{n-m} \\ HOH^* + B\text{---}HOH &\longrightarrow B\text{---}H^*OH + HOH \end{aligned} \qquad (2.16)$$

involving first a reversible ionization to produce an amine-water hydrogen-bonded complex and then an exchange of a water molecule between bulk solvent and the site adjacent to B.

The rate constant k_2 for direct proton transfer from $BH^{\oplus}$ to B decreases sharply with methyl substitution on nitrogen as seen in Table 2.5. This could be due to increased steric hindrance or to increased energy of the desolvation step that must precede proton transfer. On the other hand, the rate constant k_3 for the reaction of $BH^{\oplus}$, water, and B is not highly sensitive to methyl substitution. The small changes that are observed parallel the changes in $1/K_a$. One water molecule is involved in the proton transfer between trimethylamine and trimethylammonium ion.[18] On the basis of this and an analogous result obtained in methanol, eq. 2.14 is probably the best description of the reaction associated with k_3. A single water molecule was likewise found to be necessary for the proton exchange between dihydrogen phosphate and monohydrogen phosphate ions and between phenol and phenoxide ion in aqueous solution.[19,20]

Table 2.5
Rate Constants for Proton Transfer and Acid Dissociation Constants, Water, 25°[a]

$BH^{\oplus}$	k_1 (sec^{-1})	$k_{-1} \times 10^{-10}$ (M^{-1} sec^{-1})	$k_2 \times 10^3$ (M^{-1} sec^{-1})	$k_3 \times 10^3$ (M^{-1} sec^{-1})	$K_a \times 10^{10}$ (M)	Reference
$NH_4^{\oplus}$	25	4.3	11.7	0.9	5.68	12
$CH_3NH_3^{\oplus}$	—	—	4.0	5.3	0.242	13
$(CH_3)_2NH^{\oplus}$	—	—	0.5	9.0	0.168	14
$(CH_3)_2NH^{\oplus}$	4.7	3.0	0.0	3.4	1.57	15
$HO_2CCH_2NH_2CH_3^{\oplus}$	110	3.2	—	—	30	16

[a] The rate constants in this table apply to eqs. 2.12 to 2.14 and to dilute solutions.

Proton exchange of carboxylic acids has also been investigated by p.m.r. methods.[21] Table 2.6 lists pseudo first-order rate constants for proton ex-

Table 2.6

Analysis of Rate Constants for Proton Exchange[21]

1. Carboxylic acids with neopentyl alcohol in glacial acetic acid, 25°.

RCO_2H	k_{exch} (sec^{-1})	K_i	$1/k_i$	$1/k_{-i}{}^a$
CH_3CO_2H	6.7×10^5	5×10^{-8}	13×10^5	3×10^{13}
$NCCH_2CO_2H$	8.8×10^6	1×10^{-5}	18×10^6	2×10^{12}
Cl_3CCO_2H	2.5×10^8	3×10^{-3}	5×10^8	2×10^{11}

2. Substituted benzoic acids with methanol in methanol, 25°.

Substituent	k_{exch} (sec$^{-1} \times 10^5$)	$10^9 K_A$	$k_i{}^a$ (sec^{-1})	$k_{-i}{}^a$ (sec^{-1})
All H	0.66	0.39	3.9×10^{-9}	3×10^{13}
m-NO_2	1.8	4.5	4.5×10^{-8}	8×10^{12}
p-NO_2	2.2	4.7	4.7×10^{-8}	9×10^{12}
o-NO_2	10	27.8	2.8×10^{-7}	7×10^{12}
3,5-$(NO_2)_2$	8	42	4.2×10^{-7}	4×10^{12}

3. Acetic acid with water in water, 25°.

$k_{exch} = 4.8 \times 10^7$; $K_A = 1.75 \times 10^{-5}$; $k_i = 2.3 \times 10^{-5}$; $k_{-i}{}^a = 4 \times 10^{12}$

^{a} $k_i = 2k_{exch}$.

change between a carboxylic acid and an alcohol or water in a variety of solvents. The first section of the table lists exchange between aliphatic acids and neopentyl alcohol. In glacial acetic acid, rate constants increase systematically with the strength of the acid, the sequence being $CH_3CO_2H <$ $NCCH_2CO_2H < Cl_3CCO_2H$. The second and third sections of the table list rate constants obtained in methanol and water, with RCO_2H as solute. For reactions in glacial acetic acid, the exchange involves one molecule of acid and one of alcohol. The exchange reaction can then be written as

$$\text{R-O}\begin{array}{c}\text{H-\!-\!-O}\\ \diagup \quad \diagdown\\ \text{C-CH}_3\\ \diagdown \quad \diagup\\ \text{H}\overset{*}{-}\text{O}\end{array} \underset{k_{-i}}{\overset{k_i}{\rightleftarrows}} \text{R-O}^{\oplus}\begin{array}{c}\text{H-\!-\!-O}\\ \\ \ominus\ \text{C-CH}_3\\ \\ \text{H-\!-\!-O}\overset{*}{}\end{array} \qquad (2.17)$$

On the other hand, for reactions in methanol, the proton exchange involves 1 molecule of acid and 2 molecules of methanol; hence, the exchange must be more complicated. The most straightforward interpretation of this exchange involves an ionization to produce carboxylate ion and methyloxonium ion. A rate constant for this process may be calculated from the known ionization (equilibrium) constant for the reaction, 4×10^{-10} M, and the calculated diffusion-controlled rate constant for the reverse reaction, 6×10^{10} M^{-1} sec^{-1}, giving a dissociation rate constant of 26 sec^{-1} whereas the rate constant of exchange is of the order of 10^5 sec^{-1}. A simple dissociation process thus cannot be operative. A solution to this dilemma is the process

$$\text{(structure)} \quad \underset{k_{-i}}{\overset{k_i}{\rightleftarrows}} \quad \text{(structure)} \tag{2.18}$$

We shall now make the simplest possible assumptions concerning these proton transfers: (1) ionization (with rate constant k_i) is the rate-determining step for proton exchange; and (2) the carboxylate group in the ion pair is symmetrical; that is, the two oxygen atoms are equivalent. Using these assumptions, k_i of eq. 2.17 is related to k_{exch} according to

$$k_i = 2k_{\text{exch}} \tag{2.19}$$

while k_i of eq. 2.18 is directly related to the rate constant of exchange. The effect of structural variation on the rate constants gives an indication that eqs. 2.17 and 2.18 are correct. For both proton transfer of carboxylic acids with neopentyl alcohol in glacial acetic acid and proton transfer of benzoic acids in methanol, the stronger acid gives a faster reaction. In fact the Brønsted α, a measure of the relationship between rate constant and (equilibrium) acidity (see Chapter 4), is the same for these two sets of reactions, that is, approximately 0.54. This value is consistent with eqs. 2.17 and 2.18 since the same electronic factors that affect the equilibria of these reactions should affect the rate of the forward reaction to some extent. Incidentally, the value of 0.54 for the Brønsted α for eq. 2.18 rules out a concerted cyclic proton transfer process in which no ion pair intermediate is formed, since the latter process should show no effect of structural variation on the rate constant. The rate constants k_i of Table 2.6 are quite high, but well below the range of diffusion-controlled rate constants that characterize the reverse reactions.

2.5 CHAIN PROTON TRANSFER

The mechanism of proton transport is not fully described by the foregoing discussion. Eigen and DeMaeyer[22] have shown that ice is a protonic semiconductor. The crystal consists of a three-dimensional network of water molecules linked together by hydrogen bridges. Ice does conduct direct current: the charge carriers may be assumed to be oxonium and hydroxide ions formed by dissociation of the water molecule. Charge can be transported through the crystal by multiple proton jumps in hydrogen bridges in an electrical field.

$$
\begin{array}{c}
\text{H} \quad \text{H} \quad \text{H} \quad \text{H} \quad \text{H} \quad \text{H} \\
| \quad\ | \quad\ | \quad\ | \quad\ | \quad\ | \\
\cdots\text{H}-\overset{\oplus}{\text{O}}-\text{H}\cdots\text{O}-\text{H}\cdots\text{O}-\text{H}-\text{O}-\text{H}\cdots\text{O}-\text{H}-\text{O}-\text{H}\cdots \longrightarrow \\[6pt]
\text{H} \quad \text{H} \quad \text{H} \quad \text{H} \quad \text{H} \\
| \quad\ | \quad\ | \quad\ | \quad\ | \\
\cdots\text{H}-\text{O}\cdots\text{H}-\text{O}\cdots\text{H}-\text{O}-\text{H}-\text{O}-\text{H}-\overset{\oplus}{\text{O}}-\text{H}
\end{array}
\tag{2.20}
$$

Conduction of direct current requires that charge transport always be possible in the same direction. But after passage of a single charge, all chains of water molecules are in the state shown in the right hand side of eq. 2.20, unsuitable for further transport of positive charge from left to right. To explain conduction of direct current, the chain must be renewed for proton passage: rotation of the water molecules in the crystal will serve this purpose. Thus by a combination of proton jumps in hydrogen bridges and reorientation of water molecules, a proton can migrate as an individual through the entire ice crystal.

The rate-determining step of the conduction in ice is the proton translation in the hydrogen bond.[16,17] Here proton transfer follows a "nonclassical" mechanism, probably involving a tunnel effect. The mobility of the proton in ice (Table 2.1) is only one or two orders of magnitude less than that of the electron in metals.

Although the structure of water does not present the perfect hydrogen-bonded structure that ice does for the chain proton jump mechanism of eq. 2.20, the high mobility of the proton in water must be due to a similar process.

2.6 PROTON TRANSFERS IN NONAQUEOUS MEDIA

Although in water, the rates of proton transfer from nitromethane, phenol, p-toluenesulfonic acid, or water to a base are vastly different, in dimethyl sulfoxide proton transfers from these acids to triphenylmethyl or fluorenyl anions proceed at identical rates.[23] Furthermore, the proton exchange between dimethyl sulfoxide and its conjugate base occurs with a rate constant of 7 M^{-1} sec^{-1}, approximately a millionfold faster than the proton exchange of the carbon base fluorenyllithium with fluorene in ether solution (10^{-5} M^{-1} sec^{-1}).[25] These observations emphasize the importance of changes in hybridization on the rates of proton transfer. When rehybridization (e.g., of a carbon acid) occurs in a solvent where anionic stabilization by the solvent is important, solvent reorganization will contribute a major part of the activation energy of the reaction[24]; when rehybridization is unimportant (e.g., in the conversion of an oxyacid to its anion), less solvent reorganization is required. Hence in a solvent where anion solvation is unimportant, such as dimethyl sulfoxide, solvent reorganization should not be necessary for ionization and proton transfer rate constants should be larger and less dependent on structure.

In solvents of low dielectric constant, such as benzene, proton transfer will lead not to separated ions but rather to ion pairs, triplets, and quadruplets.[26]

The rate constants of proton transfer in nonaqueous solution may possibly be related to the equilibrium constants of these reactions. In the solvent cyclohexylamine, the acidities of a number of hydrocarbons were determined by competition with methoxide ion.[27] For a series of reactions represented by

$$\text{R—H} + \text{OMe}^{\ominus} \underset{k_r''}{\overset{k_f''}{\rightleftarrows}} \text{R}^{\ominus} + \text{MeOH} \quad (2.21)$$

a linear relationship was found between log k_f'' and log K, with a slope of 0.4. These results mean that k_r'' is a diffusion-controlled process, and is the same for all members of the series. The results further mean that the forward reaction is not diffusion-controlled.

It appears that proton transfers in hydroxylic solvents always involve hydrogen bonding to the solvent or to another molecule. They must involve at least one and possibly a very large number of solvent molecules. The rates of proton transfers, as those of other reactions, are susceptible to structural variation in the substrate, and to considerable solvent effects. In the next chapter, we will consider the participation of proton transfers in acid and base catalysis.

REFERENCES

1. M. Eigen and L. DeMaeyer, in *Techniques of Organic Chemistry*, A. Weissberger, Ed., 2nd ed., Vol. 8, Part 2, Interscience Publishing Co., New York, 1963, p. 895.
2. H. Strehlow, in *Techniques of Organic Chemistry*, A. Weissberger, Ed., 2nd ed., Vol. 8, Part 2, Interscience Publishing Co., New York, 1963, pp. 799, 865.

3. R. M. Noyes and A. Weller, in *Techniques of Organic Chemistry*, A. Weissberger, Ed., 2nd ed., Vol. 8, Part 2, Interscience Publishing Co., New York, 1963, p. 845.
4. G. Porter, in *Techniques of Organic Chemistry*, A. Weissberger, Ed., 2nd ed., Vol. 8, Part 2, Interscience Publishing Co., New York, 1963, p. 1055.
5. E. S. Caldin, *Fast Reactions in Solution*, John Wiley and Sons, New York, 1964.
6. L. K. Patterson, *Chem. in Britain*, **4**, 24 (1968).
7. M. Eigen, *Angew. Chem. Int. Ed.*, 3, 1 (1964); for measurement of very high reaction rates by Raman line broadening, see M. M. Kreevoy and C. A. Mead, *J. Amer. Chem. Soc.*, **84**, 4596 (1962).
8. R. M. Noyes, *Disc. Faraday, Soc.*, **39**, 130 (1965).
9. E. Grunwald, C. F. Jumper, and S. Meiboom, *J. Chem. Phys.*, **25**, 382 (1956).
10. E. Grunwald and M. Cocivera, *Disc. Faraday Soc.*, **39**, 105 (1965).
11. E. Grunwald and E. Price, *J. Amer. Chem. Soc.*, **86**, 2965, 2970 (1965).
12. M. T. Emerson, E. Grunwald, and R. A. Kromhout, *J. Chem. Phys.*, **33**, 547 (1960).
13. E. Grunwald, P. J. Karabatsos, R. A. Kromhout, and E. L. Purlee, *ibid.*, **33**, 556 (1960).
14. A. Loewenstein and S. Meiboom, *ibid.*, **27**, 1067 (1957).
15. E. Grunwald, *J. Phys. Chem.*, **67**, 2208 (1963).
16. M. Sheinblatt, *J. Chem. Phys.*, **36**, 3103 (1962).
17. B. Silver and Z. Luz, *J. Amer. Chem. Soc.*, **83**, 786 (1961).
18. Z. Luz and S. Meiboom, *J. Chem. Phys.*, **39**, 366 (1963).
19. Z. Luz and S. Meiboom, *J. Amer. Chem. Soc.*, **86**, 4764 (1964).
20. Z. Luz and S. Meiboom, *ibid.*, **86**, 4766 (1964).
21. E. Grunwald, *Progr. Phys. Org. Chem.*, 3, 344 (1965).
22. M. Eigen and L. DeMaeyer, *Proc. Roy. Soc.*, **A247**, 505 (1958).
23. C. D. Ritchie and R. Uschold, *J. Amer. Chem. Soc.*, **86**, 4488 (1964).
24. C. D. Ritchie, G. A. Skinner, and V. G. Badding, *ibid.*, **89**, 2063, 2960 (1967).
25. J. I. Brauman and N. J. Nelson, *J. Amer. Chem. Soc.*, **88**, 2332 (1966).
26. J. Steigman and P. M. Lorenz, *ibid.*, **88**, 2083, 2093 (1966).
27. A. Streitwieser, Jr., et al., *ibid.*, **87**, 384 (1965).

Chapter 3

OXONIUM AND HYDROXIDE IONS IN CATALYSIS

3.1 Oxonium Ion Catalysis 39
3.2 Acidity Functions 41
3.3 The Effect of Acidity on Oxonium Ion-Catalyzed Reactions 45
3.4 Mechanism of Some Oxonium Ion-Catalyzed Reactions 50
3.5 The Catalytic Function of the Proton 59
3.6 Hydroxide Ion Catalysis 62
3.7 The Catalytic Function of the Hydroxide Ion 67
3.8 Lyoxide Catalysis in Solvents Other than Water 68

Acid-base catalysis was originally identified as catalysis by strong acids or bases.[1] It is now explicitly recognized that the catalytic species present in these solutions are oxonium and hydroxide ions. Subsequently, acid-base catalysis was shown to be effected not only by those species but also by any Brønsted or Lewis acid or base as well. In the present chapter, the original concept of catalysis by oxonium and hydroxide ions, usually referred to now as specific oxonium and hydroxide ion catalysis, will be discussed. Other lyonium and lyoxide ion catalysis will also be treated here, but further generalization of acid-base catalysis will be considered later.

In water, pH ($= -\log [H_3O^\oplus]$) can affect the rate of a chemical reaction in many ways, as shown in Fig. 3.1. The logarithm of the rate constant can

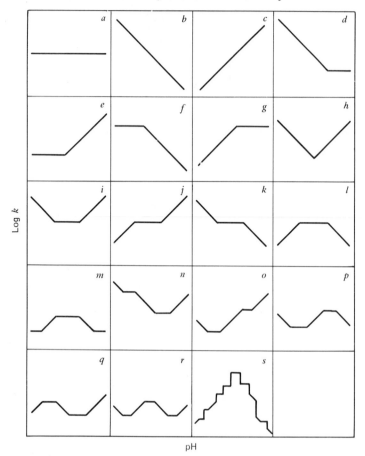

Fig. 3.1 Some hypothetical pH – log k profiles.

be independent of pH (curve a, slope 0); it can be directly dependent on the logarithm of the oxonium ion concentration (curve b, slope -1); on the logarithm of the hydroxide ion concentration (curve c, slope $+1$). It can depend on one or more combinations of these: curve i is a typical combination curve showing dependence on oxonium ion concentration, independence of pH, and dependence on hydroxide ion concentration with ascending pH. The rate constant corresponding to curve i is

$$k_{\text{obs}} = k_0 + k_{\text{H}}[\text{H}_3\text{O}^\oplus] + k_{\text{OH}}[\text{OH}^\ominus] \qquad (3.1)$$

In eq. 3.1, k_0 is a rate constant of a reaction insensitive to pH, usually called a spontaneous reaction or, more accurately, a water reaction, while k_{H} and

k_{OH} are rate constants for oxonium and hydroxide ion-catalyzed reactions. A reaction can also be dependent on one or more forms of the prototropic group(s) on one or other of the reactants. In more complicated systems, the rate can be dependent on the prototropic groups not only of the reactants, but also of the intermediates. The slopes of log k-pH profiles involving dependence on a prototropic group can also have values of $+1$, -1, and 0, leading to considerable ambiguity in the interpretation of such profiles. In this chapter we shall concentrate on reactions solely dependent on either oxonium or hydroxide ions and postpone consideration of the ambiguities to later chapters. One ambiguity, however, that must always be met directly is curve s; although such curves are meaningless, they have been published repeatedly.

3.1 OXONIUM ION CATALYSIS

Oxonium catalysis can be generally represented by an equilibrium between the substrate and oxonium ion, followed by a rate-determining transformation of the protonated substrate. The two variants of this scheme are differentiated by the fact that one represented by eq. 3.2 does not involve a nucleophile (R) in the rate-determining step, whereas the other, eq. 3.3, does. In either eq. 3.2 or 3.3, the rate of reaction depends on the *concentration* of oxonium ion, but in no way upon the *source* of this species.

$$S + H^\oplus \xrightleftharpoons{K} SH^\oplus \xrightarrow{\text{slow } (k)} \text{products}$$
$$\text{rate} = k[SH^\oplus] = kK[S][H^\oplus] \tag{3.2}$$

$$S + H^\oplus \xrightleftharpoons{K} SH^\oplus \xrightarrow[R]{\text{slow } (k)} \text{products}$$
$$\text{rate} = k[SH^\oplus][R] = kK[S][H^\oplus][R] \tag{3.3}$$

Equations 3.2 and 3.3 identify the intermediate in these reactions as the protonated substrate. All organic compounds can be considered bases, although variation in base strength of different compounds is quite profound. Essentially every organic compound can be protonated to some extent; even the extremely weak base methane can be converted to $CH_5^\oplus$ in the mass spectrometer,[2] and evidence for this species in very strong acids has been reported.[3] Protonation of organic compounds has been investigated in great depth; some results concerning the basicities of important classes of organic compounds are summarized in Table 3.1.

The structure of $SH^\oplus$ is sometimes ambiguous. Since protonation often can occur in more than one position in an organic molecule, the protonated species cannot always be specified uniquely. For example, protonation of a

Table 3.1
Basicities in Aqueous Acid of Representative Bases of Important Classes[a]

Functional Group	Compound	pK_a[b]	% H_2SO_4 to Half-Ionize
Aldehyde	Alkyl aldehydes	(ca. −8)[c]	(ca. 88)
	Benzaldehyde	−7.1	81
Amide	Acetamide	ca. 0.0	6.5
	Benzamide	ca. −2.0	34
Amine	Methylamine	10.6	
	Dimethylamine	10.6	
	Trimethylamine	9.8	
	Aniline	4.6	
Amine oxide	Trimethylamine oxide	4.7	
Aromatic hydrocarbon	Hexamethylbenzene	[d]	90.5
Carboxylic acid	Acetic acid	−6.1	74
	Benzoic acid	−7.2	82
Carboxylic ester	Ethyl acetate	(ca. −6.5)[c]	(ca. 77)
	Ethyl benzoate	−7.4	83
Ether	Diethyl ether	−3.6	52
	Tetrahydrofuran	−2.1	36
	Anisole	−6.5	77
Hydroxyl	Methanol	−2	34
	Phenol	−6.7	78
Ketone	Acetone	−7.2	82
	Acetophenone	−6.2	74
Mercaptan	Methyl mercaptan	−6.8	78
Nitro	Nitromethane	−11.9	Oleum
	Nitrobenzene	−11.13	Oleum
Olefin	1,1-Diphenylethylene	[d]	71
Phosphine	n-Butylphosphine	0.0	6.5
	Dimethylphosphine	3.9	—
	Trimethylphosphine	8.7	—
Phosphine oxide	Trimethylphosphine oxide	0	6.5
Sulfide	Dimethyl sulfide	−5.4	68
Sulfoxide	Dimethyl sulfoxide	0	6.5

[a] E. M. Arnett, *Prog. Phys. Org. Chem.*, **1**, 324 (1963).
[b] The pK_a refers to a thermodynamic equilibrium constant only in the case of weak bases that exactly obey the activity coefficient postulate. Hence most of these values only refer to the H_0 at half-ionization.
[c] No measured values are available for these compounds. Values shown are estimated by analogy from suitable compounds of known basicity.
[d] These compounds probably follow H_R instead of H_0 (see next section). Therefore no pK_a on the pH-H_0 scale is given.

carboxylic acid derivative can lead to either **1** or **2**

$$\underset{\mathbf{1}}{\underset{X}{R-C}\overset{\oplus OH}{\diagup\diagdown}}} \qquad \underset{\mathbf{2}}{\underset{XH^{\oplus}}{R-C}\overset{O}{\diagup\diagdown}}$$

where X is any group with an unshared electron pair. Usually the most abundant species is **1**. The effect of substituents on the pK_a's of *para*-substituted acetophenones is the same as that on the pK_a's of *para*-substituted benzoic acids. Since the protonation of an acetophenone leads unambiguously to **1** (X = CH$_3$), the protonation of benzoic acid[4] must lead to **1** (X = OH) also.

For amides in acidic media, n.m.r. spectroscopy provides strong evidence that the principal protonated species is **1** (X = NR$_2$).[5]

Although species such as **1** are present in the highest concentrations, other species can be present in lower concentrations; in certain cases, these less abundant species are the reactive ones. An example (the acid-catalyzed hydrolysis of esters) will be considered later in this chapter.

3.2 ACIDITY FUNCTIONS

In dilute aqueous solutions of strong acids, the proton-donating power of the medium is acceptably defined as the concentration of oxonium ion. In concentrated acid solution, however, where the solvent is no longer essentially water, the proton-donating power of the solvent cannot be expressed so simply. Many years ago Hammett and Deyrup suggested that the proton-donating power of strongly acidic solutions may be referred to a thermodynamic standard state in water by the stepwise application of progressively less basic indicators in progressively stronger acid solutions.[6] For an uncharged indicator we may write

$$\mathrm{HIn}^{\oplus} \rightleftharpoons \mathrm{In} + \mathrm{H}^{\oplus} \qquad (3.4)$$

$$K_a = \frac{a_{\mathrm{H}^{\oplus}} f_{\mathrm{In}}}{f_{\mathrm{HIn}^{\oplus}}} \frac{[\mathrm{In}]}{[\mathrm{InH}^{\oplus}]} \qquad (3.5)$$

where $a_i = f_i[i]$. Since the ratio of concentrations [In]/[HIn$^{\oplus}$] can be measured spectrophotometrically, a quantitative comparison of indicator color changes yields a direct, empirical measurement of the acidity of the medium without appreciably changing it. Neither activities nor activity coefficients can be measured directly, but the quantity $a_{H^\oplus} f_{In}/f_{HIn^\oplus}$ can be experimentally determined; it is this quantity that Hammett defined as the acidity function, h_0.

$$h_0 = a_{H^\oplus} \frac{f_{In}}{f_{HIn^\oplus}} \qquad (3.6)$$

If the ratio of activity coefficients, $f_{In}/f_{InH^\oplus}$, is independent of small structural variations in the indicator, then h_0 defines the proton-donating power of a strongly acidic medium, since eqs. 3.5 and 3.6 can be combined in logarithmic form ($H_0 = -\log h_0$) to give

$$pK_a = H_0 + \log \frac{[HIn^\oplus]}{[In]} \qquad (3.7)$$

Equation 3.7 has the same form as the equation defining pK_a in a dilute aqueous medium

$$pK_a = pH + \log \frac{[HIn^\oplus]}{[In]} \qquad (3.8)$$

Equation 3.7 approaches eq. 3.8 at infinite dilution in water, and thus the H_0 scale is an extension of the pH scale (Fig. 3.2). In proton-donating power measured by H_0, solutions of strong acids of the same titratable acidity differ from one another markedly, as seen in Fig. 3.3.

The discussion so far has been restricted to processes in aqueous solution.[6] But implicit in all ionization processes is the participation of the solvent. This participation leads to the phenomenon of the *leveling effect*.[7] Consider some reference solvent in which a group of acids behave as strong acids; in a more weakly basic solvent only some of these acids will behave as strong acids (ionize) while others will behave as weak acids; conversely, in a more strongly basic solvent other acids, which behaved as weak acids in the reference solvent, will behave as strong acids. The proton-donating power of an acid is thus moderated somewhat by the proton-accepting power of the solvent; for example, hydrogen bromide is a more powerful proton donor in acetic acid than in water. This phenomenon has in part been met in the discussion above of acidity functions; it can be extended to many other solvents as shown in Fig. 3.4.

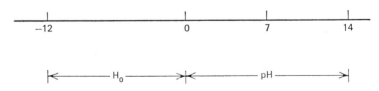

Fig. 3.2 The relationship of H_0 to pH.

Acidity Functions

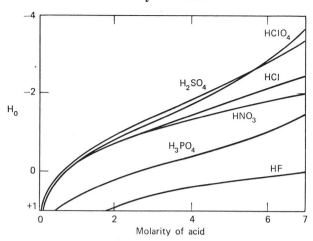

Fig. 3.3 H_0 and related acidity functions. From M. A. Paul and F. A. Long, *Chem. Rev.*, **57**, 13 (1957). © 1957 The Williams and Wilkins Co., Baltimore, Md. 21202 U.S.A.

As mentioned above, the H_0 scale is constructed by using overlapping indicators, starting in dilute acid solution where activity coefficients are equal to one and proceeding stepwise to solutions of high acidity. The assumption necessary for the validity of this procedure, that the activity coefficient ratio is independent of the particular indicator used, may be expressed by

$$\frac{f_{InH^{\oplus}}}{f_{In}} \times \frac{f_{In'}}{f_{In'H^{\oplus}}} = 1 \tag{3.9}$$

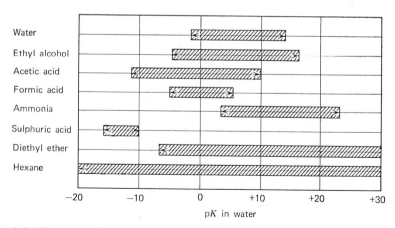

Fig. 3.4 The range of existence of acids and bases in different solvents. From R. P. Bell, *The Proton in Chemistry*, Cornell University Press, Ithaca, N.Y., 1959, p. 42.

which indicates that the two indicator ratios must cancel. By restricting his choice of indicators to a series of aniline bases, Hammett was certain that eq. 3.9 was obeyed. However, the question may be asked as to whether other types of bases obey eq. 3.9; the answer is that many compounds do not.

Variants of the Hammett acidity function have been devised to treat compounds the acidity of which cannot be described in terms of H_0. One obvious variant is an acidity function for a series of bases or indicators of a different charge type than the neutral aniline bases. A more subtle change than going to a different charge type occurs when the base or indicator belongs to a different family than the aniline bases. For prototropic equilibria involving carbonium ions, acid dependency is so markedly different from that of a Hammett base that it has been necessary to devise a new acidity function H_R,[8] (J_0).[9] The operational definition of H_R is similar to that of H_0 (see eq. 3.38).

$$H_R = pK_{R^\oplus} + \log \frac{[ROH]}{[R^\oplus]} \tag{3.10}$$

$$H_R = \log a_{H^\oplus} + \log a_{H_2O} + \log \frac{f_{R^\oplus}}{f_{ROH}}$$

The relationship of h_R to the oxonium ion activity is given by

$$h_R = \frac{a_{H^\oplus} f_{ROH}}{a_{H_2O} f_{R^\oplus}} \tag{3.11}$$

which differs from the definition of h_0 principally by the inclusion of a term involving the activity of water. Fig. 3.5 shows the relationship of various acidity functions to the sulfuric acid concentration in 0 to 100% sulfuric acid. The acidity function H_R is much more sensitive to the concentration of sulfuric acid than is H_0; in fact, the former changes nearly twice as rapidly as the latter with increasing sulfuric acid concentration. This difference in sensitivity to the medium could be due to *two* causes: (1) H_R is dependent on the activity of water in the medium whereas H_0 is not; and (2) H_R is related to carbonium ions whereas H_0 is related to ammonium or oxonium ions. Consideration of a third acidity function, $H_{R'}$, indicates that both these factors contribute to the difference between H_0 and H_R. The acidity function $H_{R'}$ is defined by

$$H_{R'} = H_R - \log a_{H_2O} = pK_{R'^\oplus} + \log \frac{[\text{olefin}]}{[R^\oplus]} \tag{3.12}$$

and applied to the protonation of olefins in strongly acidic solutions. It is obvious that $H_{R'}$ differs from H_R solely in the dependency of the latter on the activity of water. Conversely, $H_{R'}$ differs from H_0 in the effect of the

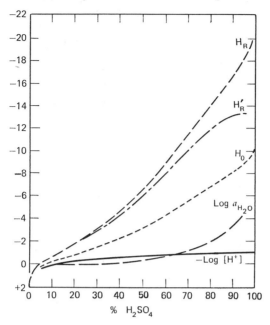

Fig. 3.5 The variation of different acidity functions with acid strength for sulfuric acid solutions. From E. M. Arnett, *Prog. Phys. Org. Chem.*, **1**, 239 (1963).

medium on carbonium ions rather than on oxonium or ammonium ions. This difference may be due to hydrogen bonding of the latter two species with the solvent.[10]

3.3 THE EFFECT OF ACIDITY ON OXONIUM ION-CATALYZED REACTIONS

The effect of the acidity of the medium on the rate of an acid-catalyzed reaction must be discussed in terms of its effect on the equilibrium protonation of organic compounds considered above. On the basis of the general eqs. 3.2 and 3.3 in 1939 Zucker and Hammett[11,12] proposed two systems to account for the effects of highly acidic media on the rates of oxonium ion-catalyzed reactions. The first of these considers a reaction analogous to eq. 3.2

$$S + H^{\oplus} \underset{}{\overset{K}{\rightleftharpoons}} SH^{\oplus} \xrightarrow{k_2 \text{ (slow)}} \text{products} \qquad (3.13)$$

Applying the Brønsted equation for the effects of changes in the solvent to the rate-determining step of eq. 3.13 (which is equivalent to assuming the transition state theory) yields

$$\text{rate} = k_2[\text{SH}^\oplus]\frac{f_{\text{SH}^\oplus}}{f_\ddagger} \qquad (3.14)$$

where f is activity coefficient and $\ddagger$ is transition state. Equation 3.14 may be transformed with the help of

$$K = \frac{[\text{H}^\oplus][\text{S}]f_{\text{H}^\oplus}f_\text{S}}{[\text{SH}^\oplus]f_{\text{SH}^\oplus}} \quad \text{thus} \quad [\text{SH}^\oplus] = \frac{1}{K}[\text{H}^\oplus][\text{S}]\frac{f_{\text{H}^\oplus}f_\text{S}}{f_{\text{SH}^\oplus}} \qquad (3.15)$$

which assumes a fast equilibrium formation of $\text{SH}^\oplus$, to give

$$\text{rate} = \frac{k_2}{K}[\text{S}]a_{\text{H}^\oplus}\cdot\frac{f_\text{S}}{f_\ddagger} \qquad (3.16)$$

If we assume that the activity coefficient ratio $f_\text{S}/f_\ddagger$ equals the activity coefficient ratio found in the Hammett acidity function $f_{\text{In}}/f_{\text{InH}^\oplus}$ then eq. 3.16 may be transformed to

$$\text{rate} = \frac{k_2}{K}[\text{S}]h_0 \qquad (3.17)$$

Since the observed rate law shows a first-order dependence on substrate eq. 3.17 leads to

$$k_{\text{obs}} = \frac{k_2}{K}h_0 \qquad (3.18)$$

Equation 3.18 may be expressed in the more usual log form

$$\log k_{\text{obs}} = -H_0 + \text{constant} \qquad (3.19)$$

which indicates that a reaction following eq. 3.18 should show a linear dependence of $\log k_{\text{obs}}$ on H_0 with a slope of -1.0.

A reaction analogous to eq. 3.3, shown in eq. 3.20, leads to a completely different dependence of the rate on acidity

$$\text{S} + \text{H}^\oplus \xrightleftharpoons{K} \text{SH}^\oplus \xrightarrow[k_2 \text{ (slow)}]{\text{H}_2\text{O}} \text{products} \qquad (3.20)$$

Applying the same Brønsted equation to the rate-determining step of eq. 3.20 yields

$$\text{rate} = k_2[\text{SH}^\oplus][\text{H}_2\text{O}]\frac{f_{\text{SH}^\oplus}f_{\text{H}_2\text{O}}}{f_\ddagger} \qquad (3.21)$$

Assuming a fast equilibrium between S and $\text{SH}^\oplus$, we may write

$$\text{rate} = \frac{k_2}{K}[\text{S}][\text{H}_3\text{O}^\oplus]\frac{f_\text{S}f_{\text{H}_3\text{O}^\oplus}}{f_\ddagger} \qquad (3.22)$$

The species S and $\text{H}_3\text{O}^\oplus$ are tantamount to the (highest) transition state of the reaction; therefore, it is reasonable to assume that the ratio of activity coefficients in eq. 3.22 is equal to unity. If this condition holds, the rate of

The Effect of Acidity on Oxonium Ion-Catalyzed Reactions

the reaction will be directly proportional to the oxonium ion concentration as shown in eq. 3.23,

$$\text{rate} = \frac{k_2}{K}[S][H_3O^\oplus] \qquad (3.23)$$

and the log of the observed rate constant will be linearly related to the log of the concentration of oxonium ion with a slope of $+1.0$:

$$\log k_{obs} = \log [H_3O^\oplus] + \text{constant} \qquad (3.24)$$

A large number of acid-catalyzed reactions approximately follow eq. 3.19 whereas a number of other acid-catalyzed reactions approximately follow eq. 3.24. (See Tables 3.3 and 3.4.)

Although the use of eqs. 3.19 and 3.24 as criteria of mechanism of acid-catalyzed reactions became widespread, difficulties arose with this treatment. By the early 1960s, both theoretical and experimental objections to the use of these equations appeared.[13] The reasons for these objections were manifold. One reason was that few reactions fit either category perfectly; even when a reasonable fit was observed, a slope of precisely 1.0 was rare. Furthermore, qualitative contradictions appeared. Consider for example, the hydration of isobutylene and the dehydration of *tert*-butyl alcohol. Since the transition state of these forward and reverse processes must be identical, the transition state (for the hydration) must consist of substrate, proton, and a water molecule but the other transition state (for the dehydration) can consist of no more than substrate plus a proton. Thus, these transition states require that one of these reactions show dependence on H_0 while the other show dependence on $\log [H_3O^\oplus]$. Both the hydration and dehydration rates, however, show dependence on H_0.

A further contradiction appeared in the protonation of aromatic compounds, which depends on H_0, although the reaction is known to be general acid-catalyzed.[14,15] Since eq. 3.13 describes specific oxonium ion catalysis, these two results are incompatible. Two obvious reasons may account for the difficulties with eqs. 3.19 and 3.24. (1) The activity coefficients of only a limited number of substrates conform to the requirements set forth earlier; that is, many substrates are not Hammett bases. (2) The dependence of the rate of reactions on the activity of water may not be correctly described by eqs. 3.19 and 3.24.

Bunnett has made an extensive empirical correlation of reactions occurring in strong aqueous mineral acid.[4,5] He discovered that many plots of $\log k + H_0$ versus $\log a_{H_2O}$ are approximately linear. He suggested that the slope, ω, of these plots categorizes the reaction mechanistically. Here 147 sets of data concerning 97 reactions were treated. Some persistent curvature was observed at high positive ω values; in these cases a linear relationship was obtained by plotting $\log k - \log [HX]$ versus $\log a_{H_2O}$, giving a slope

ω^*. The Bunnett hypothesis has been advocated as a successor to the Zucker-Hammett hypothesis, since the magnitude of the ω and ω^* values has been correlated with the function of water molecules in the mechanisms of these acid-catalyzed reactions, as shown in Table 3.2.

Table 3.2

The Mechanistic Interpretation of ω and ω^* Values[16]

ω	ω^*	Function of Water in Rate-Determining Step
For substrates protonated on nitrogen or oxygen		
-2.5 to 0	—	Is not involved
$+1.2$ to $+3.3$	< -2	Acts as a nucleophile
$> +3.3$	> -2	Acts as a proton transfer agent
For substrates protonated on carbon		
~ 0	—	Acts as a proton transfer agent

Reprinted from *J. Amer. Chem. Soc.*, **83**, 4956 (1961). © 1961 by the American Chemical Society. Reprinted by permission of the copyright owner.

Plotting $\log k + H_0$ versus $\log a_{H_2O}$ is equivalent to considering the extent to which a plot of $\log k$ versus $-H_0$ deviates from the ideal slope of 1.0, as a function of $\log a_{H_2O}$. It is an empirical fact that the deviation is often linear in $\log a_{H_2O}$. If the slope of $\log k$ versus H_0 is greater than one, ω is negative; if it is less than one, ω is positive. The ω plots thus magnify the deviations from the Zucker-Hammett line.

Whereas the treatment of Zucker and Hammett was based upon two discrete categories corresponding to eqs. 3.19 and 3.24, the range of ω values appears to be large and uniform. The mechanistic categories given in Table 3.2 are somewhat arbitrary; they depend, however, on a classification according to water participation that stems from the Zucker and Hammett treatment. Since these acid-catalyzed reactions appear to be linearly dependent on the logarithm of the activity of water, water participation in these reactions must be carefully considered. Zucker and Hammett considered that water either participated or it did not, whereas Bunnett considered that if water participates, it can do so either as a nucleophile or as a proton-transfer agent, these two functions having different sensitivities to the acidic medium. Other alternatives may be possible, but these are the major ones. Those acid-catalyzed reactions in which no water participates ($\omega = -2.5$ to 0) include the hydrolyses of acetals, of *t*-butyl acetate, and of methyl mesitoate. Those reactions in which water acts as a nucleophile ($\omega = +1.2$ to $+3.3$) include the hydrolyses of diethyl ether, ethylenimine, and carboxamides, and the *cis-trans*-isomerization of benzalacetophenone. Those reactions in which water acts as proton transfer agent include the enolization of ketones (oxygen bases; $\omega > +3.3$) and aromatic hydrogen exchange (carbon bases; $\omega \sim 0$). The subclassification (in Table 3.2) of reactions involving substrates protonated on nitrogen or oxygen, or on carbon, may reflect the division of substrates into Hammett bases and non-Hammett bases.

In summary, Bunnett has shown that the rate constants of many reactions in moderately concentrated acid solution fit the following:

$$\log k + H_0 = \omega \log a_{H_2O} + \text{constant} \qquad (3.25)$$

By considering the hydration of all species, he showed that the ω parameters can be related to the change in hydration between the reactants and the transition state. His treatment, however, assumes that the activity coefficient ratios for species of like charge are solvent independent, an assumption similar to that made in the derivation of the Hammett acidity function. It has been shown that this assumption cannot be correct for all families of organic compounds. For example, the activity coefficient ratio for an amide and its conjugate acid is clearly solvent dependent.[17] Thus ω parameters can be used only for those substrates that have been demonstrated to follow H_0 as bases; they cannot be related only to hydration changes involved in the rate-determining step. Since most organic compounds have been demonstrated not to be H_0 bases, the Bunnett treatment may not be generally valid.

Yates and Stevens[17] have developed a rigorous treatment of the effect of acidity on the rate of an acid-catalyzed reaction by considering the effect of acidity both on equilibrium protonation and on the rate of reaction of the substrate. Let us consider a generalized acid-catalyzed reaction involving hydrated species. Such a reaction is

$$\begin{align} S(H_2O)_s + H(H_2O)_n^\oplus &\overset{K_{SH^\oplus}}{\rightleftharpoons} SH(H_2O)_p^\oplus + (s+n-p)H_2O \\ SH(H_2O)_p^\oplus + rH_2O &\rightleftharpoons S^\ddagger(H_2O)_t \overset{k}{\longrightarrow} \text{products} \end{align} \qquad (3.26)$$

Assuming transition state theory and a fast preequilibrium protonation of the substrate, we can write the rate equation for an acid-catalyzed reaction

$$v = \frac{k}{K_{SH^\oplus}} [S(H_2O)_s] \frac{f_{S(H_2O)_s}}{f_{S^\ddagger(H_2O)_t}} \times a_{H(H_2O)_n^\oplus} \times a_{H_2O}^{r-(s+n-p)} = k_{obs}[S]_{st} \qquad (3.27)$$

Equation 3.28, a generalized equation for an indicator equilibrium,

$$In(H_2O)_b + H(H_2O)_n^\oplus \rightleftharpoons InH(H_2O)_c^\oplus + (b+n-c)H_2O \qquad (3.28)$$

leads to eq. 3.29, a general expression for an acidity function in aqueous solution.

$$h = \frac{a_{H(H_2O)_n^\oplus}}{a_{H_2O}^{(b+n-c)}} \times \frac{f_{In(H_2O)_b}}{f_{InH(H_2O)_c^\oplus}} \qquad (3.29)$$

Substitution for the oxonium ion activity in eq. 3.27 in terms of the relation in eq. 3.29 leads to

$$v = \frac{k}{K_{SH^\oplus}} [S_h] \frac{f_{S_{aq}} f_{InH_{aq}^\oplus}}{f_{S^\ddagger_{aq}} f_{In_{aq}}} \times h \times a_{H_2O}^{r+(b-c)-(s-p)} \qquad (3.30)$$

where the subscripts, aq, refer to the fully hydrated species. If the substrate and the indicator are of the same family of organic compounds, $f_{S_{aq}}$ and $f_{In_{aq}}$ refer to the same species and thus will cancel. Likewise the $f_{S^\ddagger_{aq}}$ and $f_{InH_{aq}^\oplus}$ contributions, which are determined chiefly by electrostatic and volume expansion terms, will cancel. Further, $(b - c) - (s - p) = 0$ since these two quantities refer to the same process. Finally $h = h_{amide}$. Thus eq. 3.31 holds for the hydrolysis of amides in strongly acidic solution.

$$v = \frac{k}{K_{SH^\oplus}} [S_{aq}] h_{amide} (a_{H_2O})^r \qquad (3.31)$$

For weakly basic amides or for a solution in which the amide is not appreciably protonated, $[S_{aq}] = [S_{aq}]_{st}$ (where st = stoichiometric) and eqs. 3.32 or 3.33 describe the dependence of the rate on the acidity.

$$k_{obs} = \frac{k}{K_{SH^\oplus}} h_{amide} (a_{H_2O})^r \qquad (3.32)$$

$$\log k_{obs} + H_A = r \log a_{H_2O} + \text{constant} \qquad (3.33)$$

Equation 3.33 has the same form as eq. 3.25 except that an acidity function strictly appropriate to the substrate is involved. Also r can be directly related to the number of water molecules needed to convert the protonated substrate to its transition state. For the acid-catalyzed hydrolyses of benzamide and p-nitrobenzamide, plots of eq. 3.33 are linear with r equal to 2.6 and 2.7, respectively. These results indicate that at least three water molecules may be involved in the transition state of the acid-catalyzed hydrolysis of benzamides. One of these water molecules is the nucleophile. The other two water molecules may be concerned with proton transfer or some process that cannot be specified other than as solvation.[18]

Although the correlation of rate constant and acidity using parallel functions is bringing this area into focus, it is discouraging that such correlations must be done on an individual basis. This lack of generalization must be overcome if the field is to expand.

3.4 MECHANISM OF SOME OXONIUM ION-CATALYZED REACTIONS

Many organic reactions exhibit oxonium ion catalysis. Several mechanistic classifications of oxonium ion-catalyzed reactions have been proposed. One such classification parallels the distinction made by eqs. 3.13 and 3.20

Mechanism of Some Oxonium Ion-Catalyzed Reactions

between the participation or nonparticipation of water in the rate-determining step of the reaction. Equations 3.13 and 3.20 have been designated A-1 and A-2 reactions by Ingold,[19] to indicate their apparent order in species other than oxonium ion. As pointed out above, another classification utilizes not only the participation or nonparticipation of water in the rate-determining step of the reaction, but also the function of the water when it does participate, either as a nucleophile or as a proton transfer agent. A third classification utilizes the number of oxonium ions participating in forming the unstable intermediate. As might be expected, these classifications overlap. We shall look at oxonium ion-catalyzed reactions from all three points of view, pointing out the overlaps.

Some acid-catalyzed reactions that conform to eq. 3.13 (and thus are A-1 reactions) are listed in Table 3.3. Essentially all the reactions of Table 3.3 show a linear dependence of $\log k$ on $-H_0$. As mentioned earlier, few of these linear relationships have the theoretical slope of 1.00. The reactions

Table 3.3
Some Acid-Catalyzed Reactions that Follow Mechanism A-1

Reaction	Reference
Hydrolysis of t-butyl acetate	C. A. Bunton and J. L. Wood, *J. Chem. Soc.*, 1522 (1955).
	P. Salomaa, *Suomen Kemi*, **328**, 145 (1959).
Hydrolysis of methyl mesitoate	C. T. Chmiel and F. A. Long, *J. Amer. Chem. Soc.*, **78**, 3326 (1956).
	M. L. Bender, H. Ladenheim, and M. C. Chen, *J. Amer. Chem. Soc.*, **83**, 123 (1961).
Hydrolysis of alkoxymethyl esters	P. Salomaa, *Acta Chem. Scand.*, **11**, 125, 132, 141 (1957).
Lactonization of γ-hydroxybutyric acid	F. A. Long, F. B. Dunkle, and W. F. McDevit, *J. Phys. Colloid Chem.*, **55**, 813, 829 (1951).
Hydrolysis of β-propiolactone	F. A. Long and M. Purchase, *J. Amer. Chem. Soc.*, **72**, 3267 (1950).
Esterification of alcohols with sulfuric acid	G. Williams and D. J. Clark, *J. Chem. Soc.*, 1304 (1956).
Hydrolysis of sucrose	P. M. Leminger and M. Kilpatrick, *J. Amer. Chem. Soc.*, **60**, 2891 (1938).
Depolymerization of paraldehydes	R. P. Bell and A. H. Brown, *J. Chem. Soc.*, 774 (1954).

(continued)

Table 3.3 (*continued*)

Reaction	Reference
Hydrolysis of tertiary ethers	R. L. Burwell, Jr., *Chem. Rev.*, **54**, 615 (1954).
Hydrolysis of epoxides	J. G. Pritchard and F. A. Long, *J. Amer. Chem. Soc.*, **78**, 2667 (1956).
Hydrolysis of acetic anhydride	V. Gold and J. Hilton, *J. Chem. Soc.*, 843, 848 (1955).
Hydrolysis of acetals, ketals, and glucosides	F. A. Long and M. A. Paul, *Chem. Rev.*, **57**, 965 (1957).
Dehydration of tertiary alcohols	R. W. Taft, Jr., *J. Amer. Chem. Soc.*, **74**, 5372 (1952).
Hydrolysis of carboxylic anhydrides	C. A. Bunton, J. H. Fendler, A. Fuller, S. Perry, and J. Rocek, *J. Chem. Soc.*, 6174 (1965).
Pinacol rearrangement	C. A. Bunton, T. Hadwick, D. R. Llewellyn, and Y. Pocker, *J. Chem. Soc.*, 402 (1958).
Hydrolysis of benzyl fluoride	C. G. Swain and R. E. T. Spaulding, *J. Amer. Chem. Soc.*, **82**, 6104 (1960).
Beckman rearrangement of acetophone oximes	L. P. Hammett and A. J. Deyrup, *J. Amer. Chem. Soc.*, **54**, 2721 (1932).
Decarbonylation of formic acid	L. P. Hammett and A. J. Deyrup, *J. Amer. Chem. Soc.*, **54**, 2721 (1932).
Decarboxylation of mesitoic acid	W. M. Schubert, *J. Amer. Chem. Soc.*, **71**, 639 (1949).
Hydrolysis of diisopropyl phosphorofluoridate	M. Kilpatrick, Jr., and M. L. Kilpatrick, *J. Phys. Colloid. Chem.*, **53**, 1371, 1385 (1949).
Hydrolysis of ethyl vinyl ether	A. J. Kresge and Y. Chiang, *J. Chem. Soc.*, B53, 58 (1967); *J. Amer. Chem. Soc.*, **89**, 4411 (1967); **90**, 5309 (1968).
Intramolecular migration during aromatic hydroxylation	D. Jerina, J. Daly, W. Landis, B. Witkop, and S. Udenfriend, *J. Amer. Chem. Soc.*, **89**, 3347 (1967). D. M. Jerina, J. W. Daly, and B. Witkop, *J. Amer. Chem. Soc.*, **89**, 5488 (1967).
Hydrolysis of neopentyl phosphate	C. A. Bunton et al., *J. Chem. Soc.*, B292 (1966).
Hydrolysis of dihydroxy and/or methoxy benzenes	W. M. Schubert and R. H. Quacchia, *J. Amer. Chem. Soc.*, **85**, 1278, 1284 (1963).

of Table 3.3 also show Bunnett ω values of -2.5 to 0, indicating that water is not involved in the rate-determining step.

The hydrolysis of acetals is probably the most thoroughly studied reaction of Table 3.3. This reaction shows the characteristics[20,21]: (1) the rate of the reaction at very low acidities is proportional to the oxonium ion concentration; (2) the rate of the reaction at high acidities is proportional to h_0; (3) ω is close to zero; (4) the entropy of activation is near zero; (5) the volume of activation is near zero; (6) the rate of reaction is unaffected by added nucleophiles, although the product is not; (7) the $\rho(\rho^*)$ (Taft) of hydrolysis is -3 to -4; (8) the rate of reaction is independent of steric effects; (9) the rate of disappearance of the methoxyl protons of a methyl ketal and the appearance of the hydroxylic protons of methanol from a methyl ketal are identical; (10) in CD_3OD, no deuterium appears in the reactant. These observations require

$$\underset{\overset{|}{OR}}{R_1 - \overset{\overset{R_2}{|}}{C} - OR} \;\overset{H^\oplus}{\rightleftharpoons}\; \underset{\overset{|}{OR}}{R_1 - \overset{\overset{R_2}{|}}{C} - OR} \;\overset{H^\oplus \text{ slow}}{\underset{ROH}{\rightleftharpoons}}\; \underset{\overset{|}{OR}}{R_1 - \overset{\overset{R_2}{|}}{C}{}^\oplus}$$

$$\overset{+H_2O}{\underset{-H^\oplus}{\rightleftharpoons}} \; \underset{\overset{|}{OR}}{R_1 - \overset{\overset{R_2}{|}}{C} - OH} \;\overset{H^\oplus}{\rightleftharpoons}\; R_1 - \overset{\overset{R_2}{|}}{C}{=}O + ROH \quad (3.34)$$

Several acid-catalyzed reactions depend on the acidity raised to a power greater than one. Such a reaction is shown schematically by

$$SOH \;\overset{H_2SO_4}{\underset{HSO_4^\ominus}{\rightleftharpoons}}\; SOH_2^\oplus \;\overset{-H_2O}{\underset{slow}{\longrightarrow}}\; S^\oplus \;\overset{fast}{\longrightarrow}\; \text{products} \quad (3.35)$$

Formally one can produce $S^\oplus$ with only one proton, but in practice, the strong acid solutions required for this process also protonate the water formed in the reaction. A classical example is the nitration of nitrobenzene:

$$C_6H_5NO_2 + HNO_3 \;\overset{H_2SO_4}{\longrightarrow}\; O_2NC_6H_4NO_2 + H_2O \quad (3.36)$$

The rate of this reaction in 80 to 90% sulfuric acid follows H_R rather than any of the other measures of acidity discussed above.[22] This is explicable in terms of the pre-equilibrium formation of nitronium ion according to eq. 3.37, which is exactly analogous to the equilibrium defining the H_R function, eq. 3.38. The dependence of the rate on H_R means a great sensitivity of the rate to the stoichiometric concentration of acid, the greatest sensitivity found in an organic reaction. Other reactions following eq. 3.35 may also show this sensitivity.

$$HONO_2 + 2H_2SO_4 \rightleftharpoons NO_2^\oplus + H_3O^\oplus + 2HSO_4^\ominus \quad (3.37)$$

$$ROH + 2H_2SO_4 \rightleftharpoons R^\oplus + H_3O^\oplus + 2HSO_4^\ominus \quad (3.38)$$

Table 3.4
Some Acid-Catalyzed Reactions Following Mechanism A-2

Reaction	Reference
Hydrolysis of diethyl ether	J. Koskikallio and E. Whalley, *Can. J. Chem.*, **37**, 788 (1959).
Hydrolysis of ethylenimine	G. J. Buist and H. L. Lucas, *J. Amer. Chem. Soc.*, **79**, 6157 (1957).
Hydrolysis of carboxamides	K. Yates and J. B. Stevens, *Can. J. Chem.*, **43**, 529 (1965) and references therein.
	J. T. Edward and S. C. R. Meacock, *J. Chem. Soc.*, 2000 (1957).
Hydrolysis of esters	C. A. Lane, *J. Amer. Chem. Soc.*, **86**, 2521 (1964) and references therein.
	R. P. Bell, A. L. Dowding, and J. M. Noble, *J. Chem. Soc.*, 3106 (1955).
cis-trans Isomerization of benzalacetophenone	D. S. Noyce, W. A. Pryor, and P. A. King, *J. Amer. Chem. Soc.*, **81**, 5423 (1959).
Hydrolysis of pyrophosphoric acid	C. A. Bunton and H. Chaimovitch, *Inorg. Chem.*, **4**, 1763 (1965).
Hydrolysis of trimethylacetic anhydride	C. A. Bunton and J. H. Fendler, *J. Org. Chem.*, **30**, 1365 (1965).
Hydrolysis of aryl phosphate	P. W. C. Barnard et al., *J. Chem. Soc.*, B227 (1966).
Hydrolysis of aryl phosphinates	P. Haake and G. Hurst, *J. Amer. Chem. Soc.*, **88**, 2544 (1966).
Hydration of carbonyl groups	M. Byrn and M. Calvin, *J. Amer. Chem. Soc.*, **88**, 1916 (1966).
	P. Greenzaid, Z. Luz, and D. Samuel, *ibid.*, **89**, 749, 756 (1967).
	Y. Pocker, J. E. Meany, and B. J. Nist, *J. Phys. Chem.*, **71**, 4509 (1967).
Hydration of alkenes	N. C. Deno, F. A. Kish, and H. J. Peterson, *J. Amer. Chem. Soc.*, **87**, 2157 (1965).

Many acid-catalyzed reactions conform to eq. 3.20 and are classified mechanism A-2 by Ingold; some examples of these reactions are shown in Table 3.4. Reactions classified in this way usually have the following characteristics: (1) the rate of reaction is proportional to the oxonium ion concentration up to fairly high acidities; (2) at higher acidities the rate may level off or diminish with increasing acidity; (3) the rate of reaction of the protonated substrate may be dependent on the activity of water; (4)

the value of ω is positive; (5) the entropy of activation is negative and usually large; (6) the volume of activation is positive; and (7) the rate of reaction is increased by added nucleophiles or bases. In addition to these general criteria of A-2 reactions, other mechanistic features can occur in individual instances.

Not included in Table 3.4 are those reactions that formally fit the A-2 mechanism of eq. 3.20, but where the chemistry, the ω value, or other criteria indicate that water acts as a proton transfer agent. These reactions are better discussed as general acid-catalyzed reactions, and will be treated in Chapter 4.

One of the reactions of Table 3.4, the hydrolysis of benzamide, has been discussed earlier in detail. It shows still other features of interest. Although in dilute acid solution the rate of hydrolysis is proportional to the oxonium ion concentration, a rate maximum occurs, as shown in Fig. 3.6, at acidities somewhat larger than the pK_a of the amide. The hypothesis of a protonated amide intermediate effectively accounts for all the experimental facts

$$RCONH_2 + H^\oplus \rightleftharpoons RC(OH)\!\!=\!\!NH_2^\oplus \xrightarrow{H_2O} \text{hydrolysis products}$$
(3.39)

Below the acid concentration necessary for maximum rate, the effect of increasing acid strength is chiefly to increase the concentration of the protonated intermediate; above the concentration for maximum rate, the effect is chiefly to decrease the concentration or activity of water. This hypothesis can account quantitatively for the changes in the rate of hydrolysis with

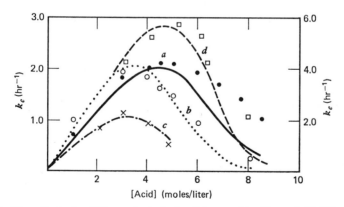

Fig. 3.6 The effect of acidity on hydrolysis of benzamide. From J. T. Edward and S. C. R. Meacock, *J. Chem. Soc.*, 2000 (1957).

		Theoretical	Experimental
Left-hand scale	benzamide in HCl	a	●
	benzamide in H_2SO_4	b	○
	p-methoxybenzamide in H_2SO_4	c	×
Right-hand scale	p-nitrobenzamide in H_2SO_4	d	□

concentration of the acid.[23] The experimental rate constant, k_{obs}, can be related to the second-order rate constant k_2 by the expression

$$k_{obs} = \frac{k_2 K(H_3O^\oplus)}{K + h_0} \qquad (3.40)$$

where K is the equilibrium constant for the protonation of the amide and h_0 is the Hammett acidity function.† For very weak bases or at low acidity, $K \gg h_0$ and eq. 3.40 simplifies to

$$k_{obs} = k_2(H_3O^\oplus) \qquad (3.41)$$

Equation 3.41 predicts a linear dependence of the experimental first-order constant on the concentration of hydronium ion; such a relationship is found in dilute acid solution. However, when $K \ll h_0$, eq. 3.40 simplifies to

$$k_{obs} = \frac{k_2 K(H_3O^\oplus)}{h_0} \qquad (3.42)$$

Since h_0 increases with acid concentration much more rapidly than $[H_3O]^\oplus$ when concentrations of mineral acid exceed about 2 M, k_{obs} decreases in high concentrations of acid. When $K \ll h_0$, protonation of the amide will be substantially complete; the decrease in rate of hydrolysis can then be regarded as due to the decreasing availability of water since $[H_2O] = K_a[H_3O]^\oplus/h_0$, where K_a is the equilibrium constant for the reaction $H_3O^\oplus \rightleftarrows H_2O + H^\oplus$.

A further corroboration of the mechanism shown for the acid-catalyzed hydrolysis of an amide stems from the effect of structure on the reaction. In the hydrolysis of substituted benzamides the effect of substituents on the overall rate constant could be manifested either in the pre-equilibrium protonation, in the subsequent rate-determining step, or both. The Hammett ρ constants for the various steps bear the relationship $\rho_{overall} = \rho_1 + \rho_2$, where 1 and 2 refer to the preequilibrium and rate-determining steps, respectively. In this system, it is possible to determine each of the three ρ constants independently of one another; the experimental data thus found closely approximate the theoretical equation given above for this relationship.[24]

Since the acid-catalyzed hydrolysis of nitriles does not show a rate maximum in concentrated acid solution as the hydrolysis of amides does, it is possible to convert a nitrile to an isolable amide in strong acid solution. Conversely, in dilute acid solutions, the hydrolysis of a nitrile will proceed directly to carboxylic acid, since under these conditions the rate of hydrolysis of the amide is faster than that of the nitrile.

The acid-catalyzed hydrolysis of esters, like other acid-catalyzed reactions,

† More rigorously h_{amide} should be used here in light of the earlier discussion, but this has not as yet been done.

proceeds via a complex pathway from reactants to products. In this instance, the scheme of eq. 3.20 must be amplified to the following to specify the five intermediates of the reaction (see Fig. 3.7):

$$\underset{RCOR}{\overset{O}{\|}} \underset{H_2O}{\overset{H_3O^{\oplus}}{\rightleftharpoons}} \underset{R-C-OR}{\overset{OH^{\oplus}}{\|}} \underset{slow}{\overset{H_2O}{\rightleftharpoons}} \underset{\underset{OH_2^{\oplus}}{|}}{\overset{OH}{\underset{|}{R-C-OR}}} \rightleftharpoons \underset{\underset{OH}{|}}{\overset{OH}{\underset{|}{R-C-OR}}}$$

$$\rightleftharpoons \underset{\underset{OH}{|}}{\overset{OH}{\underset{|}{\underset{H^{\oplus}}{R-C-O-R}}}} \underset{ROH}{\overset{slow}{\rightleftharpoons}} \underset{RCOH}{\overset{OH^{\oplus}}{\|}} \underset{H_3O^{\oplus}}{\overset{H_2O}{\rightleftharpoons}} \underset{RCOH}{\overset{O}{\|}} \quad (3.43)$$

The mechanism given for this reaction is based on (1) first-order dependence of the rate on the oxonium ion concentration; (2) acyl-oxygen bond fission; (3) spectrophotometric observation of protonated esters and acids; (4) rate insensitivity to structural change in either the alkyl or acyl portion of the molecules; and (5) the simultaneous hydrolysis and carbonyl oxygen exchange.[25]

In concentrated sulfuric acid solutions, the rate of hydrolysis of ethyl acetate decreases rapidly[26,27] and then rises again with increase in acidity.[27] The decrease in rate may be explained in terms of low activity of water in these solutions, in an analogous fashion to the hydrolysis of amides discussed above. The rise can then only be explained by a change in mechanism from A-2 to A-1 in extremely concentrated sulfuric acid. The dependence of the rate of hydrolysis of the protonated ester on the activity of water is consistent with this interpretation.

Such changes in mechanism are met in many other acid-catalyzed reactions. For example, the hydrolyses of both methyl mesitoate and methyl benzoate follow eq. 3.20 (A-2) in dilute acid solution. However, in moderately strong acid solutions, the hydrolysis of the former changes to eq. 3.13 (A-1), as reflected in dependence on H_0, lack of oxygen exchange, a Hammett rho constant of -3.7, and a greater rate of hydrolysis than unhindered methyl

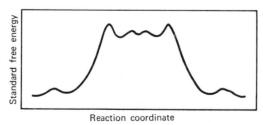

Fig. 3.7 Standard free energy versus reaction coordinate for the acid–catalyzed hydrolysis of an ester. From M. L. Bender, *Chem. Rev.*, **60**, 68 (1960). © The Williams and Wilkins Co., Baltimore, Md. 21202 U.S.A.

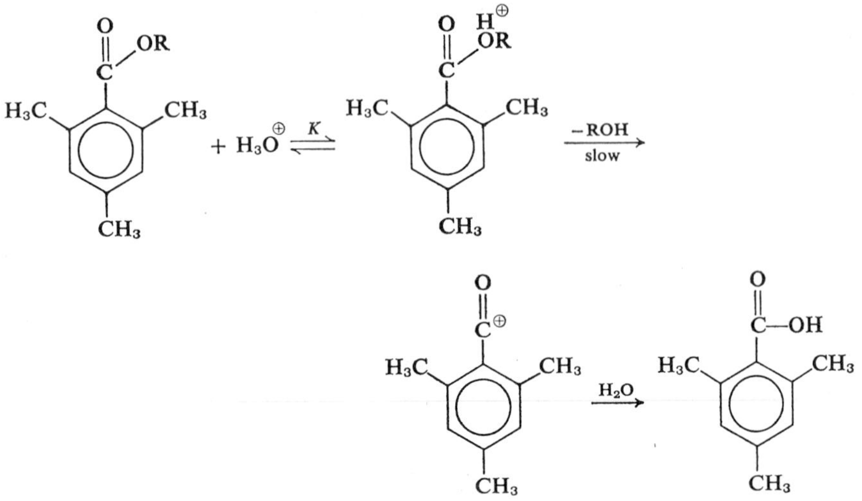

Oxonium and Hydroxide Ions in Catalysis

(3.44)

benzoate.[28,29] These data, together with the observation that methyl mesitoate ionizes to five (methyl benzoate gives two) particles in 100% sulfuric acid (acylium ion, oxonium ion, methyl hydrogen sulfate, and two bisulfate ions), suggest mechanisms 3.44 and 3.45 as those of the reactions of methyl mesitoate and methyl benzoate in strong acid solution.

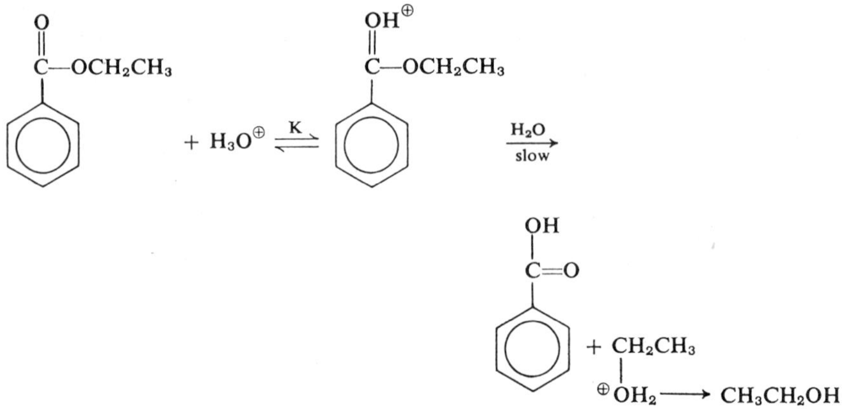

(3.45)

The hydrolysis of methyl and ethyl benzoates in 99.9% sulfuric acid illustrates another mechanistic change in acid-catalyzed hydrolysis of esters. The effect of structure on the rate constant for these reactions is shown in Fig. 3.8.[30] The Hammett rho constant for methyl benzoates is -3.7, the reaction probably following the acylium ion mechanism.

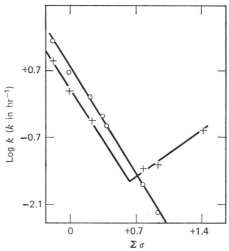

Fig. 3.8 The hydrolysis of methyl and ethyl benzoates in 99.9% sulfuric acid. From D. N. Kershaw and J. A. Leisten, *Proc. Chem. Soc.*, 84 (1960).

Ethyl benzoates appear to follow the same pattern from a Hammett sigma of -0.3 to 0.7, but above 0.7 the slope of the line changes abruptly to a reaction facilitated by electron-withdrawing substituents. Since this break occurs with an ethyl ester but not with a methyl ester, it is reasonable to assume that on the right-hand side of the break, ethyl benzoate reacts via eq. 3.45 with an acylium ion intermediate.

3.5 THE CATALYTIC FUNCTION OF THE PROTON

The proton is a convenient and powerful agent for the distortion of the electronic configuration of a substrate in order to facilitate reaction. The mechanism by which this process occurs has many variants. For example, a covalent bond may be more easily broken after protonation of one of the bonded atoms [31]

$$ROH_2^{\oplus} \longrightarrow R^{\oplus} + H_2O$$

is easier than

$$ROH \longrightarrow R^{\oplus} + OH^{\ominus} \qquad (3.46)$$

Another possibility is that addition of a nucleophile to an unsaturated linkage may occur more readily after protonation of one of the atoms (the more basic) of the double bond:

$$\underset{\substack{\|\\OH^{\oplus}}}{RCOR} + N^{\ominus} \longrightarrow R-\underset{\substack{|\\N}}{\overset{\substack{OH\\|}}{C}}-OR$$

is easier than

$$RCOR + N^\ominus \longrightarrow R-\underset{N}{\underset{|}{\overset{O^\ominus}{\overset{|}{C}}}}-OR \qquad (3.47)$$

where R—C—OR has O⁻ above and N below.

Finally, the abstraction of a proton from a molecule may be facilitated by the introduction of a proton at a different position in the molecule

$$R\overset{OH^\oplus}{\overset{\|}{C}}-CH_3 + B \longrightarrow R-\overset{OH}{\overset{|}{C}}=CH_2 + BH^\oplus$$

is easier than

$$R\overset{O}{\overset{\|}{C}}-CH_3 + B \longrightarrow R-\overset{O^\ominus}{\overset{|}{C}}=CH_2 + BH^\oplus \qquad (3.48)$$

Instead of distorting the electronic configuration, a proton can stabilize a leaving group and thus facilitate reaction. This description of protonic catalysis is similar to that in eq. 3.46 and can be illustrated by

$$RCX + H^\oplus \longrightarrow RC^\oplus + HX$$
$$\phi CH_2F + H^\oplus \longrightarrow \phi CH_2^\oplus + HF \qquad (3.49)$$

Distinguishing between these two descriptions of protonic catalysis may not be meaningful. The difference pertains to whether a pre-equilibrium protonation of the substrate occurs, for certainly in the transition state, eq. 3.46 and eq. 3.49 are identical. Thus since only the difference between the ground state and transition state of a reaction is meaningful, these two descriptions are identical. On either basis, acid catalysis of the hydrolyses of diisopropyl phosphorofluoridate and acetyl fluoride, but not of diisopropyl phosphorochloridate and acetyl chloride, is explicable,[32,33] since association of a proton with the fluorine compounds should occur more readily than with the chlorine compounds.

Finally, a proton is an electrophile that can add to a π-electron system to produce a reactive entity capable of further processes, as in eqs. 3.50 and 3.51.

$$\text{olefin} + H^\oplus \longrightarrow \text{carbonium ion} \begin{array}{l} \xrightarrow{+ ArH} \\ \xrightarrow{+ HOH} \\ \xrightarrow{+ \text{Nucleophile}} \\ \xrightarrow{- H^\oplus} \\ \xrightarrow{+ \text{olefin}} \\ \xrightarrow{t\text{-alkane}} \\ \xrightarrow{\text{rearrangement}} \end{array} \qquad (3.50)$$

$$\text{C}_6\text{H}_5\text{Y} + \text{H}^\oplus \longrightarrow [\text{C}_6\text{H}_6\text{Y}]^\oplus \qquad (3.51)$$

where Y stands for any electron-donating substituent

Equation 3.50 schematically represents an enormous number of reactions, including the polymerization and hydration of olefins and alkylation of aromatics. A particularly important subsequent reaction of the reactive carbonium ion intermediate is hydride transfer from a tertiary position of an alkane leading to a new carbonium ion. This hydride transfer is the basic mechanistic feature of the alkylation of paraffins with olefins, a reaction considered more fully in Chapter 5 under electrophilic catalysis.

Thus the rapid combination of a proton with almost any organic molecule leads to a multitude of possibilities for electronic distortion and facilitation of reaction. Probably the simplest analysis of protonic catalysis is an electrostatic one. The introduction of a proton into an organic molecule is little more than the introduction of a single positive charge. When a positive charge is introduced into an organic molecule, normal electrostatic effects must ensue. Reaction of the protonated species with a negatively charged nucleophile will be much faster because of coulombic interaction, and slower if the nucleophile is positively charged. If the nucleophile is a dipole, an ion-dipole interaction will occur. Thus, many of the qualitative observations of eqs. 3.46 and 3.51 may be quantitatively explained on the basis of electrostatic theory. In addition to simple electrostatic perturbation, the introduction of a positive charge into a π-electron system can cause a considerable electron delocalization that can also alter the reactivity profoundly.

In considering the function of the proton, the short-lived nature of protonated organic molecules must be kept in mind.[34] The mobility of the proton in the conductivity band of a substrate requires that a reaction surmount the activation barrier during the very short time in which the catalyzing proton is in the reaction complex. This restriction casts doubt on the sequence of events in eq. 3.43, for this equation (in reverse) would require that the heavy inert carboxylic acid and alcohol molecules approach to within bonding distance, rehybridize, and bond all within the short lifetime of $\text{RC(OH)}_2^\oplus$. These difficulties can be avoided if it is assumed that, simultaneously with the separation of the catalyzing proton from the reaction complex, the activation barrier of the reaction is surmounted. Initial addition of a proton to the carboxyl group of the acid (in the esterification direction) is followed by elimination of a proton from an alcohol molecule in the solvation shell of the acid with formation of an ion pair. This step involves participation of a catalyzing proton in the reaction complex and is extremely fast. The activation energy barrier of the reaction is reached and the transition state can be approximately described by the ion pair $RC(OH)_2^{\oplus}RO^{\ominus}$. Deactivation then takes place in a slower reaction step, the tetrahedral intermediate being formed in the absence of the catalyzing proton. The proton rearrangement thus proceeds so rapidly that the heavy molecular skeletons retain their initial spatial positions (the Franck-Condon principle for proton-catalyzed reactions). Decomposition of the intermediate, with formation of either the starting materials or the end products, proceeds by a mechanism similar to that for formation.

3.6 HYDROXIDE ION CATALYSIS

As in acid catalysis, two types of base catalysis are seen: (1) specific hydroxide ion (lyoxide) ion catalysis and (2) general-base catalysis. Only in the former catalysis is the hydroxide (lyoxide) ion independent of the source of the ion.

Some kinetic schemes that satisfy specific hydroxide ion catalysis are given below. Pre-equilibrium removal of a proton from a substrate by a base may occur, followed by reaction of the substrate anion to give products, with or without another reactant.

$$HS + B \underset{}{\overset{K}{\rightleftharpoons}} S^{\ominus} + BH^{\oplus} \xrightarrow[k]{\text{slow}} \text{products} \quad (3.52)$$

$$HS + B \underset{}{\overset{K}{\rightleftharpoons}} S^{\ominus} + BH^{\oplus} \xrightarrow[k]{R \text{ (slow)}} \text{products} \quad (3.53)$$

The kinetic equation for reaction 3.52 is given by

$$\text{rate} = k[S^{\ominus}] \quad (3.54)$$

Assuming a fast pre-equilibrium this can be transformed to

$$\text{rate} = \frac{kK[SH][B]}{[BH^{\oplus}]} \quad (3.55)$$

The ratio $[B]/[BH^{\oplus}]$ is, of course, related to the hydroxide ion concentration through

$$B + H_2O \overset{K_B}{\rightleftharpoons} BH^{\oplus} + OH^{\ominus} \quad (3.56)$$

Thus the rate expression for eq. 3.52 can be written as

$$\text{rate} = \frac{kK}{K_B}[SH][OH^{\ominus}] \quad (3.57)$$

The rate equation corresponding to mechanism 3.53 is analogous to eq. 3.57, with an added term in the concentration of R.

Hydroxide Ion Catalysis

In addition to the mechanisms of specific hydroxide ion catalysis involving the removal of a proton from the substrate, a mechanism involving the addition of hydroxide ion to the substrate is possible (see Chapter 6).

$$\text{HS} + \text{OH}^\ominus \underset{k_{-1}}{\overset{k_1}{\rightleftarrows}} \text{HS—OH}^\ominus \xrightarrow{k_2} \text{products} \tag{3.58}$$

The rate equation corresponding to eq. 3.58 is

$$\text{rate} = \frac{k_1 k_2}{k_{-1} + k_2} [\text{SH}][\text{OH}^\ominus] \tag{3.59}$$

For all three mechanisms, 3.52, 3.53, and 3.58, the observed rates are proportional to the hydroxide ion concentration of the solvent; other bases present do not influence the experimentally determined rates.

Corresponding to the generalization that all organic molecules are bases is the generalization that all organic molecules are weak acids. Table 3.5 presents the pK_a of a few carbon acids whose pK_a's range from 10 to 84; it presents acidities of a series of hydrocarbons based on a molecular orbital calculation. For the first three members of this series, fair agreement of the calculated values with experiment is found.

Corresponding to the acidity functions in strong acid solutions,[35] basicity functions have been devised for strongly basic solutions, including those of sodium hydroxide, potassium hydroxide, hydrazine, ethylenediamine, and ethanolamine in water. Studies of a series of weakly acidic organic indicators in these basic solutions led to the basicity function $H_\ominus$ (see eq. 3.61). The indicator equilibrium observed is

$$\text{HIn} + \text{OH}^\ominus \rightleftharpoons \text{In}^\ominus + \text{H}_2\text{O} \tag{3.60}$$

Table 3.5
Acidities of Hydrocarbons[35]

Hydrocarbon	pK_a
Fluoradene	10
Cyclopentadiene	14
Indene	23
Fluorene	31
4,5-Methylenephenanthrene	31
Cycloheptatriene	45
2-Methylanthracene	57
Toluene	59
Methane	84

Reprinted from *J. Amer. Chem. Soc.*, **85**, 1761 (1963). © 1963 by the American Chemical Society. Reprinted by permission of the copyright owner.

In analogous fashion to the definition of the Hammett acidity function, $H_\ominus$ is defined by:

$$H_\ominus = -\log \frac{a_{H_2O} f_{In^\ominus}}{a_{OH^\ominus} f_{HIn}} = -\log a_{H^\oplus} \frac{f_{In^\ominus}}{f_{HIn}} \quad (3.61)$$

In dilute solution the acidity function $H_\ominus$ becomes identical with pH, so that the pH scale can be extended on the alkaline side of neutrality as it was previously extended on the acid side. Values of $H_\ominus$ as high as 19 can be achieved in solutions of a tetraalkylammonium hydroxide in water, sulfolane-water, pyridine-water, and dimethyl sulfoxide-water solutions.[37,38] Of the media tested, the dimethyl sulfoxide-water mixture gives the highest $H_\ominus$ at given stoichiometric water and hydroxide ion concentrations.

Two basicity scales are needed to reflect the ability of a base either to remove a proton from an organic compound or to add to the organic compound to give an addition intermediate. To this end, a $J_\ominus$ basicity function has been defined (eq. 3.62) to supplement $H_\ominus$. Indicators that reflect the $H_\ominus$

$$J_\ominus = -\log \frac{f_{InOH}}{f_{In} a_{OH^\ominus}} \quad (3.62)$$

and $J_\ominus$ scales are 2,4-dinitroaniline and N,N-dimethyl-2,4,6-trinitroaniline, respectively.[39]

Compared to the many correlations between rate and acidity for acid-catalyzed reactions, there are few correlations between rate and basicity for base-catalyzed reactions. One correlation describes the rate of racemization of (+)-2-methyl-3-phenylpropionitrile in dimethyl sulfoxide-methanol solutions containing methoxide ion.[40] Increases by factors as great as 10^9 in the rate of this racemization were found on substituting dimethyl sulfoxide for methanol as solvent. The basicity function $H_\ominus$ was determined over the entire range of methanol-dimethyl sulfoxide compositions (Fig. 3.9). An excellent linear correlation was found between $\log k_{racemization}$ and $H_\ominus$ with a slope of 0.87, over a range of 10^6-fold change in rate constant.

The mechanism of this reaction may be written as

$$AH + B^\ominus \underset{k_{-a}}{\overset{k_a}{\rightleftarrows}} \underset{\substack{\text{optically}\\\text{active}}}{A^\ominus \text{---} HB} \overset{k_b}{\longrightarrow} \underset{\text{racemic}}{A^\ominus \text{---} HB} \overset{k_{-a}}{\longrightarrow} A\text{---}H + B^\ominus \quad (3.63)$$

The rate constant for eq. 3.63 is given by

$$k_{obs} = k_a \frac{k_b}{k_{-a} + k_b} \quad (3.64)$$

If k_b and $k_{-a} \gg k_a$ and $k_{-a} \gg k_b$, a pre-equilibrium followed by a rate-determining step occurs and eq. 3.64 reduces to

$$k_{obs} = K k_b \quad (3.65)$$

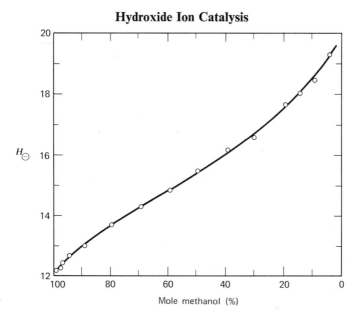

Fig. 3.9 The acidity function $H_\ominus$ in dimethyl sulfoxide-methanol solutions. From R. Stewart, J. P. O'Donnell, D. J. Cram, and B. Rickborn, *Tetrahedron*, **18**, 917 (1962).

If, however, k_b and $k_{-a} \gg k_a$ but $k_b \gg k_{-a}$ a slow proton transfer occurs and eq. 3.64 reduces instead to

$$k_{obs} = k_a \qquad (3.66)$$

The above correlation between $\log k$ and $H_\ominus$ and other independent kinetic evidence,[40] favors eq. 3.65, since the transition state of this process resembles in structure the nitrogen anions used as indicators in the establishment of the $H_\ominus$ scale. Conversely, the transition state corresponding to eq. 3.66 involves an anion whose charge is distributed on both carbon and oxygen atoms.

The above argument can be put on a quantitative basis in the following way. Equation 3.67 describes the effect of activity coefficients on the pre-equilibrium mechanism.

$$\frac{-d[\text{AH}]}{dt} = k_b[\text{A}^\ominus\text{---HB}]\frac{f_{\text{A---HB}}}{f^\ddagger} \qquad (3.67)$$

A definition of $H_\ominus$ can be given in terms of the species

$$H_\ominus = pK - \log \frac{[\text{AH}]}{[\text{A}^\ominus\text{---HB}]} \qquad (3.68)$$

Combining eqs. 3.67 and 3.68 leads to

$$\frac{-d[\text{AH}]}{dt} = \frac{k_b K[\text{AH}]}{h_\ominus} \cdot \frac{f_{\text{A}\ominus\text{---HB}}}{f\ddagger} \qquad (3.69)$$

If the nitrile, AH, is only slightly ionized, its actual concentration equals its stoichiometric concentration and thus eq. 3.70 may be written, which may be integrated to give eq. 3.71

$$k_{\text{obs}} = \frac{d[\text{AH}]}{[\text{AH}]\,dt} \frac{k_b K}{h_\ominus} \cdot \frac{f_{\text{A}\ominus\text{---HB}}}{f\ddagger} \qquad (3.70)$$

$$\log k_{\text{obs}} = H_\ominus + \log \frac{f_{\text{A}\ominus\text{---HB}}}{f\ddagger} \qquad (3.71)$$

Equation 3.71 predicts that a plot of log k_{obs} versus $H_\ominus$ should be linear with unit slope provided that the activity coefficient ratio in the equation is constant.

For the mechanism corresponding to eq. 3.66 the following rate law can be derived

$$k_{\text{obs}} = \frac{d[\text{AH}]}{[\text{AH}]\,dt} = k_a[\text{B}^\ominus]\frac{f_{\text{AH}} f_{\text{B}\ominus}}{f\ddagger} \qquad (3.72)$$

This equation will give a linear plot of log k_{obs} versus $H_\ominus$ only if the term $[\text{B}^\ominus]f_{\text{AH}}f_{\text{B}\ominus}/f\ddagger$ is a linear function of $H_\ominus$. There is no reason to believe that this would be so and therefore the linear correlation of log k with $H_\ominus$ in this racemization is reasonable evidence to favor mechanism 3.63, with the proviso that the first step is a fast pre-equilibrium and the second step is rate-determining.

Specific hydroxide (lyoxide) catalysis is seen in many classical organic reactions, including the dealdolization of diacetone alcohol, the Claisen, Michael, Perkin, and aldol condensations, and the bromination of β-disulfones and β-dinitriles. All of these reactions follow specific hydroxide ion catalysis according to eqs. 3.52 or 3.53. The alkaline hydrolysis of an ester follows specific hydroxide ion catalysis according to eq. 3.58.

Aldol condensation illustrates some of the features of specific hydroxide ion catalysis. The mechanism of this reaction may be written as

$$\text{B} + \text{CH}_3\text{CHO} \underset{k_{-1}}{\overset{k_1}{\rightleftarrows}} {}^\ominus\text{CH}_2\text{—CHO} + \text{BH}^\oplus$$

$${}^\ominus\text{CH}_2\text{CHO} + \text{CH}_3\text{CHO} \xrightarrow{k_2} \text{products} \qquad (3.73)$$

At low acetaldehyde concentrations, evidence for this mechanism includes the following. (1) The rate is proportional to the second power of the acetaldehyde concentration and to the first power of the hydroxide ion concentration. (2) When the reaction is carried out in D_2O, deuterium atoms are incorporated into the reactant at a rate greater than that of the overall reaction. (3) Variation in structure indicates that the rate of reaction is a function of both the ease of enolate ion formation and the ease of addition to various carbonyl compounds. At very high (10 M) acetaldehyde concentrations, the reaction becomes general base-catalyzed (Chapter 4). The rate becomes proportional to all bases in the medium and to the first power of the acetaldehyde concentration, and no deuterium is incorporated into the reactant in D_2O solution. This change from specific hydroxide ion catalysis to general basic catalysis is consistent with mechanism 3.73, since at some high acetaldehyde concentration, step 2 must become faster than either step 1 or its reverse and then the initial proton transfer becomes rate-determining.[42]

The dealdolization of diacetone alcohol shows another complication affecting specific hydroxide ion-catalyzed reactions. The mechanism of this reaction is given by

$$\underset{\substack{|\\CH_3}}{CH_3-\underset{|}{C}-CH_2-\overset{O}{\overset{\|}{C}}-CH_3} \;\underset{BH^{\oplus}}{\overset{B}{\rightleftharpoons}}\; \underset{\substack{|\\CH_3}}{CH_3-\underset{|}{\overset{O^{\ominus}}{C}}-CH_2-\overset{O}{\overset{\|}{C}}CH_3}$$

$$\downarrow \text{slow}$$

$$CH_3\overset{O}{\overset{\|}{C}}CH_3 \;\underset{BH^{\oplus}}{\overset{B}{\rightleftharpoons}}\; \underset{\substack{|\\CH_3}}{CH_3-\overset{O}{\overset{\|}{C}}} + {}^{\ominus}CH_2-\overset{O}{\overset{\|}{C}}-CH_3 \qquad (3.74)$$

Despite the definition of specific hydroxide ion catalysis, primary and secondary amines catalyze the reaction, while tertiary amines do not. This behavior cannot be attributed to general basic catalysis, for all bases should catalyze the reaction if that were the case. Therefore, the effect of primary and secondary amines must represent some form of specific amine catalysis. As will be discussed in Chapter 6, this special situation is explicable in terms of imine (Schiff base) formation between diacetone alcohol and the amines.

3.7 THE CATALYTIC FUNCTION OF THE HYDROXIDE ION

Although the hydroxide ion does not have the unique characteristics of small size per unit of charge that the proton has, it serves a powerful and important function in catalysis. The function of the hydroxide ion is the reversible removal of a proton from the substrate, usually an organic molecule, or alternatively, the addition of hydroxide ion to the substrate to form an unstable addition intermediate that decomposes to form products (eqs. 3.52, 3.53, and 3.58). In contrast to oxonium ion catalysis, acceleration by hydroxide ion is often not true catalysis since the hydroxide ion is usually not regenerated.

Carbanions, often stabilized by resonance, are formed in many hydroxide ion reactions. These intermediates may lead to other intermediates or give products in a variety of ways; some possibilities are shown in eqs. 3.73 and 3.75 to 3.77.

$$R_2\overset{\ominus}{C}X \xrightarrow{-X^{\ominus}} R_2C: \longrightarrow \text{products} \tag{3.75}$$

$$\underset{R_2C-\overset{\ominus}{C}R_2}{\overset{X}{|}} \xrightarrow{-X^{\ominus}} R_2C=CR_2 + X^{\ominus} \tag{3.76}$$

$$\underset{\text{benzyne}}{\text{[C}_6\text{H}_4\text{X}^{\ominus}\text{]} \xrightarrow{-X^{\ominus}} \text{[benzyne]} \longrightarrow \text{products}} \tag{3.77}$$

In these processes the function of hydroxide ion is the same, that is, the initial removal of a proton from the substrate, thereby introducing a negative charge into it. The effect of this negative charge on subsequent reactivity may be considered from the electrostatic and resonance points of view given earlier for the introduction of a positive charge into a substrate.

3.8 LYOXIDE CATALYSIS IN SOLVENTS OTHER THAN WATER

Using the glass electrode (a potentiometric method) the acidities of over 20 carbon acids were determined. The data provide an explanation of discrepancies between the relative acidities in dimethyl sulfoxide solution and in cyclohexylamine solution.[43]

Amide ion in liquid ammonia is a powerful basic catalyst whose action parallels that of hydroxide ion in water. Amide ion is both a stronger base and a more reactive nucleophile than hydroxide ion; these properties make it valuable in many of the reactions discussed above where a stronger catalytic species is needed. This system has been of considerable utility in the polymerization of styrene[46] and especially of acrylonitrile and of methyl methacrylate, whose substituents stabilize the anionic charge of the intermediate formed by addition of amide ion to the monomer of the substrate and analogous intermediates formed by the addition of this species to other monomers.

Other lyoxide ion catalysts include sodium ethoxide in ethanol and sodium t-butoxide in t-butyl alcohol.

In dimethyl sulfoxide or hexamethylphosphoramide, t-butoxide-catalyzed additions of aromatic heterocycles to conjugated hydrocarbons (homogeneous alkylation) have been achieved.[44] In addition, the base-catalyzed isomerization of alkynes in ethanol containing some potassium hydroxide is known.[45]

The proton-abstracting ability of these systems is midway between the hydroxide ion-water and the sodium amide-ammonia systems. Another proton-abstracting species is the anion of dimethyl sulfoxide.

Rearrangement of 2,2,3-triphenylpropyllithium to 1,1,3-triphenylpropyllithium in tetrahydrofuran solution in the presence of various organolithium reagents, RLi (phenyllithium, ethyllithium, n-butyllithium, isopropyllithium, and benzyllithium-α-C^{14}, occurs by cleavage into benzyllithium and 1,1-diphenylethene followed by recombination to give 1,1,3-triphenylpropyllithium. Products (RCH_2CPh_2Li and $PhCH_2Li$) of reaction of the added organolithium reagent with 1,1-diphenylethene were detected only in the case of isopropyllithium and benzyllithium: that is, of the reagents tested, only isopropyllithium was able to compete with benzyllithium for the intermediate 1,1-diphenylethene. The direct addition of organolithium reagents to 1,1-diphenylethene under the general conditions of rearrangement was examined and from these results and from those on rearrangement it was concluded that the reactivity of organolithium compounds toward 1,1-diphenylethene increases along the series PhLi < EtLi, n-BuLi ≪ i-PrLi < $PhCH_2Li$.[10] Rearrangement of 2,2,2-triphenylethyllithium to 1,1,2-triphenylethyllithium in tetrahydrofuran in the presence of phenyllithium-C^{14} or benzyllithium occurs without detectable incorporation of radioactive phenyl or benzyl in the products. Rearrangement of 2,2,2-triphenylethyllithium must therefore occur by an intramolecular mechanism. Similarly, both Stevens and Sommelet rearrangements of dibenzyldimethylammonium halide brought about by benzyllithium-α-C^{14} in tetrahydrofuran occur without detectable incorporation of radioactivity into the dimethyl(1,2-diphenylethyl)amine and dimethyl(o-methylbenzhydryl)amine produced and are therefore judged to be intramolecular rearrangements.[47]

While the use of strong bases as catalysts for organic reactions has been a powerful synthetic tool, the mechanisms of these reactions, especially in nonaqueous solution, are only now being investigated in a definitive manner.[47-49] Organometallic compounds or alkali metals, for example, have strongly basic properties. They are able to abstract allylic or benzylic protons from hydrocarbons. The carbanions thus formed are capable of undergoing profound changes such as isomerization, reaction with other olefins, and cyclizations. The future holds promise for further profound rate accelerations in this area of organic chemistry.

REFERENCES

1. W. Ostwald, *Phys. Z.*, **3**, 313 (1902).
2. F. H. Field et al., *J. Amer. Chem. Soc.*, **87**, 3294 (1965) and earlier articles.
3. G. A. Olah and R. H. Schlosberg, *ibid.*, **90**, 2726 (1968).

4. R. Stewart and K. Yates, *ibid.*, **84**, 405 (1960).
5. R. J. Gillespie and T. Birchall, *Can. J. Chem.*, **41**, 148, 2642 (1963).
6. L. P. Hammett and A. J. Deyrup, *J. Amer. Chem. Soc.*, **54**, 2721 (1932); L. P. Hammett, *Physical Organic Chemistry*, McGraw-Hill Book Co., New York, 1940, p. 263.
7. R. P. Bell, *The Proton in Chemistry*, Cornell University Press, Ithaca, N.Y., 1959, p. 41.
8. N. C. Deno, P. T. Grove, and G. Sims, *J. Amer. Chem. Soc.*, **81**, 5790 (1951); N. C. Deno, J. Jaruzelski, and A. S. Schriesheim, *ibid.*, **77**, 3044 (1955).
9. V. Gold and B. W. V. Hawes, *J. Chem. Soc.*, 2102 (1951); V. Gold, *ibid.*, 1263 (1955).
10. E. M. Arnett, *Progr. Phys. Org. Chem.*, I, 239 (1963).
11. L. Zucker and L. P. Hammett, *J. Amer. Chem. Soc.*, **61**, 2791 (1939).
12. F. A. Long and M. A. Paul, *Chem. Rev.*, **57**, 938 (1957); C. Perrin, *J. Amer. Chem. Soc.*, **86**, 256 (1964).
13. R. W. Taft, Jr., N. C. Deno, and P. S. Skell, *Ann. Rev. Phys. Chem.*, **9**, 287 (1958).
14. A. J. Kresge and Y. Chiang, *J. Amer. Chem. Soc.*, **81**, 5509 (1959); **83**, 2877 (1961).
15. F. A. Long and J. Schulze, *J. Amer. Chem. Soc.*, **83**, 3340 (1961).
16. J. F. Bunnett, *J. Amer. Chem. Soc.*, **83**, 4956, 4968, 4973, 4978 (1961).
17. K. Yates and J. B. Stevens, *Can. J. Chem.*, **43**, 529 (1965).
18. See A. J. Kresge, R. A. More O'Farrell, L. E. Hakka, and V. P. Vatullo, *Chem. Comm.*, 46 (1965), for another treatment of the relationship between kinetic and equilibrium acidity dependencies in general acid-catalyzed aromatic hydrogen exchange.
19. C. K. Ingold, *Structure and Mechanism in Organic Chemistry*, Cornell University Press, Ithaca, N.Y., 1953, Chap. XIV.
20. F. A. Long and M. A. Paul, *Chem. Rev.*, **57**, 935 (1957).
21. A. M. Wenthe and E. H. Cordes, *J. Amer. Chem. Soc.*, **87**, 3173 (1965).
22. F. H. Westheimer and M. S. Kharasch, *J. Amer. Chem. Soc.*, **68**, 1871 (1946).
23. J. T. Edward and S. C. R. Meacock, *J. Chem. Soc.*, 2000 (1957).
24. J. A. Leisten, *J. Chem. Soc.*, 765 (1959).
25. M. L. Bender, *Chem. Rev.*, **60**, 53 (1960); see, however, S. A. Shain and J. F. Kirsch, *J. Amer. Chem. Soc.*, **90**, 5848 (1968).
26. C. A. Lane, *J. Amer. Chem. Soc.*, **86**, 2521 (1964).
27. D. Jaques, *J. Chem. Soc.*, 3874 (1965).
28. C. T. Chmiel and F. A. Long, *J. Amer. Chem. Soc.*, **78**, 3326 (1956).
29. M. L. Bender, H. Ladenheim, and M. C. Chen, *ibid.*, **83**, 123 (1961).
30. D. M. Kershaw and J. A. Leisten, *Proc. Chem. Soc.*, 84 (1960); see also K. Yates and R. A. McClelland, *J. Amer. Chem. Soc.*, **89**, 2686 (1967); G. J. Harvey and V. R. Stimson, *Aust. J. Chem.*, **15**, 757 (1962).
31. For extreme examples not involving the bond that is broken see W. von E. Doering and V. Z. Pasternak, *J. Amer. Chem. Soc.*, **72**, 143 (1950); L. A. Cohen and W. M. Jones, *J. Amer. Chem. Soc.*, **82**, 1907 (1960).
32. I. Dostrovsky and M. Halmann, *J. Chem. Soc.*, 502, 516 (1953).
33. C. W. L. Bevan and R. F. Hudson, *J. Chem. Soc.*, 2187 (1953).
34. H. Zimmerman and J. Rudolph, *Angew. Chem., Int. Ed.*, **4**, 40 (1965).
35. A. Streitwieser, Jr., W. C. Langworthy, and J. I. Brauman, *J. Amer. Chem. Soc.*, **85**, 1761 (1963).
36. M. Paul and F. A. Long, *Chem. Rev.*, **57**, 1 (1957); C. H. Rochester, *Quart. Rev.*, **20**, 511 (1966).

37. C. Langford and R. L. Burwell, Jr., *J. Amer. Chem. Soc.*, **82**, 1503 (1960).
38. R. Stewart and J. P. O'Donnell, *J. Amer. Chem. Soc.*, **84**, 493 (1962).
39. M. R. Crampton and V. Gold, *Proc. Chem. Soc.*, 298 (1964).
40. R. Stewart, J. P. O'Donnell, D. J. Cram, and B. Rickborn, *Tetrahedron*, **18**, 917 (1962).
41. D. J. Cram, C. A. Kingsbury, and B. Rickborn, *J. Amer. Chem. Soc.*, **83**, 3688 (1961).
42. R. P. Bell, *The Proton in Chemistry*, Cornell University Press, Ithaca, N.Y., 1959, p. 137.
43. C. D. Ritchie and R. E. Wechold, *J. Amer. Chem. Soc.*, **89**, 1721, 2752 (1967).
44. H. Pines and W. M. Stalick, *Tetrahedron Lett.* **34**, 3726 (1963).
45. Y. Tekeuchi, Ph.D. dissertation, Northwestern University, Evanston, Ill., 1968.
46. E. Grovenstein, Jr., and G. Wentworth, *J. Amer. Chem. Soc.*, **89**, 1852 (1967).
47. F. R. Mayo and C. Walling, *Chem. Rev.*, **46**, 191 (1950).
48. D. J. Cram, *Carbanions in Organic Chemistry*, W. A. Benjamin, Inc., New York, 1965.
49. H. Pines and L. Schaap, *Adv. Catalysis*, **12**, 117 (1960).
50. E. Grovenstein and G. Wentworth, *J. Amer. Chem. Soc.*, **89**, 1852 (1967).

Chapter 4

GENERAL ACID-BASE CATALYSIS: THEORY

4.1	The Brønsted Catalysis Law	74
4.2	The Relation of the Brønsted Catalysis Law to Proton Transfer	76
4.3	The Meaning of the Brønsted Catalysis Law	82
4.4	The Meaning of α and β	85
4.5	The Relation of General Catalysis to Specific Catalysis	90

Up to this point, we have confined our discussion to catalysis of reactions by the ions characteristic of the solvent; since we usually refer to aqueous solutions, these catalytic ions are oxonium and hydroxide. The simplest examples of such catalyses are those effected by strong acids or bases, such as hydrochloric acid or potassium hydroxide; the sole function of these compounds is to provide oxonium or hydroxide ions, which are the actual catalysts. It is important to recognize that the only acidic or basic species present in these solutions are the solvent and the ions related to it.

Catalysis by weak acids or bases can, and often does, occur in the same way; the actual catalysts are again the solvent-related ions resulting from the dissociation of the acid or base, although this dissociation is incomplete. The only important consequence of the partial dissociation of these acids and bases is that, mole for mole, they do not provide as high a concentration of oxonium or hydroxide ion as do the strong electrolytes. Although the system contains acidic or basic species (the unionized acid or base) not related to the solvent, these entities are not catalytically active. In such cases, as long as the oxonium and hydroxide ion concentrations are not changed, variation in the concentration of the weak acid or base does not affect the reaction velocity; in other words, the rate is insensitive to the buffer *concentration* (at constant ionic strength—see Chapter 7), since it is the buffer *ratio* that governs the pH.

Catalysis of this kind, in which the only effective species are the ions related to the solvent, is defined as *specific* acid or base catalysis.

In numerous reactions catalyzed by weak acids or bases, however, the rates can be shown to depend not only on the concentrations of the solvent ions, but also upon the concentrations of all other proton donors or acceptors present. Classical examples of such reactions include the decomposition of nitramide, the iodination of acetone, and the mutarotation of glucose. Catalysis of this kind is defined as *general* acid or base catalysis; it is experimentally distinguished from specific catalysis by the presence in the rate expression of terms involving the concentration of every proton donor and acceptor present. The general forms of such rate expressions are shown in eqs. 4.1 and 4.2:

$$\text{rate} = \sum^{i} k_i^b [B_i][\text{substrate}] \quad \text{for general base catalysis} \quad (4.1)$$

$$\text{rate} = \sum^{i} k_j^a [HA_j][\text{substrate}] \quad \text{for general acid catalysis} \quad (4.2)$$

For example, the rate law for a general acid-base-catalyzed reaction in an acetate buffer solution is

$$\text{rate} = k_{\text{cat}}[S]$$
$$= \{k_0 + k_H[H_3O^\oplus] + k_{OH}[OH^\ominus] + k_{HOAc}[HOAc] + k_{OAc}[OAc^\ominus]\}[S] \quad (4.3)$$

The term k_0 involves the solvent water presumably acting either as an acid or a base catalyst. A complicated system such as eq. 4.3 can be analyzed experimentally in the following way. From reactions carried out in solutions sufficiently acidic that $[OH^\ominus]$ and $[OAc^\ominus]$ are negligible, a plot of the observed rate constant versus the oxonium ion concentration leads to a straight line of slope k_H and intercept k_0. From reactions carried out in solutions sufficiently alkaline that $[H^\oplus]$ and $[HOAc]$ are negligible, a plot of k_{obs} versus hydroxide ion concentration leads to a straight line of slope k_{OH} and intercept k_0. From reactions carried out in acetate buffer solutions, the remaining two rate constants, k_{HOAc} and $k_{OAc^\ominus}$ can be determined. A plot of the observed rate constants versus the acetate ion concentration for such a reaction is shown in Fig. 4.1 at two different values of r, defined as the ratio $[HOAc]/[OAc^\ominus]$. The slopes of the two lines in Fig. 4.1 are $r_1 k_{HOAc} + k_{OAc^\ominus}$

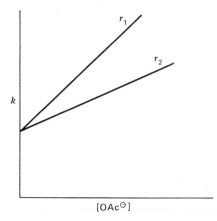

Fig. 4.1 The rate constant of a reaction catalyzed by both acetic acid and acetate ion.

and $r_2 k_{HOAc} + k_{OAc^\ominus}$. The individual rate constants can be calculated from the slopes of the two lines.

In simple systems (e.g., if k_{HOAc} or $k_{OAc^\ominus}$ is zero) the involvement of a general acid or general base catalyst in a reaction may be deduced from the pH dependence of the reaction (Fig. 3.1). The stoichiometry of a prototropic species may be expressed as $[HA]_T = [HA] + [A]$. Combining this conservation equation with that for the ionization of the acid leads to

$$[HA] = \frac{[HA_T]}{1 + \frac{K_a}{[H_3O^\oplus]}} \tag{4.4}$$

$$[A^\ominus] = \frac{[HA_T]}{1 + \frac{[H_3O^\oplus]}{K_a}} \tag{4.5}$$

where K_a is the dissociation constant of HA. Consequently a reaction rate dependent on HA, when graphed as k versus pH, will have the sigmoid shape of a titration curve for an acid; likewise a rate dependent on $A^\ominus$, graphed as k versus pH, has the sigmoid shape of a titration curve for a base. In logarithmic terms, these curves have the form of Fig. 3.1 f and g.

4.1 THE BRØNSTED CATALYSIS LAW

A fundamental question in catalysis is the relationship between the structure of the catalyst and its catalytic activity. In general acid or base catalysis, a pertinent measure of structure is the pK_a of the catalyst. It is therefore of interest to consider the relationship between the pK_a of a catalyst and its catalytic rate constant. Very early in the development of general acid-base catalysis, a relationship of this kind was proposed by Brønsted and Pedersen. The Brønsted catalysis law may be expressed either by

$$k_A = G_A K_a^{\alpha} \quad (4.6a)$$

where A and α refer to an acid and G is a constant,

$$k_B = G_B (1/K_a)^{\beta} \quad (4.6b)$$

where B and β refer to a base, and where K_a is the conventional acid dissociation constant of the acid A or of the acid conjugate to the base B, or by

$$\log k_A = -\alpha p K_a + \text{constant}$$
$$\log k_B = \beta p K_a + \text{constant} \quad (4.7)$$

Equations 4.7 indicate that a plot of $\log k_A$ versus pK_a should be linear with slope $-\alpha$, while a plot of $\log k_B$ versus pK_a should be linear with slope $+\beta$. For general acids or bases containing more than one ionizing group, it

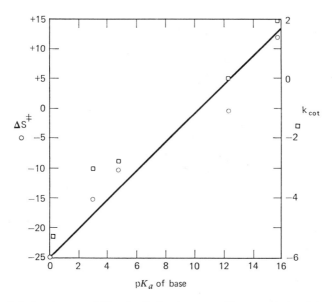

Fig. 4.2 Relation between $\Delta S^{\ddagger}$ and pK_a for the general base catalysis of the mutarotation of glucose using H_2O, $HCO_2^{\ominus}$, $CH_3CO_2^{\ominus}$, $C_6H_{11}O_6^{\ominus}$, and $OH^{\ominus}$. From H. Schmid, *Chem. Zeitung*, **90**, 35 (1966).

must be realized that the ionization constants are usually influenced by each other. When such compounds are to be treated by the Brønsted relationship, statistical corrections must be applied; details of such corrections are discussed by Brønsted.[2]

Figure 4.2 shows a plot of eq. 4.7 for some data on general base catalysis of the mutarotation of glucose. A fair linear correlation is observed over 16 pK units and approximately 7 powers of 10 in the rate constants. The slope β of the line is 0.44. (Many other Brønsted plots are known, some with considerably more data than Fig. 4.2—see later.)

This plot is shown here in order to show that further correlations of the catalytic activity may be made. For example, Fig. 4.2 indicates that a reasonable correlation exists between the pK_a of the base and the entropy of activation of the reaction.[3] This correlation implies that the enthalpy of activation is a constant independent of the pK_a of the base, since the free energy of activation (which is proportional to log k) and the entropy of activation give parallel correlations. The catalytic efficiency of a series of bases or acids reflects the entropies of activation in the iodination of acetone[4] and the hydrolysis of trifluoro-N-methylacetanilide.[5] This correlation is by no means universal, however; several other general catalyses such as the decomposition of nitramide[2] and the bromination of acetoacetic ester show no such correlation.

4.2 THE RELATION OF THE BRØNSTED CATALYSIS LAW TO PROTON TRANSFER

The Brønsted catalysis law is a relation between an equilibrium constant and a rate constant and thus constitutes a relation between the free energy of a reaction and the free energy of activation of another reaction. Since the Brønsted catalysis law describes reactions involving proton transfer, let us review the previously discussed equilibrium-rate relationships involving proton transfer in simple systems.

(1) Reactions of nitrogen or oxygen bases with oxonium ion, or reactions of similar acids with hydroxide ion, are diffusion-controlled. This behavior is illustrated in Figs. 4.3 and 4.4. (2) The reactions of nitrogen or oxygen bases or similar acids with water, show a linear relation between the log of the rate constant of proton transfer and the pK of the base or acid. As pointed out earlier, the reason for this dependence is that the reverse reaction corresponding to each of these processes is diffusion-controlled. (3) The reactions of carbon acids with hydroxide ion and the reactions of carbanions with oxonium ion do not follow the simple relationships of (1) or (2). The complications inherent in carbon acids and bases perturb the simple relation between the log of the rate constant for proton transfer and the pK of carbon acid or base

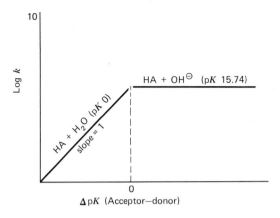

Fig. 4.3 The reactions of oxygen and nitrogen acids, HA, with water or with hydroxide ion.

for a series of ketones (Fig. 4.5). The rate of reaction of a ketone with hydroxide ion never approaches the rate of a diffusion-controlled process, no matter how acidic the ketone is. On the other hand, the rate of reaction of a carbanion with an oxonium ion to give an enol does approach the rate of a diffusion-controlled reaction. Even though a linear relation does not exist between the log of the rate constant and the pK of the ketone or carbanion, a continuous relation is seen in Fig. 4.5.[6,7] (4) The reactions of general acids with general bases of the oxygen or nitrogen variety exhibit a linear $\log k - \Delta pK$ relationship (Figs. 2.3, 2.4, and 2.5). (5) The reactions of acetylacetone with general bases, reactions characteristic of carbon acids, exhibit a relationship

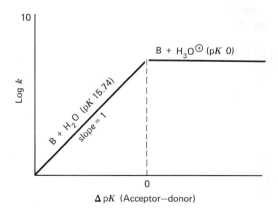

Fig. 4.4 The reactions of oxygen and nitrogen bases, B, with water or with oxonium ion.

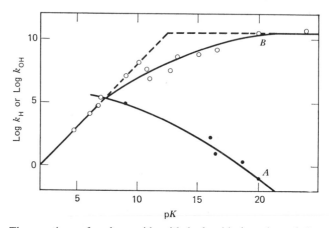

Fig. 4.5 The reactions of carbon acids with hydroxide ion, A, and the reactions of carbanions with oxonium ion to form enols, B. From M. Eigen, *Angew. Chem. Int. Ed.*, **3**, 12 (1964); R. G. Pearson and R. L. Dillon, *J. Amer. Chem. Soc.*, **75**, 2439 (1953).

between $\log k$ and ΔpK_a which shows continuous curvature over a wide range of ΔpK (see Fig. 2.6).

On the basis of these relations for proton transfer reactions exhibiting $\log k$ versus ΔpK_a curves with slopes of zero, one, zero and one, and continuous variation between zero and one, one may ask what the Brønsted catalysis law of implied constant slope really means. Consideration of the proton transfer behavior summarized above has led Eigen[6] to conclude that the Brønsted relation is only an approximation and to suggest that it is the first (linear) term of a MacLaurin's series relating $\log k$ and ΔpK, in the same way that a MacLaurin's series describes proton transfer reactions. In the general case, the, one must consider terms past the first which will lead to nonlinear Brønsted relationships. Brønsted and Pedersen in 1923 did in fact realize the possibility of a variable slope from one to zero over a range of pK values.[4,7]

An obvious reason for apparent constancies of the Brønsted slopes α and β is the limited range of both the catalysts and substrates employed in these investigations. These limitations were imposed mainly by the experimental time scale of classical kinetic investigations. The conversion of a linear $\log k$ versus ΔpK relation into a nonlinear one is seen in the reaction of nitramide with various bases. The original work on this reaction exhibited an apparently linear relation over a range of approximately 3 pK units. However, when the study was extended to a range of 17 pK units the clearly nonlinear relation of Fig. 4.6 was seen.

Let us consider the evidence for linear or nonlinear Brønsted relationships. Table 4.1 lists a number of reactions for which Brønsted relationships studied

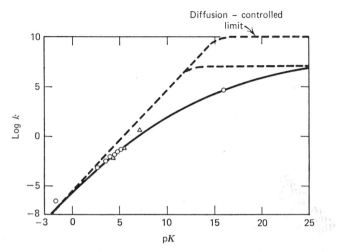

Fig. 4.6 Brønsted relationship in the general base-catalyzed decomposition of nitramide. From M. Eigen, *Angew. Chem. Int. Ed.*, **3**, 19 (1964).

over an extensive pK range show linear dependences of log k on ΔpK. For none of these reactions does the maximal catalytic rate constant approach that of a diffusion-controlled reaction, and thus it may be surmised that the diffusion-controlled rate has not been reached and only the first term of the MacLaurin's series is being observed.

Although Table 4.1 shows a number of apparently linear Brønsted relationships, others appear to be nonlinear. The decomposition of nitramide shows a Brønsted relationship that deviates from linearity at high values of pK, as pointed out by Eigen (Fig. 4.6). The only deviant point, however, is the hydroxide ion point. There are no other data near the hydroxide ion point to indicate whether the deviation is specific to hydroxide ion or is general. The maximal rate constant in the nitramide decomposition is 10^7 M^{-1} sec^{-1} (for hydroxide ion), a value close to diffusion control. If one further takes into account the equilibrium constant of the tautomerization preceding the decomposition,

$$\begin{array}{c}H\\ \diagdown\\ \end{array}\!\!N\!\!-\!\!N\!\!\begin{array}{c}\diagup O\\ \\ \diagdown O\end{array} \underset{}{\overset{K}{\rightleftharpoons}} HN\!\!=\!\!N\!\!\begin{array}{c}\diagup OH\\ \\ \diagdown O\end{array} \overset{B}{\longrightarrow} N_2O + H_2O \qquad (4.8)$$

then the rate-determining step is diffusion-controlled when ΔpK is large and positive.

A curved Brønsted plot implies a changing α or β. Although it is difficult to study a large range of catalyst pK_a's, it is possible to study a large range

Table 4.1
Some Linear Brønsted Relations

Reaction	β or α	Maximal Rate Constant (M^{-1} sec^{-1})	Range of pK_a	Strongest Acid	or	Weakest Base	Reference
$CH_3CH(OH)_2 \xrightleftharpoons{HA} CH_3CHO + H_2O$	0.54	11	10	Chloroacetic acid		Hydroquinone	9
α-Glucose $\xrightleftharpoons{B}$ β-glucose	0.40	90	17	$OH^\ominus$		H_2O	3, 4
α-Glucose $\xrightleftharpoons{HA}$ β-glucose	0.30	6.7×10^{-3}	17	$H_3O^\oplus$		H_2O	3, 4
$H_2C(OH)_2 \xrightarrow{B} H_2C=O + H_2O$	0.40	1600	17	$OH^\ominus$		H_2O	11
$H_2C(OH)_2 \xrightarrow{HA} H_2C=O + H_2O$	0.10	2.7	17	$H_3O^\oplus$		H_2O	11
$Cl_2CHCO_2Et + H_2O \xrightarrow{B} Cl_2CHCO_2^\ominus + EtOH$	0.47	13×10^{-4}	9	Imidazole		H_2O	10
4-Cl-Butanol $\xrightarrow{B}$ tetrahydrofuran	0.25	2×10^{-2}	17	$OH^\ominus$		H_2O	8

The Relation of the Brønsted Catalysis Law to Proton Transfer

of carbon acid pK_a's. For the latter, a very large variation in β is seen, as shown in Table 4.2. The variation of β in this series of ketones can be explained readily by the hypothesis that the range of catalysts studied to determine each β is rather small and thus one is observing different parts of the overall log k versus ΔpK relationship, the acetone reaction being close to the limiting slope of one and the bromoacetylacetone reaction being close to the changeover to the limiting slope of zero. The continuous relationship in Table 4.2 between β and pK in the ketone family is not obeyed when a different family, the nitro compounds, is considered.

Table 4.2
The Brønsted Relationship for the Bromination of Some Ketones[a]

Substrate	log R[b]	β	pK_a of ketone
CH_3COCH_3	-8.56	0.88	20.0
H_3CCOCH_2Cl	-7.85	0.82	16.5
$H_2CCOCHCO_2Et$ $\angle(CH_2)_3\backslash$	-1.76	0.64	13.1
$H_3CCOCH_2COO^{\ominus}$	-0.45	0.52	9.7
$H_3CCOCHBrCOCH_3$	$+0.26$	0.42	8.3

[a] Data taken from R. P. Bell, *The Proton in Chemistry*, Cornell University Press, Ithaca, N.Y., 1959, p. 172. © 1959 by the Cornell University Press. Used by permission.
[b] The quantity R is the catalytic constant (in M^{-1} sec^{-1}) of the anion of a hypothetical acid of $pK_a = 4$.

As mentioned above, the catalytic rate constant for hydroxide ion in the nitramide decomposition is abnormally low with respect to other basic catalysts. This phenomenon is also seen in iodination and bromination reactions involving acetonylacetone, monochloroacetone, monobromoacetone, and dichloroacetone. In these reactions the hydroxide ion rate constant is smaller by a factor of 10^3 to 10^4 than that calculated on the basis of other catalysts participating in the reaction. Interestingly enough, the experimental hydroxide ion rate constant agrees exactly with the prediction from other catalysts in the mutarotation of glucose and the dehydration of formaldehyde hydrate. In the last two reactions, proton transfer is not the only process in the catalysis, whereas in the other reactions it is.

In summary, linear Brønsted relationships are found in catalytic reactions involving: (1) reactions of oxygen or nitrogen acids and bases either as substrates or catalysts; (2) reactions involving chemical changes in addition to proton transfer; and (3) reactions not approaching a diffusion-controlled rate constant. On the other hand, nonlinear Brønsted relations occur among reactions (1) involving carbon acids or bases either as substrates or catalysts; (2) involving only proton transfer; and (3) approaching a diffusion-controlled rate constant. Which of these features are the crucial ones is not clear at present.

In addition to nonlinearity of the Brønsted relation, other deviations exist. In the general acid-catalyzed dehydration of acetaldehyde hydrate, several carbon acid catalysts show a negative deviation (of 1-2 logarithmic units) from the line defined by carboxylic acids and phenols while several oxime catalysts show a positive deviation of that amount.[12] Both the negative and the positive deviations probably reflect a reorganization, either within the molecule or in the solvation shell, that affects the kinetics of proton transfer, but not the equilibrium acidity. These deviations point out the arbitrary nature of the defined Brønsted line. Other deviations from the Brønsted relation include primary, secondary, and tertiary amines that require different Brønsted lines either because of solvation differences[13] or because of steric strain,[14] the presence of steric hindrance in the catalyst and/or the substrate, special solvation effects,[12] and changes in mechanism.

All the above considerations pertain to reactions in aqueous (protic) solution. Linear Brønsted plots have been found for catalysis in a number of nonaqueous systems, both protic and aprotic.[15,16] In an aprotic solvent, catalysis by acids or bases by definition must be general catalysis, so Brønsted relationships in these solvents are to be expected but they must naturally be based on equilibria and kinetics in these solvents.

4.3 THE MEANING OF THE BRØNSTED CATALYSIS LAW

Let us consider the meaning of the Brønsted relationship with regard to the rate and equilibrium of a particular catalytic step involving proton transfer. In order to probe this relationship we will assume a specific mechanism for the base-catalyzed dehydration of acetaldehyde hydrate and then show that the existence of the Brønsted relation implies the existence of a rate-equilibrium relationship for one of the individual steps of the reaction. The assumed mechanism is given by

$$CH_3\text{—}CH(OH)_2 + B \underset{}{\overset{K}{\rightleftharpoons}} CH_3\text{—}\underset{|}{\overset{O^{\ominus}}{C}}H(OH) + BH^{\oplus} \underset{k_{-c}}{\overset{k_c}{\rightleftharpoons}} CH_3\text{—}\overset{O}{\underset{}{C}}H + H_2O + B \quad (4.9)$$

The overall catalytic rate corresponding to this mechanism is given by

$$\text{overall rate} = k_c[BH^{\oplus}][CH_3\overset{O^{\ominus}}{\underset{|}{C}}H(OH)] = v \quad (4.10)$$

The Meaning of the Brønsted Catalysis Law

Equation 4.10 may be transformed to eq. 4.11 using K_b, the ionization constant of $BH^\oplus$, and K_a, the ionization constant of acetaldehyde hydrate.

$$v = k_c \frac{K_a}{K_b} [B][CH_3CH(OH)_2] \tag{4.11}$$

Since the observed rate of the reaction is $k_{obs}[B][CH_3CH(OH)_2]$, k_{obs} may be given by

$$k_{obs} = \frac{k_c K_a}{K_b} \tag{4.12}$$

Now we wish to relate the equilibrium constant, K_c, corresponding to step k_c, to the ionization constant of the catalyst, K_b. To do this, we will utilize the equilibrium definition

$$K_c = \frac{[CH_3CHO][B]}{[CH_3CH(O^\ominus)(OH)][BH^\oplus]} \tag{4.13}$$

which can be transformed to eq. 4.14 and then to eq. 4.15.

$$K_c = \frac{[CH_3CHO]}{[CH_3CH(O^\ominus)(OH)][H^\oplus]} \times \frac{[B][H^\oplus]}{[BH^\oplus]} \tag{4.14}$$

$$K_c = K_d \times K_b \tag{4.15}$$

where K_d is the equilibrium constant of the reaction: $H^\oplus + CH_3CH(OH)O^\ominus \rightleftarrows CH_3CHO + H_2O$. We can now demonstrate that the observed Brønsted relationship between k_{obs} and K_b implies the existence of a rate-equilibrium relationship between k_c and K_c, the rate and equilibrium constants of the (same) rate-determining proton transfer reaction. The experimental Brønsted relationship can be written as

$$\log k_{obs} = \beta \log K_b + c \tag{4.16}$$

The logarithmic form of eq. 4.12 is

$$\log k_{obs} = \log k_c + \log K_a - \log K_b \tag{4.17}$$

Combining eqs. 4.16 and 4.17, we obtain

$$\log k_c = (\beta + 1) \log K_b - \log K_a + c \tag{4.18}$$

Substituting eq. 4.15, we obtain

$$\log k_c = (\beta + 1)[\log K_c] - (\beta + 1)[\log K_d] - \log K_a + c \tag{4.19}$$

Since the last three terms of eq. 4.19 are constants, the equation may be written as

$$\log k_c = \beta' \log K_c + \text{constant} \tag{4.20}$$

General Acid-Base Catalysis: Theory

Equation 4.20 shows that k_c and K_c, the rate and equilibrium constants of the same step, are related to one another logarithmically. This conclusion is independent of the assumed mechanism, with the proviso that the rate and equilibrium constants refer to a rate-determining proton transfer step. In fact, the mechanism used for illustrative purposes above may be mechanistically incorrect (see later). It was used in order to provide a complicated mechanism where intuitively one cannot see any direct relationship between k_c and K_c. In simpler systems, the relationship is straightforward.

The Brønsted relation can also be interpreted in terms of molecular potential energy diagrams.[17,18] If one assumes that a proton transfer reaction is a three-center reaction, that the only important repulsion terms are those between the two heavy atoms involved in the reaction, and that the proton moves between the two heavy atom centers that remain stationary at a fixed distance, then the potential energy-reaction coordinate diagram of Fig. 4.7 can be drawn. In this diagram the resonance interaction lowering the transition state and the repulsion energies have been omitted for the sake of clarity. Curve I represents SH + B, where SH is a substrate and B a basic catalyst, while curve II represents the reaction products $S^\ominus$ + $BH^\oplus$. The activation energy is then $E°$, and the energy change in the reaction $\epsilon°$. If the basic catalyst is modified by the introduction of a substituent producing a weaker base, the change can be represented by the replacement of curve II by II′, which has the same shape and position along the reaction coordinate, but is displaced vertically in the direction of higher energy. The changes, $\delta E°$ and $\delta \epsilon°$,

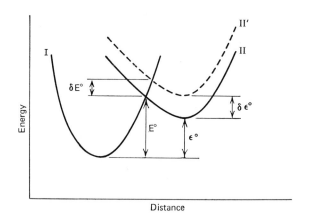

Fig. 4.7 The molecular basis of the Brønsted relation in terms of a potential energy-reaction coordinate diagram. I represents SH + B; II represents $S^\ominus$ + $BH^\oplus$; II′ represents $S^\ominus$ + $B'H^\oplus$, a slightly weaker base. From R. P. Bell, *The Proton in Chemistry*, Cornell University Press, Ithaca, N.Y., 1959, p. 169.

are related to one another by

$$\delta E^\circ = \frac{S_1}{S_1 + S_2} \delta\epsilon^\circ = \beta\, \delta\epsilon^\circ \tag{4.21}$$

where S_1 and S_2 are the slopes of the two curves at the point of intersection. Equation 4.21 is related to the Brønsted catalysis law since the potential energy differences, δE° and $\delta\epsilon^\circ$, can be related to observed velocity constants and equilibrium constants.[18]

Figure 4.7 assumes that a change in catalyst will not change the shape of the curve. This assumption is valid only if the electronic redistribution in the substrate and/or catalyst upon proton transfer is constant. Since much evidence indicates that this is not always so, especially with carbon acids and bases, Fig. 4.7 is of limited validity.

Another limitation of Fig. 4.7 is the assumption of a three-center proton transfer. In organic chemical reactions, many proton transfers are intimately connected with other changes in the substrate. Numerous examples include proton transfers in β-elimination reactions, the removal of a proton alpha to a carbonyl group, additions to carbonyl groups and so on. In such systems, the maximal free energy of the system as a whole is not necessarily identical with the maximal free energy concerned with the proton transfer. We shall discuss this point in some detail later.

An alternative treatment to the transition state theory developed above is the electrostatic theory of proton transfer.[19] This treatment offers a comparable picture of the Brønsted catalysis law but offers no significant advantages over the treatment given here.

4.4 THE MEANING OF α AND β

The rate-equilibrium relationship of the Brønsted catalysis law can be restated in terms of the familiar relation between the free energy of reaction and that of activation of the catalytically important step. For a system in which there is a linear relationship between $\Delta F^\ddagger$ and ΔF°, we may inquire about the relationship between the ΔF°s of the two ground states and the $\Delta F^\ddagger$ of the transition state. In order to carry out this inquiry, we must make two assumptions: (1) the free energy quantities can be described as linear combinations of the free energies associated with variations such as substituent changes; and (2) since a transition state has considerable resemblance to the reactant and product (both in composition and structure), any change in its free energy can be represented as a linear combination of the corresponding changes in the free energies of the reactant and product. These two assumptions lead to

$$\delta F^\ddagger = a\, \delta \bar{F}^\circ_P + b\, \delta \bar{F}^\circ_R \tag{4.22}$$

where δ is an operator representing a substituent change, and $\bar{F}_i^\circ$ is the standard partial molar free energy of the ith component.[20] Going beyond the extrathermodynamic[20] assumptions made above, a further assumption can be made. Since the reaction coordinate of a transition state lies between that of the reactant and product, we may assume that other properties of the transition state will be intermediate as well. Although the standard free energy of the transition state is at a maximum, it is likely that any changes in its value due to substituent changes will be intermediate between corresponding changes for reactant and product. This assumption allows us to convert eq. 4.22 to

$$\delta F^\ddagger = \alpha\, \delta \bar{F}_P^\circ + (1 - \alpha)\, \delta \bar{F}_R^\circ \qquad (4.23)$$

where $0 < \alpha < 1$.

Equation 4.28 may be readily transformed to

$$\delta \Delta F^\ddagger = \alpha\, \delta \Delta \bar{F}^\circ \qquad (4.24)$$

Since the value of α (or β) in the Brønsted relation is always less than one (both on the above theoretical grounds and from experiment; see, however, Chapter 5), the change in the free energy of activation must be less than the change in the free energy of reaction for the rate-determining step of the reaction. This then requires that a substituent change in the general catalyst must have a larger effect on the equilibrium constant for this step than on either k_f or k_r, the rate constants for the forward and reverse reactions. These latter rate constants thus must change in the opposite direction with change in substituent. For example, if we move to a stronger catalyst, k_f will increase and k_r will decrease, a larger change in the equilibrium constant than in either of the rate constants resulting.†

The Brønsted relationship for a Lewis acid system has recently been investigated. The reaction studied is shown by

$$S:MX_n + B \underset{k_r}{\overset{k_f}{\rightleftharpoons}} B:MX_n + S \qquad (4.25)$$

where the substrate is acetone, the Lewis acid is zinc bromide, and the bases are six substituted anilines. A linear relationship was found between $\log k_f$ and the log of the equilibrium constant for eq. 4.25. However, the slope of the line, α, was found to be 1.6, indicating that the change in the free energy of activation must be greater than the change in the free energy of reaction for this Lewis acid system. A stronger base thus increases both k_f and k_r in this system unlike the protonic systems. This result was attributed to the fact that unlike the proton the metal can carry a negative charge and thus the nucleophilic reaction in the reverse direction is affected by polarizability not present in the protonic system.[22]

---

† This is the classical treatment of the subject. For another, more recent, view see F. G. Bordwell and W. J. Boyle, Jr., *J. Amer. Chem. Soc.*, **93**, 511, 512 (1971).

The Meaning of α and β

Let us now consider the meaning of α or β in more detail. From eq. 4.23, α is seen to be a parameter measuring the extent to which the transition state resembles the product or reactant with respect to its sensitivity to structural changes (or changes in solvent). That is, when α is equal to zero the transition state resembles the reactant completely, but when α is equal to one the transition state resembles the product completely. Therefore α should approximate the fractional displacement of the transition state along the reaction coordinate from reactants to products.

The guiding principle for the a priori prediction of the fractional displacement of a transition state along the reaction coordinate is that the transition state bears a greater resemblance to the less stable of the species in the chemical reaction.[23] Thus if the free energy of reaction is positive, α should be greater than 0.5, whereas if the free energy of reaction is negative, α should be less than 0.5. For a simple proton transfer, consider

$$B_1H^\oplus + B_2 \rightleftharpoons B_1 + B_2H^\oplus \tag{4.26}$$

When $B_1H^\oplus$ is a stronger acid than $B_2H^\oplus$ (implying that B_1 is a weaker base than B_2) the proton transfer can be represented by Fig. 4.8. In the transition state of Fig. 4.8 the bond to the strong acid (to the weaker base) is largely complete. On the other hand, the position of the proton in the transition state is far from the weaker acid (stronger base).

The representation of proton transfer in terms of Fig. 4.8 creates an anomalous situation. On the one hand, there is a constancy of α or β when the pK of an acid or base is changed by 15 units in many Brønsted relations. On the other hand, changes in pK of the acid or base of that magnitude should

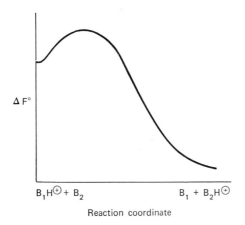

Fig. 4.8 The resemblance of the transition state of a proton transfer reaction to the stronger acid and weaker base.

change the position of the proton in the transition state according to the above argument and thus change $\alpha(\beta)$. As pointed out earlier, however, some Brønsted plots do in fact show curvature. Perhaps the missing factor in this apparently anomalous situation is that one seldom, especially in organic reactions, measures solely proton transfer reactions. When proton transfer is accompanied by other chemical processes, the position of the proton need not correspond to the remainder of the transition state.

The relationship between $\alpha(\beta)$ and the pK_a of a component of a catalytic process is seen in Table 4.2. In this table, β changes from 0.88 with acetone, a weak acid of pK_a 20, to 0.42 with bromoacetylacetone, a strong acid of pK_a 8.3. These data confirm the general prediction of eq. 4.23, but do not answer the question concerning the constancy of $\alpha(\beta)$ raised above.

The position of the proton in the transition state may be determined another way by the use of deuterium isotope effects. The picture of proton transfer given above, together with the concept that maximal isotope effects will occur when the proton is half-transferred,[24,25] predicts that when $\alpha(\beta)$ is zero or one, only a small deuterium kinetic isotope effect shoud be seen, while a maximal kinetic isotope effect should be observed when $\alpha(\beta) = 0.5$. Although such an experiment has not been attempted, an analogous experiment has been carried out by determining the magnitude of the kinetic deuterium isotope effect as a function of ΔpK_a (pK_a of the substrate—pK_a of the catalyst). This experiment, involving the general base-catalyzed proton abstraction from a series of organic substrates (Fig. 4.9), shows a maximal kinetic isotope effect when ΔpK_a approximates zero. (Notice that there are no points at the top of the curve and only four points on the left-hand side).

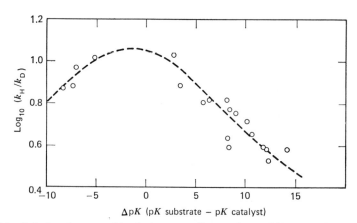

Fig. 4.9 Relation between kinetic isotope effect and pK difference of the reacting systems in the bromination of organic substrates. From R. P. Bell and D. M. Goodall, *Proc. Roy Soc.*, **A294**, 273 (1966).

This is in accord with the hypothesis that a maximal kinetic isotope effect should occur when the proton is half-transferred.* A similar maximum is seen in the proton abstraction from a series of phenonium ions of varying reactivity in the aromatic hydrogen exchange reaction.[26]

Notwithstanding the difficulty in the interpretation of magnitude of α or β, the relationship shown in Table 4.2 between the Brønsted $\alpha(\beta)$ and the pK_a of the substrate is an empirically useful one. For example, it is possible to distinguish between transition states **5** and **6** in the reaction of bases with these adducts of acetone by consideration of the Brønsted β of the process.

$$\underset{5}{\overset{\overset{\oplus}{CH_3NH_2}}{\underset{HO}{\diagup}}\underset{CH_2-H + B}{\overset{CH_3}{\diagdown}}} \qquad \underset{6}{\overset{CH_3NH^{\oplus}=C}{\underset{CH_2-H + B}{\overset{CH_3}{\diagdown}}}}$$

The experimentally determined β equals 0.4. The β predicted by transition state **5** should be very close to one, since the proton removed in transition state **5** should have a pK much greater than 20. On the other hand, transition state **6** contains a proton that is quite acidic because of its protonated nitrogen atom that can act as an electron sink and should have a β much less than one. Consideration of the Brønsted β therefore clearly eliminates **5** and strongly supports **6** as the transition state for the reaction.[27]

α and β can have a further meaning concerned with the discrimination of the reaction. When $\alpha(\beta)$ approaches zero the implication is that the substrate shows no discrimination toward acids or bases; thus, the acid or base that catalyzes the reaction is the one that is present in the largest concentration. In an aqueous system this would be water. When α or β approaches one, the implication is that the substrate has a great discrimination and only the most potent catalyst will be operative; in aqueous soluton this would be either oxonium ion or hydroxide ion. These considerations can be placed on a quantitative basis by using the relationship between selectivity and reactivity.[20] Selectivity is a measure of the probability that an encounter will result in reaction. When a reagent is modified to increase its reactivity, eventually every encounter leads to reaction. Consider a system of two substrates reacting with two reagents, one a "hot" (reactive) reagent and the other "cold." Such a system can be described by Fig. 4.10 and by eq. 4.27. Since the "hot" reagent is more reactive than the "cold" reagent the transition state of the former more nearly resembles its reactant than does the latter. Hence α_h is less than α_c. The selectivity of the "hot" reagent is less than that of the "cold"

---

* See footnote, p. 86.

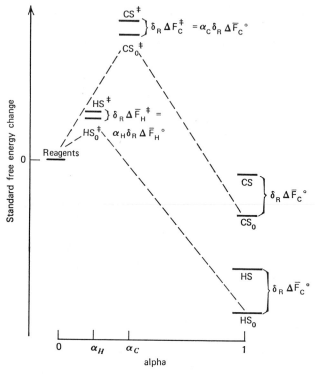

Fig. 4.10 Graphical representation of the relationship between selectivity and reactivity. From J. E. Leffler and E. Grunwald, *Rates and Equilibria of Organic Reactions*, John Wiley & Sons, Inc., New York, 1963, p. 163.

reagent if $\delta_r \Delta F_h^\circ < \delta_r \Delta F_c^\circ$ as can be seen in the definition in eq. 4.27.

$$\frac{\text{selectivity of ``hot''}}{\text{selectivity of ``cold''}} = \frac{\delta_R \Delta \bar{F}_H^\ddagger}{\delta_R \Delta \bar{F}_c^\ddagger} = \frac{\alpha_H \, \delta_R \Delta \bar{F}_H^\circ}{\alpha_c \, \delta_R \Delta \bar{F}_c^\circ} \qquad (4.27)$$

A selectivity-reactivity relationship can be seen from the data of Table 4.2; as the log of the rate constant for a given base decreases, the β for the reaction increases. This is a selectivity-reactivity relationship since β is proportional to selectivity while log k measures reactivity.

4.5 THE RELATION OF GENERAL CATALYSIS TO SPECIFIC CATALYSIS

Let us consider the differences between general and specific catalysis, using as an example general acid catalysis and specific oxonium ion catalysis. We wish to consider the question of why some reactions are general acid-catalyzed whereas other reactions are specific oxonium ion-catalyzed.

One obvious function of general catalysis is to avoid the formation of high energy intermediates and, at the same time, contribute to the polarization necessary for catalysis. For example, large concentrations of oxonium or hydroxide ion are obviously unstable in neutral solution. However, large concentrations of general acids (or general bases) are stable in neutral solution and can effect, to some extent, the same kind of electronic perturbation that the oxonium or hydroxide ion can.

A relationship between the inherent reactivity of a system and the kind of catalysis observed in the system can be advanced in several ways. Proton transfer must constitute the slow step of a reaction when the rest of the rate processes of the reaction are faster. In general acid catalysis, this phenomenon is seen in slow proton transfers to or from a carbon atom. But when proton transfers involve oxygen or nitrogen atoms rather than carbon, other aspects of the rate processes must be considered. For example, the hydrolysis of acetals involves only specific oxonium ion catalysis whereas the hydrolysis of orthoesters involves general acid catalysis. This difference can be rationalized by saying that the rate-determining step in acetal hydrolysis is a more difficult process than in orthoester hydrolysis, since the carbonium ion produced in the latter reaction is more resonance-stabilized than in the former reaction. The reaction in which general acid catalysis is seen may thus be described either as one in which the other rate processes are fast enough so that proton transfer becomes rate-determining, or alternatively as a facile reaction where a strong acid, oxonium ion, is not necessary and a weak acid, a general acid, suffices.

A somewhat different approach to this problem was taken by Bell who made a calculation of the appearance of general acid catalysis as a function of the Brønsted α of the reaction.[28] He set up hypothetical catalysis by acetic acid in a buffer containing 0.1 M acetic acid and 0.1 M acetate ion. Table 4.3 shows the relative contribution to the total velocity made by oxonium ion, water, and acetic acid, assuming three different values of α. When α is equal to 0.1 the predominant catalyst is water. Under these conditions little selectivity is seen and that catalytic species present in highest concentration, water, predominates in the reaction. On the other hand, when α is equal to 1.0, the overriding reaction is the oxonium ion-catalyzed reaction. This is reasonable under those circumstances since an α of 1.0 indicates a very high selectivity in the reaction and that catalyst which is the strongest, oxonium ion, predominates. Finally at intermediate values of α, for example, $\alpha = 0.5$, catalysis by the general acid, acetic acid, predominates. It may thus be concluded that general acid (base) catalysis will only be detectable for intermediate values of $\alpha(\beta)$. Earlier it was pointed out that the Brønsted α is a function of the ease of

Table 4.3
The Effect of α on the Occurrence of General Acid Catalysis[a]

Percent of Catalysis by Various General Acids

α	by $OH_3^\oplus$	by H_2O	by CH_3COOH
0.1	0.002	98	2
0.5	3.6	0.01	96.4
1.0	99.8	5×10^{-12}	0.2

[a] A hypothetical system containing 0.1 M acetic acid and 0.1 M acetate ion. From R. P. Bell, *Acid-Base Catalysis*, Clarendon Press, (Oxford), 1941, p. 94.

reaction. That is, $\alpha(\beta)$ can be considered to be a selectivity factor related to reactivity. This discussion leads to the conclusion that general acid (general base) catalysis will be seen for reactions of intermediate reactivity. However, the upper limit of $\alpha(\beta)$ and not the lower limit is the crucial one. The lower limit of $\alpha(\beta) = 0$ will intrude only in unlikely situations of complete indiscrimination. The upper limit of $\alpha(\beta) = 1.0$ determines whether a reaction is specific-catalyzed or general-catalyzed. The probability of detecting general catalysis is then dependent on the stability of a transition state (or intermediate) in a series of related reactions. The greater the stability of the intermediate, the closer the transition state will lie to the initial reactants, increasing the probability that the Brønsted $\alpha(\beta)$ will be less than unity, and that general catalysis will be observed.

There are many indications that an absolute distinction between general and specific catalysis does not exist. In fact, there are numerous ways in which one may be transformed into the other. Some are described here. In a multistep reaction consisting of one proton transfer step plus other steps, general catalysis will prevail if the proton transfer step is rate-determining, while specific catalysis will prevail if one of the other steps is rate-determining. Thus, one may convert a reaction from specific to general catalysis by speeding up those steps not involving proton transfer, or vice versa. There are a number of avenues for effecting such a transition. One possibility is to increase the concentration of a reagent that is involved in one of the steps not involving proton transfer. For example, in the aldol condensation, an increase in acetaldehyde concentration leads to an increase in the rate of the condensation step so that the proton transfer step becomes rate-determining. Another possibility is to increase the temperature of the reaction. If the activation energy of the steps not involving proton transfer is higher than the activation energy of the proton transfer step, a transition will be made at higher temperature from specific to general catalysis. In enzymatic catalysis, the bond making and breaking steps other than proton transfer may be entropically accelerated to such an extent that proton transfer often becomes rate-determining. For the ethylene oxide ring opening reaction of epichlorohydrin with sodium iodide in the presence of acetic acid, Swain studied the transition from specific oxonium ion catalysis to general acid catalysis.[29] From the rate constant of this reaction with oxonium ion and a Brønsted α taken from other general acid-catalyzed reactions, he calculated a theoretical rate constant for the acetic acid-catalyzed reaction, predicting that doubling of the rate would occur with 1.6 molar acetic acid. He found a doubling of the rate with 1.2 molar acetic acid, in fair agreement with the calculation. The use of such large concentrations of catalyst makes the result questionable. However, 2 M dioxane or 2 M acetamide depresses the rate whereas acetic acid increases the rate. Furthermore the ionic strength of the system was kept constant and a (protonic) indicator was not affected by this concentrated system, indicating that the pK_a of the acetic acid may not have been affected.

If there is no absolute distinction between specific and general acid catalysis, general catalysis will occur at any time that the concentration of the general acid becomes sufficiently elevated with respect to the concentration of oxonium ion. In the hydrolysis of ethyl orthoformate,[30] specific oxonium ion catalysis is found in aqueous acetic acid buffer but when the medium is changed to an aqueous dioxane solution containing acetic acid buffer, general acid catalysis by acetic acid occurs. The aqueous dioxane solution was effective since the ratio [CH_3CO_2H]/[$H_3O^{\oplus}$] is approximately 1000 times higher in such a solution than in an aqueous solution (because of the change in pK_a of acetic acid). In many intramolecular catalyses to be discussed later the effective local concentration of the intramolecular general acid (base) catalyst is increased ten to a thousandfold over the concentration of the corresponding intermolecular catalyst (Chapter 11). Thus in intramolecular reactions general catalysis can be more easily observed. The same can be said of enzymatic catalysis (Chapter 19).

REFERENCES

1. J. N. Brønsted and K. Pedersen, *Z. Physik. Chem.*, **108**, 185 (1924). This paper is worthy of careful perusal even at this date.
2. J. N. Brønsted, *Chem. Rev.*, **5**, 322 (1928).
3. H. Schmid, *Chem. Zeitung*, **90**, 351 (1966).
4. R. P. Bell, *Acid-Base Catalysis*, Oxford University Press, Oxford, 1941, p. 176.
5. R. L. Schowen, H. Jayaraman, L. Kershner, and G. W. Zuorick, *J. Amer. Chem. Soc.*, **88**, 4008 (1966).

6. M. Eigen, *Angew Chem. Int. Ed.*, **3**, 1 (1964).
7. R. G. Pearson and R. L. Dillon, *J. Amer. Chem. Soc.*, **75**, 2439 (1953), and references therein.
8. C. G. Swain, D. A. Kuhn, and R. L. Schowen, *ibid.*, **87**, 1553 (1965).
9. R. P. Bell and W. C. E. Higginson, *Proc. Roy. Soc.*, **A197**, 141 (1949).
10. W. P. Jencks and J. Carriuolo, *J. Amer. Chem. Soc.*, **83**, 1743 (1961).
11. R. P. Bell, personal communication.
12. R. P. Bell, *The Proton in Chemistry*, Cornell University Press, Ithaca, N.Y., 1959, pp. 165, 174–182.
13. R. G. Pearson, *J. Amer. Chem. Soc.*, **70**, 204 (1948).
14. H. C. Brown, H. Bartholomay, Jr., and M. D. Taylor, *ibid.*, **66**, 435 (1944).
15. R. P. Bell, *Acid-Base Catalysis*, Oxford University Press, Oxford, 1941, p. 150.
16. L. P. Hammett, *Physical Organic Chemistry*, McGraw-Hill Book Co., New York, 1940, p. 288.
17. J. Horiuchi and M. Polanyi, *Acta Physicochim. URSS*, **2**, 505 (1935).
18. R. P. Bell, *The Proton in Chemistry*, Cornell University Press, Ithaca, N. Y., 1959, p. 166.
19. R. P. Bell, *Proc. Roy. Soc.*, **A154**, 414 (1936).
20. J. E. Leffler and E. Grunwald, *Rates and Equilibria of Organic Reactions*, John Wiley and Sons, New York, 1963, p. 156.
21. For another derivation of the Brønsted relation, see J. J. Weiss, *J. Chem. Phys.*, **41**, 1120 (1964).
22. R. S. Satchel and D. P. N. Satchel, *Proc. Chem. Soc.*, 362 (1964).
23. G. S. Hammond, *J. Amer. Chem. Soc.*, **77**, 334 (1955).
24. F. H. Westheimer, *Chem. Rev.*, **61**, 265 (1961).
25. A. V. Willi and M. Wolfsberg, *Chem. Ind.*, 2097 (1964).
26. A. J. Kresge, *Disc. Faraday Soc.*, **39**, 48 (1965); J. L. Longridge and F. A. Long, *J. Amer. Chem. Soc.*, **89**, 1292 (1967).
27. M. L. Bender and A. Williams, *J. Amer. Chem. Soc.*, **88**, 2502 (1966).
28. R. P. Bell, *Acid-Base Catalysis*, Oxford University Press, Oxford, 1941, p. 93.
29. C. G. Swain, *J. Amer. Chem. Soc.*, **74**, 4108 (1952).
30. R. H. DeWolfe and R. M. Roberts, *J. Amer. Chem. Soc.*, **76**, 4379 (1954).

Chapter 5

GENERAL ACID-BASE CATALYSIS: MECHANISM

5.1 Distinguishing Mechanistic Ambiguities in General Catalysis: General Base Versus Nucleophilic Catalysis 101
5.2 Specification of General Catalysis in Stepwise Processes 107
5.3 Mechanisms of General Acid-Base-Catalyzed Reactions 115
5.4 Susceptibility to General Acid-Base Catalysis 134
5.5 The Position of the Proton in the Transition State of General-Catalyzed Reactions 138

General acid and general base catalysis can be represented by several kinetic schemes. Those for general acid catalysis follow. The simplest is one involving a rate-determining proton transfer followed by a fast decomposition step.

$$S + HA \underset{}{\overset{slow}{\rightleftarrows}} SH^{\oplus} + A^{\ominus} \overset{fast}{\longrightarrow} products \quad (5.1)$$

$$rate = k[S][HA]$$

The rate-determining proton transfer may also occur after a covalent change involving heavy (heavier than hydrogen) atoms.

$$S \underset{}{\overset{fast}{\rightleftarrows}} S'$$

$$S' + HA \underset{}{\overset{slow}{\rightleftarrows}} S'H^{\oplus} \overset{fast}{\longrightarrow} products$$

$$rate = k[S][HA] \quad (5.2)$$

The significant point about each of these equations is that the rate-determining step is solely a proton transfer. Proton transfer may also occur simultaneously with covalent changes involving heavy atoms. Two possibilities exist: (1) the proton transfer and covalent changes occur directly; and (2) the rate-determining proton and covalent transfers occur only after a preliminary prototropic pre-equilibrium.

$$S + HA \xrightleftharpoons{\text{slow}} S'H^{\oplus} + A^{\ominus} \xrightarrow{\text{fast}} \text{products}$$

where the prime means a different species

$$\text{rate} = k[S][HA] \tag{5.3}$$

$$S + HA \xrightleftharpoons{\text{fast}} SH^{\oplus} + A^{\ominus} \xrightarrow{\text{slow}} \text{products}$$

$$\text{rate} = k[S][HA] \tag{5.4}$$

The above equations do not consider a second substrate molecule that is often involved. If such is the case, prior complexing between two of the three reactants can occur before any of the steps listed in eq. 5.1 to 5.4. Other possibilities can be envisioned.

The general acid catalyses of eq. 5.1 to 5.4 have their counterparts in the general base catalyses of eq. 5.5 to 5.8. Equations 5.5 and 5.6 consist solely of rate-determining proton transfers, with or without prior covalent change involving heavy atoms.

$$SH + B \xrightleftharpoons{\text{slow}} S^{\ominus} + BH^{\oplus} \xrightarrow{\text{fast}} \text{products}$$

$$SH \xrightleftharpoons{\text{fast}} SH'$$

$$\text{rate} = k[SH][B] \tag{5.5}$$

$$S'H + B \xrightleftharpoons{\text{slow}} S'^{\ominus} + BH^{\oplus} \xrightarrow{\text{fast}} \text{products}$$

$$\text{rate} = k[SH][B] \tag{5.6}$$

Equations 5.7 and 5.8 involve a rate-determining step that includes both proton transfer and a covalent change involving heavy atoms. They differ by the prior prototropic equilibrium of eq. 5.8.

$$SH + B \xrightleftharpoons{\text{slow}} S'^{\ominus} + BH^{\oplus} \xrightarrow{\text{fast}} \text{products}$$

$$\text{rate} = k[SH][B] \tag{5.7}$$

General Acid-Base Catalysis: Mechanism

$$SH + B \rightleftharpoons S^{\ominus} + BH^{\oplus} \longrightarrow \text{products}$$
$$\text{rate} = k[SH][B] \qquad (5.8)$$

Examples of general acid and general base catalysis and general acid base catalysis are found in Tables 5.1, 5.2, and 5.3. In these tables several varieties of general catalysis appear. Two major groups are seen: (1) reactions involving proton transfer to or from carbon atoms; and (2) reactions involving proton transfer to or from nitrogen, oxygen, and sulfur atoms. Chapter 2 indicates that the first group can have solely a rate-determining proton transfer while the second group probably cannot. Thus, the first group should conform to eq. 5.1, 5.2, 5.5, and 5.6 while the second group should conform to eq. 5.3, 5.4, 5.7, and 5.8. This first approximation will be re-examined in detail later.

Equations 5.1 to 5.8 contain many mechanistic ambiguities. In all general acid mechanisms with the exception of eq. 5.4, the general acid catalyst acts as a proton donor. However, eq. 5.4 is mechanistically an oxonium ion-general base or nucleophilic reaction. Thus, ambiguities between general acid and base reactions, and between general acid and nucleophilic reactions exist.

The ambiguities in general base catalysts are much more serious. All the general base mechanisms given with the exception of eq. 5.8 can be interpreted either as general base reactions as shown, or, alternatively, as nucleophile reactions. The sole exception, eq. 5.8, is mechanistically a general acid-hydroxide ion reaction, since the pre-equilibrium in this mechanism can be shown to depend on the hydroxide ion concentration, that is, the equilibrium $B + H_2O \rightleftharpoons BH^{\oplus} + OH^{\ominus}$ leads to the kinetic equivalence of $[B][H_2O]$ and $[BH^{\oplus}][OH^{\ominus}]$ in aqueous solution.

Establishing criteria for these pathways is a major problem in analyzing the mechanisms of general acid and general base catalysis. There are essentially three kinds of ambiguities involved. (1) Is the catalyst acting as a proton transfer agent or a nucleophile? (2) In which step of a multistep process is the catalyst acting? (3) What is the position of the proton in the important transition state? Not until we have solved these problems will it be possible to speak with assurance about the mechanism of general acid-base catalysis.

In eqs. 5.1 to 5.8, all steps not considered to be rate-determining were tacitly assumed to be fast equilibria. However, to take general base catalysis as an example, if one makes no assumptions concerning eq. 5.5, the general rate expression for the process is

$$1 - \frac{[X]}{a} = \frac{\rho_2}{\rho_2 - \rho_1} e^{-\rho_1 t} - \frac{\rho_1}{\rho_2 - \rho_1} e^{-\rho_2 t} \qquad (5.9)$$

where a is the initial concentration of substrate, $[X]$ is the concentration of product at time, t, and ρ_1 and ρ_2 are the roots of the equation $\rho^2 - (k_{-1} + k_1 + k_2)\rho + k_1 k_2 = 0$,[1] where k_1 and k_{-1} refer to the forward and reverse

Table 5.1
General Acid Catalysis Where an Extensive Series Exists

1. Proton transfer to carbon
   a. Diazoacetic ester decomposition
   b. Nitromethane ion + HA
   c. Nitroethane ion + HA
   d. Azodicarbonate decomposition
   e. Acetylacetone enolate ion + HA
   f. Hydration of p-methoxy-α-methylstyrene
   g. Cleavage of C–Hg bond
   h. Hydration of mesityl oxide and crotonaldehyde
   i. Addition of H_2O to acetylenic thiolesters
   j. Aromatic hydrogen exchange
2. Proton transfer from carbon
   a. Acetone iodination
   b. Keto-enol transformation
   c. Bromination of nitromethane
   d. Bromination of acetoacetic ester
3. Carbonyl addition and reverse (proton transfer to O,N,S)
   a. Dihydroxyacetone depolymerization
   b. Dehydration of acetaldehyde hydrate
   c. Schiff base formation
   d. Hydrolysis of Schiff bases
   e. Semicarbazone formation
   f. Carbonyl compound + anilines
   g. Carbonyl compound + H_2O, ROH
   h. Carbonyl compound + thiourea
   i. Oxime formation
   j. Bisulfite addition
   k. Nitrone formation
   l. Thiol addition to carbonyl
   m. Phenylhydrazone formation
   n. Hydrolysis of ortho esters
   o. Hydrolysis of N-arylglucosamines
   p. Hydration of 1-benzyl-1,4-dihydronicotinamide
   q. Triose condensation
   r. Reactions in ice
4. Reactions of carboxylic acid and phosphoric acid derivatives
   a. Esterification of acetic acid by methanol
   b. Phosphate hydrolysis
   c. Hydroxylaminolysis of hydroxamic acid
   d. Hydrolysis of piperazine-2,5-dione
   e. Esterification
   f. Amide hydrolysis
   g. Hydroxylaminolysis of amides
5. Miscellaneous
   a. Cleavage of C–B bond
   b. Cleavage of C–Sn bond
   c. Hydrolysis of $NaBH_4$
   d. Nucleophilic aromatic substitution and elimination

General Acid-Base Catalysis: Mechanism

Table 5.2
General Base Catalysis Where an Extensive Series Exists

1. Proton transfer from carbon
    a. Bromination of ketones
    b. Hydrogen exchange of isobutyraldehyde
    c. Nitramide decomposition (from nitrogen, not carbon atoms)
    d. Halogenation, isomerization, and deuterium exchange of many organic substances containing an acidic hydrogen atom, and the racemization of such substances when the acidic hydrogen is on an asymmetric center
    e. Aldol condensation
    f. Condensation of glyceraldehyde and dihydroxyacetone
    g. Transamination
2. Carbonyl addition
    a. Hydrolysis of Schiff base
    b. Cyanohydrin formation
    c. Semicarbazone formation
    d. Phenylhydrazone formation
    e. Solvolysis of nitrostyrene
    f. Condensation of formaldehyde and tetrahydrofolic acid
3. Aromatic substitution
    a. Aromatic nucleophilic substitution
    b. Aromatic electrophilic substitution
4. Nucleophilic reactions, mainly of carboxylic acid derivatives
    a. Solvolysis of tetrabenzyl pyrophosphate
    b. Ammonolysis of phenyl acetate
    c. Hydrolysis of phosphoric carbonic anhydride
    d. Hydration of carbon dioxide
    e. Hydrolysis of Leuch's anhydride
    f. Hydrolysis of ethyl trifluorothiolacetate
    g. Hydrolysis of lactones and amides
    h. Aminolysis of esters and thiolesters
    i. Hydrolysis of glycine ester-cobalt complex
    j. Ethanolysis of ethyl trifluoroacetate
    k. Hydrolysis of γ-phenyltetronic acid enol esters
    l. Hydrolysis of phenyl acetates
    m. Hydrolysis of anhydrides
    n. Hydroxylaminolysis of esters
    o. Hydrolysis of acyl cyanides
    p. Hydrolysis of acetyl fluoride
    q. Nucleophilic reactions of N-acetylimidazole
    r. Hydrolysis of 1-acetyl-3-methylimidazolium ion
    s. Ester hydrolysis
    t. Aminolysis of phenyl acetates

(continued)

Table 5.2 (continued)

u. Imidazolysis of esters
v. Aminolysis of acetyl phosphate
w. Aminolysis of imido esters
x. Hydrolysis of methyl ethylene phosphate
y. Hydrolysis of trifluoroacetanilide
z. Hydrolysis of N-methyltrifluoroacetanilide
aa. Hydrazinolysis of esters
bb. Hydrolysis of aspirin anion
cc. S–N acyl transfer
dd. Cleavage of carbon–carbon bonds
ee. Hydrolysis of N-acylimidazoles
ff. Hydrolysis of amidines

5. Miscellaneous
   a. Cyclization of 4-chlorobutanol
   b. Hydrolysis of 4(5)-hydroxymethylimidazole
   c. Silicon–oxygen bond cleavage
   d. Addition of $HSiCl_3$ to acetylene

Table 5.3

General Acid-Base Catalysis Where an Extensive Series Exists

Reaction
a. Hydrolysis of diisopropyl phosphorofluoridate
b. Aminolysis of thiol esters
c. Mutarotation of glucose
d. Hydration of acetaldehyde
e. Enolization of ketones
f. Amide hydrogen exchange

protropic shifts. This expression, containing two exponential terms, is not a simple function of the general base. It may, however, be simplified in several ways.

1. $k_1 \ll k_{-1}$ and $k_2 \ll k_{-1}$. Under these conditions, eq. 5.9 reduces to a simple first-order reaction in hydroxide ion, that is, a specific hydroxide ion reaction, with rate constant $= k_1 k_2 / k_{-1}$.

2. $k_1 \sim k_{-1}$ and $k_2 \gg k_{-1}$ or $k_1 \ll k_{-1} \gg k_2$. These conditions convert eq. 5.9 to a general base-catalyzed reaction where the observed velocity is a linear function of the catalyst concentration, with rate constant k_1.

3. Several other simplifications of eq. 5.9 may occur, involving more complicated expressions. For example, if $k_1 \ll k_{-1}$ (and $k_2 \sim k_{-1}$),

$$k = \frac{k_2}{\left(\frac{k_{-1}}{k_2}\right) + 1} \tag{5.10}$$

If $k_1 \sim k_{-1}$ and $k_2 \ll k_{-1}$

$$k = \frac{k_1 k_2}{k_1 + k_{-1}} \tag{5.11}$$

The great majority of acid-base catalyses show either specific hydroxide ion catalysis corresponding to (1) or general catalysis corresponding to (2). There are no experimental data showing the full complications of eq. 5.9. However, many general acid or base catalyses, including many two-step reactions where the general catalysis occurs in only one of the two steps, conform to eq. 5.10. These reactions will be discussed in detail later.

5.1 DISTINGUISHING MECHANISTIC AMBIGUITIES IN GENERAL CATALYSIS: GENERAL BASE VERSUS NUCLEOPHILIC CATALYSIS

The question of determining whether a catalyst acts as a proton transfer agent or a nucleophile is a common one. This ambiguity is particularly prevalent in reactions of carboxylic acid derivatives, because of the wide occurrence of both general base and nucleophilic catalysis in these reactions. To illustrate the problem and its resolution, we will consider methods used to differentiate general base from nucleophilic catalysis in this area.

1. Catalysis by the leaving group of a reaction is evidence of general base rather than nucleophilic catalysis. The reaction shown in eq. 5.12 cannot be explained by nucleophilic attack on the carboxylic acid derivative since such a reaction could only regenerate the starting material, not lead to a chemical transformation.

$$\underset{\text{RCB}}{\overset{\text{O}}{\|}} + H_2O \xrightarrow{B} \underset{\text{RCOH}}{\overset{\text{O}}{\|}} + HB \tag{5.12}$$

Therefore, the function of B must be that of a general base. Examples of catalysis by the leaving group of a reaction include: (1) the hydrolysis of acetic anhydride by acetate ion;[2] (2) the hydrolysis and thiolation of N-acetylimidazole by imidazole;[3] and (3) the hydrolysis of acetyl fluoride by fluoride ion.[4] This mechanistic criterion is probably the most rigorous for distinguishing between general base and nucleophilic catalysis.

2. Catalysis by a second mole of an attacking nucleophile is evidence that general base catalysis occurs.[5-8] The first mole of attacking nucleophile may

be involved either as a nucleophilic catalyst or a nucleophile; this can be ascertained by product analysis. The second mole of nucleophile cannot act either as nucleophilic catalyst or nucleophile, but can act as general base catalyst. There are several complications to this interpretation. Kinetic dependence on a second mole of nucleophile is usually not seen until very high concentrations of nucleophile are introduced into the system, leading to a changing medium and all the attendant errors that may obscure the true kinetics. In addition, there may be pre-equilibria involving the nucleophile.

$$\text{BH} + \text{OH}^\ominus \rightleftharpoons \text{B}^\ominus + \text{H}_2\text{O}$$
$$\text{BH} + \text{BH} \rightleftharpoons \text{B}^\ominus + \text{BH}_2^\oplus \qquad (5.13)$$

Such reactions suggest the possibility that $\text{B}^\ominus$ rather than BH is the reactive species, or alternatively that $\text{B}^\ominus$ and $\text{BH}_2^\oplus$ are the true reactants rather than 2 moles of BH. The combination of ions can be ruled out if the calculated rate constant for the reaction of the substrate with $\text{B}^\ominus$ and $\text{BH}_2^\oplus$ is larger than that of a diffusion-controlled process.

3. Observation of a transient intermediate that can be identified as the intermediate of nucleophilic catalysis by chemical characterization and adherence of the kinetics of the overall process to a nucleophilic catalysis is proof that the catalyst is acting as a nucleophile rather than a general base. The observation may be a direct observation of the appearance and disappearance of the transient or may be a trapping experiment in which the reaction mixture is treated with a reagent to trap the hypothetical intermediate as an isolable species. For example, in the imidazole-catalyzed hydrolysis of p-nitrophenyl acetate, the formation and decomposition of N-acetylimidazole can be observed spectrophotometrically at 243 nm, the λ_{max} of this species, yielding data that rigorously fit a two-step process of nucleophilic catalysis.[9, 10]

Conversely, failure to detect an intermediate that should be observable by independent evidence, either by direct observation or by trapping experiments, can be used as evidence for general base catalysis. For example, the acetate ion-catalyzed hydrolysis of p-chlorophenyl acetate must proceed via general base catalysis since the presumed intermediate of nucleophilic catalysis, acetic anhydride, is not trapped by aniline, a proven trapping agent for this species.[11]

4. A Brønsted plot for a series of nucleophiles or bases can be characteristic of either a general base-catalyzed reaction or a nucleophile-catalyzed reaction. A Brønsted plot for a family of general bases adheres, with few exceptions, to a single line, independent of the structural characteristics of the bases making up the plot. On the other hand, a Brønsted-type plot of a series of nucleophiles of differing steric requirements and nucleophilic atoms can show considerable deviations from a single line. Plots of these two types

are shown in Fig. 5.1, a Brønsted plot for true general base catalysis, the hydrolysis of ethyl dichloroacetate, and Fig. 5.2,[12] a Brønsted-type plot for the nucleophilic reactions of p-nitrophenyl acetate, only some of which are catalytic.† To take a particular example, pyridine is equivalent to acetate ion in basicity and will fall at the same point on a Brønsted plot for general base catalysis. However, pyridine is at least a thousand times more reactive nucleophile than acetate ion, a difference common to nitrogen and oxyanion nucleophiles of the same basicity, and on a Brønsted-type plot of nucleophilic reactions, there will be a wide discrepancy in these points. Another pair of nitrogen and oxyanion nucleophiles of identical basicity, imidazole and monohydrogen phosphate ion, also show this behavior in general base and nucleophilic catalysis. Thus, general base catalysis will be characterized by Brønsted plots such as Fig. 5.1 showing identical catalytic rate constants for bases of identical strength. Contrariwise, nucleophilic catalysis is characterized by a Brønsted-type plot such as Fig. 5.2 and by large differences in catalytic rate constants for (N and O) pairs of catalysts of equal basicity.‡

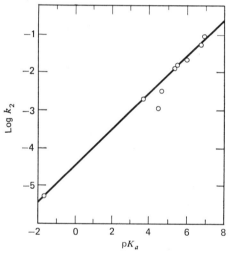

Fig. 5.1 Brønsted plot of the catalytic constants for general base-catalyzed hydrolysis of ethyl dichloroacetate. No statistical corrections have been made. From W. P. Jencks and J. Carriuolo, *J. Amer. Chem. Soc.*, 83, 1747 (1961). © 1961 by the American Chemical Society. Reproduced by permission of the copyright owner.

† The catalytic reactions show no different scatter than the noncatalytic reactions.

‡ The thousandfold difference in nucleophilicity between pyridine and acetate ion may in fact be greater since, as mentioned above, acetate ion serves as general base catalyst toward p-nitrophenyl acetate whereas pyridine serves as nucleophilic catalyst. This change in mechanism complicates the tacit assumption of a constant mechanism made above.

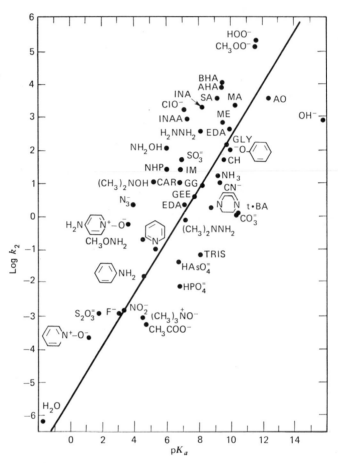

Fig. 5.2 Rates of nucleophilic reactions with *p*-nitrophenyl acetate in aqueous solution at 25° plotted against the basicity of the attacking reagent. Abbreviations: AHA, acetohydroxamic acid; AO, acetoxime; BHA, *n*-butyrylhydroxamic acid; CAR, carnosine; CH, chloral hydrate at 30°; EDA, ethylenediamine; GEE, glycine ethyl ester; GG, glycyclycine; GLY, glycine; IM, imidazole; INA, isonitrosoacetone; INAA, isonitrosoacetylacetone; MA, sodium mercaptoacetate; ME, mercaptoethanol; NHP, *N*-hydroxyphthalimide; SA, salicylaldoxime; t·BA, *t*-butylamine; TRIS, tris(hydroxymethyl)aminomethane. No statistical corrections have been made. From W. P. Jencks and J. Carriuolo, *J. Amer. Chem. Soc.*, **82**, 1779 (1960). © 1960 by the American Chemical Society. Reproduced by permission of the copyright owner.

The slope of a Brønsted plot for nucleophilic catalysis may be higher than that for general base catalysis. In the examples given here, the reaction of p-nitrophenyl acetate with nucleophiles has a slope of 0.68 while the hydrolysis of ethyl dichloroacetate with general bases has a slope of 0.3. The Brønsted plot for general base catalysis, furthermore, cannot have a slope greater than one; in a Brønsted-type plot for nucleophilic catalysis, however, no such dictum applies.[13]

5. A corollary of the arguments concerning Brønsted plots is the rule that steric hindrance is of less importance in general base catalysis than in nucleophilic catalysis.[14] A priori this should be so, since the former involves proton abstraction while the latter involves attack at carbon, the steric requirements of the former being less than the latter. Steric hindrance of nucleophilic catalysis is particularly evident in the pyridine-catalyzed hydrolysis of acetic anhydride, in which compounds such as 2-picoline, 2,6-lutidine, and 2,4,6-collidine do not catalyze hydrolysis at all, while pyridine does so efficiently.[17] On the other hand, steric hindrance is not absent in general base-catalyzed reactions such as the iodination of ketones. Negative deviations from a Brønsted plot up to fiftyfold are found in the iodination of ketones when catalyzed by 2,6-lutidine and 2,4,6-collidine. Here the kinetic effect is dependent on the steric requirements of the ketone as well as the catalyst.[14–16],†

6. Another corollary of the arguments concerning Brønsted plots is that a nucleophile exhibiting an α-effect,[18] that is, enhanced nucleophilicity due to the juxtaposition of two adjacent atoms containing unshared electron pairs will be an exceptionally good nucleophilic catalyst, but only an ordinary general basic catalyst. For example, hydroxide ion is a stronger base than hydroperoxide ion, but the latter is a stronger nucleophile by as much as 10^4 in certain nucleophilic displacement reactions. Thus, a comparison of the reactivity of these two substances with a given substrate can be used to distinguish between nucleophilic and general-base reactions, including catalytic processes.[19] This differentiation is based on the assumption that the α-effect is not seen in proton transfer rections, an assumption that is in agreement with a few facts.[20]

$$(CH_3C)_2O + \underset{N}{\bigcirc} \rightleftharpoons CH_3\overset{O}{\overset{\|}{C}}-\overset{\oplus}{N}\bigcirc \overset{\ominus}{O}\overset{O}{\overset{\|}{C}}CH_3 \xrightarrow{H_2O} CH_3\overset{O}{\overset{\|}{C}}OH + \underset{N}{\bigcirc}$$

(5.14)

† The tacit assumption is again made that hindered and unhindered bases catalyze via a common mechanism.

7. The common ion effect can be used as a criterion to distinguish between general basic and nucleophilic catalysis. In the pyridine-catalyzed hydrolysis of acetic anhydride, for example (eq. 5.14), the addition of acetate ion significantly decelerates reaction indicating nucleophilic catalysis by pyridine.[17, 21] Likewise, imidazole-catalyzed hydrolysis of trifluoroethyl acetate is inhibited by added trifluoroethanol[22] but the acetate ion-catalyzed hydrolysis of acetylimidazole is not inhibited by imidazole.[23] Such decelerations require the reversible formation of an intermediate and nucleophilic catalysis, assuming that the product of the reaction is insensitive to reversal by the addend. A corollary of this criterion is the incorporation of a radioactive (or other) tracer into a reactant using a common ion. This result, accomplished in the pyridine-catalyzed hydrolysis of acetic anhydride, also points to nucleophilic rather than general base catalysis.

8. Deuterium oxide solvent isotope effects can, with caution, be used to distinguish between general base and nucleophilic catalysis. For example, the imidazole-catalyzed hydrolysis of *p*-nitrophenyl acetate has a solvent isotope effect, k^{H_2O}/k^{D_2O}, of 1.0.[24] Contrariwise, the imidazole-catalyzed hydrolysis of ethyl dichloroacetate has a solvent isotope effect of 3.0.[25] The previous discussion presents other criteria indicating that imidazole acts as a nucleophile in the former, but as a general base in the latter reaction. The deuterium oxide solvent isotope effects are consistent with the other criteria. No such isotope effect would be expected in a nucleophilic reaction but a sizable effect would be expected in a general base reaction involving a rate-determining proton transfer because of the difference in zero point energy of vibration of proton and deuteron bonds. It must be emphasized, however, that differences in solvation between the ground and transition states can result in sizable solvent isotope effects in reactions not involving a rate-determining proton transfer.[27] Furthermore, some general catalyses involving proton transfer show little or no primary effect (see below). This criterion must therefore be used in conjunction with other arguments.

9. Product analysis can sometimes be helpful in distinguishing between general base and nucleophilic catalysis. A nucleophilic catalyst such as imidazole can yield a hydrolytic product via a nucleophilic catalysis involving reaction of the tertiary amine. However, a nucleophilic analog of imidazole, aniline, in a comparable nucleophilic reaction will give an anilide as product rather than a hydrolytic product. The use of the latter nucleophile in conjunction with the former catalyst can verify its mechanism. Conversely, if imidazole acts as a general base yielding a hydrolytic product, aniline should likewise yield a hydrolytic product.[25] The limitation in these arguments is that one must ascertain that both imidazole and aniline react by the same mechanism, a requirement difficult to fulfill.

5.2 SPECIFICATION OF GENERAL CATALYSIS IN STEPWISE PROCESSES

Another of the ambiguities met in the determination of the mechanism of general acid-base-catalyzed reactions is the problem of specifying the step in which catalysis occurs in a multistep reaction. There are many multistep reactions subject to general acid-base catalysis. The more prominent ones include: (1) additions to carbonyl and imine groups; (2) nucleophilic reactions of carboxylic acid derivatives; and (3) aromatic substitution reactions. All of these reactions have been shown to proceed through multistep pathways. The experimental basis for a multistep pathway will not be explicitly given unless it directly touches on the catalysis, but we will attempt to provide mechanistic arguments as to the step catalyzed in these reactions.

1. The reactions of nucleophiles with carbonyl compounds and imines are multistep reactions consisting of addition followed by elimination.[28] The oximation of acetone follows:

$$(CH_3)_2C=O + H_2NOH \underset{k_{-1}k'_{-1}[H^\oplus]}{\overset{k_1k'_1[H^\oplus]}{\rightleftharpoons}} (CH_3)_2C(OH)(NHOH) \xrightarrow{k[H^\oplus]} (CH_3)_2C=NOH + H_2O \quad (5.15)$$

From the observation of a pH-rate constant profile (a bell-shaped curve) together with the direct spectrophotometric observation of a transient species formed rapidly and decomposed slowly, one must conclude that the oximation of ketones occurs with the formation of a tetrahedral intermediate as shown in eq. 5.15. As the pH is raised, the rate-determining step changes from rate-determining formation of the intermediate at low pH (limited by the concentration of free hydroxylamine) to rate-determining decomposition of the intermediate at neutrality and above (limited by the oxonium ion concentration needed for decomposition). The observation of general catalysis in neutral and alkaline, but not in acid, solutions confirms the two-step process. Since the rate-determining step in alkaline solution is the decomposition of the intermediate, and since general catalysis is found in this region, the general catalysis must be associated with the decomposition of the tetrahedral intermediate. The pH-rate profile of this reaction (Fig. 5.3)[26] shows not only the bell-shaped curve at neutrality and below, but also water and hydroxide ion reactions at successively higher pH's. The presence of the water reaction (the flat in the pH rate constant profile) is in itself indicative of general catalysis.

The pH dependences of the formation of semicarbazones from carbonyl compounds and of the hydrolysis of imines, although more complicated,

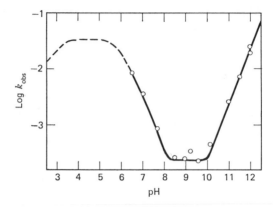

Fig. 5.3 The oximation of acetone at 25.0°, I = 0.2, 0.02 M hydroxylamine. The solid line from 7 to 8 has a slope of −1 and the line from 10 to 12.5 has a slope of +1. From A. Williams and M. L. Bender, *J. Amer. Chem. Soc.*, **88**, 2511 (1966). © 1966 by the American Chemical Society. Reproduced by permission of the copyright owner.

can also be analyzed in terms of an addition-elimination sequence. Semicarbazone formation, by analogy with oximation, may be expected to show general catalysis in the decomposition of the tetrahedral intermediate.[29] In the hydrolysis of imines, the reverse of oxime formation, the low pH region is associated with rate-determining decomposition of the adduct while the high pH region is associated with rate-determining formation of the adduct, opposite to the designation of steps in oxime formation.[27] The conclusion can then be made that the step subject to general catalysis in this reaction is the formation of the tetrahedral intermediate rather than its decomposition.

The condensation of formaldehyde with tetrahydrofolic acid to produce methylenetetrahydrofolate is a reaction similar to the above condensations in chemical nature and in pH dependence.[30] The triphasic pH-rate profile demands a two-step mechanism with a hydroxymethyl intermediate, together with a change in rate-determining step with pH. On the basis of a spectrophotometric detection of rapid formation of the intermediate in the alkaline region, followed by slow conversion of the intermediate to products, the pH dependence has been interpreted in terms of rate-determining acid-catalyzed dehydration of the hydroxymethyl intermediate in the alkaline region and general acid-catalyzed attack of tetrahydrofolic acid on formaldehyde in the acid region.

2. The nucleophilic reactions of carboxylic acid derivatives including esters[31] and amides proceed through an addition-elimination pathway with the formation of a tetrahedral intermediate, in a similar fashion to the reactions of carbonyl compounds.[32,33] The question can again be raised as to

which of these two steps, or both, is subject to general catalysis. In many instances, no data exist to answer this question. However, in a few special cases, extrakinetic arguments are available to specify the step subject to catalysis.

One of these arguments applies to the hydrolysis of ethyl trifluorothiolacetate.[34] The most interesting feature of this reaction is the pH-rate profile at zero buffer concentration (Fig. 5.4).[34] The combination of a pH-independent region near neutrality and a region at low pH dependent on pH can be interpreted as a general base catalysis by water and inhibition by oxonium ion. Catalysis by general bases and inhibition by general acids are also observed. These observations are consistent with general base-catalyzed nucleophilic attack of water on the ester bond leading to a tetrahedral intermediate that is partitioned unsymmetrically, collapsing spontaneously to products but regenerating the ester via general acid catalysis.

$$CF_3COSC_2H_5 + H_2O \underset{k_2[H_3O^\oplus]}{\overset{k_1[H_2O]}{\rightleftarrows}} CF_3\underset{OH}{\overset{OH}{\underset{|}{\overset{|}{C}}}}-SC_2H_5 \xrightarrow{k_3} CF_3COOH + C_2H_5SH$$

(5.16)

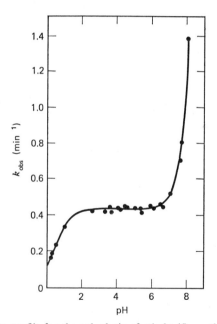

Fig. 5.4 The pH-Rate profile for the solvolysis of ethyl trifluorothiolacetate in H_2O at zero buffer concentration (30°; $I = 1.0\ M$). The solid line is theoretical and the points are experimentally determined values. From L. R. Fedor and T. C. Bruice, *J. Amer. Chem. Soc.*, **86**, 5697 (1964). © Copyright 1964 by the American Chemical Society. Reproduced by permission of the copyright owner.

An alternative to this mechanism is one involving attack of hydroxide ion on the ester to form a tetrahedral intermediate that is converted to products by oxonium ion catalysis. But this mechanism may be ruled out since the second-order rate constant for hydroxide attack would be ca. 10^{13} M^{-1} sec^{-1}, a value greater than the rate constant for a diffusion-controlled process. Tetrahedral intermediate formation and the specific mechanism of eq. 5.16 are substantiated by the observation of carbonyl oxygen exchange concurrent with hydrolysis, following the kinetics required by eq. 5.16, but not that of direct hydroxide ion attack.[35] In this reaction, one can unequivocally say that general catalysis occurs in the formation but not in the decomposition of the tetrahedral intermediate.

A different result is found in the alkaline hydrolysis of 2,2,2-trifluoro-N-methylacetanilide. This reaction was found to have both a first-order and second-order dependence on the hydroxide ion concentration.[36] Subsequently the second-order dependence on the hydroxide ion concentration was found to be a manifestation of general base catalysis superimposed on a first-order hydroxide ion reaction.[37] Thus the overall rate equation is

$$\frac{-d[A]}{dt} = [A]\{k_0 + [OH^\ominus](k_1 + \sum_i k_i[B_i])\} \qquad (5.17)$$

Since related anilide hydrolyses exhibiting second-order hydroxide ion dependence show concurrent carbonyl oxygen exchange and hydrolysis,[38] this general base catalysis must be associated with either the formation or decomposition of a tetrahedral intermediate. This conclusion is substantiated by the finding that while the hydrolysis is mixed first-order and second-order in hydroxide ion at low hydroxide ion concentrations, it changes to a strictly first-order dependence on hydroxide ion at high hydroxide ion concentrations,[39] modifying eq. 5.17 (when $B = OH^\ominus$)

$$\frac{-d[A]}{dt} = [A]\left\{k_0 + [OH^\ominus]\frac{k_a(k_1 + k_2[OH^\ominus])}{k_a + k_1 + k_2[OH^\ominus]}\right\} \qquad (5.18)$$

This finding not only requires a tetrahedral intermediate in the reaction, but also requires that the general catalysis be operative solely in the elimination of N-methylaniline from the tetrahedral intermediate. The requirement of a tetrahedral intermediate stems from the change in kinetic order of hydroxide ion that demands a two-step process with a change in rate-determining step as the hydroxide ion concentration increases. The requirement of the participation of general catalysis in the second step stems from the fact that the dependence on hydroxide ion concentration changes from second- to first-order, and not vice versa. This result is consistent only with a change in the rate-determining step from decomposition of the tetrahedral intermediate at low basicity (where general catalysis is observed) to the formation of the tetrahedral intermediate at high basicity (where general catalysis is not observed). (See the next section.) Thus the overall mechanism may be written as

$$\underset{\substack{\|\ \ \ |\\ \text{O}\ \ \text{CH}_3}}{\text{CF}_3\text{-C-N-C}_6\text{H}_5} + \text{OH}^\ominus \rightleftharpoons \underset{\substack{|\ \ \ \ \ |\\ \text{OH}}}{\underset{\substack{\text{O}^\ominus\ \text{CH}_3}}{\text{CF}_3\text{-C-NC}_6\text{H}_5}} \begin{array}{c} \xrightarrow{\text{low OH}^\ominus} \text{CF}_3\text{CO}_2^\ominus + \text{CH}_3\text{NHC}_6\text{H}_5 \\ \\ \xrightarrow{\text{high OH}^\ominus} \text{CF}_3\text{CO}_2^\ominus + \text{CH}_3\text{N}^\ominus\text{C}_6\text{H}_5 \end{array} \quad (5.19)$$

The hydrolyses of diethyl acetylmalonate and diethyl acetylethylmalonate, reactions involving the cleavage of a carbon–carbon bond, have very complicated pH-rate constant profiles exhibiting two regions in which the rate is independent of pH and two regions in which it is dependent on the hydroxide ion concentration. These results indicate a two-step process involving hydration of the carbonyl group followed by carbon–carbon cleavage, each process being catalyzed by either water or hydroxide ion,[38] as shown in the pinacolone decomposition (eq. 5.20).

$$\underset{\text{CH}_3}{\overset{\text{O}\ \ \ \text{CH}_3}{\underset{\|}{\text{CH}_3\text{-C}}}\diagdown\text{CH}_3} \underset{k_{-1},\ k_{-2}[\text{OH}^\ominus]}{\overset{k_1,\ k_2[\text{OH}^\ominus]}{\rightleftharpoons}} \underset{\text{OH}\ \ \text{CH}_3}{\overset{\text{OH}\ \text{CH}_3}{\text{CH}_3\text{-C}\diagdown\text{CH}_3}} \xrightarrow{k_3,\ k_4[\text{OH}^\ominus]} \underset{\text{CH}_3}{\overset{\text{O}}{\underset{\|}{\text{CH}_3\text{-C-O}^\ominus}}} + \text{H}\diagdown\overset{\text{CH}_3}{\underset{\text{CH}_3}{\text{-CH}_3}}$$

(5.20)

A number of extrakinetic arguments indicate that carbon–carbon cleavage is rate-determining at low pH and hydration of the carbonyl group is rate-determining at high pH, rather than the opposite. Since general base catalysis is seen at high pH but not at low pH, only the formation of this tetrahedral intermediate and not its decomposition is subject to general base catalysis. A number of other carbon–carbon cleavages, including the hydrolyses of 2-nitroacetophenone and of nitroacetone, show similar kinetic behavior[41] and probably show similar catalytic behavior. However, the alkaline hydrolysis of acetylacetone resembles the hydrolysis of 2,2,2-trifluoro-N-methylacetanilide rather than the carbon–carbon cleavages just discussed.[42] The rate expression of the acetylacetone hydrolysis has a term second-order in hydroxide ion, suggesting a rate-determining carbon–carbon cleavage. Although it has not as yet been shown that this reaction is general base-catalyzed, analogy with the hydrolysis of 2,2,2-trifluoro-N-methylacetanilide would suggest this possibility. It is of interest that those reactions with rate-determining decomposition of a tetrahedral intermediate involve reactants with a leaving group whose pK is greater than that of water (15.7): acetone (ca. 20) and N-methylaniline (ca. 20), while those reactions with rate-determining formation of a tetrahedral intermediate involve reactants with a leaving group pK equal to or less than that of water: diethyl malonate (15.2), ethanol (16), ethanethiol (9), phenol (10), acetate ion (4.7), and halide ion (ca. 0). This correlation, based on the partitioning of tetrahedral intermediates, suggests that the former set will be subject to general base catalysis but the latter set may or may not be, depending on the reactivity of the attacking nucleophile.

The aminolyses of ethyl benzimidate and ethyl m-nitrobenzimidate show a sharp pH-rate constant maximum.[43] For weak amines, including hydroxylamine, methoxylamine, and semicarbazide, general base catalysis is observed on the acidic side of the pH-rate constant profile. As in the oximation of acetone, the occurrence of a bell-shaped pH-rate constant profile implies the operation of a two-step mechanism with a change in rate-determining step from formation of the tetrahedral intermediate to its decomposition on the two sides of the bell. A number of arguments indicate that the low pH part of the bell involves rate-determining decomposition of the tetrahedral intermediate.[43] On this basis, general base catalysis is associated with the decomposition of the tetrahedral intermediate in this reaction. Likewise, the apparent general catalysis in the hydrolysis of ethyl benzimidates[43] can most easily be explained in terms of a general base-catalyzed decomposition of the conjugate acid of the tetrahedral intermediate[44]

$$\text{ArC}(\!\!=\!\!\text{NH})\text{OEt} + \text{H}^{\oplus} \rightleftharpoons \text{ArC}(\!\!=\!\!\text{NH}_2)\text{OEt}^{\oplus} \xrightleftharpoons{\text{H}_2\text{O}}$$

$$[\text{ArC}(\text{OH}_2)(\text{NH}_2)(\text{OEt})]^{\oplus} \xrightarrow[\text{slow}]{\text{B}} \text{ArCO}_2\text{Et} + \text{NH}_3 + \text{BH}^{\oplus} \quad (5.21)$$

on the following basis. The catalysis could not occur by protonation of ethyl benzimidate and formation of the tetrahedral intermediate since ethyl benzimidate is largely protonated in the acetate buffers used. In addition, general base catalysis of hydrolysis of a protonated benzimidate could not account for the Hammett rho of $+1.4$ found for substituted ethyl benzimidates,[44] since such a process should have a rho near zero. Both these pieces of information, however, are consistent with eq. 5.21. A similar argument may be made for the general catalysis of the hydrolysis of N,N'-diarylformamidinium ions.[45]

In the hydrolysis of o-carboxyphthalimide, in which a bell-shaped pH-rate constant profile is found, a similar two-step reaction must also take place. This reaction shows two pH dependencies in the region of pH 2-4, but the reactant has only one ionizable group. To account for this pH dependence, the (intramolecular) carboxyl group must be involved in *both* the formation of the tetrahedral intermediate *and* its decomposition. This reaction is one of a number in which there is a symmetrical catalytic partitioning of a tetrahedral intermediate, as opposed to the reactions listed previously, some of which may be described as reactions involving unsymmetrical catalytic partitioning of the intermediate.

The most important example of symmetrical catalytic partitioning of a tetrahedral intermediate is exemplified by

$$\underset{\text{RCX}}{\overset{O}{\|}} + YH \underset{BH^{\oplus}}{\overset{B}{\rightleftharpoons}} \underset{\underset{Y}{|}}{\overset{O^{\ominus}}{\underset{|}{RC-X}}} \underset{B}{\overset{BH^{\oplus}}{\rightleftharpoons}} \underset{\text{RCY}}{\overset{O}{\|}} + HX \qquad (5.22)$$

or its catalytic mirror image. It consists of the general base-catalyzed ethanolysis of ethyl trifluoroacetate.[47] This symmetrical reaction must show symmetrical partitioning of its tetrahedral intermediate, catalytically and otherwise. From the symmetry of the reaction, together with the requirement that the microscopic reverse of general acid catalysis is general base catalysis, eq. 5.22 (or its general acid-hydroxide ion equivalent) can be written. Other reactions approaching this symmetrical situation should conform to this scheme also. One of these is the general base-catalyzed hydrolysis of ethyl trifluoroacetate that shows a pH-independent pH-rate constant profile,[35] unlike that of ethyl trifluorothiolacetate, implying that while the latter reaction shows unsymmetrical partitioning the former reaction shows symmetrical partitioning.

An interesting property of such symmetrical reactions is that although both the formation and decomposition of the tetrahedral intermediate are catalyzed, the overall reaction is only first-order in catalyst. This conclusion follows from the fact that only one of the steps is rate-limiting (or an inequality of rate constants occurs). If the reaction proceeded with the same catalysis without an intermediate, the reaction would be second-order in catalyst, since then both the addition and decomposition processes would occur simultaneously. First-order dependence of a reaction on a general base cannot, however, serve as a criterion for a stepwise rather than a one-step process involving a single general base. Both reactions will show the same kinetic behavior.[48]

3. Aromatic substitution reactions, both nucleophilic and electrophilic, exhibit general catalysis. In aromatic nucleophilic substitution, a distinctive aspect of general catalysis emerges, which requires a two-step process with a tetrahedral intermediate and indicates that the general catalysis occurs in only one of the two steps. This phenomenon results in a curvilinear dependence of the rate constant on the catalyst concentration, for example, the reaction of p-nitrophenyl phosphate and dimethylamine (Fig. 5.5).[49] It shows a marked change in the effect of hydroxide ion and dimethylamine catalysts on the second-order rate constant from strong to essentially no catalysis at all with

increasing catalyst concentration.[49] Although this mechanism requires a two-step reaction and the appearance of general catalysis in one of the steps, it does not specify which step. The differentiation can be made by comparing the reactions of p-nitrophenyl phosphate and p-nitrochlorobenzene with the nucleophile, dimethylamine. In the latter reaction, no hydroxide ion or dimethylamine catalysis is seen, although the former reaction does show such catalysis. Assuming a two-step mechanism, the p-nitrochlorobenzene reaction must have $k_2 > k_{-1}$ as in eq. 5.23, since chloride ion is a much better leaving group than dimethylamine. Therefore, in this reaction step 1 must be rate-determining. This conclusion is supported by studies of a series of 1-substituted 2,4-dinitrobenzenes with piperidine. A large difference in the rate constants of these reactions would be expected if the expulsion of the 1-substituent is involved in the rate-determining step. However, there is essentially no difference, indicating that step 1 is rate-determining.[50] If the reaction of p-nitrochlorobenzene has k_1 as the rate-determining step, then eq. 5.23 must be the proper description of the catalysis in the reaction of dimethylamine with p-nitrophenyl phosphate, because in the chlorobenzene reaction no significant general base catalysis of this step by dimethylamine is seen.

$$R_2NH + XArNO_2 \underset{k_{-1}}{\overset{k_1}{\rightleftharpoons}} IH \xrightarrow{k_2[OH^\ominus], k_2[B]} \text{products} \qquad (5.23)$$

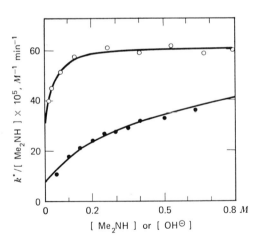

Fig. 5.5 Second-order rate constants for the nucleophilic aromatic substitution reaction of p-nitrophenyl phosphate and dimethylamine at 39°, ionic strength 1.0, as a function of hydroxide ion concentration in 0.2 M dimethylamine (○) and of dimethylamine concentration at pH 11.0 (●). From A. J. Kirby and W. P. Jencks, *J. Amer. Chem. Soc.*, **87**, 3217 (1965). © 1965 by the American Chemical Society. Reproduced by permission of the copyright owner.

where X stands for any substituent. This conclusion is consistent with the fact that the rate constant for the *p*-nitrochlorobenzene-dimethylamine reaction is similar to that for the *p*-nitrophenyl phosphate-dimethylamine reaction at very high base concentration, where a change in rate-determining step from a catalyzed to an uncatalyzed process is indicated.

Curvilinear dependence of a rate constant on the concentration of a general catalyst, changing from a strong dependence to essentially no dependence on catalyst concentration, characterizes many other reactions. Some of these are other aromatic nucleophilic substitutions such as the general base-catalyzed reactions of piperidine with 2,4-dinitroanisole and 2,4-dinitrophenyl phenyl ether.[51] Related reactions are aromatic electrophilic substitutions such as the azo coupling reaction of *p*-chlorobenzenediazonium ion with 2-naphthol-6,8-disulfonic acid, which is catalyzed by general bases.[52] Still other reactions include the general base-catalyzed hydrolysis of diethyl acetylmalonate,[40] the general acid-catalyzed reaction of hydroxylamine with amides,[53] the general base-catalyzed cyclization of glutamine and glutamate esters,[54] general acid-catalyzed semicarbazone formation,[55] and the general base-catalyzed hydrolysis of 2,2,2-trifluoro-*N*-methylacetanilide.[39]

As presented before, the change in catalyst dependence from strong dependence at low catalyst concentration to no dependence at high catalyst concentration is an argument that general catalysis is operative in the decomposition of the tetrahedral intermediate rather than its formation. The opposite is seen in semicarbazone formation.[56] If the opposite is true, no change in the catalyst dependence would occur. This argument stems mathematically from the steady-state expression for the observed rate constant for a two-step reaction.

$$k_{obs} = \frac{k_1}{\left(\frac{k_{-1}}{k_2}\right) + 1} \qquad (5.24)$$

If a general catalyst is operative in the k_1 step, its effect will always be seen, independently of the magnitude of the catalyst concentration. If, however, the general catalyst is associated with k_2, the catalyst dependence will be seen only when the quotient k_{-1}/k_2 [catalyst] is appreciable with respect to one. This analysis is in agreement with other criteria given above for reactions showing variable catalyst dependence.

5.3 MECHANISMS OF GENERAL ACID-BASE-CATALYZED REACTIONS

Having discussed mechanistic criteria for distinguishing between general base and nucleophilic catalysis and for distinguishing the particular step of a reaction in which general catalysis occurs, let us consider some examples of general acid-base catalysis, in order to specify the proton transfer and other changes occurring in these reactions. Two general classes of proton transfer reactions can be identified: (1) proton transfers to or from carbon; and (2) proton transfers to or from electronegative atoms such as nitrogen, oxygen, or sulfur. In the former reactions, rate-determining proton transfer to the exclusion of other processes is the rule. However, in the latter reactions, proton transfer is seldom solely rate-determining. We will discuss a few selected examples of each category in order to depict possible modes of behavior of proton transfer processes.

Proton Transfers to or from Carbon

A proton may be transferred from a general acid to a carbanion, an olefin,[57] or an aromatic compound in a rate-determining process. Some of these reactions are shown in Table 5.1. Since general base catalysis is the microscopic reverse of general acid catalysis, reversals of the above reactions show general base catalysis. The reverse of the first reaction is well known as a general base-catalyzed reaction; the others are not, only because the species involved are not easily available. Examples of the removal of a proton from ketones, esters, aldehydes, nitro compounds, and so on, are shown in Table 5.2.

Aromatic hydrogen exchange is an aromatic electrophilic substitution reaction in which the electrophile is either an oxonium ion or a general acid. The aromatic hydrogen exchanges of 1,3,5-trimethoxybenzene[58] and azulene[59] have been found to be subject to general acid catalysis in solutions of weak acids. In addition, in concentrated solutions of sulfuric acid or perchloric acid, proton donation to 1,3-dimethoxybenzene by both sulfuric acid and bisulfate ion appears probable on the basis of the 2.5 to 3-fold greater rate in sulfuric acid solution than in perchloric acid solution.[60]

For 1,3,5-trihydroxybenzene and 1,3,5-trimethoxybenzene[61] the rate constants of aromatic hydrogen exchange were found to be proportional to $h_0^{0.8}$ and $h_0^{1.146}$. Similar differences in equilibrium protonation of these two substances were also found, the former having a dependency on $h_0^{-1.10}$ and the latter $h_0^{-1.80}$. Thus acidity affects both the equilibrium and rate of these two reactions similarly, and this fact may be mechanistically meaningful.

The equation for this reaction may be written as

$$\text{ArH} + \text{H}^\oplus \rightleftharpoons [\ddagger] \longrightarrow \text{HArH}^\oplus \qquad (5.25)$$

From transition state theory the rate expression for eq. 5.25 and its dependence on the medium can be given by

$$\frac{v}{[\text{HAr}]} = k_{\text{obs}} = k^\circ [\text{H}^\oplus] \frac{f_{\text{H}^\oplus} f_{\text{HAr}}}{f^\ddagger} \qquad (5.26)$$

Following Leffler[62] the transition state may be assumed to be intermediate between the ground state (HAr + H$^\oplus$) and the product ion (HArH$^\oplus$). Defining α as the degree of resemblance of the transition state to the product ion HArH$^\oplus$, eq. 5.27 describing the activity coefficient of the transition state may be written

$$f^\ddagger = (f_{HAr}f_{H^\oplus})^{1-\alpha}(f_{HArH^\oplus})^\alpha \tag{5.27}$$

and thus eq. 5.26 may be transformed to

$$k_{obs} = k°[H^\oplus]\left(\frac{f_{HAr}f_{H^\oplus}}{f_{HArH^\oplus}}\right)^\alpha \tag{5.28}$$

Equilibrium protonation can be defined by eq. 5.29 for the overall equilibrium constant and eq. 5.30 for the experimentally determinable indicator ratio.

$$K = \frac{[H^\oplus]f_{H^\oplus}[HAr]f_{HAr}}{[HArH^\oplus]f_{HArH^\oplus}} \tag{5.29}$$

$$I = \frac{[HArH^\oplus]}{[HAr]} \tag{5.30}$$

Using eqs. 5.27 to 5.30, eq. 5.31 can be derived.

$$\frac{k_{obs}}{[H^\oplus]} = k_0 K\alpha\left(\frac{I}{[H^\oplus]}\right)^\alpha \tag{5.31}$$

Equation 5.31 shows a direct dependence of the observed rate constant for aromatic hydrogen exchange on the equilibrium constant for protonation of the same substance and also a direct relationship between the rate constant for aromatic hydrogen exchange and the indicator ratio of that same substance in a solution of particular acidity. When this relationship is tested by taking logarithms of eq. 5.31 and plotting the left-hand side of the equation versus log(I/[H$^\oplus$]), α is found to be 0.48 ± 0.03 for 1,3,5-trihydroxybenzene and 0.44 ± 0.03 for 1,3,5-trimethoxybenzene. The kinetic acidity dependence for these two reactions thus is consistent with a single reaction mechanism in which the transition state appears to resemble the reactant and the product equally.

The bromination of acetone, first investigated by Lapworth in 1904,[63] has been of special importance in the theories of physical organic chemistry. The fact that the base-catalyzed bromination is first-order in acetone, first-order in base, but zero-order in bromine led Lapworth to the idea of a stepwise process. In this process bromine appears not in the rate-determining step but rather in the fast product-determining step. In modern terminology the mechanism is written as

$$CH_3COCH_3 \underset{k_{-1}[BH^\oplus]}{\overset{k_1[B]}{\rightleftharpoons}} CH_3COCH_2^\ominus \xrightarrow{k_2[Br_2]} \text{product} \tag{5.32}$$

from which the following steady-state rate law can be derived.

$$k_{obs} = \frac{k_1 k_2 [Br_2][B]}{k_{-1}[BH^\oplus] + k_2[Br_2]} \quad (5.33)$$

When $k_2[Br_2] > k_{-1}[BH^\oplus]$, $k_{obs} = k_1[B]$. At high halogen concentration this formulation predicts that the rate of bromination should equal the rate of iodination, a fact demonstrated by Bartlett.[64] This scheme also predicts that halogenation, deuterium exchange, and racemization of ketones should have identical rates. These predictions have likewise been shown to be true.[65, 67] These data mean that the rate-determining step of base-catalyzed halogenation must involve a proton transfer; in other words, this is a general base-catalyzed reaction in which a proton is removed from carbon.

Acid-catalyzed halogenation of ketones follows eq. 5.34. It is subject to general acid catalysis[68] and follows the kinetic dicta[66, 67] listed above for general base catalysis.

$$CH_3COCH_3 \underset{k_{-1}[A^\ominus]}{\overset{k_1[HA]}{\rightleftarrows}} CH_3\overset{OH^\oplus}{\underset{\|}{C}}-CH_3 \underset{k_{-2}[HA]}{\overset{k_2[A^\ominus]}{\rightleftarrows}} CH_3\overset{OH}{\underset{|}{C}}=CH_2 \xrightarrow{k_3[X_2]} \text{product} \quad (5.34)$$

This mechanism is more complicated than that for general base catalysis because of the pre-equilibrium protonation of the ketone. Otherwise it is analogous.

There is, however, an ambiguity in the mechanisms for these general acid-catalyzed and base-catalyzed halogenation reactions. For example, kinetic dependence on a general base may be mechanistically described either in terms of the action of the general base (eq. 5.32) or of the action of hydroxide ion and the conjugate general acid (eq. 5.35).

$$CH_3COCH_3 \underset{k_{-1}}{\overset{k_1[HB]}{\rightleftarrows}} CH_3COCH_3 \cdot HB \xrightarrow{OH^\ominus} CH_3\overset{O^\ominus}{\underset{|}{C}}=CH_2 \quad (5.35)$$

Likewise, kinetic dependence on a general acid may be mechanistically described either in terms of general acid (and water) (eq. 5.36) or of oxonium ion and conjugate general base (eq. 5.34).

$$CH_3COCH_3 \underset{k_{-1}}{\overset{k_1[HA]}{\rightleftarrows}} CH_3COCH_3 \cdot HA \xrightarrow{H_2O} CH_3\overset{OH}{\underset{|}{C}}=CH_2 \quad (5.36)$$

This ambiguity (essentially of the position of the proton in the transition state) is met in essentially all general acid-base-catalyzed reactions. There are a number of ways in which this ambiguity will be met; some will be discussed in this section while others will be treated in the section on carbonyl reactions.

One of the methods of differentiating two kinetically indistinguishable mechanisms is to study an intermediate species that is completely protonated.

Mechanisms of General Acid-Base-Catalyzed Reactions   119

For example, in the racemization of D-a-phenylisocaprophenone in concentrated sulfuric acid, a decrease in rate constant occurs in going from 85 to 94% sulfuric acid solution. This decrease occurs at an acidity at which the ketone is essentially completely converted to its conjugate acid. It must therefore reflect the requirement for a base to abstract a proton from this conjugate acid, specifying eq. 5.34 rather than eq. 5.36 as the mechanism of the acid-catalyzed enolization of a ketone.[69]

Solvent deuterium isotope effects tend to confirm the description of eq. 5.34 for the catalysis of enolization by acetic acid.[70] For this reaction the solvent isotope effect $(k_{HOAc}/k_{DOAc}) = 1.09$ at 100°. Mechanism eq. 5.34 would lead to a solvent isotope effect described by

$$\frac{k_{HA}}{k_{DA}} = \frac{K_{DS}}{K_{HS}} \times \frac{K_{HA}}{K_{DA}} \times \frac{k_H}{k_D} \tag{5.37}$$

Since the solvent does not participate in the rate-determining step of mechanism 5.34, $k_H/k_D = 1$. Furthermore, $K_{HA}/K_{DA} = 3.3$ from experiment. Finally, it would be expected that K_{DS}/K_{HS} would be approximately 1/3.[71] Thus the ratio k_{HA}/k_{DA} should be approximately 1, as found experimentally. On the other hand, mechanism 5.36 predicts a solvent isotope effect of the order of 3 since it involves a rate-determining proton transfer. Therefore, the most reasonable explanation for the catalysis of enolization by acetic acid is mechanism 5.34.†

The identification of mechanisms 5.32 and 5.34 as the mechanisms of general base and general acid catalysis is consistent with the susceptibility of various ketones to these two kinds of catalysis. For example, the enolization of acetone in aqueous solution is strongly accelerated by general acids while the enolization of bromoacetone and of pyruvic acid is insensitive to acid catalysis.[72] The insensitivity of the latter compounds is related to a lower pK_a of the carbonyl groups in those compounds with electronegative substituents. This phenomenon has been treated quantitatively.[67]

Since the enolization of ketones can be catalyzed by both general acids and general bases, it is of interest to consider the possibility of catalysis by both species simultaneously (Chapter 10). In acetate buffer one sees not only catalysis of enolization by the various individual acid and basic species of the medium, but also a product term involving both acetic acid and acetate ion.[73] The more recent data of Bell and Jones[73] for the enolization of acetone in acetate buffers is given by

$$10^6 v = 5 \times 10^{-4}[H_2O] + 1600[H^\oplus] + 1.5 \times 10^7[OH^\ominus]$$
$$+ 5.0[HOAc] + 15[OAc^\ominus] + 20[HOAc][OAc^\ominus] \tag{5.38}$$

† Secondary solvent isotope effects are neglected, resulting in a weakened argument (see Ref. 102).

When both a general acid and a general base serve simultaneously as catalysts, two mechanistic possibilities can be discussed a priori. One can be described as catalysis in which the various acids and bases act individually.

$$HS + A \longrightarrow products$$

and (5.39)

$$HS + B \longrightarrow products$$

The other states that the acidic and basic catalysts act in a concerted manner.

$$B + HS + A \longrightarrow products \quad (5.40)$$

The product term appearing in eq. 5.38 is the crucial distinguishing feature between the consecutive processes of eq. 5.39 and the concerted process of eq. 5.40. Shortly after Dawson and Spivey's original data appeared, Pedersen[74] pointed out inconsistencies in the product term. Consider the data for each of the six terms of eq. 5.38 which are analyzed in terms of a concerted reaction by an acid and a base:

Velocity	5×10^{-4}	1600	1.5×10^7	5	15	20	
HA	H_2O	$H_3O^\oplus$	H_2O	HOAc	H_2O	HOAc	(5.41)
B	H_2O	H_2O	$OH^\ominus$	H_2O	$OAc^\ominus$	$OAc^\ominus$	

Comparison of the first and fifth terms involving water and acetate ion as bases shows a difference of 3×10^4; however, a comparison of the fourth and sixth terms again involving a comparison of water and acetate ion as bases shows a difference of only 4. Since these two ratios should be identical, the requirement of a concerted process is not met. However, Swain[75] then pointed out that Pedersen had not included some of the equivalent kinetic possibilities of the various terms and proceeded to explain the data of eq. 5.38 in terms of a concerted mechanism. The general expression for a concerted mechanism must take the form of

$$k = \sum_i \sum_j k_{ij}[A_i][B_j] \quad (5.42)$$

Swain made the assumption that eq. 5.42 could be replaced by

$$k = \sum_i k'_i[A_i] \sum_j k'_j[B_j] \quad (5.43)$$

which is equivalent to saying that the relative effectiveness of an acid is independent of the base with which it acts and vice versa. Thus eq. 5.38 can be expressed in terms of

$$k = c([H_2O] + a_1[HA] + a_2[H^\oplus])([H_2O] + b_1[A^\ominus] + b_2[OH^\ominus]) \quad (5.44)$$

The six terms of eq. 5.38 correspond to the (nine) terms of eq. 5.44. Equation

Mechanisms of General Acid-Base-Catalyzed Reactions

5.44 can be expanded and the coefficients compared with those of eq. 5.38. When this is done eq. 5.45 results, giving a fit to eq. 5.38

$$k = 10^{-5}([H_2O] + 10^2[HOAc] + 1.7 \times 10^6[H^\oplus])([H_2O] \\ + 1.4 \times 10^4[OAc^\ominus] + 3 \times 10^{10}[OH^\ominus]) \quad (5.45)$$

within a factor of 2 for any given term. However, Bell and Jones[73] pointed out that the agreement between eqs. 5.45 and 5.38 is illusory, since the term, $10^2[HOAc]$, in eq. 5.45 contributes very little to the coefficients of [HOAc] and [OAc$^\ominus$] in eq. 5.38, but enters directly in the coefficient of [HOAc][OAc$^\ominus$]. This factor can therefore be adjusted within wide limits to fit the product term. In fact, any coefficient from 0 to 1000 would suffice. In quantitative terms this criticism can be expressed in the following way: when one compares the coefficients of eq. 5.45 with those of eq. 5.38 the coefficient for acetic acid catalysis and acetate ion catalysis are given by

$$\frac{k_{HOAc}}{C} = a_1[H_2O] + \underline{a_2 b_1 K_a} \quad (5.46)$$

$$\frac{k_{OAc^\ominus}}{C} = \underline{b_1[H_2O]} + a_1 b_2 \frac{K_w}{K_a} \quad (5.47)$$

Under the experimental conditions of the enolization of acetone with acetate buffers the underlined terms of eqs. 5.46 and 5.47 are dominant. Thus in using eqs. 5.46 and 5.47 to fit eq. 5.48 the product term, a_1, can be varied within wide limits to fit the data, since k_{HOAc} and $k_{OAc^\ominus}$ are dominated as shown.

$$\frac{k_{HOAc \times OAc^\ominus}}{C} = a_1 b_1 \quad (5.48)$$

The dominant terms of eqs. 5.46 and 5.47 lead to the ratio of the catalytic coefficients for acetic acid and acetate ion catalysis shown by

$$\frac{k_{HOAc}}{k_{OAc^\ominus}} = \frac{a_2 b_1 K_a}{b_1[H_2O]} = \text{constant} \times K_a \quad (5.49)$$

The Brønsted relationship also gives a ratio of the catalytic coefficients of the acetic acid and acetate ion terms.

$$k_{HOAc} = G_{HOAc} K_{HOAc}^\alpha \quad (5.50)$$

$$k_{OAc^\ominus} = G_{OAc^\ominus} \left(\frac{1}{K_{HOAc}}\right)^\beta \quad (5.51)$$

$$\frac{k_{HOAc}}{k_{OAc^\ominus}} = \text{constant} \times K_{HOAc}^{(\alpha+\beta)} \quad (5.52)$$

The equivalence of eqs. 5.49 and 5.52 requires that $(\alpha + \beta) = 1$. However, for the enolization of acetone, $\alpha = 0.88$ and $\beta = 0.55$. The sum of the two equals 1.43, very different from 1. This inconsistency suggests that eq. 5.43, which assumes independence of the relative reactivity of acids and bases, is incorrect. For a concerted process of the kind described here either the product term must contribute a large part of the observed velocity of the catalyzed reaction or else the relationship $\alpha + \beta = 1$ must hold. Since neither of these criteria is found in the halogenation of acetone, one must conclude that a concerted process is not operative here and that a stepwise process such as shown in eq. 5.39 or its counterpart is the correct interpretation of catalysis.

This conclusion gives no explanation of the product term involving both acetic acid and acetate ion in the halogenation of acetone. One may question the meaning of this term since earlier discussion ruled out product terms involving acetic acid and hydroxide ion, a stronger base, and acetic acid and water, a weaker base. The explanation of the acetic acid-acetate ion term probably involves some special feature. One possibility is the operation of the acetic acid-acetate ion complex as a (single) general basic catalyst as has been shown.[76]

Proton Transfers to or from Electronegative Atoms

Although proton transfers to or from carbon can quite clearly be slower than other bond-making or bond-breaking processes, proton transfers to or from electronegative atoms such as oxygen, nitrogen, or sulfur, whose rate constants are often diffusion-controlled, are in general faster than other bond-making and bond-breaking processes (Chapter 3). The description of general acid base catalysis involving proton transfers to and from such atoms is therefore one of considerable interest. Several typical examples of these reactions will be discussed from the viewpoint of the relation of the proton transfers to other processes occurring in these reactions. Two general mechanisms may be advanced for these processes: either a solely rate-determining proton transfer or a rate-determining proton transfer concerted with heavy atom bond making/breaking.

The hydrolysis of ortho esters exhibits a simple version of general acid catalysis. This reaction, long the object of study,[77] has been described severally as a fast proton transfer followed by a rate-determining carbon–oxygen cleavage,

$$RC(OR)_3 \underset{H_2O}{\overset{H_3O^{\oplus}}{\rightleftharpoons}} RC(OR)_2\overset{\overset{H^{\oplus}}{|}}{\underset{}{OR}} \xrightarrow{slow} RC^{\oplus}(OR)_2 \xrightarrow[H_2O]{fast} R\overset{\overset{O}{\|}}{C}OR + ROH \quad (5.53)$$

a fast proton transfer followed by participation of the conjugate base of the catalyst in a subsequent rate-determining cleavage,

$$RC(OR)_3 \underset{H_2O}{\overset{H_3O^{\oplus}}{\rightleftharpoons}} RC(OR)_2 \begin{matrix} H^{\oplus} \\ | \\ OR \end{matrix} \xrightarrow[\text{slow}]{A^{\ominus}} RC^{\oplus}(OR)_2 \longrightarrow \text{products} \quad (5.54)$$

a slow proton transfer reaction concerted with carbon–oxygen cleavage,

$$A\text{—}H + \overset{R}{\underset{|}{O}}\text{—}CR(OR)_2 \longrightarrow [A\overset{\delta\ominus}{\cdots}H\cdots\overset{R}{\underset{|}{O}}\overset{\delta\oplus}{\cdots}CR(OR)_2] \longrightarrow$$
$$A^{\ominus} + ROH + RC^{\oplus}(OR)_2 \longrightarrow \text{products} \quad (5.55)$$

or a slow proton transfer reaction followed by a fast carbon–oxygen cleavage.

$$RC(OR)_3 + H_3O^{\oplus} \underset{}{\overset{\text{slow}}{\rightleftharpoons}} RC(OR)_2(\overset{H}{\underset{\oplus}{O}}R) + H_2O \quad (5.56)$$
$$\xrightarrow{\text{fast}} \text{products}$$

Of the mechanisms, only eqs. 5.54 through 5.56 conform to general acid catalysis, eq. 5.56 consisting solely of a rate-determining proton transfer while eqs. 5.54 and 5.55 consist of a rate-determining step involving proton transfer and other bond cleavage.

Several arguments exclude eq. 5.53. For one thing it does not explicitly take into account the possibility of general acid catalysis. Secondly, calculations of rate constants following eq. 5.53 are not consistent with general acid catalysis. Using known or calculated values of equilibrium constants for the protonation of ortho esters, the calculated rate constant for the rate-determining step of eq. 5.53 is 2×10^{11} sec^{-1} (a value greater than diffusion-control) and thus it is extremely improbable that $k_{-1} > k_2$ for this reaction.[78-80]

Equation 5.54 may be excluded since several good nucleophiles, hydroxylamine, semicarbazide, and iodide ion, have no effect on the rate of hydrolysis of ortho esters, although the first two nucleophiles can divert a considerable amount of the carbonium ion to product. Good evidence therefore exists for a carbonium ion intermediate, and at the same time eq. 5.54 can be eliminated.

The effect of substituents on the hydrolysis of ortho esters is very difficult to rationalize in terms of eq. 5.53 but is explicable in terms of mechanism 5.55. Although acetal hydrolysis shows as much as 10^7 spread in rate constant with varying stabilization of the carbonium ion, ortho ester hydrolysis is insensitive to structural changes that would normally lead to appreciable carbonium ion stabilization. A change from an orthoformate to an orthocarbonate results in a decrease in rate constant, even though a carbonium ion should be appreciably stabilized in this transformation. In addition, the

Hammett rho constant of the oxonium ion-catalyzed hydrolysis of four substituted methyl orthobenzoates in aqueous methanol correlates with σ rather than σ^+.[81] Finally, the Hammett rho plot (relating structure and reactivity) of the chloroacetic acid-catalyzed hydrolysis of these substituted methyl orthobenzoates is convexly curved.[81] These substituent effects certainly cannot be explained in terms of either eq. 5.53 or eq. 5.54. Mechanism 5.53 requires a transition state close to the carbonium ion. The implication of the substituent effects is that the transition state is not close to the carbonium ion at all. Mechanism 5.55 is consistent with these data if carbon–oxygen cleavage has not proceeded very far in the transition state.

One may calculate the forward rate constant for proton transfer from an oxonium ion to a molecule such as triethyl orthoacetate (eq. 5.56) using the estimated ionization constant of this substrate and the maximal rate constant for the reverse process 10^{11} M^{-1} sec^{-1}. On this basis, the forward rate constant of eq. 5.56 is estimated to be 10^4 sec^{-1} (as a maximal value). This value is in good agreement with the experimentally determined proton rate constant of 10^4 sec^{-1}. However, this agreement is probably fortuitous since it is difficult to conceive of this carbon–oxygen bond cleavage being faster than 10^4 sec^{-1}. In this reaction, however, no absolute refutation of eq. 5.56 can be given and one must consider both it and eq. 5.55 as possibilities for the mechanism. A carbon-14 kinetic isotope effect could distinguish between these two possibilities. If eq. 5.55 is operative, the proton would attack the electrons of the carbon–oxygen bond, but if eq. 5.56 is operative, the proton would attack the lone pair on an oxygen atom. Since the calculation given above applies only to eq. 5.56, eq. 5.55 is not ruled out.

General acid-base catalysis is widespread in additions to the carbonyl group and reverse reactions. Both general acid catalysis and general base catalysis are found in the addition of weak nucleophilic reagents to the carbonyl group. As before, general acid-base catalysis of these reactions can involve either a rate-determining proton transfer unaccompanied by any bond-making to carbon, or a concerted process in which proton transfer occurs at the same time or very nearly the same time as bond formation to carbon.[28, 29]

If a proton transfer alone is the rate-determining step two mechanisms for the reaction may be written

$$HX + R_2C{=}O \underset{}{\overset{fast}{\rightleftharpoons}} HX^\oplus{-}CR_2{-}O^\ominus \underset{A^\ominus \text{ slow}}{\overset{HA}{\rightleftharpoons}} H{-}X^\oplus{-}CR_2OH \overset{fast}{\rightleftharpoons} H^\oplus + XCR_2OH \quad (5.57)$$

$$HX + R_2C{=}O + H^\oplus \rightleftharpoons HX^\oplus{-}CR_2OH \underset{HA \text{ slow}}{\overset{A^\ominus}{\rightleftharpoons}} XCR_2OH \quad (5.58)$$

Mechanisms 5.57 and 5.58 are kinetically indistinguishable because of the equilibrium, $H_2O + HA \rightleftharpoons H_3O^\oplus + A^\ominus$, as discussed before. In the forward

Mechanisms of General Acid-Base-Catalyzed Reactions

direction, these reactions involve a general acid proton donation or a general base proton removal in the rate-determining step. The reverse reactions are described in the opposite manner. Since these reactions require proton transfer from oxygen, nitrogen, or sulfur to be slower than bond-making to a carbon atom, they are unlikely. If $HX^{\oplus}\!-\!\overset{\diagdown}{\underset{\diagup}{C}}\!-\!O^{\ominus}$ is a stronger base than $A^{\ominus}$, the free energy of proton transfer to the intermediate is negative and therefore the slow step of reaction 5.57 is diffusion-controlled. In most instances, this inequality in base strength will hold (although it is not mandatory). When it does, it is difficult to conceive of a pre-equilibrium addition of HX to the carbonyl group that can be faster than diffusion-controlled. Therefore, this mechanistic possibility cannot be taken seriously. However, when $HX^{\oplus}\!-\!\overset{\diagdown}{\underset{\diagup}{C}}\!-\!O^{\ominus}$ is a weaker base than $A^{\ominus}$, this dictum no longer applies.

Several pieces of experimental evidence make a solely rate-determining proton transfer improbable. One is that carbon-14 kinetic isotope effects have been found in the formation of semicarbazones, hydrazones, and 2,4-dinitrophenylhydrazones from carbonyl compounds containing carbon-14 in the carbonyl group.[82] Since an isotope effect can occur only if the carbonyl carbon atom participates in the rate-determining step, this fact rules out eq. 5.57 or 5.58. If the free energy of proton transfer of reaction 5.57 or 5.58 is negative, no dependence of the rate of the reaction on the pK_a of the general catalyst should be observed, a result contrary to experiment. Thus one must exclude mechanisms 5.57 or 5.58 for general acid (and general base)-catalyzed reactions when the free energy of the proton transfer is negative.

More reasonable mechanisms of these general acid and base catalyses involve concerted addition of HX and a proton transfer.

$$HX + \underset{CH_3}{\overset{CH_3}{\diagdown}}\!\!=\!\!O + HA \underset{}{\overset{slow}{\rightleftarrows}} H\!-\!X^{\oplus}\!-\!\underset{CH_3}{\overset{CH_3}{|}}\!\!C\!-\!O\!-\!H + A^{\ominus} \overset{fast}{\rightleftarrows}$$

$$H^{\oplus} + A^{\ominus} + X\!-\!\underset{CH_3}{\overset{CH_3}{|}}\!\!C\!-\!OH \quad (5.59)$$

$$A^{\ominus} + HX + \underset{CH_3}{\overset{CH_3}{\diagdown}}\!\!=\!\!OH^{\oplus} \overset{slow}{\rightleftarrows} A\!-\!H + X\!-\!\underset{CH_3}{\overset{CH_3}{|}}\!\!C\!-\!OH$$

$$A^{\ominus} + H^{\oplus} + HX + \underset{CH_3}{\overset{CH_3}{\diagdown}}\!\!=\!\!O \quad (5.60)$$

$$A^\ominus + HX + \underset{CH_3}{\overset{CH_3}{>}}\!\!=\!\!O \underset{\text{slow}}{\rightleftharpoons} A\!-\!H + X\!-\!\underset{CH_3}{\overset{CH_3}{\underset{|}{C}}}\!-\!O^\ominus \underset{\text{fast}}{\rightleftharpoons} A^\ominus + X\!-\!\underset{CH_3}{\overset{CH_3}{\underset{|}{C}}}\!-\!OH \quad (5.61)$$

$$A^\ominus + HX + \underset{CH_3}{\overset{CH_3}{>}}\!\!=\!\!O \underset{\text{fast}}{\rightleftharpoons} X^\ominus + \underset{CH_3}{\overset{CH_3}{>}}\!\!=\!\!O + HA \underset{\text{slow}}{\rightleftharpoons} X\!-\!\underset{CH_3}{\overset{CH_3}{\underset{|}{C}}}\!-\!OH + A^\ominus \quad (5.62)$$

These mechanisms are intermolecular only from a mechanistic standpoint, not from a kinetic standpoint. Presumably any two of the three reactants can form a loose hydrogen-bonded complex first and then react in a slow step to give the products. Mechanism 5.59 may be described in the forward direction as a general acid-catalyzed reaction and in the reverse direction as an oxonium ion-general base-catalyzed reaction. Mechanism 5.60 can be described in the foward direction as an oxonium ion general base-catalyzed reaction and in the reverse direction as a general acid-catalyzed reaction. Mechanisms 5.59 and 5.60 are kinetically indistinguishable for reasons given above, and are equivalent to one another in a mirror image sense. In a similar fashion, mechanism 5.61 involves a general base catalysis in the forward direction and general acid-alkoxide ion catalysis in the reverse direction, while mechanism 5.62 involves general acid-alkoxide ion catalysis in the forward direction and general base catalysis in the reverse direction. Again, mechanisms 5.61 and 5.62 are kinetically indistinguishable.

There are several methods for distinguishing between mechanistic pairs such as eqs. 5.59 and 5.60 or eqs. 5.61 and 5.62.[28] (1) A substrate containing an alkyl group can be substituted for one containing a hydrogen atom. The position of the alkyl group is, of course, fixed by synthesis; thus, the ambiguous position of the proton can be defined by the unambiguous position of the alkyl group if analogy between the alkyl group and the proton is correct. If the alkyl-substituted substrate reacts at the same rate as the protonated compound, it can be inferred that the proton is in the same position as the alkyl group. (2) The reaction can be examined under conditions of pH in which the substrate is completely protonated or deprotonated. Obviously under such conditions a catalyst cannot be used to protonate or deprotonate the substrate. (3) A mechanism can be rejected if the calculated rate constant for a given step of the reaction, usually the rate-determining step, is greater than the rate constant for a diffusion-controlled reaction. (4) The sensitivity to general acid catalysis, α, changes as the nucleophilicity of the attacking reagent increases. Opposite direction of the change is predicted when a general acid-catalyzed reaction is viewed as such or is viewed as the kinetically equivalent general base-oxonium ion reaction. Thus, one can distinguish the individuals of a pair from one another. This conclusion can be reached from structure-reactivity considerations,[83] or from the application of the "solvation rule" and the "reacting-bond rule."[84] Each of these methods will be utilized to distinguish between the above mechanisms.

Let us consider first the use of an alkyl group in place of a proton. The hydrolyses of acetals, ketals, and glycosides proceed via specific acid catalysis, involving a pre-equilibrium addition of a proton followed by a rate-determining heterolysis. The reverse of these reactions is a rate-determining addition of alcohol to the carbonyl group according to eq. 5.60, except that an alkyl group rather than a proton is on the carbonyl oxygen atom. Since the hydrolysis of acetals shows no general acid catalysis, the reverse reaction must show no general base catalysis. But such catalysis is well known for the addition of water and alcohols to carbonyl compounds. Thus the hydrolysis of acetals is not a model for carbonyl hydration, and eq. 5.60 may be ruled out. By process of elimination, eq. 5.59 can be considered the correct description for the addition of alcohols and semicarbazides to carbonyl compounds.

Oxime formation is subject to general acid catalysis. Specifically the dehydration of the carbinolamine forming the oxime is general acid-catalyzed at pH's of neutrality and above where this step is rate-determining. Oxime formation using hydroxylamine and nitrone formation using N-methylhydroxylamine show complete parallelism. Both reactions proceed through a carbinolamine intermediate whose rate-determining decomposition above pH 5 is subject to general acid catalysis with identical Brønsted α values of 0.77.[85] These data require the same mechanism for oxime and nitrone formation, which is compatible with substitution of a methyl group for a proton. The reverse of eq. 5.60 (where O = N and X = O) fulfills this requirement, but not the reverse of eq. 5.59, because proton removal from the nitrogen atom is not possible in nitrone formation.

Secondly, the decomposition of formocholine chloride, the reverse of the general base mechanism of eq. 5.62, is not subject to general base catalysis. In order to obviate this contradiction, eq. 5.61 must be preferred for general base-catalyzed addition to the carbonyl group. Some of the other approaches to specifying the position of the proton follow. The hydrolysis of benzylidineanilines is strongly inhibited by strong acids. The rate-determining step in this process is the loss of amine from the carbinolamine, the reverse of eq. 5.59 or 5.60. In strong acid solutions, there can be no shortage of acid molecules to protonate the nitrogen atom of the adduct; therefore, eq. 5.60 cannot account for the inhibition. However, proton removal from oxygen by the solvent (eq. 5.59) would decrease in strong acid.

The rate constant for addition of semicarbazide to the protonated form of p-nitrobenzaldehyde may be calculated from the observed rate constant for the oxonium ion-catalyzed reaction and the known equilibrium constant for the protonation of the aldehyde, yielding 2.7×10^{12} M^{-1} sec^{-1}. For a general base-catalyzed addition of semicarbazide to protonated benzaldehyde, an even higher value must occur. Since both these rate constants exceed that of the fastest diffusion-controlled process, eq. 5.60, the model for this calculation, must be ruled out and eq. 5.59 must be operative.

Susceptibility to general acid catalysis in the addition of nucleophiles to carbonyl groups is variable. Table 5.4 illustrates a variation in the Brønsted α from zero (no general acid catalysis) to ca. 0.7, depending on the basicity of the nucleophile. As one might expect intuitively, strong bases require no general acid catalysis whereas weak bases do. This variation in susceptibility to general acid catalysis, which occurs in the decomposition of the carbinolamine to give the product, is in agreement with the reverse of eq. 5.60 (with O = N and X = O) on the basis of both structure-reactivity[83] and "solvation rule" and "reacting bond rule"[84] considerations (see preceding information).

On the basis of the above criteria, the following generalizations can be drawn. In carbonyl addition reactions, general acid catalysis involves proton transfer to and from the carbonyl oxygen atom whereas general base catalysis involves proton transfer to and from HX, the attacking nucleophile. On the other hand, in the reactions of imines, general acid catalysis involves proton transfer to and from the attacking group. In carbonyl addition reactions, general acid and base catalyses are not simply related to one another. This behavior is in contrast to the enolization of ketones where the two catalyses are directly related to one another, differing only in the presence of a proton. In general acid catalysis of carbonyl and imine addition reactions, opposite

Table 5.4

Brønsted Slopes in General Acid Catalysis of the Addition of Nucleophiles to Carbonyl Groups[28, 83, 87, 88]

Nucleophile	pK of Nucleophile	Carbonyl Compound	α
RNH_2	~10	$X\Phi CHO^a$	0
$RS^{\ominus}$	~9	CH_3CHO	0
$CN^{\ominus}$	9.4	Several	~0
$SO_3^{2\ominus}$	7.0	ΦCHO	~0
NH_2OH	6.0	ΦCHO	Very small
$\Phi NHNH_2$	5.2	ΦCHO	0.2
ΦNH_2	4.6	$pCl\Phi CHO$	0.25
$H_2NNHCONH_2$	3.7	$X\Phi CHO$	0.25
Urea	0.2	CH_3CHO	0.45
H_2NCSNH_2	−1.3	$HCHO$	0.27
H_2O	−1.7	CH_3CHO	0.54
RSH	−7	CH_3CHO	~0.7

[a] Φ = phenyl.

catalytic behavior is seen, the former involving the carbonyl oxygen atom and the latter the attacking group, indicating the sensitivity of the catalysis to structural changes in the molecule.

The extensive data on the general acid-base-catalyzed dehydration of formaldehyde hydrate and mutarotation of glucose allows probing the mechanism of general catalysis. Using Tables 5.5 and 5.6, one can estimate the individual

Table 5.5
Acid-Catalyzed and Base-Catalyzed Dehydration of Formaldehyde Hydrate[89]

Acidic Species	p^b	q^c	K	$k_A{}^a$	$k_B{}^a$
H_2O	2	1	1.8×10^{-16}	0.0051/55.5	1600
H_3BO_3	3	3	6.2×10^{-10}	0.005	3.1
H_6TeO_6	6	1	2.0×10^{-8}	0.030	0.45
$H_2PO_4{}^{\ominus}$	2	3	6.2×10^{-8}	0.088	0.39
$HSO_3{}^{\ominus}$	1	3	6.2×10^{-8}	—	0.22
$H_2PO_3{}^{\ominus}$	1	3	1.5×10^{-7}	0.044	0.22
$H_2AsO_4{}^{\ominus}$	2	3	1.7×10^{-7}	0.079	0.29
Me_2AsOOH	1	2	5.4×10^{-7}	0.031	0.11
$NH_3OH^{\oplus}$	3	1	1.07×10^{-6}	0.020	0.032
C_6Cl_5OH	1	1	5.5×10^{-6}	—	0.022
$C_5H_5NH^{\oplus}$	1	1	6.0×10^{-6}	—	0.015
CMe_3CO_2H	1	2	8.9×10^{-6}	0.025	0.022
$MeCO_2H$	1	2	1.75×10^{-5}	0.043	0.022
$C_6H_5CH{=}CHCO_2H$	1	2	3.65×10^{-5}	—	0.017
$p\text{-}Me \cdot C_6H_4 \cdot CO_2H$	1	2	4.24×10^{-5}	—	0.015
$C_6H_5CH_2CO_2H$	1	2	4.88×10^{-5}	—	0.015
$CO_2H \cdot CO_2{}^{\ominus}$	1	4	5.41×10^{-5}	—	0.026
$C_6H_5CO_2H$	1	2	6.30×10^{-5}	—	0.013
$o\text{-}MeO \cdot C_6H_4 \cdot CO_2H$	1	2	8.06×10^{-5}	—	0.015
$o\text{-}Me \cdot C_6H_4 \cdot CO_2H$	1	2	1.24×10^{-4}	—	0.010
$CH_2OH \cdot CO_2H$	1	2	1.48×10^{-4}	—	0.012
HCO_2H	1	2	1.77×10^{-4}	0.070	0.013
$NH_2CONHNH_3{}^{\oplus}$	3	1	2.24×10^{-4}	0.059	0.0048
$p\text{-}NO_2 \cdot C_6H_4 \cdot CO_2H$	1	2	3.61×10^{-4}	—	0.0068
HF	1	1	6.71×10^{-4}	—	0.018
$C_6H_5OCH_2 \cdot CO_2H$	1	2	6.75×10^{-4}	—	0.0057
$o\text{-}HO \cdot C_6H_4 \cdot CO_2H$	1	2	1.05×10^{-3}	—	0.0052
$CH_2Cl \cdot CO_2H$	1	2	1.36×10^{-3}	0.088	0.0047
$HSO_4{}^{\ominus}$	1	4	1.03×10^{-2}	—	0.0066
$H_3O^{\oplus}$	3	1	55.5	2.7	0.0051/55.5

[a] $M^{-1} \sec^{-1}$.
[b] Number of ionizable protons in the acidic species molecule.
[c] Number of positions in the conjugate base that will accept protons.

Table 5.6

General Acid-Base Catalysis of the Mutarotation of Glucose[a, 90]

Acidic Species	K^c	$k_A{}^b$	$k_B{}^b$
H_2O	1.8×10^{-16}	4.9×10^{-6}	90.7
$C_6H_{12}O_6$	3.5×10^{-13}		1
CH_3CO_2H	1.75×10^{-5}	1.2×10^{-4}	1.48×10^{-4}
HCO_2H	1.77×10^{-4}	2.9×10^{-4}	8.6×10^{-4}
$H_3O^\oplus$	55.5		4.9×10^{-6}

[a] 25°.
[b] $M^{-1} \sec^{-1}$.
[c] M.

rate constants of catalytic processes. For these calculations, we will arbitrarily choose general base catalysis of these reactions, with water, acetate ion, and hydroxide ion acting as general bases. The basic mechanism assumed in these calculations is eq. 5.63, which conforms to our earlier conclusions concerning general base catalysis in reactions at the carbonyl group (eq. 5.61).

$$B + \underset{CH_3}{\overset{R-O}{>}}\!\!\!\!<\!\!\!\!\underset{CH_3}{\overset{OH}{}} \underset{k_2}{\overset{k_1}{\rightleftharpoons}} \underset{CH_3}{\overset{R-O}{>}}\!\!\!\!<\!\!\!\!\underset{CH_3}{\overset{O^\ominus}{}} + HB^\oplus \xrightarrow{k_3} \text{products} \quad (5.63)$$

With this equation and the over-all catalytic rate constant, one can calculate individual rate constants in one of two ways: (1) assume a fast pre-equilibrium followed by a rate-determining decomposition, as implied by our previous discussion; or (2) put no restriction on values of any of the rate constants. Eigen has followed the former practice. With an estimated value of an assumed pre-equilibrium constant and the over-all catalytic constant, he has calculated values of k_3 for the dehydration of acetaldehyde hydrate (Fig. 5.6).[91] This calculation leads to the contradictory result that some values of k_3 are larger than diffusion-controlled rates. In order to explain this anomaly, Eigen has suggested that the stepwise process of eq. 5.63 is incorrect and that the reaction is better described as a "one-encounter" reaction[85] where the carbon–oxygen bond break occurs in the same encounter as the proton transfer, although the two processes are not precisely concerted with one another. Another possible explanation of this contradiction is that the calculation itself is incorrect. In order to calculate individual rate constants on the basis of a steady-state treatment, eq. 5.64 can be used.

$$k_{obs} = \frac{k_1}{\frac{k_2}{k_3} + 1} \quad (5.64)$$

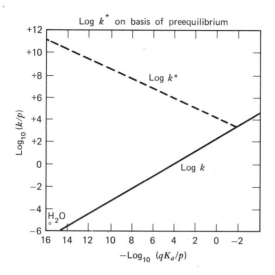

Fig. 5.6 Brønsted plot of the catalytic rate constants (M^{-1} sec^{-1}) of various acids in the dehydration of acetaldehyde hydrate according to Bell and Higginson.[125] Solid line: measured rate constant against pK_a of catalyst (including statistical correction factors p and q). This linear relationship is obeyed by more than 50 substances showing a mean deviation of 0.15 log units throughout the whole pK range. Broken line: rate constant k^* for rate-limiting step (deprotonation) assuming a stepwise mechanism.

$$\text{HS + HA} \underset{}{\overset{\text{rapid}}{\rightleftarrows}} \text{HS}^{\oplus}\text{H + B}$$
$$\text{HS}^{\oplus}\text{H + B} \longrightarrow \text{SH + HA}$$

For pK_{HS} (HS = hydrated acetaldehyde) a value of -2 has been chosen arbitrarily. From M. Eigen, *Disc. Faraday Soc.*, **39**, 13 (1965).

Values of k_1 are estimated on the basis of: (1) the relationships given in Chapter 2 between the pK_a of an acid and/or base and its rate constant with water; and (2) a diffusion-controlled rate constant of 10^{10} M^{-1} sec^{-1} for a proton transfer process with a negative free energy. Using this procedure, the values of the individual rate constants shown in Table 5.7 were calculated. The values in this table are all reasonable, in contrast to some in Fig. 5.6. Furthermore, the rate-determining step changes from k_1 for reactions catalyzed by water and acetate ion to k_3 for reactions catalyzed by hydroxide ion. Thus the prohibition against a (solely) rate-determining proton transfer given above must be softened in those cases where the free energy of the proton transfer process is positive. (The reverse of such a process must have a negative free energy and a diffusion-controlled rate constant.) The free energy-reaction coordinate profiles for all of these reactions, however, show that the highest peak over which the reaction must travel is the k_3 peak. Thus, if one defines the rate-determining step in this fashion, all the reactions have the same rate-determining step. From the above considerations, no rigorous case can be made for a "one-encounter" general catalysis. We shall take up this question again in Chapter 10 on multiple catalysis.[86]

General base catalysis of nucleophilic reactions of carboxylic acid derivatives is beset with the same ambiguities as those in carbonyl reactions with respect to the position of the proton. These reactions which include the aminolysis and ammonolysis of oxygen esters[35, 48, 92–99] and thiol esters[34, 35, 94–96, 101–105] and imidates,[44] the hydrolysis of esters, amides,[37–40] and amidines,[45, 46] and carbon–carbon cleavages,[41, 43] can be described as two-step processes with the intermediacy of a tetracovalent addition compound. As discussed earlier, either the formation or decomposition of the tetrahedral intermediate may be subject to general catalysis. Following earlier arguments we will assume that proton transfer is probably not the sole rate-determining process. General base catalysis of the formation of the tetrahedral intermediate can occur by either eq. 5.65, involving removal of a proton by a general base from the attacking nucleophile, or eq. 5.66, involving polarization of the carbonyl group by donation of a proton from the conjugate general acid. Likewise, general base catalysis of the decomposition of the tetrahedral intermediate can occur by eq. 5.67 (the reverse of eq. 5.66) or by eq. 5.68, (the reverse of eq. 5.65). Mechanisms 5.65 and 5.68 are analogous to the preferred mechanisms in carbonyl addition reactions, and are therefore preferred here.

Table 5.7

Calculated Rate Constants of Mechanism 5.64 on a Steady-State Basis[100]

Reaction	Catalyst	k_1	k_2	k_3
		\multicolumn{3}{c}{(M^{-1} sec^{-1})}		
Mutarotation of glucose[a]	Water	5×10^{-2}	10^{10}	2×10^4
Dehydration of formaldehyde hydrate[b]	Water	1×10^{-2}	10^{10}	10^8
Mutarotation of glucose	Acetate ion	0.10	10^{10}	1.4×10^5
Dehydration of formaldehyde hydrate	Acetate ion	10	10^{10}	2.2×10^7
Mutarotation of glucose	Hydroxide ion	10^{10}	10^8	1
Dehydration of formaldehyde hydrate	Hydroxide ion	10^{10}	10^8	16

[a] pK_a glucose = 12.46.
[b] pK_a formaldehyde hydrate = 13.29.

Mechanisms of General Acid-Base-Catalyzed Reactions

$$HX + R\overset{O}{\overset{\|}{C}}Y \underset{BH^\oplus}{\overset{B}{\rightleftharpoons}} R\overset{O^\ominus}{\underset{X}{\overset{|}{C}-Y}} \xrightarrow{\text{fast}} \text{products} \qquad (5.65)$$

$$R\overset{O}{\overset{\|}{C}}Y + X^\ominus \underset{B}{\overset{BH^\oplus}{\rightleftharpoons}} R\overset{OH}{\underset{X}{\overset{|}{C}-Y}} \xrightarrow{\text{fast}} \text{products} \qquad (5.66)$$

$$HX + R\overset{O}{\overset{\|}{C}}Y \rightleftharpoons R\overset{OH}{\underset{X}{\overset{|}{C}-Y}} \underset{BH^\oplus}{\overset{B}{\rightleftharpoons}} R\overset{O}{\overset{\|}{C}}\diagdown_X + YH \qquad (5.67)$$

$$HX + R\overset{O}{\overset{\|}{C}}Y \rightleftharpoons R\underset{XH^\oplus}{\overset{O^\ominus}{\overset{|}{C}-Y}} \overset{\text{fast}}{\rightleftharpoons} R\underset{X}{\overset{O^\ominus}{\overset{|}{C}-Y}} \xrightarrow[-B]{+BH^\oplus} R\overset{O}{\overset{\|}{C}}\diagdown_X + HY \qquad (5.68)$$

Preference for these mechanisms is consistent with the fact that strong bases need no general catalysis and that the fewest unstable intermediates are formed in these mechanisms. In the general base-catalyzed hydrolysis of ethyl trifluorothioacetate,[34] eq. 5.65 is preferred on the grounds that the calculated rate constant of eq. 5.66 leads to a value greater than diffusion-controlled. Although general acid catalysis of carboxylic acid reactions has not been mentioned, obviously the reverse of any reaction discussed here is subject to general acid catalysis. General acid catalysis is especially prevalent in the reactions of amines with thiol esters where the rate constant of general acid catalysis is of the same order of magnitude as the rate constant of general base catalysis.[106] In fact, in the reaction of n-butyl thiolacetate with hydroxylamine, an equilibrium between a general base pathway and a general acid pathway was suggested.[105] Experiments probing the factors underlying the occurrence of general acid catalysis in contrast to general base catalysis will undoubtedly be fruitful.

Similar to the carboxylic acid reactions are the aromatic nucleophilic substitution reactions. These processes involve the formation of a tetrahedral intermediate whose decomposition is subject to general base catalysis. Two mechanisms may be written for this catalysis, eqs. 5.69 and 5.70, assuming no proton transfer is solely rate-determining.

$$B\frown H\overset{\oplus}{N}R_2 \frown X \xrightarrow{\text{slow}} \text{products} \qquad (5.69)$$

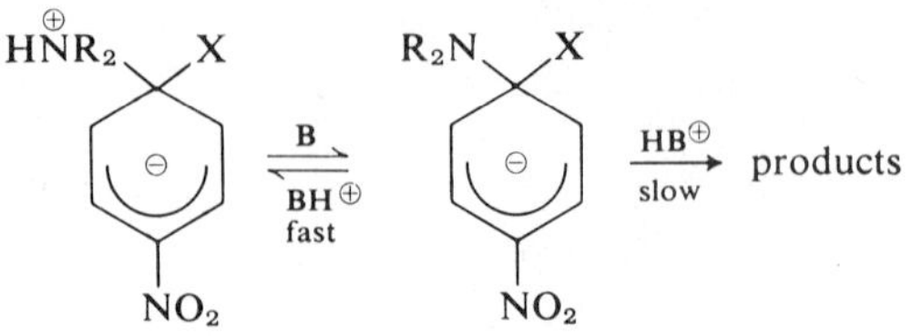

$$\underset{NO_2}{\overset{H\overset{\oplus}{N}R_2 \quad X}{\bigcirc}} \underset{\underset{\text{fast}}{BH\oplus}}{\overset{B}{\rightleftarrows}} \underset{NO_2}{\overset{R_2N \quad X}{\bigcirc}} \overset{HB\oplus}{\underset{\text{slow}}{\longrightarrow}} \text{products} \qquad (5.70)$$

Since these reactions occur in strongly basic solution, the pre-equilibrium in eq. 5.70 would be expected to take place, thus making it the best description of the catalysis in which the conjugate general acid facilitates the loss of X.

In nucleophilic substitution and elimination reactions in aprotic solvents, general acids, including phenols, carboxylic acids, and alcohols, facilitate reactions proceeding through carbonium ion intermediates.[87,88] These reactions have in common an incipient halide ion in the transition state. The simplest explanation for the catalysis involves proton transfer (or hydrogen bonding) to the halide ion, thus stabilizing the transition state.

5.4 SUSCEPTIBILITY TO GENERAL ACID-BASE CATALYSIS

General acid-base catalysis is a widespread but not universal phenomenon in organic reactions. Earlier, consideration was given to the occurrence of general or specific catalysis in a given reaction. Let us now try to specify the influence of the structure of the reagents in a related family of reactions on the susceptibility of the reaction to general catalysis. What we wish to probe is essentially a multiple structure-reactivity (linear free energy) correlation.[107] That is, we wish to determine the effect of the reactivity of the substrate or nucleophile on the Brønsted α or β values, a measure of the susceptibility of the reaction to general catalysis.

By combining the Hammett and Brønsted equations for a family of general acid-catalyzed reactions, one can obtain

$$\frac{\sigma_i}{\alpha_0 - \alpha_i} = \frac{pK_{a_2} - pK_{a_1}}{\rho_2 - \rho_1} = C \qquad (5.71)$$

where α is a Brønsted coefficient, ρ and σ refer to the Hammett equation, and the pK's are those of the general acids A_1 and A_2. For a family of base-catalyzed reactions one can obtain the similar expression

$$\frac{\sigma_1}{\beta_i - \beta_0} = \frac{pK_{a_1} - pK_{a_2}}{\rho_1 - \rho_2} = C \qquad (5.72)$$

Susceptibility to General Acid-Base Catalysis

where the symbolism is similar. Equation 5.71 indicates that electron withdrawing substituents on the substrate will result in a smaller value of α for general acid catalysis; that is, as the reactivity of the substrate increases, the value of α will decrease. Equation 5.72 is a similar relationship in which increased reactivity of the substrate results in a lowered sensitivity to general base catalysis, that is, a decreased value of β.

In a similar manner, a relationship between the Brønsted parameters, α or β, and the reactivity of a nucleophile participating in a general acid or base reaction can be derived from the Brønsted and Swain-Scott equations. For general acid catalysis the relationship is

$$\frac{pK_{a_2} - pK_{a_1}}{S_1 - S_2} = \frac{n_k}{\alpha_k - \alpha_0} = C \tag{5.73}$$

and for general base catalysis, it is

$$\frac{pK_{a_2} - pK_{a_1}}{S_1 - S_2} = \frac{\beta_0 - \beta_i}{n_i} = C \tag{5.74}$$

in which n is a measure of nucleophilic reactivity, often proportional to the basicity of the nucleophile, and S is a measure of the sensitivity to nucleophilic reactivity. These relationships can be expressed in two different ways. Equation 5.73 states that if a substrate is made more reactive because of catalysis by a stronger acid, its sensitivity to the reactivity of the nucleophile will decrease; alternatively, eq. 5.73 states that as the nucleophilicity (basicity) of the attacking reagent increases, the sensitivity to the strength of the catalyzing acid, α, decreases. Similar statements can be made for eq. 5.74.[28,88,108] These equations can be considered to be quantitative statements of the considerations of Hammond,[109] Leffler,[63] and Swain and Thornton.[111]

Using basicity as a measure of nucleophilicity eq. 5.73 can be transformed[88] to

$$\frac{\alpha_2 - \alpha_1}{pK_{a_1}^N - pK_{a_2}^N} = C \tag{5.75}$$

where pK_a^N refers to the pK_a of the conjugate acid of the nucleophilic reagent. A test of eq. 5.74, using the data of Table 5.4, is shown in Fig. 5.7. Of the 10 points, only one is deviant. This deviation may be associated with the fact that this point refers to reaction with formaldehyde, a more reactive aldehyde than any of the others used in this correlation (the correlation, of course, really requires constancy of substrate which has not been adhered to because of lack of data). This correlation leads to the conclusion that the general acid stabilizes the carbonyl oxygen, and not the nitrogen, sulfur, or other nucleophilic atom in the transition state, since an increase in pK of the

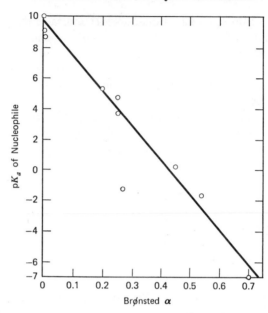

Fig. 5.7 Relationship between reactivity (pK_a) of the nucleophile and the susceptibility to general acid catalysis in a series of additions to the carbonyl group. Data of Table 5.4.

nucleophile should require less stabilization of the oxyanion formed (since the transition state should be closer to the reactants). Further, the correlation is incompatible with a mechanism in which the conjugate base of the acid removes a proton from the nucleophile since β ($=1-\alpha$) should decrease with increasing pK_a of the nucleophile (again since the transition state is closer to reactants), which requires that α increase with increasing pK_a of the nucleophile, a situation not found experimentally.

The correlations given above indicate that general catalysis will be seen to a greater extent with substrates and nucleophiles of low reactivity. Since these substances represent the majority of reactants in enzyme-catalyzed reactions, it is not surprising that general catalysis is seen frequently in such reactions.

Thus far we have discussed the effect of structural variation of the reactants on general catalysis. Let us now discuss the effect of structural variations of intermediates on general catalysis. A blanket statement can be made that reactions proceeding through stable intermediates need no general catalysis, but reactions proceeding through unstable intermediates need general catalysis. This statement is a corollary of the generalization that catalysis is seen and becomes important when it is most needed.[28] Since the preferred reaction pathway is that which requires the stablest transition state [and thus the

Susceptibility to General Acid-Base Catalysis

stablest intermediate(s)], that catalytic pathway will be favored that leads to the stabilization of that particular group in a transition state that is inherently the most unstable. For example, reactions at the carbonyl group are aided most by the avoidance of the formation of the highly unstable fully protonated carbonyl group through the transfer of a proton to the carbonyl oxygen atom from a general acid catalyst. On the other hand, since protonation of a nitrogen atom occurs more readily than an oxygen atom, catalysis of the reactions of imines occurs through the stabilization of the incipient hydroxide ion in the transition state by a transfer of a proton to or from the oxygen atom by a catalyst molecule.

Comparison of various possible catalyzed and uncatalyzed pathways can lead to a clearer picture of the function of a general catalyst in stabilizing transition states and, relatedly, unstable intermediates. Some hypothetical pathways for the hydration of a cationic imine or the dehydration of a carbinolamine are shown in eq. 5.76.[85]

$$
\begin{array}{c}
 {\scriptstyle \delta\oplus} {\scriptstyle \delta\ominus} \\
{>}N{=\!\!=}C{\cdots}OH \rightleftharpoons {>}\overset{\oplus}{N}{=}C{<} + {}^{\ominus}OH \\
 1 \\
\end{array}
\quad \pm H^{\oplus}
$$

$$
{>}N{-}\underset{|}{C}{-}OH \overset{\pm HA}{\rightleftharpoons} {>}\overset{\delta\oplus}{N}{=\!\!=}\underset{|}{C}{\cdots}O{\cdots}H{\cdots}\overset{\delta\ominus}{A} \overset{\pm A^{\ominus}}{\rightleftharpoons} {>}\overset{\oplus}{C}{=}N{<} \quad (5.76)
$$

$$
\pm HA^{\oplus} \qquad H \qquad H_2O
$$

$$
{>}N{-}\underset{|}{\overset{\oplus}{C}}{-}O{\overset{H}{\diagdown}} \rightleftharpoons {>}\overset{\delta\oplus}{N}{=\!\!=}\underset{|}{\overset{\delta\oplus}{C}}{\cdots}O{\overset{H}{\diagdown}} \quad
$$

The upper and lower pathways are for the attack and expulsion of hydroxide ion and water, respectively, while the middle path shows general acid catalysis. In the course of this reaction, there is a large change in the acidity and basicity of the reacting groups. In the absence of proton transfer, the basicity of the oxygen atom of the carbinolamine intermediate will increase by some 18 pK units if it becomes hydroxide ion (upper path); a comparable change will occur in the acidity of water if it becomes the conjugate acid of the carbinolamine (lower path). The transition state for the upper path should occur near the products since hydroxide ion is a poor leaving group.[109] Consequently a considerable basicity should have developed in the incipient hydroxide ion, and significant stabilization should occur by interaction with a general acid. Likewise, the transition state for the lower path should occur near the reactants since water is a poor nucleophile.[109] Thus the protonated carbinolamine will be highly acidic and stabilization will be expected by interaction with a catalyzing base. In both these hypothetical situations, interaction with the catalyst alters the transition state toward the center of the reaction coordinate as shown in Fig. 5.8.[110]

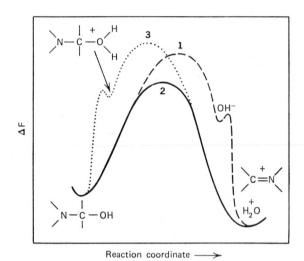

Fig. 5.8 Schematic transition-state diagram for the reaction mechanisms shown in the upper (- - - -), center (———), and lower (······) lines of eq. 5.76 at neutrality. From J. E. Reimann and W. P. Jencks, *J. Amer. Chem. Soc.*, **88**, 3981 (1966). © 1966 by the American Chemical Society. Reprinted by permission of the copyright owner.

These considerations are consistent with the generalizations given above that catalysis occurs where it is most needed and that reactions proceed by pathways that avoid the formation of unstable intermediates. These generalizations have been stated in the form of a rule: in general-catalyzed reactions involving proton transfer to or from nitrogen, oxygen, or sulfur, bases will react with atoms which become more acidic in the transition state (and product).[85]

5.5 THE POSITION OF THE PROTON IN THE TRANSITION STATE OF GENERAL-CATALYZED REACTIONS

General catalysis in the overall sense involves the transfer of a proton, preceded by hydrogen bonding (Chapter 2). On this basis, the efficiency of a general catalysis may depend on several factors: (1) the equilibrium constant of formation of the hydrogen-bonded complex; (2) the degree of bond polarization induced by complex formation; and (3) the rate of proton transfer.[108] However, evaluation of such factors depends on the intimate description of the proton transfer process occurring in general catalysis.

The Position of the Proton

One of the incisive methods of probing the position of the proton in proton transfer reactions is the kinetic isotope effect, using either deuterium or tritium. In proton transfers to or from carbon, a kinetic isotope effect roughly reflecting the difference in zero point energies of carbon–hydrogen versus carbon–deuterium (or carbon–tritium) bonds is seen.[112-117] However, as indicated in Chapter 4, the kinetic isotope effect is dependent on the relation between the pK of the substrate and the pK of the general catalyst, a maximal value of the isotope effect occurring when the difference in the pK's of the proton donor and acceptor is zero. This phenomenon is consistent with the prediction of a maximal kinetic isotope effect when the proton is exactly half-transferred from donor to acceptor in the transition state.[118] Thus in general catalyses of proton transfer to or from carbon, one can define the position of the proton with some degree of assurance.

One exception must be made to the above statement. In several proton transfer reactions involving carbon, in particular, the removal of protons from nitromethane by general bases, the deuterium kinetic isotope effect ($k(CH_3NO_2)/k(CD_3NO_2)$) with hindered bases such as 2,6-lutidine and 2,4,6-collidine is much larger than the maximum predicted from a difference in the zero-point energies of vibration of carbon–hydrogen and carbon–deuterium bonds.[119] This result together with curvature in the Arrhenius plots of certain proton transfer reactions[120] has led to the suggestion of the occurrence of quantum mechanical tunnelling that is more facile for a proton than a deuteron.[121]

Parallel kinetic isotope effect studies have been carried out in general-catalyzed reactions in which proton transfer occurs only between electronegative atoms. These isotope effects are necessarily carried out by observing deuterium oxide solvent isotope effects since any proton on an electronegative atom will very rapidly equilibrate with a protic solvent. Unfortunately in such solvent isotope effect studies, so-called secondary isotope effects, arising from solvation differences, complicate the primary isotope effects associated with proton transfer.[122] A few attempts have been made to disentangle the primary and secondary isotope effects in such systems.[24, 27, 123, 124] Probably the most interesting to date is the work of Schowen and co-workers.[124] On the basis of the assumption that the secondary contributions to the solvent isotope effect are given by the following:

$$\frac{k^{H_2O}}{k^{D_2O}} = \left[\left(\frac{k^{H_2O}}{k^{D_2O}}\right)_{max}\right]^{\beta} \tag{5.77}$$

where β is the Brønsted slope and $(k^{H_2O}/k^{D_2O})_{max}$ is estimated according to a number of simple rules,[123] the primary isotope effects in the general base-catalyzed hydrolysis of 2,2,2-trifluoro-N-methylacetanilide and in the general base-catalyzed mutarotation of glucose have been calculated. In the former reaction, the primary isotope effect is calculated to be 4.6, in agreement with that calculated from the β of 4.3 for the reaction.[124] In the latter reaction a primary isotope effect of 2.4 is found whereas a primary isotope effect of 6.8 is calculated from the Brønsted β of 0.4. On this basis the authors conclude that the anilide hydrolysis is a reaction showing a true rate-determining proton transfer (without nitrogen–carbon cleavage) while the mutarotation of glucose is one in which the transfer of the proton occurs in a fast step after the rate-determining elimination of alkoxide ion from the carbonyl group. Thus the former reaction is one that is normally associated with a rate-determining proton transfer. The function of the general catalyst in the latter reaction, still an important catalytic process, is the specific solvation of the alkoxide ion.[123] These two variants of general catalysis, which may be viewed as extremes, are shown in Fig. 5.9.[123] The difference in stability of the ammonium ion (of the anilide) and oxonium ion (of glucose) with respect to the corresponding anionic forms is suggested as the fundamental reason for the differing behavior.

Undoubtedly, more probes of this kind to specify general acid-base catalysis more closely will be seen in the future. In viewing proton transfer it should be kept in mind that the transition state for large atom movement and the transition state for proton transfer need not coincide: one may precede or follow the other. Certainly the potential energy of the proton is affected by the change in the properties of the groups between which it is transferred and vice versa. In addition, the proton is highly mobile with respect to the large atoms (Chapter 2). Thus the true position of the proton may be indeterminate and it may be improper to specify its position in the manner described above.[83] In the transition state, the proton has almost certainly moved toward the atom to which it is transferred; in this sense, the reaction is a concerted proton transfer and large atom movement.

The previous discussion has been concerned with stepwise mechanisms for proton transfers, with no discrete mention of the participation of the protic solvent. However, proton transfers through intervening water molecules may have to be considered. In many of the reactions described here, the proton initially on the nucleophile may be accepted by the conjugate base

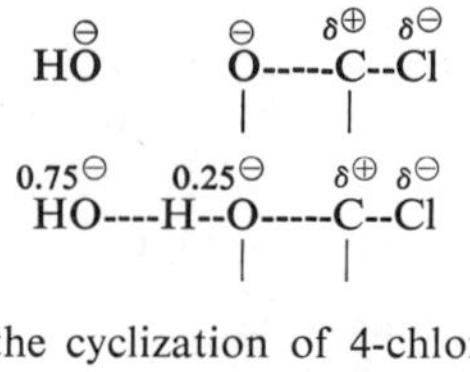

$$\underset{}{\text{H}\overset{\ominus}{\text{O}}} \quad \underset{|}{\overset{\ominus}{\text{O}}}\text{-----}\underset{|}{\overset{\delta\oplus}{\text{C}}}\text{--}\overset{\delta\ominus}{\text{Cl}}$$

$$\underset{}{\overset{0.75\ominus}{\text{HO}}}\text{----}\text{H}\text{--}\underset{|}{\overset{0.25\ominus}{\text{O}}}\text{-----}\underset{|}{\overset{\delta\oplus}{\text{C}}}\text{--}\overset{\delta\ominus}{\text{Cl}}$$

Fig. 5.9 Transition states in the cyclization of 4-chlorobutanol. From C. G. Swain, D. A. Kuhn, and R. L. Schowen, *J. Amer. Chem. Soc.*, **87**, 1556 (1965). © by the American Chemical Society. Reprinted by permission of the copyright owner.

of the general acid catalyst, possibly through the intermediacy of water molecule(s), before the diffusion-controlled separation of the products. This "one-encounter" mechanism[85] will be discussed more in Chapter 10.

REFERENCES

1. R. P. Bell, *The Proton in Chemistry*, Cornell University Press, Ithaca, N.Y., 1959, p. 134.
2. M. Kilpatrick, Jr., *J. Amer. Chem. Soc.*, **50**, 2891 (1928); A. R. Butler and V. Gold, *J. Chem. Soc.*, 2305 (1961).
3. W. P. Jencks and J. Carriuolo, *J. Biol. Chem.*, **234**, 1272, 1280 (1959).
4. C. A. Bunton and J. H. Fendler, *J. Org. Chem.*, **31**, 2307 (1966).
5. G. DiSabato and W. P. Jencks, *J. Amer. Chem. Soc.*, **83**, 4393 (1961).
6. M. Caplow and W. P. Jencks, *Biochemistry*, **1**, 883 (1962).
7. T. C. Bruice and S. J. Benkovic, *J. Amer. Chem. Soc.*, **86**, 418 (1964).
8. J. F. Kirsch and W. P. Jencks, *J. Amer. Chem. Soc.*, **86**, 833 (1964).
9. M. L. Bender and B. W. Turnquest, *J. Amer. Chem. Soc.*, **79**, 1652 (1957).
10. S. A. Bernhard and H. Gutfreund, *Proc. Int. Symp. Enzyme Chem.*, Tokyo and Kyoto, Maruzen, Tokyo, 1958, p. 124.
11. A. R. Butler and V. Gold, *J. Chem. Soc.*, 1334 (1962).
12. W. P. Jencks and J. Carriuolo, *J. Amer. Chem. Soc.*, **82**, 1778 (1960); B. M. Anderson and W. P. Jencks, *ibid.*, **82**, 1773 (1960); D. G. Oakenfull, T. Riley, and V. Gold, *Chem. Comm.*, 365 (1966).
13. Brønsted coefficients larger than one and less than zero exist for proton removal from carbon acids, F. G. Bordwell et al., *J. Amer. Chem. Soc.*, **91**, 4002 (1969).
14. J. A. Feather and V. Gold, *Proc. Chem. Soc.*, 306 (1963).
15. F. Covitz and F. H. Westheimer, *J. Amer. Chem. Soc.*, **85**, 1773 (1963).
16. J. A. Feather and V. Gold, *J. Chem. Soc.*, 1752 (1965).
17. V. Gold and E. G. Jefferson, *ibid.*, 1409 (1953); A. R. Butler and V. Gold, *ibid.*, 4362 (1961).
18. J. O. Edwards and R. G. Pearson, *J. Amer. Chem. Soc.*, **84**, 16 (1962).
19. R. G. Pearson and D. N. Edgington, *ibid.*, **84**, 4607 (1962); see Chapter 4 of this book.
20. M. Eigen, *Angew. Chem. Int. Ed.*, **3**, 1 (1964).
21. S. L. Johnson, *J. Phys. Chem.*, **67**, 495 (1963).
22. J. F. Kirsch and W. P. Jencks, *J. Amer. Chem. Soc.*, **86**, 833 (1964).
23. W. P. Jencks, F. Barley, R. Barnett, and M. Gilchrist, *ibid.*, **88**, 4464 (1966).
24. M. L. Bender, E. J. Pollock, and M. C. Neveu, *J. Amer. Chem. Soc.*, **84**, 595 (1962); B. M. Anderson, E. H. Cordes, and W. P. Jencks, *J. Biol. Chem.*, **236**, 455 (1961).
25. W. P. Jencks and J. Carriuolo, *J. Amer. Chem. Soc.*, **83**, 1743 (1961).
26. A. Williams and M. L. Bender, *ibid.*, **88**, 2508 (1966).
27. C. A. Bunton and V. J. Shiner, Jr., *J. Amer. Chem. Soc.*, **83**, 3214 (1961).
28. W. P. Jencks, *Prog. Phys. Org. Chem.*, **2**, 63 (1964).
29. There is not much general acid catalyses here because of the difference in basicity of the species. W. P. Jencks, personal communication. For a different example see G. J. Buist, C. A. Bunton, and J. Lomas, *J. Chem. Soc.*, B, 1099 (1966); the aldolization reaction has been investigated in detail: C. D. Gutsche et al., *J. Amer. Chem. Soc.*, **89**, 1235 (1967).

30. R. G. Kallen and W. P. Jencks, *Fed. Proc.*, **24**, 541 (1965); *J. Biol. Chem.*, **241**, 5851 (1966).
31. S. B. Tove, *Biochim. Biophys. Acta*, **57**, 230 (1962); B. Birks and P. Boyer, *Cereal Chem.*, **28**, 483 (1951).
32. M. L. Bender, *Chem. Rev.*, **60**, 53 (1960).
33. L. do Amaral, K. Koehler, D. Bartenbach, T. Pletcher, and E. H. Cordes, *J. Amer. Chem. Soc.*, **89**, 3537 (1967); S. O. Eriksson, *Acta Pharm. Suec.*, **6**, 121 (1969); J. Brown, S. C. K. Su, and J. A. Shafer, *J. Amer. Chem. Soc.*, **88**, 4468 (1966); S. C. K. Su and J. A. Shafer, *J. Org. Chem.*, **34**, 926, 2911 (1969).
34. L. R. Fedor and T. C. Bruice, *J. Amer. Chem. Soc.*, **86**, 5697 (1964).
35. M. L. Bender and H. d'A. Heck, *J. Amer. Chem. Soc.*, **89**, 1211 (1967).
36. S. S. Biechler and R. W. Taft, Jr., *J. Amer. Chem. Soc.*, **79**, 4927 (1957). See also M. J. Gregory and T. C. Bruice, *ibid.*, **89**, 2121 (1967).
37. R. L. Schowen and G. W. Zuorick, *Tetrahedron Lett.*, 3839 (1965); *J. Amer. Chem. Soc.*, **88**, 1223 (1966).
38. M. L. Bender and R. J. Thomas, *J. Amer. Chem. Soc.*, **83**, 4183 (1961).
39. R. L. Schowen, H. Jayaraman, and L. Kershner, *Tetrahedron Lett.*, 497 (1966); *J. Amer. Chem. Soc.*, **88**, 3373 (1966).
40. G. E. Lienhard and W. P. Jencks, *J. Amer. Chem. Soc.*, **87**, 3855 (1965).
41. R. G. Pearson, D. H. Anderson, and L. L. Alt, *J. Amer. Chem. Soc.*, **77**, 527 (1955).
42. R. G. Pearson and E. A. Mayerle, *J. Amer. Chem. Soc.*, **73**, 926 (1951).
43. E. S. Hand and W. P. Jencks, *J. Amer. Chem. Soc.*, **84**, 3505 (1962).
44. R. H. DeWolfe and F. B. Augustine, *J. Org. Chem.*, **30**, 699 (1965).
45. R. H. DeWolfe, *J. Amer. Chem. Soc.*, **82**, 1585 (1960); R. H. DeWolfe and R. M. Roberts, *J. Amer. Chem. Soc.*, **75**, 2942 (1953).
46. B. Zerner and M. L. Bender, *J. Amer. Chem. Soc.*, **83**, 2267 (1961).
47. S. L. Johnson, *J. Amer. Chem. Soc.*, **86**, 3819 (1964).
48. T. C. Bruice and S. J. Benkovic, *Bioorganic Mechanisms*, Vol. I, W. A. Benjamin, Inc., New York, 1966, p. 85.
49. A. J. Kirby and W. P. Jencks, *J. Amer. Chem. Soc.*, **87**, 3217 (1965).
50. J. F. Bunnett, E. W. Garbisch, Jr., and K. M. Pruitt, *J. Amer. Chem. Soc.*, **79**, 385 (1957).
51. J. F. Bunnett and R. H. Garst, *J. Amer. Chem. Soc.*, **87**, 3879 (1965), and references therein; J. F. Bunnett and C. Bernasconi, *J. Amer. Chem. Soc.*, **87**, 5209 (1965).
52. H. Zollinger, *Helv. Chim. Acta*, **38**, 1597, 1617, 1623 (1955).
53. W. P. Jencks and M. Gilchrist, *J. Amer. Chem. Soc.*, **86**, 5616 (1964).
54. R. B. Martin, A. Parcell, and R. I. Hedrick, *J. Amer. Chem. Soc.*, **86**, 2406 (1964).
55. E. H. Cordes and W. P. Jencks, *J. Amer. Chem. Soc.*, **84**, 4319 (1962).
56. W. P. Jencks, personal communication.
57. An elegant series of reactions have been carried out in aprotic medium. A. Schriesheim et al., *J. Amer. Chem. Soc.*, **84**, 3160 (1962); **85**, 2111 (1963); A. Zwierzak and H. Pines, *J. Org. Chem.*, **27**, 4084 (1962); H. Pines and W. M. Stalick, *Tetrahedron Lett.*, **34**, 3723 (1968); N. J. Turro and W. B. Hammond, *J. Amer. Chem. Soc.*, **87**, 3258 (1965); D. J. Cram and R. D. Guthrie, *ibid.*, **88**, 5760 (1966).
58. A. J. Kresge and Y. Chiang, *J. Amer. Chem. Soc.*, **81**, 5509 (1959); **83**, 2877 (1961); **89**, 4471 (1967).
59. S. Colapietro and F. A. Long, *Chem. Ind.*, 1056 (1960); B. C. Challis and F. A. Long, *J. Amer. Chem. Soc.*, **85**, 2524 (1963); J. Schulze and F. A. Long, *J. Amer. Chem. Soc.*, **86**, 331 (1964); R. J. Thomas and F. A. Long, *J. Amer. Chem. Soc.*, **86**, 4770 (1964).

References

60. A. J. Kresge, L. E. Hakka, S. Mylonakis, and Y. Sato, *Disc. Faraday Soc.*, **39**, 75 (1965).
61. A. J. Kresge, R. A. More O'Farrell, L. E. Hakka, and V. P. Vitullo, *Chem. Comm.*, 46, (1965).
62. J. E. Leffler, *Science*, **117**, 340 (1953).
63. A. Lapworth, *J. Chem. Soc.*, **85**, 30 (1904).
64. P. D. Bartlett, *J. Amer. Chem. Soc.*, **56**, 967 (1934).
65. C. K. Ingold and C. L. Wilson, *J. Chem. Soc.*, 773 (1934); S. K. Hsu and C. L. Wilson, *J. Chem. Soc.*, 623 (1936).
66. S. K. Hsu, C. K. Ingold, and C. L. Wilson, *J. Chem. Soc.*, 78 (1938); O. Reitz, *Z. Physik. Chem.*, **A179**, 119 (1937).
67. P. D. Bartlett and C. H. Stauffer, *J. Amer. Chem. Soc.*, **57**, 2580 (1935).
68. H. M. Dawson and J. S. Carter, *J. Chem. Soc.*, 2282 (1926); L. Zucker and L. P. Hammett, *J. Amer. Chem. Soc.*, **61**, 2785 (1939).
69. C. G. Swain, E. C. Stivers, J. F. Reuwer, Jr., and L. J. Schaad, *J. Amer. Chem. Soc.*, **80**, 5885 (1958).
70. C. G. Swain, A. J. DiMilo, and J. P. Cordner, *J. Amer. Chem. Soc.*, **80**, 5983 (1958).
71. E. Högfeldt and J. Bigeleisen, *J. Amer. Chem. Soc.*, **82**, 15 (1960).
72. E. D. Hughes, H. B. Watson, and E. D. Yates, *J. Chem. Soc.*, 3318 (1931); H. B. Watson and E. D. Yates, *J. Chem. Soc.*, 1207 (1932).
73. H. M. Dawson and E. Spivey, *J. Chem. Soc.*, 2180 (1930); R. P. Bell and P. Jones, *J. Chem. Soc.*, 88 (1953).
74. K. J. Pedersen, *J. Phys. Chem.*, **38**, 581 (1934).
75. C. G. Swain, *J. Amer. Chem. Soc.*, **72**, 4578 (1950).
76. F. J. C. Rossotti, *Nature*, **188**, 936 (1960).
77. J. N. Brønsted and W. F. K. Wynne-Jones, *Trans. Faraday Soc.*, **25**, 59 (1929).
78. C. A. Bunton and R. H. DeWolfe, *J. Org. Chem.*, **30**, 1371 (1965).
79. J. G. Fullington and E. H. Cordes, *J. Org. Chem.*, **29**, 970 (1964).
80. A. J. Kresge and R. J. Preto, *J. Amer. Chem. Soc.*, **87**, 4593 (1965).
81. H. Kwart and M. B. Price, *J. Amer. Chem. Soc.*, **82**, 5123 (1960).
82. G. A. Ropp and V. F. Raaen, *J. Chem. Phys.*, **22**, 1223 (1954).
83. G. E. Lienhard and W. P. Jencks, *J. Amer. Chem. Soc.*, **88**, 3982 (1966).
84. C. G. Swain and J. C. Worosz, *Tetrahedron Lett.*, 3199 (1965).
85. J. E. Reimann and W. P. Jencks, *J. Amer. Chem. Soc.*, **88**, 3973 (1966).
86. General base catalysis has been found in reactions of azomethines, D. S. Auld and T. C. Bruice, *ibid.*, **89**, 2090, 2098 (1967); a termolecular mechanism is suggested by the experiments of J. Warkentin and O. S. Tee, *Chem. Comm.*, 190 (1966).
87. E. H. Cordes and W. P. Jencks, *J. Amer. Chem. Soc.*, **84**, 4319 (1962).
88. L. do Amaral, W. A. Sandstrom, and E. H. Cordes, *ibid.*, **88**, 2225 (1966).
89. R. P. Bell, personal communication.
90. H. Schmid, *Chem. Zeitung*, **90**, 351 (1966), and references therein.
91. M. Eigen, *Disc. Faraday Soc.*, **39**, 17 (1965).
92. R. L. Betts and L. P. Hammett, *J. Amer. Chem. Soc.*, **59**, 1568 (1937).
93. W. H. Watanabe and L. R. DeFonso, *J. Amer. Chem. Soc.*, **78**, 4542 (1956).
94. J. F. Bunnett and G. T. Davis, *J. Amer. Chem. Soc.*, **82**, 665 (1960).
95. W. P. Jencks and J. Carriuolo, *J. Amer. Chem. Soc.*, **82**, 675 (1960).
96. T. C. Bruice and M. F. Mayahi, *J. Amer. Chem. Soc.*, **82**, 3067 (1960).
97. T. C. Bruice and R. G. Willis, *J. Amer. Chem. Soc.*, **87**, 531 (1965).
98. W. P. Jencks and M. Gilchrist, *J. Amer. Chem. Soc.*, **88**, 104 (1966).
99. T. C. Bruice and S. J. Benkovic, *J. Amer. Chem. Soc.*, **86**, 418 (1964).

100. R. P. Bell and P. T. McTigue, *J. Chem. Soc.*, 2983 (1960).
101. P. J. Hawkins and D. S. Tarbell, *J. Amer. Chem. Soc.*, **75**, 2982 (1953).
102. D. S. Tarbell and D. P. Cameron, *J. Amer. Chem. Soc.*, **78**, 2731 (1956).
103. K. A. Connors and M. L. Bender, *J. Org. Chem.*, **26**, 2498 (1961).
104. T. C. Bruice, J. J. Bruno, and W.-S. Chou, *J. Amer. Chem. Soc.*, **85**, 1659 (1963).
105. T. C. Bruice and L. R. Fedor, *ibid.*, **86**, 738, 739, 4117, 4886 (1964).
106. T. C. Bruice and S. J. Benkovic, *Bioorganic Mechanisms*, Vol. I, W. A. Benjamin, Inc., New York, 1966, p. 291.
107. S. I. Miller, *J. Amer. Chem. Soc.*, **81**, 101 (1959).
108. E. H. Cordes and W. P. Jencks, *J. Amer. Chem. Soc.*, **84**, 4319 (1962).
109. G. S. Hammond, *J. Amer. Chem. Soc.*, **77**, 334 (1955).
110. J. E. Reimann and W. P. Jencks, *J. Amer. Chem. Soc.*, **88**, 3973 (1966).
111. C. G. Swain and E. R. Thornton, *J. Amer. Chem. Soc.*, **84**, 817 (1962).
112. O. Reitz, *Z. Physik. Chem.*, **A176**, 363 (1936); **A179**, 119 (1937); O. Reitz and J. Kopp, *Z. Physik. Chem.*, **A184**, 429 (1939).
113. R. P. Bell, *The Proton in Chemistry*, Cornell University Press, Ithaca, N.Y., 1959, p. 201.
114. R. P. Bell and J. E. Crooks, *Proc. Roy. Soc.*, **A286**, 285 (1965).
115. R. P. Bell and D. M. Goodall, *Proc. Roy. Soc.*, **A294**, 273 (1966).
116. R. P. Bell, *Disc. Faraday Soc.*, **39**, 16 (1965); H. Shechter, et al., *J. Amer. Chem. Soc.*, **84**, 2905 (1962).
117. C. G. Swain, E. C. Stiver, S. J. F. Reuwer, Jr., and L. J. Schaad, *J. Amer. Chem. Soc.*, **80**, 5885 (1958).
118. F. H. Westheimer, *Chem. Rev.*, **61**, 265 (1961).
119. L. Funderburk and E. S. Lewis, *J. Amer. Chem. Soc.*, **86**, 2531 (1964).
120. J. R. Hulett, *Proc. Roy. Soc.*, **A251**, 274 (1959); E. F. Caldin and M. Kasparian, *Disc. Faraday Soc.*, **39**, 25 (1965).
121. R. P. Bell, *Trans. Faraday Soc.*, **55**, 1 (1959).
122. C. A. Bunton and V. J. Shiner, Jr., *J. Amer. Chem. Soc.*, **83**, 3207 (1961).
123. C. G. Swain, D. A. Kuhn, and R. L. Schowen, *J. Amer. Chem. Soc.*, **87**, 1553 (1965).
124. R. L. Schowen, H. Jayaraman, L. Kershner, and G. W. Zuorick, *J. Amer. Chem. Soc.*, **88**, 4008 (1966).
125. R. P. Bell and W. C. E. Higginson, *Proc. Roy. Soc.*, **A197**, 141 (1949).

Part Two

Organic and Inorganic Catalysis

Chapter 6

NUCLEOPHILIC AND ELECTROPHILIC CATALYSIS

6.1	Introduction	147
6.2	Nucleophilic Catalysis	148
	6.2.1 Nucleophilicity	148
	6.2.2 Mechanisms of Nucleophilic Catalysis	153
	Halide Ions as Catalysts	153
	Oxyanions as Catalysts	155
	Tertiary Amines as Catalysts	159
	Catalysis by Primary and Secondary Amines	165
	Catalysis by Cyanide Ion and Thiazolium Ion	172
	Catalysis by Carbanions	175
	6.2.3 The Transition between Nucleophilic and General Base Catalysis .	176
6.3	Electrophilic Catalysis	179
	6.3.1 Electrophilicity	179
	6.3.2 Mechanisms of Electrophilic Catalysis	181
	Catalysis by Lewis Acid Metal Halides, Hydrides, and Carbonium Ions	181
	Catalysis by Halogens and Other Electrophiles	185
	Catalysis by Carbonyl Compounds	186
6.4	Summary	189

6.1 INTRODUCTION

    Nucleophilic and electrophilic catalysis resemble ordinary chemical reactions more than any other kind of catalysis. The reason for this resemblance is that these catalyses occur via several partial reactions, each of which is the same as a noncatalytic reaction. What then is the difference between a nucleophilic reaction and a nucleophilic catalysis, or between an electrophilic reaction and an electrophilic catalysis? The answer to a first approximation is that in the catalyses, the initial product of reaction, although sometimes isolable, is unstable toward further reaction. Thus these catalyses, like general acid-base catalyses, form unstable *intermediates* that decompose to give products and regenerate the catalyst.

But in order for a nucleophilic or electrophilic catalysis to occur at a significant rate, several requirements must be met. The first is that the catalyst, whether it be nucleophile or electrophile, must react with the substrate faster than does the final acceptor, as shown in eq. 6.1 for nucleophilic catalysis. The intermediate must, of course, react with the final acceptor or otherwise decompose faster than does the original substrate.

$$\text{S-Q} + \text{N} \rightleftarrows \text{S-N} \xrightleftharpoons{[R]} \text{P} + \text{N} \qquad (6.1)$$
$$+\text{Q}$$

That is, the catalyst must be an unusually effective nucleophile (electrophile) and the intermediate unusually susceptible to nucleophilic (electrophilic) attack or other decomposition. Finally, the equilibrium between the catalyst and substrate yielding the intermediate must not be as favorable as between the reactant and the eventual acceptor.

Nucleophilic and electrophilic catalysis have been recognized for many years, although these names are of more recent origin. Nucleophilic catalysis was exploited by Langenbeck in the 1930s[1] and analyzed by Baker and Rothstein.[2] Examples of nucleophilic and electrophilic catalysis are numerous; we attempt to assess their importance and to delineate some selected mechanisms.

6.2 NUCLEOPHILIC CATALYSIS

6.2.1 Nucleophilicity

Nucleophilic catalysis ultimately depends on nucleophilicity. For this reason, we briefly consider the relative efficacy of nucleophiles. Unfortunately, there is no simple correlation between structure and nucleophilic (to a carbon atom) reactivity, but we examine some of the factors involved. Edwards has suggested two components of nucleophilic character, one related to classical basicity (toward a proton) and the other related to polarizability or oxidizability of the nucleophile[3]

$$\log \frac{k}{k_0} = \alpha P + \beta H \qquad (6.2)$$

where P is defined in terms of the relative molar refractivity, (eq. 6.3), while H (eq. 6.4) is a function of the basicity.

$$P = \log \frac{R_N}{R_{H2O}} \tag{6.3}$$

where R = refractive index and N = nucleophile.

$$H = pK_a + 1.74 \tag{6.4}$$

Equation 6.2 correlates a number of disparate nucleophiles in several families of reactions. It is superior to using basicity or polarizability alone when the nucleophilic atom is changed (e.g., from oxygen to sulfur). If attention is confined to a single nucleophilic atom, a rough correlation between basicity and nucleophilicity is noted. In fact, within a limited range of structural variation, a quantitative logarithmic correlation between basicity and nucleophilicity holds, as seen in a family of amines or thiols in Fig. 5.1. Such a quantitative correlation may be viewed as a manifestation either of a Brønsted or of a Hammett relation, both of which are linear free energy correlations.

When the nucleophilic atom is altered, all correlation between basicity and nucleophilicity vanishes. For example, sodium butylmercaptide is about as basic as sodium phenoxide but is 10^3 times as nucleophilic toward saturated carbon; phenoxide and bromide ions are of comparable nucleophilicity toward saturated carbon, but differ by a factor of about 10^{17} in basicity. Even when atoms of the same column of the periodic table are compared, there is no correlation between basicity and nucleophilicity. Instead, in water one finds the orders of decreasing nucleophilicity toward saturated carbon of

$$I^\ominus > Br^\ominus > Cl^\ominus > F^\ominus$$
$$SeR^\ominus > SR^\ominus > OR^\ominus$$

The best nucleophiles in these series form the weakest bonds to carbon, indicating no correlation between thermodynamics and kinetics in these reactions. It has been suggested many times that the electronic charge cloud of the nucleophile must be distorted by the electrophilic center of the substrate in the transition state. Thus such polarizability may explain the sequences given above, although orbital overlap, steric hindrance, and solvation energies may also need to be considered. Solvation is of considerable importance in determining the nucleophilicity of anions. In fact, the order of nucleophilicity of the halide ions can be reversed by changing the solvent (Chapter 7).

Another factor determining nucleophilicity is seen in nucleophiles possessing the structure YN where N is a nucleophilic atom and Y is an electronegative atom containing one or more pairs of unshared electrons. Examples of such nucleophiles are hydroxylamine, hydrazine, hydroperoxide ion, hypochlorite ion, oximate ions, and hydroxamate ions. These nucleophiles show exceptional nucleophilicity not accounted for on the grounds of either polarizability or basicity. The exceptional reactivity of these nucleophiles has been attributed to the stabilization of the transition state of the nucleophilic reaction by the neighboring lone pair electrons in the same way that the transition state in the solvolysis of an α-chloroether is stabilized. In the latter reaction the carbonium ion-like transition state is stabilized as in structure 6.5. In a similar manner, a nucleophilic reaction by hypochlorite ion may be stabilized by structure 6.6, an extreme form of a nucleophilic substitution in which a pair of electrons

$$ROCH_2^{\oplus} \longleftrightarrow RO^{\oplus}=CH_2 \tag{6.5}$$

$$Cl-O^{\oplus} \longleftrightarrow Cl^{\oplus}=O \tag{6.6}$$

has been *completely* donated from nucleophile to substrate ($ClO^{\ominus} - 2\epsilon$). In reality complete donation does not occur, but partial donation does. On this basis structure 6.6 explains the exceptional nucleophilicity found. This phenomenon has been termed the alpha effect.[4]

The order of nucleophilic reactivity is dependent not only on the nucleophile but also on the nature of the substrate as well. Analysis of the vast amount of data on substitution reactions has indicated that for some substrates the rates are sensitive chiefly to the ordinary proton basicity of the nucleophile, whereas for other substrates the rates are sensitive chiefly to the polarizability of the nucleophile. The properties of the electrophilic center determine which type of behavior is found. If the properties of the electrophilic center are those that make it a hard (nonpolarizable) acid, then basicity is the dominant factor. If the electrophilic site is a soft (polarizable) center, then polarizability is the most important factor in the rates. For example, electrophilic centers such as $RCO^{\oplus}$, $H^{\oplus}$, $RSO_2^{\oplus}$, $(RO)_2PO^{\oplus}$, and $(RO)_2B^{\oplus}$ react rapidly with nucleophiles that are strongly basic to the proton and not very polarizable, such as $OH^{\ominus}$ and $F^{\ominus}$. Other electrophilic centers such as $RCH_2^{\oplus}$, $R_2P^{\oplus}$, $RS^{\oplus}$, $Br^{\oplus}$, $R_2N^{\oplus}$, $RO^{\oplus}$, and $Pt^{2\oplus}$ react rapidly with highly polarizable nucleophiles such as $I^{\ominus}$ and R_3P. These facts have been generalized by Pearson in terms of the principle of "hard and soft acids and bases."[5] A general rule can be stated: hard electrophilic centers (acids) react rapidly with hard nucleophiles (bases), and soft electrophilic centers react rapidly with soft nucleophiles. The rule refers to S_N2 or S_E2 mechanisms.

We can explore relative nucleophilicity toward the carbonyl carbon atom, the tetrahedral phosphorus atom, and the saturated carbon atom by means of the data in Tables 6.1 and 6.2. Table 6.1 indicates that the important

Table 6.1
Nucleophilic Reactivity Toward p-Nitrophenyl Acetate and Isopropyl Methylphosphonofluoridate[a,b]

Nucleophile	pK_{HA}	p-Nitrophenyl Acetate	Isopropyl Methyl-phosphonofluoridate
$HOO^\ominus$	11.5	2×10^5	1.0×10^5
Acetoximate	12.4	3×10^3	—
Salicylaldoximate	9.2	3.2×10^3	1.5×10^3
$OH^\ominus$	15.7	9×10^2	1.6×10^3
$\phi O^\ominus$	10.0	1×10^2	34
NH_2OH	6	1×10^2	1.3
$OCl^\ominus$	7.2	1.6×10^3	7×10^{-3}
$CO_3^{2\ominus}$	10.4	1.0	—
NH_3	9.2	16	—
$CN^\ominus$	10.4	11	—
$\phi S^\ominus$	6.4	—	7.4×10^{-3}
ϕNH_2	4.6	1.5×10^{-2}	—
C_5H_5N	5.4	0.1	—
$NO_2^\ominus$	3.4	1.3×10^{-3}	—
$CH_3CO_2^\ominus$	4.8	5×10^{-4}	—
$F^\ominus$	3.1	1×10^{-3}	Very reactive
$S_2O_3^{2\ominus}$	1.9	1×10^{-3}	Unreactive
H_2O	−1.7	6×10^{-7}	1×10^{-6}

[a] Rate constants in $M^{-1}\ min^{-1}$.
[b] From J. O. Edwards and R. G. Pearson, *J. Amer. Chem. Soc.*, **84**, 16 (1962). © 1962 by the American Chemical Society. Reprinted by permission of the copyright owner.

component of the nucleophile with respect to p-nitrophenyl acetate is its basicity. There are some obvious exceptions to this statement pointing up the fact that those nucleophiles containing adjacent electronegative atoms with unshared electron pairs are exceptionally good nucleophiles. Table 6.1 also points up a fact (Chapter 5) that nitrogen nucleophiles are considerably more reactive than oxygen nucleophiles of the same pK. The behavior of nucleophiles toward a tetrahedral phosphorus atom, shown in Table 6.1, is quite similar. The reactive centers of both p-nitrophenyl acetate and the phosphonofluoridate can thus be described as hard acids or electrophiles.

On the other hand, the data of Table 6.2, pertaining to reactions at saturated carbon, show a different order of nucleophilicity. The most important nucleophilic component appears to be polarizability rather than basicity, although both contribute something to the overall nucleophilicity. The importance of polarizability in reactions at saturated carbon may be seen from a comparison of the reactivities of malonate and ethoxide ions toward ethyl

Table 6.2
Nucleophilic Reactivity Toward Methyl Chloride[a]

Nucleophile	$k(M^{-1}\text{ sec}^{-1})$
$SO_3^{2\ominus}$	2.3×10^{-4}
$S_2O_3^{2\ominus}$	1.7×10^{-4}
$SC(NH_2)_2$	2.5×10^{-5}
$I^\ominus$	1.2×10^{-5}
$CN^\ominus$	1×10^{-5}
$SCN^\ominus$	3.2×10^{-5}
$NO_2^\ominus$	1.8×10^{-5}
$OH^\ominus$	1.2×10^{-6}
$N_3^\ominus$	8×10^{-7}
$Br^\ominus$	5×10^{-7}
NH_3	2.2×10^{-7}
$Cl^\ominus$	1.1×10^{-7}
C_5H_5N	9×10^{-8}
H_2O	1×10^{-10}

[a] From J. O. Edwards and R. G. Pearson, J. Amer. Chem. Soc., **84**, 16 (1962). © 1962 by the American Chemical Society. Reprinted by permission of the copyright owner.

bromide. The former reaction is two times faster than the latter, although ethoxide ion is some 500 times more basic than diethyl malonate ion.

Toward aromatic carbon the following order of nucleophilicity is seen.

$$C_6H_5S^\ominus,\ CH_3O^\ominus > C_5H_{10}NH > C_6H_5O^\ominus$$
$$> N_2H_4 > OH^\ominus > C_6H_5NH_2 > Cl^\ominus > CH_3OH$$

There is a large spread in these rate constants: methoxide ion is 10^4 times faster than aniline, which in turn is 10^9 times faster than methanol. Nucleophilicity toward aromatic carbon apparently involves both polarizability and basicity.

Saville has used the softness-hardness principle as a guide for selecting catalysts in substitution reactions.[6] Substitution reactions of a hard-soft acid-base combination are suggested to be the easiest to catalyze since then the matching of hardness and softness is most straightforwardly carried out. The rules are simply that if the leaving group is a hard base, then a hard electrophile is used as catalyst (or vice versa). Likewise, when the electrophilic center is hard, a hard nucleophile is used as catalyst (or vice versa).

6.2.2 Mechanisms of Nucleophilic Catalysis

Catalysis by a base or a nucleophile can be interpreted mechanistically as a general base or a nucleophilic catalysis (Chapter 5). The principal criteria for nucleophilic catalysis include the following: (1) observation of a transient intermediate whose structure and kinetics indicate that it is an intermediate in the pathway of reaction; (2) trapping of the intermediate as an isolable species; (3) dependence of relative catalytic power on nucleophilicity rather than basicity, characterized by the nucleophilic reactivity orders discussed above, considerable steric hindrance, and importance of nucleophiles exhibiting the alpha effect; (4) occurrence of a common ion effect by the leaving group of the substrate (this amounts to trapping the intermediate as the reactant); (5) no primary deuterium oxide solvent isotope effect; and (6) product analysis of noncatalytic nucleophiles analogous to alleged catalytic nucleophiles.

Halide Ions as Catalysts. An example of nucleophilic catalysis is the iodide ion-catalyzed hydrolysis of methyl bromide[7,8]:

$$CH_3Br + I^{\ominus} \longrightarrow CH_3I + Br^{\ominus}$$
$$CH_3I + H_2O \longrightarrow CH_3OH + I^{\ominus}$$
(6.7)

In nucleophilic substitution reactions methyl bromide reacts with iodide ion faster than it does with water; furthermore, methyl iodide reacts with water faster than does the starting material. Thus the principal requirements of nucleophilic catalysis are met although none of the rigorous mechanistic criteria have been demonstrated. It might seem anomalous that iodide ion could be both a good nucleophile and leaving group at the same time; but, considering microscopic reversibility, this behavior can occur when factors other than basicity are involved. Nucleophilicity and leaving tendency depend on different aspects of structure (polarizability vs bond strength) and thus one suitably constituted group may contain both properties. The free energy-reaction coordinate diagram for the catalyzed process is compared in Fig. 6.1 with the corresponding uncatalyzed reaction. The catalyzed reaction bears the relationship to the uncatalyzed reaction predicted by the simplest theory of catalysis: the pathway is changed to one of greater complexity and the height of any individual peak in the catalyzed reaction is less than the height of the single peak in the uncatalyzed reaction. The relative heights of the peaks in the catalyzed reaction are not known because a thermodynamic analysis of the system has not been carried out. Halide and acetate ions also catalyze the solvolysis of sulfinylsulfones.[9]

The fluoride ion-catalyzed hydrolysis of ethyl chloroformate[10] probably

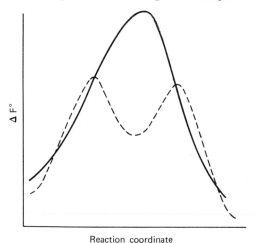

Fig. 6.1 The hydrolysis of methyl bromide. ——— uncatalyzed; - - - - catalyzed by iodide ion.

proceeds through a similar nucleophilic catalysis,

$$C_2H_5OCCl \underset{}{\overset{+F^\ominus}{\rightleftarrows}} C_2H_5OCF \overset{H_2O}{\longrightarrow} C_2H_5OH + CO_2 + HF \qquad (6.8)$$

(with C=O on both ester structures)

since the chloroformate may be readily transformed to the fluoroformate in nonaqueous medium and since the hydrolysis of the latter is about 30 times faster than that of the former. These arguments do not rigorously differentiate nucleophilic from general base catalysis, but seem preferable at present.

Halide ion catalysis is also seen in acid-catalyzed reactions. For example, the perchloric acid-catalyzed hydrolyses of methyl *p*-toluenesulfinate and sulfite esters are accelerated by the addition of sodium chloride, and even more by the addition of sodium bromide.[11] These reactions, which occur with sulfur–oxygen fission, can be explained most readily by postulating that catalysis occurs by the conversion of the ester into the readily hydrolyzable sulfinyl halide. More effective catalysis by bromide ion than chloride ion is consistent with this mechanism since the former is a better nucleophile (and HBr is a stronger acid).[11] Enhanced catalysis by halogen acids over perchloric acid in the hydrolysis of glucose-6-phosphate may be explained in a similar manner.[12] This reaction, which follows an A-2 (acid-catalyzed, second-order) mechanism with carbon–oxygen cleavage, probably proceeds through nucleophilic attack of the halide ion on the anomeric carbon atom, leading to a glucosyl halide that rapidly hydrolyzes to products.

Nucleophilic Catalysis

The relationship of nucleophilic catalysis to ordinary nucleophilic reactions is nowhere seen as clearly as in halide ion-catalyzed racemizations, such as the iodide ion-catalyzed racemization of 2-octyl iodide (eq. 6.9)[13] and the fluoride ion-catalyzed racemization of isopropyl methylphosphonofluoridate (eq. 6.10).[14]

$$(+)\text{ 2-octyl iodide} + I^\ominus \rightleftharpoons (-)\text{ 2-octyl iodide} + I^\ominus \qquad (6.9)$$

$$(+)\text{CH}_3\underset{\underset{\text{OCH(CH}_3)_2}{|}}{\overset{\overset{O}{\|}}{P}}\!\!-\!\!F \;+\; F^\ominus \rightleftharpoons (-)\text{CH}_3\underset{\underset{F}{|}}{\overset{\overset{O}{\|}}{P}}\!\!-\!\!\text{OCH(CH}_3)_2$$

$$\xrightarrow{H_2O} F^\ominus + \text{CH}_3\underset{\underset{OH}{|}}{\overset{\overset{O}{\|}}{P}}\!\!-\!\!\text{OCH(CH}_3)_2 \qquad (6.10)$$

The former is, of course, one of the basic experiments delineating nucleophilic substitution reactions at saturated carbon (S_N2 reactions).

Oxyanions as Catalysts. The saponification of an ester may be described as a nucleophilic catalysis[2] or more properly a nucleophile-promoted hydrolysis since the catalyst is not regenerated.

$$\text{R}\!-\!\overset{\overset{O}{\|}}{C}\!-\!\text{OR}' + \text{OH}^\ominus \rightleftharpoons \text{R}\!-\!\underset{\underset{OH}{|}}{\overset{\overset{O^\ominus}{|}}{C}}\!-\!\text{OR}' \longrightarrow \text{R}\!-\!\text{CO}_2^\ominus + \text{R}'\text{OH} \qquad (6.11)$$

The mechanism of this reaction,[15] the addition of hydroxide ion to the carbonyl group to form an unstable intermediate that collapses to give products, is of the form of nucleophilic catalysis even though the catalyst is not regenerated. The alkoxide ion-catalyzed alcoholysis of an ester is a true nucleophilic catalysis, following the same mechanism as the saponification reaction.

$$\text{R}\!-\!\overset{\overset{O}{\|}}{C}\!-\!\text{OR}' + \text{OR}''^\ominus \rightleftharpoons \text{R}\!-\!\underset{\underset{OR''}{|}}{\overset{\overset{O^\ominus}{|}}{C}}\!-\!\text{OR}' \rightleftharpoons \text{R}\!-\!\overset{\overset{O}{\|}}{C}\text{OR}'' + \text{OR}'^\ominus \qquad (6.12)$$

Since the alkoxide ion produced in the alcoholysis is in equilibrium through

$$\text{OR}'^\ominus + \text{R}''\text{OH} \rightleftharpoons \text{HOR}' + \text{R}''\text{O}^\ominus \qquad (6.13)$$

with the alkoxide ion of the attacking nucleophile, the alcoholysis is a true catalytic process.

Since alcohol is a better nucleophile than water and alkoxide ion is a better nucleophile than hydroxide ion toward carboxylic acid derivatives,[16] alcohol or alkoxide ion should serve as a nucleophilic catalyst for the hydrolysis of a carboxylic acid derivative such as an amide whose inherent hydrolytic rate is less than that of an ester. An interesting special case of catalysis by an alcoholate ion is the chloralate ion-catalyzed hydrolysis of *p*-nitrophenyl acetate.

$$
\begin{array}{c}
CCl_3\overset{O^{\ominus}}{\underset{OH}{C}}-H + CH_3CO\phi NO_2 \longrightarrow CCl_3\overset{O-CCH_3}{\underset{OH}{-CH}} \longrightarrow {}^{\ominus}OCCH_3 \\
\qquad\qquad\qquad\qquad\qquad\qquad\qquad\qquad\qquad + \\
\qquad\qquad\qquad\qquad\qquad\qquad\qquad\qquad\qquad CCl_3CH + H^{\oplus} \\
\updownarrow \qquad\qquad\qquad\qquad\qquad\qquad + NO_2\phi O^{\ominus} \qquad \parallel \\
\qquad\qquad\qquad\qquad\qquad\qquad\qquad\qquad\qquad O \\
CCl_3-\overset{OH}{\underset{OH}{CH}} \longleftarrow \underset{H_2O}{}
\end{array}
\qquad (6.14)
$$

Since the pK of chloral hydrate is 9.7, an appreciable amount of chloralate ion will exist near neutrality, producing a hemiacetal intermediate that can decompose to give acetate ion and regenerate chloral, which is capable of ready hydration.[17]

Carboxylate ions catalyze the hydrolysis of several carboxylic acid derivatives. As mentioned in Chapter 5, some of these reactions proceed via general base catalysis while others proceed via nucleophilic catalysis. The formate ion-catalyzed hydrolysis of acetic anhydride and the formate and acetate ion-catalyzed hydrolysis of propionic anhydride[18,19] are best explained by nucleophilic catalysis (eq. 6.15).

$$
\underset{CH_3COCCH_3}{\overset{O\;\;O}{\parallel\;\;\parallel}} + HCO_2^{\ominus} \rightleftarrows \underset{CH_3COCH}{\overset{O\;\;O}{\parallel\;\;\parallel}} + CH_3CO_2^{\ominus} \xrightarrow{H_2O} \underset{HCOH}{\overset{O}{\parallel}} + \underset{CH_3COH}{\overset{O}{\parallel}}
\qquad (6.15)
$$

The postulated intermediate in eq. 6.15, acetic formic anhydride, would be expected to be more susceptible to hydrolysis than the reactant acetic anhydride since formic acid is a stronger acid than acetic. Furthermore, catalysis by formate ion is greater than that by acetate opposite to the order of their base strengths. On these grounds, the formate ion catalyses conform to eq. 6.15. However, since the acetate ion-catalyzed hydrolysis of acetic anhydride is a general base catalysis, and since the intermediate acetic propionic anhydride should not be exceptionally reactive, the acetate ion catalysis of hydrolysis of propionic anhydride could equally well be a general base catalysis.

Nucleophilic Catalysis

In the acetate ion-catalyzed hydrolysis of 2,4-dinitrophenyl esters, evidence for nucleophilic catalysis has been demonstrated both by an isotopic tracer experiment[21] (see eq. 6.16) and by experiments using aniline as a trapping agent.[22]

$$\text{PhCO-O-C}_6\text{H}_3(\text{NO}_2)_2 + \text{CH}_3\text{C}(=O)\text{-O}^{18\ominus} \rightarrow \text{PhC}(=O)\text{-O}^{18}\text{-C}(=O)\text{CH}_3 \rightarrow \text{PhCO}^{18}\text{H} + \text{CH}_3\text{CO}^{18}\text{H}$$

(6.16)

Nucleophilic catalysis requires the observed isotopic result since the fission of the intermediate acetic benzoic anhydride should take place preferentially at the acetic carbon atom, from the known reactivities of acetic and benzoic derivatives. Experimentally 75% of the oxygen-18 derived from one atom of the original acetate ion is found in the benzoic acid product, indicating that the major pathway of this catalysis is indeed a nucleophilic one. The trapping by aniline of the anhydride intermediate from the acetate ion-catalyzed hydrolysis of 2,4-dinitrophenyl acetate accounts for essentially 100% of the reaction product, indicating this reaction proceeds only by nucleophilic catalysis.[22]

Phosphate ion is a catalyst for a number of hydrolyses including those of p-nitrophenyl acetate,[23] esters of thiocholine,[24] N-acetylimidazole,[25] chloramphenicol,[26,27] acetic anhydride,[28] methyl acetate,[29] tetraethyl pyrophosphate,[30] and the reaction of dialkyl sulfides with iodine.[31] The kinetics of some of these reactions indicate that phosphate ion, probably in the form of monohydrogen phosphate ion, participates in the reaction. In the hydrolysis of p-nitrophenyl acetate, the catalytic constant for monohydrogen phosphate is ca. 1000 times less than that for imidazole, implying a nucleophilic order, as suggested in Chapter 5. Evidence is unknown concerning the mechanism of phosphate catalysis but is known for imidazole catalysis. The intermediate in the former nucleophilic catalysis would be acetyl phosphate, which hydrolyzes in a spontaneous reaction that is moderately fast[32]; the corresponding acyl arsenate decomposes much more rapidly. Since the nucleophilicities of the arsenate and phosphate ions are comparable, arsenate ion would appear to be a better catalyst than phosphate ion, other things being equal. In the hydrolysis of tetraethyl pyrophosphate, both trapping and isotopic tracer experiments[29] indicate the occurrence of nucleophilic catalysis through the intermediacy of $(C_2H_5O)_2P(=O)OPO_3^{2\ominus}$, which should more readily decompose to metaphosphate than the neutral reactant would.

$$(EtO)_2\overset{O}{\underset{\|}{P}}-O-\overset{O}{\underset{\|}{P}}(OEt)_2 \xrightarrow[\substack{O^\ominus \\ -O\overset{\|}{P}(OEt)_2}]{+HPO_4{}^{2\ominus}} (EtO)_2\overset{O}{\underset{\|}{P}}-O\overset{O}{\underset{\substack{\| \\ OH}}{P}}-O^\ominus$$

$$\xrightarrow[-(EtO)_2\overset{O}{\underset{\|}{P}OH}]{} PO_3{}^\ominus - \begin{cases} \xrightarrow{HPO_4{}^{2\ominus}} \text{polymetaphosphate} \\ \xrightarrow{MeOH} MeOPO_3H^\ominus \end{cases} \quad (6.17)$$

Catalysis of anhydride formation by sulfite ion can be explained by a nucleophilic mechanism.[33]

Several nucleophiles exhibiting the alpha effect serve as catalysts for carboxylic and phosphoric acid derivatives. These catalysts include nitrite, hypochlorite, and oximate ions. Nitrite ion catalyzes the hydrolyses of acetic anhydride and bis-(dimethylamino)-phosphorochloridate.[34] Nucleophilic catalytic mechanisms are suggested.

$$CH_3\overset{O}{\underset{\|}{C}}O\overset{O}{\underset{\|}{C}}CH_3 + NO_2{}^\ominus \longrightarrow CH_3\overset{O}{\underset{\|}{C}}ONO + CH_3CO_2{}^\ominus$$

$$CH_3\overset{O}{\underset{\|}{C}}ONO \begin{cases} \xrightarrow{H_2O} CH_3\overset{O}{\underset{\|}{C}}OH + NO_2{}^\ominus \\ \xrightarrow{\alpha\text{-naphthylamine}} \text{4-amino-1,1'-azonaphthalene} \end{cases} \quad (6.18)$$

$$\underset{Me_2N}{\overset{Me_2N}{>}}\overset{O}{\underset{}{\overset{\|}{P}}}-Cl \xrightarrow[-Cl^\ominus]{+ONO^\ominus} \underset{Me_2N}{\overset{Me_2N}{>}}\overset{O}{\underset{}{\overset{\|}{P}}}-ONO \xrightarrow[-ONO^\ominus]{+H_2O} \underset{Me_2N}{\overset{Me_2N}{>}}\overset{O}{\underset{}{\overset{\|}{P}}}-OH \quad (6.19)$$

Evidence for the postulated intermediate, acetyl nitrite (eq. 6.18) has been obtained by the addition of α-naphthylamine to the system. This diverts acetyl nitrite from its usual hydrolytic path to form 4-amino-1,1′-azonaphthalene. No evidence exists for eq. 6.19. Hypochlorite ion is an efficient catalyst for the hydrolysis of isopropyl methylphosphonofluoridate (Sarin).[35] This reaction shows first-order dependence on both substrate and hypochlorite ion, but the latter is not consumed. Anions such as chloride ion have no effect. On this basis, mechanism 6.20 has been proposed.

$$\underset{RO}{\overset{RO}{>}}P\underset{F}{\overset{O}{\nearrow}} \xrightarrow[-F^{\ominus}]{+OCl^{\ominus}} \underset{RO}{\overset{RO}{>}}P\underset{OCl}{\overset{O}{\nearrow}} \xrightarrow[-OCl^{\ominus}]{+H_2O} \underset{RO}{\overset{RO}{>}}P\underset{OH}{\overset{O}{\nearrow}} \quad (6.20)$$

In the second step, cleavage of the chlorine–oxygen bond by analogy with the decomposition of an alkyl hypochlorite is suggested. Toward Sarin, hypochlorite ion shows approximately the same nucleophilicity as hydroxide ion, even though its basicity is eight powers of ten less than that of hydroxide ion. This fact together with the instability of acyl hypochlorites leads to a facile nucleophilic catalysis. 2-Pyridinealdoxime methiodide is a nucleophilic catalyst for the thiolation of proteins with thioparaconic acid or *N*-acetylhomocysteine thiolactone.[36] On the basis of its high nucleophilicity (presumably due to the alpha effect) toward tetrahedral phosphorus[37] and the known susceptibility of the products toward nucleophilic attack,[38] a nucleophilic catalysis may be suggested.

The chlorination of aromatic compounds in aqueous solution is catalyzed by acetic acid. This catalysis can be explained by the conversion of hypochlorous acid to acetyl hypochlorite (eq. 6.21), the eventual chlorinating agent; the acetic acid thus acts as nucleophilic catalyst, forming an unstable intermediate more reactive than the original substrate.[39]

$$HOCl + CH_3\overset{O}{\overset{\|}{C}}OH \rightleftharpoons CH_3\overset{O}{\overset{\|}{C}}OCl + H_2O$$

$$CH_3\overset{O}{\overset{\|}{C}}OCl + ArH \rightleftharpoons ArCl + CH_3CO_2H \quad (6.21)$$

Likewise, the diazotization of aromatic amines by nitrous acid is accelerated in the presence of acetic acid, excess nitrous acid,[40,41] chloride ion, or bromide ion.[42] These catalyses are probably related to the mechanism of eq. 6.21.

Tertiary Amines as Catalysts. Of great importance in nucleophilic catalysis is catalysis by tertiary amines. The synthetic utility of pyridine in acylation reactions has long been recognized.[43] The acylations of amines, alcohols, and phenols with an acyl chloride or anhydride in pyridine solution or with catalytic amounts of pyridine in inert solvents are well-known synthetic procedures. These catalyses can be explained on the same basis as the pyridine-catalyzed hydrolysis of acetic anhydride,[44] a reaction intensively investigated from many points of view. (1) A series of substituted pyridines of constant steric requirement catalyzes with relative rate constants conforming to a Brønsted-type relation.[44] Compounds such as 2-picoline,

2,6-lutidine, and 2,4,6-collidine do not catalyze at all.[45] (2) The addition of acetate ion significantly decelerates the reaction.[45-47] (3) Added hydroxylamine traps an intermediate as a hydroxamic acid.[32] (4) The isolation of the postulated intermediate, acetylpyridinium chloride, from the reaction of acetyl chloride and pyridine under anhydrous conditions has been reported.[32] These data are consistent with eq. 6.22.

$$CH_3\overset{O}{\overset{\|}{C}}O\overset{O}{\overset{\|}{C}}CH_3 + \underset{N}{\bigcirc} \rightleftharpoons CH_3\overset{O}{\overset{\|}{C}}N^{\oplus}\underset{}{\bigcirc} + \overset{O^{\ominus}}{\overset{|}{O}}\overset{}{\overset{\|}{C}}CH_3 \quad \Bigg\downarrow H_2O \quad CH_3\overset{O}{\overset{\|}{C}}OH + \underset{N}{\bigcirc} + H^{\oplus}$$

(6.22)

The intermediate acylammonium ion is probably the most reactive acylating agent known. A member of this family has been prepared in stable form as the antimony hexachloride salt.[48] This species, however, is a highly reactive one from both resonance and inductive considerations; it is not to be confused with the protonated form of an amide where protonation occurs on oxygen with resultant conservation of resonance stabilization.

The hydrolysis of acetyl phosphate is catalyzed by tertiary amines, and in the reactions catalyzed by pyridine, 4-methylpyridine, triethylenediamine, and trimethylamine, the reaction occurs with P–O fission. This is nucleophilic catalysis since an intermediate phosphorylated tertiary amine may be trapped by fluoride ion to give phosphorofluoridate.[49]

$$CH_3\overset{O}{\overset{\|}{C}}O-\underset{\underset{O^{\ominus}}{|}}{\overset{\overset{O^{\ominus}}{|}}{P}}=O \xrightarrow[-CH_3CO_2^{\ominus}]{+NR_3} O=\underset{\underset{O^{\ominus}}{|}}{\overset{\overset{\ominus O}{|}}{P}}-\overset{\oplus}{N}R_3 \begin{array}{c} \xrightarrow{H_2O} H_2PO_4^{\ominus} \\ \xrightarrow{F^{\ominus}} HOPO_2F^{\ominus} \end{array}$$

(6.23)

The tertiary amine, imidazole, has achieved prominence in nucleophilic catalysis. It has many special characteristics: (1) it is a constituent of essentially every enzyme, as the side chain of the amino acid histidine; (2) it is an amine nucleophile of pK 7.0 and therefore can operate efficiently at neutrality both as a base and nucleophile; (3) acylimidazoles and phosphorylimidazoles, the initial products of nucleophilic attack on acyl and phosphoryl derivatives, are reactive species. Nucleophilic reactions of many carboxylic acid derivatives (although not all) are catalyzed by imidazole. In many (although not all), catalysis occurs via a nucleophilic mechanism. The imidazole-catalyzed hydrolysis of *p*-nitrophenyl acetate is illustrative of the process. Kinetic experiments with substrate concentration much larger than imidazole concentration show a catalytic process depending on the unionized form of the nucleophile.[50] Complementary experiments with imidazole concentration much larger than substrate concentration also show the same kinetic dependency; and, in addition, the formation and decomposition of an unstable intermediate.[51] The intermediate can be isolated from the reaction in nonaqueous medium and shown to be N-acetylimidazole.[52] Furthermore, the kinetics of the overall reaction are completely accounted for by the formation and decomposition of N-acetylimidazole.[53-55] On this basis, the mechanism must be described by the nucleophilic catalysis of eq. 6.24, where the first step is ordinarily rate-determining.

$$CH_3CO\text{-}C_6H_4\text{-}NO_2 + \text{Im-NH} \rightarrow CH_3C(O)\text{-Im-NH}^{\oplus} \xrightarrow{OH^{\ominus}} CH_3COH + \text{Im-NH} \quad (6.24)$$

Nucleophilic catalysis by imidazole combines the properties of a good, although not exceptional, nucleophile with an intermediate that is quite unstable. In nucleophilicity toward p-nitrophenyl acetate, imidazole is approximately 10^5 times better than acetate ion. However, in the (hydrolytic) decomposition of the intermediates formed by nucleophilic attack, N-acetylimidazole is only about tenfold slower than acetic anhydride. Thus the requirements of an efficient nucleophilic catalysis are met: high nucleophilicity and an unstable intermediate.

The reason for the simultaneous occurrence of these two properties lies in the unique nature of the intermediate that is subject to nucleophilic attack by neutral nucleophiles such as water through the ready protonation of the N-acylimidazole (pK ca. 5) to a species resembling an acyltrialkylammonium ion, which is markedly susceptible to nucleophilic attack. A superficial way to account for the nucleophilic instability of N-acylimidazoles is to view the compounds as nitrogen analogs of anhydrides.

Many other imidazole and histidine derivatives serve as catalysts in the hydrolysis of p-nitrophenyl acetate, including N-methylimidazole, benzimidazole derivatives, and histidine-containing peptides, as shown in Table 6.3. Catalysis by N-methylimidazole approximately equivalent to that by

Table 6.3
Catalysis by Imidazoles of the Hydrolysis of p-Nitrophenyl Acetate[a,g]

Catalyst	pK_a	k'_2
Imidazoles		
1. Imidazole	6.95	20.2
2. 2-Methylimidazole	7.75	2.7
3. 4-Methylimidazole	7.45	25.1
4. N-Methylimidazole[c]	7.05	0.5
5. 4-Bromoimidazole	3.7	0.28
6. 4-Hydroxymethylimidazole	6.45	5.6
7. 4-Nitroimidazole	1.5 (9.1)	35.5
Histidines		
1. Histidine[59,61]	6.0	d
2. N-Acetylhistidine	7.05	11.2
3. Histidine methyl ester	5.2	5.6
4. N-Benzoyl-L-histidine methyl ester		
5. 1-Methylhistidine	6.5	d
6. β-Aspartylhistidine	6.9	d
7. Histidylhistidine	6.8	d
8. Histamine	6.0	7.0
9. Carnosine	6.8	d
10. Anserine	7.0	d
11. Carbobenzoxy-L-histidyl-L-tyrosine ethyl ester	6.25	8.9
12. 8:1 copolymer of alanine + histidine[b]		6.0
13. Poly-L-histidine		e
14. Bacitracin[60]	~6	<20
15. Gly-L-His-L-Ser[59 f]		15
16. Copoly L-His-L-Ser[59,61,62 f]		9.7
17. L-Ser-L-His-L-Asp[59 f]		45
18. L-Thr-L-Ala-L-Ser-L-His-L-Asp[59 f]		92
19. L-Ser-γ-NH$_2$Bu-L-His-γ-NH$_2$Bu-L-Asp[59 f]		147
20. Cyclo Gly-L-His-L-Ser-Gly-L-His-L-Ser[59]		7
21. Copoly L-His-L-Asp[61,62 f]		~6
22. Copoly L-His-L-Lys[61,62 f]		~6
23. Copoly L-His-L-Gln[61,62 f]		~6
Benzimidazoles		
1. Benzimidazole	5.4	0.96
2. 2-Methylbenzimidazole	6.1	0.0375
3. 6-Aminobenzimidazole	6.0	2.95
4. 6-Nitrobenzimidazole	3.05	4.8
5. 4-Hydroxybenzimidazole	5.3	2.8

(continued)

Table 6.3 (continued)

6. 4-Methoxybenzimidazole	5.1	0.31
7. 4-Hydroxy-6-nitrobenzimidazole	3.05	3.75
8. 4-Hydroxy-6-aminobenzimidazole	5.9	6.15
9. 2-Methyl-4-hydroxy-6-nitrobenzimidazole	3.9	1.1
10. 2-Methyl-4-hydroxy-6-aminobenzimidazole	6.65	1.5
11. 4-(2′,4′-Dihydroxyphenyl)imidazole	6.45	9.4

[a] In 28.5% ethanol-water, at pH 8.0, ionic strength = 0.55 M, and 30°C, unless otherwise noted.
[b] In 5% aqueous ethanol, pH 7.0.
[c] In 5% dioxane-water, pH 7.0, 25°C.
[d] Zero-order rates reported. The relative rates of these six compounds vary from histidine (relative rate = 1) to anserine (relative rate = 3.6).
[e] Reported to be five to ten times more effective than histidine (per residue).
[f] 40% (v/v) dioxane-water, pH 7.76, 28°.
[g] From M. L. Bender, *Chem. Rev.*, **60**, 53 (1960). © 1960 by the Williams and Wilkins Co. Reprinted by permission of the copyright owner.

imidazole points out the fact that a neutral N-acylimidazole is not a necessary intermediate in this catalysis. The catalytic rate constants of many imidazole derivatives of constant steric requirement are dependent on the pK of the catalyst, as in the pyridine family.[56] Thus in comparing catalysis by various imidazole derivatives, the relative basicity of the catalysts must be considered first. A favorite occupation has been to attempt to synthesize polypeptides showing a greater catalytic activity than imidazole toward p-nitrophenyl acetate. Table 6.3 indicates limited success so far in this pursuit. The most effective peptide is L-seryl-γ-aminobutyryl-L-histidyl-γ-aminobutyryl-L-aspartic acid,[59] which has a catalytic rate constant toward p-nitrophenyl acetate sevenfold greater than that of imidazole; the pK of this peptide is unknown. This peptide, as well as several others,[60] exhibits partial stereoselectivity in hydrolysis of optically active phenylalanine esters. Thermally prepared polyanhydro-α-amino acids containing eighteen amino acids common to protein accelerate the hydrolysis of p-nitrophenyl acetate.[63] Histidine residues play a key role in the hydrolysis, the activity sometimes being more than 10 times that of monomeric histidine. A polymer of ethylenimine containing histidine moieties has catalytic activity toward p-nitrophenyl acetate (See Chapter 11).

Imidazole may also serve as a nucleophilic catalyst for the hydrolysis of phosphate derivatives. Thus although 2,6-lutidine acts as a general base catalyst in the solvolysis of tetrabenzyl pyrophosphate,[64] imidazole and N-methylimidazole serve as nucleophilic catalysts in this reaction.[65] The catalysis results from nucleophilic attack by the amine on phosphorus,

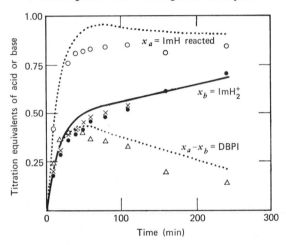

Fig. 6.2 Effect of added guanidinium dibenzyl phosphate (0.03 M) on the solvolysis of tetrabenzyl pyrophosphate (0.02 M) in propanol at 10°; imidazole, 0.04 M. ○, ●, and △ represent corrected experimental points for x_a, x_b, and [DBPI], respectively; x = uncorrected x_b values. Dotted and solid lines were computed with $k_1 = 0.70$, $k_{-1}K = 0.65$, and $k_2K = 0.50$ M^{-1} min^{-1}. From R. Blakeley, F. Kerst, and F. H. Westheimer, *J. Amer. Chem. Soc.*, **88**, 112 (1966). © 1966 by the American Chemical Society. Reproduced by permission of the copyright owner.

forming N-(dibenzylphosphoryl)imidazolium ion, by displacing dibenzyl phosphate ion. Further, imidazole catalyzes the hydrolysis of cytidine cyclic phosphate, probably by a general base mechanism.[65] The kinetics of the overall process with imidazole are in accord with, and demand, the intermediate N-(dibenzylphosphoryl)imidazole, the kinetics of whose reactions can be measured independently. Figure 6.2[65] shows a kinetic analysis of this reaction in the presence of added guanidinium dibenzyl phosphate, which should exert a common-ion effect, taking into account mechanism 6.25, as well as the equilibrium (K) between the protonated intermediate and its neutral form.

(6.25)

Nucleophilic Catalysis

Many other reactions show catalysis by tertiary amines, phosphines, sulfides, tertiary amine oxides, and phosphine oxides. Some examples are the dialkyl sulfide catalysis of the disulfide-sulfinic acid reaction,[66] the tertiary phosphine-catalyzed transesterification of phosphates, phosphonates, and phosphinates with alkyl halides,[67] the phosphine oxide-catalyzed conversion of isocyanates to carbodiimides,[68] the pyridine N-oxide-catalyzed reaction of isocyanates with alcohols,[69] the triethylamine-catalyzed alcoholysis of N-benzoylphosphoramidate,[70] and the dimethylformamide-catalyzed reactions of phosphorochloridates with alcohols, acids, and amines.[71] Presumably all these reactions involve nucleophilic catalysis with the formation of unstable intermediates. Only in the last reaction has the formation of a reactive intermediate been rigorously demonstrated.

Catalysis by Primary and Secondary Amines. Many reactions of carbonyl compounds show specific catalysis by ammonia and primary and secondary amines. These reactions include the dealdolization of diacetone alcohol, the decarboxylation of β-keto acids, the Knoevenagel condensation, semicarbazone formation, and the enolization of acetone. Catalysis of this specific kind must reflect some common property of these systems. The common property is the ability of the catalyst and the substrate to form, in a facile nucleophilic reaction, an imine or protonated imine that can serve as an intermediate for further reaction.

The equilibrium between carbonyl compounds and imines is known to be established rapidly under relatively mild conditions. Furthermore, the properties of imines are conducive to service as intermediates in many reactions. If one compares related carbonyl and imine species in condensation and enolization reactions, imines are found to be relatively reactive. In Table 6.4, a protonated carbonyl compound, a carbonyl compound, and a protonated imine are compared in reactivity toward the removal of a proton. Of these three species the protonated carbonyl compound is the fastest but is not present in appreciable concentration unless extremely high acidity is used, since the pK of a carbonyl compound is ca. -6. On the other hand, immonium ions have pK's near neutrality. Therefore, if one considers only those species in appreciable concentration near neutrality, only the protonated imine, imine, and carbonyl compound need be considered. Of these compounds, the protonated imine is much more susceptible to enolization than the carbonyl compound. The apparent reason for this difference in reactivity is that the protonated imine is a much more efficient "electron sink" than the carbonyl group. The neutral imine, for which there are little data, is probably a better "electron sink" than the neutral carbonyl compound, since in reactions of these compounds, general acid catalysis results in proton donation to the electronegative atom resulting in neutral products rather than anionic ones. On this basis, the relative reactivity of the three

Table 6.4*
Rate Constants of Base-Catalyzed Enolization of Some Unsaturated Substrates

Substrate pK_a of	$\rightarrow (CH_3)_2C{=}OH^\oplus$	$(CH_3)_2C{=}N^\oplus HGly$	$(CH_3)_2C{=}NH^\oplus CH_3$	$(CH_3)_2C{=}O$
Substrate	$\rightarrow -2^a$	10^b	10^b	16^c

Base	$k_2\ (M^{-1}\ \text{sec}^{-1})$			
Water	1.53^d	2.04×10^{-3}	$2.62 \times 10^{-2\,f}$	$0.84 \times 10^{11\,c}$
Acetate ion	$1.32 \times 10^{4\,e}$	7.2		2.5×10^{-7}
Methylamine				1.87×10^{-2}
Glycine		5.5×10^2		1.49×10^{-3}

[a] Approximate pK_a of $ROH_2^\oplus$.
[b] Approximate pK_a of $RNH_3^\oplus$.
[c] Approximate pK_a of ROH.
[d] $k_{H_2O} = k_{H_3O^\oplus} + K_s/[H_2O]$ where $k_{H_3O^\oplus}$ is the rate constant of the hydronium ion-catalyzed enolization of acetone and K_s is the ionization constant of protonated acetone.
[e] $k_{acetate} = k_{acetic\ acid} K_s/K_a$, where $k_{acetic\ acid}$ is the rate constant of the acetic acid catalyzed enolization of acetone, K_s is the ionization constant of protonated acetone, and K_a is the ionization constant of acetic acid.
[f] $k_{H_2O} = k_e$, which is defined in Chapter 2.

* From M. L. Bender and A. Williams, *J. Amer. Chem. Soc.* **88**, 2507 (1966). © 1966 by the American Chemical Society. Reprinted by permission of the copyright owner.

species are protonated imine > imine > carbonyl compound. The enhanced reactivity of the protonated imine over the carbonyl compound would also be expected to be involved in nucleophilic additions to these unsaturated linkages, since the same properties are involved in both reactions.

Anilinium ions catalyze the formation of semicarbazones from benzaldehydes.[73] This catalysis is ten to a thousandfold more efficient than catalysis by other acids of comparable strength, including the conjugate acids of secondary amines whose catalytic constants are close to those predicted from the Brønsted plot obtained for carboxylic acids. Many independent pieces of evidence indicate an imine intermediate in this catalysis. (1) N-Benzylidineanilines in dilute solutions of semicarbazide partition to aldehyde and semicarbazone, the fraction of semicarbazone formed being dependent on the concentration of semicarbazide. (2) The rate of the anilinium ion-catalyzed semicarbazone formation is independent of semicarbazide concentration except in very dilute solution; in fact, the rate of imine formation accounts quantitatively for the rate of the anilinium ion-catalyzed semicarbazone formation. (3) The rate of anilinium ion-catalyzed semicarbazone formation is identical to the rate of anilinium ion-catalyzed oxime formation. These results taken together strongly suggest that the catalysis by anilinium ion must proceed by nucleophilic catalysis.

$$\begin{array}{c} H_3C \\ H_3C \end{array}\!\!\!\!\!\!\!\!=\!O + ArNH_3^{\oplus} \rightleftharpoons \begin{array}{c} H_3C \\ H_3C \end{array}\!\!\!\!\!\!\!\!=\!NH^{\oplus} Ar + H_2O \quad \text{slow}$$

$$\begin{array}{c} H_3C \\ H_3C \end{array}\!\!\!\!\!\!\!\!=\!NH^{\oplus} Ar + NH_2NH\overset{O}{\overset{\|}{C}}NH_2 \longrightarrow \begin{array}{c} H_3C \\ H_3C \end{array}\!\!\!\!\!\!\!\!=\!NNH\overset{O}{\overset{\|}{C}}NH_2 + ArNH_3^{\oplus}$$

(6.26)

The three requirements of nucleophilic catalysis are well satisfied here: (1) aniline is more reactive than semicarbazide toward the original carbonyl compound; (2) the rate of attack of semicarbazide on the (protonated) imine is greater than on the original carbonyl compound; and (3) the equilibrium constant of semicarbazone formation is more favorable than that of imine formation by approximately 10^5.

In the conversion of pyridoxal and pyridoxal phosphate to the corresponding semicarbazones, nucleophilic catalysis by both primary and secondary amines is found.[74] The species that can form semicarbazones include the aldehyde, the imine, the hydrogen-bonded imine, and the protonated imine. Of these species the protonated imine gives by far the highest rate of semicarbazone formation. With the primary amines (methylamine, glycine, and ornithine), the rate of semicarbazone formation is independent of the catalyst concentration, indicating that conversion of the imine to the semicarbazone is the rate-determining step. The overall reaction with these catalysts is approximately 50 times faster than with the aldehyde itself. The secondary amines morpholine and proline can give only cationic imines. These catalyses show kinetic dependence on the catalyst concentration; consequently, the iminium ion intermediate must react as fast as it is formed. The conclusion drawn from these results is that the iminium ion must be extremely reactive compared to the imine. On the basis of the above experiments a protonated imine was suggested as the structure of enzyme-bound pyridoxal phosphate.

Ammonium ion catalysis is also seen in the iodination of acetone[72] and in the α-hydrogen exchange of isobutyraldehyde-2-d.[75] The deuterium exchange of isobutyraldehyde-2-d in the presence of free methylamine, where the aldehyde exists largely as the imine, depends on the methylammonium ion concentration; an important reaction path may therefore involve the rate-controlling removal of deuterium by methylamine from the conjugate acid of the imine. In the iodination of acetone, although amine catalysis adheres to the Brønsted plot defined by other general bases (acids), ammonium ion catalysis is as much as 10^6-fold faster than that predicted on the basis of carboxylic acids of comparable acidity (Fig. 6.3).[72] On this basis, the ammonium ion catalysis of iodination proceeds via imine formation.

This hypothesis is supported by the observation that the rate constant of the methylammonium ion-catalyzed iodination is not a linear function of the buffer ratio, but is explainable in terms of a two-step reaction involving a protonated imine intermediate, with a change in the rate-determining step

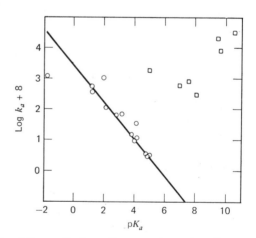

Fig. 6.3 Catalysis of the enolization of acetone by carboxylic acids, ○, and ammonium ions, □. The line is an arbitrary line through the carboxylic acid points, giving $\alpha = 0.66$. From M. L. Bender and A. Williams, *J. Amer. Chem. Soc.*, **88**, 2504 (1966). © 1966 by the American Chemical Society. Reproduced by permission of the copyright owner.

$$\underset{H_3C}{\overset{H_3C}{>}}=O + NH_3R^\oplus \rightleftharpoons \underset{H_3C}{\overset{H_3C}{>}}\underset{NH_2^\oplus R}{\overset{OH}{<}} \underset{H_2O}{\rightleftharpoons}$$

(6.27)

$$\underset{H_3C}{\overset{H_3C}{>}}=NH^\oplus R \underset{BH^\oplus}{\overset{B}{\rightleftharpoons}} \underset{H_2C}{\overset{H_3C}{>}}-NHR \xrightarrow[\text{fast}]{I_2} \text{products}$$

as the buffer ratio changes. This mechanism is further supported by the observation of kinetic terms involving both the ammonium ion and a base.

Primary and secondary, but not tertiary, amines catalyze the decarboxylation of β-keto acids such as dimethylacetoacetic acid, acetonedicarboxylic acid, and oxaloacetic acid.[76-78] The postulation of imine formation in this reaction[77] is supported by the observation that the rate constant of imine formation from aniline and ethyl oxaloacetate is identical to the rate constant of the aniline-catalyzed decarboxylation of oxaloacetic acid in ethanol solution.[78] This kinetic identity requires both that an imine be formed and that the decarboxylation of the β-imino acid be more facile than the decarboxylation of the β-keto acid. On the basis of this evidence, eq. 6.28 can be formulated.

$$R-\overset{O}{\underset{\|}{C}}-CH_2-\overset{O}{\underset{\|}{C}}OH \underset{H_2O}{\overset{NH_2R}{\rightleftharpoons}} R-\overset{HN^\oplus R}{\underset{\|}{C}}-CH_2CO_2^\ominus \xrightarrow{-CO_2} R-\overset{HNR}{\underset{|}{C}}=CH_2 \xrightarrow[+H_2O]{-NH_2R} R-\overset{O}{\underset{\|}{C}}CH_3$$

(6.28)

In the aniline-catalyzed decarboxylation of oxaloacetic acid in water, however, chemical and kinetic evidence indicates that the carbinolamine of the aniline and the substrate, rather than the imine, may decarboxylate to give the products (eq. 6.29).

$$^\ominus CO_2-\overset{O}{\underset{\|}{C}}-CH_2-\overset{O}{\underset{\|}{C}}O^\ominus \underset{H_2O}{\overset{NH_2R}{\rightleftharpoons}} {}^\ominus CO_2-\underset{\underset{R}{\overset{|}{NH_2^\oplus}}}{\overset{CH_2}{\underset{|}{\overset{|}{\underset{HO}{C}}}}}\overset{}{\diagdown}C=O \xrightarrow{}$$

(6.29)

$$^\ominus CO_2-\underset{\|}{\overset{}{C}}-CH_3 \longleftarrow {}^\ominus CO_2-\underset{\underset{H}{\overset{|}{O}}}{\overset{}{C}}=CH_2 + CO_2 + RNH_2$$

Primary amines also catalyze the decarboxylation of α-keto acids. This work, originated by Langenbeck[1,79] and extended by others,[80] may be most easily explained in terms of intermediate imine formation. Catalysis by simple amines such as ethylamine or glycine can be explained by eq. 6.30.

$$\text{R—CO—CO}_2\text{H} + \text{R'NH}_2 \longrightarrow \underset{\substack{\|\\ \text{N}\\ | \\ \text{R'}}}{\text{R—C—CO}_2\text{H}} \longrightarrow \underset{\substack{\|\\ \overset{\oplus}{\text{N}}\\ \diagup \ \diagdown \\ \text{R'} \quad \text{H}}}{\text{R—C—CO}_2^{\ominus}} \longrightarrow$$

$$\left[\underset{\substack{\|\\ \overset{\oplus}{\text{N}}\\ \diagup \ \diagdown \\ \text{R'} \quad \text{H}}}{\text{R—C}^{\ominus}} \longleftrightarrow \underset{\substack{|\\ \text{N}\\ \diagup \ \diagdown \\ \text{R'} \quad \text{H}}}{\text{R—C:}} \right] \longrightarrow \underset{\substack{\|\\ \text{N}\\ | \\ \text{R'}}}{\text{R—C—H}} \xrightarrow{\text{H}_2\text{O}} \text{R—CHO} + \text{R'NH}_2 \quad (6.30)$$

The important transition state in this reaction resembles its product, a resonance-stabilized zwitterion. The most efficient of Langenbeck's catalysts are the 3-amino-2-oxindoles, which are as much as 650 times more effective than ethylamine. The imine formed from an α-keto acid and such an amine has an active hydrogen atom which may tautomerize. The resulting tautomer is of the vinylogous β-keto acid type; thus it decarboxylates readily as β-keto acids are known to do.

[Structure A with –NH₂] + CH₃COCO₂H $\xrightarrow{-\text{H}_2\text{O}}$ [Structure with –N=C(CH₃)–CO₂H] ⟶

[Structure with N–CH(CH₃)–CO₂H] $\xrightarrow[-\text{H}^{\oplus}]{-\text{CO}_2}$ [Structure with N=CH–CH₃, enolate] ⟶ (6.31)

[Structure with –N=CH–CH₃] $\xrightarrow{\text{H}_2\text{O}}$ A + CH₃CHO

Historically, catalysis of carbon–carbon condensations by primary and secondary amines was the first of this class of nucleophilic catalyses to be recognized. These reactions include aldol, Michael, and Knoevenagel condensations. They proceed by the intermediate formation of an imine or immonium ion. The catalysis of aldol condensations by ammonia and primary and secondary amines observed by Knoevenagel is of historical interest;[81] but the mechanistic interpretation of this reaction was developed

Nucleophilic Catalysis

much later, starting with observations of the dealdolization of diacetone alcohol. Although primary and secondary amines [82] as well as amino acids [84] serve as catalysts, other bases such as phenoxide ion and tertiary amines do not.[83] As mentioned earlier, this reaction conforms to specific hydroxide ion catalysis rather than general basic catalysis. On this basis, the amine catalysis must reflect a specific amine reaction. By analogy with our previous discussion, we may write the following to explain this specific catalysis.[83,85]

$$
\begin{array}{c}
\underset{\substack{|\\ \text{OH}}}{(CH_3)_2C}-CH_2-\underset{\substack{\|\\ O}}{C}-CH_3 \xrightleftharpoons{R_2NH} \underset{\substack{|\\ \text{OH}}}{(CH_3)_2C}-CH_2-\underset{\substack{\|\\ N^\oplus R_2}}{C}-CH_3 \longrightarrow \\
\underset{\substack{|\\ O^\ominus}}{(CH_3)_2C}-CH_2-\underset{\substack{\|\\ N^\oplus R_2}}{C}-CH_3 \\
\updownarrow \\
2CH_3\underset{\substack{\|\\ O}}{C}CH_3 + NR_2H \xrightleftharpoons{H_2O} (CH_3)_2\underset{\substack{\|\\ O}}{C} + CH_2=\underset{\substack{|\\ NR_2}}{C}-CH_3
\end{array}
\tag{6.32}
$$

An aniline-catalyzed Michael condensation is seen in the formation of the compound Warfarin.[86] A plausible mechanism is

$$
\phi-CH=CH-\underset{\substack{\|\\ O}}{C}-CH_3 \rightleftharpoons \underset{\substack{|\\ NH_2^\oplus\phi}}{\phi-CH-CH_2}-\underset{\substack{\|\\ N^\oplus H\phi}}{C}-CH_3 \rightleftharpoons \phi-CH=CH-\underset{\substack{\|\\ N^\oplus H\phi}}{C}-CH_3 \rightleftharpoons \tag{6.33}
$$

A more direct demonstration of the intermediacy of an imine in a condensation reaction is the *n*-butylammonium acetate-catalyzed condensation of piperonal with nitromethane.[87] In this reaction the rate of piperonylidenebutylamine formation and the rate of attack of nitromethane on the imine account for the overall rate of appearance of 3,4-methylenedioxy-β-nitrostyrene.

$$\underset{\underset{\text{CH}_2}{\overset{\text{O}}{\bigcirc}}}{\overset{\text{CHO}}{\bigcirc}} \xrightarrow[\text{BuNH}_2]{\underset{\text{2 steps}}{k_1}} \underset{\underset{\text{CH}_2}{\overset{\text{O}}{\bigcirc}}}{\overset{\text{CH=NBu}}{\bigcirc}} \xrightarrow[\text{CH}_2\text{NO}_2^{\ominus}]{\underset{\text{2 steps}}{k_2}} \underset{\underset{\text{CH}_2}{\overset{\text{O}}{\bigcirc}}}{\overset{\text{CH=CHNO}_2}{\bigcirc}} \quad (6.34)$$

These catalyses by primary and secondary amines have a common feature: the substrate is a carbonyl compound. The catalysis almost always proceeds through an imine or immonium ion intermediate, formed in a facile reaction under mild conditions. The presence of the imino group in the molecule facilitates subsequent reaction whether it be a condensation, decarboxylation, or enolization, by stabilization of the transition state through a resonance interaction or electrostatic effect or both. Finally, the amine catalyst is regenerated by a facile reversion of imine formation. Although the mechanisms are by no means rigorously proven, they form a consistent pattern from which to proceed.

Catalysis by Cyanide Ion and Thiazolium Ion. These catalysts are two of the most specific nucleophilic catalysts known, promoting only the acyloin condensation and the decarboxylation of α-keto acids, and related reactions. The essential features of the acyloin condensation were proposed by Lapworth[88] (eq. 6.35).

$$\text{PhCHO} \underset{}{\overset{\text{CN}^{\ominus}}{\rightleftarrows}} \underset{\text{CN}}{\overset{\text{O}^{\ominus}}{\text{Ph}-\overset{|}{\underset{|}{\text{C}}}-\text{H}}} \underset{\text{B}}{\overset{\text{BH}^{\oplus}}{\rightleftarrows}} \underset{\text{CN}}{\overset{\text{OH}}{\text{Ph}-\overset{|}{\underset{|}{\text{C}}}-\text{H}}} \underset{\text{BH}^{\oplus}}{\overset{\text{B}}{\rightleftarrows}} \left[\underset{\text{CN}}{\overset{\text{OH}}{\text{Ph}-\overset{|}{\underset{|}{\text{C}}}^{\ominus}}} \leftrightarrow \underset{\underset{\text{N}^{\ominus}}{\overset{\|}{\text{C}}}}{\overset{\text{OH}}{\text{Ph}-\overset{|}{\underset{}{\text{C}}}}} \right] \overset{\text{PhCHO}}{\rightleftarrows}$$

$$\underset{\text{NC H}}{\overset{\text{HO O}^{\ominus}}{\text{Ph}-\overset{|}{\underset{|}{\text{C}}}-\overset{|}{\underset{|}{\text{C}}}\text{Ph}}} \rightleftarrows \underset{\text{NC H}}{\overset{\text{O}^{\ominus}\text{ OH}}{\text{Ph}-\overset{|}{\underset{|}{\text{C}}}-\overset{|}{\underset{|}{\text{C}}}-\text{Ph}}} \overset{-\text{CN}^{\ominus}}{\rightleftarrows} \underset{\text{H}}{\overset{\text{O OH}}{\text{Ph}-\overset{\|}{\text{C}}-\overset{|}{\underset{|}{\text{C}}}-\text{Ph}}}$$

$$(6.35)$$

This condensation must involve the activation of the hydrogen atom attached to a carbonyl group, but the carbanion formed by the removal of such a hydrogen has no obvious stabilizing feature. However, if cyanide

Nucleophilic Catalysis

ion is added to benzaldehyde, for example, to form an anion of mandelonitrile, equilibrium between this anion and a tautomer produces a species equivalent to the carbanion formed by removal of the hydrogen attached to the carbonyl group. This carbanion, stabilized by resonance interaction with the cyano group, can participate in the condensation reaction. The cyanide ion can then leave the product by reversal of the original carbonyl addition reaction. The unique characteristics of the cyanide ion are its ready reversible addition to carbonyl groups and its ability to stabilize a carbanionic charge on an adjacent carbon atom. Similar reactions catalyzed by cyanide ion are the cleavage of α-diketones to esters and aldehydes[89] and the reaction of ninhydrin to yield hydrindantin and phthalonic acid.[90]

Cyanide ion catalyzes both the decarboxylation and acyloin condensation of α-keto acids.[91] Catalysis of decarboxylation by cyanide ion is shown in eq. 6.36.

$$CH_3-\underset{O}{\underset{\|}{C}}-CO_2^\ominus \underset{}{\overset{CN^\ominus}{\rightleftarrows}} CH_3-\underset{CN}{\underset{|}{\overset{O^\ominus}{\overset{|}{C}}}}-CO_2^\ominus \overset{BH^\oplus}{\underset{B}{\rightleftarrows}} CH_3-\underset{CN}{\underset{|}{\overset{OH}{\overset{|}{C}}}}-\underset{O}{\overset{\nearrow O^\ominus}{C}} \underset{-CO_2}{\rightleftarrows}$$

$$\left[CH_3-\underset{\underset{N}{\overset{|}{\overset{\|}{C}}}}{\overset{OH}{\overset{|}{C^\ominus}}} \leftrightarrow CH_3-\underset{\underset{N^\ominus}{\overset{\|}{C}}}{\overset{OH}{\overset{|}{C}}}\right] \underset{B}{\overset{BH^\oplus}{\rightleftarrows}} CH_3-\underset{CN}{\underset{|}{\overset{OH}{\overset{|}{CH}}}} \overset{B}{\underset{BH^\oplus}{\rightleftarrows}} CH_3-\underset{CN}{\underset{|}{\overset{O^\ominus}{\overset{|}{CH}}}} \underset{CN^\ominus}{\rightleftarrows} CH_3\overset{O}{\overset{\|}{CH}}$$

(6.36)

Like catalysis of the acyloin condensation, it depends on the capacity of cyanide ion to add rapidly and reversibly to the carbonyl group to give a species that can stabilize a negative charge. Ordinarily the product of this reaction is an acyloin. However, when the reaction is carried out in aqueous dioxane solution of dimedone, the aldehyde can be trapped as a derivative.

Although cyanide ion is an efficient catalyst for decarboxylation of α-keto acids and the acyloin condensation, it is not unique. The coenzyme thiamine catalyzes the acyloin condensation in the absence of enzyme. Likewise other thiazolium ions catalyze this reaction in *mildly basic solution*.[92] The decarboxylation of α-keto acids is accelerated (nonenzymically) by thiamine. Breslow[93] discovered that thiazolium salts are in rapid protropic equilibrium with the zwitterion.

(6.37)

Nucleophilic and Electrophilic Catalysis

The relative stability of the zwitterion is at first sight surprising since the carbanion is not stabilized by enolate ion resonance as are most stable carbanions, but on closer scrutiny it is apparent that the ion is closely related to cyanide ion, being a double bond analog of the triply bonded cyanide carbanion. Both are strongly stabilized by resonance involving two canonical forms, one of which has a six-electron carbon atom, symbolized in eq. 6.37. The thiazolium zwitterion is a member of a wide class of substances in which this resonance occurs.

Given that such a zwitterion as eq. 6.37 exists, it is possible to formulate a mechanism by which it might catalyze the acyloin condensation and the decarboxylation of α-keto acids as cyanide ion does.[94] Equation 6.38 gives the mechanism of the thiazolium ion (thiamine)-catalyzed decarboxylation and acyloin condensation of pyruvic acid.

(6.38)

Nucleophilic Catalysis

Both the intermediate formed between pyruvate and the thiazolium ion and the adduct of acetaldehyde and the thiazolium ion have been isolated from reactions of the enzyme carboxylase,[95,96] and the former has been isolated from the native enzyme.[97] The thiazolium ion also catalyzes the transketolase reaction both enzymically and nonenzymically.[95,98] A typical example is the reaction of fructose-6-phosphate with glyceraldehyde-3-phosphate to give a tetrose phosphate and xylulose-5-phosphate. The reaction, given in eq. 6.39, consists of a reverse aldol condensation, leading to the release of one aldehyde group, followed by the aldol condensation of the acetol adduct of the thiazolium ion with a new aldehyde to form the product.

$$
\begin{array}{c}
\text{CH}_2\text{OH} \\
| \\
\text{C}=\text{O} \\
| \\
\text{HOCH} \\
| \\
\text{R} \\
\text{Fructose} \\
\text{6-phosphate}
\end{array}
+ \text{thiazolium}^{\ominus} \underset{B}{\overset{BH^{\oplus}}{\rightleftharpoons}}
\begin{array}{c}
\text{HOCH}_2 \\
| \\
\text{HOC—thiazolium} \\
| \\
\text{HOCH} \\
| \\
\text{R}
\end{array}
\rightleftharpoons
\begin{bmatrix}
\text{HOH}_2\text{C} \\
\diagdown \\
\text{C}^{\ominus}\text{—thiazolium}^{\oplus} \\
\diagup \\
\text{HO}
\end{bmatrix}
\updownarrow
\begin{bmatrix}
\text{HOH}_2\text{C} \\
\diagdown \\
\text{C}=\text{thiazolium} \\
\diagup \\
\text{HO}
\end{bmatrix}
$$

$-\text{RCHO (aldotetrose phosphate)}$

$+\text{R}'\text{CHO (glyceraldehyde phosphate)}$

$$
\begin{array}{c}
\text{CH}_2\text{OH} \\
| \\
\text{C}=\text{O} \\
| \\
\text{HOCH} \\
| \\
\text{R}' \\
\text{Xylulose} \\
\text{5-phosphate}
\end{array}
+ \text{thiazolium}^{\ominus} \underset{B}{\overset{BH^{\oplus}}{\rightleftharpoons}}
\begin{array}{c}
\text{HOCH}_2 \\
| \\
\text{HOC—thiazolium} \\
| \\
\text{HOCH} \\
| \\
\text{R}'
\end{array}
$$

$R = -\text{CHOHCHOHCH}_2\text{OPO}_3\text{H}^{\ominus}$
$R' = -\text{CHOHCH}_2\text{OPO}_3\text{H}^{\ominus}$

(6.39)

In this catalysis, the carbanionic intermediate is again resonance stabilized by its attachment to the thiazolium group with consequent interaction with the cationic nitrogen "electron sink"[99] (see Chapter 17).

Catalysis by Carbanions. Initiation of the polymerization of olefins in organic solvents can be effected by carbanions in the form of organometallic compounds such as phenylsodium and butyllithium. These highly reactive substances add to an olefin leading to a new organometallic compound that can repeat the process until the olefin is depleted or until termination occurs, usually through interaction with a Lewis acid.

$$CH_3CH_2CH_2CH_2Li \xrightarrow{H_2C=CHR} CH_3(CH_2)_4CHRLi$$
$$CH_3(CH_2)_4CHRLi \xrightarrow{H_2C=CHR} CH_3(CH_2)_4CHRCH_2CHRLi$$
$$CH_3(CH_2)_4CHRCH_2CHRLi \xrightarrow{H_2C=CHR} \text{ and so on}$$

(6.40)

The stereochemistry of carbanionic vinyl polymerization is, however, dependent on the counterion. For example, if isoprene is polymerized in a hydrocarbon solvent with a lithium counterion, the polymer has a stereochemistry almost exclusively cis-1,4. If a sodium counterion is used in the same solvent, the polymer will possess not cis-1,4, but rather trans-1,4 stereochemistry. These results indicate that the active catalyst is not the free carbanion, rather it is the ion pair. In this sense, the catalyst (rigorously the initiator) is not truly a nucleophile but a combination of a nucleophile and an electrophile.[100]

6.2.3 The Transition between Nucleophilic and General Base Catalysis

We have just considered related reactions, some of which are subject to nucleophilic catalysis and some to general base catalysis. Which species will act as nucleophilic catalysts and which will act as general base catalysts? Some generalizations may be put forth. Those bases or nucleophiles that are polarizable, or soft, will tend to act as nucleophilic catalysts. On the other hand, those bases or nucleophiles that are not polarizable, but hard, will tend to act as general base catalysts. If Brønsted plots for nucleophilic catalysis have higher slopes than for general base catalysis, stronger bases will tend to act as nucleophiles whereas weaker bases will tend to react as general base catalysts. One might then expect a break in a Brønsted plot, weaker bases exhibiting general base catalysis with a lower slope and stronger bases exhibiting nucleophilic catalysis with a higher slope, as shown in Fig. 6.4. Figure 6.4 is not entirely hypothetical. It represents in a schematic way the hydrolysis of ethyl dichloroacetate catalyzed by a series of bases. With bases weaker than ammonia, hydrolysis occurs via general basic catalysis. With bases of the strength of ammonia or higher, a nucleophilic reaction occurs that may lead either to hydrolysis or to other reactions.

The question can also be asked as to why a particular member of a substrate family is susceptible to general base catalysis while another is susceptible to nucleophilic catalysis. This dichotomy of mechanism is seen in imidazole catalysis of ester hydrolysis when the structure of the ester is varied. Esters activated in the acyl group and containing a poor leaving group show general base catalysis by imidazole. On the other hand, esters with a

Nucleophilic Catalysis

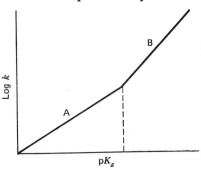

Fig. 6.4 The Brønsted plot of the general base (A) and nucleophilic (B) catalysis of the hydrolysis of ethyl dichloroacetate.

good leaving group are subject to nucleophilic catalysis by imidazole.[101] Likewise, substituted phenyl acetates containing highly electron-withdrawing substituents are subject to nucleophilic catalysis by acetate ion while those with other substituents are subject to general base catalysis by acetate ion.[102] Kirsch and Jencks[101] have analyzed the change from general base to nucleophilic catalysis by imidazole, using a structure-reactivity correlation involving a comparison of the rate constants of the imidazole-catalyzed and hydroxide ion-catalyzed reactions of a series of esters, as shown in Fig. 6.5.[101] The hydroxide ion rate constants are assumed to follow a common mechanism for all esters, and are used as an empirical measure of structural change, rather than a Hammett sigma or Taft sigma* constant. By comparing the rate constants of imidazole catalysis to the rate constants of hydroxide ion catalysis, the structural changes affecting reactivity will automatically be taken into account, and thus one can determine structural variations important to the change in catalysis. In this analysis, another factor must also be considered in addition to the change in the catalytic mechanism; that is, the rate-determining step of the two-step ester hydrolysis can be changed by structural variation.

The reactions on the upper line of Fig. 6.5 are identified as nucleophilic catalyses on the basis of the demonstration of an N-acetylimidazole intermediate. The reactions on the lower line are assumed to be general base-catalyzed reactions. A number of these reactions are discussed in Chapter 5. The ethyl acetate point requires special comment since the imidazole-catalyzed hydrolysis of ethyl acetate is a very slow process indeed. It may be ascribed to general base catalysis on the basis of: (1) the rate constant for imidazole catalysis is equal to that for phosphate ion catalysis; (2) the deuterium oxide isotope effect conforms to general base catalysis; and (3) the point lies on a line extrapolated from more reactive acetates that follow general base catalysis.

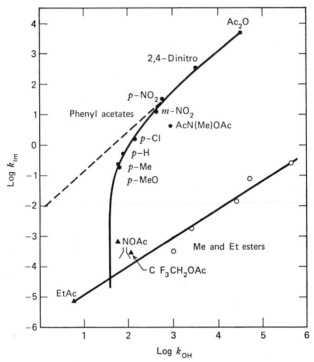

Fig. 6.5 Rates of imidazole-catalyzed ester hydrolysis as a function of the rate of alkaline hydrolysis: nucleophilic reactions of acetates; ●; general base catalysis of acetates, ▲; general base catalysis of methyl and ethyl esters,[5] ○ (ionic strength 1.0; 25°). Trifluoroethyl acetate rate measured with N-methylimidazole. From J. F. Kirsch and W. P. Jencks, *J. Amer. Chem. Soc.*, **86**, 843 (1964). © 1964 by the American Chemical Society. Reproduced by permission of the copyright owner.

At the top of Fig. 6.5 lie a series of acetate derivatives with excellent leaving groups. As we proceed down the upper line of the graph, the leaving groups become progressively poorer. With good leaving groups, essentially every tetrahedral intermediate formed by the addition of imidazole to the carbonyl group is partitioned to products; there will be little effect of structural change of the leaving group on the rate constant and thus the imidazole and hydroxide ion reactions parallel one another well (slope 1.2). As the leaving group gets poorer, the energy barrier for the decomposition of the intermediate to products (Fig. 6.6) becomes larger than the energy barrier for the regeneration of the starting material from the intermediate, and thus the former becomes rate-determining. Thus a given structural change in the leaving group will be directly reflected in the rate constant. This leads to the downward curvature in the upper line of Fig. 6.5. Finally, as the leaving

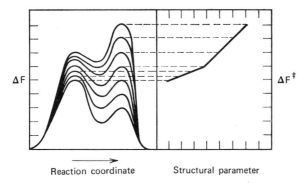

Fig. 6.6 Transition state diagram for a two-step reaction showing how changes in structure that affect principally the second step result in a nonlinear structure reactivity correlation. From J. F. Kirsch and W. P. Jencks, *J. Amer. Chem. Soc.*, **86**, 845 (1964). © 1964 by the American Chemical Society. Reproduced by permission of the copyright owner.

group becomes still worse, the rate constant of the nucleophilic reaction of imidazole with ester becomes so small that it is not observed: a general base-catalyzed reaction that is less sensitive to the nature of the leaving group takes over, bringing us to the ethyl acetate reaction at the bottom of the sigmoid curve. Two breaks are thus seen in this curve, one reflecting the change in rate-determining step from the formation of the tetrahedral intermediate to its decomposition, and the second reflecting the transformation from nucleophilic to general base catalysis.

Other changes in catalytic mechanism with changes in the leaving group are seen in reactions of imidazole with acetate derivatives.[103] Those compounds, such as acetic anhydride and *p*-nitrophenyl acetate, that have very good leaving groups have imidazole reactions that are first-order in imidazole. However, acetate esters of less acidic phenols have imidazole reactions second-order in imidazole; presumably the second imidazole is a general base catalyst for the first, the nucleophile. In the imidazole reactions of acetates of acetoxime and trifluoroethanol, still worse leaving groups, hydroxide ion is needed for catalysis. Finally ethyl acetate shows no nucleophilic reaction with imidazole at all.

6.3 ELECTROPHILIC CATALYSIS

6.3.1 Electrophilicity

Electrophilic catalysis, the mirror image of nucleophilic catalysis, consists of electrophilic attack by the catalyst on the substrate producing an unstable intermediate that decomposes to give the product and regenerate the catalyst. The broadest definition of an electrophilic catalyst would include metal ions as well as many compounds capable of increasing their coordination number usually spoken of as Lewis acids. We arbitrarily relegate metal ion catalysis to Chapter 8, because of its somewhat special character, and concentrate here on reactions catalyzed by electrophilic metal halides, halogens, and carbonyl compounds. No ambiguities exist between electrophilic and general acid catalysts since we arbitrarily define electrophiles as Lewis acids while general acids are proton donors.

On the basis of an earlier classification of metal ions[104] Pearson classified other generalized electrophiles.[5] The main criterion used was whether the electrophile forms its most stable complex with the first ligand atom of each group of the periodic table (class a hard electrophiles), or with the second or subsequent member of each group (class b soft electrophiles). When equilibrium data were not available, other criteria were used. One was that class b electrophiles complex readily with a variety of bases of negligible proton basicity. These include carbon monoxide, olefins, and aromatic hydrocarbons. Rate data were also used. Table 6-5 shows a variety of electrophiles for which available data allow an assignment of class a or b behavior. Some borderline and duplicative cases are also shown. The listing in Table 6.5 can be used in conjunction with the data of Table 6.2 by the use of the rule: hard electrophiles tend to associate with hard nucleophiles and soft electrophiles tend to associate with soft nucleophiles.

An illustration of this principle is given by a comparison of $CH_3Hg^\oplus$, a typical soft electrophile, and $H^\oplus$, a hard electrophile.[105] Both form stable complexes with $OH^\ominus$, a hard base, and with $S^{2\ominus}$, a soft base. However, the stability constants are such that the competition reaction,

$$H^\oplus + CH_3HgOH \rightleftharpoons H_2O + CH_3Hg^\oplus \qquad (6.41)$$

has an equilibrium constant of $10^{6.3}$. The competition reaction,

$$H^\oplus + CH_3HgS^\ominus \rightleftharpoons HS^\ominus + CH_3Hg^\oplus \qquad (6.42)$$

on the other hand, has a constant of $10^{-8.4}$. The preferences of the proton for the hard base and $CH_3Hg^\oplus$ for the soft base are dramatically demonstrated.

The distinguishing features of class a electrophiles are small size, high positive oxidation state, and the absence of any outer electrons that are easily excited to higher states. In addition, the hardness of a given acceptor atom is a function of the other groups attached to it. Thus BF_3 is a class a electrophile, but BH_3 is a typical class b electrophile, forming complexes such as BH_3CO. Class b electrophiles in general have one or more of the following properties: low or zero positive charge, large size, and several

Electrophilic Catalysis

TABLE 6.5
Classification of Electrophiles[a,b]

Hard (Class a)	Soft (Class b)
$H^\oplus$, $Li^\oplus$ $Na^\oplus$, $K^\oplus$	$Cu^\oplus$, $Ag^\oplus$, $Au^\oplus$, $Tl^\oplus$, $Hg^\oplus$, $Cs^\oplus$
$Be^{2\oplus}$, $Mg^{2\oplus}$, $Ca^{2\oplus}$, $Sr^{2\oplus}$, $Mn^{2\oplus}$	$Pd^{2\oplus}$, $Cd^{2\oplus}$, $Pt^{2\oplus}$, $Hg^{2\oplus}$, $CH_3Hg^\oplus$
$Al^{3\oplus}$, $Sc^{3\oplus}$, $Ga^{3\oplus}$, $In^{3\oplus}$, $La^{3\oplus}$	$Tl^{3\oplus}$, $Au^{3\oplus}$, $Te^{4\oplus}$, $Pt^{4\oplus}$
$Cr^{3\oplus}$, $Co^{3\oplus}$, $Fe^{3\oplus}$, $As^{3\oplus}$, $Ce^{3\oplus}$	$Tl(CH_3)_3$, BH_3, $Co(CN)_5^{2\ominus}$
$Si^{4\oplus}$, $Ti^{4\oplus}$, $Zr^{4\oplus}$, $Th^{4\oplus}$, $Pu^{4\oplus}$	$RS^\oplus$, $RSe^\oplus$, $RTe^\oplus$
$Ce^{4\oplus}$, $Ge^{4\oplus}$, $VO^{2\oplus}$	$I^\oplus$, $Br^\oplus$, $HO^\oplus$, $RO^\oplus$
$UO_2^{2\oplus}$, $(CH_3)_2Sn^{2\oplus}$	I_2, Br_2, ICN, and so on
$BeMe_2$, BF_3, BCl_3, $B(OR)_3$	Trinitrobenzene, and so on
$Al(CH_3)_3$, $Ga(CH_3)_3$, $In(CH_3)_3$, AlH_3	Chloranil, quinones, and so on
$RPO_2^\oplus$, $ROPO_2^\oplus$	Tetracyanoethylene, and so on
$RSO_2^\oplus$, $ROSO_2^\oplus$, SO_3	O, Cl, Br, I, N
$I^{7\oplus}$, $I^{5\oplus}$, $Cl^{7\oplus}$, $Cr^{6\oplus}$, $Se^{6\oplus}$	M° (metal atoms)
$RCO^\oplus$, CO_2, $NC^\oplus$	Bulk metals
HX (hydrogen bonding molecules)	

[a] The following are in a borderline class between class a and b: $Fe^{2\oplus}$, $Co^{2\oplus}$, $Ni^{2\oplus}$, $Cu^{2\oplus}$, $Zn^{2\oplus}$, $Pb^{2\oplus}$, $Sn^{2\oplus}$, $Sb^{3\oplus}$, $Bi^{3\oplus}$, $Rh^{3\oplus}$, $Ir^{3\oplus}$, $B(CH_3)_3$, SO_2, $NO^\oplus$, $Ru^{2\oplus}$, $Os^{2\oplus}$, $R_3C^\oplus$.
[b] From R. G. Pearson, *J. Amer. Chem. Soc.*, **85**, 3533 (1963). © 1963 by the American Chemical Society. Reprinted by permission of the copyright owner.

easily excited outer electrons. For metals these outer electrons are *d*-orbital electrons.

Hydrogen-bonded complexes are class a, so that stronger bonds are formed to N, O, and F donors than to P, S, and I donors. A similar analysis may be made of the difference in rate constants of proton transfers from oxygen and nitrogen acids versus carbon acids. As mentioned earlier, the softness-hardness principle may be used as a guide to selecting electrophilic and nucleophilic catalysts.

6.3.2 Mechanisms of Electrophilic Catalysis

Catalysis by Lewis Acid Metal Halides, Hydrides, and Carbonium Ions. Electrophilic catalysis common to the experience of practically all organic chemists is the use of metal halides of the Lewis acid variety for accelerating reactions proceeding through carbonium ion intermediates. These reactions include the mercuric chloride-catalyzed hydrolysis of α-phenylethyl, *t*-butyl, and benzhydryl chlorides.[106] This complex reaction is considered in Chapter 8. In these catalyses, assistance of carbonium ion formation is the important factor. Assistance of the generation of carbonium ions from alkyl halides is also seen in catalysis of the racemization of optically active α-phenylethyl chloride by many electrophilic metal chlorides.[107,108]

Catalysis of the alkylation and acylation of aromatic and aliphatic compounds by metallic halides from every group of the periodic table except group IA is an important electrophilic catalysis. In these reactions, usually called Friedel-Crafts reactions, the most common catalysts include halides of aluminum, tin, antimony, iron, zinc, boron, and gallium.[109] In the boron halide family, the order of Lewis acid strength toward pyridine and possibly the catalytic effectiveness is

$$BBr_3 > BCl_3 > BH_3 > BF_3$$

Considerations of electronegativity only would predict an order in which BF_3 is the strongest electrophile and BH_3 the weakest. It has therefore been suggested that the formally empty $2p$ orbital on boron is at least partially occupied by electron pairs donated from the halogen atoms. Toward dimethyl ether, the order of the halides of group 3A is

$$\text{gallium} > \text{indium} > \text{thallium}$$

Toward hydrogen fluoride, the order of the halides of group 5A is

$$\text{antimony} > \text{arsenic} > \text{phosphorus}$$

The function of the metal halide catalyst in the Friedel-Crafts reaction is to loosen or break the carbon–halogen bond in the alkyl or acyl halide, making the latter reagent more electrophilic. In order to probe this process, the interactions between the metal halide and various components of these reactions have been investigated.[110] Aluminum bromide is readily soluble in alkyl bromides. Vapor pressure measurements demonstrate the presence of a stable complex, $MeBr \cdot AlBr_3$, in solvent methyl bromide at 0°, and a solid of this kind has been obtained by partial removal of the solvent at $-78°$. Some solutions of this kind, although not all, conduct electricity. In many instances, exchange of halogen between alkyl and aluminum halides occurs readily. These observations may be explained by the equilibria

$$R-X: + MeX_3 \rightleftharpoons R-X:MeX_3 \rightleftharpoons R^{\oplus}MeX_4^{\ominus} \qquad (6.43)$$

In addition to these complexes, complexes between the aromatic hydrocarbon and metal halide, hydrogen halide and metal halide, aromatic hydrocarbon-hydrogen halide-metal halide, and aromatic hydrocarbon-alkyl halide-metal halide have been investigated. The last is the most interesting, for presumably it is the intermediate in a Friedel-Crafts alkylation reaction. Complexes of alkylbenzenes, alkyl fluorides, and boron fluoride have been observed and isolated at low temperatures.[111-113] These complexes have been described as either sigma-complexes[111] or oriented pi-complexes.[113] These complexes, together with other aspects of the mechanism of these reactions such as substituent effects, can be accounted for in terms of the following mechanism.

$$RX + BF_3 \underset{k_{-1}}{\overset{k_1}{\rightleftharpoons}} {}^{\delta\oplus}R\text{—}X\cdots{}^{\delta\ominus}BF_3$$

$$\bigcirc + {}^{\delta\oplus}R\text{—}X\cdots{}^{\delta\ominus}BF_3 \underset{k_{-2}}{\overset{k_2}{\rightleftharpoons}} \bigcirc\!\!\begin{array}{c}H\\ \end{array}\!\!\cdots{}^{\delta\oplus}R\text{—}X\cdots{}^{\delta\ominus}BF_3 \xrightarrow{k_3}$$

$$\left[\bigcirc\!\!\begin{array}{c}H\\R\end{array}\right]^{\oplus} BF_3X^{\ominus} \xrightarrow{k_4} \bigcirc\!\!\!-R + HX + BF_3 \quad (6.44)$$

where $k_1, k_{-1}, k_2, k_{-2}, k_4 \gg k_3$.

In the Friedel-Crafts acylation reaction, similar processes occur. The available ultraviolet spectra, heats of formation, and infrared spectra are best in accord with the undissociated oxonium ion structure for the complex of aluminum chloride and benzoyl chloride. Of course, a difference exists in the alkylation and acylation reactions in that the latter require stoichiometric amounts of the electrophilic promoter because the product is basic and forms a stable complex with it.[114]

In the alkylation reaction, a significant difference in product distribution is seen between ferric chloride and aluminum or zirconium chloride catalysts. When ferric chloride is used as catalyst with a t-hexyl chloride as alkylating agent, no rearrangement of the carbon skeleton occurs, whereas rearrangement to a secondary isomer predominates with the others.[115] Solvent effects are also seen.[115] This difference may be associated with differences in the degree of ionization of the alkyl halide by the catalyst according to eq. 6.43.

Catalyst concentration has an unexpected effect in the rearrangement of alkylated aromatics. In the system m-xylene-hydrogen fluoride-boron trifluoride, the product of migration of the methyl group is markedly dependent on the concentration of boron trifluoride, as shown in Table 6.6.[116] When the ratio [BF_3]/[xylene] is much less than one, a mixture of xylenes is produced in which the m-xylene content of 60% is in agreement with the thermodynamically calculated value of 57%, whereas at higher concentrations of boron trifluoride, the m-xylene content rises, finally approaching 100% when [BF_3]/[xylene] = 3. This seemingly anomalous situation can be explained if one takes into account the fact that the reaction system really must be considered in terms of two interlocking equilibria. At low boron trifluoride concentrations, a true thermodynamic equilibrium among the

Table 6.6
Isomerization of Xylenes with HF and BF_3[a,b]

Xylene	Meta-	Meta-	Ortho-	Meta-	Meta-	Ortho-	Para-
Moles BF_3/mole xylene	0.09	0.06	0.10	0.13	0.65	0.72	3.0
Temperature (°C)	82	100	100	121	82	82	3
Contact time (min)	30	30	30	3	30	30	1380
Yield of C_8 product (%)	98	91	83	91	94	84	100
Composition of C_8 product, % o-xylene	15	19	19	18	11	11	0
m-xylene	64	60	61	60	74	75	100
p-xylene	22	21	20	22	15	14	0

[a] 6 moles HF per mole of xylene.
[b] From D. A. McCaulay and A. P. Lein, *J. Amer. Chem. Soc.*, **74**, 6246 (1952). © 1952 by the American Chemical Society. Reprinted by permission of the copyright owner.

xylene isomers (in the hydrocarbon phase) will be found. However, at high boron trifluoride concentrations, because of the conversion of the xylenes to their conjugate acids, a new (true) thermodynamic equilibrium among the various *protonated* xylenes is obeyed, the compounds now being found in the acid phase. Since the resonance stabilization of the conjugate acid of *m*-xylene is so much higher than that of the other xylenes, at high boron trifluoride concentration only *m*-xylene is found upon removal of the catalyst. This double equilibrium is schematically shown by

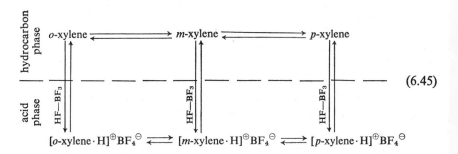

(6.45)

A similar phenomenon is seen in the rearrangement of methylated phenols, where a trace of hydrogen chloride produces a small amount of rearrangement but a swamping amount produces a larger amount.[117]

Many metal halide-catalyzed alkylation reactions require a co-catalyst. In fact, rigorously purified anhydrous aluminum chloride neither isomerizes cyclohexane nor polymerizes olefins. But in the presence of trace amounts of water, alcohols, alkyl halides, or Brønsted acids, these reactions proceed smoothly and sometimes even explosively. This general concept was discovered in working with heterogeneous systems employing solid aluminum halides in a vacuum line,[118] but it is equally applicable to homogeneous catalyses. For example, catalysis of the copolymerization of isobutene and isoprene using diethylaluminum chloride, which is completely miscible with hydrocarbons, is accelerated by Brønsted acids in the order HCl, HBr > HF, H_2O > CCl_3COOH > CH_3OH > CH_3COCH_3. HCl and HBr give explosive polymerization rates at concentrations as low as 10^{-4} M at $-50°$. These catalyses can be explained straightforwardly by postulating that the combination of Brønsted and Lewis acids promotes the formation of the carbonium ions needed to initiate the polymerization. On this basis, any other carbonium ion source should also be effective. t-Butyl chloride fulfills this requirement, and indeed it is effective as a co-catalyst for the polymerization of isobutene with diethylaluminum chloride in methyl chloride solution.[119]

Catalysis by Halogens and Other Electrophiles. The halogenation of aromatics is catalyzed by many Lewis acids such as metal halides and the halogens themselves. For example, the bromination of aromatic compounds by bromine alone and in the presence of iodine follows the kinetic equations

$$k_{obs} = k[Br_2] + k'[Br_2]^2 + k''[Br_2]^3 \tag{6.46}$$

$$k_{obs} = k[Br_2]^2[IBr] + k'[Br_2][IBr]^2 \tag{6.47}$$

The higher powers of bromine and the first and second moles of iodine bromide act as electrophilic catalysts in facilitating the cleavage of the bromine–bromine bond of the halogen directly involved to the substitution reaction.[120] Iodine also catalyzes the dehydration of tertiary alcohols.[121] This catalysis occurs by the conversion of the alcohol to the corresponding hypoiodite, which forms a carbonium ion readily, leading to the olefinic product. Carbonium ion formation from the hypoiodite is more facile than from the alcohol, because hypoiodite ion is a better leaving group than is hydroxide ion. The overall reaction is a catalytic process, since the product hypoiodous acid and hydriodic acid re-form molecular iodine in an equilibrium process.

$$\begin{array}{c}
\underset{\underset{\text{OH}}{|}}{\overset{\overset{\text{CH}_3}{|}}{\text{CH}_3-\text{C}-\text{CH}_2-\text{CH}_3}} + I_2 \rightleftarrows \underset{\underset{\text{OI}}{|}}{\overset{\overset{\text{CH}_3}{|}}{\text{CH}_3-\text{C}-\text{CH}_2-\text{CH}_3}} + \text{HI} \\
\underset{\underset{\text{OI}}{|}}{\overset{\overset{\text{CH}_3}{|}}{\text{CH}_3-\text{C}-\text{CH}_2-\text{CH}_3}} \xrightarrow{-\text{OI}^\ominus} \overset{\overset{\text{CH}_3}{|}}{\text{CH}_3-\text{C}^\oplus-\text{CH}_2-\text{CH}_3} \xrightarrow{-\text{H}^\oplus} \overset{\overset{\text{CH}_3}{|}}{\text{CH}_3-\text{C}=\text{CH}-\text{CH}_3} \\
\text{HOI} + \text{HI} \rightleftarrows I_2 + H_2O
\end{array} \tag{6.48}$$

The nitration of activated aromatic compounds is catalyzed by nitrous acid, the rate of nitration being dependent on the concentrations of aromatic compound and of nitrous acid at constant nitric acid concentration. Because this catalysis occurs only with activated aromatics and because of its kinetic dependence, the catalysis probably proceeds through electrophilic attack by $NO^{\oplus}$ forming a nitrosoaromatic intermediate which is then oxidized by nitric acid, yielding the nitroaromatic product and regenerating the nitrous acid catalyst.[120,122]

Bromine catalyzes the hydration of carbon dioxide. This has recently been shown to be due to hypobromous acid.[125]

Catalysis by Carbonyl Compounds. Since primary and secondary amines are nucleophilic catalysts for many reactions of carbonyl compounds, it might be predicted that carbonyl compounds can serve as electrophilic catalysts in the reactions of amines. An example of electrophilic catalysis by an aldehyde is seen in the hydrolysis of phosphoramidate ion. A rapid condensation of formaldehyde and phosphoramidate occurs, dependent on the first and second power of the formaldehyde concentration. The formation of the formaldehyde-phosphoramidate adduct is not complete, however, even at 0.8 M formaldehyde. On the basis of these observations and the observation that the overall hydrolysis is accelerated further by the addition of amines to the solution, eq. 6.49 has been suggested for this process.[123]

$$H_3\overset{\oplus}{N}PO_3{}^{2\ominus} + HCHO \underset{}{\overset{fast}{\rightleftarrows}} HOCH_2\overset{\oplus}{N}H_2PO_3{}^{2\ominus} \underset{}{\overset{fast}{\rightleftarrows}} (HOCH_2)_2\overset{\oplus}{N}HPO_3{}^{2\ominus}$$

$$RN\overset{\oplus}{H_3}\diagup$$

$$R\overset{\oplus}{N}H_2CH_2\overset{\oplus}{N}H_2PO_3{}^{2\ominus}$$

$$-H_2NCH_2NH_2R \diagdown$$

$$H_2PO_4{}^{\ominus} + H_2NCH_2OH \qquad H_2PO_4{}^{\ominus} + HN(CH_2OH)_2$$

(6.49)

The overall reaction is catalytic because the formaldehyde-amine products either decompose or react further. The hydrolysis of phosphoramidate is also catalyzed by hypochlorite ion.

The Strecker degradation of amino acids catalyzed by α-dicarbonyl compounds may be explained by an electrophilic catalysis of similar mechanism. In this reaction, the intermediate imine is subject to facile decarboxylation, followed by hydrolytic cleavage, yielding carbon dioxide, ammonia, and an aldehyde from the original α-amino acid. Sugar ketoses can effect a similar change important in browning reactions.[124]

The most important electrophilic catalyst of the aldehyde variety is pyridoxal, vitamin B_6. It serves as coenzyme for many enzymatic reactions of α-amino acids such as transamination, decarboxylation, racemization, elimination, and condensation. Most of these reactions are catalyzed by pyridoxal alone, albeit less efficiently and less specifically than in the presence of the enzyme.

Early work on the reaction of α-amino acids and α-keto acids indicated a transamination reaction.[126] With glyoxylic acid the reaction proceeds at room temperature and no decarboxylation ensues.[127] These reactions are most easily rationalized in terms of the intermediate formation of an imine that can form a tautomeric imine by prototropic shift. Cleavage of the resultant imine will lead to products of transamination. Both pyridoxal and pyridoxamine participate in transamination reactions with glutamic acid and α-ketoglutaric acid, respectively.[128] These reactions, as well as some of the earlier reactions, were found to be catalyzed by cupric, ferric, and aluminum salts.[129] On the basis of these studies, mechanism 6.50 was proposed for the pyridoxal-catalyzed transamination reaction.[130]

$$(6.50)$$

Since, as mentioned above, pyridoxamine, the product of reaction, can react with another α-keto acid producing another amino acid and regenerating pyridoxal, the sum of two such reactions constitutes an overall catalysis. Chelated imines corresponding to **1** and **3** have been spectrally identified, and in some cases isolated.[131,132] Both 1:1:1 and 2:2:1 pyridoxal-amino acid-metal ion chelates have been identified. The metal ion functions in the catalytic mechanism by stabilizing the intermediate imine and thus helping to provide a planar conjugated system, and by lowering the electron density at the α-carbon atom. The removal of a proton, converting imine **1** to imine **2**, is the key step in this process, depending for its efficiency on the absorption of the electronic charge produced on proton loss by the pyridinium nitrogen electron sink. This process depends on the unique pi electron system that appears on imine formation. In the conversion of imine **2** to imine **3**, the dihydropyridine moiety serves as an electron source, and in the final step the isomeric imine, **3**, is hydrolytically cleaved. Thus, the essential parts of the pyridoxal catalyst are: (1) the aldehyde group for imine formation; (2) the pyridinium nitrogen atom to serve as an electron sink (and source) through a conjugated pi electron system; and (3) the *ortho*-hydroxyl group to stabilize the imine chelate. This mechanistic description is confirmed by the replacement of pyridoxal in these catalyses by 2-hydroxypyridine-4-carboxaldehyde, indicating that the methyl and hydroxyl group of pyridine are not required in the catalysis, and by 2-hydroxy-4-nitrobenzaldehyde (but not by 2-hydroxy-3-nitrobenzaldehyde) indicating that any suitably oriented electron sink can replace the pyridinium ion.[130]

Pyridoxal serves as catalyst for the racemization of optically active α-amino acids.[133] The reaction is accelerated by metal ions. As predicted by eq. 6.51, this reaction accompanies the transamination reaction. In the pyridoxal-alanine-aluminum ion system, racemization predominates at pH 9.6 while transamination is the major reaction at pH 5, although both reactions are competitive at all pH's. The racemization reaction may be viewed simply as the formation of imine **2**, which of course no longer possesses the original asymmetric center, and reversion to starting materials, which in a symmetric (nonenzymatic) environment should lead to a racemic mixture.

Pyridoxal also serves as catalyst for the decarboxylation of α-amino acids.[134-136] For example, when α-aminoisobutyric acid is heated with pyridoxal, carbon dioxide, isopropylamine, acetone, and pyridoxamine are produced.[136] This reaction may be interpreted in terms of the cleavage of the α-carbon–carboxylate bond in imine **1** rather than the α-carbon–hydrogen bond. Mechanistically, the rest of the process is the same.

In enzymic reactions, pyridoxal also catalyzes eliminations and condensations (see Chapter 17). Some of these include the conversion of serine and cysteine to pyruvate, internal oxidation-reduction reactions of the pinacol-pinacolone rearrangement type, the formation of tryptophan from serine and indole, the decomposition of threonine to glycine and acetaldehyde, and the conversion of *O*-phosphohomoserine to threonine. Only partial success has been achieved in effecting these reactions using pyridoxal alone. Vanadium salts in combination with pyridoxal phosphate catalyze a highly specific elimination of hydrogen sulfide from cysteine. The reaction occurs in two stages: ring fission of a thiazolidine intermediate and elimination of hydrogen sulfide.[137] Tryptophan is formed in low yield (0.6–0.8%) from serine, indole, pyridoxal, and alum.[138] Threonine may be cleaved by pyridoxal and metal ions yielding glycine and acetaldehyde in 64% yield, ammonia and α-ketobutyric acid in 6% yield, and β-pyridoxal serine in 8% yield.[139] The latter reactions may also proceed through imine intermediates.

6.4 SUMMARY

Nucleophilic and electrophilic catalysis pervade organic chemistry. The diversity of organic chemical reactions is mirrored by the diversity of these catalyses. It may be reasonably predicted that many important further catalyses of this kind will be found by looking at still more organic reactions. One of the striking characteristics of these catalyses is their specificity. Since they depend on specific chemical interactions between substrate and catalyst, chemical complementarity must exist between the substrate and catalyst, leading to a first approximation of catalytic specificity. Although many different nucleophiles and electrophiles have been utilized as catalysts, by no means have all possibilities been investigated. Particularly absent from view at the moment are thiolate anion nucleophiles, which possess good basicity and high polarizability. These substances and others will be utilized as catalysts when the reactions are found that lead to reactive intermediates. For this is the key to nucleophilic and electrophilic catalysis. Can a given nucleophile or electrophile easily produce a reactive intermediate that will lead to a wanted product, and at the same time regenerate the catalyst?

REFERENCES

1. W. Langenbeck, *Die Organischen Katalysatoren*, J. Springer, Berlin, 1935; *Adv. Enzymol.*, **14**, (1952).
2. J. W. Baker and E. Rothstein in *Handbuch der Katalyse*, Vol. II, G. M. Schwab, Ed., J. Springer, Vienna, 1940, p. 45.
3. J. O. Edwards, *J. Amer. Chem. Soc.*, **76**, 1540 (1954); **78**, 1819 (1956).
4. J. O. Edwards and R. G. Pearson, *J. Amer. Chem. Soc.*, **84**, 16 (1962); see also T. C. Bruice et al., *J. Amer. Chem. Soc.*, **89**, 2106 (1967).
5. R. G. Pearson, *J. Amer. Chem. Soc.*, **85**, 3533 (1963); *Science*, **151**, 172 (1966); R. G. Pearson and J. Songstad, *J. Amer. Chem. Soc.*, **89**, 1827 (1967), R. G. Pearson, *J. Chem. Ed.*, **45**, 581, 643 (1968).
6. B. Saville, paper presented, *Symp.* on *SHAB*, Cyanamid European Research Institute, Geneva, May 1965; *Chem. Eng. News*, **43**, 100 (1965); *Angew. Chem. Int. Ed.*, **6**, 928 (1967).

7. E. A. Moelwyn-Hughes, *Proc. Roy. Soc.*, **A164**, 295 (1938); *J. Chem. Soc.*, 779 (1938).
8. R. E. Robertson, *Progr. Phys. Org. Chem.*, **4**, 213 (1967).
9. J. Kice and G. Guaraldi, *J. Amer. Chem. Soc.*, **89**, 4113 (1967).
10. R. F. Hudson and M. Green, *J. Chem. Soc.*, 1055 (1962).
11. C. A. Bunton, P. B. D. DeLaMare, and J. G. Tillett, *J. Chem. Soc.*, 4754 (1958); C. A. Bunton and B. N. Hendy, *J. Chem. Soc.*, 2562 (1962); *Chem. Ind.*, 466 (1960); J. G. Tillett, *J. Chem. Soc.*, 5138 (1960).
12. C. A. Bunton and H. Chaimovich, *J. Amer. Chem. Soc.*, **88**, 4082 (1966); C. A. Bunton, D. Kellerman, K. G. Oldham, and C. A. Vernon, *J. Chem. Soc.*, B 292 (1966).
13. E. D. Hughes, F. Juliusberger, S. Masterman, B. Topley, and J. Weiss, *J. Chem. Soc.*, 1525 (1935).
14. P. J. Christen and J. A. C. M. Van den Muysenberg, *Biochim. Biophys. Acta*, **110**, 217 (1965).
15. M. L. Bender, *Chem. Rev.*, **60**, 53 (1960).
16. M. L. Bender, G. E. Clement, C. R. Gunter, and F. J. Kézdy, *J. Amer. Chem. Soc.*, **86**, 3697 (1964).
17. F. Gawron and F. Draus, *J. Amer. Chem. Soc.*, **80**, 5392 (1958).
18. M. Kilpatrick, Jr., *J. Amer. Chem. Soc.*, **50**, 2891 (1928).
19. M. Kilpatrick, Jr., *J. Amer. Chem. Soc.*, **52**, 1410 (1930).
20. M. Kilpatrick, Jr., and M. L. Kilpatrick, *J. Amer. Chem. Soc.*, **52**, 1419 (1930).
21. M. L. Bender and M. C. Neveu, *J. Amer. Chem. Soc.*, **80**, 5388 (1958).
22. D. G. Oakenfull, T. Riley, and V. Gold, *Chem. Comm.*, 385 (1966).
23. T. C. Bruice and R. Lapinski, *J. Amer. Chem. Soc.*, **80**, 2265 (1958).
24. E. Heilbronn, *Acta Chem. Scand.*, **12**, 1492 (1958).
25. E. R. Stadtman, in *The Mechanism of Enzyme Action*, W. D. McElroy and B. Glass, Eds., Johns Hopkins Press, Baltimore, 1954, p. 581.
26. T. Higuchi et al., *J. Amer. Pharm. Assoc.*, **43**, 129, 530 (1954).
27. C. Trolle-Lassen, *Arch. Pharm. Chemi.*, **60**, 632 (1953).
28. A. W. D. Avison, *J. Chem. Soc.*, 732 (1955).
29. J. M. Holland and J. G. Miller, *J. Phys. Chem.*, **65**, 463 (1961).
30. D. M. Brown and N. K. Hamer, *J. Chem. Soc.*, 1155 (1960).
31. D. Samuel and B. Silver, *J. Chem. Soc.*, 4321 (1961).
32. D. E. Koshland, Jr., *J. Amer. Chem. Soc.*, **74**, 2286 (1952).
33. T. Higuchi, J. D. McRae, and A. C. Shah, *J. Amer. Chem. Soc.*, **88**, 4015 (1966).
34. E. B. Lees and B. Saville, *J. Chem. Soc.*, 2262 (1958); E. W. Crunden and R. F. Hudson, *Chem. Ind.*, 748 (1959).
35. J. Esptein, V. E. Bauer, M. Saxe, and M. M. Demek, *J. Amer. Chem. Soc.*, **78**, 4068 (1956).
36. I. M. Klotz and S. G. Elfbaum, *Biochim. Biophys. Acta*, **86**, 100 (1964).
37. I. B. Wilson and S. Ginsberg, *Biochim. Biophys. Acta*, **18**, 168 (1955).
38. S. Ginsberg and I. B. Wilson, *J. Amer. Chem. Soc.*, **79**, 481 (1957); B. E. Hackley, Jr., G. M. Steinberg, and J. C. Lamb, *Arch. Biochem. Biophys.*, **80**, 211 (1959).
39. P. B. D. De la Mare, I. C. Hilton, and C. A. Vernon, *J. Chem. Soc.*, 4039 (1960); P. B. D. De la Mare, I. C. Hilton, and S. Varma, *J. Chem. Soc.*, 4044 (1960).
40. C. A. Bunton and M. Masui, *J. Chem. Soc.*, 304 (1960).
41. J. H. Ridd, *Quart. Rev.*, **15**, 418 (1961).
42. H. Schmid, *Chem. Zeitung*, **90**, 351 (1966).
43. H. Adkins and Q. E. Thompson, *J. Amer. Chem. Soc.*, **71**, 2242 (1949).

References

44. V. Gold and E. G. Jefferson, *J. Chem. Soc.*, 1409, 1416 (1953).
45. A. R. Butler and V. Gold, *J. Chem. Soc.*, 4362 (1961).
46. S. L. Johnson, *J. Phys. Chem.*, **67**, 495 (1963).
47. C. A. Bunton, N. A. Fuller, S. G. Perry, and V. J. Shiner, *Tetrahedron Lett.*, **14**, 458 (1961).
48. F. Klages and E. Zange, *Ann.*, **607**, 35 (1957).
49. G. Di Sabato and W. P. Jencks, *J. Amer. Chem. Soc.*, **83**, 4393 (1961); the accumulation of a phosphorylpyridinium ion has been demonstrated in W. P. Jencks and M. Gilchrist, *J. Amer. Chem. Soc.*, **87**, 3199 (1965).
50. T. C. Bruice and G. L. Schmir, *Arch. Biochem. Biophys.*, **63**, 484 (1956); *J. Amer. Chem. Soc.*, **79**, 1663 (1957).
51. M. L. Bender and B. W. Turnquest, *J. Amer. Chem. Soc.*, **79**, 1652, 1656 (1957).
52. W. Langenbeck and R. Mahrwald, *Chem. Ber.*, **90**, 2423 (1957).
53. D. M. Brouwer, M. J. vander Vlugt, and E. Havinga, *Koninkl. Ned. Akad. Wetenschap. Proc.*, **60B**, 275 (1957).
54. S. A. Bernhard and H. Gutfreund, *Proc. Int. Symp. Enzyme Chemistry*, Tokyo–Kyoto, Maruzen, Tokyo, 1958, p. 124.
55. See also W. P. Jencks et al., *Biochim. Biophys. Acta*, **24**, 227 (1957); *J. Biol. Chem.*, **234**, 1272, 1280 (1959).
56. T. C. Bruice and G. L. Schmir, *J. Amer. Chem. Soc.*, **80**, 148 (1958).
57. J. C. Sheehan and D. N. McGregor, *J. Amer. Chem. Soc.*, **84**, 3000 (1962).
58. P. Cruickshank and J. C. Sheehan, *J. Amer. Chem. Soc.*, **86**, 2070 (1964).
59. J. C. Sheehan, G. B. Bennett, and J. A. Schneider, *J. Amer. Chem. Soc.*, **88**, 3455 (1966).
60. D. T. Elmore and J. J. Smyth, *Biochem. J.*, **94**, 563 (1965).
61. E. Katchalski, G. D. Fasman, E. Simons, E. R. Blout, F. R. N. Gurd, and W. L. Koltun, *Arch. Biochem. Biophys.*, **88**, 361 (1960).
62. J. Noguchi and T. Saito, *Polyamino Acids, Polypeptides and Proteins*, M. A. Stahmann, Ed., University of Wisconsin Press, Madison, Wis., 1962, p. 313.
63. D. L. Rohlfing and S. W. Fox, *Arch. Biochem. Biophys.*, **118**, 122 (1967).
64. G. O. Dudek and F. H. Westheimer, *J. Amer. Chem. Soc.*, **81**, 2641 (1959).
65. R. Blakeley, F. Kerst, and F. H. Westheimer, *J. Amer. Chem. Soc.*, **88**, 112 (1966).
66. J. L. Kice and E. H. Morkved, *J. Amer. Chem. Soc.*, **86**, 2270 (1964).
67. H. G. Henning and G. Busse, *Angew Chem., Int. Ed.*, **4**, 951 (1965).
68. J. J. Monagle, *J. Org. Chem.*, **27**, 3851 (1962); T. W. Campbell and J. J. Monagle, *J. Amer. Chem. Soc.*, **84**, 1493 (1962).
69. J. Burkus, *J. Org. Chem.*, **27**, 474 (1962).
70. C. Zioudrou, *Tetrahedron*, **18**, 197 (1962); see also A. S. Shawali and S. S. Biechler, *J. Amer. Chem. Soc.*, **89**, 3020 (1967).
71. F. Cramer and M. Winter, *Chem. Ber.*, **94**, 989 (1961).
72. M. L. Bender and A. Williams, *J. Amer. Chem. Soc.*, **88**, 2502 (1966).
73. E. H. Cordes and W. P. Jencks, *J. Amer. Chem. Soc.*, **84**, 826 (1962).
74. E. H. Cordes and W. P. Jencks, *Biochemistry*, **1**, 773 (1962); see also K. Nagano and D. E. Metzler, *J. Amer. Chem. Soc.*, **89**, 2891 (1967).
75. J. Hine et al., *ibid.*, **87**, 5050 (1965); **88**, 3367 (1966); **89**, 1205, 2264, 3085 (1967).
76. E. O. Wiig, *J. Phys. Chem.*, **32**, 961 (1928).
77. K. J. Pedersen, *J. Amer. Chem. Soc.*, **51**, 2098 (1929); **60**, 595 (1938); *J. Phys. Chem.*, **38**, 559 (1932).
78. R. W. Hay, *Austral. J. Chem.*, **18**, 337 (1965).
79. W. Langenbeck, *Ergeb. Enzymforsch.*, **2**, 314 (1933).

80. K. G. Stern and J. L. Melnick, *J. Biol. Chem.*, **131**, 597 (1939); R. E. Schachat, E. I. Becker, and A. D. McLaren, *J. Org. Chem.*, **16**, 1349 (1951); *J. Phys. Chem.*, **56**, 722 (1952); A. S. Endler and E. I. Becker, *J. Amer. Chem. Soc.*, **77**, 6608 (1955); *J. Phys. Chem.*, **61**, 747 (1957).
81. E. Knoevenagel, *Ann.*, **281**, 25 (1894).
82. J. G. Miller and M. Kilpatrick, Jr., *J. Amer. Chem. Soc.*, **53**, 3217 (1931).
83. F. H. Westheimer and H. Cohen, *J. Amer. Chem. Soc.*, **60**, 90 (1938).
84. J. C. Speck, Jr. and A. A. Forist, *J. Amer. Chem. Soc.*, **79**, 4659 (1957).
85. F. H. Westheimer, *Ann. N. Y. Acad. Sci.*, **39**, 401 (1940).
86. C. Schroder, S. Preis, and K. P. Link, *Tetrahedron Lett.*, **13**, 23 (1960).
87. T. I. Crowell and D. W. Peck, *J. Amer. Chem. Soc.*, **75**, 1075 (1953).
88. A. Lapworth, *J. Chem. Soc.*, **83**, 995 (1903); **85**, 1206 (1904).
89. H. Kwart and M. M. Baevsky, *J. Amer. Chem. Soc.*, **80**, 580 (1958).
90. T. C. Bruice and F. M. Richards, *J. Org. Chem.*, **23**, 145 (1958).
91. V. Franzen and L. Fikentscher, *Ann.*, **613**, 1 (1958).
92. T. Ugai, S. Tanaka, and S. Dokawa, *J. Pharm. Soc. Japan*, **63**, 269 (1943).
93. R. Breslow, *J. Amer. Chem. Soc.*, **79**, 1762 (1957).
94. R. Breslow, *ibid.*, **80**, 3719 (1958). This elegant paper is well worth reading.
95. L. O. Krampitz, I. Suzuki, and G. Guell, *Fed. Proc.*, **20**, 971 (1961).
96. H. Holzer and K. Beaucamp, *Angew. Chem.*, **71**, 776 (1959); *Biochim. Biophys. Acta.*, **46**, 225 (1961).
97. G. L. Carlson and G. M. Brown, *J. Biol. Chem.*, **236**, 2099 (1961).
98. K. W. Bock, L. Jaenicke, and H. Holzer, *Biochem. Biophys. Res. Comm.*, **9**, 472 (1962).
99. R. Breslow, *J. Cell. Comp. Physiol.*, **54**, suppl. 1, 89 (1959).
100. J. K. Stille, *Introduction to Polymer Chemistry*, John Wiley & Sons, Inc., New York, 1962.
101. J. F. Kirsch and W. P. Jencks, *J. Amer. Chem. Soc.*, **86**, 837 (1964).
102. D. G. Oakenfull, T. Riley, and V. Gold, *Chem. Comm.*, 385 (1966).
103. J. F. Kirsch and W. P. Jencks, *J. Amer. Chem. Soc.*, **86**, 833 (1964).
104. S. Ahrland, J. Chatt, and N. R. Davies, *Quart. Rev. London*, **12**, 265 (1958).
105. G. Schwarzenbach, paper presented at *Symp. on SHAB*, Cyanamid European Research Institute Geneva, May 1965, quoted in Ref. 6.
106. C. K. Ingold, *Structure and Mechanism in Organic Chemistry*, Cornell University Press, Ithaca, N.Y., 1953.
107. K. Heald and G. Williams, *J. Chem. Soc.*, 362 (1954).
108. R. S. Satchell, *J. Chem. Soc.*, 5464 (1964); 797 (1965).
109. R. J. Gillespie in *Friedel Crafts and Related Reactions*, Vol. I, G. A. Olah, Ed., Interscience Publishers, New York, p. 194, 1953.
110. G. Baddeley, *Quart. Rev.*, **8**, 355 (1954).
111. G. A. Olah and S. J. Kuhn, *J. Amer. Chem. Soc.*, **80**, 6541 (1958).
112. G. A. Olah, S. J. Kuhn, and S. H. Flood, *ibid.*, **83**, 4571 (1961); **84**, 1688 (1962).
113. R. Nakane, A. Natsubori, and O. Kurihara, *J. Amer. Chem. Soc.*, **87**, 3597 (1965).
114. H. C. Brown and F. R. Jensen, *J. Amer. Chem. Soc.*, **80**, 2291 (1958).
115. L. Schmerling and J. P. West, *J. Amer. Chem. Soc.*, **76**, 1917 (1954).
116. D. A. McCaulay and A. P. Lein, *J. Amer. Chem. Soc.*, **74**, 6246 (1952).
117. L. A. Fury, Jr., and D. E. Pearson, *J. Org. Chem.*, **30**, 2301 (1965).
118. H. Pines and R. C. Wackher, *J. Amer. Chem. Soc.*, **68**, 595, 599 (1946).
119. J. P. Kennedy, *Chem. Eng. News*, 56 (September 19, 1966).
120. P. B. D. De la Mare and J. H. Ridd, *Aromatic Substitution*, Academic Press, Inc., New York, 1959, p. 116.

References

121. D. J. Cram and G. S. Hammond, *Organic Chemistry*, McGraw-Hill Book Co., New York, 1960, p. 403.
122. R. M. Schramm and F. H. Westheimer, *J. Amer. Chem. Soc.*, **70**, 1782 (1948); C. A. Bunton, E. D. Hughes, C. K. Ingold, D. I. H. Jacobs, M. H. Jones, G. J. Minkoff, and R. I. Reed, *J. Chem. Soc.*, 2628 (1950) and references therein.
123. W. P. Jencks and M. Gilchrist, *J. Amer. Chem. Soc.*, **86**, 1410 (1964).
124. J. E. Hodge, B. E. Fisher, and E. C. Nelson, *Proc. Amer. Soc. Brew. Chemists*, 84 (1963).
125. M. Caplow, *J. Amer. Chem. Soc.*, **93**, 230 (1971).
126. R. M. Herbst, *Advan. Enzymol.*, **4**, 76 (1944).
127. H. I. Nakada and S. Weinhouse, *J. Biol. Chem.*, **204**, 831 (1953).
128. E. E. Snell, *J. Amer. Chem. Soc.*, **67**, 194 (1945).
129. D. E. Metzler and E. E. Snell, *J. Amer. Chem. Soc.*, **74**, 979 (1952).
130. D. E. Metzler, M. Ikawa, and E. E. Snell, *J. Amer. Chem. Soc.*, **76**, 648 (1954). This is an important paper.
131. G. L. Eichorn and J. W. Dawes, *J. Amer. Chem. Soc.*, **76**, 5663 (1954).
132. L. Davis, F. Roddy, and D. E. Metzler, *J. Amer. Chem. Soc.*, **83**, 127 (1961).
133. J. Olivard, D. E. Metzler, and E. E. Snell, *J. Biol. Chem.*, **199**, 669 (1952).
134. E. Werle and W. Koch, *Biochem. Z.*, **319**, 305 (1949).
135. J. Thanassi and J. S. Fruton, *Biochemistry*, **1**, 975 (1962).
136. G. D. Kalyanker and E. E. Snell, *Biochemistry*, **1**, 594 (1962).
137. F. Bergel, K. H. Harrup, and A. M. Scott, *J. Chem. Soc.*, 1101 (1962).
138. A. E. Braunstein, *Advan. Protein Chem.*, **3**, 1 (1947).
139. D. E. Metzler, J. B. Longenecker, and E. E. Snell, *J. Amer. Chem. Soc.*, **75**, 2786 (1953).

Chapter 7

CATALYSIS BY FIELDS

7.1 Introduction 194
7.2 Salt Effects 196
    7.2.1 The Effect of Ionic Strength on Rate Constants 196
    7.2.2 Electrolyte Catalysis 198
7.3 Solvent Effects 202
    7.3.1 The Effect of the Dielectric Constant on Kinetics 202
    7.3.2 Special Solvent Effects on Reactions of Ionic Species 205
    7.3.3 Other Special Solvent Effects 208

7.1 INTRODUCTION

In this chapter, we discuss catalysis of chemical reactions by substances whose interaction cannot be specified as discretely as those given earlier. For this reason the loose description, catalysis by fields, is used. The substance of this chapter is in fact catalysis by salts and solvents. Both of these substances are external to the substrate of the reaction and both lead to important rate accelerations. Neither salts nor solvents appear simply in the rate expression, as do other catalysts of the general acid-base or nucleophilic-electrophilic variety. However, they do affect the standard free energy of the ground state and/or the transition state, and therefore can profoundly affect the rate constant of a reaction. (We will not consider effects on equilibria.) In general, the pathway of reaction is not changed by salts and solvents, in contrast to many of the catalyses described earlier. But from both a practical and theoretical point of view, these effects must be considered in describing the acceleration of chemical reaction.

Introduction

Sometimes they are called medium effects while at other times they are called catalytic effects. The gradation between medium and catalytic effects is a subtle one, but the magnitude of the effects, whatever they may be called, is far from subtle. If the rate accelerations are significant, we refer to them as catalyses.

The fundamental basis of the catalyses discussed here can be seen in a comparison of the rates of chemical reactions in solution compared to the gas phase, using the transition state theory of reaction rates.[1] Consider the reaction

$$A + B \rightleftharpoons M^{\ddagger} \tag{7.1}$$

where A and B are reactants and $M^{\ddagger}$ is the activated complex; the specific rate constant for this reaction in the gas phase is $K_g = (RT/Nh)K_g^{\ddagger}$, where $K_g^{\ddagger}$ is the "equilibrium constant" between complex and reactants, assuming ideal behavior of the gases. In solution the analogous equation must be corrected to account for deviations from ideal behavior. The thermodynamic "equilibrium constant" between the ground and transition state should be defined as a ratio of activities.

$$K_s = \frac{a^{\ddagger}}{a_A a_B} = \frac{C^{\ddagger}}{C_A C_B} \frac{f^{\ddagger}}{f_A f_B} \tag{7.2}$$

Consequently if the *rate* of reaction is proportional to the *concentration* of the activated complex, the *rate constant* is dependent on the ratio of *activity coefficients*.

$$\text{rate} = \left(\frac{RT}{Nh}\right) C^{\ddagger} = \left(\frac{RT}{Nh}\right) K_s^{\ddagger} C_A C_B \frac{f_A f_B}{f^{\ddagger}} \tag{7.3}$$

$$k_{\text{obs}} = \left(\frac{RT}{Nh}\right) K_s^{\ddagger} \frac{f_A f_B}{f^{\ddagger}} \tag{7.4}$$

The activity coefficients can be referred to any convenient standard state, the usual one being that of infinite dilution of the solutes. The rate constant in the standard state will then be equal to $(RT/Nh)K_s^{\ddagger}$, and can be defined as k_0, yielding

$$k_{\text{obs}} = k_0 \frac{f_A f_B}{f^{\ddagger}} \tag{7.5}$$

Equation 7.5 was derived by Brønsted[2] and by Bjerrum[3] with special reference to the effects of salts on the rates of chemical reactions, although the transition state theory was not used in the original derivations. This treatment can, of course, be used to analyze the effect of other external variables on the rate constant of a chemical reaction.

7.2 SALT EFFECTS

7.2.1 The Effect of Ionic Strength on Rate Constants

The most important application of eq. 7.5 occurs when one or more of the reactants are ions. According to the Debye-Hückel theory, the relation between the activity coefficient of an ion and the ionic strength is given for dilute solutions (less than 0.01 M) by

$$-\ln f_i = \frac{Z_i^2 \alpha \sqrt{\mu}}{1 + \beta r_i \sqrt{\mu}} \tag{7.6}$$

where μ is the ionic strength, r_i is the distance of closest approach of an ion to the ith ion, α and β are constants, for a given solvent and temperature, and Z is the charge. Using eq. 7.6, 7.5 can be written as

$$\ln k_{obs} = \ln k_0 - \frac{Z_A^2 \alpha \sqrt{\mu}}{1 + \beta r_A \sqrt{\mu}} - \frac{Z_B^2 \alpha \sqrt{\mu}}{1 + \beta r_B \sqrt{\mu}} + \frac{(Z_A + Z_B)^2 \alpha \sqrt{\mu}}{1 + \beta r_\ddagger \sqrt{\mu}} \tag{7.7}$$

or, if we assume a mean value of r for the distance of closest approach,

$$\ln k_{obs} = \ln k_0 + \frac{2 Z_A Z_B \alpha \sqrt{\mu}}{1 + \beta_r \sqrt{\mu}} \simeq \ln k_0 + 2 Z_A Z_B \alpha \sqrt{\mu} \tag{7.8}$$

Equation 7.8 predicts a linear relationship between log k_{obs} and the square root of the ionic strength, with a slope proportional to the product of charges $Z_A Z_B$. Figure 7.1 shows excellent quantitative agreement of a number of ionic reactions with this equation. This figure indicates that reactions between ions of the same sign are accelerated by an increase of ionic strength, whereas reactions between ions of opposite sign are retarded. The former may be viewed as catalyses, the latter as inhibitions.

Equation 7.8 has been much abused by attempting to apply it to concentrated solutions where it is not valid. Even in dilute solutions, complex formation between ionic compounds of opposite sign can invalidate the relationship. Certainly at higher concentrations specific effects of added ions will be found.

If one of the reactants is a neutral molecule so that $Z_A Z_B = 0$, eq. 7.8 predicts no effect of ionic strength. This appears true for very dilute solutions. But at higher ionic concentration, the rate constants may change because of changes in the activity coefficients not given by the Debye-Hückel theory, and because the activity coefficients of neutral molecules are affected by higher ionic strength. For example, the acid-catalyzed hydrolysis of γ-butyrolactone is affected by the presence of salts. The logarithm of the rate constant is a linear function of the first power of the ionic strength for a number of salts; sodium sulfate and sodium chloride increase the rate constant while sodium iodide and sodium perchlorate decrease it.

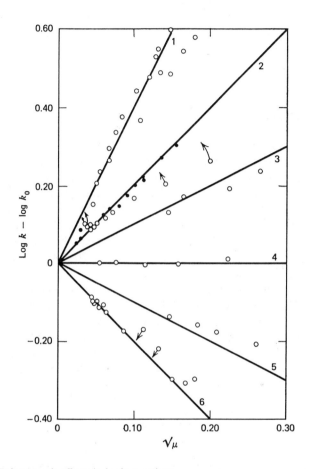

Fig. 7.1 Primary salt effects in ionic reactions.
1. $2[Co(NH_3)_5Br]^{2\oplus} + Hg^{2\oplus} + 2H_2O \rightarrow 2[Co(NH_3)_5H_2O]^{3\oplus} + HgBr_2$.
2. Circles. $CH_2BrCOO^\ominus + S_2O_3^{2\ominus} \rightarrow CH_2S_2O_3COO^{2\ominus} + Br^\ominus$. Sodium salt: no foreign salt added.
   Dots. $S_2O_8^{2\ominus} + 2I^\ominus \rightarrow I_2 + 2SO_4^{2\ominus}$ (bimolecular).
3. $[NO_2{=}N{-}COOC_2H_5]^\ominus + OH^\ominus \rightarrow N_2O + CO_3^{2\ominus} + C_2H_5OH$.
4. Inversion of cane sugar, catalyzed by $OH^\ominus$ ions.
5. $H_2O_2 + 2H^\oplus + 2Br^\ominus \rightarrow 2H_2O + Br_2$ (first order with respect to $H^\oplus$ and $Br^\ominus$).
6. $[Co(NH_3)_5Br]^{2\oplus} + OH^\ominus \rightarrow [Co(NH_3)_5OH]^{2\oplus} + Br^\ominus$.

From R. P. Bell, *Acid Base Catalysis*, Clarendon Press, (Oxford), 1941, p. 33.

Likewise eq. 7.8 predicts no effect of ionic strength on reactions of two neutral molecules, except when the activities of the neutral molecules are affected by higher ionic strength. But when neutral molecules react to form oppositely charged ions, as in the hydrolysis of an alkyl halide, the transition state may be considered a strong dipole, and salt effects will again occur.

7.2.2 Electrolyte Catalysis

In the solvolysis of secondary and tertiary halides, reactions that proceed through carbonium ion intermediates, the dipolar nature of the transition state has been elucidated. A theoretical treatment suggests dependence of the logarithm of the rate constant on the ionic strength, with a slope dependent on the square of the dipolar charge and on the distance separating the assumed point dipoles.[6] An experimental study in 90% acetone-water solvent bears out these predictions semiquantitatively, but the rate enhancements are quite small, 0.1 M salts leading to only 30–100% rate increases.[6] However, in aprotic solvents, salt effects on reactions proceeding through carbonium ion intermediates lead to rate enhancements greater than a millionfold. The rate of ionization of p-methoxyneophyl p-toluenesulfonate in diethyl ether is accelerated by a factor of 10^5 by 0.1 M lithium perchlorate, while the same reaction in acetic acid is accelerated by a factor of only 2.5, as seen in Fig. 7.2.[7] The pattern of salt effects in the range shown in the figure is different for the two solvents: it is linear for acetic acid (eq. 7.9), whereas it is complex for diethyl ether (eq. 7.10).

$$k_{obs} = k_0\{1 + b[\text{LiClO}_4]\} \qquad (7.9)$$

$$k_{obs} = k_0\{1 + b[\text{LiClO}_4] + c[\text{LiClO}_4]^n\} \qquad (7.10)$$

The occurrence of large salt effects on ionization may drastically alter the relative ionizing power of solvents, as seen in Fig. 7.2. Whereas the rate constant of ionization in acetic acid in the absence of lithium perchlorate exceeds that in ether by a factor of 2×10^4, ether becomes a better ionizing medium than acetic acid at concentrations of lithium perchlorate above 0.036 M.

Even larger salt effects are seen in the ionization of a spirodienyl p-nitrobenzoate in ether solution, where 0.05 M lithium perchlorate accelerates the rate by $10^{8.7}$. The effect of lithium perchlorate on the rate constant again fits an equation of the form of eq. 7.10. With both this substrate and the p-toluenesulfonate, different salts show different accelerations, cations showing the order, Li > Na > Bu$_4$N, identical with the order of increasing ionic

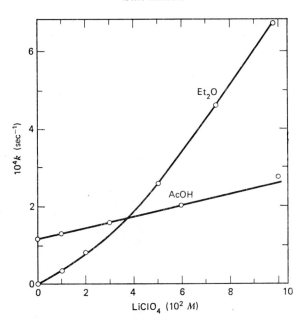

Fig. 7.2 The effect of lithium perchlorate on the ionization of *p*-methoxyneophyl *p*-toluenesulfonate. From S. Winstein, S. Smith, and D. Darwish, *J. Amer. Chem. Soc.*, **81**, 5511 (1959). © 1959 by the American Chemical Society. Reprinted by permission of the copyright owner.

radius. Considerable anion specificity is also evident.† The magnitude of the salt effects and the observed cation order suggest that the chief role of the salt in this catalysis is to provide specific electrophilic assistance to ionization. Assistance to ionization by an ion pair $M^\oplus Y^\ominus$, may be

$$RX + M^\oplus Y^\ominus \rightleftarrows R^\oplus X^\ominus M^\oplus Y^\ominus \longrightarrow \text{products} \qquad (7.11)$$

Whereas uncatalyzed ionization leads to an ion pair in a medium of low dielectric, electrolyte-catalyzed ionization leads to an ion quadruplet. Estimated interaction energies between two ion pairs are consistent with very large assistance to ionization provided by an ion pair in this medium.

The catalytic effect of uncharged bases alone in the base-catalyzed mutarotation of tetramethylglucose and tetraacetylglucose in nitromethane solution is very low. This effect is greatly enhanced by the addition of a wide variety of salts.[9,10] The mutarotation of tetraacetylglucose in pyridine solution is accelerated tenfold by 0.02 M lithium perchlorate. Other rate accelerations vary widely with both the anionic and cationic component of the salt

† A nucleophilic salt effect by chloride and bisulfate ions was found in the hydrolysis of diazoketones in perchloric acid. J. G. Tillett and S. Aziz, *J. Chem. Soc.*, B 1302 (1968).

and are smaller than the lithium perchlorate acceleration. The mechanism of catalysis by an uncharged base requires the intermediate formation of an ion pair

$$\underset{\underset{|}{O}}{\overset{\underset{|}{O-H}}{\underset{H}{\diagup}\hspace{-6pt}\diagdown}} + B \longrightarrow \underset{\underset{|}{O^{\ominus}}}{\overset{\underset{}{O}}{\underset{H}{\diagup}\hspace{-6pt}\diagdown}} \quad BH^{\oplus} \qquad (7.12)$$

since the reaction proceeds through the aldehydic form of glucose. The electrolyte catalysis can then be explained by the stabilization of the transition state leading to this charge-separated intermediate by means of anion quadruplet interaction of the kind proposed in eq. 7.11. This suggestion is consistent with the relative efficacy of different cations in promoting reactions that follow the stability of their oxonium salts.

$$\underset{\underset{|}{O^{\ominus}M^{\oplus}X^{\ominus}}}{\overset{\underset{}{O\text{---}HB^{\oplus}}}{\underset{H}{\diagup}\hspace{-6pt}\diagdown}} \qquad (7.13)$$

Although the electrolyte catalyses described above show specificity with respect to both the cation and the anion of the salt, they must be described in terms of catalysis by the salt as a whole rather than by either of the component parts. In addition to these electrolyte catalysts, some salts act catalytically as anions while others act catalytically as cations. One can therefore define anionic catalysis and cationic catalysis. Many, although not all, of the metal ion catalyses described in Chapter 8 may be viewed in this context. For example, alkaline earth cations catalyze the hydrolysis of many phosphate esters and anhydrides. They do so by stabilization of the transition state of the hydrolytic reaction through an electrostatic interaction made favorable by the chelation possible with these metal ions. Likewise the lithium counterion may be viewed similarly in the carbenoid reactions of $LiCCl_3$ in aprotic solvents.†

Cationic or metal ion catalysis has its counterpart in anionic catalysis. The analogy is not complete because the usual monovalent anions cannot form chelates to a polyvalent metal ion in the same sense that the polyvalent

---

† Both cations and anions catalyze the peroxide oxidation of sugars to keturonic acids. M. R. Everett and F. Sheppard, *Oxidation of Carbohydrates to Keturonic Acids*, Times-Journal Pub. Co., Oklahoma City, 1944.

metal ion can coordinate with the polyanionic substrate. Some interesting possibilities in the use of polyanions as catalysts have therefore not as yet been probed.

In electrophilic substitution reactions, "one" and "two anion" catalyses have been found[11-13] (Fig. 7.3).[12] The isotopic exchange reaction of a simple alkylmercuric halide with mercuric halide, labeled with radiomercury, is a second-order reaction, proceeding with retention of configuration.[11] Since increasing ionic character of the mercuric substituting agent leads to increasing rate, the anion of the substituting agent is presumably not involved in the reaction. In ethanol solution, the reaction is catalyzed by lithium halides, the order of catalytic effectiveness being $I^\ominus > Br^\ominus > Cl^\ominus > CH_3CO_2^\ominus > NO_3^\ominus$.[12] A plot of the second-order rate constant for the exchange reaction against the added salt concentration is biphasic (Fig. 7.3) with a linear dependence on the salt concentration up to the equivalence point of the salt and the substituting agent, followed by a change in slope to a new linear dependence on the salt concentration. The relative slopes of the first and second portions of the curve vary, depending on the anion of the salt and also on the alkyl group of the substrate. These two slopes imply two

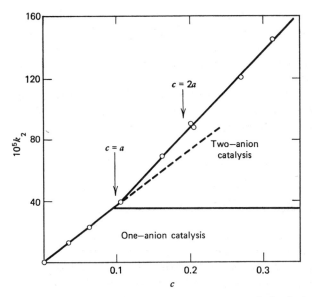

Fig. 7.3 Dependence of second-order rate constants on one-alkyl substitution, $k_{2c}^{(1)}$, by mercuric bromide in methylmercuric bromide in ethanol at 59.8°, on the concentration (c) of added lithium bromide, the concentrations (a) and (b) of the mercury-exchanging substances being 0.095 M. From H. B. Charman, E. D. Hughes, Sir C. Ingold, and H. C. Volger, *J. Chem. Soc.*, 1142 (1961).

processes, one depending on a higher order of the salt concentration than the other. Since the slope changes at the point of 1:1 stoichiometry of salt and substituting agent, these two processes can be identified as "one" and "two-anion" catalysis, respectively.

The one and two anion catalyses [14–16] may be described as

$$CH_3HgBr \xrightleftharpoons{Br^\ominus} CH_3Hg\begin{matrix}Br\\\\Br^\ominus\end{matrix} \xrightharpoonup{\overset{*}{HgBr_2}}$$

$$\left[\begin{matrix}HgBr_2\\CH_3 \\ \overset{*}{HgBr_2}\end{matrix}\right]^\ominus \xrightleftharpoons{HgBr_2} CH_3\overset{*}{HgBr_2}{}^\ominus \xrightleftharpoons{Br^\ominus} CH_3\overset{*}{HgBr} \quad (7.14)$$

$$CH_3HgBr \xrightleftharpoons{Br^\ominus} CH_3HgBr_2{}^\ominus \xrightharpoonup{\overset{*}{Br^\ominus,HgBr_2}}$$

$$\left[\begin{matrix}HgBr_2\\CH_3 Br\\ \overset{*}{HgBr_2}\end{matrix}\right]^\ominus \xrightleftharpoons{HgBr_2,Br^\ominus} CH_3\overset{*}{HgBr_2}{}^\ominus \xrightleftharpoons{Br^\ominus} CH_3\overset{*}{HgBr} \quad (7.15)$$

which can be designated S_E1 mechanisms because of their cyclic nature in contrast to the uncatalyzed S_E2 mechanism, which is noncyclic. In both S_E1 and S_E2 mechanisms, the anionic catalyst stabilizes the transition state through its ability to coordinate with the mercury atom, either through single coordination or bridging.

A suggested case of electrostatic facilitation involves the reaction of the anion of o-nitrophenyl oxalate with the conjugate acid of 2-aminopyridine [18] but this has been disputed. [19]

7.3 SOLVENT EFFECTS

7.3.1 The Effect of the Dielectric Constant on Kinetics

The influence of the dielectric constant of the medium on the rate constants of reactions between two ions can be considered in terms of eq. 7.5. The electrostatic free energy for bringing two ions from infinite separation to the equilibrium distance in the activated complex ($r^\ddagger$) is

$$\Delta F^\ddagger_{el} = \frac{Z_A Z_B e^2}{D r^\ddagger} \quad (7.16)$$

where D is the dielectric constant of the medium and e is the charge on the electron. From eq. 7.16, it is possible to write an expression for the dependence of the rate constant on the dielectric constant.[20]

$$\ln k = \ln k_0' - \frac{N Z_A Z_B e^2}{DRTr^{\ddagger}} \qquad (7.17)$$

where k_0' is a specific rate constant in the medium of infinite dielectric constant, N Avogadro's number, R the gas constant and T the temperature. This equation predicts a linear plot of log k against $1/D$, with negative slope if the charges on the ions are of the same sign and a positive slope if the charges are of opposite sign. Figure 7.4 shows data on the alkaline fading of bromphenol blue in ethanol–water mixtures and on the reaction between oxonium ion and the doubly negative azodicarboxylate ion in mixtures of water and dioxane. The first reaction is between univalent negative and divalent negative ions, while the second is between a univalent positive ion and a divalent negative ion. In agreement with predictions of theory, the slope of log k versus $1/D$ is negative for the former reaction and positive for the latter. The slopes lead to values of r of 2.81 Å for the former and 3.42 Å for the latter reaction, values of the expected order of magnitude. In the reaction of oxonium ion and the azodicarboxylate ion in dioxane–water mixtures, the enthalpy of activation remains sensibly constant while the entropy of activation increases with increasing dioxane content. A similar but exaggerated phenomenon is seen in the hydrolysis of alkyl hydrogen sulfates in dioxane–water mixtures. The rate constant of this reaction in

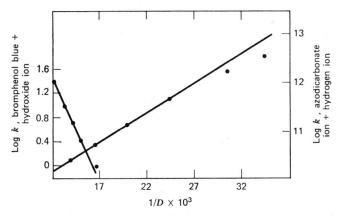

Fig. 7.4 Influence of dielectric constant on rate of reaction between two ions of same sign (left axis) and of opposite sign (right axis). From E. S. Amis and V. K. La Mer, *J. Amer. Chem. Soc.*, **61**, 905 (1939); C. V. King and J. J. Josephs, *J. Amer. Chem. Soc.*, **66**, 767 (1944). © 1939 and 1944 by the American Chemical Society. Reprinted by permission of the copyright owner.

98% dioxane–water is 10^7 times larger than in water alone solely because of a difference in entropy of activation of ca. 30 entropy units.[22] The logarithm of the rate constant is again linearly dependent on $1/D$; the slope of this line leads to a value of r of 16 Å, an extraordinarily large value whose meaning is not clear. The latter two reactions, which involve oppositely charged ions, have transition states that are less highly solvated than the ground states. This phenomenon should lead to an increase in entropy. When the solvent is changed to one of lower dielectric, this effect is markedly enhanced.

A corollary of this treatment is that when neutral molecules form a transition state in which separation of charge occurs, an equal but opposite solvent effect should occur. The most familiar example of this phenomenon is in reactions following the $S_N 1$ mechanism in which an ionization leads to a transition state resembling a carbonium ion–anion pair. For these and other reactions, a useful qualitative theory was expounded long ago: "an increase in the ion-solvating power of the medium will accelerate the creation and concentration of charges and inhibit their destruction and diffusion."[24] The solvolytic rates of several secondary and tertiary halides increase by approximately a thousandfold over the solvent range from 60% ethanol–water to water. This qualitative theory is probably good for small changes in solvent, for example from methanol to ethanol, but not for gross changes in solvent composition, for example from methanol to dimethyl sulfoxide. This conclusion is indicated by a quantitative correlation of the rates of solvolysis, based on a quantity Y, a measure of the "ionizing" power of the medium, as manifested by its effect on the rate of solvolyses occurring by the $S_N 1$ mechanism.[25] Y was defined by the relation

$$Y = \log \frac{k_{\text{BuCl}}}{k_{0_{\text{BuCl}}}} \tag{7.18}$$

where k_{BuCl} and $k_{0_{\text{BuCl}}}$ are the rate constants for the solvolysis of t-butyl chloride in the given solvent and in the reference solvent (80% ethanol). The following equation was suggested to correlate the rates of solvolysis of compounds reacting via the $S_N 1$ mechanism

$$\log \frac{k}{k_0} = mY \tag{7.19}$$

where k and k_0 are the solvolysis rates of a compound in a given solvent and in 80% ethanol, Y is characteristic of the given solvent, and m is a constant characteristic of the compound solvolyzed. Equation 7.19 gives good correlations, but only when restricted to one binary solvent mixture.[26]

Both theoretically and experimentally the polarity of the solvent per se has little effect on other reactions such as the reactions of ions with neutral (polar) molecules. The theoretical prediction that the rate constant of reaction is higher in a medium of lower dielectric constant is sometimes obeyed (alkyl halides + hydroxide ion), but other times disobeyed (esters + hydroxide ion).

7.3.2 Special Solvent Effects on Reactions of Ionic Species

The solvent can exert enormous effects on the rate constants of reactions of ions with organic molecules, whether they are nucleophiles or bases. For example, the change from water to acetone solvent multiplies the second-order rate constant of the chloride ion-methyl iodide reaction by ca. 10^7.[27] In addition, the rate of racemization of optically active 2-methyl-3-phenylpropionitrile by methoxide ion is 10^9 times faster in dimethyl sulfoxide than in methanol.[28] The dielectric of the medium can cause some small part of these effects, but the major cause must certainly be some specific solvent effect. As implied above, the most profound differences are found when comparing protic and aprotic solvents. A solvent change of this kind can produce two effects, one concerned with perturbation of the equilibrium between ion pairs and free ions, the other concerned with specific solvation of ions, usually by the protic solvent. We will consider these effects in turn.

Second-order rate constants for S_N2 reactions of lithium halides with alkyl halides and toluenesulfonates in acetone solution correspond to an apparent nucleophilicity order, $I^\ominus > Br^\ominus > Cl^\ominus$.[31] The importance of ion association in controlling apparent nucleophilicity is evident when reactivities of lithium and tetra-*n*-butylammonium halides are compared. In the reaction of *n*-butyl *p*-bromobenzenesulfonate with 0.04 M halide salts in acetone, the relative reactivity pattern is

$$\begin{array}{ccc}
\text{LiI, 6.2} & > & \text{LiBr, 5.7} & > & \text{LiCl, 1.0} \\
\vee & & \wedge & & \wedge \\
(n\text{-Bu})_4\text{NI, 3.7} & < & (n\text{-Bu})_4\text{NBr, 18} & < & (n\text{-Bu})_4\text{NCl, 68}
\end{array}$$

Thus lithium and tetra-*n*-butylammonium halides show exactly opposite orders. Tetra-*n*-butylammonium halides are known to be more dissociated in acetone solution than lithium halides. Hence these data suggest that the nucleophilicity order of anions in acetone is actually $Cl^\ominus > Br^\ominus > I^\ominus$, the one observed with the former salts, and that the reactivity pattern of lithium halides is governed largely by the extreme variation in the dissociation constants of the salts in acetone. In fact, when the dissociation constants are taken into account, assuming the reactivity of the ion pair to be negligibly small, good agreement between the reactivities of the lithium and tetra-*n*-butylammonium salts is found. This analysis has been employed for many other reactions of similar nature.[30-35]

When association of ions to ion pairs (and higher complexes) is taken into account, it is then possible to evaluate the reactivity of dissociated ions in different solvents. These reactivities appear to be intimately associated with specific solvations of the ions. In protic solvents, anions are solvated by ion-dipole interactions, on which is superimposed a strong hydrogen bonding component, greatest for small anions on electrostatic grounds. This solvation by protic solvents decreases strongly in the series $OH^\ominus$, $F^\ominus \gg Cl^\ominus > Br^\ominus > N_3^\ominus > I^\ominus > SCN^\ominus >$ picrate$^\ominus$.[36] In dipolar aprotic solvents, the solvation of anions occurs to a much smaller extent due only to ion-dipole interactions. There is no significant contribution to solvation by hydrogen bonding. Solvation of anions by dipolar aprotic solvents is thus relatively insensitive to the structure of the anion.

Two important effects are noted in comparing the rate constants of S_N2 reactions of halides in aprotic and protic solvents. The order of nucleophilicity changes from $Cl^\ominus > Br^\ominus > I^\ominus$ in solvents such as dimethylformamide[37] and acetone[29] to $I^\ominus > Br^\ominus > Cl^\ominus$ in solvents such as water. Secondly, the absolute rate and constants of chloride ion reactions are of the order of 10^6–10^7 faster and those of iodide ion reactions are of the order of 10^4 faster in aprotic solvents than in water.[30,38] These effects are related to one another because the solvent has a profound effect on the nucleophilicity of the different halide ions. The small chloride ion with its high charge density is retarded 24-fold, whereas the large iodide ion with its lower charge density is retarded only twofold.[37]

It is thus concluded that both solvation of anions and ion-pairing have a retarding effect on bimolecular substitution reactions. Interestingly, both retardations are most effective with small anions with their high charge density. It is a matter of choice whether one considers aprotic solvents to have an accelerating effect on bimolecular substitution reactions or whether one considers protic solvents to have a retarding effect. The choice is really which solvent is the standard state. Since this discussion is oriented toward aqueous systems, the changes occurring in aprotic solvents are regarded as accelerations. The stereospecificity of polymerization may also be affected by the solvent.[38]

The rate of any reaction involving a metal cation-carbanion ion pair in an aprotic solvent is increased by the addition of a solvent that will preferentially solvate the cation and thus lead to a higher concentration of dissociated anions. The alkylation of enolate ions is approximately a hundredfold faster in diglyme than in diethyl ether.[39] The reaction of sodium diethyl n-butylmalonate with alkyl halides is 1420 times faster in dimethyl sulfoxide than in benzene; the addition of even 5% dimethylformamide to a benzene solution results in a twentyfold rate enhancement of this reaction.[41]

Anion solvation through hydrogen bonding to protic solvents is most effectively reduced by the addition of solvents such as dimethyl sulfoxide. For example, increasing the dimethyl sulfoxide concentration of an aqueous solvent increases the rate constant of the saponification of esters, markedly at higher dimethyl sulfoxide concentrations. However, adding comparable concentrations of acetone to the aqueous solvent depresses the rate (Fig. 7.5).[36]

A very large kinetic effect of a solvent change occurs in the methoxide ion-catalyzed racemization of 2-methyl-3-phenylpropionitrile in methanol–dimethyl sulfoxide solutions, as seen in Fig. 7.6.[28] From pure methanol to 98.5% dimethyl sulfoxide-methanol, the rate of racemization increases by a factor of 5×10^7. Extrapolation of the curve of Fig. 7.6 to 100% dimethyl sulfoxide gives a rate increase over that in methanol of about nine powers of ten. These two solvents have dielectric constants that do not differ widely from one another (34 for methanol and 49 for dimethyl sulfoxide), and the bulk of evidence supports the hypothesis that at low base concentrations the metal alkoxides are dissociated. The big difference in activity of methoxide anion in the two solvents is attributed to the presence of solvent-anion hydrogen bonds in methanol ($CH_3OH \cdots OCH_3^{\ominus}$) that are absent in dimethyl sulfoxide. Chapter 3 points out that the kinetic data of Fig. 7.6 give

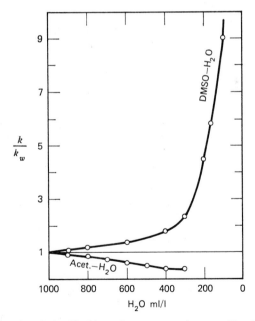

Fig. 7.5 Effect of dimethyl sulfoxide and acetone on the saponification of esters. From E. Tommila and M.-L. Murto, *Acta Chem. Scand.*, **17**, 1947 (1963).

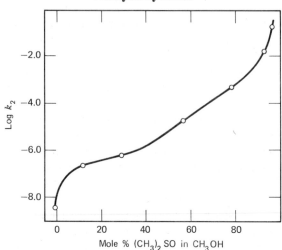

Fig. 7.6 Plot of mole % $(CH_3)_2CO$ in CH_3OH against logarithm of k_2 (M^{-1} sec^{-1}) for CH_3OM-catalyzed racemization of 2-methyl-3-phenylpropionitrile at 25.0°. From D. J. Cram, B. Rickborn, C. A. Kingsbury, and P. Haberfield, *J. Amer. Chem. Soc.*, **83**, 3678 (1961). © 1961 by the American Chemical Society. Reprinted by permission of the copyright owner.

an excellent correlation with the equilibrium data of Fig. 3.9 for these same solvents. Dimethyl sulfoxide may be thought of as a medium containing a species of higher basicity: while methoxide ions are in a sense "buffered" by methanol solvent, they are less solvated in dimethyl sulfoxide In methanol, the driving force for dissociation of a metal methoxide derives from solvation of methoxide anion by the hydrogen and of the metal cation by the oxygen of the hydroxyl group. In dimethyl sulfoxide, positive sulfur solvates the methoxide anion and negative oxygen solvates the metal cation. Since the charge on sulfur is shielded by three shells of electrons and is also sterically hindered, the main impetus for dissociation probably derives from solvation of the metal cation that leaves the methoxide anion relatively poorly solvated and highly reactive. This is schematically shown in the following:

$$CH_3O^\ominus \text{---} HOCH_3 + 2(CH_3)_2SO \rightleftarrows$$
$$CH_3O^\ominus \text{---} S^\oplus(CH_3)_2 + (CH_3)_2S^\oplus \text{---} O^\ominus \text{---} HOCH_3 \quad (7.20)$$
$$\underset{O^\ominus}{|}$$

7.3.3 Other Special Solvent Effects

Implicit in the previous discussion was the assumption that solvent effects on reactions involving ions are primarily associated with solvent effects on the ion in the ground or transition state, but not on neutral molecules in the ground or transition state. This assumption in general must be incorrect. But since the effects on ions so far outweigh other effects, the latter may ordinarily be neglected.

A closer analysis of reactions in dimethyl sulfoxide solutions reveals the following points.[28,36,43] (1) Rate increases occur even at low dimethyl sulfoxide concentration, with no sharp change ordinarily noted as it passes from the minor to the major solvent species. (2) The rate increase per increment of dimethyl sulfoxide is due mainly to a change in $\Delta H^\ddagger$. (3) Addition of dimethyl sulfoxide to a system increases the rate irrespective of the charge character of the reaction, with either a negatively charged or a neutral nucleophile, and with either a neutral or positively charged substrate. These data do not support the simple picture of eq. 7.18 for the action of dimethyl sulfoxide. The observation of large rate increases in solvent mixtures where there is sufficient methyl alcohol present to form hydrogen bonds with both dimethyl sulfoxide and the nucleophile shows that some other factor must be influencing the reaction rates. Moving from solvent mixtures in which dimethyl sulfoxide is the minor component to mixtures where it predominates results in only a moderate increase; if the anion were poorly solvated at high dimethyl sulfoxide concentrations, a much larger increase might have been expected. Also expected would be a difference in the sensitivity of oxygen and sulfur anions to dimethyl sulfoxide concentration, since the former smaller anion has a considerably higher solvation energy. Changes in solvation should be reflected in $\Delta S^\ddagger$, but were observed in $\Delta H^\ddagger$. Changes in the state of solvation of the nucleophile apparently do not become important until the concentration of the hydroxylic species in the solvent is very low and the concentration of dimethyl sulfoxide is very high. Also, desolvation of halide anions may occur more readily than desolvation of alkoxides in going to highly concentrated dimethyl sulfoxide solutions.

At low dimethyl sulfoxide concentrations, the solvent must lower the energy of the transition state. One attractive interpretation suggests enhanced dipolar character of the substrate upon interaction with a random dimethyl sulfoxide molecule.[43] Attack of a nucleophile upon such a dipolar species may well be facile compared to an unpolarized substrate molecule.

REFERENCES

1. A. A. Frost and R. G. Pearson, *Kinetics and Mechanism*, 2nd ed., John Wiley & Sons, Inc., New York, 1961, p. 127.
2. J. N. Brønsted, *Z. Phys. Chem.*, **102**, 169 (1922); **115**, 337 (1925).
3. N. Bjerrum, *Z. Phys. Chem.*, **108**, 82 (1924); **118**, 251 (1925).
4. R. P. Bell, *Acid Base Catalysis*, Oxford University Press, Oxford, 1941, p. 33.

5. F. A. Long, W. F. McDevit, and F. B. Dunkle, *J. Phys. Colloid Chem.*, **55**, 813 (1951).
6. L. C. Bateman, M. G. Church, E. D. Hughes, C. K. Ingold, and N. A. Taher, *J. Chem. Soc.*, 979 (1940).
7. S. Winstein, S. Smith, and D. Darwish, *J. Amer. Chem. Soc.*, **81**, 5511 (1959).
8. S. Winstein, E. C. Friedrich, and S. Smith, *J. Amer. Chem. Soc.*, **86**, 305 (1964).
9. A. M. Eastham, E. L. Blackall, and G. A. Latremouille, *ibid.*, **77**, 2182 (1955); E. L. Blackall and A. M. Eastham, *ibid.*, **77**, 2184 (1955).
10. Y. Pocker, *Chem. Ind.*, 968 (1960).
11. H. B. Charman, E. D. Hughes, and Sir C. Ingold, *J. Chem. Soc.*, 2523 (1959).
12. H. B. Charman, E. D. Hughes, Sir C. Ingold, and H. C. Volger, *ibid.*, 1142 (1961).
13. E. D. Hughes, C. K. Ingold, and R. M. G. Roberts, *ibid.*, 3900 (1964).
14. C. K. Ingold, *Eleventh Conference on Reaction Mechanisms*, Hamilton, Ont., July 1966.
15. O. A. Reutov, B. Praisnar, I. P. Beletskaya, and V. I. Sokolov, *Izv. Akad. Nauk SSSR Otd. Khim. Nauk.*, 970 (1963).
16. R. E. Dessy and F. E. Paulik, *J. Amer. Chem. Soc.*, **85**, 1812 (1963).
17. R. E. Dessy, T. Hieber, and F. Paulik, *J. Amer. Chem. Soc.*, **86**, 28 (1964).
18. M. L. Bender and Y.-L. Chow, *J. Amer. Chem. Soc.*, **81**, 3929 (1959).
19. T. C. Bruice and B. Holmquist, *J. Amer. Chem. Soc.*, **89**, 4028 (1967).
20. G. Scatchard, *Chem. Rev.*, **10**, 229 (1932).
21. C. V. King and J. J. Josephs, *J. Amer. Chem. Soc.*, **66**, 767 (1944).
22. B. D. Batts, *J. Chem. Soc.*, B547 (1966).
23. E. D. Hughes and C. K. Ingold, *J. Chem. Soc.*, 244 (1935).
24. C. K. Ingold, *Structure and Mechanism in Organic Chemistry*, Cornell University Press, Ithaca, N.Y., 1953, p. 349.
25. E. Grunwald and S. Winstein, *J. Amer. Chem. Soc.*, **70**, 846 (1948).
26. A. H. Fainberg and S. Winstein, *J. Amer. Chem. Soc.*, **79**, 1597, 1602 (1957).
27. E. R. Swart and L. J. le Roux, *J. Chem. Soc.*, 406 (1957).
28. D. J. Cram, B. Rickborn, C. A. Kingsbury, and P. Haberfield, *J. Amer. Chem. Soc.*, **83**, 3678 (1961).
29. S. Winstein, L. G. Savedoff, S. Smith, I. D. R. Stevens, and J. S. Gall, *Tetrahedron Lett.*, **9**, 24 (1960).
30. E. R. Swart and L. J. le Roux, *J. Chem. Soc.*, 2110 (1956).
31. E. A. Moelwyn-Hughes, *Trans. Faraday Soc.*, **45**, 167 (1949).
32. N. N. Lichtin and K. N. Rao, *J. Amer. Chem. Soc.*, **83**, 2417 (1961).
33. S. D. Bowers, Jr., and J. M. Sturtevant, *J. Amer. Chem. Soc.*, **77**, 4903 (1955).
34. E. A. S. Cavell and J. A. Speed, *J. Chem. Soc.*, 1453 (1960).
35. C. C. Evans and S. Sugden, *J. Chem. Soc.*, 270 (1949).
36. J. Miller and A. J. Parker, *J. Amer. Chem. Soc.*, **83**, 117 (1961).
37. W. M. Weaver and J. D. Hutchison, *J. Amer. Chem. Soc.*, **86**, 261 (1964).
38. A. J. Parker, *Quart. Rev.*, **16**, 163 (1962); *Chem. Rev.*, **69**, 1 (1969).
39. H. D. Zook and T. J. Russo, *J. Amer. Chem. Soc.*, **82**, 1258 (1960).
40. H. E. Zaugg, B. W. Horrom, and S. Borgwaldt, *ibid.*, **82**, 2895, (1960); H. E. Zaugg, *ibid.*, **82**, 2903 (1960); **83**, 837 (1961).
41. C. Bernasconi and H. Zollinger, *Helv. Chim. Acta*, **49**, 103 (1966).
42. E. Tommila and M.-L. Murto, *Acta Chem. Scand.*, **17**, 1947 (1963).
43. C. A. Kingsbury, *J. Org. Chem.*, **29**, 3262 (1964).

Chapter 8

METAL ION CATALYSIS

8.1	Introduction	211
8.2	Superacid Catalysis	212
	8.2.1 Heterolytic Cleavage of Bonds to Electronegative Atoms	214
	8.2.2 Cleavage of Carbon–Carbon and Carbon–Hydrogen Bonds	215
	8.2.3 Additions to Carbon–Oxygen and Carbon-Nitrogen Double Bonds	221
	8.2.4 Cleavage of Phosphates	228
	8.2.5 Factors in Superacid Catalysis	232
8.3	Catalysis via Redox Reactions	235
	8.3.1 Metal Ion Redox Reactions	235
	8.3.2 Reactions Involving Copper	236
	8.3.3 Decomposition of Peroxides	237
	8.3.4 Autoxidation	242
8.4	Catalysis by Organometallic Compounds	242
	8.4.1 Transition Metal Complexes	243
	8.4.2 Hydrogenation and Related Reactions	248
	8.4.3 Carbon–Carbon Bond Formation	258
	8.4.4 Reactions Involving Metal Carbonyls	266
	8.4.5 Oxidation of Hydrocarbons	270

8.1 INTRODUCTION

Metal ions can produce powerful accelerations of reactions in homogeneous solution. Metal ions catalyze (or promote) a large number of organic and inorganic reactions. Ultimately catalysis occurs because the metal ion has the ability to coordinate with the substrate and thereby stabilize the transition state. However, the mode of action of metal ions in the stabilization of transition states is varied. In some reactions, the metal ion acts as a "superacid" catalyst; that is, it serves as a proton of magnified charge. In other reactions, the metal ion serves as a carrier of electrons. In still other reactions, the metal ion forms a metal–carbon bond and participates directly in the covalent changes. In many instances, the metal ion acts as a template on which reactants can bind and react.

Although distinctions among the various modes of metal ion catalysis are somewhat arbitrary and overlapping, this discussion will be divided into three parts: (1) superacid catalysis; (2) catalysis involving redox reactions; and (3) catalysis involving metal–carbon bond formation. Metal ions form bonds with many ligands (nucleophiles). Metal ions may have, in general, square planar, tetrahedral, or octahedral coordination. These complexes may be of the low or high spin variety (as determined by esr). Some examples of different coordination are seen in Fig. 8.1.

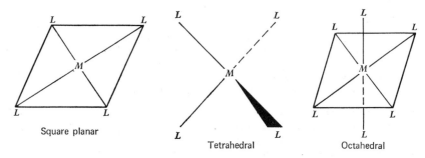

Fig. 8.1 Some examples of metal (M)-ligand (L) coordination.

8.2 SUPERACID CATALYSIS

Metal ions catalyze many organic reactions by means of a mechanism akin to acid or electrophilic catalysis. Since an oxonium ion contains a formal charge of plus one while a metal ion can carry a charge of up to four units, catalysis by metal ions analogous to that seen with oxonium ions has been dubbed "superacid" catalysis.[1] This type of metal ion catalysis occurs in many nucleophilic reactions of organic compounds. Some of these are described and the factors underlying this catalysis analyzed.

In oxonium ion catalysis, the high rate of proton transfer with respect to other covalent changes is one of the factors ensuring efficient catalysis. It is therefore of interest to ascertain the speed of transfer of metal ions from one ligand to another. Since much of our discussion centers around reactions in aqueous solution, we consider the rate constants for water substitution in the inner coordination sphere of metal ions. Figure 8.2 shows rate constants of water substitution for some alkali metal, alkaline earth, transition metal, and other ions.[2] As with proton transfer, rate constants of water substitution vary greatly. For ions of alkali metals and alkaline earth metals, the rate constants are high with all ligands used, but there is some variation from one ligand to another. These results indicate that the loss of water molecules from the coordination sphere of the ions is easy, so that there is fast exchange with the solvent, and the rate-determining step is the attachment of the ligand. There is a group of cations for which the rate constants are less than 10^7 sec^{-1} and practically independent of the nature of the ligand. For example, with magnesium ion, the rate constant is about 10^5 sec^{-1} for reaction with sulfate, thiosulfate, and chromate ions. This result indicates that the removal of water from the coordination shell of the cation is the rate-determining step. Besides magnesium ion, this group includes most of the divalent ions of the first series of transition elements. In a third group of cations, the prototropic ionization of a water molecule in the coordination shell appears to be rate-determining. The rates are comparatively low and vary with the ligand. Examples of such cations are ferric and aluminum ions. The rate constants for the reactions of ferric ion with various anions parallel the basicity of the anion, suggesting a process analogous to general base catalysis for substitution.

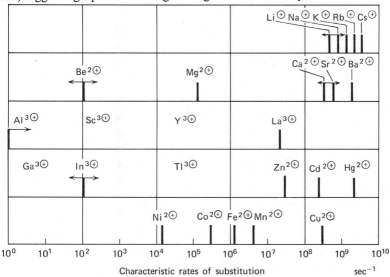

Fig. 8.2 Characteristic rate constants for H_2O substitution in the inner coordination sphere of metal ions (abscissa in sec^{-1}). From M. Eigen, *Pure Appl. Chem.*, **6**, 105 (1963), by permission of the International Union of Pure and Applied Chemistry and Butterworths Scientific Publications.

Although the rate constants of ligand substitution are below 10^7 sec^{-1}, indicating that some other step may be rate-determining, there are very few rate constants below 10^4 sec^{-1}, indicating that, in general, ligand substitution will be rapid compared to other changes.

8.2.1 Heterolytic Cleavage of Bonds to Electronegative Atoms

A classical metal ion catalysis is the silver or mercuric ion-catalyzed cleavage of carbon–halogen bonds in alkyl halides. The compounds susceptible to this catalysis are all capable of forming carbonium ions. In this sense, these catalyses are identical with the electrophilic catalyses by Lewis acids such as aluminum, tin, and iron halides (Chapter 6). As in other reactions proceeding through carbonium ion intermediates, there is no direct relationship between the effect of a mercuric ion catalyst on the total rate and on the relative distribution of products, a result indicative of the catalytic mechanism:[3]

$$RX + M^{\oplus} \longrightarrow R^{\oplus} + MX$$
$$R^{\oplus} + H_2O \longrightarrow products \tag{8.1}$$

The relative reactivity pattern in silver ion catalysis is likewise consistent with eq. 8.1.[4] The interaction of silver and mercuric ions with halogen atoms is a relatively specific one. This is due to the great affinity of these metal ions for halide ion, as evidenced by the small ionization of mercuric halides, the insolubility of silver halides, and the formation of complexes such as $AgX_2^{\ominus}$ from silver halides and excess halide ion. A kinetic analysis of the mercuric bromide-catalyzed solvolysis of alkyl halides indicates that only mercuric ion, but not bromomercuric ion or mercuric bromide, catalyzes the solvolysis.[5]

In acetonitrile solution, reaction of alkyl iodides with silver nitrate is faster than reaction with silver perchlorate. Furthermore, neopentyl iodide (steric hindrance in S_N2) has a low reactivity. These and other data point to a mechanism in which both silver cations and the accompanying anions participate in the rate-determining step of the reactions. The intermediate can thus be described as an ion pair of an anion and a carbonium ion.[6]

A further kinetic analysis of the reaction of alkyl bromides with silver nitrate indicates that the rate is proportional to the first power of the alkyl bromide and the 1.5 power of the silver nitrate. Although reaction with tetraethylammonium nitrate gives only substitution products, reaction with silver nitrate gives elimination products as well; the rate constant with silver nitrate is 112-fold larger than with tetraethylammonium nitrate.[7] These results demand the direct intervention of silver ion in the reaction. Since the fraction of elimination is twice as high in the reaction of silver nitrate with 2-octyl bromide as with 2-octyl chloride, the halide ion must still be present in the transition state. The substitution reaction under kinetically controlled conditions proceeds with almost complete inversion of configuration. On the basis of these results, a mechanism involving an ion quadruplet intermediate may be written (see Chapter 1).

$$RX + Ag^{\oplus}NO_3^{\ominus} \longrightarrow NO_3^{\ominus}R^{\oplus}X^{\ominus}Ag^{\oplus} \begin{array}{c} \nearrow O_2NOR + AgX \\ \longrightarrow HNO_3 + \text{alkene} + AgX \\ \searrow_{CH_3CN} \\ \text{acetonitrile reaction} \end{array} \quad (8.2)$$

The meaning of the 1.5 kinetic order in silver nitrate can be explained on the basis of a combination of silver nitrate with either silver ion or nitrate ion. Since the reaction is one-half order in the sum of stoichiometric silver nitrate and tetraethylammonium nitrate, the second hypothesis appears to be the correct one, and thus one may suggest an "anionic" catalysis of this electrophilic reaction (Chapter 7).[7,8]

Other metal ions such as plumbous, aluminum, manganous, and ferric ions also facilitate carbonium ion formation.[9] Furthermore, the solvolysis of sulfides and disulfides is induced by metal ions including silver and cupric ions,[10,11] and the S_N2 reaction of thiosulfate and bromoacetate ions is accelerated by lanthanum chloride.[12]

Catalysis of the trans-cis isomerization of bis(oxalato)diaquochromate(III)[13] and of tris(oxalato)chromate(III)[14] is brought about by oxonium ion and by many divalent and trivalent metal ions, the most effective being cerium(III) and lanthanum(III), which are much more efficient than oxonium ion. The logarithm of the catalytic rate constant of the divalent ions is directly proportional to the formation constants of the monooxalato complexes of the metal ions, indicating that the transition state may be described by such a complex. A "one-ended dissociation" mechanism may be proposed for the isomerization. In this mechanism, attack of the cation on the chelated oxalate forms a five-coordinated intermediate with only one end of the oxalate bound to the chromium. Subsequently the released end of the oxalate becomes rebound to the chromium with simultaneous release of the cation. In this reaction, as well as in reactions involving carbonium ion intermediates, the metal ion stabilizes the leaving group in the transition state.

8.2.2 Cleavage of Carbon–Carbon and Carbon–Hydrogen Bonds

Polyvalent metal ions catalyze the decarboxylation of oxaloacetic acid[15] and many other β-keto acids containing a second carboxylic acid group adjacent or nearly adjacent to the ketonic function, including dihydroxyfumaric acid, acetonedicarboxylic acid, oxalosuccinic acid, and dihydroxytartaric acid.[16] On the other hand, neither monocarboxylic acids nor β-keto acids containing no second carboxylic acid group are decarboxylated by metal ions. Of the many metal ions tested as catalysts in the decarboxylation of oxaloacetic acid, aluminum, ferric, ferrous, and cupric ions were the most efficient; sodium, potassium, and silver ions were inactive.

In the aluminum ion-catalyzed decarboxylation of oxaloacetic acid, an intermediate with intense absorption at 252 nm was observed.[17] A similar intermediate was also observed in the metal ion-catalyzed decarboxylation of dimethyloxaloacetic acid, a related compound incapable of forming an enolate ion.[18] This intermediate must therefore be the enolate ion of the product, rather than of the reactant. The effect of pH on the rate constant of decarboxylation shows that the metal ion complex of the dinegative ion reacts, but not the complex of the singly charged anion or of the dissociated acid. Furthermore, the monoester of dimethyloxaloacetic acid, in which the carboxylic acid group adjacent to the ketone is esterified, is not subject to metal ion catalysis while the diion is. This evidence indicates that the metal ion is coordinated to the carboxylate ion gamma to the one that is lost.[18] Cupric ion, which usually shows square planar coordination, gives excellent catalysis. Hence the metal ion is probably not coordinated with the carbonyl oxygen atom and with both of the carboxylate ion groups simultaneously. On this basis, the coordination of the metal ion is postulated to involve the γ-carboxylate ion and the ketonic oxygen atom, as in

(8.3)

Since aluminum ion is an excellent catalyst and since it does not undergo valency change, it appears that no valence change of the metal ion occurs in this reaction. The absence of catalysis by such highly charged ions as $Co(NH_3)_6^{3+}$ indicates that the cation acceleration described here is due to an interaction of a specific short-range character, that is, a chelate, and not to a purely electrostatic interaction. The following visual observations in the decarboxylation catalyzed by ferric ion are strongly indicative of eq. 8.3.

bright yellow ⟶ green ⟶ blue ⟶ deep blue ⟶ colorless products

DMOAA–Fe^{3+} Fe^{3+}-enolate complex

(8.4)

During the decarboxylation an electron pair initially associated with the carboxylate ion group is transferred to the rest of the molecule. A metal ion associated with the carbonyl group should assist this transfer because of its position and positive charge. Thus, mechanism 8.3[18] assigns an electrophilic function to the metal ion catalysis.

A corollary of this description is that the higher the charge and the more readily the metal ion coordinates with the carboxyl group, the greater should be the catalytic activity of the ion. This prediction is borne out by Table 8.1, as far as charge is concerned. The rate constants of the manganous and nickelous ion-catalyzed decarboxylation of dimethyloxaloacetate are three orders of magnitude larger in high dioxane-water solution than in water owing to an increased association constant for the catalyst-substrate complex.[19]

Many attempts have been made in this and other metal ion-catalyzed reactions to correlate catalytic rate constants of different metal ions with association constants of the metal ions with either the reactant or the product.[19] Two such attempts in metal ion-catalyzed decarboxylation showed a linear free energy correlation between the rate constants and the product association constants but not with the reactant association constants.[20, 21] If one assumes

Table 8.1

Effect of Metal Ions on Rate of Decarboxylation of Dimethyloxaloacetic Acid[c]

Metal Ion	Concentration (M)	pH	k_1 (min^{-1})
None	—	4.6	0.0024
Cu^{2+}	0.001	4.6	0.143
Al^{3+}	0.001	4.6	0.128
Ni^{2+}	0.01	4.6	0.0216
Mn^{2+}	0.01	4.6	0.0058
None	—	2.4[a]	0.0032
Fe^{2+} [b]	0.002	2.4	0.0102
Fe^{3+}	0.002	2.3[a]	0.301
None	—	0	0.00032
Pd^{2+}	0.01	0	0.00061

[a] Solution unbuffered.
[b] Experiment conducted under nitrogen.
[c] From R. Steinberger and F. H. Westheimer, *J. Amer. Chem. Soc.*, **73**, 429 (1951). © 1951 by the American Chemical Society. Reprinted by permission of the copyright owner.

that the transition state of the decarboxylation reactions resembles the product, that is, that it has considerable enolate ion character, the data of Fig. 8.3[20] are meaningful.

The effects of various coordinating agents upon cupric ion catalysis of dimethyloxaloacetic acid and acetonedicarboxylic acid decarboxylations are informative. Negative ions such as citrate and acetate ions diminish the catalytic activity of cupric ion; the amount of diminution is much greater for citrate ion than for acetate ion.[18] In the decarboxylation of acetonedicarboxylic acid, the species $Cu^{2\oplus}A^{2\ominus}$ and $CuOAc^{\oplus}A^{2\ominus}$ are catalytically active but the species $Cu(OAc)_2A^{2\ominus}$ is completely inactive (Fig. 8.4).[21] These facts are in agreement with the hypothesis that any ligand reducing the effective charge of the metal ion-substrate complex reduces the catalytic effectiveness of the metal ion. Conversely, a complex-forming agent that does not destroy the charge on the cupric ion does not destroy the catalytic activity. Thus pyridine, which readily forms complexes with cupric ion, promotes the cupric ion-catalyzed decarboxylation of dimethyloxaloacetic acid.[18] Likewise, o-phenanthroline enhances the catalytic activity of manganous ion sixteenfold.[19] In an enzymic process involving metal ion catalysis, the enzyme protein may have three functions: (1) the usual function of imparting specificity toward

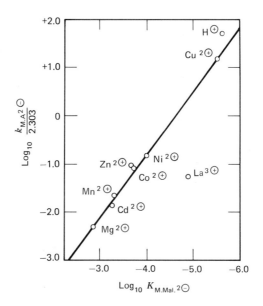

Fig. 8.3 Linear free energy relationship in the decarboxylation of acetonedicarboxylic acid by various metal ions. The log of the rate constant is plotted against the association constant of the same ions with malonate ion. From J. E. Prue, *J. Chem. Soc.*, 2337 (1952).

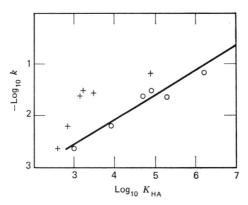

Fig. 8.4 Linear free energy relationship between log k of the metal ion-catalyzed decarboxylation of oxaloacetic acid and the association constants of the metal ion with the reactant (+) or product (○). From E. Gelles and A. Salama, *J. Chem. Soc.*, 3689 (1958).

the substrate; (2) specific complexation of the metal ion in a manner enhancing its activity; and (3) binding (of subunits or substrates) (see Chapter 17).

Metal ions do not catalyze the decarboxylation of β-keto monocarboxylic acids because metal ion complexation with the ground state is more stable than complexation with the transition state. Reversal of this argument suggests that metal ions can catalyze carboxylation of substances containing active hydrogen atoms since the metal ion complex should be more stable in the transition than in the ground state. For example, magnesium methyl carbonate may be used to carboxylate both ketones (eq. 8.5) and aliphatic nitro compounds, an elegant synthetic method as well as an interesting catalytic phenomenon.[22, 23] In addition the magnesium complexes of the enolate ions of the β-keto acids may be further alkylated (eq. 8.5) to form ketones.[24]

$$R \cdot \overset{O}{\overset{\|}{C}} CH_2 R' + (CH_3 O \overset{O}{\overset{\|}{C}} O^{\ominus})_2 Mg^{2\oplus} \longrightarrow R - \underset{R'}{\overset{\overset{Mg^{2\oplus}}{\overset{\ominus O}{\diagup} \overset{O^{\ominus}}{\diagdown}}}{\underset{\|}{C}}} \underset{C=O}{\overset{}{C}} + 2CH_3OH + CO_2 \quad (8.5)$$

$$\downarrow \begin{array}{l} 1 \ R''X \\ 2 \ H_3O^{\oplus}, -CO_2 \end{array}$$

$$R - \underset{\underset{R''}{|}}{\overset{O}{\overset{\|}{C}}} - CHR'$$

In addition to cleavage of carbon–carbon bonds, the cleavage of carbon–hydrogen bonds is facilitated by metal ions in systems where the metal ion can coordinate with the substrate in the proper position. Many multivalent cations such as cupric, nickelous, lanthanum, zinc, manganous, cadmium, magnesium, and calcium ions are efficient catalysts in the bromination of ethyl acetoacetate and 2-carbethoxycyclopentanone.[25] Likewise, zinc ion catalyzes the iodination of pyruvate[26] and o-carboxyacetophenones.[27] The rate-determining step in these ketone halogenations is the formation of an enol through a proton transfer to a general base. The metal ion can catalyze the reaction in a manner similar to that for decarboxylation by stabilizing the anionic charge generated in the cleavage of the carbon–hydrogen bond. The relative catalytic efficiency of the series of metal ions listed above parallels their stability order in complexes with salicylaldehyde.[25] Since this complex resembles an enolate ion transition state, these data are consistent with this description of the metal ion catalysis.

Copper and nickel chelates of imines formed from salicylaldehyde and esters of optically active α-amino acids undergo rapid racemization, ester exchange, and oxidative deamination.[28] Analysis of these reactions was a forerunner of the explanation of pyridoxal-catalyzed reactions. In the latter (nonenzymic) reactions, a metal ion complex of the imine formed from an amino acid and pyridoxal undergoes racemization, decarboxylation, and transamination reactions, among others. In the salicylaldehyde chelates, the metal ion can facilitate carbon–hydrogen bond cleavage leading to racemization, as well as facilitate ester exchange by polarization of the carbonyl group. In the pyridoxal imine chelates, the metal ion can likewise facilitate carbon–hydrogen or carbon–carbon bond cleavage in conjunction with the conjugated pi electron system in the molecule, as described in Chapter 6 (eq. 6.50).

Multivalent metal ions catalyze aldol condensation reactions. For example, in basic solution cupric diglycinate condenses with formaldehyde, producing cupric complexes of serine, or with acetaldehyde, producing complexes of threonine and allothreonine.[29-31] These reactions occur readily at pH 11 at 100°. The catalytic activity of the metal ion in these reactions stems from the facilitation of carbon–hydrogen bond cleavage yielding the enolate ion necessary for condensation. The labilization of an α-hydrogen atom by chelation of an α-amino acid to a metal ion has been demonstrated in isolation by an nmr technique.[33] On this basis, the mechanism may be

$$(8.6)$$

Superacid Catalysis

The effectiveness of the metal ion catalysis in this reaction can be markedly improved by the use of a cobalt(III) bis(ethylenediamine) complex, which combines with only *one* molecule of glycine producing a complex with an overall charge of $\oplus 2$ rather than the charge of zero in eq. 8.6.[34]

8.2.3 Addition to Carbon–Oxygen and Carbon–Nitrogen Double Bonds

The hydrolysis of many esters is subject to catalysis by metal ions.[32] Structurally, the esters have a common feature: they contain a secondary functional group that can serve as ligand for the metal ion. For instance, metal ions effectively catalyze the hydrolysis of α-amino acid esters.[35] As with all amines, these substances would be expected to form coordination compounds with heavy metal ions. The hydrolysis of amino acid esters complexed with cobaltous, cupric, manganous, calcium, and magnesium ions is extremely rapid even at pH 7–8 where the esters are ordinarily stable (Table 8.2). When the concentration of metal ion is varied, the rate constant of the hydrolysis varies, reaching a maximal value as the [metal ion]: [ester] ratio approaches unity, indicating that the most active species is a complex of *one* metal ion and *one* ester molecule. With the five metal ions utilized, ester hydrolysis increases with the increasing tendency of the metal ion to coordinate with amines. On the basis of this evidence, a 1:1 complex between the metal ion and the α-amino acid ester may be postulated in which the metal ion chelates with the α-amino group and the carbonyl oxygen atom of the ester. There is direct evidence in the solid state for the 1:1 complexes of certain α-amino acid esters with the halides of $Cu^{2\oplus}$, $Ni^{2\oplus}$ and $Cd^{2\oplus}$.[36]

Table 8.2

Effect of Several Metal Ions on Hydrolysis of Glycine Methyl Ester[a,b]

Metal Ion	pH	$k_{obsd}(\sec^{-1})$
$Cu^{2\oplus}$	7.3	0.0425
$Co^{2\oplus}$	7.9	0.0156
$Mn^{2\oplus}$	7.9	0.00351
$Ca^{2\oplus}$	7.9	0.0007

[a] Temperature 25.4°; metal ion concentration 0.016 M; ester concentration 0.016 M.
[b] From H. Kroll, *J. Amer Chem. Soc.*, **74**, 2036 (1952). © 1952 by the American Chemical Society. Reprinted by permission of the copyright owner.

In glycine buffer at pH 7.3, glycine methyl ester and phenylalanine ethyl ester undergo a facile hydrolysis catalyzed by cupric ion.[37] Under these conditions, the reactions follow first-order kinetics in the substrate, and it is thus possible to compare the rate constants of hydrolysis of DL-phenylalanine ethyl ester catalyzed by oxonium, hydroxide, and cupric ions at pH 7.3 and 25°: $H_3O^{\oplus}$, 1.46×10^{-11} sec^{-1}; $OH^{\ominus}$, 5.8×10^{-9} sec^{-1}; $Cu^{2\oplus}$ (0.0775 M), 2.67×10^{-3} sec^{-1}. Although the last constant is a complex constant not directly comparable to the former two, it clearly shows the facile catalysis by cupric ion. Further, measurements of the metal ion-ester equilibrium indicate that this cupric ion rate constant is a reasonable approximation to the true catalytic rate constant.[38]

The greater catalytic effectiveness of cupric ion relative to hydroxide or oxonium ion in this hydrolysis cannot be due solely to the electrostatic effect of a positively charged α-amino ester toward hydroxide ion. The introduction of a positive charge two atoms away from the carbonyl group of an ester can increase the rate constant of alkaline hydrolysis by a factor of up to 10^3 whereas here there is a difference of approximately 10^6 between the cupric ion-catalyzed and alkaline hydrolyses of DL-phenylalanine ethyl ester. Since the effective charge on the cupric ion:glycine(buffer):ester complex is $\oplus 1$, the factor of 10^6 is not due to an increase in charge over that present in a singly charged ester. Thus, the rapid hydrolysis of α-amino esters by cupric ion appears to be due to a direct interaction of the metal ion with the reaction center, the ester group.

This conclusion is emphasized by work in which the relative second-order rate constants for reactions of hydroxide ion with various substrates have been determined: glycine ethyl ester, 1; protonated glycine ethyl ester, 41; and the cupric ion complex of glycine ethyl ester, 3200×41.[39] The rate constant ratio of protonated to unprotonated glycine ethyl ester reflects the electrostatic interaction between the cationic charge and the hydroxide ion. The 41-fold difference in rate constants is related to the 74-fold difference in the pK_a's of glycine with and without a negative charge. These differences are somewhat smaller than the 1000-fold factor mentioned above. However, the effect of the cupric ion is definitely larger than that of the positive charge and demands an additional effect beyond a simple electrostatic one. This effect corresponds to superacid catalysis. With histidine, cysteine, and aspartic acid esters, the rate of cupric ion-catalyzed hydrolysis is only 100-fold greater than that of the neutral substrate. That is, the rate of the cupric ion catalysis with these molecules is equivalent to the rate of the protonated substrate. In these systems, the metal ion can chelate at two sites *other* than the ester bond; hence a catalytic effect is seen that corresponds only to an electrostatic effect. But when one of the two sites with which the metal ion can complex is the reactive ester linkage itself, superacid catalysis occurs.

Metal ion-catalyzed hydrolysis of α-amino esters thus can be described by the interaction of the metal ion with the ester linkage. This interaction can take place in two ways, in which the metal ion complexes with the carbonyl oxygen atom (eq. 8.7) or alternatively the alkoxyl oxygen atom (eq. 8.8) of the ester.

$$\begin{array}{c}
H_2C\text{---}C\!\!=\!\!O \\
| \quad\quad\quad | \\
H_2N\diagdown\quad O^\ominus \\
\quad Cu^{2\oplus} \\
H_2N\diagup \quad O^{18} \\
| \quad\quad\quad \| \\
RCH\text{---}C\text{---}OCH_3
\end{array} \xrightarrow{OH^\ominus} \begin{array}{c}
H_2C\text{---}C\!\!=\!\!O \\
| \quad\quad\quad | \\
H_2N\diagdown\quad O^\ominus \\
\quad Cu^{2\oplus} \\
H_2N\diagup \quad O^{18} \\
| \quad\quad\quad | \\
RCH\text{---}C\text{---}OCH_3 \\
\quad\quad\quad | \\
\quad\quad\quad OH
\end{array} \xrightarrow{+H^\oplus} \begin{array}{l} \xrightarrow{-H_2O} \text{exchange} \\ \xrightarrow{-CH_3OH} \text{hydrolysis} \end{array}$$

(8.7)

$$\begin{array}{c}
H_2C\text{---}C\!\!=\!\!O \\
| \quad\quad\quad | \\
H_2N\diagdown\quad O^\ominus \\
\quad Cu^{2\oplus}\diagup CH_3 \\
H_2N\diagup \quad O \\
| \quad\quad\quad | \\
RCH\text{---}C\!\!=\!\!O^{18}
\end{array} \xrightarrow{OH^\ominus} \begin{array}{c}
H_2C\text{---}C\!\!=\!\!O \\
| \quad\quad\quad | \\
H_2N\diagdown\quad O^\ominus \\
\quad Cu^{2\oplus}\diagup CH_3 \\
H_2N\diagup \quad O \\
| \quad\quad\quad | \\
RCH\text{---}C\text{---}^{18}O^\ominus \\
\quad\quad\quad | \\
\quad\quad\quad OH
\end{array} \xrightarrow{+H^\oplus} \begin{array}{l} \xrightarrow{-H_2O} \text{exchange} \\ \xrightarrow{-CH_3OH} \text{hydrolysis} \end{array}$$

(8.8)

In either case, polarization of the carbonyl group ensues, facilitating nucleophilic attack. A tetrahedral addition intermediate is postulated in these reactions because carbonyl oxygen exchange accompanies metal ion-catalyzed hydrolysis.[37]

An alternative explanation can also account for the metal ion catalysis. Since cupric ion complexes hydroxide ion around neutrality,[25,40] the catalysis can consist of a template-like reaction in which the cupric ion binds both the substrate and hydroxide ion leading to an intracomplex reaction of hydroxide ion with the ester. This mechanism was originally suggested for the calcium, barium, and thallous ion-catalyzed hydrolyses of potassium ethyl oxalate and of potassium ethyl malonate.[41] The oxalate ester is more susceptible to metal ion catalysis than the malonate ester which in turn is catalyzed to a greater extent than the adipate ester (Table 8.3). Alkali metal ions, on the other hand, have only a small negative salt effect on the hydrolysis of potassium ethyl malonate. On the basis of these structural and metal ion effects, the transition state of the calcium, barium, and thallous ion catalyses was postulated to have the chelate structure

$$\begin{array}{c}
\quad\quad OC_2H_5 \\
\quad\quad | \\
O\!\!=\!\!C\text{-------}OH^\ominus \\
\quad\quad | \quad\quad\quad\vdots \\
O\!\!=\!\!C\diagdown_{O}^{\ominus}\diagup M^{n\oplus}
\end{array}$$

(8.9)

Table 8.3

Metal Ion Catalysis of Hydrolysis of Esters Containing Free Carboxylate Ion[41]

Base	Ester		
	EtOx$^\ominus$	EtMal$^\ominus$	EtAd$^\ominus$
OH$^\ominus$	29.2	0.778	2.01
TlOH	1330	5.65	4.37
BaOH$^\oplus$	9200	38.0	—
CaOH$^\oplus$	—	39.8	7.94
Co(NH$_3$)$_5$OH$_2{}^{2\oplus}$	1740	16.0	17.1

A particularly effective metal ion-catalyzed hydrolysis is seen in the following[42]:

$$\text{(8.10)}$$

At pH 7 and 25°, the half-life for this reaction is approximately 0.5 min. The stereochemistry of this reaction dictates that the metal ion bound to the phenanthroline nitrogen atoms can coordinate with the transition state of the reaction resembling a trigonal amide but not with the ground state which is a linear nitrile. Thus a metal ion or other complexing agent can facilitate reaction. The rate constant for this reaction is approximately 10^{10} times larger than that of benzonitrile plus hydroxide ion.[43] At pH 7, approximately 50% of the complex ion will be present in the form of a hydroxy derivative. Again internal hydroxide ion attack on the nitrile cannot be distinguished from external hydroxide ion attack. Notwithstanding this ambiguity, this reaction is an elegant example of metal ion catalysis. A similar reaction is seen in the metal ion-catalyzed reaction of isocyanates and alcohols.[44,45]

Mechanism 8.9, depicting interaction of a bound hydroxyl ion with the ester linkage, and mechanisms 8.7 and 8.8, depicting polarization of the carbonyl linkage, are kinetically indistinguishable. Extrakinetic arguments, however, indicate that eqs. 8.7 and 8.8 are preferable. As pointed out earlier, no catalysis other than an electrostatic effect is seen in the cupric ion-catalyzed hydrolysis of histidine methyl ester. This result can be readily explained in terms of mechanism 8.7 or 8.8 since the cupric ion cannot complex with the ester linkage directly. However, this result is not compatible with mechanism 8.9 since eq. 8.11 shows that internal hydroxide ion attack is still feasible.

$$\text{(structure showing MeO-C(=O) with }^{\ominus}\text{HO, N=, NH, H}_2\text{N, Cu}^{2\oplus}\text{)} \tag{8.11}$$

The mechanistic ambiguity in metal ion catalysis between polarization of the ester linkage and attack by an internally bound hydroxide ion is resolved in the hydrolysis of α-amino ester coordinated with a cobalt complex ion. Amine ligands of cobalt(III) ions are not exchanged readily and the molecule may be considered to have the permanent structure in the following:

$$\text{[Co complex with Cl, NH}_2\text{, CH}_2\text{COR]} \xrightarrow{\text{Hg}^{2\oplus}} \text{[Co complex with O=C(OR)CH}_2\text{, NH}_2\text{]} \xrightarrow{\text{OH}^{\ominus}}$$

$$\text{[Co complex with }^{\ominus}\text{O-C(=O)CH}_2\text{, NH}_2\text{]} + \text{ROH} \tag{8.12}$$

When the cobalt complex of eq. 8.12 is treated with mercuric ion, the chloride ligand is removed and hydrolysis occurs, although the ester is stable in the absence of mercuric ion. Infrared spectrophotometric studies show that the hydrolysis of the ester occurs in two steps, with the intermediate formation of a cobalt complex containing the ester group bound both through the amino and carbonyl groups. Hence, the metal ion catalysis of the hydrolysis occurs by polarization of the carbonyl group through direct interaction with the metal ion. This special situation may not of course be applicable to all esters.[46]

Metal ions including transition metal ions[47] and rare earth ions[48] catalyze the hydrolysis of a variety of amides including acylamino acids, dipeptides, tripeptides, and amino acid amides. In these compounds the metal ion can complex with one or more amine or carboxylate ion ligands, in addition to the amide group. Thus, the structural prerequisites for metal ion catalysis of amide hydrolysis parallel those for ester hydrolysis. In the hydrolysis of glycinamide and phenylalanylglycinamide,[47] cupric, cobaltous, and nickelous ions catalyze the hydrolysis in that order of effectiveness. However, metal ion catalysis in amide hydrolysis is not nearly as striking as in ester hydrolysis. For example, hydrolysis of glycinamide in the presence of 0.02 M cupric ion is only twentyfold faster than the spontaneous hydrolysis. This result is at first surprising since most of the infrared evidence for the interaction of metal ions with carboxylic acid derivatives shows stronger interaction with amides than esters, presumably because of the greater basicity of the former. But this argument is fallacious, since stronger complexing of the metal ion with the ground state than with the transition state must necessarily lead to a slower rather than faster reaction. This point is clearly seen in the pH dependence of the cupric ion-catalyzed hydrolysis of glycylglycine.[49] This reaction shows a bell-shaped pH-rate constant profile with a maximum at pH 4.2 that can be correlated with the pH dependence of the ionizations of (cupric-glycylglycine)$^\oplus$. At lower pH's no cupric ion complex would be expected since the coordinating ligands become protonated. However, at higher pH's, complex formation should still occur, and indeed it does, but in the form of (cupric-glycylglycine)^0 which has lost a proton from the amide group. The latter metal ion complex is exceedingly stable, and quite unreactive.[50]

The hydrolytic cleavage of N-terminal peptide bonds is smoothly accomplished by the complex Co(trien)H$_2$O(OH)$^{2\oplus}$, similar to the species in eq. 8.12.[51] The N-terminal amino acid residue is selectively hydrolyzed and simultaneously converted to the inert Co(III) complex as shown by

$$\text{Co(trien)H}_2\text{O(OH)}^{2\oplus} + \text{H}_2\text{NCHRCNHP} \longrightarrow \text{Co(trien)(H}_2\text{NCHRCOO)}^{2\oplus} + \text{H}_2\text{NP} \quad (8.13)$$

The reaction takes place rapidly in aqueous solution at 65° and pH 7–8. Although stoichiometric rather than catalytic, this process is perhaps the best model at present for the action of exopeptidases containing metal ions.*
If the N-terminal amino group of the substrate is blocked by a benzyloxycarbonyl residue, the reaction does not take place, presumably because binding by the amino group to the metal ion is essential. Assuming that eq. 8.12 is

* See Chapter 17.

Superacid Catalysis

followed, the carbonyl oxygen of the amide link to be cleaved also becomes coordinated, leading to the facilitation of the hydrolytic process. The high efficiency of this reaction compared to other metal ion-catalyzed amide hydrolyses probably depends on the stable, highly charged complex ion used, containing only two replaceable positions.

Thiol esters are particularly susceptible to cleavage by heavy metal ions such as mercuric, lead, and silver ions. Since these cations will cleave simple esters containing no secondary ligand groups, chelation does not appear to be essential in these reactions as it was in the hydrolysis of oxygen esters and amides. Presumably the coordination of the sulfur atom with the heavy metal ion is the principal driving force of this reaction.[52-56]

Reactions of carbonyl compounds and imines are also catalyzed by metal ions. The hydrolyses of some imines are catalyzed by divalent metal ions such as cupric or nickelous ions. Spectrophotometric evidence indicates the formation of a metal ion: substrate complex with subsequent facile cleavage of the complex.

$$\text{(8.14)}$$

The hydrolyses of imines derived from salicylaldehyde are, however, retarded by metal ions, presumably because the reactant complex is stabler than the transition state (product) complex. The facile cleavage (eq. 8.14) may be described in terms of the polarization of the carbon–nitrogen double bond by the metal ion.[57]

In two common organic reactions, the Meerwein-Ponndorf reduction and its partner, the Oppenauer oxidation, carbonyl compounds and their corresponding alcohols are interconverted under the catalytic influence of certain aluminum alkoxides. The mechanism of this reaction almost certainly involves hydride transfer in a cyclic complex involving aluminum as a coordinating atom (eq. 8.15).[58]

$$(8.15)$$

The function of the catalyst is twofold. First of all, it acts as a strong base so that the alcohol, a weak hydride donor, is converted to the alkoxide ion, a much more powerful reducing agent; secondly, the metal ion coordinates with the carbonyl oxygen atom, making the carbonyl carbon a better electrophile, and at the same time, holding the carbonyl compound in an optimal position with respect to the hydride donor. This description, while satisfying, is undoubtedly oversimplified since it has been shown that an aluminum alkoxide trimer is the reactive entity. Furthermore, hydride transfer is apparently not the rate-determining step of the reaction, since the alcohol product is formed more slowly than the acetone product.[59] The Tischenko reaction, the aluminum alkoxide-catalyzed disproportionation of an aldehyde to the ester of the corresponding acid and alcohol, follows a similar mechanistic pathway in which the aluminum ions carry out the functions described above.[60] In a similar fashion, the lithium aluminum hydride reduction of ketones may be catalyzed by polarization of the carbonyl group by the lithium ion, or alternatively by the aluminum trihydride molecule. In the addition of a Grignard reagent to a ketone, two molecules of Grignard are found kinetically. One hypothesis is that one molecule coordinates with the carbonyl oxygen atom, polarizing the carbonyl group and thus facilitating the addition of the alkyl group of the second Grignard molecule to the carbonyl carbon atom.[61]

Polymerization of propylene oxide and cyclic ethers and acetals in general is known.[62] An epoxide can be viewed as an analog of a carbon oxygen double bond.

8.2.4 Cleavage of Phosphates

Metal ions catalyze the hydrolysis of phosphoric and phosphonic acid halides, phosphate esters, and various phosphoric acid anhydrides including acyl phosphates, pyrophosphates, and triphosphates.

The hydrolyses of diisopropyl phosphorofluoridate (DFP) and isopropyl methylphosphonofluoridate (Sarin) are susceptible to catalysis by many metal salts and chelates such as $MoO_4^{2\ominus}$, Ce(III), Mn(II), Cu(II), $WO_4^{2\ominus}$, and $CrO_4^{2\ominus}$.[63-66] In the hydrolysis of DFP by cupric ion, the magnitude of the catalysis is affected by the structure of the ligands on the metal ion. Maximum activity is found with metal ion complexes meeting the following requirements: (1) maximum positive charge; (2) minimum number of ligands; and (3) minimum formation of μ-dihydroxodinuclear species.[62] In the hydrolysis of Sarin, catalysis by cerous, cupric, and manganous ions in the form of bifunctional species containing a nucleophilic center (hydroxide ion) and an electrophilic center (metal ion) is particularly effective.[65] Although the hydroxymetallic ions are considerably weaker bases than hydroxide ion itself, they are catalytically more active by a factor of 10. Although this activity can be explained by this bifunctionality, other mechanisms may be operative (see earlier).

The solvolysis of tetrabenzyl pyrophosphate catalyzed by the general base lutidine is additionally catalyzed by calcium ion. In the presence of 0.02 M calcium ion and 0.2 M lutidine, the rate of solvolysis of tetrabenzyl pyrophosphate (with cleavage of the P–OP bond) is increased by a factor close to a millionfold over the uncatalyzed reaction, each catalyst making approximately a thousandfold contribution.[68] The divalent cation can chelate with two oxygen atoms of the pyrophosphate ester, making the phosphorus atom more susceptible to nucleophilic attack.

$$(PhCH_2O)_2\overset{O}{\overset{\|}{P}}O\overset{O}{\overset{\|}{P}}(OCH_2Ph)_2 + Ca^{2\oplus} \longrightarrow (PhCH_2O)_2P\underset{O}{\overset{O\cdots Ca^{2\oplus}\cdots O}{\|}}P(OCH_2Ph)_2 \quad (8.16)$$

$$\downarrow \text{lutidine, 2-propanol}$$

$$\text{products}$$

The hydroxides of lanthanum, cerium, and thorium promote the hydrolysis of α-glyceryl phosphate from pH 7 to 10.[69,70] At pH 8.5, lanthanum hydroxide gel accelerates the alkaline hydrolysis of α-glyceryl phosphate by more than a thousandfold. The rate of the metal ion-catalyzed reaction is considerably increased by nitrogen- or oxygen-containing substituents involving the carboxylate ion[71] in the β-position of the ester. The reaction can therefore be suggested to proceed through the complex:

$$(8.17)$$

In this complex, the lanthanum ion is coordinated with a β-substituent as well as with an oxyanion or an oxygen atom of the leaving group. Since dilution of the lanthanum hydroxide gel with water increases the rate of reaction, the catalytically active species is the metal ion in solution in equilibrium with the gel, and not the metal ion of the gel. The hydrolyses of diesters of phosphoric acid are also susceptible to metal ion catalysis, in particular by multivalent cations such as barium, stannous, and cupric ions. The diesters that undergo metal ion-catalyzed hydrolyses include open chain diesters and cyclic diesters containing both five- and six-membered rings.[72]

Nucleophilic reactions of acetyl phosphate are catalyzed by cations such as magnesium, calcium, cobalt, manganese, nickel, zinc, and lithium.[73–76] Calcium ion catalyzes a neutral as well as a basic hydrolysis, and also catalyzes the reactions of glycine and mercaptoacetate with the carbonyl group of the anhydride. At pH 7.7, magnesium ion catalyzes the hydrolysis of acetyl phosphate markedly, the reaction being first-order in both magnesium ion and acetyl phosphate concentration. The catalysis by metal ion is greater at pH 7.7, where acetyl phosphate exists as dinegative ion, than at pH 0.63, where it exists as a neutral molecule. The chelate shown in eq. 8.18 may therefore be postulated as an intermediate in the metal ion-catalyzed hydrolysis.

$$\underset{\substack{\text{Mg}^{2\oplus}\\ \diagup\quad\diagdown}}{\underset{CH_3\overset{\displaystyle\|}{C}\diagdown_O\diagup\overset{\displaystyle P=O}{}\diagdown_{O^\ominus}}{}} \longrightarrow CH_3-\overset{\overset{\displaystyle Mg^{2\oplus}}{\diagup}\ \ \ \ }{\underset{\displaystyle O}{\overset{\displaystyle O}{C}}} + \underset{\displaystyle O\ \ \ O}{\overset{\displaystyle O^\ominus}{P}} \xrightarrow[\text{fast}]{H_2O} H_2PO_4^\ominus \qquad (8.18)$$

In reactions at the carbonyl group (with carbon–oxygen fission) the metal ion may facilitate reaction by decreasing the electrostatic repulsion between the negatively charged nucleophile and negatively charged substrate, or by stabilizing the departing phosphate group. In reactions proceeding with phosphorus–oxygen fission (eq. 8.18) the metal ion may facilitate metaphosphate formation by stabilizing the leaving group in the same manner as a proton.

Enzymic phosphorylation reactions in which adenosine triphosphate (ATP) participates require magnesium ion. Although the metal ion may act as a chelating agent between nucleotide and enzyme in these reactions, several arguments suggest that part of the metal ion function may be associated with superacid catalysis of the kind described here. Certainly in nonenzymic hydrolysis of ATP,[77,78] as well as simpler polyphosphates[79] and triphosphates,[80,81] the facilitation by various divalent metal ions such as calcium, magnesium, manganous, cupric, and cadmium ions must be attributed to superacid catalysis. For instance, the rate of hydrolysis of ATP is accelerated tenfold by calcium ion at pH 9 and sixtyfold by cupric ion at pH 5. Divalent metal ions also catalyze the nonenzymic transphosphorylation of ATP[82] and γ-phenylpropyl triphosphate[81] with inorganic phosphate and carboxylate ions. In the (uncatalyzed) hydrolysis of γ-phenylpropyl diphosphate, the monoprotonated species hydrolyzes approximately 2000 times faster than the fully ionized species.[83] Consequently, a metaphosphate intermediate can be invoked here. On this basis, the mechanism of metal ion catalysis of hydrolysis of the terminal phosphate bond may be postulated to be similar to that for the hydrolysis of acetyl phosphate. Again, it is suggested that metaphosphate ion formation is facilitated by the superacid properties of the metal ion chelate.

$$ (8.19) $$

The identical rate constants of γ-phenylpropyl triphosphate and ATP in metal ion-catalyzed reactions with inorganic phosphate[81] indicate that the adenine moiety is not involved in (kinetically important) complexing with the metal ion. Nmr measurements, however, show that magnesium, calcium, and zinc ions bind predominantly to the β- and γ-phosphates of ATP, that cupric ion binds predominantly to the α- and β-phosphates, whereas manganous ion binds to all three.[84] Therefore, mechanism 8.19 is an arbitrary designation of the position of the complexing of the metal ion. It is also arbitrary with respect to the assumption of a 1:1 complex. If these reactions do not proceed through a metaphosphate intermediate, the metal ion catalysis may be described as one in which the metal ion coordinates both the substrate and the nucleophilic agent needed for the bimolecular process in a stereochemistry favorable for reaction.

The disproportionation of adenosine diphosphate to adenosine triphosphate and adenosine monophosphate in dimethyl sulfoxide solution is catalyzed by sodium, potassium, rubidium, and cesium ions, but not by hydrogen, lithium, ammonium, magnesium, or zinc ions. At constant sodium or potassium ion concentration, the initial rate is proportional to the square of the adenosine diphosphate concentration. When the sodium ion concentration is varied, the rate is proportional to the sodium-adenosine diphosphate complex concentration, indicating possible mechanistic similarity to the above reactions.[85]

8.2.5 Factors in Superacid Catalysis

In many of the reactions described previously, the metal ion is consumed stoichiometrically. Consequently, the term "metal ion catalysis" used continually in the preceding dicussion and elsewhere usually should be replaced by the more rigorous term, "metal ion promotion."

The most important characteristic of a metal ion in these reactions, acting either as catalyst or promoter, is its positive charge. In many reactions, variation in catalytic effectiveness can be directly correlated with variation in the magnitude of cationic charge. Since this charge consists of the *effective* charge on the metal ion complex, and not on the metal ion alone, the electrostatic nature of the ligands attached to the metal ion is of equal importance to the inherent charge on the metal ion. In several reactions cited above, catalysis by a polyvalent metal ion is reduced to nil by complexation of the metal ion with anionic ligands. Furthermore, charge density may be more important than net charge. The force between a charge and another charge or a dipole is calculated from simple electrostatic theory to depend on the inverse second, or third to fifth, power of the distance. With this restriction on the field of the metal ion, it is necessary, for maximal effect, to introduce the metal ion directly into the substrate molecule to be catalyzed. More specifically, it must be introduced directly into the reactive bond of the substrate to be broken. Thus stereospecific coordination of the metal ion is of utmost importance. On this basis, the effectiveness of a metal ion should increase with increasing ionic radius of the ion. However, with transition metal ions, the electrostatic effect of the ion is also affected by the shielding of the ligand from the nuclear charge of the metal ion by the d electrons and the ligand field.[86]

All polar organic reactions are influenced by electronic changes within a molecule. The principal function of the metal ion catalyst is to effect such changes; the metal ion can bring them about probably to a greater extent than any other chemical species. The simplest manifestation of electronic distortion by a metal ion is seen in the fact that the acidity of a water molecule coordinated with a cupric ion in the hydrated cupric ion is 10^7 times greater than the acidity of a free water molecule. The acidity of an organic substrate complexed with cupric ion might therefore also be 10^7 times greater than that of the organic substrate alone. This hypothesis means that a considerable shift of electron density can be brought about by coordination of a metal ion with an organic substrate; therefore, reactions dependent on electronic movement should be effectively facilitated by a metal ion.

Many reactions require the presence of an "electron sink" in the molecule to absorb electron density produced by the reaction. The introduction of a suitably positioned metal ion in the substrate undergoing such a reaction can serve this purpose. Other reactions require the neutralization of negative charge to reduce electrostatic repulsion during reaction. A metal ion will also serve this purpose. Still other reactions require the polarization of a particular bond to effect reaction. Again, introduction of a metal ion in a specific position in the substrate molecule will accelerate reaction in this way. Finally the stabilization of leaving groups will often facilitate reaction. When the leaving groups are halide, phosphate, mercaptide, or other anions, metal ions can facilitate reaction.

Either a proton or a metal ion can introduce a positive charge into a substrate molecule, producing the electronic changes discussed above. A metal ion, however, is superior to a proton on several grounds: a metal ion can introduce a multiple positive charge into an organic molecule whereas a proton can introduce only a single positive charge; a metal ion can operate in neutral solution whereas a proton cannot; a metal ion can coordinate several donor atoms whereas a proton can coordinate only one.

In listing the ways in which metal ions may promote organic reactions, the requirement that the metal ion be suitably positioned within the substrate molecule was emphasized. Specific complexation or chelation of the metal ion with the substrate appears to be an absolute requirement of metal ion catalysis of nucleophilic reactions. In many reactions chelation appears to be the rule; this requirement means that the substrate must contain either one or two donor atoms, in addition to the reactive center, with which the metal ion must coordinate. Many attempts have been made to correlate the effectiveness of catalysis by a series of metal ions with the relative formation constants of the complexes. Such correlations have been successful in a number of reactions, unsuccessful in others. The successful analyses involve a complex for the correlation closely approximating the transition state of the reaction rather than the ground state. This result indicates that the metal ion complex must stabilize the transition state of the reaction in order to assist the reaction effectively, and that metal ion complex formation in the ground state depresses the reaction rate. Not only must the metal ion be strongly chelated to the substrate, it must also be bound in the correct stereochemical fashion for entering into reaction.

Since coordination compounds of metal ions may involve a large number of ligands, it is possible to form not only a metal ion complex with the substrate but also a complex with both the substrate and the nucleophilic agent simultaneously. The metal ion can thus serve as a focal point for both components of a bimolecular reaction and assist reaction by making the entropy of activation more positive. This phenomenon is seen in the enhanced reactivity in eq. 8.20 with respect to eq. 8.21

$$(8.20)$$

$$(8.21)$$

Whereas the pyridine aldoximate ion reacts 10 times faster with *p*-nitrophenyl acetate than with the ester of eq. 8.20, the rate constants for eq. 8.20 and 8.21 are equivalent to one another, indicating that complex formation has facilitated eq. 8.20 by ten fold.[87]

A metal ion can enhance the reaction of a substrate and nucleophile by forming a complex with both, if both the substrate and nucleophile contain negative charges, for in such a complex, electrostatic repulsion between these two species is decreased and entropy enhanced. The most efficient complex of metal ion, substrate, and nucleophile can be envisioned as one in which both the substrate and the nucleophile are attached to the metal ion in positions *other* than those at which eventual covalent interactions takes place, as seen in eq. 8.20. However, this appears not to be an absolute requirement, for the reaction of the thiolate ion (eq. 8.22) with methyl iodide occurs with an energy of activation which is considerably less than that of a system not containing a metal ion.[89]

$$(8.22)$$

A complex of metal ion, substrate, and nucleophile must have the proper stereochemistry for the covalent interaction between the two ligands to occur. A metal ion complex containing hydroxide ion as a ligand serves as a carrier of hydroxide ion in neutral solution, just as the metal ion itself can be considered to approximate a superproton in neutral solution. Thus, a metal ion complex containing hydroxide ion (or another nucleophile) can be considered to be a bifunctional catalyst, the metal ion serving as a general acid and the hydroxide ion serving as general base or nucleophile.

In conclusion, a metal ion can serve as a superacid when introduced in the correct stereochemical position in a substrate molecule. This superacid catalysis can be manifested in many ways. Probably its most important future progress will be in "template" reactions where two reactants will be coordinated simultaneously to a metal ion prior to reaction. The effectiveness of catalysis in simple systems parallels the strength of binding to models of the transition state, but not the strength of binding to the reactant. In many instances, the transition state model is in fact the product, making the metal ion reaction not a catalysis but a promotion as pointed out above. In true catalyses such as enzymic catalysis the strongest coordination must be to the transition state, and not to the reactant or product. In this connection, it should be noted that the weakest chelators such as manganous and magnesium ion are commonly found in enzymic reactions.[90]

8.3 CATALYSIS VIA REDOX REACTIONS

The role of metal ions in superacid catalysis is that of an electrophile or general acid and depends on the magnitude of the cationic charge as well as complex-forming ability. Except for these factors, there is no distinction between transition, inner transition, and other metal ions. In catalysis of oxidation-reduction reactions, however, only transition metal ions play a role since the key feature in these reactions is the ability of the metal ion to exist in solution in more than one oxidation state.[91] The ability to complex plays a secondary, but still necessary, role.

8.3.1 Metal Ion Redox Reactions

If given oxidizing and reducing agents have the proper redox potentials (or standard free energies) for reaction with each other, the reaction may still be slow. This is particularly true of organic reducing agents. In such a case, a metallic ion of variable valence may greatly accelerate the rate by providing an easier reaction path.

Cupric and silver ions, as well as other cations exhibiting two or more stable oxidation states, can serve to "transport" electrons in redox reactions through a chain mechanism in which the catalytic metal ion is successively oxidized and reduced. For example, reaction 8.23 is catalyzed by Cu(II) since the rate of the reaction shows a dependence on the vanadium(III) and copper(II) concentrations but not the iron(III) concentration.[92]

$$V(III) + Fe(III) \longrightarrow V(IV) + Fe(II) \qquad (8.23)$$

The mechanism of this catalysis is given by

$$\begin{align} V(III) + Cu(II) &\longrightarrow V(IV) + Cu(I) \\ Cu(I) + Fe(II) &\xrightarrow{fast} Fe(II) + Cu(II) \end{align} \qquad (8.24)$$

When an electron transfer reaction

$$Tl(I) + 2Ce(IV) \longrightarrow Tl(III) + 2Ce(III) \qquad (8.25)$$

involves either a three-body collision or an unavailable intermediate valency state (either thallium(II) or cerium(II)), reaction will proceed slowly. However, in the presence of manganous ion, both of these difficulties are avoided by the introduction of an alternate catalytic reaction path (Chapter 1).

$$\begin{align} Ce(IV) + Mn(II) &\longrightarrow Ce(III) + Mn(III) \\ Mn(III) + Ce(IV) &\longrightarrow Ce(III) + Mn(IV) \\ Mn(IV) + Tl(I) &\longrightarrow Mn(II) + Tl(III) \end{align} \qquad (8.26)$$

Coordination often greatly increases the ease with which an electron transfer can occur.[93] An example is the catalytic effect of Mn(III) on the reaction between chlorine and oxalate ion. The key step is the internal oxidation reduction of $MnC_2O_4^{\oplus}$, followed by reaction with chlorine.[94]

$$\begin{align} MnC_2O_4^{\oplus} &\longrightarrow Mn^{2\oplus} + CO_2 + CO_2^{\ominus} \\ CO_2^{\ominus} + Cl_2 &\longrightarrow CO_2 + Cl\cdot + Cl^{\ominus} \\ Cl\cdot + Mn^{2\oplus} &\longrightarrow Cl^{\ominus} + Mn^{3\oplus} \end{align} \qquad (8.27)$$

In electron transfer between metal ions, ligands can serve as bridges. Hence, their structure is of importance to the rate of electron transfer.[95]

8.3.2 Reactions Involving Copper

The copper(0)-copper(I)-copper(II) system has long been a favorite catalytic system for reactions that may be accelerated by electron transfer, because of favorable redox properties of this system.[32] Both the Sandmeyer reaction (the substitution of an aromatic diazonium ion by halide ion) and the Meerwein reaction (the reaction of an aromatic diazonium salt with molecules containing activated double bonds such as acrylonitrile) are catalyzed by copper(I) salts. These reactions have been postulated to proceed via the phenyl free radical on the basis of the following observations: (1) although copper (II) does not promote polymerization, the aromatic diazonium ion and cuprous chloride system promotes the polymerization of acrylonitrile or methyl methacrylate; and (2) the effect of substituents on the rate of the Sandmeyer reaction is $p\text{-}NO_2 > p\text{-}Cl > H > p\text{-}MeO$.[96] On the basis of the intermediacy of a phenyl radical in these reactions, mechanism 8.28 may be suggested.

$$\text{Ph-}N^{\oplus}{\equiv}N + Cu(I) \longrightarrow \text{Ph-}N{=}N\cdot + Cu(II)$$

$$\text{Ph-}N{=}N\cdot \xrightarrow{-N_2} \text{Ph}\cdot$$

Cu(II) / Sandmeyer — Meerwein — $CH_2{=}CHCN$

$$\text{Ph}^{\oplus} \quad Cu(I) \text{ (or organocopper)} \qquad \text{PhCH}_2\dot{\text{C}}HC{\equiv}N \qquad (8.28)$$

↓ $Cl^{\ominus}$ ↓ Cu(II), $Cl^{\ominus}$

PhCl + Cu(I)   PhCH₂CHClC≡N + Cu(I)

Copper(II) salts catalyze the oxidation of aromatic compounds by oxygen, plus hydrosilation,[97] and copper(I) salts catalyze the reduction of aromatic halides by HX.[98] Organocopper intermediates may be involved in some of the reactions discussed above.[99] Copper(0) promotes the decomposition of benzenesulfonylazide.[100]

Cupric ion catalyzes the oxidation of o-aminophenol, benzenesulfonyl azide, acetophenone, and many other organic compounds.[101]

8.3.3 Decomposition of Peroxides

The decomposition of hydrogen peroxide is accelerated by many metal ions.[102–104] This reaction, which may be described by

$$\begin{aligned} Fe(II) + H_2O_2 &\longrightarrow Fe(III)(OH^{\ominus}) + HO\cdot \\ Fe(III) + H_2O_2 &\longrightarrow Fe(II) + H^{\oplus} + HO_2\cdot \\ Fe(III) + HO_2\cdot &\longrightarrow Fe(II) + H^{\oplus} + O_2 \end{aligned} \qquad (8.29)$$

is catalyzed not only by ferrous ion but also other transition metal ions with two stable valence states. The overall efficiency of the decomposition is greatly increased beyond that shown in eq. 8.29 because of chain decomposition by the HO and HO$_2$ radicals as shown in the following:

$$\text{HO} \cdot + \text{H}_2\text{O}_2 \longrightarrow \text{H}_2\text{O} + \text{HO}_2 \cdot$$
$$\text{HO}_2 \cdot + \text{H}_2\text{O}_2 \longrightarrow \text{O}_2 + \text{H}_2\text{O} + \text{HO} \cdot \tag{8.30}$$

Because of the formation of hydroxyl and hydroperoxyl radicals this system is capable of oxidizing organic compounds such as primary alcohols (Fenton reagent). A more complicated system, involving hydrogen peroxide (or molecular oxygen) in the presence of ferrous ions, ascorbic acid, and ethylenediaminetetraacetic acid, is capable of hydroxylating aromatic compounds.[105] Since this system bears some resemblance to enzymic hydroxylations of aromatic compounds by oxygen or hydrogen peroxide, it has been the subject of several investigations.[106-109]

Before discussing the latter reactions, let us consider the possible catalytic functions of metal ions in redox reactions. In reactions discussed so far, metal ions have been reversibly oxidized and reduced. Metal ions, however, can perform another catalytic function in redox reactions: a bridging, electron transfer agent between one substrate and another. For example, ferric ion catalyzes the oxidation of ascorbic acid by hydrogen peroxide.[110] This process cannot be easily understood in terms of successive oxidations and reductions of the metal ion because hydroxyl radicals formed in such a process should lead to aromatic hydroxylation, but do not. It can be understood in terms of the following:

(8.31)

This mechanism suggests that ferric ion forms a mixed complex with the enediol and hydrogen peroxide. This mixed complex can then undergo both an acid-base reaction, perhaps involving water molecules of the solution, and an electron transfer reaction in which the ferric ion serves as a bridging group between ascorbic acid and hydrogen peroxide so that ascorbic acid becomes oxidized and hydrogen peroxide reduced.[111] Mechanism 8.31 is an example of a metal ion-catalyzed oxidation-reduction reaction consisting of both proton and electron transfer. A simple hypothetical model is as follows:

$$(8.32)$$

In this intramolecular process, the catechol moiety is oxidized while the azo linkage is reduced, the overall process involving both proton transfer and electron transfer in a conjugated system. Equation 8.32 simply says that a metal ion can carry out the function of bringing two molecules together so that the process of proton transfer and electron flow between the oxidizing and reducing parts of the system can proceed.[111]

The mechanism of aromatic hydroxylation by aqueous hydrogen peroxide may be probed most easily when ferric ion and catechol are used as catalysts.[112-114] Ferric ion cannot be replaced by Cr(III), Co(III), Zn(II), Mn(II), Al(III), or Mg(II), but cupric ion may act as an inefficient catalyst for the reaction. 1,2-Dihydroxy or 1,4-dihydroxy aromatic compounds are catalysts for the reaction, whereas monohydroxy or 1,3-dihydroxy aromatic compounds are not. The rate of the reaction depends on the concentration of ferric ion and the concentration of hydrogen peroxide to the first power, but is inhibited by high concentrations of enediol catalyst. The kinetic results are not consistent with free-radical chain mechanisms but rather suggest that the hydroxylating agent is a complex of ferric ion, the enediol, and hydrogen peroxide similar to eq. 8.31. The isomer distribution of the phenols formed in the reaction and the relative reactivity of the aromatic substrates indicate that the oxidizing agent is very nonselective, and confirm that the hydroxyl radical is not the hydroxylating species. It is proposed that the actual oxidizing agent is complexed iron oxide, formed by the elimination of a molecule of water from the intermediate containing ferric ion, hydrogen peroxide and the enediol catalyst. The catalytic nature of the reaction is easily seen in the following:

$$(8.33)$$

The only reactants consumed in the cycle are hydrogen peroxide and the aromatic compound; the products are substituted phenols and water. This mechanism, which is related to eqs. 8.31 and 8.32, is consistent with all the experimental information listed above. The oxidation of saturated and unsaturated aliphatic hydrocarbons by ascorbic acid, ferric or ferrous ion, and oxygen can be described in a similar fashion.[109]

The catalytic decomposition of hydrogen peroxide to oxygen and water in the presence of manganous or ferric chelates of triethylenetetramine* is probably related to the above mechanism. In the manganous ion reactions, studied as possible models for the enzyme catalase,[115-117] the rate of reaction depends on the hydrogen peroxide concentration to greater than the first but less than the second power. This result indicates that an equilibrium between two molecules of hydrogen peroxide and the metal ion complex is involved. Thus, the mechanism of eq. 8.34 may be proposed.[111]

In this mechanism, the metal ion again serves as a bridging agent for the two substrate molecules facilitating proton transfer and electron flow that will lead to the product of reaction. In ferric chelate catalysis, the rate depends on less than the first power of the peroxide concentration, a result not consistent with eq. 8.34 unless one molecule of hydrogen peroxide is bound very

* Triethylenetetramine = TETA.

$$\text{(8.34)}$$

tightly. The structure of the ligand in these systems is crucial, the catalytic efficiency of the metal ions with ethylenediamine and diethylenetriamine being much less than that with triethylenetetramine. While iron(III) and manganese-(II) serve as efficient catalysts, Mg(II), Al(III), Ca(II), Cr(III), Co(III), Ni(II), Cu(II), Zn(II), Sr(II), Ag(I), Cd(II), Ba(II), Tl(I), and Pb(II) do not. Obviously neither size nor charge of the metal ion can be the cause of this specificity, since many of the above ions have size and charge similar to Fe(III) or Mn(II). It seems likely that the specificity is due to some special requirement of the detailed electronic structure of the metal ion. In the catalyst, the distribution of electrons around the metal nucleus must be such as to make the compounds of eq. 8.34 have appropriate stabilities. While highly unstable reaction intermediates (or activated complexes) mean high activation energies for the mechanism, extremely stable reaction intermediates will prevent the rapid regeneration of the original catalyst. The second-order rate constant for hydrogen peroxide decomposition for $(\text{TETA})\text{Fe}(\text{OH}_2)_2{}^{3\oplus}$ is 1.2×10^3 M^{-1} sec^{-1} and the value for $(\text{TETA})\text{FeF}_2{}^{\oplus}$ is about twice this value. The corresponding value for the enzyme catalase is about 1.6×10^6 M^{-1} sec^{-1} mole^{-1} of hematin iron, but that for hemoglobin is 10^{-1} M^{-1} sec^{-1}. Thus the (TETA)Fe complexes are remarkable catalysts.

Related to metal ion catalysis of the decomposition of hydrogen peroxide is the metal ion catalysis of the decomposition of alkyl hydroperoxides[118-120] and diacyl peroxides.[121] In many of these decompositions, reaction is not catalytic in the metal ion, but in some reactions it is. Cupric, cobaltous, and ferrous salts have been used in these reactions, often used to initiate polymerization. Some evidence for metal ion-hydroperoxide complexes has been found,[119] as well as evidence of organometallic and free-radical intermediates.[121]

8.3.4 Autoxidation

The autoxidation of organic compounds, that is, the reaction of oxidizable materials with molecular oxygen unaccompanied by the phenomena of flame and high temperature, is catalyzed by transition metal ions and inhibited by easily oxidized materials such as phenols, aromatic amines, and secondary alcohols. Metal ions of variable valence are effective in these reactions. Thus copper, cobalt, iron, and manganese salts are good catalysts, whereas aluminum, magnesium, zinc, and lead salts are inactive or very poor catalysts. The oxidation of hydrocarbons, probably of most practical interest, is carried out commonly with metal naphthenates, resinates, or stearates. The mechanism of the reaction for the cobalt case is given by[122]

$$\begin{aligned} &\text{Co(II)} + \text{ROOH} \longrightarrow \text{Co(III)} + \text{OH}^{\ominus} + \text{RO}\cdot \\ &\text{Co(III)} + \text{ROOH} \longrightarrow \text{Co(II)} + \text{H}^{\oplus} + \text{ROO}\cdot \\ &\left. \begin{array}{l} \text{ROO}\cdot + \text{RH} \longrightarrow \text{ROOH} + \text{R}\cdot \\ \text{R}\cdot + \text{O}_2 \longrightarrow \text{ROO}\cdot \end{array} \right\} \text{chain propagation} \quad (8.35) \\ &\text{RO}\cdot + \text{RH} \longrightarrow \text{ROH} + \text{R}\cdot \\ &2\text{ROO}\cdot \longrightarrow \text{inactive products} \end{aligned}$$

Cobalt(II), copper(II), nickel(II), and iron(II) salts are the preferred catalysts. Small amounts of organic peroxides or hydroperoxides are usually added to prevent induction periods that are sometimes encountered.[123,124] The start of the reaction coincides with the oxidation of Co(II) to Co(III). Starting the reaction with Co(III) or Cu(II) may eliminate the need for added peroxides.[124,125] The rate of the reaction expressed as $-dO_2/dt$ is generally independent of the oxygen pressure or concentration, and may be between first- and second-order in the hydrocarbon, and between zero- and first-order in the catalyst concentration. Under steady-state conditions, the rate should depend on the square root of the catalyst concentration.[126] In eq. 8.35, the first two initiating steps also regenerate the catalyst. The next two steps are the main chain-carrying steps leading to the major initial product, hydroperoxide. Further oxidation and decomposition of the hydroperoxide produce ketone and carboxylic acid.

8.4 CATALYSIS VIA ORGANOMETALLIC COMPOUNDS

Among the most significant developments in the field of homogeneous catalysis in recent years have been the discovery and elucidation of a variety of new, and often unusual, catalytic reactions of transition metal ions and coordination compounds. Examples of such reactions include the hydrogenation of olefins catalyzed by complexes of ruthenium, rhodium, cobalt,

Catalysis by Organometallic Compounds

platinum, and other metals; the hydroformylation of olefins catalyzed by complexes of cobalt, nickel, palladium, and rhodium (Oxo process); the dimerization of ethylene and polymerization of dienes catalyzed by complexes of rhodium, cobalt, chromium, and nickel; double bond migration in olefins catalyzed by complexes of rhodium, palladium, cobalt, platinum, and other metals; the oxidation of olefins to aldehydes, ketones, and alkenyl esters, catalyzed by palladium chloride (Wacker process); and the hydration of acetylenes catalyzed by ruthenium chloride. The field is a relatively new, but rapidly developing one for several reasons: its potential practical applications and the possible relevance of some of the catalytic reactions as model systems for related heterogeneous and enzymic processes.

8.4.1 Transition Metal Complexes

The majority of catalytic reactions discussed here involve as catalysts coordination compounds of the metals near the end of each transition series, comprising iron, ruthenium, osmium, cobalt, rhodium, iridium, nickel, palladium, and platinum, notably the platinum groups. The electron configurations of the metals in these groups are usually in the range d^6 to d^{10}, with the configuration d^8 especially widely represented. The catalytic complexes in these groups are usually, although not invariably, the spin-paired or low-spin type, that is, complexes in which the ligand field splittings are sufficiently large so that the d electrons first fill up the most stable orbitals available to them, with pairing if necessary, before occupying those of higher energy. First let us consider some aspects of the electronic structures and chemical reactivities of this general class of complexes, because of their relevance to an understanding of the catalytic properties in which we are interested.

The stable coordination numbers of low-spin complexes of transition metals[127] range from eight to two, and exhibit a systematic inverse dependence on the number of d electrons of the metal atom. This inverse dependence has its origin in the fact that, in general, the higher the coordination number, the fewer the d electrons that can be accommodated in stable (bonding or nearly nonbonding) orbitals of the complex. In the case of an octahedral complex, for example, the three stable t_{2g} orbitals (non-σ-bonding or possibly slightly π-bonding in the case of π-acceptor ligands such as CO or $CN^\ominus$) can accommodate up to six d electrons. Any additional electrons are forced to occupy the e_g^* orbitals, which are strongly anti-σ-bonding in the complexes we are discussing. This generally results in the destabilization of the coordination number 6, in favor of a lower coordination number that permits a larger number of d electrons to be accommodated in stable orbitals (Fig. 8.5).

The loss of a $CN^\ominus$ ligand when an electron is added to the very stable d^6

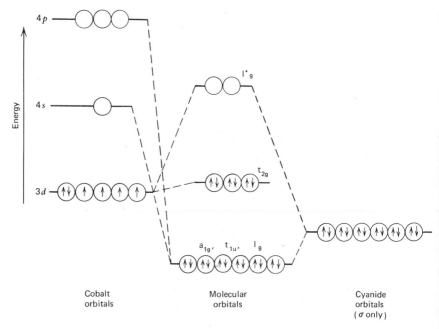

Fig. 8.5 Simplified molecular orbital diagram for $Co(CN)_6^{3\ominus}$.

complex, $Co(CN)_6^{3\ominus}$, to give the pentacoordinated low-spin cobalt(II) complex, $Co(CN)_5^{3\ominus}$, can be understood in these terms (eq. 8.36).

$$Co(CN)_6^{3\ominus} \xrightarrow{+e^{\ominus}} [Co(CN)_6^{4\ominus}] \longrightarrow Co(CN)_5^{3\ominus} + CN^{\ominus}$$
$$d^6 \quad\quad\quad d^7 \quad\quad\quad d^7$$
$$\text{Very stable} \quad \text{Unstable} \quad \text{Stable} \tag{8.36}$$

This process is analogous to that which accompanies the addition of an electron to a saturated carbon compound CX_4, for example, CCl_4, represented by the simplified molecular orbital diagram, Fig. 8.6. In CX_4, like $Co(CN)_6^{3\ominus}$ because all the stable bonding orbitals are filled, an extra electron is forced into a strongly antibonding orbital. The result is that the coordination number 4 is destabilized, and a species of lower coordination number (a free radical or, in some cases, a carbanion, depending on the relative electron affinities of X and CX_3) is generated:

$$CX_4 \xrightarrow{+e^{\ominus}} [CX_4^{\ominus}] \longrightarrow CX_3{\cdot} + X^{\ominus} \tag{8.37}$$
$$\text{Very stable} \quad \text{Unstable} \quad \text{Stable}$$

The purpose of developing this analogy is to provide a basis for the expectation—and, indeed, this turns out to be the case—that the chemical reactivities of $Co(CN)_5^{3\ominus}$ and related pentacoordinated d^7 complexes will resemble those of typical organic free radicals. This free-radical-like

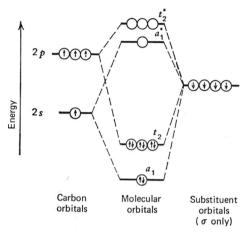

Fig. 8.6 Simplified molecular orbital diagram for CX_4.

behavior is shown, for example, by the general reaction (eq. 8.38) which $Co(CN)_5^{3\ominus}$ undergoes with organic halides to form stable organocobalt complexes.[128]

$$Co(CN)_5^{3\ominus} + RX \longrightarrow Co(CN)_5X^{3\ominus} + R\cdot$$
$$\underline{Co(CN)_5^{3\ominus} + R\cdot \longrightarrow Co(CN)_5R^{3\ominus}} \quad (8.38)$$
$$2Co(CN)_5^{3\ominus} + RX \longrightarrow Co(CN)_5X^{3\ominus} + Co(CN)_5R^{3\ominus}$$

The basis for this marked parallelism between $Co(CN)_5^{3\ominus}$ and an organic free radical is that in each case the characteristic reactivity pattern is dominated by the tendency to go from an open-shell configuration to the more stable closed-shell configuration of the species (CX_4 in the case of the organic compound and the octahedral d^6 configuration in the case of the coordination compound). The types of reactions which achieve this closed-shell configuration are the same in both cases. One such reaction which is of particular importance for the catalytic applications discussed later is that involving the reductive homolytic cleavage of single bonds:

$$2Co(II)(CN)_5^{3\ominus} + Y-Z \longrightarrow Co(III)(CN)_5Y^{3\ominus} + Co(III)(CN)_5Z^{3\ominus}$$
$$Y-Z = H-H, HO-H, HO-OH, CH_3I, \text{ and so on} \quad (8.39)$$

The analogy between pentacoordinated d^7 complexes and organic free radicals is capable of extension to coordination compounds having other electron configurations and coordination numbers. Thus, analogous reasoning can lead us to expect similarities between the reactivity patterns of tetracoordinated d^8 complexes and carbenes, pentacoordinated d^8 complexes and carbanions, and pentacoordinated d^6 complexes and carbonium ions. In each case the stoichiometries of the reactions that restore the stable closed-shell

Table 8.4

Species of Related Configurations and Reactivities in Organic and Coordination Chemistry[a,b]

	Organic species			Transition Metal Counterpart		
Species	Coordination Number	Nonbonding Electrons	Characteristic Reactions	Coordination Number	Nonbonding Electrons	Examples	Reactions
Saturated molecule (R_3C-X)	4	0	Substitution	6	6	$RhCl_3^{\ominus}$	$RhCl_3^{\ominus} + H_2O \rightleftharpoons RhCl_5OH_2^{2\ominus} + Cl^{\ominus}$ $RhCl_6^{3\ominus} + H_2 \rightleftharpoons RhHCl_5^{3\ominus} + HCl$
Free radical ($R_3C\cdot$)	3	1	Dimerization Abstraction	5	7	$Co(CN)_5^{3\ominus}$	$2Co(CN)_5^{3\ominus} \rightleftharpoons Co_2(CN)_{10}^{6\ominus}$ $Co(CN)_5^{3\ominus} + CH_3I \rightarrow Co(CN)_5I^{3\ominus} + CH_3\cdot$
Carbene ($R_2C:$)	2	2	Addition Insertion	4	8	$IrI(CO)(PPh_3)_2$	$2Co(CN)_5^{3\ominus} + CH \equiv CH \rightarrow (NC)_5CoCH=CHCo(CN)_5^{3\ominus}$ $IrI(CO)(PPh_3)_2 + C_2H_4 \rightleftharpoons IrI(CO)(C_2H_4)(PPh_3)_2$ $IrI(CO)(PPh_3)_2 + H_2 \rightleftharpoons IrH_2I(CO)(PPh_3)_2$
Carbonium ion ($R_3C^{\oplus}$)	3	0	Addition of nucleophile	5	6	*$Co(CN)_5^{2\ominus}$	$Co(CN)_5^{2\ominus} + I^{\ominus} \rightleftharpoons Co(CN)_5I^{3\ominus}$
Carbanion ($R_3C:^{\ominus}$)	3	2	Addition of electrophile	5	8	$Mn(CO)_5^{\ominus}$	$Mn(CO)_5^{\ominus} + H^{\oplus} \rightleftharpoons Mn(CO)_5H$

[a] The asterisk (*) indicates the intermediate in S_N1 substitution reactions of $Co(CN)_5OH_2^{2\ominus}$. Note that the change in coordination number and in the number of nonbonding electrons in going from one species to the next is the same in each series. This results in correspondingly similar changes in the reactivity patterns of the two series of compounds, since the reactivity pattern in each case is dominated by the tendency to return to the stable closed-shell configuration of the first member of the series.

[b] From J. Halpern, Chem. Eng. News, 68, Oct. 31, 1966. © 1966 by the American Chemical Society. Reprinted by permission of the copyright owner.

configurations are the same for both species. Hence, the reactivity patterns are similar, as shown in Table 8.4.[128]

Many of the novel reactions of d^7 and d^8 complexes that contribute to the widespread roles of these complexes in catalysis, and which reflect reactivity patterns that are relatively unfamiliar in inorganic chemistry, are thus seen to be closely related to reactivity patterns that have, in fact, long been familiar in organic chemistry. In the case of d^8 complexes in particular, the following general classes of reactions should be recognized:

$$L_4M + X \rightleftarrows L_4MX$$
$$d^8$$
$$X = CO \text{ or olefin}$$

$$L_4M + YZ \rightleftarrows L_4M\begin{matrix}Y\\ \diagdown\\ Z\end{matrix}$$
$$d^8 \qquad\qquad d^6 \qquad\qquad (8.40)$$
$$YZ = H_2, CH_3-I, Br_2, \text{ and so on}$$

$$L_4MX + YZ \rightleftarrows L_4M\begin{matrix}Y\\ \diagdown\\ Z\end{matrix} + X$$
$$d^8 \qquad\qquad d^6$$

The first two of these reactions (i.e., addition to an unsaturated molecule, and insertion into a single bond) are also characteristic of carbenes. The last reaction is essentially a combination of the first two.

The insertion reaction is a reaction of considerable generality with metal ion complexes.[129] The insertion reaction, in this sense, is the addition of the organometallic, M—X, to an unsaturated molecule, :Y, to form a new complex with Y inserted between M and X.

$$M-X + :Y \longrightarrow M-Y-X \qquad (8.41)$$

The molecule :Y may be CO, an alkene, a diene, acetylene, RCHO, RCN, SO_2, O_2, or other unsaturated system. The ligand X may be $H^\ominus$, $R^\ominus$ or $OR^\ominus$, $NR_2^\ominus$ or NR_3, $OH^\ominus$, or H_2O, halide ion, or another metal atom. The name insertion reaction, while describing accurately the overall result, may nevertheless be mechanistically misleading. The reactions described here are not free radical reactions, but rather reactions in which the group :Y must first coordinate to the metal M before reaction can occur. The group X migrates onto the group Y after coordination, in what might better be called a ligand migration reaction.

$$\underset{L-L}{\overset{L-X}{\diagup M\diagup}}+Y\longrightarrow\underset{L-L}{\overset{L+X}{\diagup M\diagup}}\xrightarrow[\text{(solvent)}]{S}\underset{L-L}{\overset{X}{\underset{Y}{\diagup M\diagup}}}+S \qquad (8.42)$$

8.4.2 Hydrogenation and Related Reactions

Many transition metal ions and complexes have the ability to catalyze reactions of molecular hydrogen in homogeneous solution.[140] The complexes include those of Cu^{II}, Cu^{I}, Ag^{I}, Hg^{II}, Hg^{I}, Co^{I}, Co^{II}, Pd^{II}, Pt^{II}, Rh^{I}, Rh^{III}, Ru^{II}, Ru^{III}, and Ir^{I}. In each case H_2 is split by the catalyst with the formation of a reactive transition metal hydride complex (which may or may not be detected) as an intermediate. A partial list of these intermediates is given in Table 8.5. The evidence for the formation of the intermediate metal hydride complexes consists in part of the observation of the stable analogs given in the table. In addition, kinetic evidence to be discussed shortly points to these postulated intermediates. The hydrogen ligand in the metal hydride complexes has anionic character in general and can act as a reducing agent. The stable analogs of the metal hydride complexes are characterized both by thermodynamic and by kinetic stability, brought about in part by the introduction of π-bonding ligands such as PR_3, $CN^{\ominus}$, CO, and olefins. The catalytic activity of the metal hydride complexes depends in general on the balance between the ease of formation of these materials and their further reactivity. This description implies a compromise between stability and lability of the hydrido complexes for optimal catalytic action.[32]

The first homogeneous catalysis of hydrogenation to be found was the reduction by dissolved hydrogen of Cu(II) salts and benzoquinone, in quinoline solution, brought about by Cu(I) salts of organic acids.[132] The reaction is homogeneous, is independent of the concentration of the oxidizing agent, and follows the rate law, rate $= k[CuA]^2[H_2]$, where A is the anion of the organic acid.[133,124] Molecular hydrogen was suggested to be split by a dimer of Cu(I), or in a termolecular process, to a hydride-like species which is an active reducing agent. The reversible formation of such an intermediate is consistent with the ability of the system to catalyze the ortho-para hydrogen conversion,[35] and the exchange of deuterium with a hydrogen donor in solution.[136]

Three distinct mechanisms have been recognized by which metal ion complexes can form hydride complexes:[136] (1) heterolytic splitting; (2) homolytic splitting; and (3) insertion, that is, the addition of H_2 to form a dihydride. The characteristic reactivities of hexacoordinated, pentacoordinated, and

Table 8.5
The Reaction Products of Molecular Hydrogen and Various Metal Ion Complexes[130]

Electron Configuration	Metal Ion	Catalytic Complex	Postulated Hydride Intermediate	Stable Analogs
d^5	Ru^{III}	$RuCl_6^{3\ominus}$	$HRuCl_5^{3\ominus}$	
d^6	Ru^{II}	$RuCl_n^{2\ominus n}$	$HRuCl_{n-1}^{2\ominus n}$	$HRuCl(Et_2PC_2H_4PEt_2)$
				$HRuCl(CO)(PPh_3)_3$
	Rh^{III}	$RhCl_6^{3\ominus}$	$HRhCl_5^{3\ominus}$	$HRhCl(trien)^{3\oplus}$
d^7	Co^{II}	$Co(CN)_5^{3\ominus}$ [a]	$HCo(CN)_5^{3\ominus}$ [a]	
d^8	Pd^{II}	$PdCl_4^{3\ominus}$	$HPdCl_3^{2\ominus}$	$HPdCl(PEt_3)_2$
	Pt^{II}	$PtCl_2$—$SnCl_2$	?	$HPtCl(PEt_3)_2$
	Ir^I	$IrCl(CO)(PPh_3)_2$	$H_2IrCl(CO)(PPh_3)_2$ [a]	
d^9	Cu^{II}	$Cu_{aq}^{2\oplus}$	$CuH_{aq}^{\oplus}$	
d^{10}	Cu^I	$CuOAc$, $CuOHp$ [b]	$CuH^{\oplus}$, CuH	
	Ag^I	$Ag_{aq}^{\oplus}$	$AgH^{\oplus}$, AgH_{aq}	
	Hg^{II}	$Hg_{aq}^{2\oplus}$	?	
	Rh^I	$RhH(CO)(PPh_3)_3$ [c]	$RhH_3(CO)(PPh_3)_3$	
		$RhCl(PPh_3)_3$		

[a] Stable species.
[b] Hp = heptanoate.
[c] Ref. 131.

tetracoordinated complexes illustrated by these reactions follow readily from the general principles already discussed.

Heterolytic splitting, which occurs widely, involves essentially a substitutional process (replacement of a chloride ligand by a hydride derived from H_2) without change in the formal oxidation number of the metal. Reactivity is thus governed by the substitution lability of the complex, by the stability of the hydride formed, and by the presence of a suitable base (which may be the solvent or the displaced ligand) to stabilize the released proton.

In this heterolytic fashion, Cu(II) and Hg(II) catalyze the reduction by hydrogen of a number of oxidizing agents such as Cr(VI).[138] The kinetic order of Cu(II) in the activation of H_2 by $Cu(H_2O)_6{}^{2\oplus}$ in CCl_4 solution changes from one to two as the acidity increases.[139] At the same time a decrease in rate occurs. The following mechanism can account for these facts:

$$Cu^{2\oplus} + H_2 \underset{k_2}{\overset{k_1}{\rightleftarrows}} CuH^{\oplus} + H^{\oplus}$$

$$CuH^{\oplus} + Cu^{2\oplus} \xrightarrow{k_3} 2Cu^{\oplus} + H^{\oplus} \quad (8.43)$$

$$2Cu^{\oplus} + \text{substrate} \xrightarrow{\text{fast}} \text{products} + 2Cu^{2\oplus}$$

Application of the steady-state method, assuming $CuH^{\oplus}$ is a reactive intermediate, gives the rate equation:

$$\text{rate} = \frac{k_1 k_3 [H_2][Cu^{2\oplus}]^2}{k_3 [Cu^{2\oplus}] + k_2 [H^{\oplus}]} \quad (8.44)$$

This mechanism can explain a number of phenomena such as the effect of solvent and added anions on the rates and kinetic order. Anions such as carboxylate ions produced high rates of reduction of substrates such as $Cr_2O_7{}^{2\ominus}$, $IO_3{}^{\ominus}$, and $Ce^{4\oplus}$, presumably because of decrease of k_2. Very similar results are seen in the activation of H_2 by $Ag^{\oplus}$.[140] In this reaction, the rate is accelerated by a factor of 10^4 if fluoride ion is incorporated in the reactions,[141] implying a four-center transition state.

Heterolytic fission of H_2 is also seen in the reaction with ruthenium(III) chloride. The kinetics of the reaction, the effect of ligands on reactivity, the effect of the medium on reactivity, and the observation of isotopic exchange between deuterium and water point to this mechanism. For example, the ruthenium(III) chloride-catalyzed reduction of ferric ion by H_2 may be represented by[142]

$$RuCl_6{}^{3\ominus} + H_2 \longrightarrow HRuCl_5{}^{3\ominus} + H^{\oplus} + Cl^{\ominus} \quad \text{(slow)}$$

$$HRuCl_5{}^{3\ominus} + 2Fe^{3\oplus} \xrightarrow{Cl^{\ominus}} RuCl_6{}^{3\ominus} + 2Fe^{2\oplus} + H^{\oplus} \quad \text{(fast)}$$

(8.45)

In the absence of ferric ion, there is no uptake of hydrogen. The first step must therefore be an equilibrium, and thus isotopic exchange between deuterium and water should be catalyzed by ruthenium chloride, as found experimentally. In the isotopic exchange, the HD/H_2 ratio increases linearly with the hydrochloric acid concentration, in accord with eq. 8.46,

$$D_2 + RuCl_6^{3\ominus} \longrightarrow D^{\oplus} + Cl^{\ominus} + DRuCl_5^{3\ominus} \begin{array}{c} \xrightarrow{HCl} HD + RuCl_6^{3\ominus} \\ \xrightarrow{H_2O} HRuCl_5^{3\ominus} \longrightarrow H_2 \end{array}$$
(8.46)

which predicts that the initial product of the exchange is HD rather than D_2, and that the HD concentration will be dependent on the concentration of hydrochloric acid. This mechanism parallels the early suggestion that the enzyme hydrogenase splits hydrogen heterolytically.[143] This suggestion was based on a similar observation, namely, that the initial product of the enzyme-catalyzed H_2–D_2O exchange is predominantly HD rather than D_2. Since the rate of the heterolytic splitting is dependent on the basicity of the group leaving the metal ion as well as the bond strength between the metal ion and the leaving group, the reaction may again have a four-center transition state in which the hydrogens are bonded both to the metal ion and the leaving group.

In homolytic splitting and insertion, hydride formation is accompanied by formal oxidation of the metal and reactivity is closely linked to the susceptibility of the metal ion to oxidation. Thus, the high reactivity of $Co(CN)_5^{3\ominus}$ toward H_2, compared with that of $Co(NCCH_3)_5^{2\oplus}$, reflects in part the tendency of $CN^{\ominus}$ to stabilize preferentially the higher oxidation state, and CH_3CN the lower oxidation state, of cobalt. In the case of square planar d^8 complexes, the expected order of the reactivity toward H_2 is $Os^0 > Ru^0 > Fe^0 > Ir^I > Rh^I > Co^I$, $Pt^{II} > Pd^{II} \gg Ni^{II}$, Au^{III}. Homolytic splitting of hydrogen is seen in the reactions of silver ion and cobalt(II) complexes, as shown in

$$2Co^{II}(CN)_5^{3\ominus} + H_2 \longrightarrow 2HCo^{III}(CN)_5^{3\ominus} \qquad (8.47)$$

The stoichiometry of this process, together with the structural analysis given above dictates the description of this process as a homolytic splitting.

Addition of hydrogen to form a dihydride is seen in the reactions of iridium (I) compounds.

$$Ir^ICl(CO)(PPh_3)_2 + H_2 \longrightarrow Ir^{III}H_2Cl(PPh_3)_2 \qquad (8.48)$$

An increase in coordination number from a square planar d^8 to an octahedral d^6 coordination occurs. The stereochemistry of this process can be specified as[144]:

$$\begin{array}{c} Ph_3P\text{———}Cl \\ \diagup \ Ir \ \diagup \\ CO\text{———}PPh_3 \end{array} + H_2 \longrightarrow \begin{array}{c} \quad\quad H \\ Ph_3P\text{—|—}H \\ \diagup \ Ir \ \diagup \\ CO\text{—|—}PPh_3 \\ \quad\quad Cl \end{array} \qquad (8.49)$$

252 Metal Ion Catalysis

Heterolytic, homolytic, and additive splitting of hydrogen by metal ion complexes are important examples of the analogy between reactivity of metal ion complexes and organic reactions of alkyl halides, free radicals, and carbenes seen in Table 8.4.

Homogeneous hydrogenation of organic substrates such as olefins occurs by transfer of hydrogen from the hydrido-transition metal complex to the substrate. As seen in Fig. 8.7[145] these reactions can compete favorably with more classical methods of hydrogenation using heterogeneous catalysts. The following examples, all discovered within the past few years, illustrate how hydrogenation of olefins is realized for each of the three mechanisms of splitting of hydrogen.

Ruthenium(II) complexes catalyze the hydrogenation of olefins such as fumaric and maleic acids in aqueous hydrochloric acid.[146] A Ru(II)-olefin complex is formed first, as determined spectrophotometrically, and is then hydrogenated in a slow step according to the rate law: rate = k[Ru(II)-olefin][H_2]. Reduction of fumaric acid with D_2 in H_2O yields undeuterated succinic acid. The hydrogen which enters the double bond thus comes from the solvent. Reduction with H_2 or D_2 in D_2O yields chiefly ($\pm$)-2,3-dideuterosuccinic acid, indicating that addition of hydrogen is *cis*. A mechanism that explains these observations is

$$\text{(8.50)}$$

The key step in this mechanism, as in all hydrogenations with transition metal ions, is the insertion of a π-bonded alkene, alternatively described as ligand migration of hydride ion. This mechanism has many features in common with that proposed for heterogeneous hydrogenation on chromia gel catalyst.[147]

The hydrogenation of olefins by pentacyanocobaltate ion occurs with the intermediate formation of the hydrido complex described in eq. 8.47. Conjugated olefins as well as α, β-unsaturated acids and aldehydes, 1,2-diketones, nitrobenzene, and benzaldehyde are reduced by this catalyst. The penta-

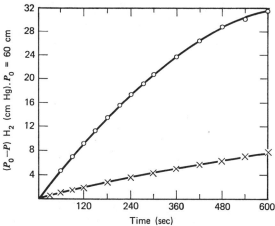

Fig. 8.7 Comparison of the rates of homogeneous and heterogeneous hydrogenation. From G. Wilkinson, *Proc. R. A. Welch Found. Conf. Chem. Res.*, **9**, 139 (1966).

cyanocobaltate ion also catalyzes the exchange between deuterium and water, hydrogen and hydrogen peroxide, and the reaction between hydrogen and ferricyanide ion. The hydrogenation of butadiene proceeds by the following mechanism[148]:

$$2Co(CN)_5^{3\ominus} + H_2 \longrightarrow 2HCo(CN)_5^{3\ominus}$$
$$HCo(CN)_5^{3\ominus} + C_4H_6 \longrightarrow C_4H_7Co(CN)_5^{3\ominus} \quad (8.51)$$
$$C_4H_7Co(CN)_5^{3\ominus} + HCo(CN)_5^{3\ominus} \longrightarrow C_4H_8 + 2Co(CN)_5^{3\ominus}$$

Although eq. 8.51 explains the overall chemistry in a straightforward fashion, it does not explain the dependence of the butene product distribution on the $CN^\ominus$/cobalt ratio. At low values of this ratio, the product is predominantly *trans*-2-butene, whereas at high values, the product is predominantly 1-butene. If one postulates that there can be more than one butenyl metal complex, these results may be explained. Equation 8.52 indicates an equilibrium between a sigma-bonded butenyl complex and a π-allyl complex formed by rearrangement of the butenyl complex with loss of cyanide ion.

$$CH_3\text{—}CH\text{=}CH\text{—}CH_2\text{—}Co(CN)_5^{3\ominus} \rightleftharpoons \left[\begin{array}{c} CH_2 \\ H\text{—}C \quad Co(CN)_4 \\ C \\ H \quad CH_3 \end{array}\right]^{3\ominus} + CN^\ominus$$

$$\downarrow [H] \qquad\qquad\qquad\qquad \downarrow [H] \qquad\qquad (8.52)$$

$$\text{1-butene} \qquad\qquad\qquad\qquad trans\text{-2-butene}$$

If the sigma complex forms 1-butene while the π-allyl complex forms 2-butene, the dependence of the product distribution on the $CN^\ominus$/cobalt ratio may be rationalized. The proposed sigma bonded intermediate $(C_4H_7)Co(CN)_5^{3\ominus}$ has been prepared by the reaction of crotyl bromide with $Co(CN)_5^{3\ominus}$. Studies using nmr prove its structure, and also demonstrate its rearrangement as in eq. 8.52 to the π-allyl complex with loss of $CN^\ominus$.[149]

The homogeneous hydrogenations of olefins by $IrCl(CO)[P(C_6H_5)_3]_2$[150, 152] and $RhCl[P(C_6H_5)_3]_3$ probably involve dihydride intermediates of the kind shown in eq. 8.48. A plausible mechanism for the hydrogenation involving steps of the type already described, is shown in the following scheme.

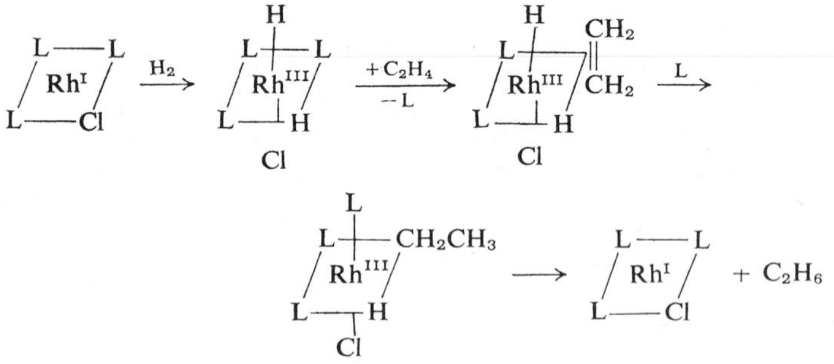

$$\begin{array}{c} \text{(scheme 8.53)} \end{array} \qquad (8.53)$$

L = PPh$_3$ or solvent

The trigonal bipyramidal d^8 complex, $IrH(CO)(P(C_6H_5)_3)_3$, is an even more efficient catalyst for the hydrogenation of ethylene.[153] Both H_2 and C_2H_4 reversibly coordinate with the iridium complex. It appears that a seven-coordinated species is formed, since two atoms of hydrogen are added with no loss of the original ligands.

Other catalysts for hydrogenation in homogeneous solution include the monohydride of an osmium complex ion,[153] a number of soluble Ziegler-type catalysts consisting of an aluminum trialkyl and some complex of Co(II) or other metal ion,[154] and lithium aluminum hydride.[155]

Similar to the catalysis of olefin hydrogenation is catalysis of the addition of silanes[156-158] and germanes to olefins by rhodium(I) and platinum(II) complexes. Olefin π-complexes have been suggested to be intermediates in these reactions,

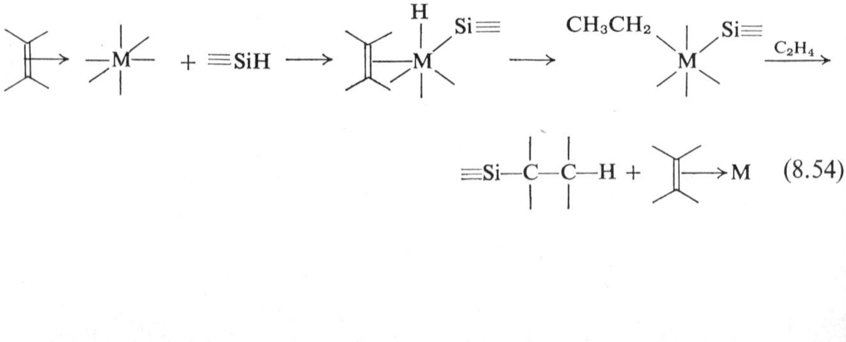

$$\qquad (8.54)$$

The hydration of acetylene catalyzed by ruthenium(III) is also related to these catalyses.[159] The rate of this reaction is proportional to both the acetylene and ruthenium(III) concentrations. Dependence on the hydrochloric acid concentration indicates that the reaction proceeds only with a ruthenium species containing at least one water molecule. That is, at high chloride ion concentration, no reaction occurs, since water is excluded from the metal ion complex. On this basis, the hydration may be written as

$$RuOH_2 \xrightarrow[-H^\oplus]{RC\equiv CR'} \underset{RR'}{\overset{RuOH}{C=C}} \xrightarrow{H^\oplus} RCH_2\overset{O}{\overset{\|}{C}}R' + Ru(?) \quad (8.55)$$

The hydration of acetylene by mercuric salts, although not fully resolved, may be related to eq. 8.55.[160]

Because of historical circumstances, heterogeneous hydrogenation catalysts were well known before any homogeneous ones were recognized. For this reason, the view has often been taken that catalytic activity depends critically on some property characteristic of the solid state (e.g., crystal lattice geometry or electronic band structure) which cannot be realized in species that exist in solution. The numerous examples cited here of homogeneous activation of hydrogen by metal ions and complexes in solution clearly force a modification of this view. The importance of chemical structure rather than electronic band properties in catalytic hydrogenation is also seen in heterogeneous polymeric models of hydrogenation catalysts. A polymer containing quinoid rings is catalytic but one having only conjugated double bonds is not. A correlation is found between the concentration of free spins as measured by esr and activity toward dehydrogenation, but no correlation is found with the semiconducting properties of the materials.[161]

Perhaps the most important general requirement for catalytic activity that has emerged from the great variety of systems examined is the availability of low-lying filled orbitals, preferably with a high degree of d-character, which can donate electrons to hydrogen and form weak bonds with it. This property can be realized in a variety of metal ions and complexes, as well as in solids, and in each case is most pronounced just prior to the filling of a d-shell or d-band.

The higher activity that metallic catalysts sometimes display, relative to metal ions in solution, may be associated in part with stronger metal–hydrogen bonding. Thus, chemisorption of hydrogen on catalytic metals is usually exothermic, whereas splitting of hydrogen by metal ions in solution appears, in every case, to be endothermic. Connected with this is the observation that while the rate-determining process in heterogeneously catalyzed reactions is frequently desorption, it always appears to be the hydrogen splitting step in homogeneous systems.[162]

Complexes of many transition metals, including cobalt, iridium, iron, rhodium, palladium, and platinum catalyze double bond migration of terminal olefins to internal olefins.[163] Two families of plausible mechanisms have been suggested for these catalyses: one invokes an intermediate in which the olefinic residue has assumed π-allylic character; the other, an intermediate in which the olefin has assumed the character of an alkyl group.

$$\underset{M}{RCH_2CH\!=\!\!=\!CH_2} \rightleftarrows \underset{MH}{RCH\diamond\!\!\diamond CH_2} \rightleftarrows \underset{M}{RCH\!=\!\!=\!CH-CH_3} \quad (8.56)$$

$$\underset{MH}{RCH_2CH\!=\!\!=\!CH_2} \longrightarrow \underset{M}{RCH_2-CH-CH_3} \longrightarrow \underset{M}{RCH\!=\!\!=\!CH-CH_3} \quad (8.57)$$

Several isotopic tracer experiments require eq. 8.57. The isomerization of $C_5H_{11}CD_2CH=CH_2$ by a palladium catalyst yields little if any $C_5H_{11}CD=CHCH_2D$, the predicted product of the π-allyl mechanism.[164] The isomerization of 1-butene in CH_3OD using rhodium(I) compounds leads to displacement of approximately one olefin proton by deuterium per molecule of 1-butene isomerized. The product also contains substantial amounts of deuterated 1-butene and nondeuterated 2-butenes.[165,166]

Many transition metal compounds become catalysts for isomerization only when they are converted to hydrides. Thus no isomerization of 1-butene is observed at 25° with any of the following: $Ni[P(OC_2H_5)_3]_4$, $[(C_2H_4)_2RhICl]_2$, H_2PtCl_6–$SnCl_2$, Li_2PdCl_4, or (at 0°) $RhCl_3$. However, a variety of hydride-generating reactions is available.[167] Cocatalysts may produce hydride by oxidation of a carbon compound (eq. 8.58), disproportionation or oxidation with hydrogen (eq. 8.59), oxidation of a metal with hydrogen chloride (eq. 8.60) or proton (eq. 8.61), or displacement of hydrogen from coordinated olefin (eq. 8.62). Although any of these reactions furnishes initiating hydride ion, the exchange of olefin protium and solvent deuterium,

$$CH_3CH_2OH + OH^\ominus + ClPt^{II} \longrightarrow CH_3CHO + HPt^{II} + Cl^\ominus + H_2O \quad (8.58)$$

$$H_2 + ClRh^{III} \longrightarrow HRh^{III} + Cl^\ominus + H^\oplus$$
$$H_2 + Rh^I \longrightarrow H_2Rh^{III} \quad (8.59)$$

$$HCl + Rh^I \longrightarrow HRh^{III}Cl \quad (8.60)$$

$$\begin{array}{c}\diagdown\!CH\\ \|\\ C\diagup\diagdown\end{array}\!\!-\!Pd^{II}\!-\!Cl \longrightarrow \begin{array}{c}\diagdown\!CCl\\ \|\\ C\diagup\diagdown\end{array}\!\!-\!Pd\!-\!H \quad (8.61)$$

which accompanies isomerization in CH_3OD or CH_3COOD, indicates that hydride ion is regenerated from solvent proton; accordingly the working cocatalyst is a proton.

The isomerization reaction involving the complexes listed above thus proceeds by mechanism 8.57 after formation of a hydrido complex. Accompanying the isomerization is a rapid, reversible olefin exchange, as determined by deuterium tracer experiments.[167] A mechanism applying to a number of rhodium, palladium, platinum, nickel, and iron catalysts is illustrated in eq. 8.62.

$$\begin{array}{c} L{-}L \\ /Rh^I/ \\ L{-}L \end{array} \xrightleftharpoons[L]{RCH_2CH=CH_2} \begin{array}{c} L{-}L \\ /Rh^I/\overset{CH_2R}{\underset{CH_2}{\overset{|}{CH}}} \\ L{-}L \end{array} \xrightleftharpoons{H^\oplus + L} \begin{array}{c} H \\ L{+}\overset{CH_2R}{\underset{CH_2}{\overset{|}{CH}}} \\ /Rh^{III}/ \\ L{-}L \\ L \end{array}$$

where $L = Cl^\ominus$ or solvent.

$+L \updownarrow -L$

$$\begin{array}{c} L \\ L{+}\overset{CH_2R}{\underset{}{CH}} \\ /Rh^{III}/{\diagdown}CH_3 \\ L{-}L \\ L \end{array}$$

$-L \updownarrow +L$

$$RCH{=}CHCH_3 + \begin{array}{c} L{-}L \\ /Rh^I/ \\ L{-}L \end{array} \xrightleftharpoons{L} \begin{array}{c} L{-}L \\ /Rh^I/\overset{CHR}{\underset{CH_3}{\overset{\|}{CH}}} \\ L{-}L \end{array} \xleftarrow{-(H^+ + L)} \begin{array}{c} H \\ L{+}\overset{CHR}{\underset{CH_3}{\overset{\|}{CH}}} \\ /Rh^{III}/ \\ L \end{array}$$

(8.62)

where $L = Cl^\ominus$ or solvent.

Some isotopic tracer results give a different picture of the isomerization process. For example, in isomerization with bis(benzonitrile)-dichloropalladium, the reaction occurs mainly by a 1,3 shift of deuterium or hydrogen. Although intermolecular deuterium exchange occurs, the extent is only about half that observed with the above catalysts, and very little of the exchanging deuterium becomes attached to doubly bonded carbon atoms.[166] In the isomerization of allylbenzene to propenylbenzene with cobalt hydrotetracarbonyl, no deuterium appears in the product when $DCo(CO)_4$ is used and, furthermore, rates of isomerization using $HCo(CO)_4$ and $DCo(CO)_4$ are identical.[168] These observations may be explained most readily by an internal 1,3-hydrogen shift, possibly through a π-allyl intermediate if the hydrogen atom that is transferred retains its integrity, distinct from the hydrogen originally on the cobalt.

8.4.3 Carbon–Carbon Bond Formation

The conversion of acetylene to oligomeric products was extensively investigated by Reppe.[169] Acetylene in nonaqueous solvents under the influence of nickel complexes such as $Ni(CN)_2$ or $Ni(acac)_2$, for example, cyclizes to cyclooctatetraene.[170] For maximal catalytic activity, the nickel complex must be octahedral (or square planar, which is easily transformed to octahedral) rather than tetrahedral and contain two stable ligands such as those described above. These requirements suggest that four moles of acetylene coordinated to octahedral nickel form cyclooctatetraene by an electronic rearrangement with little movement of atoms.[171]

$$\tag{8.63}$$

This mechanism describes the action of the nickel both as an activator and as a template for the proper arrangement of reacting groups. A corollary of this hypothesis predicts that if one further position of the octahedral nickel is blocked by a stable ligand, only three acetylene molecules will be able to complex with nickel, and therefore only benzene can form. This is indeed the case. Finally, if four positions are blocked on the equatorial plane of the

$$\tag{8.64}$$

nickel complex by the use of o-phenanthroline, no reaction should occur, as is found experimentally. Thus nickel serves as a selective template.

$$\text{(structure: B}_2\text{Ni(N-N)(≡)}_2\text{ complex)} \quad \xrightarrow{\quad\not\quad} \quad (8.65)$$

Other template reactions on nickel are seen in reactions of dienes. Butadiene, for example, can be cyclized by the use of so-called bare nickel to give cyclododecatriene.[172] A single nickel atom with its outer electron shell is both the activator and the matrix for a process that requires extremely high stereoregulation. The models of cyclododecatriene nickel (Fig. 8.8),[173] based on results obtained by X-ray analysis, demonstrate how accurately the matrix fits the cyclododecatriene molecule. The nickel atom is situated exactly in the center of the ring, the distances between the nickel atom and the carbon atoms of the three double bonds being about 2 Å each. This picture represents a "lock and key" fit very precisely.[173]

The nickel complex used as catalyst in the cyclotrimerization reaction must be coordinated to weakly bonded ligands so that they can be displaced by the reactant groups. One catalyst used is the nickel complex of cyclododecatriene, whose reaction with butadiene is shown by

$$\text{CDT-Ni} \xrightarrow[\substack{-\text{CDT} \\ -40°}]{+3\text{C}_4\text{H}_6} [\text{Ni(butadiene)}_3] \longrightarrow \text{Ni}(C_{12}\text{-chain}) \quad (8.66)$$

The cyclododecatriene ligand is displaced by three butadiene molecules at $-40°$, presumably to give an octahedral complex of nickel with the three butadiene molecules, which then undergoes an electronic rearrangement on the nickel template forming a C_{12} chain, the ends of which are bound to the nickel atom in π-allyl bonds. The complex of nickel with the twelve carbon chain has been isolated, and shown to give *n*-dodecane and nickel on hydrogenation at 20°, proving the linearity of the carbon chain. The suggestion that the product complex is a bis(π-allyl)nickel compound is strengthened by the fact that a simple analog, bis(π-allyl)nickel may be synthesized from allylmagnesium chloride and nickel bromide. In fact, bis(π-allyl)nickel can function as an active catalyst system, that is, as a source of

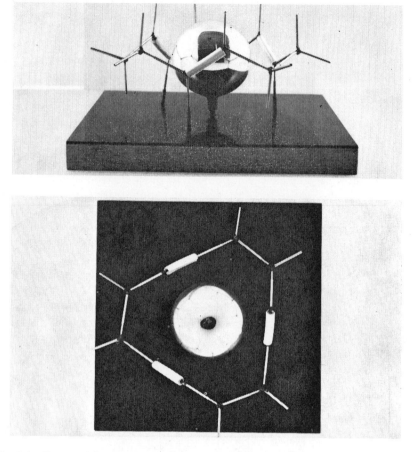

Fig. 8.8 Two models of cyclododecatriene nickel. From G. Wilke, *Proc. R. A. Welch Found. Conf. Chem. Res.*, **9**, 166 (1966).

bare nickel. In that case, the displaced allyl groups combine to form 1,5-hexadiene, while the butadiene itself forms the trimer.[174]

The product of eq. 8.66 may be treated with carbon monoxide at 20° to give cyclododecatriene plus nickel tetracarbonyl, or may be treated with a strong ligand such as trialkylphosphine to give a complex of cyclododecatriene and the phosphine with nickel. Ring closure occurs in both of these reactions.

Transition elements having partly filled d-orbitals form stable π-allyl complexes; those with nearly empty d-orbitals form labile π-allylic systems; transition elements with completely filled d-orbitals form reactive π-allylic systems or under low temperature conditions σ-allyl systems. The decrease in

$$\text{[Ni complex]} \xrightarrow[20°]{CO} \text{[cyclododecatriene]} + Ni(CO)_4$$

$$\xrightarrow{PR_3} \text{[Ni(PR_3)_2 complex]} \quad (8.67)$$

the ability for back donation of the transition elements may be responsible for this sequence.

π-Allyl metal complexes catalyze a variety of dimerization and polymerization reactions besides the cyclization reaction mentioned above. Tris(π-allyl)cobalt, an extremely labile complex, catalyzes the dimerization of butadiene to 3-methylheptatriene. A very different catalyst is obtained if one π-allyl group of tris(π-allyl)cobalt is replaced by an iodine atom. This compound, bis(π-allyl)cobalt iodide, also an extremely labile compound, catalyzes the polymerization of butadiene to a product that contains mainly a cis-1,4-structure. The mechanism of the polymerization can be formulated as follows, consistent with the cyclotrimerization reaction.

$$\text{[Co complex]} + 2C_4H_6 \longrightarrow \text{[diene]} + \text{[Co complex]} \xrightarrow{C_4H_6}$$

$$\text{[Co complex]} \xrightarrow{C_4H_6} \text{and so on} \quad (8.68)$$

Tris(π-allyl)chromium also catalyzes the polymerization of butadiene, but the product is substantially 1,2-polybutadiene. One π-allyl group of tris-(π-allyl)chromium can be replaced by an iodine atom, forming bis(π-allyl)-chromium iodide, which reacts with butadiene to form cyclododecatriene as in eq. 8.66. Formally the chromium iodide acts as a pseudo bare nickel. (π-Allyl)nickel aluminum tetrachloride produces a mixture of 1,4-*trans*- and 1,2-polybutadiene from butadiene. Addition of trialkylphosphine (1:1) yields a complex which polymerizes butadiene rapidly at 0° to 1,4-*cis*-polybutadiene. Apparently the strongly basic phosphine ligand increases the stereospecificity of the catalyst. Using a catalyst prepared from (π-allyl)nickel chloride, aluminum chloride, and triphenylphosphine in the ratio 1:2:1, propylene can be dimerized, yielding a mixture of hexenes. Again, the structure of the hexene is a function of the basicity of the phosphine ligand, the most basic ligand producing the most branched structure.[173,174]

The effect of a ligand is also seen in the cyclodimerization of butadiene on nickel blocked by a stable donor. A bis(π-allyl)nickel complex of 2 molecules of butadiene and such a nickel atom can be isolated. This complex on treatment with hydrogen gives n-octane, but on treatment either with an excess of the donor atom attached to nickel or with carbon monoxide leads to eight-carbon products which may be either cyclooctadiene or vinylcyclohexene.

$$L:Ni \xrightarrow{2C_4H_6} [Ni] \begin{cases} \xrightarrow{H_2} n\text{-octane} \\ \xrightarrow{L:} \text{cyclooctadiene} + NiL_4 \\ \xrightarrow[CO]{-80°} [Ni] \xrightarrow{CO} \text{vinylcyclohexene} + LNi(CO)_3 \end{cases} \quad (8.69)$$

A cyclooligomerization on nickel can be carried out using both ethylene and butadiene simultaneously. When these two olefins are present initially, the original bis(π-allyl) compound will not necessarily cyclize as in eq. 8.69 before it reacts with an ethylene molecule. Thus two products can result, either a cyclooctadiene or a cyclodecadiene.[172]

$$L:Ni \xrightarrow{2C_4H_6} [Ni] \begin{cases} \xrightarrow{spontaneous} [Ni] \longrightarrow \text{cyclooctadiene} \\ \xrightarrow{C_2H_4} [Ni] \longrightarrow \text{cyclodecadiene} \end{cases} \quad (8.70)$$

Several factors appear to be responsible for the facile catalyses by transition metal atoms in these oligomerizations. One is the facile change of the oxidation state of the catalyst, for example $Ni^0 \rightleftarrows Ni^{II}$ in eq. 8.66 or $Cr^I \rightleftarrows Cr^{III}$

Catalysis by Organometallic Compounds 263

in a comparable reaction. Secondly, the outer electron shells of transition metal ions can form matrices for selective catalytic processes, selectivity that can be strongly influenced either by anionic substituents or by electron donors of different basicity. Finally, the transition metal atoms can form π-allyl compounds of high reactivity.

Ziegler discovered that a mixture of a transition metal compound and a metal alkyl derived from a strongly electropositive metal would catalyze the polymerization of ethylene or propylene.[175–177] These components, for example, titanium tetrachloride and a trialkylaluminum, form an insoluble complex in the inert hydrocarbon solvents used. The heterogeneity of these catalysts is usually assumed to be responsible for the stereospecific polymerization found in these reactions by Natta and co-workers,[178] although the stereospecific polymerization of t-butyl methacrylate has been carried out in homogeneous solution.[179] Whether or not stereospecific polymerization can be carried out in homogeneous solution, the polymerizations themselves can certainly be carried out in homogeneous solution, by catalytic systems such as VCl_4–$Sn(C_6H_5)_4$[180] and by $(C_2H_5)_2TiCl_2$–$(CH_3)_2AlCl$.[181] In these mixed catalyst systems, the question to be immediately asked is whether one or both metals participate in catalysis, and if one, which one. This question was answered by an investigation of the copolymerization of ethylene and propylene with different metal alkyls in combination with different transition metal compounds. When a common transition metal compound VCl_4 is used in conjunction with a series of different metal alkyls such as CH_3TiCl_3, AlR_3, or R_2Zn, the same reactivity ratio in the copolymer is found in all cases. However, when the copolymerization is carried out using a common metal alkyl, AlR_3, in combination with a series of different transition metal compounds such as $VOCl_3$, VCl_4, $TiCl_4$, $ZrCl_4$, or $HfCl_4$, different reactivity ratios are found in every case. Hence, the growing polymer chain must be attached to the transition metal center and not the metal alkyl.[182]

On the basis of the experiments specifying the transition metal ion as the site of attachment of the growing chain and by analogy with the hydrogenation of olefins described above, one may postulate the mechanism given below. In this mechanism, the key step is an insertion reaction involving rearrangement of an olefin from π-bonding to σ-bonding to the transition metal ion with simultaneous migration of the growing chain.[183]

$$\overset{\delta\oplus \;\; \delta\ominus}{M-R} + CH_2 = CH_2 \longrightarrow \overset{\delta\oplus \;\; \delta\ominus}{\underset{\underset{CH_2=CH_2}{\uparrow}}{M-R}} \longrightarrow MCH_2CH_2R \qquad (8.71)$$

where M = transition metal and R = alkyl, aryl, or hydrogen.

Experimental evidence indicates that lower valence compounds of titanium or vanadium are necessary for reaction.[184,185] The role of the aluminum alkyl is to alkylate the transition metal ion for the initiation of chain transfer, and to act as a scavenger in the system. The active site on the catalyst surface is then a transition metal ion of lower valence state in octahedral coordination, having a growing polymer chain, R, at one position, and one position vacant or occupied only by solvent. The other sites are occupied by halide ions, for example, in $TiCl_3$. The olefin is π-bonded to a vacant site of the transition metal ion, and an insertion reaction occurs by a rate-determining migration of the group R. A new vacant site, also on the surface, has now been created so that another molecule of olefin can be adsorbed. The chain can continue to grow by switching back and forth between the two sites. Octahedral rather than tetrahedral coordination is required since only in this way can the internal rearrangement and insertion occur. The driving force for this rearrangement is due to the labilization of the alkyl group by the bonding of the olefin to the metal ion. A corollary to this mechanism is that the metal alkyl bond must not be too stable, which in effect means that the coligands such as chlorine should not be too soft.

The layer-lattice structure of the transition metal halides is ideally[183] suited to provide two or three surface coordination sites. Figure 8.9 shows the layer-lattice structure with active sites containing an R group and a coordinated solvent molecule. These sites correspond to rigid complexes of the type cis-$TiCl_4AB$. Hence they exist as a racemic mixture of asymmetric sites. The adsorption of an α-olefin can occur in two ways to give essentially diastereomeric systems of different energy.[186] If one form is favored sufficiently, an isotactic polymer is formed.

Nontransition metal alkyls catalyze the polymerization of olefins, but usually to materials of low molecular weight. Examples are the aluminum,[187] gallium, beryllium, and indium[188] alkyl growth reactions. Lithium alkyls are effective particularly for the polymerization of dienes.[189] These are presumably examples of the insertion reactions of olefins. In the insertion reaction of R_2AlH, cis-addition to the olefin occurs, as required by a mechanism in which the olefin is first coordinated to the metal.[190]

Many olefins and diolefins can be polymerized, or dimerized, in aqueous or alcoholic solution. The catalysts are salts of group VIII metals. The first example was Rh(III) chloride or nitrate, which polymerizes butadiene to crystalline trans-1,4-polybutadiene.[191] Salts of Pd(II), Ir(III), Ru(III), and Co(II) are also effective.[192] The reduction of Rh(III) to Rh(I) in the presence of ethylene, forming $[Rh(C_2H_4)_2Cl]_2$, provides a rationale for labile insertions.[193] The dimerization of ethylene to 1-butene[194] with $RhCl_3$ shows the following characteristics.[195] An induction period occurs in which the $RhCl_3$ is presumably converted to the true catalytic form. The rate of dimerization

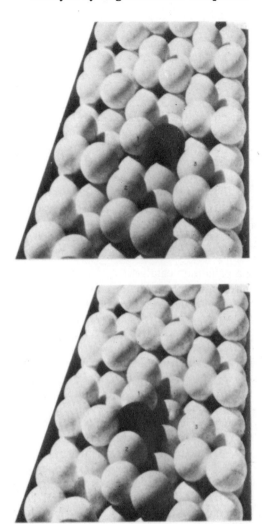

Fig. 8.9 Surface coordination sites for stereospecific polymerization of α-olefins. Black sphere represents an alkyl group. From E. J. Arlman and P. Cossee, *J. Catalysis*, **3**, 99 (1964).

is first-order in each of the components, $H^{\oplus}$, $Cl^{\ominus}$, and C_2H_4 at low concentration, but becomes independent of the concentration at high concentration. For chloride ion, the rate goes to zero at ratios of Cl/Rh of less than 3. The reduction of $RhCl_3$ to $[Rh(C_2H_4)_2Cl]_2$ together with the kinetics suggest the composition of $Rh(C_2H_4)(C_2H_5)Cl_3s^{\ominus}$ as the composition of the transition state, and eq. 8.72 as the pathway.

$$\text{"Rh}^{\text{III}}\text{Cl}_3 \cdot 3\text{H}_2\text{O"} \quad\quad \text{C}_2\text{H}_4 + \text{`C}_2\text{H}_4\text{Rh}^{\text{III}}\text{Cl}_3 s\text{'}^{\ominus}$$

$$\downarrow \text{C}_2\text{H}_4 \quad\quad\quad\quad +s\uparrow\downarrow -s$$

$$\text{L}_2\text{Rh}^{\text{I}}(\text{C}_2\text{H}_4)_2 \xrightarrow{\text{H}^{\oplus} + \text{Cl}^{\ominus}} [\text{Cl}_2\text{Rh}^{\text{I}}(\text{C}_2\text{H}_4)_2]^{\ominus} \xrightarrow{+\text{H}^{\oplus} + \text{Cl}^{\ominus}} [\text{C}_2\text{H}_5\text{Rh}^{\text{III}}\text{Cl}_3(\text{C}_2\text{H}_4)s]^{\ominus}$$

$$\uparrow \text{C}_2\text{H}_4 \quad\quad\quad\quad\quad\quad\quad\quad\quad\quad\quad\quad \downarrow$$

$$[\text{Cl}_2\text{Rh}^{\text{I}}s(\text{CH}_3\text{CH}_2\text{CH}{=}\text{CH}_2)]^{\ominus} \xrightarrow{-\text{H}^{\oplus} - \text{Cl}^{\ominus}} [\text{CH}_3\text{CH}_2\text{CH}_2\text{CH}_2\text{Rh}^{\text{III}}\text{Cl}_3 s_2]^{\ominus}$$

where L_2 = acetylacetonyl or $(\text{C}_2\text{H}_4)_2\text{Rh}\begin{smallmatrix}\text{Cl}\\\diagup\diagdown\\\diagdown\diagup\\\text{Cl}\end{smallmatrix}$ (8.72)

s = solvent.

This mechanism involves the reversible transformation between planar Rh(I) and octahedral Rh(III) complexes postulated in the isomerization of olefins brought about by this catalyst.

Carbon–carbon bonds may also be formed from an arylmercuric halide and an olefin using a palladium catalyst.[196] The process occurs by the initial formation of an aryl-palladium bond followed by insertion of the olefin and finally cleavage of the alkyl-palladium bond. The catalyst may be regenerated by the use of cupric chloride and oxygen. This general catalytic process is widely applicable for the preparation of organic compounds.

$$\text{ArHgCl} + \text{Li}_2\text{PdCl}_4 \longrightarrow [\text{ArPdCl}] + 2\text{LiCl} + \text{HgCl}_2$$
$$[\text{ArPdCl}] + \text{C}_2\text{H}_4 \longrightarrow \text{ArCH}_2\text{CH}_2\text{PdCl}$$
$$\text{ArCH}_2\text{CH}_2\text{PdCl} \longrightarrow \text{ArCH}{=}\text{CH}_2 + \text{HPdCl}$$
$$\text{HPdCl} + 2\text{CuCl}_2 \longrightarrow 2\text{CuCl} + \text{PdCl}_2 + \text{HCl}$$
$$2\text{CuCl} + 2\text{HCl} + \tfrac{1}{2}\text{O}_2 \longrightarrow 2\text{CuCl}_2 + \text{H}_2\text{O}$$
$$\underline{\text{PdCl}_2 + 2\text{LiCl} \longrightarrow \text{Li}_2\text{PdCl}_4}$$
$$\text{ArHgCl} + \text{C}_2\text{H}_4 + \text{HCl} + \tfrac{1}{2}\text{O}_2 \longrightarrow \text{ArCH}{=}\text{CH}_2 + \text{HgCl}_2 + \text{H}_2\text{O}$$
(8.73)

The cocatalyst system tungsten hexachloride–ethylaluminum dichloride (WCl_6–$\text{C}_2\text{H}_5\text{AlCl}_2$) will simultaneously aid alkylation of aromatics with olefins, and metathesis, or disproportionation of olefins.[230]

8.4.4 Reactions Involving Metal Carbonyls

An important homogeneous catalysis is the oxo or hydroformylation reaction.[197] This and related reactions are catalyzed by cobalt hydrotetracarbonyl and dicobalt octacarbonyl, as well as nickel tetracarbonyl. A system composed of cobalt salts, carbon monoxide, molecular hydrogen, and olefin can undergo a variety of reactions in the liquid state including hydrogenation of the olefin, hydrogenolysis of carbon–oxygen bonds in alcohols, homologation of alcohols by carbon monoxide and hydrogen, and hydroforymlation, that is, the addition of carbon monoxide and hydrogen to an olefin producing an aldehyde containing one more carbon atom than the olefin.

Since both cobalt hydrotetracarbonyl and dicobalt octacarbonyl form under the conditions of the oxo reactions[198] and since cobalt hydrotetracarbonyl plus the organic substrate will at room temperature and pressure give the same products as are formed in the oxo reaction, $HCo(CO)_4$ is the active catalyst to consider. This catalyst is an interesting substance. On the one hand, it is an unstable liquid that decomposes readily to give hydrogen and dicobalt octacarbonyl. On the other hand, the tetracarbonyl cobaltate anion is stable in aqueous solution. That is, this hydrocarbonyl acts as a strong acid in aqueous solution, although the nmr of the parent compound exhibits a hydride-like environment surrounding the hydrogen atom.

The hydroformylation reaction is first-order in olefin and approximately first-order in the amount of cobalt present.[199] The rate rises with increasing hydrogen pressure and falls with increasing CO pressure.[200]

Because of the overall complexity of the oxo reaction, the mechanism is best approached from studies of probable individual steps. For example, alkylcobalt tetracarbonyls, as do alkylmanganese pentacarbonyls, undergo an insertion reaction with carbon monoxide.[201, 202]

$$RCo(CO)_4 + CO \rightleftharpoons RCOCo(CO)_4 \qquad (8.74)$$

When radioactive *CO was used in the corresponding manganese reaction, a coordinated inactive CO rather than the incoming radioactive CO is inserted between the alkyl group and the manganese atom.

$$RMn(CO)_5 + {}^*CO \rightleftharpoons RCOMn(CO)_4({}^*CO) \qquad (8.75)$$

In keeping with this observation it is also possible to use other ligands, L, such as amines and phosphines to cause the insertion reaction.[203]

$$RCo(CO)_4 + L \rightleftharpoons RCOCo(CO)_3L \qquad (8.76)$$

These observations suggest that the insertion reaction is better considered as a migration by analogy with familiar examples from organic chemistry such as methyl migration.[204] One of the mechanisms consistent with the insertion process viewed as a ligand migration involves a dissociative process in which elimination of a ligand to a coordinately unsaturated complex occurs before or after migration. Postulation of a dissociation mechanism is required by the inhibition by carbon monoxide of reactions of acylcobalt tetracarbonyls with $HCo(CO)_4$, hydrogen, olefins, dienes, and acetylenes.[205]

Olefins readily react with cobalt hydrotetracarbonyl by the insertion mechanism.[205] The product is a σ-bonded alkyl group.

$$\text{HCo(CO)}_4 \underset{\text{inhibition}}{\rightleftarrows} \text{HCo(CO)}_3 + \text{CO}$$
$$1$$

$$\text{CH}_2=\text{CH}_2 + \text{HCo(CO)}_3 \longrightarrow (\text{CO})_3\overset{\text{H}}{\text{Co}} \longleftarrow \| \text{CH}_2\text{CH}_2$$
$$2$$

$$(\text{CO})_3\overset{\text{H}}{\text{Co}} \longleftarrow \| \text{CH}_2\text{CH}_2 \longrightarrow (\text{CO})_3\text{Co}-\text{CH}_2\text{CH}_3$$
$$3$$

$$(\text{CO})_3\text{Co}-\text{CH}_2\text{CH}_3 + \text{CO} \longrightarrow (\text{CO})_4\text{CoCH}_2\text{CH}_3$$
$$4$$

(8.77)

$$(\text{CO})_4\text{Co}-\text{CH}_2\text{CH}_3 + \text{CO} \longrightarrow (\text{CO})_4\text{Co}-\overset{\text{O}}{\underset{\|}{\text{C}}}\text{CH}_2\text{CH}_3$$
$$5$$

$$(\text{CO})_4\text{Co}\overset{\text{O}}{\underset{\|}{\text{C}}}\text{CH}_2\text{CH}_3 + \text{H}_2 \longrightarrow \text{HCo(CO)}_4 + \text{CH}_3\text{CH}_2\text{CHO}$$
$$\text{or HCo(CO)}_4 \longrightarrow \text{Co}_2(\text{CO})_8 + \text{CH}_3\text{CH}_2\text{CHO}$$
$$6$$

Further insertion of olefin with the alkylcobalt tetracarbonyl product is slow and incomplete: if 1-pentene is used as the olefin, a mixture of 1-pentyl- and 1-methylbutylcobalt tetracarbonyls is produced. Both insertion and isomerization are inhibited by carbon monoxide. This inhibition again requires a reversible dissociation mechanism with a coordinately unsaturated tricarbonyl as an active intermediate.[205] The coordinative unsaturation allows the addition of olefin, presumably as a π-complex. This complex then undergoes a hydride ion migration reaction to give the alkyl derivative. Pickup of a CO molecule completes the process (eq. 8.77).

The formation of an alkylcobalt tetracarbonyl is followed by a CO insertion reaction (eq. 8.77-5) under oxo conditions. The mechanism of the final reduction of the acyl carbonyl compound to aldehyde is not very well known. H_2 or $HCo(CO)_4$ may function as the reducing agent (eq. 8.77-6).[206] The species actually reduced may be $CH_3CH_2COCo(CO)_3$ since the reduction is completely inhibited by high CO pressure.[201]

A homogeneous model for the heterogeneous Fischer–Tropsch catalysis of steam and carbon monoxide involves iron pentacarbonyl as a catalyst.[207] In basic solution, the anion, $HFe(CO)_4^{\ominus}$ is formed, which in dimeric form can serve as catalyst for this reaction. The dimer, $H_2Fe_2(CO)_8^{2\ominus}$, loses molecular hydrogen forming $Fe_2(CO)_8^{2\ominus}$ which can then split water, and in the presence of CO, regenerate the anion $H_2Fe_2(CO)_8^{2\ominus}$.[208, 209]

Catalysis by Organometallic Compounds

The reactions of cobalt carbonyls are manifold.[32, 210] Esters, for example, may be prepared from olefins, carbon monoxide, and alcohols. This reaction, which takes place through the corresponding acylcobalt derivative, can lead to isomeric mixtures of esters and is accelerated by base. A plausible mechanism is given in eq. 8.78.

$$RCH=CH_2 \xrightarrow{+HCo(CO)_4} \begin{bmatrix} \rightarrow RCHCo(CO)_4 \\ \quad\quad | \\ \quad\quad CH_3 \\ \rightarrow RCH_2CH_2Co(CO)_4 \end{bmatrix} \xleftarrow[-X^\ominus]{+Co(CO)_4^\ominus} RCH_2CH_2X$$

$$\downarrow +CO$$

$$RCH_2CH_2\overset{O}{\overset{\|}{C}}X \xrightarrow[-X^\ominus]{+Co(CO)_4^\ominus} RCH_2CH_2\overset{O}{\overset{\|}{C}}Co(CO)_4 \xrightarrow[B]{R'OH} RCH_2CH_2\overset{O}{\overset{\|}{C}}OR'$$

$$R'NH_2 \Big| B$$
$$\quad\quad\rightarrow RCH_2CH_2\overset{O}{\overset{\|}{C}}NHR'$$

(8.78)

Esters have likewise been made from alkyl halides, carbon monoxide, and alcohols in basic medium in the presence of cobalt tetracarbonyl anion. The same acylcobalt intermediate is again formed in these reactions. Finally acyl halides plus cobalt tetracarbonyl anion form acylcobalt derivatives, eventually yielding esters.[201, 211] Cobalt octacarbonyl catalyzes the reaction of dialkylmercury compounds with carbon monoxide to give (symmetrical) dialkyl ketones.[229]

Nickel carbonyls catalyze analogous reactions such as the carboxylation or carboxyalkylation of olefins and acetylenes by carbon monoxide.[212] Triphenylphosphine, iodide ion, and hydrogen halides are activators of nickel carbonyl, suggesting that some CO must be replaced before reaction proceeds.[213, 214] A mechanism for the hydrogen halide activation is given.

$$HX + Ni(CO)_4 \rightleftharpoons HNi(CO)_2X + 2CO$$
$$C_2H_4 + HNi(CO)_2X \longrightarrow CH_3CH_2Ni(CO)_2X$$
$$CH_3CH_2Ni(CO)_2X + CO \longrightarrow CH_3CH_2\overset{O}{\overset{\|}{C}}Ni(CO)_2X$$
$$CH_3CH_2\overset{O}{\overset{\|}{C}}Ni(CO)_2X + 2CO \longrightarrow CH_3CH_2\overset{O}{\overset{\|}{C}}X + Ni(CO)_4$$
$$\Big| ROH$$
$$\quad\rightarrow CH_3CH_2\overset{O}{\overset{\|}{C}}OR$$

or

$$+ ROH \longrightarrow CH_3CH_2\overset{O}{\overset{\|}{C}}OR + HNi(CO)_2X$$

(8.79)

Palladium and rhodium complexes also serve as catalysts for the hydroformylation reaction.[215] Both conjugated and nonconjugated dienes in the presence of carbon monoxide and alcohols give unsaturated esters in the presence of diiodobis(tributylphosphine)palladium(II).[216] Likewise, 1-hexene with hydrogen and carbon monoxide in the presence of 1% $RhCl_3(PPh_3)_3$ gives n-heptaldehyde at low temperature.[217]

8.4.5 Oxidation of Hydrocarbons

The oxidation of ethylene by aqueous palladium(II) chloride and cupric chloride leads to a catalytic process for the production of acetaldehyde. In the absence of cupric chloride, the reaction gives metallic palladium.

$$C_2H_4 + PdCl_2 + H_2O \longrightarrow CH_3CHO + Pd^0 + 2HCl \qquad (8.80)$$

Since cupric chloride oxidizes palladium metal to palladium(II), and since the cuprous chloride produced can be air-oxidized to the cupric state, a true homogeneous catalytic process is established.[218, 219] The reaction occurs in two stages, an initial rapid uptake of ethylene by palladium chloride, followed by a slow reaction yielding acetaldehyde. The initial rapid uptake of ethylene can be described by

$$PdCl_4^{2\ominus} + C_2H_4 \underset{}{\overset{K}{\rightleftharpoons}} \begin{bmatrix} Cl & Cl \\ & Pd \\ Cl & \end{bmatrix}^{\ominus} Cl^{\ominus} \qquad (8.81)$$

The volume of gas initially absorbed decreases as the chloride ion concentration is increased. This dependency is in accord with equilibrium 8.81. The product of eq. 8.81 is written as a π-complex between ethylene and palladium chloride because the equilibrium is readily reversible. The rate of the slow product-forming reaction is[219]

$$\frac{-d[C_2H_4]}{dt} = \frac{k[PdCl_3C_2H_4^{\ominus}]}{[Cl^{\ominus}][H^{\oplus}]} \qquad (8.82)$$

Only a very small isotope effect is seen when deuteroethylene is oxidized ($k_H/k_D = 1.07$). In deuterium oxide solution, no deuterium is incorporated into the acetaldehyde product.[220] These kinetic and isotopic results lead to the mechanism of eq. 8.83.

The two pre-equilibrium steps in eq. 8.83 are consistent with the inverse dependence of the rate on both chloride ion and hydrogen ion. The combination of the lack of an isotope effect and an intramolecular process, dictated by the tracer experiment in deuterium oxide, demands that the final decomposition of the hydroxyl complex be a two-step process in which the first step,

$$\left[\begin{array}{c}\text{Cl}\diagdown\quad\diagup\text{Cl}\\\text{Pd}\\\text{Cl}\diagup\quad\diagdown\text{\hspace{-2mm}}\diagup\hspace{-2mm}\diagdown\end{array}\right]^{\ominus} + \text{H}_2\text{O} \underset{}{\overset{K}{\rightleftharpoons}} \begin{array}{c}\text{Cl}\diagdown\quad\diagup\text{Cl}\\\text{Pd}\\\text{H}_2\text{O}\diagup\quad\diagdown\text{\hspace{-2mm}}\diagup\hspace{-2mm}\diagdown\end{array} + \text{Cl}^{\ominus}$$

$$\begin{array}{c}\text{Cl}\diagdown\quad\diagup\text{Cl}\\\text{Pd}\\\text{H}_2\text{O}\diagup\quad\diagdown\end{array} \underset{}{\overset{K}{\rightleftharpoons}} \left[\begin{array}{c}\text{Cl}\diagdown\quad\diagup\text{Cl}\\\text{Pd}\\\text{HO}\diagup\quad\diagdown\end{array}\right]^{\ominus} + \text{H}^{\oplus}$$

$$\left[\begin{array}{c}\text{Cl}\diagdown\quad\diagup\text{Cl}\\\text{Pd}\\\text{HO}\diagup\end{array}\right]^{\ominus} \xrightarrow{\text{slow}} \text{ClPd}\!-\!\text{CH}_2\text{CH}_2\text{OH} + \text{Cl}^{\ominus}$$

$$\text{(complex)} \longrightarrow \text{Cl}^{\ominus} + \text{PdCH}_3\overset{\text{H}}{\text{C}}\!\!=\!\!\text{O} + \text{H}^{\oplus}$$

(8.83)

an insertion reaction, is rate-determining. The experimental data are inconsistent with a vinyl alcohol intermediate. Palladium catalyzes this reaction by coordinating the ethylene and hydroxo groups, and also by serving as an electron sink and carrier agent for the internal hydrogen transfer in the last step.

If acetic acid is used as a solvent, vinyl acetate is formed.[221] The final breakdown step in this reaction must take a different course from that given in eq. 8.83, such as the transfer of hydrogen from carbon to palladium giving a π-complex of vinyl acetate which then dissociates.

Other oxidizing metal ions such as Hg(II),[222] Pb(IV),[223] and Tl(III),[224,225] produce carbonyl compounds from olefins. In these reactions, glycols are formed as well. Kinetic information on the thallium(III) reaction is best explained by a mechanism in which a π-complex is rapidly formed and a slow insertion reaction occurs in which water is the migrating ligand.[215]

The extraordinary catalytic activity of metal ions has been investigated in a serious manner only for the last decade. Many important problems remain to be solved. Among these are the homogeneous catalytic activation of saturated hydrocarbons and of molecular nitrogen. The recent success in fixing nitrogen with transition metal catalysts of the Ziegler-Natta type[226] as well as the recent discovery of several transition metal complexes containing coordinated nitrogen, $Ru(NH_3)_5N_2^{2\oplus}$ [227] and $IrCl(N_2)(PPh_3)_2$,[228] indicate that attainment of at least one of these objectives may be close. There is also reason to hope that the emerging understanding of catalytic mechanisms in these simple metal ion systems will make a contribution to our understanding of related catalytic phenomena in enzymic systems.

REFERENCES

1. F. H. Westheimer, *Trans. N. Y. Acad. Sci.*, **18**, 15 (1955).
2. M. Eigen, *Pure Appl. Chem.*, **6**, 105 (1963).
3. I. Roberts and L. P. Hammett, *J. Amer. Chem. Soc.*, **59**, 1063 (1937).
4. E. D. Hughes, C. K. Ingold, and S. Masterman, *J. Chem. Soc.*, 1236 (1937).
5. O. T. Benfey, *J. Amer. Chem. Soc.*, **70**, 2163 (1948).
6. G. S. Hammond, M. F. Hawthorne, J. H. Waters, and B. M. Graybill, *J. Amer. Chem. Soc.*, **82**, 704 (1960).
7. Y. Pocker and D. N. Kevill, *J. Amer. Chem. Soc.*, **87**, 4760, 4771, 4778 (1965).
8. G. D. Parfitt, A. L. Smith, and A. G. Walton, *J. Phys. Chem.*, **69**, 661 (1965).
9. M. R. V. Sahyun, *Nature*, **206**, 788 (1965).
10. B. Saville, *J. Chem. Soc.*, 4062 (1962).
11. I. M. Klotz and B. J. Campbell, *Arch. Biophys. Biochem.*, **96**, 92 (1962).
12. H. Kunnap and A. Parts, *Apophoreta Tartu. Soc. Lit. Eston. in Svecia* (Stockholm), 377 (1949); *Chem. Abstr.*, **44**, 7629i (1950).
13. K. R. Ashley and R. E. Hamm, *Inorg. Chem.*, **4**, 1120 (1965).
14. N. W. D. Beese and C. H. Johnson, *Trans. Faraday Soc.*, **31**, 1632 (1935).
15. H. A. Krebs, *Biochem. J.*, **36**, 303 (1942).
16. R. W. Hay, *Rev. Pure Appl. Chem.*, **13**, 157 (1963).
17. A. Kornberg, S. Ochoa, and A. H. Mehler, *J. Biol. Chem.*, **174**, 159 (1948).
18. R. Steinberger and F. H. Westheimer, *J. Amer. Chem. Soc.*, **73**, 429 (1951).
19. J. V. Rund and R. A. Plane, *J. Amer. Chem. Soc.*, **86**, 367 (1964).
20. J. E. Prue, *J. Chem. Soc.*, 2331 (1952).
21. E. Gelles and A. Salama, *J. Chem. Soc.*, 3683, 3689 (1958).
22. M. Stiles, *J. Amer. Chem. Soc.*, **81**, 2598 (1959).
23. M. Stiles and H. L. Finkbeiner, *J. Amer. Chem. Soc.*, **81**, 505 (1959).
24. H. Finkbeiner, *J. Amer. Chem. Soc.*, **87**, 4588 (1965).
25. K. J. Pedersen, *Acta Chem. Scand.*, **2**, 252 (1948).
26. A. Schellenberger and G. Hubner, *Chem. Ber.*, **98**, 1938 (1965).
27. A. Schellenberger, G. Oehme, and G. Hubner, *Chem. Ber.*, **98**, 3578 (1965); J. Halpern, personal communication.
28. P. Pfeiffer, W. Offermann, and H. Werner, *J. Prakt. Chem.*, **159**, 313 (1941).
29. M. Sato, K. Okawa, and S. Akabori, *Bull. Chem. Soc., Japan*, **30**, 937 (1957).
30. S. Akabori, T. T. Otani, R. Marshall, M. Winitz, and J. P. Greenstein, *Arch. Biochem. Biophys.*, **83**, 1 (1959).
31. T. Otani and M. Winitz, *Arch. Biochem. Biophys.*, **102**, 464 (1963).
32. *Adv. Chem. Series*, no. 37, Washington, 1963; no. 70, Washington, 1968.
33. D. H. Williams and D. H. Busch, *J. Amer. Chem. Soc.*, **87**, 4644 (1965).
34. M. Murakami and K. Takahashi, *Bull. Chem. Soc., Japan*, **32**, 308 (1959).
35. H. Kroll, *J. Amer. Chem. Soc.*, **74**, 2036 (1952).
36. M. P. Springer and C. Curran, *Inorg. Chem.*, **2**, 1270 (1963).
37. M. L. Bender and B. W. Turnquest, *J. Amer. Chem. Soc.*, **79**, 1889 (1957).
38. W. A. Connor, M. M. Jones, and D. L. Tuleen, *Inorg. Chem.*, **4**, 1129 (1965).
39. H. L. Conley, Jr., and R. B. Martin, *J. Phys. Chem.*, **69**, 2914, 2923 (1965).

References

40. K. J. Pedersen, *Acta Chem. Scand.*, **2**, 385 (1948).
41. J. E. Prue, *J. Chem. Soc.*, 2331 (1952).
42. R. Breslow, R. Fairweather, and J. Keana, *J. Amer. Chem. Soc.*, **89**, 2135 (1967).
43. Y. Ogata and M. Okano, *J. Chem. Soc., Japan*, **70**, 32 (1949).
44. R. S. Bruenner and A. E. Oberth, *J. Org. Chem.*, **31**, 887 (1966).
45. See also Y. Pocker and J. E. Meany, *J. Phys. Chem.*, **71**, 3113 (1967); **72**, 655 (1968) for the general acid-base and metal ion catalysis of the hydration of pyridinecarboxaldehydes.
46. M. D. Alexander and D. H. Busch, *J. Amer. Chem. Soc.*, **88**, 1130 (1966).
47. L. Meriwether and F. H. Westheimer, *J. Amer. Chem. Soc.*, **78**, 5119 (1956).
48. E. Bamann, J. G. Haas, and H. Trapmann, *Arch. Pharm.*, **294**, 569 (1961). E. Bamann, A. Rother, and H. Trapmann, *Naturwiss.*, **43**, 326 (1956); E. Bamann, H. Trapmann, and A. Rother, *Chem. Ber.*, **91**, 1744 (1958).
49. I. J. Grant and R. W. Hay, *Austral. J. Chem.*, **18**, 1189 (1965); see also R. W. Hay and P. J. Morris, *Chem. Comm.*, 23 (1967).
50. B. R. Rabin, *Trans. Faraday Soc.*, **52**, 1130 (1956).
51. J. P. Collman and D. A. Buckingham, *J. Amer. Chem. Soc.*, **85**, 3039 (1963); D. A. Buckingham, L. G. Marzilli, and A. M. Sargeson, *ibid.*, **89**, 2772 (1967); D. A. Buckingham et al., *ibid.*, **89**, 1082 (1967); D. A. Buckingham, L. G. Marzilli, and A. M. Sargeson, *ibid.*, **89**, 4539 (1967).
52. G. Sachs, *Ber.*, **54**, 1849 (1921).
53. R. Schwyzer, *Helv. Chim., Acta*, **36**, 414 (1953).
54. J. R. Stern, *J. Biol. Chem.*, **221**, 33 (1956).
55. F. Lynen, E. Reichert, and L. Rueff, *Ann.*, **574**, 1 (1951).
56. R. Benesch and R. E. Benesch, *Proc. Nat. Acad. Sci., U.S.*, **44**, 848 (1958).
57. G. L. Eichhorn and J. C. Bailar, Jr., *J. Amer. Chem. Soc.*, **75**, 2905 (1953); G. L. Eichhorn and I. M. Trachtenberg, *J. Amer. Chem. Soc.*, **76**, 5183 (1954).
58. R. B. Woodward, N. L. Wendler, and F. J. Brutschy, *J. Amer. Chem. Soc.*, **67**, 1425 (1945).
59. V. J. Shiner, Jr., and D. Whittaker, *J. Amer. Chem. Soc.*, **85**, 2337 (1963).
60. J. Hine, *Physical Organic Chemistry*, 2nd ed., McGraw-Hill Book Co., New York, 1962, p. 269.
61. C. G. Swain and H. B. Boyles, *J. Amer. Chem. Soc.*, **73**, 870 (1951).
62. H. Tani et al., *J. Amer. Chem. Soc.*, **89**, 173 (1967); *Chem. Eng. News*, February 12, 1968, p. 43.
63. T. Wagner-Jauregg, B. F. Hackley, Jr., T. A. Lies, O. O. Owens, and R. Proper, *J. Amer. Chem. Soc.*, **77**, 922 (1955).
64. L. Larsson, *Acta Chem. Scand.*, **12**, 1226 (1958).
65. J. Epstein and D. H. Rosenblatt, *J. Amer. Chem. Soc.*, **80**, 3596 (1958).
66. R. C. Courtney, R. L. Gustafson, S. J. Westerback, H. Hyytiainen, S. C. Chaberek, Jr., and A. E. Martell, *J. Amer. Chem. Soc.*, **79**, 3030 (1957); see also F. J. Farrell, W. A. Kjellstrom, and C. S. Spiro, *Science*, **164**, 320 (1969).
67. A. E. Martell, *Adv. Chem. Series*, **37**, 161 (1963).
68. F. H. Westheimer, *Spec. Publ. Chem. Soc.*, **8**, 1 (1957).
69. E. Bamann and E. Nowotny, *Chem. Ber.*, **81**, 451, 455, 463 (1948).
70. W. W. Butcher and F. H. Westheimer, *J. Amer. Chem. Soc.*, **77**, 2420 (1955).
71. R. Hofstetter, Y. Murakami, G. Mont, and A. E. Martell, *J. Amer. Chem. Soc.*, **84**, 3041 (1962).
72. M. Smith, G. I. Drummond, and H. G. Khorana, *J. Amer. Chem. Soc.*, **83**, 698 (1961).
73. F. Lipmann and L. Tuttle, *J. Biol. Chem.*, **153**, 571 (1944).

74. D. E. Koshland, Jr., *J. Amer. Chem. Soc.*, **74**, 2286 (1952); J. L. Kurz and C. D. Gutsche, *J. Amer. Chem. Soc.*, **82**, 2175 (1960).
75. G. DiSabato and W. P. Jencks, *J. Amer. Chem. Soc.*, **83**, 4393 (1961).
76. C. H. Oestreich and M. M. Jones, *Biochemistry*, **5**, 2926, 3151 (1966).
77. D. Lipkin, R. Markham, and W. H. Cook, *J. Amer. Chem. Soc.*, **81**, 6075 (1959); R. A. Alberty, *J. Biol. Chem.*, **243**, 1337 (1968).
78. M. Tetas and J. M. Lowenstein, *Biochem.*, **2**, 350 (1963); see also R. A. Alberty, *J. Biol. Chem.*, **243**, 1343 (1968).
79. J. van Steveninck, *Biochem.*, **5**, 1998 (1966).
80. P. Schneider, H. Brintzinger, and H. Erlenmeyer, *Helv. Chim. Acta*, **47**, 992 (1964).
81. D. L. Miller and F. H. Westheimer, *J. Amer. Chem. Soc.*, **88**, 1514 (1966).
82. J. M. Lowenstein, *Biochim. Biophys. Acta*, **28**, 206 (1958); J. M. Lowenstein, *Biochem. J.*, **70**, 222 (1958).
83. D. L. Miller and F. H. Westheimer, *J. Amer. Chem. Soc.*, **88**, 1507 (1966).
84. M. Cohn and T. R. Hughes, Jr., *J. Biol. Chem.*, **235**, 3250 (1960); **237**, 176 (1962).
85. E. A. Hopkins and J. H. Wang, *J. Amer. Chem. Soc.*, **87**, 4391 (1965).
86. L. E. Orgel, *Biochem. Soc. Symp.*, **15**, 8 (1958); charge can also have an effect on rate constant.
87. R. Breslow and D. Chipman, *J. Amer. Chem. Soc.*, **87**, 4195 (1965).
88. D. S. Auld and T. C. Bruice, *J. Amer. Chem. Soc.*, **89**, 2083, 2090, 2098 (1967).
89. D. H. Busch, J. A. Burke, Jr., D. C. Jicha, M. C. Thompson, and M. L. Morris, *Adv. Chem. Series*, **37**, 125 (1963).
90. I. M. Klotz in *The Mechanism of Enzyme Action*, W. D. McElroy and B. Glass, Eds., Johns Hopkins Press, Baltimore, 1954, p. 275.
91. M. F. Ansell and B. C. L. Weedon, *Chem. Britain*, **3**, 306 (1967).
92. W. C. E. Higginson, D. R. Rosseinsky, J. B. Stead, and A. G. Sykes, *Disc. Faraday Soc.*, **29**, 49 (1960).
93. F. R. Duke, *J. Amer. Chem. Soc.*, **69**, 3054 (1947); F. R. Duke and A. A. Forist, *J. Amer. Chem. Soc.*, **71**, 2790 (1949).
94. H. Taube, *J. Amer. Chem. Soc.*, **69**, 1418 (1947); F. R. Duke, *ibid.*, **69**, 2885 (1947).
95. H. Taube and E. L. King, *J. Amer. Chem. Soc.*, **76**, 4053 (1954).
96. D. C. Nonhebel and W. A. Waters, *Adv. Catalysis*, **9**, 353 (1957).
97. T. Saegusa, Y. Ito, S. Kabayashi, and K. Hirota, *J. Amer. Chem. Soc.*, **89**, 2240 (1967).
98. R. G. R. Bacon and H. A. O. Hill, *Quart. Rev.*, **19**, 95 (1965).
99. A. H. Lewin and T. Cohen, *Tetrahedron Lett.*, 4531 (1965).
100. H. Kwart and A. A. Kahn, *J. Amer. Chem. Soc.*, **89**, 1950, 1951, (1967).
101. C. Wasmuth, personal communication, H. Kwart and A. A. Kahn, *J. Amer. Chem. Soc.*, **89**, 1950, 1951 (1967); K. Wiberg and W. G. Nigh, *ibid.*, **87**, 3849 (1965); J. K. Kochi, *Science*, **155**, 415 (1967); H. Erlenmeyer, U. Muller, and H. Sigel, *Helv. Chim. Acta*, **49**, 681 (1966).
102. F. Haber and J. Weiss, *Proc. Roy. Soc.*, **A147**, 332 (1934).
103. N. Uri, *Chem. Rev.*, **50**, 375 (1952).
104. H. Erlenmeyer, U. Muller, and H. Sigel, *Helv. Chim. Acta*, **49**, 681 (1966); U. Muller and H. Sigel, *Helv. Chim. Acta*, **49**, 671 (1966).
105. S. Udenfriend, C. T. Clark, J. Axelrod, and B. B. Brodie, *J. Biol. Chem.*, **208**, 731 (1954).
106. R. Breslow and L. N. Lukens, *J. Biol. Chem.* **235**, 292 (1960).
107. R. R. Grinstead, *J. Amer. Chem. Soc.*, **82**, 3472 (1960).
108. R. O. C. Norman and G. K. Radda, *Proc. Chem. Soc.*, 138 (1962).
109. G. A. Hamilton, *J. Amer. Chem. Soc.*, **86**, 3391 (1964).

References

110. R. R. Grinstead, *J. Amer. Chem. Soc.*, **82**, 3464 (1960).
111. G. A. Hamilton, personal communication.
112. G. A. Hamilton and J. P. Friedman, *J. Amer. Chem. Soc.*, **85**, 1008 (1963).
113. G. A. Hamilton, J. P. Friedman, and P. M. Campbell, *ibid.*, **88**, 5266 (1966).
114. G. A. Hamilton, J. W. Hanifin, Jr., and J. P. Friedman, *ibid.*, **88**, 5269 (1966).
115. J. H. Wang, *J. Amer. Chem. Soc.*, **77**, 822 (1955).
116. J. H. Wang, *J. Amer. Chem. Soc.*, **77**, 4715 (1955).
117. R. C. Jarnagin and J. H. Wang, *J. Amer. Chem. Soc.*, **80**, 6477 (1958).
118. J. K. Kochi and P. E. Mocaldo, *J. Org. Chem.*, **30**, 1134 (1965).
119. W. H. Richardson, *J. Amer. Chem. Soc.*, **87**, 247, 1096 (1965); **88**, 975 (1966); *J. Org. Chem.*, **30**, 2804 (1965).
120. T. J. Wallace, R. M. Skomoroski, and P. J. Lucchesi, *Chem. Ind.*, 1764 (1965).
121. J. K. Kochi and R. D. Gilliom, *J. Amer. Chem. Soc.*, **86**, 5251 (1964).
122. G. A. Russell, *J. Amer. Chem. Soc.*, **79**, 3871 (1957); *J. Chem. Ed.*, **36**, 111 (1959).
123. J. P. Wibaut and A. Strang, *Koninkl. Ned. Akad. Wetenschap. Proc.*, **B54**, 102 (1951).
124. T. J. Wallace and R. M. Skomoroski, *Chem. Ind.*, 348 (1965).
125. R. Lombard and L. Rammert, *Bull. Soc. Chim. France*, **23**, 36 (1956).
126. H. Boardman, *J. Amer. Chem. Soc.*, **84**, 1376 (1962).
127. L. G. Sillen and A. E. Martell, *Stability Constants*, The Chemical Society, London, 1964.
128. J. Halpern, *Chem. Eng. News*, 68, October 31, 1966.
129. For a review, see R. F. Heck, *Adv. Chem. Ser.*, **49**, 181 (1965).
130. J. Halpern, *Proc. 3rd Int. Cong. Catalysis*, J. Wiley and Sons, New York, 1965, p. 146.
131. F. H. Jardine, J. A. Osborn, G. Wilkinson, and J. F. Young, *Chem. Ind.*, 560 (1965); *Chem. Comm.*, 131 (1965); M. C. Baird, J. T. Mague, J. A. Osborn, and G. Wilkinson, *J. Chem. Soc. (A)*, 1347 (1967).
132. M. Calvin, *Trans. Faraday Soc.*, **34**, 1181 (1938).
133. M. Calvin, *J. Amer. Chem. Soc.*, **61**, 2230 (1939); M. Calvin and W. K. Wilmarth, *J. Amer. Chem. Soc.*, **78**, 1301 (1956).
134. W. K. Wilmarth and M. K. Barsh, *J. Amer. Chem. Soc.*, **78**, 1305 (1956).
135. W. K. Wilmarth and M. K. Barsh, *J. Amer. Chem. Soc.*, **75**, 2237 (1953).
136. S. Weller and G. A. Mills, *J. Amer. Chem. Soc.*, **75**, 769 (1953).
137. J. Halpern, *Ann. Rev. Phys. Chem.*, **16**, 103 (1965).
138. A. H. Webster and J. Halpern, *J. Phys. Chem.*, **60**, 280 (1956); G. J. Korinek and J. Halpern, *J. Phys. Chem.*, **60**, 285 (1956).
139. J. Halpern and E. Peters, *J. Chem. Phys.*, **23**, 605 (1955).
140. A. H. Webster and J. Halpern, *J. Phys. Chem.*, **61**, 1239 (1957).
141. M. T. Beck, I. Gimesi, and J. Farkas, *Nature*, **197**, 73 (1963).
142. J. F. Harrod, S. Ciccone, and J. Halpern, *Can. J. Chem.*, **39**, 1372 (1961), J. Halpern and B. R. James, *Can. J. Chem.*, **44**, 671 (1966).
143. D. Rittenberg and A. Krasna, *Disc. Faraday Soc.*, **20**, 185 (1955); see also N. C. Deno et al., *J. Amer. Chem. Soc.*, **84**, 4713 (1962); H. B. Charman, *Nature*, **212**, 278 (1966); J. Chatt, *Science*, **160**, 723 (1968).
144. L. Vaska and J. W. DiLuzio, *J. Amer. Chem. Soc.*, **84**, 679 (1962).
145. G. Wilkinson, *Proc. Robert A. Welch Found. Conf. Chem. Res.*, No. **9**, 139 (1965).
146. J. Halpern, J. F. Harrod, and B. R. James, *J. Amer. Chem. Soc.*, **83**, 753 (1961); **88**, 5150 (1966).
147. R. L. Burwell, Jr., A. B. Littlewood, M. Cardew, G. Pass, and C. T. H. Stoddart, *J. Amer. Chem. Soc.*, **82**, 6272 (1960).
148. J. Kwiatek, I. L. Mador, and J. K. Seyler, *ibid.*, **84**, 304 (1962), *Adv. Chem. Ser.*, **37**, 201 (1963).

149. J. Kwiatek and J. K. Seyler, *Proc. 8 Int. Conf. Coord. Chem.*, Springer-Verlag, Vienna, 1964, p. 308.
150. L. Vaska and R. E. Rhodes, *J. Amer. Chem. Soc.*, **87**, 4970 (1965).
151. This compound also is a catalyst for the polymerization of trioxane and acetals in general. *Chem. Eng. News*, 43 (February, 1968).
152. F. H. Jardine, J. A. Osborn, G. Wilkinson, and J. F. Young, *Chem. Ind.*, 560 (1965), *Chem. Comm.*, 131 (1965); G. Wilkinson, *Proc. Robert A. Welch Found. Conf. Chem. Res.*, **9**, 139 (1965).
153. L. Vaska, *Inorg. Nuclear Chem. Lett.*, **1**, 89 (1965).
154. M. F. Sloan, A. S. Matlack, and D. S. Breslow, *J. Amer. Chem. Soc.*, **85**, 4014 (1963).
155. L. H. Slaugh, *Tetrahedron*, **22**, 1741 (1966).
156. J. C. Saam and J. L. Speier, *J. Amer. Chem. Soc.*, **80**, 4104 (1958); **83**, 1351 (1961).
157. A. J. Chalk and J. F. Harrod, *J. Amer. Chem. Soc.*, **87**, 16 (1965).
158. H. G. Kuivila and C. R. Warner, *J. Org. Chem.*, **29**, 2845 (1964); R. H. Fish and H. G. Kuivila, *ibid.*, **31**, 2445 (1966); H. A. Tayim and J. C. Bailar, Jr., *J. Amer. Chem. Soc.*, **89**, 4330 (1967); L. H. Sommer, K. W. Michael, and H. Fujimoto, *ibid.*, **89**, 1519 (1967).
159. J. Halpern, B. R. James, and A. L. W. Kemp, *J. Amer. Chem. Soc.*, **83**, 4097 (1961).
160. W. L. Budde and R. E. Dessy, *J. Amer. Chem. Soc.*, **85**, 3964 (1963).
161. J. Manassen and S. Khalif, *J. Amer. Chem. Soc.*, **88**, 1943 (1966).
162. J. Halpern, *Adv. Catalysis*, **11**, 301 (1959).
163. M. Orchin, *Adv. Catalysis*, **16**, 1 (1966); also isotopic exchange, J. L. Garnett and R. J. Hodges, *J. Amer. Chem. Soc.*, **89**, 4546 (1967).
164. N. R. Davies, *Nature*, **201**, 490 (1964); N. R. Davies, *Australian J. Chem.*, **17**, 212 (1964).
165. R. Cramer, *J. Amer. Chem. Soc.*, **88**, 2272 (1966).
166. J. F. Harrod and A. J. Chalk, *J. Amer. Chem. Soc.*, **88**, 3491 (1966).
167. R. Cramer and R. V. Lindsey, Jr., *ibid.*, **88**, 3534 (1966); see also R. Cramer, *ibid.*, **89**, 1633 (1967).
168. L. Roos and M. Orchin, *J. Amer. Chem. Soc.*, **87**, 5502 (1965).
169. J. W. Reppe, *Acetylene Chemistry*, C. A. Meyer and Co., New York, 1949.
170. J. W. Reppe et al., *Ann.*, **560**, 1 (1948).
171. G. N. Schrauzer and S. Eichler, *Chem. Ber.*, **95**, 550 (1962); G. N. Schrauzer, *Angew. Chem. Int. Ed.*, **3**, 185 (1964).
172. G. Wilke, *Angew. Chem.*, **75**, 10 (1963); *Angew. Chem. Int. Ed.*, **2**, 105 (1963); palladium catalysis is apparently similar, *Chem. Eng. News*, April 21, 1969, p. 48.
173. G. Wilke, *Proc. Robert A. Welch Found. Conf. Chem. Res.*, **9**, 165 (1965).
174. G. Wilke et al., *Angew Chem.*, **78**, 157 (1966); *Angew Chem. Int. Ed.*, **5**, 151 (1966).
175. K. Ziegler, E. Holzkamp, H. Breil, and H. Martin, *Angew. Chem.*, **67**, 541 (1955).
176. For reviews, see J. K. Stille, *Chem. Rev.*, **58**, 541 (1958); A. D. Ketley and F. X. Werber, *Science*, **145**, 667 (1964); F. Dawans and P. Teyssie, *Bull. Soc. Chim. France*, **10**, 2376 (1963).
177. See also R. G. Miller, J. T. Kealy, and A. L. Barney, *J. Amer. Chem. Soc.*, **89**, 3756 (1967).
178. G. Natta, *Atti accad. nazl. Lincei, Mem., Classe sci. fis., mat. e nat.*, ser. *VIII*, **4**, sez. II, 61 (1955); G. Natta, P. Pino, G. Mazzanti, and P. Longi, *Gazz. Chim. Ital.*, **87**, 549 (1957). L. A. M. Rodriguez and H. M. VanLooy, *J. Polymer Sci.*, **14**, pt. A-1 1951, 1971 (1966).
179. G. Smets, personal communication.
180. W. L. Carrick et al., *J. Amer. Chem. Soc.*, **82**, 1502, 3887, 5319 (1960); **83**, 2654 (1961).

References

181. W. P. Long and D. S. Breslow, *J. Amer. Chem. Soc.*, **82**, 1953 (1960); J. C. W. Chien, *ibid.*, **81**, 86 (1959); D. S. Breslow and N. R. Newburg, *ibid.*, **81**, 81 (1959).
182. F. J. Karol and W. L. Carrick, *J. Amer. Chem. Soc.*, **83**, 2654 (1961).
183. P. Cossee, *J. Catalysis*, **3**, 80 (1964); *Tetrahedron Lett.*, **17**, 12, 17 (1960); E. J. Arlman and P. Cossee, *J. Catalysis*, **3**, 99 (1964).
184. W. L. Carrick, A. G. Chasar, and J. J. Smith, *J. Amer. Chem. Soc.*, **82**, 5319 (1960).
185. D. B. Ludlum, A. W. Anderson, and C. E. Ashby, *J. Amer. Chem. Soc.*, **80**, 1380 (1958).
186. G. Natta, *Angew. Chem.*, **68**, 393 (1956).
187. K. Ziegler, *Angew. Chem.*, **64**, 323 (1952); **68**, 721 (1956).
188. K. Ziegler and H. G. Gellert, U. S. Patent, 2,699,453 (1955).
189. K. Ziegler, F. Crossman, H. Kleiner, and O. Shafter, *Ann.*, **473**, 1 (1929); H. Sinn and F. Patat, *Angew. Chem. Int. Ed.*, **3**, 93 (1964).
190. G. Wilke and H. Muller, *Ann.*, **618**, 267 (1958).
191. R. E. Rinehart, H. P. Smith, H. S. Witt, and H. Romeyn, Jr., *J. Amer. Chem. Soc.*, **83**, 4864 (1961); **84**, 4145 (1962).
192. A. J. Canale, W. A. Hewett, T. M. Shryne, and E. A. Youngman, *Chem. Ind.*, 1054 (1962); A. J. Canale and W. A. Hewett, *J. Polymer Sci.*, **B2**, 1041 (1964).
193. R. Cramer, *Inorg. Chem.*, **1**, 722 (1962).
194. T. Alderson, E. L. Jenner, and R. V. Lindsey, Jr., *J. Amer. Chem. Soc.*, **87**, 5638 (1965).
195. R. Cramer, *J. Amer. Chem. Soc.*, **87**, 4717 (1965).
196. R. F. Heck, *J. Amer. Chem. Soc.*, **90**, 317 (1968); for the reductive coupling of alcohols to hydrocarbons by metal ions see E. E. van Tamelen and M. A. Schwartz, *J. Amer. Chem. Soc.*, **87**, 3277 (1965).
197. H. W. Sternberg and I. Wender, *Spec. Publ. Chem. Soc.*, **13**, 35 (1959); C. W. Bird, *Chem. Rev.*, **62**, 283 (1962).
198. I. Wender, H. W. Sternberg, and M. Orchin, *J. Amer. Chem. Soc.*, **75**, 3041 (1953); M. Orchin, L. Kirch, and I. Goldfarb, *J. Amer. Chem. Soc.*, **78**, 5450 (1956).
199. G. Natta and R. Ercoli, *Chim. ind. (Milan)*, **34**, 503 (1952).
200. G. Natta, R. Ercoli, S. Castellano, and F. H. Barbieri, *J. Amer. Chem. Soc.*, **76**, 4049 (1954).
201. D. S. Breslow and R. F. Heck, *Chem. Ind.*, 467 (1960); *J. Amer. Chem. Soc.*, **83**, 4023 (1961); **84**, 2499 (1962).
202. T. H. Coffield, J. Kozikowski, and R. D. Closson, *J. Org. Chem.*, **22**, 598 (1957).
203. R. J. Mawby, F. Basolo, and R. G. Pearson, *J. Amer. Chem. Soc.*, **86**, 3994 (1964).
204. F. Basolo and R. G. Pearson, *Mechanisms of Inorganic Reactions*, 2nd ed., J. Wiley, and Sons, New York, 1967.
205. R. F. Heck, *J. Amer. Chem. Soc.*, **86**, 5138 (1964).
206. C. L. Aldridge and H. B. Jonassen, *J. Amer. Chem. Soc.*, **85**, 886 (1963).
207. J. W. Reppe, *Ann.*, **582**, 116 (1953).
208. P. Krumholz and H. M. A. Stettiner, *J. Amer. Chem. Soc.*, **71**, 3035 (1949).
209. H. W. Sternberg, R. Markby, and I. Wender, *J. Amer. Chem. Soc.*, **79**, 6116 (1957); W. Hieber and G. Brendel, *Z. Anorg. Allgem. Chem.*, **289**, 324 (1957).
210. L. Watts, J. D. Fitzpatrick, and R. Pettit, *J. Amer. Chem. Soc.*, **87**, 3253 (1965); J. D. Fitzpatrick, L. Watts, G. F. Emerson, and R. Pettit, *ibid.*, **87**, 3254 (1965); G. F. Emerson and R. Pettit, *ibid.*, **84**, 4591 (1962); M. L. H. Green and P. L. I. Nagy, *ibid.*, **84**, 1310 (1962); J. Tsuji and K. Ohno, *Tetrahedron Lett.*, 4713 (1966); E. J. Corey and E. K. W. Wat, *J. Amer. Chem. Soc.*, **89**, 2757 (1967); G. N. Schrauzer et al., *ibid.*, **88**, 4890 (1966); A. J. Chalk and J. F. Harrod, *ibid.*, **89**, 1640 (1967); Shell Int. Res. Maat. N. V. Belg. Patent 626, 407 (June 21, 1963), French Patent 1,352,206 (January 1, 1964), U.S. Patent, 3,258,502.

211. R. F. Heck and D. S. Breslow, *J. Amer. Chem. Soc.*, **85**, 2779 (1963).
212. W. Reppe and H. Kroper, *Ann.*, **582**, 38 (1953).
213. J. D. Rose and F. S. Statham, *J. Chem. Soc.*, 69 (1950).
214. R. F. Heck, *J. Amer. Chem. Soc.*, **85**, 2013 (1963).
215. J. Tsuji, J. Kiji, S. Imamura, and M. Morikawa, *J. Amer. Chem. Soc.*, **86**, 4350 (1964); J. Tsuji, M. Morikawa, and J. Kiji, *ibid.*, **86**, 4851 (1964); J. Tsuji and S. Hosaka, *ibid.*, **87**, 4075 (1965); J. Tsuji and K. Ohno, *J. Amer. Chem. Soc.*, **88**, 3452 (1966), see also *Chem. Eng. News*, 50 (September 18, 1967); K. B. Yatsimirskii and A. P. Filippov, *Kin. Cat.*, **6**, 599 (1965).
216. S. Brewis and P. R. Hughes, *Chem. Comm.*, 489 (1965).
217. J. A. Osborn, G. Wilkinson, and J. F. Young, *Chem. Comm.*, 17 (1965).
218. J. Smidt et al., *Angew Chem.*, **71**, 176 (1959); *Angew. Chem. Int. Ed.*, **1**, 80 (1962); J. Smidt, *Chem. Ind.*, 54 (1962).
219. P. M. Henry, *J. Amer. Chem. Soc.*, **86**, 3246 (1964).
220. I. I. Moiseev, M. N. Vargaftik, and Y. K. Syrkin, *Izv. Akad. Nauk SSSR Otd. Khim. Nauk*, 1144 (1963).
221. I. I. Moiseev, M. N. Vargaftik, and Y. K. Syrkin, *Dokl. Akad. Nauk SSSR*, **133**, 377 (1960); E. W. Stern and M. L. Spector, *Proc. Chem. Soc.*, 370 (1961); E. W. Stern, *ibid.*, 111 (1963).
222. G. F. Wright, *Ann. N.Y. Acad. Sci.*, **65**, 436 (1957).
223. R. Criegee, *Angew. Chem.*, **70**, 173 (1958).
224. R. R. Grinstead, *J. Org. Chem.*, **26**, 238 (1961).
225. P. M. Henry, *J. Amer. Chem. Soc.*, **87**, 990, 4423 (1965).
226. M. E. Volpin and V. B. Shur, *Nature*, **209**, 1236 (1966); H. Brintzinger, *J. Amer. Chem. Soc.*, **88**, 623, 4305, 4307, (1966).
227. A. D. Allen and C. D. Senoff, *Chem. Comm.*, 621 (1965); S. Ikeda et al., *J. Amer. Chem. Soc.*, **90**, 1089 (1968).
228. J. P. Collman and J. W. Kang, *J. Amer. Chem. Soc.*, **88**, 3459 (1966); H. A. Scheidegger, J. N. Armor, and H. Taube, *J. Amer. Chem. Soc.*, **90**, 3263 (1958); A. Sacco and M. Rossi, *Chem. Comm.*, 316 (1967); A. Misono, Y. Uchida, T. Saito, and M. Song, *ibid.*, 419 (1967); A. Yamamoto, S. Kitazume, L. S. Pu and S. Ikeda, *ibid.*, 79 (1967); E. E. van Tamelen et al., *J. Amer. Chem. Soc.*, **91**, 1551 (1969); G. W. Parshall, *J. Amer. Chem. Soc.*, **89**, 1822 (1967).
229. D. Seyferth, personal communication.
230. *Chem. Eng. News*, September 28, 1970, p. 39.

Part Three

Bridging Nonenzymic and Enzymic Catalysis

Chapter 9

INTRAMOLECULAR CATALYSIS

9.1 Intramolecular General Acid-Base Catalysis 283
    9.1.1 Catalysis by the Carboxylate Ion and Carboxyl Groups 283
    9.1.2 Catalysis by the Phosphoric Acid Group 287
    9.1.3 Catalysis by the Alcohol and Alkoxide Groups 290
9.2 Intramolecular Nucleophilic-Electrophilic Catalysis. 294
    9.2.1 Catalysis by Alkoxyl, Alkylthio, and Halogen Groups 295
    9.2.2 Catalysis by Carboxylate Ion and Carboxyl Groups 297
    9.2.3 Catalysis by Tertiary Amine Groups 304
    9.2.4 Catalysis by Amide Groups 306
    9.2.5 Catalysis by Hydroxyl and Primary Amino Groups 309
    9.2.6 Catalysis by Electrophilic Groups 312
9.3 Comparison of Intermolecular and Intramolecular Catalysis 312

A commonplace generalization is that intramolecular reactions occur more readily than corresponding intermolecular reactions. This phenomenon is widely seen in organic reactions. Nowhere has it been documented more extensively than in reactions exhibiting neighboring group participation. A multitude of substitution reactions at saturated carbon centers are subject to assistance by a wide variety of nucleophiles, delineated in large part by Winstein and co-workers. These processes have been attributed to anchimeric[1] or synartetic[2] assistance. Another view of these reactions is as intramolecular nucleophilic-catalyzed reactions for those reactions in which the overall stoichiometry is not affected by the participating group. The nucleophiles that participate in these reactions are diverse species such as the iodine atom, the alkylthio, methoxy, and ester groups, and various forms of carbon atoms including both unsaturated and saturated systems. These reactions have been the subject of much discussion[3] and considerable controversy.[4,5]

Although classical neighboring group participation has been restricted mainly to interactions by nucleophiles, there is no reason why intramolecular catalysis should not cover the entire gamut of catalysis, from general acid-base catalysis to nucleophilic-electrophilic catalysis. The problem of elucidating the mechanism of intramolecular catalysis is the problem of differentiating general base from nucleophilic catalysis, general base catalysis from a combination of general acid-hydroxide ion catalysis, and general acid catalysis from general base-oxonium ion catalysis. In other words, all the ambiguities of intermolecular systems are also found in intramolecular systems. The classical example of intramolecular catalysis (by internal carboxylate ion) is found in the hydrolysis of aspirin.[51]

In recent years considerable advances in intramolecular catalysis have been made, stimulated in large part by the hypothesis that intramolecular catalysis can serve as a simple model of the intracomplex catalysis exhibited by enzymes, and fostered by the hope that modes of catalysis seen in enzymic systems but not in intermolecular systems would appear in intramolecular systems.

Many catalyses not observed before have been observed in intramolecular catalysis, chiefly in general catalysis. This form of catalysis is quite important in intramolecular systems for several reasons. An intramolecular catalyst must have a high effective local concentration. If its stereochemistry is right, its reaction should occur readily. A corollary to this argument is that the ratio of the effective concentration of the intramolecular general catalyst to the concentration of external oxonium or hydroxide ion must be large, and thus the intramolecular catalyst should effectively compete with external catalysts.

Intramolecular catalysis is often detected when prototropic groups of the substrate become of kinetic importance. The pH-rate constant profile is thus indicative of such dependence, as seen in several of the graphs of Fig. 3.1. A sigmoidal rate constant versus pH curve that is maximal at high pH indicates a general base or nucleophilic catalysis while a sigmoidal curve that is maximal at low pH indicates a general acid catalysis. Of course, the ambiguities mentioned above must be considered. One experimental method of differentiating these ambiguities is to determine the effect of ionic strength on the profile (although this is not a general method). Consider a sigmoidal curve that is maximal at high pH, nominally a general base catalysis. If at low pH, ionic strength greatly affects the rate constant, the reaction is general base-catalyzed, since at low pH the amount of the free base form is minute and subject to a large percentage change by a slight perturbation of the pK

Intramolecular General Acid-Base Catalysis

by the ionic strength. If, however, the reaction is general acid-catalyzed, little effect of ionic strength on the rate constant will be seen at low pH since most of the species will be in the protonated form. A general acid catalysis can be distinguished from its general base-oxonium ion counterpart in like manner. Other methods of distinguishing between mechanistic alternatives follow those outlined in Chapters 5 and 6.

9.1 INTRAMOLECULAR GENERAL ACID-BASE CATALYSIS

9.1.1 Catalysis by the Carboxylate Ion and the Carboxyl Groups

Intermolecular general base catalysis of ketone enolization has its counterpart in intramolecular catalysis.[1-9] The iodinations of pyruvic acid and levulinic acid show dependence on both the ionized and unionized forms of the substrate but independence of external buffer concentration.[7,9] These reactions may be interpreted in terms of catalysis by the internal carboxylate ion and carboxylic acid, respectively. In levulinic acid, iodination takes place at the methylene group, and thus, the stereochemistry of the cyclic transition states in the two reactions is the same. A larger catalytic contribution by the carboxyl group of the more acidic pyruvic acid and a larger catalytic contribution of the carboxylate ion of the more basic levulinate ion are consistent with this interpretation.

The iodination of o-isobutyrylbenzoic acid exhibits the pH-rate constant profile shown in Fig. 9.1.[8] This profile can be kinetically described in terms of four components: an oxonium ion-catalyzed reaction, a hydroxide ion-catalyzed reaction, and reactions of the anionic form and the neutral form of the substrate, as shown in eq. 9.1.

$$k_{obs} = k_H[H^{\oplus}] + k_{OH}[OH^{\ominus}] + k_{RCO_2^{\ominus}} \left[\begin{array}{c} \text{O} \\ \parallel \\ \text{C6H4(CCHMe}_2\text{)(CO}_2^{\ominus}) \end{array} \right] + k_{RCO_2H} \left[\begin{array}{c} \text{O} \\ \parallel \\ \text{C6H4(CCHMe}_2\text{)(CO}_2\text{H)} \end{array} \right] \quad (9.1)$$

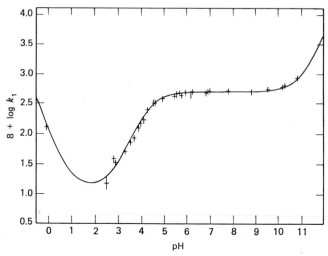

Fig. 9.1 The kinetics of iodination of *o*-isobutyrylbenzoic acid at 25.0° in aqueous solution, I = 0.50. The curve is a calculated curve. From E. T. Harper and M. L. Bender, *J. Amer. Chem. Soc.*, **87**, 5625 (1965). © 1965 by the American Chemical Society. Reprinted by permission of the copyright owner.

The principal kinetic term in the region pH 2.5 to 9 is the third term, involving carboxylate ion. This term, which is at least 50 times more important than the carboxylic acid term, is the one of mechanistic interest. Two mechanisms can be written for the carboxylate ion term.

Equation 9.2 depicts internal carboxylate ion catalysis whereas eq. 9.3 depicts external hydroxide ion catalysis facilitated by the internal carboxylic acid group. Mechanism 9.3 is ruled out on grounds that: (1) such catalysis is not seen in intermolecular reactions; (2) the rate constant of such a reaction would be too close to diffusion control for proton transfer; and (3) the rate constant of intramolecular iodination is higher for the isopropyl derivative described here than for the corresponding methyl derivative, whereas in intermolecular reactions the reverse is true. The only reasonable explanation of this inversion of reactivity is through mechanism 9.2, which postulates that stereochemically freezing the rotation of the molecule increases the reaction rate of the isopropyl over that of the methyl compound.

Although the hydrolysis of acetals via external catalysis is specific acid-catalyzed, general acid catalysis occurs when the catalyst is internal. The hydrolyses of o-carboxyphenyl β-D-glucoside (eq. 9.4),[10] of acid polysaccharides such as the alginates,[11] and of salicyl methyl formal[12] have pH-rate constant profiles showing dependence of an acid group of pK 3–4. At pH 3.5, the rate of hydrolysis of o-carboxyphenyl β-D-glucoside is 10^4 times faster than the rate of hydrolysis of the p-carboxyphenyl compound. On the basis of the profile and relative rates, the *ortho*-carboxyl group must be participating directly in the hydrolysis.[9]

$$\text{(structures)} \quad (9.4)$$

A priori, an intramolecular general acid catalysis or the kinetically equivalent combination of nucleophilic attack by internal carboxylate ion on the protonated substrate could occur. These mechanisms cannot be distinguished in the glucoside hydrolysis, but can in the similar hydrolysis of salicyl methyl formal. The relative rates of hydrolysis of this material and two related compounds (eq. 9.5) indicate that the *ortho*-carboxy compound hydrolyzes at an exceptionally high rate.

$$\text{(structures)} \quad (9.5)$$

(Relative rates)

Three mechanisms have been considered for this process.[12]

$$\text{(scheme)} \quad (9.6)$$

$$\text{(scheme)} \quad (9.7)$$

$$\text{(scheme)} \quad (9.8)$$

Equations 9.6 and 9.7 involve electrophilic-nucleophilic mechanisms in which the *ortho*-carboxylate ion reacts with the protonated substrate, forming an intermediate that can decompose to products. However, the postulated intermediate in eq. 9.6 is stable under the reaction conditions, whereas the postulated intermediate in eq. 9.7 leads to the products at a rate faster than the overall reaction. At the isosbestic point between reactant and product, the intermediate does not appear when the kinetics and its extinction coefficient demand it. By process of elimination, eq. 9.8, the classical description of general acid catalysis, must be the correct description of the reaction.

In the conversion of 2,2'-tolandicarboxylic acid to 3-(2-carboxybenzylidene)phthalide, only one of the two *ortho*-carboxylic acid groups participates as a nucleophile; the other must be a catalyst[14] (eq. 9.9). 2-Tolancarboxylic acid and 2,4'-tolandicarboxylic acid undergo this reaction.

The reaction of 2,2'-tolancarboxylic acid takes place approximately 10^4 times faster than either. In aqueous buffer solution, a bell-shaped pH-rate constant profile is seen for the 2,2' isomer with inflection points of approximately 3.2 and 4.8. This profile implies that the active species is the monoacid-monoanion. The rate constant of this species is approximately 6000 times larger than that for the anion of 2-tolancarboxylic acid under the same conditions. On this basis, a carboxylic acid group of the dicarboxylic acid must catalyze the addition of the carboxylate ion to the triple bond, possibly by proton donation to the unsaturated system.

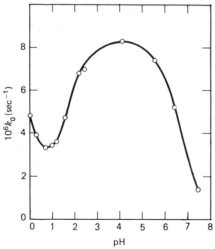

(9.9)

9.1.2 Catalysis by the Phosphoric Acid Group

Monoanions of phosphate monoesters hydrolyze with phosphorus–oxygen fission at rates much larger than those exhibited by either the dianion or the diacid. The pH-rate constant profile for the hydrolysis of a typical compound, methyl hydrogen phosphate, is shown in Fig. 9.2.[18] This

Fig. 9.2 The hydrolysis of methyl hydrogen phosphate at 100°. *A*—experimental; *B*—calculated on the basis of the monoanion. From C. A. Bunton, D. R. Llewellyn, K. G. Oldham, and C. A. Vernon, *J. Chem. Soc.*, 3574 (1958).

profile can be analyzed in terms of the sole reactivity of the monoanion, with a small contribution from an oxonium ion-catalyzed reaction of the neutral diacid. Reactions of the dianion and diacid, when they do occur, proceed with carbon–oxygen fission. In addition, the monoanion of methyl phosphate hydrolyzes approximately 10,000 times faster than the monoanion of dimethyl phosphate. Another mechanism is operative with phosphate esters of pK less than 5 and with p-nitrophenyl phosphate,[13] which differ by less than 100-fold in rate constant whereas they differ in leaving group basicity by up to 10^9. Five suggestions have been made to account for the former data.[15–18]

$$(9.10)$$

The first three transition states of eq. 9.10 portray the rate-determining formation of monomeric *meta*phosphate ion that would presumably add water in a fast step to give *ortho*phosphate. The last two transition states of eq. 9.10 indicate the direct formation of *ortho*phosphate ion. No matter what variant occurs, a phosphoric acid hydroxyl group acts as an internal general acid, facilitating the cleavage of the alcohol group. The lack of deuterium isotope effect in this reaction (a dangerous criterion) indicates that the proton transfer is not concerted with bond-breaking but must either precede or follow it.

*Meta*phosphate ion rather than *ortho*phosphate ion is probably the initial product of many of these reactions, as deduced from the small effect of the alkyl group on the rate of the reaction. Supporting evidence for this conclusion comes from a study of the hydrolysis of acetyl phosphate monoanion.[19,20] This species also contains an internal acidic hydroxyl group. Both the entropy of activation and the volume of activation are quite close to zero in this reaction while these quantities are large and negative in the hydrolysis of acetyl phenyl phosphate. The rate of hydrolysis of acetyl phosphate monoanion (involving carbon–oxygen cleavage) is considerably higher than that of simple phosphates. The effect of substituents on the hydrolysis of substituted phenyl phosphate monoanion is very small, and suggests that both proton transfer and stability of the anion are important factors leading to an overall cancellation of the effect of structure on reactivity. Thus the mechanism for the hydrolysis of acetyl phosphate monoanion may be written as eq. 9.11. The decomposition of carbamyl phosphate monoanion has been suggested to proceed in a similar fashion.[21]

$$\text{CH}_3-\overset{\overset{\text{O}}{\|}}{\text{C}}\overset{\text{H}}{\underset{\text{O}}{\diagdown}}\overset{\text{O}}{\underset{\diagdown}{\text{P}}}=\text{O} \longrightarrow \text{CH}_3-\overset{\overset{\text{O}}{\|}}{\text{C}}\overset{\text{H}}{\underset{\text{O}}{\diagdown}} + \overset{\text{O}}{\underset{\text{O}^\ominus}{\|}}\text{P}=\text{O}$$

$$\overset{\text{O}}{\underset{\text{O}^\ominus}{\diagdown}}\text{P}\overset{\text{O}}{\diagup} + \text{H}_2\text{O} \longrightarrow \text{H}_2\text{PO}_4^\ominus$$

(9.11)

Many other reactions of phosphorus derivatives occur by similar internal catalytic interactions. The monoanion of pyrophosphoric acid hydrolyzes much faster than the dianion or trianion and slightly faster than the neutral molecule.[22] The rate of hydrolysis of monoprotonated γ-phenylpropyl pyrophosphate is 2000 times as great as that for P,P'-bis-(γ-phenylpropyl) pyrophosphate.[23] These results can be interpreted in terms of internal catalysis by a phosphoric acid hydroxyl group, and also in terms of a monomeric *meta*phosphate intermediate in the hydrolytic reactions. The pH-rate constant profile for the hydrolysis of N-arylphosphoramidates indicates a mechanism related to eq. 9.10-2 or -5.[24,25] The hydrolysis of S-butylthiophosphate shows a bell-shaped pH-rate constant profile similar to that for the oxygen esters, again implying a similar mechanism.[26]

The implication of the above discussion is that any suitably situated acidic group in a phosphate molecule, no matter what its composition, should facilitate hydrolysis. The hydrolysis of salicyl phosphate exemplifies this conclusion.[27] The facile hydrolysis of this compound shows a bell-shaped pH dependence which can be interpreted in terms of the hydrolysis of a dianion, suggested to be the phosphate dianion.[27] However, there are three possible species exhibiting formal dianionic structures.

Species **1** can lead to catalysis through nucleophilic attack by carboxylate

$$\underset{1}{R\diagdown\overset{\text{OPO}_3\text{H}^\ominus}{\diagup}\diagdown\text{CO}_2^\ominus} \qquad \underset{2}{R\diagdown\overset{\text{OPO}_3{}^{2\ominus}}{\diagup}\diagdown\text{CO}_2\text{H}} \qquad \underset{3}{R\diagdown\overset{\overset{\overset{\text{PO}_3^\ominus}{|}}{\text{O}}}{\diagup}\diagdown\text{CO}_2^\ominus}\text{H}$$

(9.12)

ion with the formation of salicoyl phosphate. However, no salicoyl phosphate intermediate can be formed, since no oxygen-18 is found in the salicylic acid product when the hydrolysis is carried out in H_2O^{18} and no hydroxamic acid is formed when hydroxylamine is present. Species 2 implies that internal catalysis can take place by nucleophilic attack of phosphate dianion on the protonated carboxyl group. However, the cyclic phosphate intermediate of such a reaction hydrolyzes through salicyl phosphate and not vice versa; this then cannot be the path of the hydrolysis. A general-acid catalysis of the hydrolysis of salicyl phosphate by the *ortho*-carboxylic acid group is therefore postulated, the catalyst facilitating reaction by protonating the leaving group.

$$(9.13)$$

In eq. 9.13 the *ortho*-carboxylic acid group serves the same function as the internal hydroxyl groups of methyl hydrogen phosphate and acetyl phosphate. In no case is proton transfer concerted with bond-breaking since no deuterium oxide isotope effects are seen.[28]

A large effect of an *ortho*-carboxy group occurs in the hydrolysis of diethyl *o*-carboxyphenyl phosphate. The hydrolytic rate constant of the *ortho*-isomer exceeds the rate constant of the *para*-isomer by a factor of almost 10^8. Presumably proton donation by the *ortho*-carboxyl group is also important here.[29]

9.1.3 Catalysis by the Alcohol and Alkoxide Groups

Hydroxyl groups exist as substituents in many organic molecules. Their normal (electronic) substituent effects are not of great importance. However, as intramolecular catalysts, hydroxyl groups exert important effects.

Although simple tertiary alcohols on treatment with acid produce olefins containing the greatest alkyl substitution around the double bond, dehydration of the tertiary alcohol of eq. 9.14 produces the opposite result.

$$(9.14)$$

Intramolecular General Acid-Base Catalysis

This anomalous situation can be readily explained by an inspection of the carbonium ion intermediate formed in this reaction. It contains a hydroxyl group situated perfectly for selective removal of a proton from the terminal carbon but not from an internal carbon. Thus, the hydroxyl group may be viewed as an internal general base, facilitating this selective reaction.[30,31]

At 100° the dianion of glucose 6-phosphate is approximately five times more reactive in hydrolysis than the monoanion. This observation, which is contrary to the discussion given in the previous section, indicates that some special structural feature of glucose 6-phosphate must be involved. The 1-hydroxyl group of glucose is relatively acidic, with a pK of 10.8 at 100°. Since β-glucose 6-phosphate can assume an unfavorable conformation in which the hydrogen atom of the 1-hydroxy group can interact with the phosphate group, a plausible mechanism to explain this result involves proton transfer from the 1-hydroxyl group to the 6-oxygen atom, assisting cleavage of the phosphorus–oxygen bond.[32,33]

(9.15)

In the hydrolysis of salicylate esters and amides, the *ortho*-hydroxyl group is an important participant. The complex pH-rate constant profile for the hydrolysis of *p*-nitrophenyl 5-nitrosalicylate, shown in Fig. 9.3, can be interpreted in terms of three reactions: (1) reaction of the neutral substrate with water in the left-hand valley; (2) reaction of the anionic substrate with water in the central region, including the rising portion and the central plateau; and (3) reaction of the anion of the substrate with hydroxide ion in the right-hand region dependent on the concentration of hydroxide ion.[34]

Similar observations have been made for methyl salicylate,[35] phenyl salicylate,[36] and salicylamide.[37] The rate constant of hydrolysis of the imide from *N*-carbobenzyloxy-β-benzylaspartyl-*o*-hydroxyanilide is dependent on the

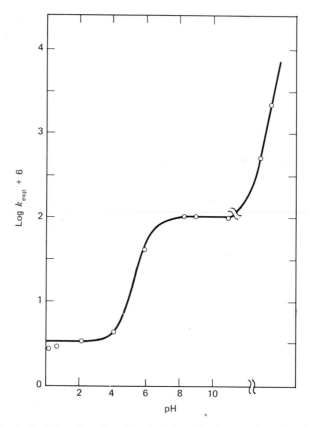

Fig. 9.3 The hydrolysis of *p*-nitrophenyl 5-nitrosalicylate, *p*-nitrophenyl 2-methoxy-5-nitrobenzoate, and *p*-nitrophenyl 3-nitrobenzoate in 34.4% dioxane-water at 25°. From M. L. Bender, F. J. Kézdy and B. Zerner, *J. Amer. Chem. Soc.*, **85**, 3017 (1963). © 1963 by the American Chemical Society. Reprinted by permission of the copyright owner.

phenolate ion form of the substrate and is 10^4 times greater than that for the *p*-isomer.[38] The second-order hydroxide ion rate constant of the *ortho*-isomer is $3.3 \times 10^3 \ M^{-1} \ \text{sec}^{-1}$, one of the largest reported for the hydrolysis of ordinary esters.

Of the three pH regions in the salicylate hydrolyses, the middle one is the most interesting. The reaction in this region was designated above as one of water with the anion of the substrate. However, as noted in eq. 9.16, it can be viewed in that fashion or alternatively as the kinetically equivalent reaction of the neutral substrate with hydroxide ion.

$$\text{(o-O}^{\ominus}\text{)C}_6\text{H}_4\text{C(O)OR} + \text{H}_2\text{O} \rightleftharpoons (\text{o-OH})\text{C}_6\text{H}_4\text{C(O)OR} + \text{OH}^{\ominus} \quad (9.16)$$

↓ products ↓ products

The hydrolysis of *p*-nitrophenyl 5-nitrosalicylate is exceptionally rapid. The rate constant calculated in terms of the reaction of hydroxide ion with the neutral substrate is approximately 2000 times larger than that of the corresponding compound containing an *ortho*-methoxyl group. However, reactions of *p*-nitrophenyl 5-nitrosalicylate with external nucleophiles such as imidazole, sulfite, and azide ion, whose concentration can be varied independently of the substrate species, are not exceptionally rapid when compared to the corresponding *o*-methoxyl compound. These reactions therefore must be described not as reactions of hydroxide ion with the neutral substrates but rather as reactions of the anionic species with water. This description implies intramolecular general base catalysis by the *ortho*-phenoxide ion.

$$(9.17)$$

Similar rate enhancements are seen in the hydrolysis of monocatecholate esters, including the acetate, chloroacetate, cinnamate, and benzoate. Reactions of these esters with hydroxide ion show rate enhancements with respect to the corresponding methoxy compound or the corresponding hydrogen compound of 200-fold to 800-fold.[36,39–41]. These rate enhancements, which are obviously not due to a nucleophilic interaction of the hydroxyl group, are meaningful only if the reaction takes place by means of the neutral substrate and hydroxide ion rather than by the kinetically equivalent anion and water. Only in one instance has this ambiguity been resolved experimentally: the hydrolysis of catechol monobenzoate. The reaction of this ester with imidazole is no faster than the reaction of the corresponding *ortho*-methoxy compound.[36] Therefore, for this process reaction of the neutral substrate with hydroxide ion must be ruled out and the reaction of the anion with water must be operative.

An aliphatic hydroxyl group located in proximity to an ester bond may facilitate its alkaline hydrolysis. In a series of cholestane 3-acetoxy-5-hydroxy steroids, the axial esters solvolyze faster than the equatorial esters, when the ester bond is *cis* to the hydroxyl group.[42] Similar acceleration of the base-catalyzed methanolysis of other cyclohexane-1,3-diol monoacetates also occurs.[43] These effects are small: the relative rate constants of alkaline hydrolysis of cholestane-3β-ol acetate and cholestane-3β,4β-diol 3-monoacetate are 1:9.[44] Rate accelerations of approximately equal magnitude occur in the alkaline hydrolysis of monoglycyl derivatives of *cis*-tetrahydrofuran-3,4-diol relative to 3-hydroxytetrahydrofuran; somewhat smaller accelerations are found in the hydrolysis of the *trans*-derivatives.[45] The largest effect of a hydroxyl group was found in comparing the alkaline hydrolysis of cyclopentyl acetate with *cis*-2-hydroxycyclopentyl acetate[46]; here a difference of 33-fold between the *cis*-2-hydroxyl and unsubstituted compounds was found, but the *cis* compound is only two-fold faster than the *trans*-compound. The 30-fold greater rate of hydrolysis of leucyl-RNA compared to leucine ethyl ester may also be due to an interaction with the neighboring hydroxyl group.[47] If these rate enhancements are in fact attributable to a direct interaction of the hydroxyl group in the transition state of the reaction, they can be described as hydrogen bonding to the carbonyl oxygen[42] or the ethereal oxygen[45] or as a microscopic solvent change.[46] However, the presence of the vicinal hydroxyl group leads to a considerable lowering of pK of the leaving group, which could account for the small rate enhancements noted here.[48] Therefore, only those reactions where a 1,3- rather than a 1,2-hydroxyl interaction occurs, where a significant difference in the rates between the hydroxy and unsubstituted compounds occurs, and where there is a significant difference between the *cis*- and *trans*-isomers can be considered to truly show a hydroxyl interaction.

9.2 INTRAMOLECULAR NUCLEOPHILIC-ELECTROPHILIC CATALYSIS

Nucleophilic interactions within molecules are manifold. It is therefore not surprising that intramolecular nucleophilic catalysis is of great importance. Since nucleophiles by definition react with electrophiles, the same reactions could also be thought of as electrophilic catalysis. However, the usual fixation on the nucleophilic reagent rather than on the electrophilic substrate leads to the result that in intramolecular systems nucleophilic catalysis is more important than electrophilic catalysis. This result is amplified by the fact that it is not easy to introduce electrophiles of the type discussed in Chapter 6 into a molecule.

9.2.1 Catalysis by Alkoxyl, Alkylthio, and Halogen Groups

As mentioned earlier, the phenomenon of neighboring group participation, especially in substitution reactions at saturated carbon, was the forerunner of much of the intramolecular nucleophilic catalysis to be discussed here. These reactions may be described as ones in which a group in a molecule stabilizes the transition state, not by ordinary electronic and steric effects transmitted through a field, but by direct or partial bonding to the reaction center. The implication of this description is that the interaction of the neighboring group results in a new pathway proceeding through an unstable intermediate. The acceleration due to the participation of a neighboring group has been defined as anchimeric or synartetic assistance. The acceleration must be defined in terms of a ratio of the rate in the presence of the neighboring group to the rate of some arbitrarily defined standard state. Many neighboring groups have both steric and electronic effects on the rate in addition to their direct participation in the reaction, especially in reactions at a saturated carbon atom where the neighboring group resides directly adjacent to the reaction center; hence, estimation of acceleration has been a knotty problem, even one of emotional involvement. Electronic effects can sometimes be calculated or otherwise approximated; steric effects are harder to evaluate. When the neighboring group is not situated directly adjacent to the reaction center, estimation of the acceleration becomes more straightforward, for both steric and electronic effects diminish.

Of the vast number of neighboring group participation reactions occurring in substitutions at saturated carbon involving catalysis by alkoxyl, alkylthio, halogen, ester, amide, hydroxyl, saturated carbon, unsaturated carbon, cyclopropane rings, and hydrogen groups, a few examples of catalysis by the first three groups will be discussed. More extensive discussion should be obtained elsewhere.[1-6]

The relative rates of solvolysis of a series of ω-methoxyalkyl p-bromobenzenesulfonates (Table 9.1) require that an effect of the methoxyl groups be other than electronic, because the maximal effect is found when the number of carbon atoms in the chain is four. The large rate constants of compounds with four (and five) carbon atoms suggest nucleophilic participation proceeding through cyclic oxonium ions (eq. 9.18).[1]

Although participation by the methoxyl group occurs through a five- or

(9.18)

Table 9.1
The Relative Rates of Solvolyses of ω-Methoxyalkyl
p-Bromobenzenesulfonates[a]

Compound	Relative Rate		
	EtOH (75°)	AcOH (25°)	HCO$_2$H (75°)
Me·[CH$_2$]$_3$·OBs	1.00	1.00	1.00
MeO·[CH$_2$]$_2$·OBs	0.25	0.28	0.10
MeO·[CH$_2$]$_3$·OBs	0.67	0.63	0.33
MeO·[CH$_2$]$_4$·OBs	20.4	657.0	461.0
MeO·[CH$_2$]$_5$·OBs	2.8	123.0	32.6
MeO·[CH$_2$]$_6$·OBs	1.19	1.16	1.13

[a] From S. Winstein et al., *Tetrahedron*, **3**, 1 (1958).

six-membered cyclic oxonium ion intermediate, but not when this intermediate would have seven, three, or four members, participation by a neighboring methylthio group in substitution reactions occurs through a three-membered cyclic intermediate. Many examples of this phenomenon occur, including the participation of a neighboring methylthio group in acetal hydrolysis.[51] The rate constants of the oxonium ion-catalyzed hydrolysis of CH$_3$CH(OEt)$_2$, CH$_3$OCH$_2$CH(OEt)$_2$, and CH$_3$SCH$_2$CH(OEt)$_2$ in 50% dioxane-water at 25° are 0.25, 2 × 10^{-4}, and 2.33 × 10^{-2} M^{-1} sec^{-1}. If the oxygen compound is used as a measure of the inductive effect of the sulfur atom on the rate constant, then the 100-fold higher reaction rate of the sulfur compound must be attributed to some specific interaction.

$$\text{CH}_3\text{SCH}_2\text{—CH(OC}_2\text{H}_5)_2 \xrightleftharpoons{\text{H}^\oplus} \text{CH}_3\text{S—CH}_2\text{—}\overset{\overset{\text{H}^\oplus}{|}\overset{\text{OC}_2\text{H}_5}{|}}{\text{CH}}\underset{\underset{\text{OC}_2\text{H}_5}{|}}{} \xrightarrow{-\text{C}_2\text{H}_5\text{OH}}$$

$$\underset{\underset{\text{OC}_2\text{H}_5}{|}}{\overset{\overset{\text{CH}_3}{|}}{\text{S}^\oplus}\diagdown\diagup}_{\text{CH}_2\text{—CH}} \xrightarrow[\text{fast}]{\text{H}_2\text{O}} \text{products} \quad (9.19)$$

Evidence for nucleophilic participation by halogen groups depends both on kinetic measurements and on the configuration of reaction products. The rate of formation of titratable acid in acetic acid solutions of *trans*-2-bromocyclohexyl p-bromobenzenesulfonate is appreciably higher than expected for its unassisted acetolysis as estimated by the rate for the *cis*-isomer, and from the correlation for other compounds of the same kind.[50] *trans*-2-Iodocyclohexyl *p*-toluenesulfonate is unstable and liberates titratable acid about 1000 times faster than the unsubstituted compound in acetic acid.[50] Although the products of these reactions were not determined, the results at least suggest participation by the bromo- and iodo-groups proceeding according to eq. 9.20. Mechanism 9.20 is consistent with a large body of stereochemical evidence indicating retention of configuration in reactions of *trans*-2-halocyclohexyl derivatives and with the fact that nucleophilic participation by halogen groups increases in the order Cl < Br < I.

$$\underset{I}{\diagup\!\!\!\diagdown}\text{OSO}_2\text{Ar} \rightleftarrows \underset{I^{\oplus\ominus}\text{OSO}_2\text{Ar}}{\diagup\!\!\!\diagdown} \xrightarrow{\text{HOAc}} \text{products} \qquad (9.20)$$

9.2.2 Catalysis by Carboxylate Ion and Carboxyl Groups

The pH-rate constant profile of the hydrolysis of aspirin consists of an acid-catalyzed reaction, a base-catalyzed reaction, and a reaction independent of pH near neutrality. The latter reaction, usually called a water reaction, can of course be a spontaneous reaction of the substrate. Closer inspection of Fig. 9.4[51] indicates that the pH-independent region is dependent on the ionization of a basic group of pK 4, such as the carboxylate ion. On

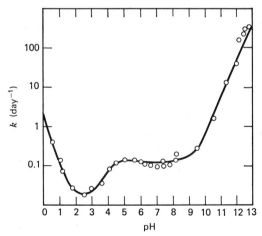

Fig. 9.4 The hydrolysis of aspirin at 25°C. From L. J. Edwards, *Trans. Faraday Soc.*, **46**, 723 (1950).

the basis of these arguments, together with the attractive stereochemical possibilities for intramolecular participation, this reaction was originally formulated as eq. 9.21a.[52] This mechanism is supported by the fact that the hydrolysis of aspirin in H_2O^{18} apparently gave a small amount of salicylic acid-O^{18} [53] and by the fact that the hydrolysis of aspirin is fifty times faster than that of its *para*-isomer.[54]

$$\text{(structures)} \quad (9.21a)$$

$$\text{(structures)} \quad (9.21b)$$

More recent evidence indicates a general base mechanism on the basis of further H_2O^{18} evidence (eq. 21b) although with certain substituents nucleophilic catalysis is observed.[55]

Monothiol esters of succinic acid hydrolyze readily in neutral solution[56] unless a nitro group is present in the molecule. A pH-rate constant profile of this reaction shows a maximum at about pH 6, indicating the participation of the carboxylate ion in the reaction and the importance of an acidic group of pK ca. 8.[57] This profile may be explained by postulating that carboxylate ion acts as a nucleophilic catalyst to displace the thiol group. The succinic anhydride thus formed can be partitioned in two directions. In the reverse direction, the thiolate ion (but not the thiol) regenerates the reactant. In the forward direction, water reacts to give the product.

Imides can apparently be hydrolyzed by neighboring carboxylate ion.[58] See also the hydrolysis of *o*-carboxyphthalimide.[117]

The hydrolysis of monoaryl succinates and glutarates has provided much information concerning intramolecular nucleophilic catalysis. Discovery of

Intramolecular Nucleophilic-Electrophilic Catalysis

catalysis by the internal carboxylate ion in these hydrolyses provided a large impetus to enlarge the study of intramolecular catalysis[59,60] beyond substitutions at saturated carbon. The pH-rate constant profiles of these reactions show dependence on a basic group of pK ca. 4.5.[59-62] The absolute rate constants of hydrolysis are quite high compared to comparable acetate esters. For example, mono-p-nitrophenyl glutarate hydrolyzes about 10^5 times faster than p-nitrophenyl acetate at pH 5.

Kinetic evidence indicates that the hydrolysis of mono-p-bromophenyl *exo*-3,6-endoxo-4-tetrahydrophthalate proceeds in two steps. On the basis of the pH dependence of the individual steps, the reaction may be most easily identified as the formation and decomposition of an anhydride intermediate.[63] In the hydrolysis of a series of mono-p-bromophenyl esters of dicarboxylic acids, rates of p-bromophenol release are in general lower than the rate of hydrolysis of the corresponding anhydride, with the exception of the ester mentioned above and the ester of maleic acid.[63,64] The relative rates of p-bromophenol loss and of anhydride hydrolysis for this series of esters are shown in Table 9.2. As one proceeds down through the table, the rate of p-bromophenol release increases markedly while no dramatic change is seen in the rate of anhydride hydrolysis. Thus a changeover occurs from rate-limiting formation of the anhydride to rate-limiting hydrolysis of the intermediate as one proceeds down the table; all of the reactions presumably follow:

$$\begin{array}{c}\text{C-O}\phi\text{Br} \\ \text{CO}^\ominus\end{array} \xrightarrow[-\text{HO}\phi\text{Br}]{k_1} \begin{array}{c}\text{C-O}\\\text{C=O}\end{array} \xrightarrow[+\text{H}_2\text{O}]{k_2} \begin{array}{c}\text{CO}_2\text{H}\\\text{CO}_2\text{H}\end{array} \qquad (9.22)$$

Some of the compounds in Table 9.2 involve succinic and glutaric acid monoesters which exist in more than one conformation as seen in eq. 9.23 for a substituted glutaric acid ester.[65]

$$^\ominus\text{O}_2\text{C}\diagdown\!\!\!\diagup\text{CO}_2\phi\text{X} \xrightleftharpoons{K} \diagdown\!\!\!\diagup\begin{array}{c}\text{CO}_2\phi\text{X}\\\text{CO}_2{}^\ominus\end{array} \qquad (9.23)$$

$$\downarrow$$
product

Table 9.2
Intramolecular Nucleophilic Catalysis in the Hydrolysis of Some Mono-p-Bromophenyl Esters[b]

Compound	$k_{\text{anhydride formation}}$ (relative)	$k_{\text{anhydride hydrolysis}}$ (relative)
cyclopentane-COOR, COO⁻	1[a]	1[a]
gem-dimethyl cyclopentane-COOR, COO⁻	20	0.07
cyclobutane-COOR, COO⁻	230	1.46
cyclobutene-COOR, COO⁻	10,000	11.2
bicyclic-COOR, CO₂⁻	53,000	5.2

[a] The two columns cannot be compared with one another quantitatively.
[b] From T. C. Bruice and U. K. Pandit, *Proc. Natl. Acad. Sci. U.S.*, **46**, 402 (1960).

In an intramolecular reaction leading to an anhydride intermediate, only one of the conformational isomers can react. Therefore, the Hammett-Curtin rule that the relative rates of two conformers are identical does not apply here. The observed rate constant for eq. 9.23 and related reactions, assuming that the reaction proceeds through a tetrahedral intermediate and from there to product, is given by eq. 9.24, where K is the conformational equilibrium constant and the k's refer to the formation and decomposition of the tetrahedral intermediate.

$$k_{\text{obs}} = \frac{K k_1}{(k_2/k_3) + 1} \tag{9.24}$$

The observed rate constants for a series of glutarate monoesters varies with

Intramolecular Nucleophilic-Electrophilic Catalysis

the structure of the substituent in the 3-position. A logarithmic plot of the observed rate constants versus the Taft steric constants for the 3-substituents shows a linear relationship, as indicated in Fig. 9.5. The effect of two individual substituents of *gem*-substitution is additive. Thus, intramolecular catalysis in flexible systems must take into account not only the catalyst but also substituents of the *gem*-dialkyl type which lead to the most favorable conformational equilibrium.

The importance of conformation is dramatically told in Table 9.2. Although the rate constants of hydrolysis of the anhydride intermediates are relatively insensitive to structure, the rate constants for anhydride formation are very sensitive to structural effects. For example, in proceeding from a glutarate monoester to a succinate monoester (a loss of one carbon atom) the rate constant increases by 230-fold; proceeding from the succinate monoester to the bicyclic system results in a further increase of 230-fold. The rate increases correspond to free energy differences of ca. 3 kcal/mole, approximately twice the difference expected from entropy considerations by removing one rotational degree of freedom in each change. Certainly, rotational entropy must account for a substantial part of these extraordinary rate differences; but other factors, including possibly the relief of steric strain in the last compound of Table 9.2, cannot be overlooked as appreciable contributors to the rate constants.

In the hydrolysis of monohydrogen phthalate esters, participation by both neighboring carboxylic acid and carboxylate ion occurs.[66-68] When the ester contains a poor leaving group such as a methoxy, ethoxy, or chloroethoxy group, the (neutral) carboxylic acid group participates in the reaction, as determined by pH-rate constant profiles. However, when the ester contains

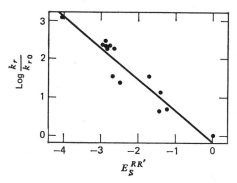

Fig. 9.5 Plot of log (k_r/K_{r0}) versus $E_s^{RR'}$ correlating the relative rate of ring closure with Taft's steric substituent constants. From T. C. Bruice and W. C. Bradbury, *J. Amer. Chem. Soc.*, **87**, 4846 (1965). © 1965 by the American Chemical Society. Reprinted by permission of the copyright owner.

a good leaving group, such as a trifluoroethoxy or phenoxy group, the anionic carboxylate ion participates in the reaction. Between these two extremes, both forms may participate simultaneously, as in propargyl and N-acetylserinamide esters. In the hydrolysis of phenyl hydrogen phthalate, the formation and the decomposition of the phthalic anhydride intermediate are seen spectrophotometrically.[66] The rate constant for the decay of the intermediate is equal to the rate constant for the hydrolysis of phthalic anhydride, proving that this hydrolysis has a nucleophilic catalysis mechanism.

$$\text{[phenyl hydrogen phthalate]} \xrightarrow{-\phi O^{\ominus}} \text{[phthalic anhydride]} \xrightarrow{+H_2O} \text{[phthalic acid]} \quad (9.25)$$

Several mechanisms have been suggested for reactions involving an intramolecular carboxylic acid catalyst. These will be discussed in the section on the hydrolysis of phthalamic acid.

A neighboring carboxylic acid group has a profound effect on the rates of hydrolysis of β-amic acids and β-cyano acids. These molecules contain both a hydrolyzable amide or nitrile linkage and a carboxylic acid group appropriately situated with respect to one another for internal interaction. The effect of the neighboring carboxylic acid group shows up in the acid hydrolysis of glycyl- and L-leucyl-L-asparagine;[69] in the release of aspartic acid from insulin, ribonuclease, and glucagon under acid conditions;[70] in the hydrolysis of succinamic acid,[71] succinanilic acid,[72] and phthalamic acid;[73] and in the hydrolysis of β-cyanopropionic acid, β-cyanobutyric acid, and o-cyanobenzoic acid.[74] Several of these hydrolyses have been shown to be independent of the external oxonium ion concentration and dependent on the unionized form of the carboxyl group. All of the reactions are quite rapid compared to the hydrolysis of ordinary amides or nitriles. These data point to the involvement of the internal carboxylic acid group in the reaction.

In order to specify the kind of effect that the carboxyl group exerts in these hydrolyses, one should consider the hydrolysis of phthalamic acid.[73] This reaction shows dependence on an acidic group of pK 3.5. At approximately pH 3, the hydrolysis of phthalamic acid is 10^5 times faster than the hydrolysis of benzamide. More importantly, at the same pH, the hydrolysis of phthalamic acid is 10^6 times faster than the hydrolysis of o-nitrobenzamide, a compound containing an ortho-substituent of similar electronic and steric properties. (The hydrolysis of o-cyanobenzoic acid may show an even greater rate enhancement.[74]) This large acceleration indicates that the carboxylic acid group cannot be affecting the reaction in the usual electronic manner, but must be participating in some covalent manner. Five mechanisms can be suggested a priori for this participation. Three of these involve the formation of phthalic anhydride intermediate either directly from the substrate through the intermediacy of a zwitterion, or indirectly by the reaction of the anion of the substrate with an external oxonium ion. The other two involve the action of the *o*-carboxylic acid group as a general acid catalyst, facilitating the addition of water to the amide group, or as a general base catalyst, facilitating the addition of water to a protonated amide group. The latter two mechanisms may be ruled out by a double label tracer experiment.

$$\text{(9.26)}$$

Decarboxylation of the phthalic acid formed via path **1** or path **2** leads to different isotopically labelled carbon dioxides, distinguishable by mass spectrometry. Results of such experiments are compatible only with path **2**. Thus, the reaction must proceed by one of the possible routes to phthalic anhydride and then to the phthalic acid product. By analogy with other general acid catalyses, the mechanism proceeding through zwitterion formation is probably the most attractive.

The microscopic reverse of these hydrolyses should also proceed by the same mechanism. This truism has been demonstrated in the reverse of the hydrolysis of succinanilic acid, the reaction of succinic acid with aniline. The formation of succinanilic acid from these components is much faster than the formation of an anilide from a monocarboxylic acid and aniline.[75] The rate-determining formation of succinic anhydride as intermediate in this reaction is supported by the pH-rate constant profiles of conversion of succinic acid to the anhydride and the reverse reaction, and of the hydrolysis of succinanilic acid, as well as the independently determined pH profile of the equilibrium constant between acid and anhydride.

9.2.3 Catalysis by Tertiary Amine Groups

Since tertiary amines are better nucleophiles than carboxylate ions, one might expect better intramolecular nucleophilic catalysis with the former than the latter.[76] Aryl esters of 4-(4'-imidazolyl)butyric acid are hydrolyzed at greatly enhanced rates in neutral solution in a reaction dependent on the concentration of free imidazole species.[77] By analogy with intermolecular nucleophilic catalysis by imidazole and because of the stereochemistry of the system, the reaction was thought to proceed through a lactam intermediate as shown in eq. 9.27. The loss of p-nitrophenol from the p-nitrophenyl ester, which presumably reflects lactam formation, has a half-life of 0.2 sec in 50% aqueous ethanol.

$$(9.27)$$

The reaction thus possesses a rate constant roughly equivalent to that for the conversion of a chymotrypsin-p-nitrophenyl acetate complex to the corresponding acyl-enzyme. A tetrahedral intermediate in the formation of the lactam was postulated because the pH dependency showed small deviations from the expected ionization curve, and because the apparent pK determined from the kinetics varied with the substituent on the aryl group.[78]

Although the hydrolysis of the methyl ester of 4-(4'-imidazolyl)butyric acid is not subject to intramolecular imidazole participation, the corresponding propylthiol ester does show such participation.[79] In fact, the reaction of the propylthiol ester to form the lactam intermediate is approximately 10^6–10^7 times as fast as the hydrolysis of a thiol ester with hydroxide ion at neutrality. In this reaction, addition of external thiol reduces the rate of disappearance of the thiol ester and therefore requires the presence of a lactam

Intramolecular Nucleophilic-Electrophilic Catalysis 305

intermediate. Furthermore, the amide of 4-(4'-imidazolyl)butyric acid is hydrolyzed with participation by the protonated imidazole group. This interaction is reminiscent of the participation of the protonated carboxyl group, rather than the carboxylate ion, in the hydrolysis of amic acids. The amic acid hydrolyses may be interpreted in terms of protonation of the amide group followed by attack by the neighboring nucleophile. The hydrolysis of 4-(4'-imidazolyl)butyramide may be similarly interpreted. This process, of course, is kinetically indistinguishable from the unlikely one involving attack by a protonated nucleophile.[80] An internal pyridine ring probably interacts in a similar way with the protonated amide group in the hydrolysis seen in eq. 9.28.[82] In this reaction, the cleavage of acylpyridinium ion would be expected to be rapid. The intermediacy of an acylpyridinium ion intermediate has been firmly established in pyridine-catalyzed acyl transfer.[81]

(9.28)

Related to intramolecular imidazole catalyses are intramolecular dimethylamino catalyses. Although imidazole is known to be an excellent nucleophile, especially toward activated esters, the dimethylamino group is a sterically hindered nucleophile. Its participation must be considered in terms of three mechanistic possibilities: (1) nucleophilic catalysis by the dimethylamino group; (2) general basic catalysis by the dimethylamino group; (3) or general acid catalysis by the dimethylammonium ion together with hydroxide ion attack.

The alkaline hydrolysis of acetylthiolcholine is surpassed in rate constant by the corresponding dimethylamino derivative by 240-fold.[83] Likewise, the relative hydroxide ion rate constants for ethyl acetate, acetylcholine, and protonated dimethylaminoethyl acetate are 1, 15, and 400.[83]

These rate differences can be explained as nucleophilic or general base participation by the dimethylamino group, or as general acid participation by the dimethylammonium ion. From a chemical point of view, the two most likely mechanisms involve either nucleophilic catalysis by the dimethylamino group or general acid catalysis by the dimethylammonium ion. Hansen favors the latter since the rate constants increase with increasing size of the alkyl group, a result not expected in a nucleophilic reaction.[84] Bruice and Benkovic favor the former, by analogy of the many O → N shifts involving comparable primary amines.[85]

In the reaction of trimethylamine with *p*-nitrophenyl acetate, nitrophenol release was interpreted to occur via nucleophilic catalysis by analogy with the corresponding imidazole reaction. The lack of a deuterium oxide effect was in agreement with this hypothesis.[86] The intramolecular analog of this reaction was investigated in the hydrolysis of *p*-substituted-phenyl 4-(*N*,*N*-dimethylamino)butyrates and valerates. Although no evidence for an acyltrialkylammonium ion intermediate was found in these reactions, they probably proceed by the same nucleophilic pathway as the corresponding intermolecular reaction since: (1) the Hammett rho constant had the same high value (+2.2–2.5) for both intermolecular and intramolecular processes; (2) the enthalpy of activation is identical for both intermolecular and intramolecular processes; (3) all steric factors in both intermolecular and intramolecular processes are reflected in the entropy of activation; and (4) $T\Delta S^{\ddagger}$ is 4–5 kcal/mole higher for the intramolecular process than for the intermolecular process.[87] Confirmation of a nucleophilic mechanism for this process comes from the fact that the rate constant for the intramolecular reaction calculated on the basis of a general acid-hydroxide ion reaction is approximately diffusion-controlled ($10^7 \ M^{-1} \ \text{sec}^{-1}$).[87]

Tertiary amines also catalyze reactions between cellulose and epoxides[88] and acylation of alcohols.[89]

9.2.4 Catalysis by Amide Groups

Intramolecular amide groups can exert profound nucleophilic catalysis. The amide anion is an ambident nucleophile; that is, either the oxygen or nitrogen atom may participate as the nucleophilic atom. A variety of mechanisms are seen when the oxygen atom is the nucleophile. These reactions proceed through oxazolone, oxazoline, or benzoylanthranil intermediates. When the nitrogen atom of the amide is the nucleophile, the intermediate is an imide.

The alkaline hydrolyses of succinamide, maleamide, phthalamide and 1,2-*cis*-cyclohexanedicarboxamide are considerably faster than the hydrolysis of acetamide. An imide intermediate was detected spectrophotometrically

Intramolecular Nucleophilic-Electrophilic Catalysis

in the hydrolysis of the diamides.[90,91] On this basis, eq. 9.29 may be written:

$$\text{(9.29)}$$

Earlier the alkaline hydrolysis of N-benzyloxycarbonyl-L-asparagine methyl ester was suggested to proceed in the same fashion because the imide intermediate was isolated and its formation and decomposition were observed polarimetrically.[92] Extraordinary rate constants are seen in these reactions. For example, the rate constant of the conversion of methyl N-methylphthalamate to the corresponding imide with hydroxide ion is 12,400 M^{-1} sec^{-1}. However, the imide hydrolyzes to the final product 200–300 times more slowly than does the parent ester with hydroxide ion.[93,94] The rate constant of these reactions is very sensitive to the structure of the alkyl group on the amide. The most significant rate increase with alkyl substitution on the amide was found when the amide group contains a serinamide moiety (eq. 9.30).

$$\text{(9.30)}$$

The rate constant for the first step of this reaction is approximately 10^7 times higher than that for the hydrolysis of benzyl propionate. This acceleration implicates the hydroxyl group in addition to the amido group.[95]

The racemization of optically active acylamino acid derivatives has been interpreted for some time in terms of the formation of oxazolone intermediates, through intramolecular interaction of the amido group with the carboxylic acid center. When it was found that both the hydrolysis of *p*-nitrophenyl *N*-benzyloxycarbonyl-glycyl-L-phenylalanine and the accompanying racemization, which is faster than hydrolysis, are dependent on the hydroxide ion concentration, it was suggested that both proceed through an oxazolone intermediate which can be partitioned to reactant and consequent racemization, or to products.[96] An oxazolone intermediate has been elegantly demonstrated in the hydrolysis of *p*-nitrophenyl hippurate on the basis of the following evidence. (1) The postulated oxazolone intermediate in this facile reaction, 2-phenyloxazolin-5-one, has a spectrum identical to the intermediate observed in alkaline solutions. (2) The rate of hydrolysis of the synthetic compound is identical to that for the intermediate. (3) The pK of the intermediate as determined kinetically is identical to the pK of the oxazolone. (4) The hydrolysis of *p*-nitrophenyl hippurate is general base catalyzed whereas the hydrolysis of *p*-nitrophenyl benzoylsarcosine, which cannot form an oxazolone, is not. (5) The appearance and disappearance of the intermediate quantitatively accounts for the entire reaction of the ester. On this basis, eq. 9.31 may be written.[97]

$$\text{(9.31)}$$

The hydrolysis of the diamide shown in eq. 9.32 proceeds approximately 10^4 times faster than that of *N,N*-dicyclohexylbenzamide in acetic acid solution.

$$\text{(9.32)}$$

Intramolecular Nucleophilic-Electrophilic Catalysis

The *ortho*-benzamido group probably functions as an intramolecular nucleophilic catalyst on the basis of: (1) the isolation of the intermediate benzoylanthranil from the reaction mixture in dry dioxane with hydrogen chloride; and (2) the formation of the hydrolysis product upon rapid decomposition of the intermediate in acetic acid solution.[98]

The alkaline hydrolysis of phosphoric acid triesters containing amido groups proceeds via a Δ^2-oxazoline intermediate.

$$\underset{\underset{NH-CH_2}{|}}{ArC} \overset{O}{\underset{CH_2-OP(OR)_2}{\Big\|}} \longrightarrow ArC \overset{O-CH_2}{\underset{N-CH_2}{\Big\langle}} \overset{H_2O}{\longrightarrow}$$

$$Ar-C \overset{O}{\underset{NH-CH_2}{\Big\langle}} \overset{CH_2OH}{|} + Ar-C \overset{O}{\underset{O-CH_2-CH_2NH_2}{\Big\langle}} \quad (9.33)$$

In aqueous solution around neutrality, an intermediate was detected spectrophotometrically and shown to be identical to an authentic sample of the Δ^2-oxazoline. In addition, the rate of decomposition of the intermediate was shown to be identical to that of the hydrolysis of the synthetic Δ^2-oxazoline.[99]

Intramolecular nucleophilic catalysis by amido groups is seen to be widespread and to occur through diverse pathways. A comparison of amide catalysis via imide or oxazolone formation indicates that the former is in general much faster than the latter assuming the same ring size, although in each case the ambident amide anion is presumably functioning. The reason for this difference is not apparent. However, the facile interaction of intramolecular amido groups must be kept in mind in analyzing the reactions of enzymes, all of which are polyamides.

9.2.5 Catalysis by Hydroxyl and Primary Amino Groups

Although an intramolecular hydroxyl group may participate as a general acid-base catalyst (see Section 9.1.3), in favorable circumstances it may also participate as an intramolecular nucleophilic catalyst. In Chapter 5, the greater nucleophilicity of alcohols than of water, and of alkoxides than of hydroxide ion, was mentioned. Since the reactivity of intramolecular versus intermolecular groups is enhanced and, since a number of cyclic oxygen groups such as five- and six-membered lactones, five-membered cyclic phosphates, and epoxides are unstable toward nucleophiles, it is not surprising that intramolecular catalysis by hydroxyl groups occurs.

The hydrolysis of amides containing a hydroxyl group in the δ-position occurs through the intermediate formation of a δ-lactone.[100,101] In N HCl, the acidic hydrolysis of δ-hydroxyvaleramide and γ-hydroxybutyramide are much faster than the hydrolysis of the corresponding unsubstituted amides.[103] Likewise, the hydrolysis of aldonamides (related to aldose) is much faster than the hydrolysis of unsubstituted amides. The significant pH independent reaction and the enhanced rate of the alkaline hydrolysis of γ-hydroxybutyramide indicate that hydroxyl participation occurs at all pH's, not just in acid.[100] The neutral and alkaline reactions can be described as internal alkoxide ion reactions with the protonated amido and neutral amido groups, respectively. The acid reaction can be described as the reaction of internal alcohol with the protonated amide.

$$(9.34)$$

Diesters of phosphoric acid containing a hydroxyl group on the β-carbon atom are particularly labile to hydrolysis compared to unsubstituted phosphate diesters. This phenomenon has been observed in the hydrolysis of glyceryl methyl phosphates,[104] glycerylphosphorylcholine,[105] alkoxyethyl methyl phosphates,[106] benzyl adenosine-3′-phosphate, and other compounds.[107] The alkaline hydrolysis of methyl 2-hydroxycyclohexyl phosphate, for example, is approximately 10^3 times faster than the hydrolysis of dimethyl phosphate.[108] Furthermore, the cis-ester hydrolyzes considerably faster than the trans-ester. These results suggest nucleophilic participation by the neighboring hydroxyl groups, forming a five-membered cyclic phosphate intermediate, which is subsequently decomposed.

Intramolecular Nucleophilic-Electrophilic Catalysis

$$\text{(9.35)}$$

The simplest five-membered cyclic phosphate, ethylene phosphate, hydrolyzes 10^7 times faster than dimethyl phosphate in alkaline solution. The latter reaction, however, involves mostly carbon–oxygen cleavage; hence, the rate differential for cleavage at phosphorus is 10^8.[109] This facile reaction, due to the strained nature of the cycle,[110] accounts for the rapid hydrolysis of the intermediate in these intramolecular catalyses, but it also introduces another problem: if the ring is strained, how can it readily be formed? This question can be answered by considering that the driving force of the cyclization is the entropy gained when two molecules are formed from one. In all fairness, however, it should be pointed out that this entropy cannot be as large as in a simple cleavage because degrees of freedom are lost on cyclization. Alternatively, the entropic gain in cyclization can be expressed in terms of the entropy due to the proximity of the reactants in the intramolecular process.

The alkaline hydrolyses of p-nitrophenyl β-glycosides exhibit greatly enhanced rates with respect to the corresponding 2-O-methyl derivatives.[111] Substitution of hydroperoxide anion for hydroxide ion reduces the rate of hydrolysis of the glycosides, indicating that the hydroxide ion acts as a base, rather than as a nucleophile. These observations suggest eq. 9.36 involving an epoxide intermediate.

$$\text{(9.36)}$$

Product analysis and the effect of hydroperoxide anion on the 2-O-methyl derivatives indicate a change in mechanism to an aromatic nucleophilic substitution reaction.

The alkaline hydrolysis of diphenyl β-aminoethyl phosphate is a rapid process yielding two moles of phenol. In contrast, the alkaline hydrolysis of diphenyl alkyl phosphates is a slow reaction yielding one mole of phenol since the anion of the diester is resistant to further hydrolysis. In addition, the first mole of phenol is produced more rapidly than the second. These observations lead to [112,113]

$$\phi O\underset{\phi O}{\overset{O}{\underset{\|}{P}}}\underset{H_2N}{\overset{O}{\diagup}} \xrightarrow{-\phi OH} \underset{\phi}{\overset{O}{\underset{\|}{P}}}\underset{N}{\overset{O}{\diagup}}\underset{H}{} \xrightarrow[-\phi OH]{H_2O}$$

$$\underset{HOP}{\overset{O}{\underset{\|}{}}}\underset{N}{\overset{O}{\diagup}}\underset{H}{} \xrightarrow{H_2O} (HO)_2\overset{O}{\underset{\|}{P}}OCH_2CH_2NH_2 \qquad (9.37)$$

9.2.6 Catalysis by Electrophilic Groups

In contrast to the multitude of intramolecular catalyses by nucleophiles, very few are brought about by electrophiles defined in the sense of Chapter 6. As mentioned earlier, this is somewhat a matter of convention. For example, hydride transfer may be viewed as a nucleophilic interaction by hydride ion or an electrophilic interaction by the acceptor (carbonium ion).

One reaction does not seem to be clouded by this ambiguity. It involves the hydrolysis of the diazonium salt of *o*-aminophenyl 2,6-dimethylbenzoate. Although the corresponding *para*-diazonium salt is stable, the *ortho*-derivative is unstable, decomposing to 2,6-dimethylbenzoic acid and *ortho*-hydroxybenzenediazonium ion.[114] This reaction can be explained by electrophilic attack of the diazonium ion on the ethereal oxygen of the ester. The acylium ion of 2,6-dimethylbenzoic acid, a reasonably stable species from other evidence, thus formed, might then decompose to the product carboxylic acid.

9.3 COMPARISON OF INTERMOLECULAR AND INTRAMOLECULAR CATALYSIS

The sparse data that exist on intramolecular catalysis indicate that the same laws governing the relative efficacy of intermolecular general acid-base and nucleophilic-electrophilic catalysts hold in intramolecular systems. Thus, general bases of greater basicity, general acids of greater acidity, nucleophiles of greater nucleophilicity and electrophiles of greater electrophilicity are expected to be better intramolecular catalysts. However, this generalization must be tempered by the realization that intramolecular interactions place much more stringent requirements on the stereochemistry of the system than do intermolecular interactions. Therefore, exceptions to the above generalizations should be expected.

The proximity of the catalyst to the reaction center leads to a predominance of nucleophilic over general base catalyses. For example, there is no well-documented case of general base catalysis by imidazole in an intramolecular reaction whereas there are several in intermolecular situations. In addition, there are many more examples of catalysis by a carboxylic acid group acting mechanistically as an oxonium ion-nucleophilic catalyst in intramolecular systems than in intermolecular systems. The list could be extended considerably to indicate the increased importance of nucleophilic catalysis in intramolecular systems.

A kinetic comparison of corresponding intramolecular and intermolecular catalyses reveals some profound differences. These catalyses correspond to first-order and second-order processes, respectively, and therefore are not amenable to straightforward comparison. Nevertheless, a comparison can be made by calculating what concentration of the intermolecular catalyst is necessary for equivalent rates of reaction of the intramolecular and intermolecular catalyses (assuming equivalent concentrations of the two substrates). This calculation leads to a concentration of external catalyst, which may be equated to the effective (local) concentration of the internal catalyst.

For example, the iodination of o-isobutyrylbenzoate ion may be compared with the benzoate ion-catalyzed iodination of acetophenone. This comparison indicates that the internal carboxylate ion is equivalent to 50 M of external carboxylate ion, this factor making the second-order rate constant of the intermolecular reaction equal the first-order rate constant of the intramolecular reaction.[8] This result can be alternatively expressed by saying that the effective concentration of the o-carboxylate ion in o-isobutyrylbenzoate is 50 M. In the less rigid aliphatic system involving levulinic acid, the efficiency of intramolecular catalysis is less, and the internal carboxylate ion is equivalent to only 1 M of external carboxylate ion.[7]

One of the plagues of this comparison is the possibility that the change from intermolecular to intramolecular catalysis will result in a change in mechanism. For example, both intramolecular catalysis by carboxylate ion in the hydrolysis of monoaryl hydrogen glutarates and intermolecular catalysis by carboxylate ion in the hydrolysis of aryl acetates with acetate ion were once thought to be nucleophilic in character. But subsequent investigation has indicated that, whereas the intramolecular catalyses are nucleophilic,[63] some of the intermolecular catalyses are general basic, some nucleophilic, and some mixed.[115]

The intramolecular-intermolecular comparison is further confused by the fact, discussed earlier, that the conformation of a molecule undergoing intramolecular catalysis can have a large effect on the catalytic rate constant. For example, if one were to compare an intermolecular catalysis of acetate ion and phenyl acetate with the intramolecular catalysis in monophenyl hydrogen glutarate, should one use the unsubstituted glutarate or the β,β-dimethyl derivative, which has a more favorable conformation for intramolecular interaction?[65]

Notwithstanding these problems, we shall sail into the muddy waters of this comparison. Comparison of the intermolecular catalysis in the hydrolysis of p-nitrophenyl acetate with acetate ion and the intramolecular hydrolysis of mono-p-nitrophenyl glutarate indicates that the internal catalyst is equivalent to ca. 600 M of the external catalyst. Since the intermolecular reaction has been shown to proceed by nucleophilic catalysis to the extent of 50%,[115] this comparison is good to a factor of 2. Table 9.3 shows calculations comparing intermolecular and intramolecular reactions in nucleophilic catalyses by imidazole and the dimethylamino group. The table indicates that an internal dimethylamino catalyst is equivalent to 5370 M of the corresponding intermolecular catalyst in the hydrolysis of a p-nitrophenyl ester. On the other hand, an internal imidazole catalyst is equivalent to only 9.4 M of the corresponding intermolecular catalyst in the same reaction. These differences may reflect differences in conformation of the two systems,

Table 9.3
Ratios of Intra:Intermolecular Catalytic Rate Constants for the Hydrolysis of m- and p-Substituted Phenyl Esters[d]

Substituent	Me_2N-[a,c] (M)	Imidazole-[b]
H	1260	24
p-Cl	1080	23
m-NO$_2$	1700	32
p-NO$_2$	5370	9.4

[a] Intramolecular reaction: γ-(N,N-dimethylamino)butyrates.
[b] Intramolecular reaction: γ-(4-imidazolyl)butyrates.
[c] The valerates are the same within a factor of 2.
[d] From T. C. Bruice and S. J. Benkovic, *J. Amer. Chem. Soc.*, **85**, 1 (1963). © 1963 by the American Chemical Society. Reprinted by permission of the copyright owner.

since the former was carried out in water while the latter was performed in 50% ethanol, assuming partitioning of the tetrahedral intermediate to be the same. The conformation of the intramolecular dimethylamino group in water may be favorable for reaction while the conformation of the intramolecular imidazole group in 50% ethanol-water may be unfavorable for reaction. This hypothesis may be tested by carrying out these reactions in the same solvent or by determining the effect of *gem*-dimethyl substitution on these two reactions. The former should have no effect while the latter should have an appreciable effect if this hypothesis is correct. Nonetheless, these values provide a range in which the comparison between intermolecular and intramolecular catalysis may be found. Certainly, intramolecular catalysis is generally much more powerful than intermolecular catalysis.

The thermodynamic reason for the difference between intermolecular and intramolecular catalysis has been determined for catalyses by the dimethylamino group. Table 9.4 shows that the enthalpies of activation of the intermolecular and intramolecular reactions are roughly constant, but that the entropies of activation for the intramolecular catalyses are approximately 4–5 kcal/mole more favorable than for the intermolecular catalyses. This result would be expected on the basis that intramolecular processes do not lose the translational degrees of freedom in going from the ground to transition state that intermolecular catalyses do. In general, intramolecular reactions may be superior because the ground state is raised or the transition state lowered.

Table 9.4

Activation Parameters for Nucleophilic Displacement by the Dimethylamino Group in Intermolecular and Intramolecular Catalyses in the Hydrolysis of *m*- and *p*-Substituted Phenyl Esters[c]

Substituent	Intermolecular (kcal mole^{-1})		Intramolecular (kcal mole^{-1})[a]		(kcal mole^{-1})[b]	
	$\Delta H^\ddagger$	$T\Delta S^\ddagger$	$\Delta H^\ddagger$	$T\Delta S^\ddagger$	$\Delta H^\ddagger$	$T\Delta S^\ddagger$
p-NO$_2$	12.3	−6.3	11.9	−1.9	11.5	−2.6
m-NO$_2$	12.1	−8.0	11.5	−4.3	11.8	−4.4
p-Cl	12.5	−9.1	15.9	−2.2	13.8	−4.1
H	12.9	−9.4	12.5	−5.7	12.3	−6.4
p-CH$_3$			13.7	−5.1	14.4	−5.5

[a] Butyrates.
[b] Valerates.
[c] From T. C. Bruice and S. J. Benkovic, *J. Amer. Chem. Soc.*, **85**, 1 (1963). © 1963 by the American Chemical Society. Reprinted by permission of the copyright owner.

Some catalyses seen in intramolecular systems do not exist in intermolecular systems. For example, catalysis is exceedingly powerful in phthalamic acid hydrolysis and related reactions. But in order to find a comparable intermolecular reaction, one must look to the reaction of *N*-butylacetamide and acetic acid at 220° C;[116] if it is truly an analogous situation, it is almost infinitely slower than the intramolecular catalysis. Thus, intramolecular catalysis not only surpasses intermolecular catalysis quantitatively but perhaps also qualitatively.

The emphasis in the previous discussion has been that proximity is of overriding importance. This is certainly so, but together with this concept, the concept of correct orientation for facile catalysis in intramolecular systems must also be considered. The catalyst must not only have an effective local concentration of 50 *M* but it must have correct stereochemical orientation. Correct orientation of the catalyst with respect to the substrate is clearly seen in the phthalamic acid molecule (Fig. 9.6).[117] Stereochemical restrictions force the two *ortho* substituents perpendicular to one another in the ground state. No degrees of rotational freedom are lost in going from this ground state to a transition state of very similar structure, and thus the reaction rate is favorable not only because of proximity but also because of orientation.

The fact that intramolecular catalysis is more powerful than intermolecular catalysis implies that its accelerations with respect to oxonium or hydroxide ion catalysis must be large. Three representative comparisons are given for intramolecular general acid catalysis, intramolecular general base catalysis, and intramolecular nucleophilic catalysis. Intramolecular general acid catalysis in the hydrolysis of *o*-carboxyphenyl β-D-glucoside leads to a rate constant 10^4 times greater than the oxonium ion-catalyzed hydrolysis of a glucoside at pH 3.5 and above.[10] Intramolecular general base catalysis in the hydrolysis of *p*-nitrophenyl 5-nitrosalicylate leads to rate constants 10^3 times greater than the hydroxide ion-catalyzed hydrolysis of *p*-nitrophenyl

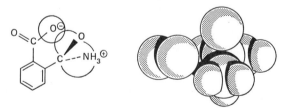

Fig. 9.6 Two representations of the phthalamic acid molecule, showing perpendicular attack of the carboxylate ion on the adjacent protonated amide. From B. Zerner and M. L. Bender, *J. Amer. Chem. Soc.*, **83**, 2267 (1961). © 1961 by the American Chemical Society. Reprinted by permission of the copyright owner.

2-methoxy-5-nitrobenzoate at pH 6 and below.[37] Intramolecular nucleophilic catalysis by the dimethylamino group in the hydrolysis of p-nitrophenyl 4-(N,N-dimethylamino)butyrate leads to rate constants 10^5 times greater than the hydroxide ion-catalyzed hydrolysis of p-nitrophenyl acetate at pH 8 and below. These large rate enhancements around neutrality are partial answers to the rate enhancements of the order of 10^{10} which enzymic catalysts show over oxonium and hydroxide ion rates around neutrality.

REFERENCES

1. S. Winstein and E. Grunwald, *J. Amer. Chem. Soc.*, **70**, 828 (1948); S. Winstein, E. Allred, R. Heck, and R. Glick, *Tetrahedron*, **3**, 1 (1958).
2. C. K. Ingold, *Structure and Mechanism in Organic Chemistry*, Cornell University Press, Ithaca, N.Y., 1953, p. 511.
3. B. Capon, *Quart. Rev.*, **18**, 45 (1964).
4. P. D. Bartlett, *Nonclassical Ions*, W. A. Benjamin, Inc., New York, N.Y., 1965; J. L. Coke, *J. Amer. Chem. Soc.*, **89**, 135 (1967).
5. G. D. Sargent, *Quart. Rev.*, **20**, 301 (1966); T. G. Traylor and C. L. Perrin, *J. Amer. Chem. Soc.*, **88**, 4934 (1966); A. Nickon and J. L. Lambert, *ibid.*, **88**, 1905 (1966); H. C. Brown et al., *J. Amer. Chem. Soc.*, **89**, 370 (1967).
6. See also, E. L. Allred and S. Winstein, *J. Amer. Chem. Soc.*, **89**, 3998, 3991, 4008, 4012 (1967).
7. R. P. Bell and M. A. D. Fluendy, *Trans. Faraday Soc.*, **59**, 1623 (1963); W. J. Albery, R. P. Bell and A. L. Powell, *J. Amer. Chem. Soc.*, **61**, 1194 (1965).
8. E. T. Harper and M. L. Bender, *J. Amer. Chem. Soc.*, **87**, 5625 (1965).
9. A. Schellenberger and G. Hübner, *Chem. Ber.*, **98**, 1938 (1965).
10. B. Capon, *Tetrahedron Lett.*, 911 (1963).
11. O. Smidsrød, A. Haug, and B. Larsen, *Acta Chem. Scand.*, **20**, 1026 (1966).
12. B. Capon and M. C. Smith, *Chem. Comm.*, 523 (1965).
13. A. J. Kirby and A. G. Varvoglis, *J. Amer. Chem. Soc.*, **89**, 415 (1967); C. A. Bunton, E. J. Fendler, and J. H. Fendler, *J. Amer. Chem. Soc.*, **89**, 1221 (1967).
14. R. L. Letsinger, E. N. Oftedahl, and J. R. Nazy, *J. Amer. Chem. Soc.*, **87**, 742 (1965); a different example is given in S. J. Benkovic, *ibid.*, **88**, 5511 (1966).
15. W. W. Butcher and F. H. Westheimer, *J. Amer. Chem. Soc.*, **77**, 2420 (1955).
16. J. Kumamoto and F. H. Westheimer, *J. Amer. Chem. Soc.*, **77**, 2515 (1955).
17. C. A. Vernon, *Spec. Publ. Chem. Soc.*, **8**, 17 (1957).
18. C. A. Bunton, C. A. Vernon, et al. *J. Chem. Soc.*, 3574 (1958); 3293 (1960); 1636 (1961).
19. G. Di Sabato and W. P. Jencks, *J. Amer. Chem. Soc.*, **83**, 4393 (1961).
20. W. P. Jencks, *Brookhaven Symp. Biol.*, **15**, 134 (1962).
21. C. M. Allen, Jr., and M. E. Jones, *Biochemistry*, **3**, 1238 (1964).
22. D. O. Campbell and M. L. Kilpatrick, *J. Amer. Chem. Soc.*, **76**, 893 (1954).
23. D. L. Miller and F. H. Westheimer, *J. Amer. Chem. Soc.*, **88**, 1507 (1966).
24. J. D. Chanley and E. Feageson, *J. Amer. Chem. Soc.*, **80**, 2686 (1958).
25. T. C. Bruice and S. J. Benkovic, *Bioorganic Mechanisms*, Vol. 2, W. A. Benjamin, Inc., New York, N.Y. 1966, p. 74.
26. E. B. Herr, Jr. and D. E. Koshland, Jr. *Biochim. Biophys. Acta*, **25**, 219 (1957).

27. J. D. Chanley, E. M. Gindler, and H. Sobotka, *J. Amer. Chem. Soc.*, **74**, 4347 (1952).
28. M. L. Bender and J. M. Lawlor, *J. Amer. Chem. Soc.*, **85**, 3010 (1963).
29. M. Gordon, V. A. Notaro, and C. E. Griffin, *J. Amer. Chem. Soc.*, **86**, 1898 (1964).
30. R. T. Arnold, *Helv. Chim. Acta*, **32**, 134 (1949).
31. P. F. G. Praill and B. Saville, *Chem. and Ind.*, 495 (1960).
32. C. A. Bunton and H. Chaimovich, *J. Amer. Chem. Soc.*, **88**, 4082 (1966).
33. Similar results have been found in the hydrolysis of glycero-1,2-hydrogen phosphate [L. Kugel and M. Halmann, *J. Amer. Chem. Soc.*, **89**, 4125 (1967)], and in the fragmentation of β-halophosphates [G. L. Kenyon and F. H. Westheimer, *J. Amer. Chem. Soc.*, **88**, 3561 (1966)].
34. M. L. Bender, F. J. Kézdy, and B. Zerner, *J. Amer. Chem. Soc.*, **85**, 3017 (1963).
35. M. L. Bender and F. L. Killian, unpublished results.
36. B. Capon and B. Ch. Ghosh, *J. Chem. Soc.*, B472 (1966).
37. T. C. Bruice and D. W. Tanner, *J. Org. Chem.*, **30**, 1668 (1965).
38. Y. Shalitin and S. A. Bernhard, *J. Amer. Chem. Soc.*, **86**, 2291 (1964).
39. B. Hansen, *Acta Chem. Scand.*, **17**, 1375 (1963).
40. E. J. Fuller, *J. Amer. Chem. Soc.*, **85**, 1777 (1963).
41. Y. Shalitin and S. A. Bernhard, *J. Amer. Chem. Soc.*, **86**, 2291 (1964).
42. H. B. Henbest and B. J. Lovell, *J. Chem. Soc.*, 1965 (1957).
43. S. M. Kupchan, W. S. Johnson, and S. Rajagopalan, *Tetrahedron*, **7**, 47 (1959); S. M. Kupchan and W. S. Johnson, *J. Amer. Chem. Soc.*, **78**, 3864 (1956); S. M. Kupchan and C. R. Narayanan, *ibid.*, **81**, 1913 (1959); S. M. Kupchan, J. H. Block, and A. C. Isenberg, *J. Amer. Chem. Soc.*, **89**, 1189 (1967).
44. S. M. Kupchan, P. Slade, and R. J. Young, *Tetrahedron Lett.*, **24**, 22 (1960).
45. H. G. Zachau and W. Karau, *Chem. Ber.*, **93**, 1830 (1960).
46. T. C. Bruice and T. H. Fife, *J. Amer. Chem. Soc.*, **84**, 1973 (1962).
47. R. Wolfenden, *Biochemistry*, **2**, 1090 (1963).
48. The logarithms of rate constants of alkaline hydrolysis of acetate esters are linearly related to the pK's of the alcoholic leaving groups with slope -0.32; J. F. Kirsch and W. P. Jencks, *J. Amer. Chem. Soc.*, **86**, 837 (1964).
49. J. C. Speck, Jr., D. J. Rynbrandt, and I. H. Kochevar, *ibid.*, **87**, 4979 (1965).
50. S. Winstein, E. Grunwald, and L. L. Ingraham, *J. Amer. Chem. Soc.*, **70**, 821 (1948).
51. L. J. Edwards, *Trans. Faraday Soc.*, **46**, 723 (1950); **48**, 696 (1952).
52. E. R. Garrett, *J. Amer. Chem. Soc.*, **79**, 3401 (1957); **80**, 4049 (1958); **82**, 711 (1960).
53. M. L. Bender, F. Chloupek, and M. C. Neveu, *J. Amer. Chem. Soc.*, **80**, 5384 (1958).
54. G. L. Schmir and T. C. Bruice, *J. Amer. Chem. Soc.*, **80**, 1173 (1958).
55. A. R. Fersht and A. J. Kirby, *ibid.*, **89**, 4857 (1967); **90**, 5826, 5833 (1968).
56. E. J. Simon and D. Shemin, *J. Amer. Chem. Soc.*, **75**, 2520 (1953).
57. M. Stiles, personal communication.
58. P. D. Hoagland and S. W. Fox, *J. Amer. Chem. Soc.*, **89**, 1389 (1967).
59. H. Morawetz and P. E. Zimmering, *J. Phys. Chem.*, **58**, 753 (1954).
60. H. Morawetz and E. W. Westhead, *J. Polymer Sci.*, **16**, 273 (1955).
61. H. Morawetz and E. Gaetjens, *J. Polymer Sci.*, **32**, 526 (1958).
62. E. Gaetjens and H. Morawetz, *J. Amer. Chem. Soc.*, **82**, 5328 (1960).
63. T. C. Bruice and U. K. Pandit, *J. Amer. Chem. Soc.*, **82**, 5858 (1960).
64. T. C. Bruice and U. K. Pandit, *Proc. Natl. Acad. Sci., U.S.*, **46**, 402 (1960); see also D. R. Storm and D. E. Koshland, Jr., *Proc. Natl. Acad. Sci., U.S.*, **66**, 445 (1970).

References

65. T. C. Bruice and W. C. Bradbury, *J. Amer. Chem. Soc.*, **87**, 4846 (1965); see also A. K. Herd, L. Eberson, and T. Higuchi, *J. Pharm. Sci.*, **55**, 162 (1966).
66. J. W. Thanassi and T. C. Bruice, *J. Amer. Chem. Soc.*, **88**, 747 (1966).
67. Å. Agren, U. Hedsten, and B. Jonsson, *Acta Chem. Scand.*, **15**, 1532 (1961).
68. L. Eberson, *Acta Chem. Scand.*, **16**, 2245 (1962); **18**, 2015 (1964).
69. S. J. Leach and H. Lindley, *Trans. Faraday Soc.*, **49**, 921 (1953).
70. J. Schultz, H. Allison, and M. Grice, *Biochemistry*, **1**, 694 (1962); T. Vajda, *Chem. Ind.*, 197 (1959).
71. A. Bruylants and F. J. Kézdy, *Rec. Chem. Progr.*, **21**, 213 (1960).
72. T. Higuchi, L. Eberson, and A. K. Herd, *J. Amer. Chem. Soc.*, **88**, 3805 (1966); T. Higuchi, L. Eberson, and J. D. McRae, *J. Amer. Chem. Soc.*, **89**, 3001 (1967).
73. M. L. Bender, *J. Amer. Chem. Soc.*, **79**, 1258 (1957); M. L. Bender, Y.-L. Chow, and F. Chloupek, *J. Amer. Chem. Soc.*, **80**, 5380 (1958).
74. S. Wideqvist, *Arkiv. fur Kemi*, **3**, 147, 281, 289 (1951); **2**, 383 (1950).
75. T. Higuchi, T. Miki, A. C. Shah, and A. K. Herd, *J. Amer. Chem. Soc.*, **85**, 3655 (1963); for a similar reaction see T. Higuchi, G. L. Flynn, and A. C. Shah, *ibid.*, **89**, 616 (1967).
76. *Chem. Eng. News*, April 3, 1967.
77. 1,2-Dimethyl-5,7-dinitrobenzimidazolyl-4-alanylglycine hydrolyzes 150,000 times as fast as 2,4-dinitrophenylalanylglycine; K. L. Kirk and L. A. Cohen, *J. Org. Chem.*, **34**, 395 (1969). This has been generalized; *J. Org. Chem.*, **34**, 390 (1969).
78. T. C. Bruice and J. M. Sturtevant, *J. Amer. Chem. Soc.*, **81**, 2860 (1959). Cf. G. L. Schmir and T. C. Bruice, *J. Amer. Chem. Soc.*, **80**, 1173 (1958).
79. T. C. Bruice, *ibid.*, **81**, 5444 (1959). See also S.-H. Chu and H. G. Mautner, *J. Org. Chem.*, **31**, 308 (1966).
80. U. K. Pandit and T. C. Bruice, *J. Amer. Chem. Soc.*, **82**, 3386 (1960).
81. A. R. Fersht and W. P. Jencks, *J. Amer. Chem. Soc.*, **91**, 2125 (1969).
82. A. Signor and E. Bordignon, *J. Org. Chem.*, **30**, 3447 (1965).
83. B. Hansen, *Acta Chem. Scand.*, **12**, 324 (1958).
84. B. Hansen, *Svensk. Kemisk Tidskrift*, **75**, 10 (1963).
85. T. C. Bruice and S. J. Benkovic, *Bioorganic Mechanism*, Vol. I, W. A. Benjamin, New York, 1966, p. 135.
86. M. L. Bender and B. W. Turnquest, *J. Amer. Chem. Soc.*, **79**, 1656 (1957); M. L. Bender, M. C. Neveu, and E. J. Pollack, *J. Amer. Chem. Soc.*, **84**, 595 (1962).
87. T. C. Bruice and S. J. Benkovic, *J. Amer. Chem. Soc.*, **85**, 1 (1963).
88. *Chem. Eng. News*, April 3, 1967, p. 52.
89. S. M. Kupchan, et al., *J. Amer. Chem. Soc.*, **89**, 1189 (1967).
90. M. B. Vigneron, P. Crooy, F. J. Kézdy, and A. Bruylants, *Bull. Soc. Chim. Belg.*, **69**, 616 (1960).
91. H. Morawetz and P. S. Otaki, quoted by T. C. Bruice, *Brookhaven Symp. Biol.*, **15**, 80 (1963).
92. E. Sondheimer and R. W. Holley, *J. Amer. Chem. Soc.*, **76**, 2467 (1954); **79**, 3767 (1957).
93. J. A. Schafer and H. Morawetz, *J. Org. Chem.*, **28**, 1899 (1963).
94. M. T. Behme and E. H. Cordes, *J. Org. Chem.*, **29**, 1255 (1964).
95. S. A. Bernhard, A. Berger, J. H. Carter, E. Katchalski, M. Sela, and Y. Shalitin, *J. Amer. Chem. Soc.*, **84**, 2421 (1962).
96. M. Goodman and K. C. Steuben, *J. Org. Chem.*, **27**, 3409 (1962).
97. J. de Jersey, A. A. Kortt, and B. Zerner, *Biochem. Biophys. Res. Comm.*, **23**, 745 (1966).
98. T. Cohen and J. Lipowitz, *J. Amer. Chem. Soc.*, **83**, 4866 (1961); **86**, 5611 (1964).

99. C. Zioudrou and G. L. Schmir, *J. Amer. Chem. Soc.*, **85**, 3258 (1963); other examples include catalysis by a quinoline nitrogen in the ammonolysis of an ester [J. H. Jones and G. T. Young, *Chem. Comm.*, 35 (1967)], catalysis by an aromatic amide in the hydrolysis of an ester [R. M. Topping and D. E. Tutt, *Chem. Comm.*, 698 (1966)], and cleavage of peptides at cysteyl residues [Y. Degani, A. Patchornik, and J. A. McLaren, *J. Amer. Chem. Soc.*, **88**, 3460 (1966)].
100. M. L. Wolfrom, R. B. Bennett, and J. D. Crum, *J. Amer. Chem. Soc.*, **80**, 944 (1958).
101. H. Zahn and L. Zürn, *Ann.*, **613**, 76 (1958).
102. T. C. Bruice and F. H. Marquardt, *J. Amer. Chem. Soc.*, **84**, 365 (1962).
103. L. Zürn, *Ann.*, **631**, 56 (1960).
104. O. Bailly and J. Gaumé, *Bull. Soc. Chim. France*, **2**, 354 (1935); M. C. Bailly, *Compt. Rend.*, **206**, 1902 (1938); **208**, 443 (1939).
105. E. Baer and M. Kates, *J. Biol. Chem.*, **175**, 79 (1948).
106. D. M. Brown and A. R. Todd, *J. Chem. Soc.*, 52 (1952); D. M. Brown, C. A. Dekker, and A. R. Todd, *J. Chem. Soc.*, 2715 (1952).
107. D. M. Brown and A. R. Todd, in *The Nucleic Acids*, E. Chargaff and J. N. Davidson, Eds., Vol. 1, Academic Press, New York, 1955, p. 409; L. Kugel and M. Halmann, *J. Amer. Chem. Soc.*, **88**, 3566 (1966).
108. D. M. Brown and H. M. Higson, *J. Chem. Soc.*, 2034 (1957).
109. J. Kumamoto and F. H. Westheimer, *J. Amer. Chem. Soc.*, **77**, 2515 (1955).
110. J. R. Cox, Jr., R. E. Wall, and F. H. Westheimer, *Chem. Ind.*, 929 (1959); P. C. Haake and F. H. Westheimer, *J. Amer. Chem. Soc.*, **83**, 1102 (1961).
111. R. C. Gasman and D. C. Johnson, *J. Org. Chem.*, **31**, 1830 (1966).
112. H. A. C. Montgomery, J. H. Turnbull, and W. Wilson, *J. Chem. Soc.*, 4603 (1956); G. Riley, J. H. Turnbull, and W. Wilson, *J. Chem. Soc.*, 1373 (1957); G. J. Durant, J. H. Turnbull, and W. Wilson, *Chem. Ind.*, 157 (1958).
113. Other examples include the participation of the anilino group in peptide bond cleavage (K. L. Kirk and L. A. Cohen, *J. Org. Chem.*, **34**, 395 (1969)), the participation of an amino group in sulfur–sulfur bond cleavage [E. S. Wagner and R. E. Davis, *J. Amer. Chem. Soc.*, **88**, 7 (1966)], and a peptide rearrangement [L. Benoiton and H. N. Rydon, *J. Chem. Soc.*, 3328 (1960)].
114. D. J. Triggle and S. Vickers, *Chem. Comm.*, 544 (1965).
115. D. G. Oakenfull, T. Riley, and V. Gold, *Chem. Comm.*, 385 (1966).
116. K. G. Wyness, *J. Chem. Soc.*, 2934 (1958).
117. B. Zerner and M. L. Bender, *J. Amer. Chem. Soc.*, **83**, 2267 (1961).

Chapter 10

MULTIPLE CATALYSIS

10.1	Introduction	321
10.2	Intermolecular Multiple Catalysis	322
	10.2.1 General Acid-Base Catalysis	322
	10.2.2 Nucleophilic-General Base Catalysis	326
	10.2.3 Nucleophilic-Electrophilic Catalysis	327
10.3	Multiple Catalysis Involving Bifunctional Molecules	328
	10.3.1 Bifunctional Nucleophiles	329
	10.3.2 Bifunctional Catalysts	330
	10.3.3 Bifunctional Catalysis Consisting of One Intermolecular and One Intramolecular Catalyst	335
	10.3.4 Bifunctional Metal Ion Catalysis	336
	10.3.5 Bifunctional Intramolecular Catalysis	337
10.4	Evaluation of Multiple Catalysis	345

10.1 INTRODUCTION

The maxim that if a little is good, a lot is better has been applied to catalysis for many years in the hypothesis that multiple catalysis should be superior to a single catalysis. Many terms have been applied to catalysis involving more one than one catalytic entity: concerted catalysis, bifunctional catalysis, and multiple catalysis. Since the former terms imply specific mechanistic or structural descriptions while multiple catalysis does not, we prefer to use multiple catalysis as the generic name for this area.

Multiple catalysis can combine all the individual kinds of catalysis discussed previously. All possible combinations and permutations are theoretically possible in terms of two-catalyst, three-catalyst systems, or higher.

At this stage in our sophistication, many synthetic two-catalyst systems are known. For higher systems we must look to the enzymes. The most common multiple catalysis involves either general acid-base catalysis or electrophilic-nucleophilic catalysis. To these natural combinations one can add nucleophilic-general base catalysis, but apparently not electrophilic-general acid catalysis.

One important catalytic contributor is excluded from discussion here: binding, covalent, or otherwise, which introduces a catalyst into the substrate molecule or into a complex with it. Although it might be considered a second contributor to a multiple catalysis, its special character makes it worthy of special treatment (Chapter 11).

The identification of multiple catalysis and the simultaneity of multiple catalytic action are vexing mechanistic questions that reappear with considerable frequency. The identification of multiple catalysis depends on kinetic observations requiring the presence of two catalytic species. Unfortunately the appearance of two catalytic species in the rate law does not always mean the action of two independent catalysts. Often kinetically indistinguishable possibilities occur in which two species associate to form one catalytically important species. Even when two catalytic species do operate, they need not act in the same step of the mechanism. If a general base catalyzes one step of a reaction and its (conjugate) general acid catalyzes a second step of the reaction, this process is usually spoken of as a single, rather than a multiple catalysis, because at any one time only one catalytic species is operative. However, if two catalysts act in two steps of a reaction in such a way that their cumulative action is synergistic in some sense, then the process must be considered as a multiple, rather than a single catalysis. If the catalytic interactions are stepwise, there nevertheless must be some transition state in the reaction containing both species simultaneously. This requirement does not of course specify the timing of attainment of the transition state.

10.2 INTERMOLECULAR MULTIPLE CATALYSIS

10.2.1 General Acid-Base Catalysis

In Chapter 5, evidence pointing to multiple catalysis by a combination of intermolecular acids and bases in the enolization of ketones, additions to carbonyl compounds, and the reverse reaction was considered, but rejected. We now return to this problem which was in effect initiated by the investigations of Lowry on the mutarotation of tetramethyl-D(+)-glucose in aprotic solvents.[1] The rate of the mutarotation of tetramethylglucose in chloroform, ethyl acetate, pyridine, or cresol solutions is negligible. In cresol-pyridine mixtures, however, the rate is appreciable. In fact, between 55 and 92% cresol, the rate constant is so large that it is not observable (Table 10.1). The suggestion was made on the basis of these data that both a general acid and base are simultaneously necessary for the mutarotation reaction, as exemplified by

$$\begin{array}{c} \text{(structure with } \text{—O, HA, OH B)} \rightleftharpoons \text{(structure with —OH, A}^{\ominus}\text{, =O, HB}^{\oplus}\text{)} \end{array} \quad (10.1)$$

This interpretation has been questioned, however, since tetra-n-butylammonium phenoxide, containing no ionizable protons whatsoever, is a powerful catalyst for the mutarotation of tetramethylglucose in benzene solution, and since pyridine-cresol mixtures may contain appreciable quantities of pyridinium cresolate.[2] If either the pyridinium ion or the cresolate ion is a better catalyst than pyridine or cresol, the data of Table 10.1 can be explained as a single rather than a multiple catalysis.

The most likely solvent for the appearance of general acid-base catalysis would be expected to be an aprotic solvent, since in aqueous solution water can serve as either general acid or base catalyst. If the two groups act at the same time, as proposed by Swain and Brown[3] for the mutarotation of tetramethylglucose in the presence of 2-hydroxypyridine (or its tautomer, 2-pyridone) in benzene, the process is called a concerted reaction. However, this catalysis may be a "tautomeric catalysis" being brought about by

Table 10.1

The Mutarotation of Tetramethyl-D(+)-Glucose in Cresol-Pyridine Mixtures[b]

Percent Cresol	$k \times 10^4$ (min^{-1})[a]
100	3
95	820
92–55	too fast to measure
52.5	1800
21	168
0	3

[a] 25°.
[b] From T. M. Lowry and I. J. Faulkner, J. Chem. Soc., **127**, 1883 (1925).

molecules that can furnish and receive hydrogens in a tautomeric system whether it be in this or another reaction. Of course, the stereochemistry of the process must be suitable so that hydrogen bonding, proton transfer, and reaction take place. Thus, 2-pyridone but not 4-pyridone is a catalyst for the mutarotation of tetramethylglucose.[3] It is interesting to note that whereas 2-pyridone is a catalyst, the comparable compound 2-aminophenol is not.[4] Eigen has argued that catalysis of the dehydration of formaldehyde hydrate in aqueous solution may occur by one-encounter processes. Hypothetical examples involving arbitrary numbers of water molecules are shown in eqs. 10.2 and 10.3 for water and hydroxide ion catalysis.

$$\text{(structures)} \quad (10.2)$$

$$\text{(structures)} \quad (10.3)$$

These processes can be considered to involve general acid and base catalysis, since one species (water) serves as proton donor while the other (water or hydroxide ion) serves as proton acceptor. However, an alternative explanation involving single (stepwise) rather than multiple catalysis also adequately accounts for the facts of this reaction (Chapter 5). The will-o-the-wisp of multiple catalysis is difficult to pin down.

An attempt has been made to determine whether in fact more than one molecule of water is involved in the hydration of *sym*-dichloroacetone by varying the concentration of water in aprotic solvents such as dioxane or acetone. In the water-catalyzed hydration reaction, a plot of log k versus water concentration has a slope of 2.95 while in the dehydration reaction the slope of the plot is 1.98, in conformity with eq. 10.2.[6] However, investigations in largely apolar media suffer from the fact that association of polar species must be exaggerated over the norm of aqueous solution.

The optimal method for detecting simultaneous general acid-base catalysis is to observe kinetic dependence on each species. Some reactions, including the halogenation of ketones,[7] the hydration of dichloroacetone,[8] and the ketonization of oxaloacetate ion,[9] show kinetic dependence on the product of general acid and general base concentrations. As pointed out in Chapter 5, those reactions involving apparent kinetic dependence on both carboxylic acid and carboxylate ion cannot be rigorously attributed to a multiple catalysis since these reactions may be due to catalysis by the kinetically indistinguishable dimeric (basic) species, $HA_2^\ominus$, whose existence is known.[10] This argument does not apply to the ketonization of oxaloacetate ion since kinetic dependence on the product of general acid and general base is seen with imidazole and triethanolamine buffers, 0.1 M of the latter buffer showing a product term that is 15–23% of the total rate constant. Since these catalysts are not subject to the ambiguity of the carboxylate catalysts, this process may then be a true multiple catalysis whose mechanism involves the simultaneous action of general acid and general base donating and removing protons, respectively.

The imidazole catalysis of the hydration of *sym*-dichloroacetone in 95% dioxane–5% water (v/v) may also be a multiple catalysis. Kinetically the reaction is both first-order and second-order in imidazole in the range of 0 to 0.6 M imidazole. Although the imidazole reaction shows a second-order dependence, catalysis by other secondary and tertiary amines including *N*-methylimidazole shows only a first-order dependence. Since this reaction is catalyzed by general bases,[8] the second-order dependence may mean a general base catalysis of general base catalysis. Although association of imidazole in this medium of low dielectric clouds the issue, the mechanism of the reaction may be[11]

$$\overset{\delta\ominus}{O}=\!\!\!\!\!\!\!\!\!\!\!\!\!\!\!\!\!\overset{H}{\underset{|}{C}}\!\!\!\!\!\!\!\!\!\!\!\!\cdots O\cdots H\cdots N\overset{\delta\oplus}{\underset{\cdots}{\diagdown}}N\cdots H\cdots N\overset{\delta\oplus}{\underset{\cdots}{\diagdown}}NH \qquad (10.4)$$

A reaction once thought to proceed via multiple catalysis is the tautomerization of some methylene-azomethines.[12] This conclusion was based on an identity of the initial rate constants of isomerization, isotopic exchange, and loss of optical activity for ethoxide ion-catalyzed conversions such as

$$CH_3\!-\!\overset{*}{\underset{\underset{H}{|}}{\overset{C_6H_5}{\overset{|}{C}}}}\!-\!N\!=\!\overset{C_6H_4Cl\text{-}p}{\underset{|}{C}}\!-\!C_6H_4Cl\text{-}p \rightleftharpoons CH_3\!-\!\overset{C_6H_5}{\underset{|}{C}}\!=\!N\!-\!\overset{C_6H_4Cl\text{-}p}{\underset{\underset{H}{|}}{C}}\!-\!C_6H_4Cl\text{-}p \qquad (10.5)$$

* stands for an optically active center.

in solvents such as 2:1 dioxane-ethanol-O-*d* or ethanol-O-*d*. The discovery of an intramolecular component in the base-catalyzed rearrangement of 3-phenyl-1-butene to *cis*- and *trans*-2-phenyl-2-butene, coupled with the fact that the rearrangement competes with simple isotopic exchange,[13] led to a reexamination of the problem. When the optically active *reactant* of eq. 10.5 is allowed to undergo about 8% isomerization, the α-phenylethylamine portion of the molecule shows neither racemization nor isotopic exchange which one would predict since $k_{\text{isomerization}} = k_{\text{isotopic exchange}} = k_{\text{loss of optical activity}}$. However, when the *product* of eq. 10.5 is 10% isomerized, the methyl and benzhydryl hydrogens of the product are >95% exchanged, indicating that $k_{\text{exchange}} \gg k_{\text{isomerization}}$. This apparent anomaly may be explained by postulating that indeed a carbanion intermediate is formed in this reaction but the collapse of the carbanion intermediate to reactant and product is not equal, but rather favors the product even though the overall equilibrium constant is 1.2.[14]

The studies described above employ strongly basic species (alkoxide ions) and very weakly acidic species (alcohols), so that mechanisms involving both acidic and basic species would not be as probable as if both of them were of comparable strength. The latter situation is anticipated in enzymic reactions. Therefore, catalysis of the transamination of pyridoxal and α-aminophenylacetic acid by the combination of imidazole and imidazolium ion is of interest.[15] Since this reaction occurs via complex formation between catalysts and substrate, it will be discussed in Chapter 11.

10.2.2 Nucleophilic-General Base Catalysis

Although nucleophilic catalysis of *p*-nitrophenyl acetate hydrolysis by imidazole proceeds without further catalysis, nucleophilic catalysis by imidazole of the hydrolysis of esters with poorer leaving groups such as *p*-methylphenol and *p*-methoxyphenol is subject to a further (general base) catalysis by imidazole. Esters with even poorer leaving groups such as phenyl acetate, trifluoroethyl acetate, and acetoxime acetate show nucleophilic catalysis by imidazole which is subject to further general (base) catalysis by hydroxide ion.[16] The reaction second-order in imidazole is defined as a general base catalysis superimposed on nucleophilic catalysis, since it shows a deuterium isotope effect, although the first-order reaction of imidazole and *p*-nitrophenyl acetate shows none. Catalysis by *N*-methylimidazole does not exhibit second-order catalysis by *N*-methylimidazole indicating that in the former reaction one imidazole may assist the other by proton abstraction. Alternative mechanisms involving general base or general acid assistance to the tetrahedral intermediate are also possible (Chapter 5). The imidazole reaction assisted by hydroxide ion can either occur in these manners or can consist of preequilibrium formation of imidazole anion that then serves as the nucleophilic entity. Indeed the reaction showing second-order dependence on imidazole concentration could occur by a pre-equilibrium autoprotolysis to imidazolium ion and imidazole anion, which could then serve as nucleophile. Although the imidazole-hydroxide ion equilibrium is a probable pathway, the imidazole-imidazolium ion equilibrium is not, since the low value of the equilibrium constant for this process leads to an unreasonably high rate constant for the subsequent nucleophilic attack on the ester. Hence, the second-order imidazole catalysis appears to be a true multiple catalysis. This catalysis involving nucleophilic and general base catalysis need not show simultaneous interaction of the two catalytic species and the substrate.

Nucleophilic catalysis by imidazole occurs in the formation of a tetrahedral intermediate; but general base catalysis assistance by imidazole can occur either in the formation or the decomposition of the tetrahedral intermediate. Multiple catalysis requires only that the nucleophile still be present when the general base reacts, so that a transition state containing both species occurs.

10.2.3 Nucleophilic-Electrophilic Catalysis

The amino acid serine is cleaved by the enzyme hydroxymethylase in the presence of the coenzymes tetrahydrofolate and pyridoxal phosphate to glycine and a derivative of formaldehyde. The same reaction can be performed at pH 5.5 nonenzymatically by a mixture of an N,N'-diarylethylenediamine, pyridoxal phosphate, and metal ion.[17] In this reaction, pyridoxal phosphate acts as an electrophilic catalyst while the N,N'-diarylethylenediamine acts as a nucleophilic catalyst. Omission of either of these reagents from the reaction mixture negates the reaction. A plausible mechanism for the reaction, although not proven by any means, is

(10.6)

The key step in eq. 10.6 is the one in which the diamine reacts with the imine of the amino acid. In this reaction, the diamine acts as nucleophile and the pyridoxal moiety as an electron sink, simultaneously facilitating the cleavage of the carbon–carbon bond under very mild conditions.

Our search for multiple catalysis, while not completely barren, has not been amply repaid. In a few instances a combination of general acid and general base appears to be superior to either alone. Likewise, a combination of nucleophilic and general basic catalysis, or nucleophilic and electrophilic catalysis, appears to be superior to either alone. Determination of the rate enhancement brought about by the introduction of the second catalyst is not easy because quantitative data are not always available and because the reference state for such calculations is not completely apparent. Let us then look to a more fruitful field, multiple catalysis involving bifunctional molecules.

10.3 MULTIPLE CATALYSIS INVOLVING BIFUNCTIONAL MOLECULES

In 1950 Swain pointed out that if a reaction is catalyzed by two components, such as nucleophilic and electrophilic catalysts, enhanced catalysis should result if two of the components, either the nucleophilic catalyst and the substrate, the electrophilic catalyst and the substrate, or the nucleophilic and electrophilic catalyst, are joined together (Fig. 10.1). He further suggested that enzymes may owe a large share of their catalytic powers to the fact that in enzymes nucleophilic and electrophilic catalysts are combined in the same molecule in the proper stereochemical arrangement for optimum interaction with the substrate.[18] Reactions of two of these combinations are

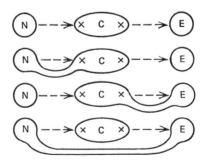

Fig. 10.1 Possible kinds of nucleophilic-electrophilic catalysis. From C. G. Swain, *J. Amer. Chem. Soc.*, **72**, 4583 (1950). © 1950 by the American Chemical Society. Reprinted by permission of the copyright owner.

Multiple Catalysis Involving Bifunctional Molecules

now known, those involving an electrophilic (general acid) catalyst and substrate, or nucleophilic and electrophilic (general acid and general base) catalysts. In addition, systems not envisioned by Swain, involving all three components, substrate, electrophilic catalyst, and nucleophilic catalyst, are known. The question we wish to address here is whether these catalytic systems are in fact superior to multiple catalysis involving separate entities and/or whether they are superior to single catalysis. In essence we are looking at a higher order of complexity of intramolecular catalysis.

10.3.1 Bifunctional Nucleophiles

Before approaching the question of bifunctional catalysis, let us briefly consider the reactions of two species, one of which is bifunctional in nature. The simplest of these reactions involves bifunctional nucleophiles and monofunctional substrates, although the definition is somewhat arbitrary.

A bifunctional nucleophile whose reactions have been intensively studied is catechol monoanion. This species shows exceptional nucleophilicity in aqueous solution toward phosphoryl halides,[19,20] sulfonyl and acyl halides,[21] and phenyl chloroacetate.[22] The catechol monoanion is the reactive species in each reaction. Not only is catechol monoanion more reactive than the dianion or diprotonated species, it is also more reactive than monoanions of phenol, resorcinol, or hydroquinone. On the other hand, phenols with three vicinal hydroxyl groups are five times more reactive than catechol toward diisopropyl phosphorofluoridate.[19] The nucleophilicity of catechol monoanion (pK 9.8) is equal to that of hydroxide ion in this reaction, indicating its enhanced reactivity. Catechol forms a complex with isopropyl methylphosphonofluoridate.[23] In the reaction of isopropyl methylphosphonofluoridate with a series of substituted catechols, kinetic evidence of complexing prior to reaction is known.[24] On this basis, as well as the pH dependence and relative reactivity studies, the following mechanism has been proposed:

$$(10.7)$$

For carbonyl and sulfonyl compounds a similar mechanism will presumably apply. In the latter reactions, however, no evidence for prior complexing exists.

The reactions of phenyl acetate with aliphatic primary amines and with diamines of structures $NH_2(CH_2)_nNH_2$ or $NH_2(CH_2)_nNH_3^{\oplus}$ (where $n = 2$–6) show kinetic dependence on the first power of the ester and the amine.[25] The values of the rate constants of the diamines are larger than predicted from the Brønsted plot of the monoamines by a factor of up to tenfold, suggesting a possible contribution from the terminal amino group or ammonium ion acting as general base or acid. Alternatively, the diamines may form a separate Brønsted series.

Although n-butylamine reacts with p-nitrophenyl acetate in chlorobenzene solution by means of a process second-order in amine, reaction with benzamidine is first-order in amine. The second-order rate constant of the benzamidine reaction is at least 15,000 times larger than the second-order rate constant of the reaction of n-butylamine. On the basis of these extraordinary results, the following mechanisms for the n-butylamine and benzamidine reactions are postulated (excluding a solvent effect).

$$\text{(structures)} \longrightarrow RNH-\underset{\underset{OH}{|}}{\overset{\overset{R}{|}}{C}}-OR + RNH_2 \longrightarrow \text{products}$$

$$\text{(structures)} \longrightarrow Ph-C\underset{\underset{H}{|}}{\overset{\overset{H}{|}}{\underset{N}{\overset{N}{\diagdown}}}} \underset{OH}{\overset{\overset{H}{|}\;\overset{R}{|}}{N-C-OR}} \longrightarrow \text{products} \tag{10.8}$$

Large concentrations of tertiary amine have little effect on the rate of the n-butylamine reaction, in agreement with the cyclic proton transfer nature of eq. 10.8.[26]

10.3.2 Bifunctional Catalysts

The classical experiment in bifunctional catalysis is catalysis of the mutarotation of α-D-tetramethylglucose in benzene solution by 2-pyridone

(2-hydroxypyridine).[27] This reaction shows second-order kinetics (at low concentrations of catalyst) whereas comparable catalysis by a mixture of phenol and pyridine shows third-order kinetics.[28] This result was first interpreted to mean that both the nitrogen and hydroxyl functions of 2-hydroxypyridine act in place of the separated pyridine and phenol catalysts. This conclusion is supported by the observation that 0.1 M phenol or 0.1 M pyridine have no effect on the rate of the reaction catalyzed by 0.0001 M 2-hydroxypyridine. The rate of the 2-hydroxypyridine reaction compares favorably with that of the pyridine-phenol reaction even though the basicity and acidity of 2-hydroxypyridine are 10^4 and 10^2 less than the basicity and acidity of pyridine and phenol, respectively. For example, at 0.05 M concentration of all catalysts, the rate of the 2-hydroxypyridine reaction is approximately 50 times that of the combination of phenol and pyridine. As expected when comparing second- and third-order reactions, when the concentration of catalysts is reduced, the rate differential is even greater: at 0.001 M catalyst concentrations, the 2-hydroxypyridine rate is 7000 times the pyridine-phenol rate.

This catalysis has stringent specificity requirements, as seen in a comparison of other similar catalysts. Benzoic acid, picric acid, and 2-aminopyridine also possess catalytic activity; the rates of these reactions are first-order in the catalyst. 3-Hydroxypyridine and 4-hydroxypyridine, however, show smaller rates of catalysis, second-order in the catalyst. Thus, the stereochemistry of the bifunctional process is important. 2-Pyridone, benzoic acid, pyrazole, and 1,2,4-triazole have been interpreted as tautomeric catalysts for enzyme-like reactions on the basis of their high reactivity and their structure.[27]

In benzene solution, complexing of materials capable of hydrogen bonding is expected. Also, the tetra-n-butylammonium, 4-nitrophenoxide ion pair catalyzes no more than diethylammonium 4-nitrophenoxide.[29] Thus there is no compelling reason to postulate a general acid-base mechanism in pyridine. If 2-hydroxypyridine is capable of self-dimerization, it should also be capable of complexing with a substrate containing the same kind of functional groups. The abnormally high initial specific rotations of solutions of tetramethyl-D(+)-glucose containing 2-hydroxypyridine point to complex formation. The substrate also complexes with picric acid, as indicated by a dynamic vapor pressure method.[27] The pyranose-like hemiacetal, 2-tetrahydropyranol, partially inhibits the mutarotation catalyzed by 2-hydroxypyridine, although phenol and pyridine do not. This may be caused by competitive complexing with the catalyst. On the basis of these data, the mechanism of catalysis by 2-hydroxypyridine and related bifunctional catalysts was given by

$$(10.9)$$

This mechanism is quite similar to that proposed for the reaction of benzamidine with *p*-nitrophenyl acetate in chlorobenzene solution (eq. 10.8). Any system having two electronegative atoms in a 1,3-juxtaposition, having an internal angle between the two electronegative atoms of less than 180°, and possessing one double bond, such as carboxylic acids, amides, amidines, and so on (rather than greater than 180° as in imidazole) appear to satisfy the requirements for bifunctional catalysts for reactions with carbonyl substrates.

The mutarotation of α-D-tetramethylglucose in nitromethane solution shows a pattern similar to that in benzene solution. Carboxylic acid catalysts are more effective than phenols of the same acidity by a factor of approximately 400. A bifunctional catalysis by the carboxylic acid is indicated.[30] A similar suggestion has been made for the acceleration by carboxylic acids in the acylation of aromatic amines with acyl chlorides in benzene or nitrobenzene solution.[31] No rate enhancements are seen with similar compounds having only one functionality; furthermore, carboxylic acids have no effect on the reaction of picryl chloride with the same aromatic amines. On this basis, it was suggested that the carboxylic acid donates a proton to the leaving group and abstracts a proton from the amine at some stage of the reaction. While this mechanism is pictorially attractive, the small rate accelerations (a factor of 2 with 0.5 *M* carboxylic acid in nitrobenzene solution) do not present a strong case for this mechanism, since medium effects of a nonspecific variety could also explain the data.

Although bifunctional catalysis of the acid-base variety occurs in aprotic solution, its occurrence in hydroxylic solution is not known with certainty. One possible occurrence is in the hydrolysis of the iminolactone, *N*-phenyl-iminotetrahydrofuran.[32] This reaction yields aniline and butyrolactone at acid pH and γ-hydroxybutyranilide in alkaline solution. The extent of conversion of iminolactone to aniline at constant pH in the region of pH 7–9 depends upon the nature and concentration of the buffer. Low concentrations of phosphate or bicarbonate buffers direct the reaction to aniline formation, at the expense of the yield of hydroxyanilide, without affecting the rate of iminolactone disappearance. The change in product ratio without change in rate may be simply interpreted in terms of a mechanism involving the rate-controlling formation of an intermediate, whose breakdown controls product formation. The only reasonable intermediate in the reaction is a tetrahedral intermediate whose breakdown must therefore be controlled by the buffer. The buffer effects have the following characteristics: (1) Tris, imidazole, or p-nitrophenol buffers do not affect aniline yields, but phosphate, bicarbonate, arsenate, acetate to a lesser extent, and dicarboxylic acid buffers do; (2) the effectiveness of phosphate buffer in increasing aniline yield falls off with increasing pH in the region pH 7–9 implying that the dianion is better than the trianion; (3) the effect of phosphate, acetate, or bicarbonate concentration on the aniline yield shows a saturation effect at 0.01, 0.1, and 0.1 M concentrations, respectively, implying some kind of complex formation; and (4) phosphate buffer is 240 times more effective than imidazole buffer in promoting the conversion of iminolactone to aniline. The last point rules out both nucleophilic and classical general base catalysis by phosphate ion since in the former, imidazole far excels phosphate ion (although their pK's are about the same) and in the latter imidazole is equal to phosphate ion (Chapters 6 and 5, respectively). However, bifunctional catalysis by phosphate ion should be superior to that by imidazole on steric grounds. Thus, if the phosphate ion can accelerate the transformation of the neutral tetrahedral to the zwitterionic intermediate in eq. 10.10, the aniline yield should increase since the former leads to anilide while the latter leads to aniline.

$$\text{(10.10)}$$

where R = alkyl or phenyl.

Although this mechanism is very attractive, it is extremely difficult to visualize a complexing in aqueous solution having a dissociation constant of 2×10^{-3} M between dihydrogen phosphate and the electronegative atoms of the substrate molecule. This apparent saturation phenomenon may reflect instead a change in rate-determining step which also can exhibit a similar kinetic saturation.

Analogous mechanisms have been advanced to explain carboxylic acid catalysis of the cyclization of glutamic acid esters to derivatives of 2-pyrrolidone-5-carboxylic acid[33] and the specific catalytic effects of carboxylic, sulfuric, and phosphoric acids on the rearrangement of N,N'-diacylhydrazines.[34] Unlike the iminolactone work, however, these reactions were carried out in nonaqueous solution. Effective catalyst concentrations were in the range of 0.1–2 M, and no comparison was made with a general acid of the strict monofunctional type. Unexpected catalytic effects of phosphate monoanion and bicarbonate ion on the reaction of urea with formaldehyde in water have been reported.[35] The conversion of glutamine to pyrrolidonecarboxylic acid in water is accelerated by high concentrations of phosphate, arsenate, and bicarbonate ions.[36] Insufficient information on these systems is available to permit assignment of mechanism; plausible hypotheses include concerted proton transfer and simple medium effects.

The occurrence of bifunctional general acid-general base catalysis may follow the same rules as monofunctional general acid-base catalysis. If this analogy is correct, bifunctional catalysis will operate to obviate the formation of unstable intermediates, such as oxonium ions, but will not be necessary when ammonium ion intermediates are formed.

Nucleophilic catalysts such as imidazole and pyridine may be altered to introduce a second functionality such as a general acid. Such bifunctional nucleophilic catalysts would be expected to be more effective than corresponding monofunctional nucleophiles. Indeed, p-nitrophenyl acetate is hydrolyzed by N-(diethylaminoethyl)imidazole 36 times faster than by imidazole at pH 7.2,[37] and is hydrolyzed by 2-hydroxypyridine (2-pyridone) at approximately the same rate as by pyridine, although the basicity of the former is about 10^4 times less than the latter.[38] The pH-rate constant profiles would be in order for these reactions, to determine whether both functions of the catalyst interact with the substrate. The former catalysis is reported to be most effective at pH 7.2 which is in qualitative agreement with such a conclusion. Peptide syntheses from both activated (cyanomethyl) esters[39] and low-energy (methyl) esters[35] and amines are accelerated by triazoles, such as 1,2,4-triazole, and 2-hydroxypyridine, in aprotic medium. Imidazole is somewhat effective as a nucleophilic catalyst with activated esters, but it fails completely with non-activated esters. Although these conclusions were reached by product analysis rather than kinetic determination, the results probably indicate some special property of 1,2,4-triazole and 2-hydroxypyridine which may be represented by the following nucleophilic-general acid catalysis mechanism.

$$\text{(10.11)}$$

10.3.3 Bifunctional Catalysis Consisting of One Intermolecular and One Intramolecular Catalyst

In theory bifunctional catalysis consisting of one intermolecular and one intramolecular catalyst should also show enhanced catalysis over two intermolecular catalysts. This phenomenon may occur in the transamination of glutamic acid with 3-hydroxypyridine-4-carboxaldehyde.[41] The kinetics of the reaction indicates that the step involving tautomerization of one imine to another is dependent on the first power of the imidazole concentration. Conversely, catalysis of the same reaction in the transamination of pyridoxal and α-aminophenylacetic acid is brought about by two moles of imidazole.[15] One explanation of this difference is that in the former reaction, the (internal) phenolic group and one (external) imidazole group are acting as a general acid-base catalytic pair, whereas in the latter reaction two (external) imidazole molecules are acting as the general acid-base catalytic pair. It is not clear at the moment why the internal phenolic group in pyridoxal does not act as an internal general acid(base) catalyst. Comparison with the reaction of pyridine-4-carboxaldehyde, a molecule not containing any internal phenolic group, would be helpful.

A multiple catalysis combining one intermolecular and one intramolecular catalyst occurs in the solvolysis of diaxial 1,3-dihydroxy acetates.[42] The three acetate esters, coprostanol acetate (1), coprostane-3β,5β-diol 3-monoacetate (2) and strophanthidin 3-acetate (3) are solvolyzed in methanol-water-chloroform solution with quite different rates for two reasons: first, solvolysis of the 3-acetate groups of compounds 2 and 3 is facilitated by the intramolecular catalysis of the 5-hydroxyl groups in these molecules; secondly, not only are compounds 2 and 3 subject to 1,3-diaxial hydroxyl interaction but also are subject to external catalysis by trimethylamine, pyridine, or N-methylimidazole buffers. The external catalysis is probably due to general base rather than nucleophilic catalysis since esters of aliphatic alcohols are not subject to nucleophilic catalysis, and since catalysis by N-methylimidazole is much poorer than catalysis by trimethylamine (the former is a better nucleophilic catalyst but the latter is a better general base catalyst). These catalyses lead to considerable differences in rate constant: for example, at 3:1 triethylamine:triethylammonium acetate buffer (0.21 M) at 40°, the relative rate constants of **1**, **2**, and **3** are 1:300:1200. The solvolysis of **2** can be pictured as

(10.12)

In this mechanism one can perceive a general acid-base catalysis of ester solvolysis, the general acid being internal and the general base being external. The fourfold rate enhancement of **3** over **2** may be due to the presence of the 19-aldehyde group which in the form of a hemiacetal could provide an additional hydroxyl interaction, leading to a multiple internal general acid catalysis. However, this effect is not large. A significant observation in this series is that when the internal general acid catalysis is not present, the external general base catalysis is likewise not present (compound **1**).

10.3.4 Bifunctional Metal Ion Catalysis

Some of the metal ion catalyses discussed in Chapter 8 involve multiple catalysis consisting of catalysis both by metal ion and by another catalyst such as a general base or nucleophile. A particularly clear example is the solvolysis of tetrabenzyl pyrophosphate in propanol solution. In the presence of 0.02 M calcium ion and 0.2 M lutidine, the reaction rate is increased by a factor close to one millionfold over the uncatalyzed reaction, each catalyst making approximately a thousandfold contribution.[43] This multiple catalysis is an additive function of the two components, indicating that each catalyst is independent of the other.

An interesting catalysis involving a metal ion plus another catalyst is the oxidative deamination of amino acids to keto acids, ammonia, and hydrogen peroxide, catalyzed by pyridoxal and manganic ion at room temperature. This reaction serves as a model for the action of some amine oxidases. In the oxidation of alanine, pyruvate is found in low yield, ammonia in high yield, and hydrogen peroxide not at all, since it is decomposed rapidly when formed. Salicylaldehyde and pyridoxine cannot replace pyridoxal but pyridoxal phosphate can. α-Methylalanine, N-methylalanine, and lactic acid are not oxidized under conditions where alanine reacts readily. Other amino acids and amino acid esters and amides can replace alanine but simple amines react slowly if at all. The rate of O_2 uptake is decreased by ethylenediaminetetraacetic acid, but is unaffected by light or by free radical inhibitors such as phenols (thus, the oxidation presumably does not occur by a free radical chain mechanism). Glycine is oxidized five to six times more rapidly than α,α-dideuteroglycine. These results are consistent with eq. 10.13. Intermediates **1** and **2** are similar to the intermediates proposed for other pyridoxal-catalyzed reactions of amino acids.

It is suggested that **2**, or some intermediate like **2**, can complex with O_2 to give **3**. The transfer of a proton through the solvent and electrons through the complex (Chapter 8) can lead to **4**, in which oxygen has been reduced to hydrogen peroxide and the rest of the complex has lost two electrons. Compound **4** would be expected to be in equilibrium with pyruvic acid, $M^{3\oplus}$, H_2O_2, and **5**. Compound **5** is a tautomer of the imine of ammonia with pyridoxal and would be expected to give pyridoxal and ammonia readily.[44]

10.3.5 Bifunctional Intramolecular Catalysis

The ultimate in combining multiple catalysts with the substrate is to attach all catalysts covalently to the substrate molecule. This in essence is intramolecular catalysis (Chapter 9) utilizing multiple rather than single catalysts. Reactions susceptible to such catalysis include hydration and dehydration reactions and reactions of carboxylic acid derivatives.

Probably the first such catalysis to be found was the hydrolysis of "succinylaspirin," **6**. Whereas aspirin, **7**, and the monomethyl ester, **8**, are hydrolyzed in the pH range of 3 to 9 at rates proportional to the ionized carboxylate ion, **6** is hydrolyzed at a rate proportional to the concentration of a species containing both a carboxylate ion and a free carboxylic acid, assuming pK's of 3.6 and 4.5 for the salicylic and succinic carboxylic acid groups,

(10.13)

respectively (Fig. 10.2).[45] Two kinetically indistinguishable pathways are possible, each involving attack by carboxylate ion and electrophilic assistance from an unionized carboxylic acid group. If the reaction proceeds as in eq. 10.14, the compound is calculated to hydrolyze 24,000 times as fast as aspirin anion, which lacks only the second unionized carboxylic acid group. If the reaction proceeds through eq. 10.15, however, the compound is calculated to hydrolyze 66 times as fast as **7**. The lower ratio must be accepted because of the lack of definitive evidence to distinguish these possibilities.

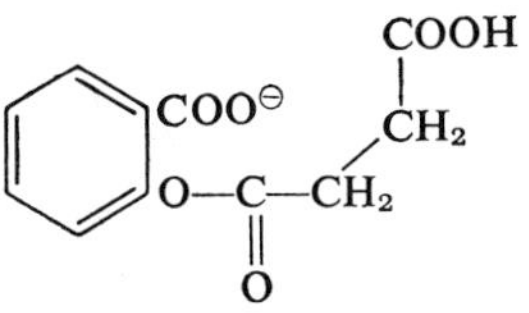

6 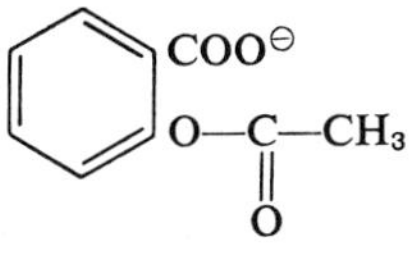   **7** 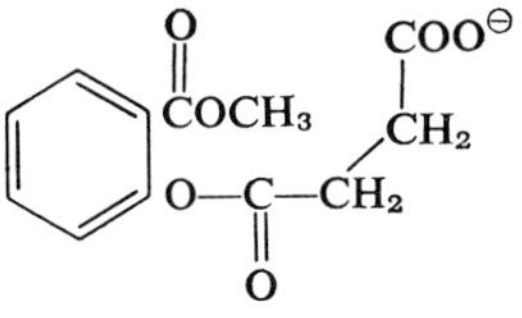   **8**

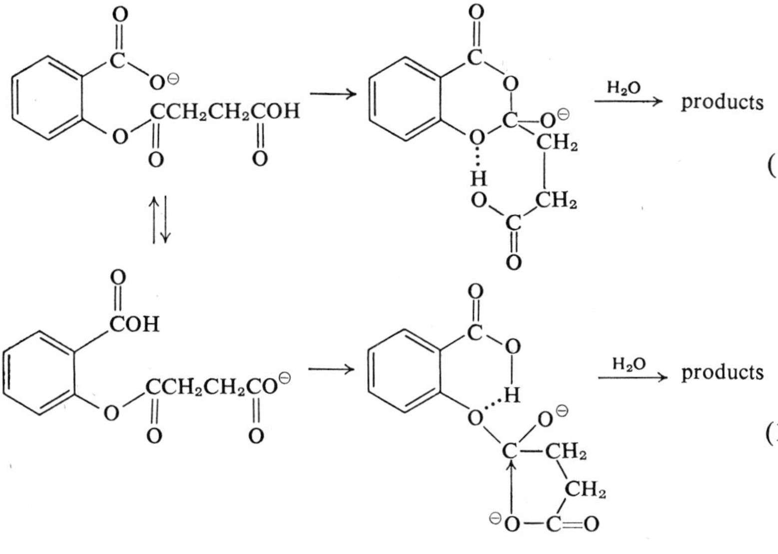

(10.14)

(10.15)

The pH-rate constant profiles for the hydrolysis of the phthalamic acid derivatives **9–12** show a similar rate difference. Compound **9** hydrolyzes with the largest rate constant of the group and exhibits a bell-shaped pH-rate constant profile with a maximum at a pH corresponding to the maximum concentration of the singly ionized species. Compounds **9** and **12** of this series show sigmoid curves, similar to the behavior of phthalamic acid (Chapter 9). Compound **11** is unreactive under the conditions investigated

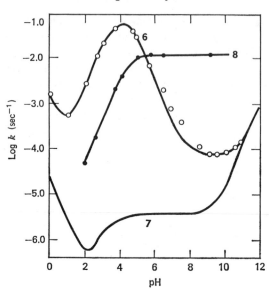

Fig. 10.2 Hydrolysis of aspirin and related compounds at 25°C. From H. Morawetz and I. Oreskes, *J. Amer. Chem. Soc.*, **80**, 2591 (1958). © 1958 by the American Chemical Society. Reprinted by permission of the copyright owner.

while compound **10** possesses reactivity similar to phthalamic acid. Assuming that the reaction proceeds through an anhydride intermediate, the carboxylate ion of **9a** rather than **9b** is the reactive nucleophile. On this basis, the rate enhancement of the multiple catalysis over the single catalysis can be calculated as the rate constant ratio $k_9/k_{10} = 80$. Correction for inductive effect differences brings the ratio to 40, very similar to the result found in the ester hydrolysis discussed above.[46]

The molecules catechol salicylate and salicyl salicylate both contain two phenolic hydroxyl groups, each of which is capable of acting individually as an intramolecular catalyst. When these two catalysts are placed in the same molecule, they might be expected to act in concert. Such is not the case, however.[47,48] Two reasons may be suggested for these failures: the conformation of these two hindered molecules does not allow simultaneous interaction of both catalytic functions and/or the loss of a good leaving group does not require an acid catalysis (Chapter 9). A molecule in which simultaneous interaction of two catalysts can occur is methyl 2,6-dihydroxybenzoate.[48] The hydrolysis of this ester shows a bell-shaped pH-rate constant profile depending on two groups of pK 8.2 and 11.6 (Fig. 10.3).[48] The rate constant at the maximum of the bell is approximately equal to the rate constant of methyl salicylate in its completely ionized form. Since the introduction of a

[Structures 9a, 10, 9, 11, 9b, 12 shown]

methyl group into methyl *o*-toluate depresses the hydrolytic rate by approximately 10^5-fold because of steric hindrance, one would expect the same rate depression when a 6-hydroxyl group is introduced into methyl salicylate. Further, the spectral properties of the system completely parallel the kinetic properties. Since the rate depression does not occur, the steric hindrance of the 6-hydroxyl group must be overcome by its catalysis. On this basis and since methyl salicylate shows intramolecular general base catalysis,[48] the hydrolysis of methyl 2,6-dihydroxybenzoate may be described as a powerful intramolecular general acid-base catalysis.

An earlier section described multiple catalysis of ester solvolysis by an internal general acid and an external general base. If both the catalysts are placed within the substrate molecule, a still more profound catalysis might be expected. The example above shows it. The solvolyses of cevadine orthoacetate diacetate (**13**) and cevadine diacetate (**14**) also show such a phenomenon. The solvolysis of these esters in methanol-water-chloroform solution have rate constants that show a sigmoidal dependence on the external buffer ratio, indicating dependence on the ionization of an internal basic group. Furthermore, the solvolytic rate constants for these esters are significantly

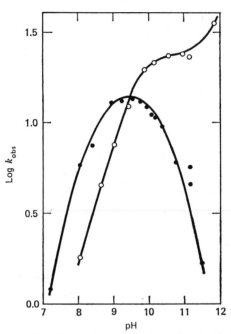

Fig. 10.3 pH-Rate profile for the kinetics of hydrolysis of methyl 2,6-dihydroxybenzoate (●) and methyl salicylate (○) in 1.15% acetonitrile-water at 60°C. From F. L. Killian and M. L. Bender, *Tetrahedron Lett.*, **16**, 1255 (1969).

larger than for the esters dehydrocevadine orthoacetate diacetate (**13**) and the formamido ketone from cevadine orthoacetate diacetate (**14**). The relative rates are **13**, 1000; **15**, 1; and **16**, 40. Compound **13** has neither the possibility of a 1,3-diaxial hydroxyl interaction nor an interaction from an internal nitrogen of reasonable basicity since it is in the weakly basic carbinolamine form. Compound **16** has internal catalysis through the 1,3-diaxial interaction of a hydroxyl group with the ester, but the nitrogen in the molecule is in the form of a weakly basic amide. However, compounds **13** and **14** have both a 1,3-diaxial hydroxyl interaction and a basic nitrogen atom for catalysis.[49] A possibility is shown in eq. 10.16.

(10.16)

[Structures 13, 14, 15, 16 shown]

Ang = angeloyl

The threefold difference in rate between **13** and **14** may be due to a difference between the chair and twist-boat conformations of the D-ring in these two compounds which may result in a small difference in the effectiveness of the 1,3-interaction between the hydroxyl and acetate groups in these two molecules. The solvolysis of cevadine diacetate (**14**—showing intramolecular general acid-base catalysis) is one to two orders of magnitude faster than the solvolysis of coprostane-3β,5β-diol 3-monoacetate [42] (**2**—showing intramolecular general acid and intermolecular general base catalysis).

In the terminology of the intramolecular-intermolecular comparison, 10–100 M external catalyst is necessary for the rate of the intermolecular reaction to be equivalent to the rate of the intramolecular reaction, a value not far different from that found in single catalyst systems (Chapter 9).

The hydration of fumaric acid to malic acid may be brought about by external oxonium ion or hydroxide ion catalysts, but, as implied in eq. 10.17, hydration of the monoanion of fumaric acid needs no assistance from external catalysts.

$$v = k^0[H_2F][H^\oplus] + k^\ominus[HF^\ominus] + k^{2\ominus}[F^{2\ominus}][OH^\ominus] \qquad (10.17)$$

This unusual phenomenon is seen in best perspective in Fig. 10.4, a pH-rate constant profile for the reaction in which both the observed rate constants and the extrapolated oxonium and hydroxide ion catalyses are depicted. At pH 4.5, which is close to neutrality at 175° where these reactions were carried out, reaction of the monoanion is approximately one million-fold faster than the extrapolations of the hydroxide and oxonium ion reactions predict. Neither the diacid nor the dianion of fumaric acid shows this uncatalyzed reaction; nor does crotonic acid, a similar molecule containing only one carboxylic acid group adjacent to the double bond. Hence, both a protonated carboxylic acid group and an ionized carboxylate ion must participate in the hydration of fumaric acid monoanion. Two possible mechanisms for this multiple intramolecular catalysis are apparent: either a general acid-base catalysis or a general acid-nucleophilic catalysis (eqs. 10.18 and 10.19). Distinction between these mechanisms may be made on the basis of experiments on the reverse reaction.

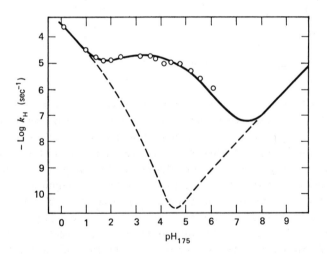

Fig. 10.4 The hydration of fumaric acid at 175°: ———, calculated curve assuming reaction of monoanion; — — — —, calculated assuming no reaction of monoanion. From M. L. Bender and K. A. Connors, *J. Amer. Chem. Soc.*, **84**, 1980 (1962). © 1962 by the American Chemical Society. Reprinted by permission of the copyright owner.

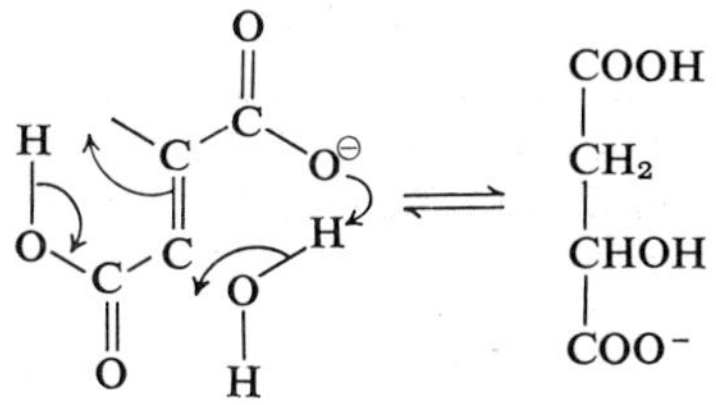

Multiple Catalysis Involving Bifunctional Molecules

(10.18)

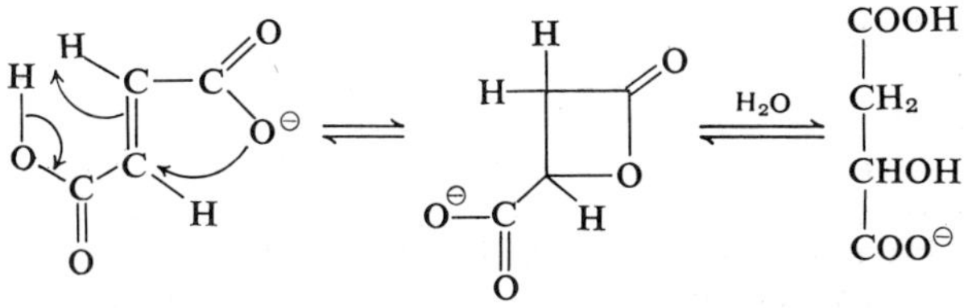

(10.19)

Using optically active malate in the reverse reaction allows one to make both polarimetric and spectrophotometric measurements of the rate constant. Equation 10.18 requires that the two rate constants be identical, but this is not necessary for eq. 10.19 since the intermediate present in this reaction may be partitioned in both directions. The observation that the polarimetric rate constant is larger than the spectrophotometric rate constant is incompatible with eq. 10.18 but consistent with eq. 10.19. Therefore, this intramolecular multiple catalysis is brought about by the combination of general acid and nucleophilic catalysts.[50] For a heterogeneous example see Ref. 51.

10.4 EVALUATION OF MULTIPLE CATALYSIS

The search for enhanced catalysis in two-catalyst versus one-catalyst systems has not been fruitful as mentioned above. The reason for this failure is clearly evident in Tables 10.2 and 10.3. The rate constants of several sets

Table 10.2
Comparison of Second- and Third-Order Rate Constants of Some Imadazole-Catalyzed Reactions

Reaction	Second-Order Rate Constant $k_{Im}(M^{-1}\,min^{-1})$	Third-Order Rate Constant $k_{Im}{}^2\,(M^{-2}\,min^{-1})$
Hydrolysis of p-methylphenyl acetate[16]	0.21	0.17
Hydrolysis of p-methoxyphenyl acetate[16]	0.20	0.19
Hydration of sym-dichloroacetone[11]	2.36	3.55

Table 10.3
Rate Constants of the Hydrazinolysis of Substituted Phenyl Acetates: Uncatalyzed, General Acid-Catalyzed and General Base-Catalyzed Reactions[a]

Substituent	k_u (M^{-1} min^{-1})	k_{ga} (M^{-2} min^{-1})	k_{gb} (M^{-2} min^{-1})
p-NO$_2$	327		
m-NO$_2$	39.7		
H	0.245	2.62	10.75
p-CH$_3$	0.130	1.98	7.86
p-OCH$_3$	0.097	1.82	8.20

[a] From T. C. Bruice and S. J. Benkovic, *J. Amer. Chem. Soc.*, **86**, 418 (1964). © 1964 by the American Chemical Society. Reprinted by permission of the copyright owner.

of singly and doubly catalyzed reactions (Table 10.2) as well as the rate constants of some sets of uncatalyzed and singly catalyzed reactions (Table 10.3) are given. In the reactions of Table 10.2, the third-order rate constant has approximately the same numerical value as the second-order rate constant. In the hydrazinolysis reactions the value of the third-order constant is an order of magnitude larger than the second-order constant. These values mean that the third-order reaction is observable only when the (last) catalyst concentration is of the order of 1 molar for reactions of Table 10.2 and 0.1 molar for reactions of Table 10.3.

What are the underlying causes of the lack of appreciable acceleration by an additional catalyst? The answer can be seen in an analysis of the thermodynamic factors in the hydrazinolysis reaction. Figure 10.5 shows the data for a series of hydrazinolysis reactions with and without catalysis (second-order and third-order reactions). The catalytic reactions have considerably lower enthalpy of activation than the uncatalyzed reactions, but the entropy of activation of the catalyzed reactions is considerably higher, the two nearly compensating one another so that the free energies of activation of the two are not appreciably different.

These and similar data can be analyzed in terms of the ratio, $T\Delta S^{\ddagger}$/kinetic order. Table 10.4 shows a surprisingly constant value of 4–6 kcal/mole/kinetic order for this ratio. This value compares favorably with a value of 5.4 kcal/mole suggested as a reasonable value for the loss of entropy from incorporation of a water molecule into a transition state.[53] A somewhat smaller value (2–3 kcal/mole) is calculated by the Sackur-Tetrode equation[54] for the loss in translational entropy of two molecules in forming a complex. In conclusion, the introduction of any additional external catalyst must

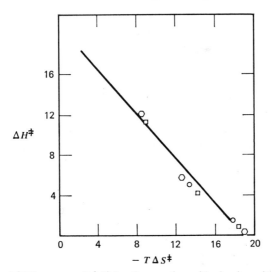

Fig. 10.5 Plot of $\Delta H^{\ddagger}$ versus $-T\Delta S^{\ddagger}$ for the reaction of hydrazine with (□) p-methyl-, (○) p-methoxy-, and (◊) unsubstituted phenyl acetates depicting compensation. From T. C. Bruice and S. J. Benkovic, *J. Amer. Chem. Soc.*, **86**, 418 (1964). © 1964 by the American Chemical Society. Reprinted by permission of the copyright owner.

Table 10.4
A Comparison of the Value of $T\Delta S^{\ddagger}$ (kcal/mole) to the Kinetic Order of Displacement Reactions on the Phenyl Ester Bond[a]

Reaction	Kinetic Order	$T\Delta S^{\ddagger}/$ Kinetic Order[b]	Number of Esters Investigated
$(CH_3)_3N$ + Ph esters	2	4	4
γ-(N,N-Dimethylamino)butyrate Ph esters	1	4	5
δ-(N,N-Dimethylamino)valerate Ph esters	1	4–5	5
$OH^{\ominus}$ + Ph esters	2	3	4
$AcO^{\ominus}$ + Ph esters	2	4	2
Monophenyl glutarates	1	4	4
H_2NNH_2 + Ph esters	2	4	4
$2(H_2NNH_2)$ + Ph esters	3	4–5	3
$H_2NNH_2 + H_2NNH_3^{\oplus}$ + Ph esters	3	6	3
Imidazole + Ph esters	2	6–7	6
2(Imidazole) + Ph esters	3	5	1

[a] From T. C. Bruice and S. J. Benkovic, *J. Amer. Chem. Soc.*, **86**, 418 (1964). © 1964 by the American Chemical Society. Reprinted by permission of the copyright owner.
[b] Values rounded to one significant figure.

necessarily lead to unfavorable entropy of activation, even though the enthalpy of activation is made more favorable.

How then is it possible to accelerate a reaction by going to a higher order catalytic process? The answer is to go to a catalytic reaction which increases the catalytic functionalities but not the kinetic order of the reaction, such as catalysis by a bifunctional molecule. If the same decrease in enthalpy of activation brought about by changing from one to two catalysts is not accompanied by a compensating activation entropy loss, the catalysis should be considerably more efficient. The mutarotation of α-D-tetramethylglucose in benzene solution by 2-pyridone was originally attributed to 2-pyridone, but is now thought to be probably due to the corresponding ion. The data are shown in Table 10.5.

Multiple catalysis can be advantageous over single catalysis if the multiple catalyst is a bifunctional or polyfunctional molecule. In such a case, the entropy of activation should remain constant for the two systems, while the activation enthalpy should be more favorable for the bifunctional catalyst than for the monofunctional catalyst (a prediction yet to be tested).

Both intramolecular catalysis (Chapter 9) and bifunctional catalysis gain their efficacy versus separated reagents through favorable activation entropies. If intramolecular and bifunctional catalysis are combined with one another, the activation entropy advantage should be the sum of the individual entropy advantages and could amount to a very significant kinetic factor. Such a summation is probably best seen in the hydration of fumaric acid monoanion to malate monoanion. Activation parameters were not determined for this reaction; nor were comparisons made with a reaction in which the two catalysts were external, for such a reaction does not exist. But the

Table 10.5

Activation Parameters in the Mutarotation of α-D-Tetramethylglucose in Benzene Solution[a,b]

Catalyst	Activation Enthalpy (kcal/mole)	Activation Entropy (e.u./mole)	Activation Free Energy (25°) (kcal/mole)
Pyridine-phenol	-8.7	-35.0 ± 3	19.4 ± 0.4
2-Pyridone	10.8 ± 1	23.1 ± 3	17.7 ± 0.4
Benzoic acid	10.8 ± 1	21.5 ± 3	17.3 ± 0.2

[a] The data refer to the chemical change from the free catalyst species and reactant to the transition state complex.

[b] From. P. R. Rony, *J. Amer. Chem. Soc.*, **90**, 2824 (1968). © 1968 by the American Chemical Society. Reprinted by permission of the copyright owner.

power of the system may be seen by comparison of the rate constant of this intramolecular bifunctional catalysis with hydroxide ion or oxonium ion catalysis of the same reaction. The intramolecular bifunctional catalysis is up to one million times faster than the external catalysis, a factor which can account for a considerable fraction of the enhancement of enzymic reactions over hydroxide or oxonium ion reactions, usually of the order of 10^{10}–10^{12}.

REFERENCES

1. T. M. Lowry and I. J. Faulkner, *J. Chem. Soc.*, **127**, 1883 (1925); T. M. Lowry, *ibid.*, 2554 (1927).
2. Y. Pocker, *Chem. Ind.*, 968 (1960).
3. C. G. Swain and J. F. Brown, Jr., *J. Amer. Chem. Soc.*, **74**, 2534, 2538 (1952).
4. P. R. Rony, W. E. McCormack, and S. W. Wunderly, *J. Amer. Chem. Soc.*, **91**, 4244 (1969); P. R. Rony, *ibid.*, **91**, 6090 (1969); the latter proposes a new form of catalysis, Tautomeric Catalysis.
5. M. Eigen, *Disc. Faraday Soc.*, **39**, 17 (1965).
6. R. P. Bell, Eleventh Conference on Reaction Mechanisms, Hamilton, Ont., 1966.
7. C. G. Swain, A. J. Di Milo, and J. P. Cordner, *J. Amer. Chem. Soc.*, **80**, 5983 (1958).
8. R. P. Bell and M. B. Jensen, *Proc. Roy. Soc.*, **A261**, 38 (1961).
9. B. E. C. Banks, *J. Chem. Soc.*, 63 (1962); 5043 (1961).
10. F. J. C. Rossotti, *Nature*, **188**, 936 (1960).
11. E. H. Cordes and M. Childers, *J. Org. Chem.*, **29**, 968 (1964).
12. C. K. Ingold, *Structure and Mechanism in Organic Chemistry*, Cornell University Press, Ithaca, N.Y., 1953, p. 572; C. K. Ingold and C. L. Wilson, *J. Chem. Soc.*, 1493 (1933); C. K. Ingold and C. L. Wilson, *ibid.*, 93 (1934); S. K. Hsu, C. K. Ingold, and C. L. Wilson, *ibid.*, 1778 (1935); R. P. Ossorio and E. D. Hughes, *J. Amer. Chem. Soc.*, 426 (1952).
13. D. J. Cram and R. T. Uyeda, *J. Amer. Chem. Soc.*, **84**, 4358 (1962); D. J. Cram and R. T. Uyeda, *J. Amer. Chem. Soc.*, **86**, 5466 (1964).
14. D. J. Cram and R. D. Guthrie, *J. Amer. Chem. Soc.*, **87**, 397 (1965).
15. T. C. Bruice and R. M. Topping, *J. Amer. Chem. Soc.*, **85**, 1480, 1488, 1493 (1963).
16. J. F. Kirsch and W. P. Jencks, *ibid.*, **86**, 833 (1964); T. C. Bruice and S. J. Benkovic quoted in this paper.
17. E. Brode and L. Jaenicke, *Biochem. Z.*, **332**, 259 (1960).
18. C. G. Swain, *J. Amer. Chem. Soc.*, **72**, 4578 (1950).
19. B. J. Jandorf, T. Wagner-Jauregg, J. J. O'Neill, and M. A. Stolberg, *J. Amer. Chem. Soc.*, **74**, 1521 (1952); K.-B. Augustinsson, *Acta Chem. Scand.*, **6**, 959 (1952).
20. J. Epstein, D. H. Rosenblatt, and M. M. Demek, *J. Amer. Chem. Soc.*, **78**, 341 (1956).
21. J. W. Churchill, M. Lapkin, F. Martinez, and J. A. Zaslowsky, *ibid.*, **80**, 1944 (1958).
22. E. J. Fuller, *J. Amer. Chem. Soc.*, **85**, 1777 (1963).
23. L. Larsson, *Arkiv Kemi*, **13**, 259 (1958).
24. T. Higuchi, Progress report submitted to Armed Forces Chemical Center, Maryland, May 20, 1959, quoted by T. C. Bruice and S. J. Benkovic in *Bioorganic Mechanisms*, Vol. 2, W. A. Benjamin Co., New York, 1966, p. 143.

25. T. C. Bruice and R. G. Willis, *J. Amer. Chem. Soc.*, **87**, 531 (1965).
26. F. M. Menger, *J. Amer. Chem. Soc.*, **88**, 3081 (1966).
27. C. G. Swain and J. F. Brown, Jr., *J. Amer. Chem. Soc.*, **74**, 2538 (1952).
28. C. G. Swain and J. F. Brown, Jr., *J. Amer. Chem. Soc.*, **74**, 2534 (1952).
29. P. R. Rony, personal communication.
30. E. L. Blackall and A. M. Eastham, *J. Amer. Chem. Soc.*, **77**, 2184 (1955).
31. L. M. Litvinenko and N. M. Oleinik, *J. Gen. Chem. USSR*, **33**, 2227 (1963) and references therein.
32. B. A. Cunningham and G. L. Schmir, *J. Amer. Chem. Soc.*, **88**, 551 (1966).
33. A. J. Hubert, R. Buyle, and B. Hargitay, *Helv. Chim. Acta*, **46**, 1429 (1963).
34. M. Brenner and W. Hofer, *ibid.*, **44**, 1794 (1961); W. Hofer and M. Brenner, *ibid.*, **47**, 1625 (1964).
35. B. Glutz and H. Zollinger, *Angew. Chem. Int. Ed.*, **4**, 440 (1965).
36. P. B. Hamilton, *J. Biol. Chem.*, **158**, 375 (1945); J. B. Gilbert, V. E. Price, and J. P. Greenstein, *ibid.*, **180**, 209 (1949); A. Meister, *ibid.*, **210**, 17 (1954).
37. V. Franzen, *Angew. Chem.*, **72**, 139 (1960).
38. E. Sacher and K. J. Laidler, *Can. J. Chem.*, **42**, 2404 (1964).
39. H. C. Beyerman and W. M. van den Brink, *Proc. Chem. Soc.*, 266 (1963).
40. H. C. Beyerman, W. M. van den Brink, and H. S. Tan, *Abst. Meeting Fed. Europ. Biochem. Soc.*, 1964, p. 31.
41. J. W. Thanassi, A. R. Butler, and T. C. Bruice, *Biochem.*, **4**, 1463 (1965).
42. S. M. Kupchan, S. P. Eriksen, and M. Friedman, *J. Amer. Chem. Soc.*, **84**, 4159 (1962); **88**, 343 (1966).
43. F. H. Westheimer, *Spec. Publ. Chem. Soc.*, **8**, 1 (1957).
44. G. A. Hamilton and A. Revesz, *J. Amer. Chem. Soc.*, **88**, 2069 (1966).
45. H. Morawetz and I. Oreskes, *J. Amer. Chem. Soc.*, **80**, 2591 (1958).
46. H. Morawetz and J. Shafer, *J. Amer. Chem. Soc.*, **84**, 3783 (1962).
47. B. Capon and B. Ch. Ghosh, *J. Chem. Soc.*, B472 (1966).
48. F. L. Killian and M. L. Bender, *Tetrahedron Lett.*, **16**, 1255 (1969).
49. S. M. Kupchan, S. P. Eriksen, and Y.-T. Shen, *J. Amer. Chem. Soc.*, **85**, 350 (1963); S. M. Kupchan, S. P. Eriksen, and Y.-T. S. Liang, *ibid.*, **88**, 347 (1966).
50. M. L. Bender and K. A. Connors, *J. Amer. Chem. Soc.*, **84**, 1980 (1962).
51. H. Pines and C. N. Pillai, *J. Amer. Chem. Soc.*, **82**, 2401 (1960).
52. T. C. Bruice and S. J. Benkovic, *J. Amer. Chem. Soc.*, **86**, 418 (1964).
53. L. L. Schaleger and F. A. Long, *Advances in Physical Organic Chemistry*, V. Gold, Ed., Academic Press, Inc., New York, 1963, p. 26.
54. I. Z. Steinberg and H. A. Scheraga, *J. Biol. Chem.*, **238**, 172 (1963).
55. P. R. Rony, *J. Amer. Chem. Soc.*, **90**, 2824 (1968).

Chapter 11

CATALYSIS BY COMPLEXATION

11.1	Covalent Complexes	352
	11.1.1 Carbonyl Complexes	353
	11.1.2 Amine Complexes	356
	11.1.3 Nucleophiles Containing a Catalytic Group	357
	11.1.4 Electrophiles Containing a Catalytic Group	359
11.2	Noncovalent Complexes	360
	11.2.1 Ionic and Hydrogen Bonding Complexes	362
	11.2.2 Apolar Complexes	364
	Pi Molecular Complexes	364
	Complexes of Aliphatic Hydrocarbons	368
	Inclusion Complexes	373
	11.2.3 Polymeric Complexes	382

Many kinds of complexes can be formed between organic molecules. These complexes include: (1) covalent complexes—defined here as rapidly and reversibly formed compounds; (2) electrostatic or ionic complexes; (3) hydrogen-bonded complexes; (4) metal ion complexes; and (5) apolar complexes. Many of these interactions can have an effect on the rates of reaction. Catalysis by metal ion complexes has been discussed in detail in Chapter 8. The effect of some other complexes on catalysis has been alluded to. In this chapter, we consider explicitly the effect of complexing on catalysis.

The essential role of complexing on catalytic action has been noted many times. In all of heterogeneous catalysis, adsorption of the substrate(s) on the catalytic surface appears to be an essential act. Likewise in enzymic catalysis, formation of an enzyme-substrate complex is usually the primary step. In many of these reactions, the catalyst-substrate complex is more stable than the uncomplexed materials. This fact is difficult to reconcile with a facilitation of reaction in which the free energy of activation must be lowered. The explanation of course is that complexing must lower the free energy of the transition state by an even greater amount than the ground state in order for catalysis to occur. This lowering can occur either through a change in reaction pathway by the complexation, or simply through a lowering of the transition state without change in pathway.

Complexing can juxtapose substrate and catalyst. When this occurs, the subsequent catalytic act should have the characteristics of an intramolecular reaction because of more favorable activation entropy. The initial complexing of substrate and catalyst will have an unfavorable translational entropy, however, unless compensated by other favorable energy factors such as solvation entropy. These favorable factors occur with sufficient frequency to make complexation an important catalytic pathway.

In addition to bringing catalyst and substrate into proximity with one another, complexation must perform one further function if it is to be effective. The complexation must be stereospecific so that the catalytic group(s) are oriented correctly with respect to the reactive center of the substrate. This requirement is important because random complexing can lead to many more nonproductive complexes than productive (catalytically active) complexes. The formation of stereospecific complexes depends on complementarity of structure between the catalyst and substrate in some sense. As we learn more about the general laws of complementarity, we will undoubtedly be able to design more and more specific catalysts for specific reactions. In fact, the chemistry of complementarity will undoubtedly be one of the exciting branches of chemistry of the near future.[1]

11.1 COVALENT COMPLEXES

Many of the nucleophilic and electrophilic catalyses described in Chapter 6 proceed through the formation of intermediates, for example, imines, which facilitate reaction. The catalyses described here operate in the same general fashion, but differ in one respect: with the intermediate a functional group is introduced, which serves as an intramolecular catalyst to complete the reaction. Thus the difference between these catalyses and nucleophilic-electrophilic catalyses is the direct versus indirect interaction of an introduced catalytic entity at the reaction center. This distinction is of course an arbitrary one with a gradation that is sometimes hard to classify.

11.1.1. Carbonyl Complexes

Rapid and reversible additions to carbonyl groups make substrates containing such groups susceptible to catalyses occurring through adducts of carbonyl compounds. Nitrogen and oxygen nucleophiles in particular add to carbonyl groups readily to form tetrahedral adducts and/or imines. If such an adduct or imine contains a functional group that can interact with the reaction center, a specific and powerful catalysis can occur.

The hydrolysis of esters containing carbonyl substituents is susceptible to this catalysis. The alkaline hydrolysis of methyl 2-benzoyl-6-methylbenzoate, for example, proceeds at a very high rate compared to the hydrolysis of other 2,6-substituted benzoate esters.[2] Likewise, the alkaline hydrolysis of methyl o-formylbenzoate is 10^5 times faster than the hydrolysis of methyl o-nitrobenzoate, a compound of similar steric and electronic properties.[3] These reactions, which are first-order in hydroxide ion concentration, may be explained by the addition of hydroxide ion to the carbonyl group of the substrate, producing an adduct whose oxyanion can function as an intramolecular nucleophilic catalyst.

$$\text{(11.1)}$$

A similar mechanism can be invoked to explain the easy saponification of methyl 21-oxo-oleanolate, which has a keto group in the *gamma* position to a hindered ester in a polycyclic system,[4] the ready saponification of compound **1** compared to the corresponding compound containing no γ-keto group,[5] and the hydrolysis of 1-benzyl-2-acetyl-6-oxo-9-carbomethoxy-1-hydroxydecahydroisoquinoline which is 10^4 times faster than that of the corresponding ester in which the 6-(γ)-keto group is replaced by a methylene group.[6]

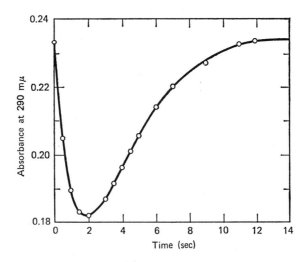

1

The alkaline hydrolyses of cinnamoylsalicylaldehyde[7] and *o*-acetoxybenzaldehyde[8] are likewise extremely fast. The nucleophilic catalysis postulated for the intramolecular alkoxide ion in eq. 11.1 is not a unique possibility. Alternative mechanisms involving general base catalysis by the alkoxide ion or general acid catalysis by the alcohol group may be written, but are less probable on the basis of arguments to be presented here.

Nucleophiles other than hydroxide ion which are capable of addition to the carbonyl group should also catalyze the hydrolysis of the esters described above. This is indeed the case. Cyanide ion increases the rate constant of

Fig. 11.1 The morpholine-catalyzed (0.085 M) hydrolysis of methyl *o*-formylbenzoate. The reaction was followed using a spectrophotometer equipped with a stopped-flow mixing device. From M. L. Bender, J. A. Reinstein, M. S. Silver, and R. Mikulak, *J. Amer. Chem. Soc.*, **87**, 4549 (1965). © 1965 by the American Chemical Society. Reprinted by permission of the copyright owner.

hydrolysis of cinnamoylsalicylaldehyde, 0.002 M leading to a ten-fold rate enhancement.[7] Likewise, morpholine catalyzes the hydrolysis of methyl o-formylbenzoate. Figure 11.1[3] illustrates a typical spectrophotometric curve of the facile morpholine-catalyzed hydrolysis of methyl o-formylbenzoate, using a stopped-flow mixing device. Since the reaction is followed at the isosbestic point of the ester reactant and acid product, the decrease followed by an increase in absorbance during the course of reaction demonstrates the formation and subsequent decomposition of an unstable intermediate. 3-Morpholinophthalide can be isolated from the reaction mixture after a few seconds of contact time and is hydrolyzed to o-formylbenzoate ion at exactly the rate at which the intermediate in the overall reaction hydrolyzes. This combination of isolation and kinetic evidence clearly indicates the following pathway for this catalysis[3,9]:

$$\text{(11.2)}$$

The intramolecular nucleophilic catalysis by the alkoxide ion in the morpholine catalysis lends support to the mechanism of eq. 11.1.

The alkaline hydrolysis of dimethyl phosphoacetoin is 2×10^6 times faster than the hydrolysis of trimethyl phosphate.[10] Likewise alkaline hydrolysis of p-nitrophenyl phenacyl methylphosphonate is many orders of magnitude faster than hydrolysis of p-nitrophenyl methyl methylphosphonate.[11] By analogy with the ester reactions described above, one may postulate addition of hydroxide ion (or water) to the carbonyl group followed by an interaction of this adduct with the phosphate ester center. Nucleophilic attack by this adduct would lead to a five-membered cyclic phosphate ester, whose ready cleavage is known (Chapter 9). It is thus reasonable to postulate a mechanism similar to eq. 11.1 for these reactions, although it is by no means proven.

The transformation of phenylglyoxal to mandelic acid (or methyl mandelate) can be catalyzed in a highly specific manner by using a catalyst that adds to the aldehyde group and then facilitates the hydride transfer necessary for this internal oxidation-reduction reaction. The catalyst is 2-dimethylaminoethanethiol.[12] This catalysis occurs at room temperature in the presence of 5 moles of catalyst per mole of substrate. The catalysis does not operate for the intermolecular analog of this reaction. Furthermore, a mixture of ethyl mercaptan and diethylamine or triethylamine catalyzes the transformation much less effectively than 2-diethylaminoethanethiol. When the reaction is carried out in deuterium oxide, no carbon-bound deuterium is found in the product, indicating the occurrence of an intramolecular hydride transfer. When the reaction is carried out at 0° in methanol, an intermediate adduct of the catalyst and substrate can be isolated, which contains no free thiol group, indicating the thiol group of this bifunctional (ambident) catalyst adds to the substrate. On the basis of these experiments, the following mechanism may be postulated.

$$(11.3)$$

In an enzymic transformation, glutathione serves as coenzyme in a reaction which may be similar mechanistically.

11.1.2. Amine Complexes

If amines can catalyze the hydrolysis of esters containing carbonyl groups, then carbonyl compounds should be able to catalyze the hydrolysis of esters containing amino groups. This prediction was confirmed in the benzaldehyde-catalyzed hydrolysis of p-nitrophenyl leucinate.[13] The rate constant for this catalysis is six times larger than the rate constant for the corresponding imidazole catalysis, even though imidazole is more basic than benzaldehyde by some 14 powers of 10! As predicted, benzaldehyde has no catalytic effect on an ester containing no amino group such as p-nitrophenyl acetate. By analogy with eq. 11.2, the mechanism of this catalysis can be postulated to proceed through the addition of the amino ester to benzaldehyde, forming an adduct capable of intramolecular catalysis of ester hydrolysis, as shown in

$$NH_2-CHR-CO\phi NO_2 + \phi CHO \longrightarrow \phi-CH\begin{array}{c}H_2\overset{\oplus}{N}\diagdown CHR \\ | \quad | \\ O^{\ominus}\quad C-O\phi NO_2 \\ \| \\ O\end{array} \xrightarrow{-O_2N\phi OH}$$

$$\phi-CH\begin{array}{c}HN\diagdown CHR \\ / \quad | \\ O-C \\ \| \\ O\end{array} \xrightarrow{H_2O} \phi CHO + NH_2CHRCO_2H \qquad (11.4)$$

Although the details of this process are still speculative, the analogy between eqs. 11.4 and 11.2 is a reasonable one, and certainly the rate enhancement is profound.

11.1.3 Nucleophiles Containing a Catalytic Group

A facile nucleophilic reaction, followed by a second (catalytic) reaction performed by a group on the original nucleophile, and finally decomposition to give the products and regenerate the nucleophile can lead to effective catalysis.

The 2-diethylaminoethanethiol catalysis of the phenylglyoxal-mandelic acid transformation is an example of such a catalytic series. The catechol monoanion-catalyzed hydrolysis of phenyl chloroacetate is another.[14] The transesterification of one aryl ester to another is known. In such a reaction the catechol monoanion, a bifunctional nucleophile (Chapter 10) would be expected to react rapidly. The facile hydrolysis of catechol monochloroacetate is also known (Chapter 9). In fact, it is faster than the overall hydrolytic rate for the catechol monoanion-catalyzed hydrolysis of phenyl chloroacetate, as it must be. These analogies together with the failure of phenol, guaiacol, resorcinol, and hydroquinone to exhibit positive catalysis are consistent with

Catalysis by Complexation

$$\text{ClCH}_2\text{CO}\phi + \underset{\text{O}^\ominus}{\text{OH}} \longrightarrow \underset{\text{HO}}{\text{ClCH}_2\text{CO}}$$

$$\underset{\ominus\text{O}}{\text{ClCH}_2\text{C}-\text{O}} \xrightarrow{\text{H}_2\text{O}} \text{ClCH}_2\text{CO}^\ominus + \underset{\text{O}^\ominus}{\text{OH}} \qquad (11.5)$$

When the original ester contains a very good leaving group, the intermediate in the reaction can be isolated.[15]

A similar reaction in which each step may be observed directly is the reaction of p-nitrophenyl acetate with o-mercaptobenzoic acid.[16] The dianion of o-mercaptobenzoic acid reacts with the ester producing thioaspirin. Thioaspirin then hydrolyzes in a pH-independent reaction (from pH 6.2 to 7.9) to give acetate ion and regenerate o-mercaptobenzoic acid. These two steps constitute an overall catalysis of ester hydrolysis brought about by o-mercaptobenzoate dianion, the rate-determining step being the second step. Whereas catalysis by catechol monoanion involves nucleophilic attack followed by intramolecular general basic catalysis, catalysis by o-mercaptobenzoate dianion involves nucleophilic attack followed by intramolecular catalysis by the *ortho*-carboxylate ion, similar to the hydrolysis of aspirin (Chapter 9).

$$\underset{\text{S}^\ominus}{\text{CO}_2^\ominus} + \underset{\text{NO}_2}{\overset{\text{CH}_3}{\underset{|}{\text{C}=\text{O}}}} \xrightarrow{-^\ominus\text{O}\phi\text{NO}_2} \underset{\text{CO}_2^\ominus}{\overset{\text{CH}_3}{\underset{\text{S}-\text{C}=\text{O}}{}}} \xrightarrow{\text{OH}^\ominus} \underset{\text{S}^\ominus}{\text{CO}_2^\ominus} + \text{CH}_3\text{CO}_2^\ominus \qquad (11.6)$$

The sluggish hydrolysis of thioaspirin makes eq. 11.6 a relatively inefficient process.

11.1.4 Electrophiles Containing a Catalytic Group

By analogy with nucleophilic reactions, a facile electrophilic reaction, followed by a second (catalytic) reaction performed by a group on the original electrophile, and finally decomposition to give the products and regenerate the electrophile should also lead to effective catalysis.

The electrophile 8-quinolineboronic acid is a catalyst for the hydrolysis of chloroethanol and 3-chloro-1-propanol in dimethylformamide solutions containing water and collidine.[17] In the absence of 8-quinolineboronic acid, the chloroalcohols undergo slow solvolysis to products that are not glycols, whereas in the presence of the catalyst, only glycols are formed. Hence, in the presence of the catalyst, alcoholic portions of these substrates are not available for cyclization, a fact that can be rationalized with the formation of borate esters from the alcohols. Inhibition of the catalytic reaction by both water and ethylene glycol support this suggestion. Hydrolysis of stereoisomeric chlorohydrins including *trans*-2-chloro-1-indanol and *erythro*-2-chloro-1,2-diphenylethanol show that the reaction occurs with inversion of configuration, ruling out direct participation by the tertiary amine. In addition, *cis*-2-chloro-1-indanol does not undergo carbon–chlorine fission when treated with 8-quinolineboronic acid. These results are consistent with a mechanism for the catalysis by 8-quinolineboronic acid in which reversible esterification of

$$\text{quinoline-B(OH)}_2 + \text{HOCH}_2\text{CH}_2\text{Cl} \rightleftharpoons \text{[cyclic borate ester]} \xrightarrow{-\text{Cl}^\ominus}$$

$$\text{[quinolinium borate]} \xrightleftharpoons[-\text{OH}^\ominus]{\text{H}_2\text{O}} \text{quinoline-B(OH)}_2 + \text{HOCH}_2\text{CH}_2\text{OH} \quad (11.7)$$

the boronic acid group by the alcoholic groups of the substrate occurs followed by displacement of the halogen atom of the substrate by a water molecule that may or may not be bonded covalently to the boron atom.[18,19] This catalyst is in essence a bifunctional catalyst, but instead of consisting of two functions that interact with the reaction center, it consists of two functions of which one binds the substrate stereospecifically and the other carries out the actual chemical transformation. As implied by this description, neither a boronic acid nor a nitrogen base alone will carry out this catalysis.

Phenyl salicylate is hydrolyzed abnormally rapidly in borate buffers.[20] The catalytic constants for the hydrolysis of phenyl salicylate in borate buffers are more than a hundredfold greater than those for phenyl o-methoxybenzoate and phenyl benzoate, while in imidazole buffers they are only two to threefold greater. Borate ion normally has only a very weak catalytic effect on ester hydrolysis, and with p-nitrophenyl acetate, no effect at all is observed. Catalysis does not occur with phenyl salicylate in phosphate buffers, nor with catechol monobenzoate in borate buffers. The most likely explanation of these results is that complex formation of the kind shown in eq. 11.8 facilitates the hydrolytic reaction.

$$\text{salicylate-O}\phi \xrightarrow[H_3BO_3]{-H_2O} \text{complex} \xrightarrow[-\phi OH]{+HO^-} \text{intermediate} \rightleftharpoons_{H_2O} \text{salicylic acid} + H_3BO_3 \quad (11.8)$$

Since the complex from a salicylate involves a six-membered ring while the corresponding complex from catechol monobenzoate would involve a less favored seven-membered ring, the catalytic specificity can be explained.

11.2 NONCOVALENT COMPLEXES

One of the special features of enzymic catalysts is their ability to form adsorptive, usually noncovalent complexes. Ideally, model reactions for enzymic processes should include such an association. The forces holding

Noncovalent Complexes

these complexes together consist of the various kinds of complexing mentioned earlier, including ionic, hydrogen-bonded, and apolar complexes. Of these the last is probably the most important. Apolar complexes take many forms such as pi molecular complexes, micellar complexes, and inclusion complexes. We shall discuss each of these.

When complexing occurs between substrate and catalyst, two pathways can be written:

$$C + S \xrightleftharpoons{K} C \cdot S \xrightarrow{k_{cat}} C + P \qquad (11.9)$$

$$C \cdot S \xrightleftharpoons{K} C + S \xrightarrow{k_{cat}'} C + P \qquad (11.10)$$

These two equations correspond to complexing that leads to catalysis and complexing that prevents catalysis. The complex in eq. 11.9 is referred to as a productive complex while that in eq. 11.10 is referred to as a nonproductive or blind-alley complex. The former is, of course, the object of prime attention.

The minimal experimental evidence for complex formation in a catalytic (or other) process is adherence to a kinetic scheme showing a "saturation" phenomenon. Kinetic analysis of eq. 11.9 may be carried out in the simplest fashion by assuming a fast pre-equilibrium formation of the complex and assuming $[S]_0 \gg [C]_0$.† Under these conditions

$$K = \frac{[C][S]}{[C \cdot S]} \qquad (11.11)$$

and

$$[C]_0 = [C] + [C \cdot S] \quad \text{and} \quad [S]_0 = [S] \qquad (11.12)$$

On substitution

$$[C]_0 = \frac{K[C \cdot S]}{[S]_0} + [C \cdot S] = [C \cdot S]\left(\frac{K}{[S]_0} + 1\right) \qquad (11.13)$$

Therefore

$$[C \cdot S] = \frac{[C]_0}{\frac{K}{[S]_0} + 1} \qquad (11.14)$$

$$v = k_{cat}[C \cdot S] = \frac{k_{cat}[C]_0}{\frac{K}{[S]_0} + 1} = \frac{k_{cat}[C]_0[S]_0}{K + [S]_0} \qquad (11.15)$$

Equation 11.15 shows a direct dependence of the velocity on $[S]_0$ at values of $[S]_0 \ll K$, but a zero-order dependence on $[S]_0$ at values of $[S]_0 \gg K$, thus showing a "saturation" of the catalyst by the substrate. The equilibrium

† See Chapter 15 for a more complete derivation of catalysis involving complexing.

constant of complex formation, K, is experimentally equal to the concentration of substrate at which the rate is half its maximal value. The rate constant, k_{cat}, is the maximal rate constant. This derivation is, of course, a complete description of Michaelis-Menten kinetics first developed for enzymic systems,[22] using the equilibrium assumption.

Equation 11.10 leads to a kinetic equation of equivalent form. Using the same assumptions as above, the velocity corresponding to equation 11.10 is

$$v = \frac{k'_{cat}K[C]_0[S]_0}{K + [S]_0} \tag{11.16}$$

Thus, the perennial problem of whether an intermediate is present in the reaction pathway, or only as a side reaction, is also a vexing problem in consideration of reactions involving complexing. In general the resolution of such a problem requires the isolation of the intermediate and determination of the kinetics of its decomposition. Obviously, the isolation of such an ephemeral substance as a catalyst-substrate complex is an exceedingly difficult task. As a consequence, the only arguments for distinguishing between eqs. 11.9 and 11.10 are extrakinetic arguments.

In addition to the simple dichotomy of eq. 11.9 and 11.10, there exists the more probable simultaneous combination of the two: several complexes may be formed between catalyst and substrate, not all of which are productive complexes. This possibility stems directly from the fact that only the complex of correct stereochemistry will lead to catalysis, in the same way that in intramolecular systems only those molecules possessing correct stereochemistry exhibit catalysis.

11.2.1 Ionic and Hydrogen Bonding Complexes

On the basis of electrostatic theory, ionic interactions are expected to be of importance in media of low dielectric constants.[22, 23] The occurrence of ion pairs mentioned in Chapter 7 and the unfavorable free energy of ion pair formation in aqueous solutions of butylammonium isobutyrates bear out this statement.[24] Since we are particularly concerned with aqueous solutions rather than apolar media of low dielectric constant, the possibility of forming simple ionic complexes of catalytic importance is remote. (See, however, the sections on apolar and polymeric complexes.)

Hydrogen bond interactions between solutes are expected to be of importance in apolar media containing no hydrogen bond donors or acceptors. The importance of the solvent is vividly seen in Fig. 11.2 which depicts the hydrogen-bonded association of N-methylacetamide.[25]

Catalyses occurring in aprotic media containing no electron donors, for example, benzene, cyclohexane, or carbon tetrachloride, may then occur

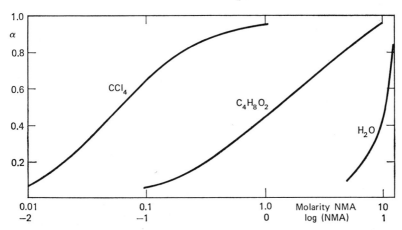

Fig. 11.2 Variation of degree of association, α, with concentration of N-methylacetamide in carbon tetrachloride, dioxane, and water, respectively. From I. M. Klotz and J. S. Franzen, *J. Amer. Chem. Soc.*, **84**, 3461 (1962). © 1962 by the American Chemical Society. Reprinted by permission of the copyright owner.

through prior hydrogen-bonded complexes. This catalysis will be of special importance for those reactions that are susceptible to general acid-base catalysis. Of particular importance will be those reactions in aprotic media that are susceptible to multiple general acid-base catalysis and have good analogy in the self-association of 2-hydroxypyridine, which shows this phenomenon.

In aqueous solution, however, hydrogen-bonded complexes should not be stable, and therefore, their occurrence in catalytic reactions has low probability, although an intermediate need never attain a high concentration. In Chapter 10, several tentative suggestions of hydrogen-bonded complexes were put forth. For example, the reaction of isopropyl methylphosphonofluoridate with monocatecholate ion and its derivatives was postulated to proceed through the initial formation of a complex between the substrate and nucleophile involving a hydrogen bond and an ion-dipole interaction. Certainly the kinetic evidence is consistent with complex formation. Furthermore, the structure of this complex can explain the enhanced reactivity of the bifunctional monocatecholate nucleophile. However, nonproductive complex formation via eq. 11.10 can occur, just as well as productive complex formation via 11.9. Since complexes of the type pictured in eq. 11.7 are improbable in aqueous solution and since the ability of phenols to associate with organophosphorus compounds related to the above substrate is inversely proportional to their reactivity toward nucleophilic reagents,[26] it appears more likely that the complex between isopropyl methylphosphonofluoridate and monocatecholate ion is a nonproductive complex, involving the interaction of the hydrocarbon portion of the molecules rather than the hydrogen bond, ionic, or dipolar parts of the molecules.

The hydrogen-bonded complex postulated between the monohydrogen phosphate ion and the tetrahedral adduct in the hydrolysis of N-phenyliminotetrahydrofuran must be viewed with caution. Although this picture is a convenient mental scaffolding to explain the data, it must be probed further before assigning an apparent dissociation constant of 2×10^{-3} M to a hydrogen-bonded complex in aqueous solution.[27]

11.2.2 Apolar Complexes

In aqueous solution, complexes form between lyophobic, usually hydrocarbon-containing, compounds. We consider the effects of some of these complexes on the kinetics of chemical reactions, with particular emphasis on catalysis. Several different forms of apolar complexes can be discussed: pi molecular complexes; complexes of aliphatic hydrocarbons, including micelles; and inclusion complexes. Although this categorization is an arbitrary one, it will allow us to discuss some interesting catalytic phenomena.

Pi Molecular Complexes. The formation of complexes between aromatic, lyophobic compounds, and other compounds such as heterocyclics, tetracyanoethylene, or other aromatic compounds is a well-established if perhaps not well-understood phenomenon.[28] These complexes, which have been called pi-complexes, donor acceptor complexes, and charge transfer complexes, have been also called pi molecular complexes on the basis that this is a descriptive but neutral name, which does not imply that the charge transfer spectrum sometimes observed at the same time that the complex is observed is necessarily the basis of complex formation.[29]

If one is to make any generalization on the effect of pi molecular complexes on the rates of organic reactions, it would be that such complexes decelerate rather than accelerate reaction. This "negative catalysis" can be viewed in terms of eq. 11.10, namely that the complexed substrate does not react to give products while the uncomplexed substrate does. This description of course means that the complex is a nonproductive one of the highest order. The simplest meaning of inhibition of reaction through complex formation is that complex formation prevents access of some reagent to the substrate and thus prevents reaction.

Inhibition by pi molecular complex formation includes the inhibition of the alkaline hydrolysis of ethyl p-aminobenzoate by caffeine.[30] Likewise, the alkaline hydrolysis of methyl *trans*-cinnamate is inhibited by several imidazole, purine, and xanthine compounds such as imidazole, benzimidazole, purine, uracil, theophylline, caffeine, and guanine.[31] Finally, the hydrolysis and aminolysis of *trans*-indoleacryloylimidazole and p-nitrophenyl *trans*-indoleacrylate are inhibited by formation of a pi molecular complex with 3,5-dinitrobenzoate ion.[32] The kinetics of all three sets of reactions are consistent with a scheme in which only the uncomplexed ester can be hydrolyzed. In each case, the binding constant of the complex determined kinetically agrees with that determined by independent physical measurement such as solubility or spectrophotometric measurements, although some systems are more complicated than simple 1:1 complexes. The apparent stability constants of the methyl cinnamate complexes range from 1.0 M^{-1} for the imidazole complex to 36 M^{-1} for caffeine (which, however, does not form only a 1:1 complex).[31] The stability constant of the *trans*-indoleacryloylimidazole-3,5-dinitrobenzoate ion complex is 22 M^{-1}.[32] The unreactivity of the complexed cinnamate and indoleacrylate esters can be explained only if the complexing agent interferes with the entry of the hydroxide ion (or amine) to the ester linkage. Thus the complexing agent must interact with these esters at a place closer to the ester linkage than the aromatic portion of the substrate.

A similar inhibition is seen in the reaction between aniline and 1-chloro-2,4-dinitrobenzene. The second-order rate constant for this reaction decreases with increasing concentration of the reactants, presumably because of the formation of an unreactive pi molecular complex. This phenomenon does not occur in the corresponding reaction with an aliphatic amine.[33]

These inhibitions can in general be attributed to the conversion of the substrate through pi molecular complex formation to a form unreactive to a third reagent. A charge transfer description of the indoleacrylate ester-3,5-dinitrobenzoate ion complex would predict that the complexed ester would be more reactive toward nucleophiles because of the loss of electron density to the complexing agent. The fact that the opposite is the case can be explained in steric terms as mentioned above or by saying that the charge transfer description is incorrect. A slight rate enhancement has been found in the alkaline hydrolysis of 4-nitrophthalimide when acenaphthene is used as complexing agent.[34] The charge transfer description of this complex would predict a higher electron density in the 4-nitrophthalimide complex and thus a lower rate constant of reaction. These two contrary predictions mean some important factor has been overlooked.

A charge transfer description of complex formation, however, affords a ready explanation for the enhancement of acetolysis of the tosylate

(11.17)

by the addition of aromatic donor molecules such as anthracene or phenanthrene.[35] The 1:1 complex with phenanthrene is 21–27 times more reactive than the uncomplexed tosylate at various temperatures. On the other hand, hexaethylbenzene shows no rate enhancement, presumably because it cannot form a complex. If the carbonium ion-like transition state of this reaction is complexed to a greater extent than is the ground state, the reaction rate should be higher. This attractive possibility, however, is difficult to reconcile with the fact that the rate enhancement is due solely to the activation entropy. The charge transfer description could be tested by investigating the kinetics of acetolysis of a tosylate in which the tosylate and complexing agent were donor and acceptor, rather than acceptor and donor, as they are in the present example. Another interesting aspect of the acetolysis of a complexed tosylate is the question of its stereochemistry: since the complexing agent prevents solvent access from the back side of the carbonium ion, retention rather than inversion or racemization should be the stereochemical result of such reactions.

Pi molecular complex formation apparently leads to rate enhancement in the following nucleophilic displacement reactions.

$$\begin{array}{lc} & \text{Relative rate} \\ \Phi O^\ominus + \Phi CH_2 S^\oplus Me_2 \longrightarrow & 2.7 \\ HO^\ominus + \Phi CH_2 S^\oplus Me_2 \longrightarrow & 0.8 \\ \hline \Phi O^\ominus + MeBr \longrightarrow & 1 \\ HO^\ominus + MeBr \longrightarrow & 1.5 \end{array} \quad (11.18)$$

Phenoxide ion is a threefold better nucleophile than hydroxide toward benzyldimethylsulfonium ion, although it is a slightly poorer nucleophile toward methyl bromide. This inversion of relative nucleophilicity is attributed to complex formation between phenoxide ion and benzyldimethylsulfonium ion which stabilizes the transition state of that reaction.[36] These data can be alternatively explained on the basis of the theory of soft and hard acids and bases (see Chapter 6).

Imidazole catalyzes the reversible transamination of pyridoxal by α-aminophenylacetic acid in aqueous solution near neutrality.[38] The catalysis is associated with the tautomerism between the two imines (aldimine and ketimine). The most interesting aspect of the catalysis is the change in the dependence of catalytic rate from the square of the total imidazole concentration at low imidazole concentrations (ca. 0.1 M) to independence of the total imidazole concentration at high imidazole concentrations (>1 M). Although the possibility of a solvent effect at the high imidazole concentrations cannot be discounted, this kinetic behavior is most easily interpreted as the binding of two molecules of imidazole. Of the several mechanisms that can be suggested to explain this binding, the only tenable one is binding of the imidazoles to the imines. Another mechanism involving rate-determining formation of the aldimine by two imidazole molecules, with a change in the rate-determining step to explain the dependence on imidazole concentration, is ruled out on the basis that the kinetics of the imidazole catalysis using the morpholine imine of pyridoxal is similar to that for the reaction using pyridoxal. Since imines react at greater rate with amines than do aldehydes, a difference in kinetics should be observed, but is not. A third mechanism involving rate-controlling complex formation between pyridoxal and two imidazole molecules is ruled out on the basis that the overall rate should be independent of amino acid concentration, which is not the case.

The pH dependence of the imidazole catalysis of the transamination reaction, shown in part in Fig. 11.3,[38] indicates dependence of the catalytic rate constant on the product of imidazole and imidazolium ion concentrations.[40] Plots such as Fig. 11.3 at various pH's show constancy of the binding constant calculated in terms of a complex of aldimine (and ketimine) with one molecule of imidazole and one molecule of imidazolium ion. The reactant α-aminophenylacetic acid also complexes with imidazole, as shown by

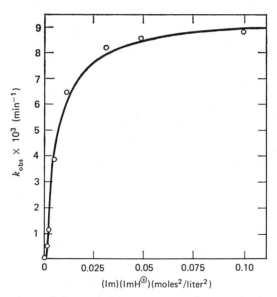

Fig. 11.3 Dependence of the catalytic rate constant of transamination on the product of imidazole and imidazolium ion concentrations. From J. C. Bruice and R. M. Topping, *J. Amer. Chem. Soc.*, **85**, 1488 (1963). © 1963 by the American Chemical Society. Reprinted by permission of the copyright owner.

solubility studies. The formation constants of this complex are quite similar to the constants determined kinetically for the imine complexes if calculated on the basis of the formation of a complex composed of one imidazole molecule, one imidazolium ion, and one amino acid molecule in the form of the zwitterion. On the basis of these experiments, a multiple catalysis (Chapter 10) of the prototropic shift by imidazole and imidazolium ion has been suggested.[40]

$$(11.19)$$

The stereochemical requirements of complexing of the imidazole moieties with the hydrocarbon portions of the imines appear not to be compatible with the stereochemical requirements of the catalysis, as postulated in eq. 11.19. Therefore, the complexes are probably nonproductive ones.[39]

Complexes of Aliphatic Hydrocarbons. The transfer of alkanes from water to a hydrocarbon solvent results in a small, positive enthalpy change and a large, positive entropy change. The addition of each additional methylene group in an alkane causes a decrease of ca. 160 cal/mole in the free energy of this change.[41] These results suggest that apolar molecules will tend to associate in aqueous solution and that the magnitude of the association constant depends on the magnitude of the interacting apolar residues. Emphasis on the entropy change involved has led to the suggestion of a "hydrophobic bond," which in fact is not a bond at all but rather a description of the postulate that a substantial entropy gain will result from the liberation of water of solvation surrounding two alkane molecules when they associate with one another.[42] This point of view leads naturally to association of hydrocarbon molecules, and thus to micelle formation in the case of detergents, and inclusion complexes in the case of such molecules as the cyclodextrins. We discuss the effect of each of these systems on various catalytic processes.

Not only do hydrocarbons associate in aqueous solution, but also ions do if the size of the interacting apolar groups on the ions is large enough. Although the free energy of formation of an ion pair from butylammonium isobutyrate is unfavorable,[24] ion pair formation from decyltrimethylammonium ion and azobenzene-4-sulfonate ion is quite favorable.[43] The general rule appears to follow that the association constant for ion pair formation increases as the size of the interacting apolar groups increases.

The enolization of a ketone by a carboxylate ion catalyst is enhanced if both the ketone and the catalyst contain sizable apolar residues that can interact with one another. For example, the rate constant for the β-phenylpropionate-catalyzed bromination of benzoylacetone is threefold higher than that predicted on the basis of the pK of the carboxylate ion catalyst.[44] Likewise, the hydrolysis of egg albumin is catalyzed to a one hundredfold greater extent by strong monobasic acids of high molecular weight than by an equivalent concentration of hydrochloric acid.[45] As expected, a longer chain acid such as cetanesulfonic acid is more efficient than a shorter chain acid such as decanesulfonic acid. These results must be explained in terms of an increased stabilization of the transition state of these reactions, which may be brought about by the apolar interactions described above.

Similar results are seen in the reactions of straight chain amines with p-nitrophenyl esters of straight chain carboxylic acids[46] and with p-nitrophenyl polyuridylate.[47] The rate constant of the reaction of decylamine with p-nitrophenyl decanoate is approximately four hundredfold larger than the rate constant of reaction of ethylamine with this ester. Likewise the rate constant of the reaction of the C_{10} amine with p-nitrophenyl polyuridylate is 600 times larger than the rate constant of the C_2 amine with this polymeric ester. These reactions show a potentiality not yet realized in corresponding catalytic processes dependent on apolar complexing.

The hydrolysis of the amide, chloramphenicol, is catalyzed by a large number of dicarboxylic acids and hydroxybenzoic acids. The pH-rate constant profiles of these reactions show that the monoanion possesses the greatest catalytic activity, with relatively little or no catalytic effect being exhibited by either the undissociated or totally dissociated species.[48] The most interesting of the dicarboxylic acids studied are the *cis*- and *trans*-cyclohexane-1,2-dicarboxylic acids. Although the catalytic rate constant is a linear function of the concentration of the monoanion of the *trans*-compound, the *cis*-isomer exhibits a limiting rate or "saturation" phenomenon, indicating binding of the catalyst with the substrate, possibly of apolar character. This is confirmed by phase solubility measurements. At 0.01 M catalyst, the rate constant of catalysis by the *cis*-isomer is approximately tenfold larger than that by the *trans*-isomer, although at concentrations above 0.3 M the reverse order begins, because of the difference in concentration dependencies. Complexing of catalyst and substrate can conceivably lead to multiple catalysis even in aqueous solution. On this basis, such catalysis is possible in enzymic reactions.

Reactions between two long chain compounds bring forth the possibility of micelle formation and the question of its effect on rate constants. Indeed the rates of many reactions are affected by micelle formation, either of the reactants themselves or by the addition of a micellar agent. Some of these effects are accelerations, others decelerations. At least three factors can account for alteration of the rate when the reactant or reactants are incorporated into or onto a micelle: proximity, electrostatic, and medium effects. Proximity effects arise if reactants are concentrated or diluted by incorporation into the micellar phase. Alterations in stability of reactants or transition states by electrostatic effects due to the micellar charges or by short range interactions involving the molecules that constitute the micelle can also affect reaction rates.

Medium and electrostatic effects are seen in the effect of salts on the rate constant of the alkaline hydrolysis of methyl 1-naphthoate in 50% dioxane-water.[49] Alkali halides show only a small, colligative retardation. However, salts containing organic anions have large and specific retardations whereas salts containing organic cations show slight to fairly large accelerations. Organic detergent salts have effects similar to those of other organic salts. For example, lauryltrimethylammonium chloride exhibits a slight acceleration whereas sodium lauryl sulfate produces a substantial deceleration. Hydrocarbons such as naphthalene or *cis*-decalin show a substantial decelerating effect. Let us consider the hydrocarbon effects first. In a system involving methyl 1-naphthoate and a hydrocarbon, the probability of finding a hydrocarbon or an aromatic compound adjacent to the ester is significantly greater than the corresponding mole fraction. Therefore, the microscopic environment of the ester is more nonpolar in the presence of the hydrocarbon than in its absence. That is, the dielectric constant of the microscopic environment is lower. Under these conditions the attack of hydroxide ion on the ester would be expected to be impeded. When the added hydrocarbon bears an electric charge, the same sorting of solvent molecules will persist, and therefore lower dielectric constant will again surround the ester substrate. But in addition, the electric charge will have an effect on the hydroxide ion, a negative charge repelling and a positive charge attracting hydroxide ion. Thus with salts of organic anions, both the environmental effect and the electrostatic effect work in the same direction to produce a substantial deceleration, while with organic cations, the environmental and electrostatic effects work in opposite directions, leading to slight to large increases if the latter effect is more prominent than the former.

On the basis of electrostatic effects, the following generalizations may be made. Cationic micelle-forming detergents accelerate the reactions of neutral organic molecules with anionic reagents but decelerate the reactions of neutral organic molecules with cationic reagents. On the other hand, anionic micelle-forming detergents accelerate the reactions of neutral organic molecules with cationic reagents but decelerate the reactions of neutral organic molecules with anionic reagents. These generalizations hold in a surprisingly large number of instances. For example, the cationic detergent cetyltrimethylammonium bromide accelerates: the reaction of glyclglycine with 1-fluoro-2,4-dinitrobenzene at pH 8 up to twentyfold[50]; the reactions of dyes with hydroxide ion four to fiftyfold[51]; and the alkaline hydrolysis of p-nitrophenyl hexanoate up to fivefold.[52] However, the (acid) hydrolysis of methyl orthobenzoate at pH 4.76 is inhibited by cetyltrimethylammonium bromide (Fig. 11.4).[52] On the other hand, anionic detergents such as sodium lauryl

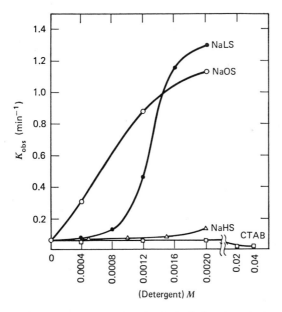

Fig. 11.4 First-order rate constants for the hydrolysis of methyl orthobenzoate in aqueous solution at 25 and pH 4.76 plotted as a function of the concentration of sodium lauryl sulfate (●), sodium oleyl sulfate (○), sodium heptadecyl sulfate (△), and cetyltrimethylammonium bromide (□). From M. T. A. Behme, J. G. Fullington, R. Noel, and E. H. Cordes, *J. Amer. Chem. Soc.*, **87**, 266 (1965). © 1965 by the American Chemical Society. Reprinted by permission of the copyright owner.

sulfate or sodium oleyl sulfate accelerate the acid hydrolysis of methyl orthobenzoate (Fig. 11.4) up to eightyfold at extremely low concentrations of detergent.[52] However, sodium dodecyl sulfate leads to a small decrease in the rate of reaction of glycylglycine with 1-fluoro-2,4-dinitrobenzene,[50] and sodium lauryl sulfate decreases the alkaline hydrolysis of p-nitrophenyl hexanoate[52] and the reation of hydroxide ion with certain dyes.[51]

Effects of substrate structure are understandable on the basis of variable micelle interactions. For example, the hydrolysis of p-nitrophenyl hexanoate is accelerated by cetyltrimethylammonium bromide to a greater extent than the hydrolysis of p-nitrophenyl acetate, presumably because of the greater incorporation of the former ester into the micelle.[52] Likewise the hydrolysis of ethyl orthovalerate and ethyl orthopropionate are accelerated in descending order by sodium lauryl sulfate, but not ethyl orthoformate.[52] The effect of structure is beautifully illustrated by the rate constants of acid hydrolysis of the straight chain alkyl sulfates given in Table 11.1. Although the lower esters from methyl to amyl have essentially identical rate constants, the higher esters from decyl to octadecyl increase markedly with chain length. This reaction is interesting for another reason, the micelle former is not an added reagent, but is the reactant itself. The increase in rate of this hydrolysis with micelle formation cannot be attributed to a medium effect, which is known to be extremely important in this reaction (Chapter 7) because the enhanced reactivity is due to a lowered activation enthalpy here whereas it is strictly an activation entropy effect in the case of medium effects. Therefore, the explanation must lie in the difference of the electrostatic potential of the

Table 11.1

The Rate Constants of the Acid Hydrolysis of Some Sodium Alkyl Sulfates[a,b]

Sodium Alkyl Sulfate	$k_H \times 10^5 (M^{-1} \sec^{-1})$
Methyl	5
Ethyl	5
Amyl	5.25
Decyl	41
Dodecyl	198
Tetradecyl	280
Hexadecyl	347
Octadecyl	505

[a] 90°, [sodium alkyl sulfate] = [HClO$_4$] = 0.04 M.
[b] From J. L. Kurz, *J. Phys. Chem.*, **66**, 2239 (1962). © 1962 by the American Chemical Society. Reprinted by permission of the copyright owner.

micelle with respect to the bulk solution. In essence, the sulfate moieties become stronger bases due to the presence of a negative potential on the micelle.[53]

The concentration of micelle-forming detergent has an effect on the rate of reaction, both in accelerations and declerations. The rate constant of the hydrolysis of sodium decyl sulfate, for example, is independent of substrate concentration and equivalent to the rate constant of hydrolysis of sodium amyl sulfate below the critical micelle concentration of sodium decyl sulfate. But above the critical micelle concentration, the hydrolytic rate constant increases markedly (Fig. 11.5).[53] In the sodium lauryl sulfate-methyl orthobenzoate system, however, the hydrolytic rate constant is proportional to the fourth power of the detergent below the critical micelle concentration and increases only slowly above that point. In the region below the critical micelle concentration, sodium lauryl sulfate is presumably a monomer. The great sensitivity of the rate constant to the detergent concentration suggests that the substrate induces the formation of micelles, containing a 4:1 ratio of sodium lauryl sulfate: substrate.

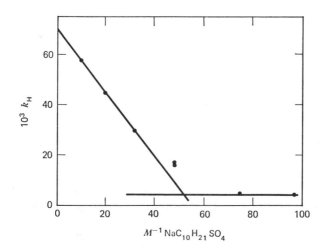

Fig. 11.5 k_H at 90° for sodium decyl sulfate as a function of its reciprocal concentration; the horizontal line is defined by the value for k_H for sodium amyl sulfate under the same conditions (I = 0.51, (HClO$_4$) = 0.02 M). From J. L. Kurz, *J. Phys. Chem.*, **66**, 2239 (1962). © 1962 by the American Chemical Society. Reprinted by permission of the copyright owner.

Inclusion Complexes. In aqueous solution, where complexation must necessarily involve largely apolar binding, the most powerful form of complexation, other than enzymic complexation, is the formation of inclusion complexes. Inclusion complexes possess high stability constants and potentially high stereospecificity as well, since the binding involves multiple interactions. Inclusion complexes resemble enzymic complexes in these important respects, and, therefore, it is worthwhile to consider in detail the catalytic properties of inclusion complexes.

Many forms of inclusion complexes exist, such as clathrate complexes, canal complexes, layer complexes, molecular sieves, intramolecular hollow space complexes, and linear polymer complexes.[54-57] Of the many inclusion compounds such as the urea and thiourea clathrates, the complexes of tri-o-thymotide and hydroquinone, the hydrates of hydrocarbons, and many others, the most pertinent are the complexes of cyclodextrins, sometimes called Schardinger dextrins, cycloamyloses, or cycloglucans. The cycloamyloses consist solely of D(+)-glucose units linked via α-1,4 links in a cyclic array. Cyclohexaamylose, cycloheptaamylose, and cyclooctaamylose (α- β- and γ-cyclodextrins) contain six, seven, or eight glucose residues per molecule. These nonreducing sugars yield only glucose on acid hydrolysis, and yield only 2,3,6-trimethylglucose on methylation followed by hydrolysis. X-ray crystallographic studies in conjunction with these chemical studies firmly establish the structure[58] and stereochemistry[59] of the cycloamyloses and their complexes. They are doughnut-shaped molecules with the glucose units in the C-1 conformation. The primary hydroxyl groups (carbon 6 of the glucose unit) are located on one side of the torus while the secondary hydroxyls (carbons 2 and 3 of the glucose units) are located on the other side of the torus. The interior of the cavity contains a ring of C–H groups, a ring of glycosidic oxygens, and another ring of C–H groups. As a consequence the interior of the cycloamylose torus is relatively apolar compared to water. The diameter of the cavity in α, β, and γ cyclodextrins is approximately 5, 7, and 10 Å.

The cycloamyloses form solid inclusion complexes with a variety of molecules and ions.[55-61] The most common guest molecules are apolar in nature. On the basis of the X-ray analysis of the cyclohexaamylose-potassium acetate complex, the guest resides in the cavity of the cycloamylose.[59] The solid complexes, however, are often not stoichiometric. A significant advance was made when it was found that the cycloamyloses also form inclusion complexes in aqueous solution.[62] Fortunately, the solution complexes are in the main stoichiometric compounds containing 1 molecule of host and 1 molecule of guest, that is, monomolecular inclusion complexes.[63]

The formation of inclusion complexes by cycloamyloses in aqueous solution has led to their utilization as model enzymes. For example, they act as asymmetric catalysts in the saponification of mandelic acid esters, although the rate effects and optical yields are small.[64] The cycloamyloses catalyze the decarboxylation of substituted cyanoacetic acids and β-keto acids, accelerations of up to fifteenfold being observed.[65] In neither of these studies was the relationship between complexing and catalysis demonstrated, although in the latter reactions, compensating activation enthalpy and entropy changes, as opposed to solution reactions, were noted. Cycloheptaamylose inhibits the basic hydrolysis of ethyl *p*-aminobenzoate, the complexed ester being completely unreactive (within experimental error) toward 0.04 N barium hydroxide solution.[66] The extent of inhibition due to added cycloheptaamylose is consistent with the formation of a 1:1 complex having a dissociation constant of 2.34×10^{-3} M. On the other hand, cycloheptaamylose catalyzes the cleavage of the β-lactam ring in penicillins.[94] The presence of a bulky group adjacent to the point of attachment was shown to be important in increasing apolar binding within the cavity. The cleavage of this strained amide is the only known example of hydrolysis of an amide by a cycloamylose.

Cycloamyloses accelerate the decomposition of diaryl phosphates in alkaline solution with concomitant phosphorylation of the amylose.[67] The acceleration of the release of phenol from bis-*p*-chlorophenyl pyrophosphate is ca. 200-fold with cycloheptaamylose, smaller accelerations being observed with different hosts such as cyclohexaamylose and cyclooctaamylose or with different guests such as diphenyl pyrophosphate. Furthermore, the acceleration was reduced when phenolic products were present in the reaction mixture. When an unsymmetrical pyrophosphate such as *p*-chlorophenyl ethyl pyrophosphate is treated with cycloheptaamylose, cycloheptaamylose phosphate, *p*-chlorophenol, and ethyl phosphate are produced. This set of products is compatible only with the mechanism of eq. 11.20, which assumes that the *p*-chlorophenyl group is selectively included in the cycloamylose ring.

(11.20)

The most searching experiments involving catalysis via cycloamylose inclusion complexes have been performed by studying the hydrolysis of phenyl esters.

The hydrolysis of a series of *meta*- and *para*-substituted phenyl acetates at pH 10.6 follows a normal Hammett relationship between the logarithm of the rate constant and the Hammett substituent constants. Likewise, when 1% methyl glucoside, a monomolecular analog of the cycloamyloses, is added to the solution, essentially the same linear Hammett relation is seen, the maximum perturbation of the rate constants being about 20%. However, when 1% cyclohexaamylose or cycloheptaamylose is added to the hydrolysis mixture, very large and variable accelerations of the rate constants are found. The accelerations do not follow a Hammett relationship (Fig. 11.6); in fact, they are the antithesis of such a relation, being independent of electronic effects. But there is an order that can be discerned in the chaos of Fig. 11.6; it is that all *meta*-substituted esters show large accelerations while all *para*-substituted esters show small accelerations.

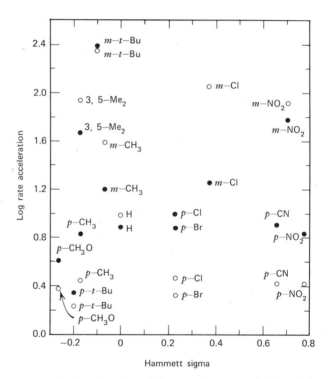

Fig. 11.6 Rate accelerations by 1% cyclohexaamylose, ○, and 1% cycloheptaamylose, ●, in the liberation of phenols from substituted phenyl acetates at pH 10.6 in 1% acetonitrile-water at 25°. From R. L. Van Etten, J. F. Sebastian, G. A. Clowes, and M. L. Bender, *J. Amer. Chem. Soc.*, **89**, 3242 (1967). © 1967 by the American Chemical Society. Reprinted by permission of the copyright owner.

Noncovalent Complexes

This differentiation between *meta-* and *para-* substituted esters independent of electronic effects must be the manifestation of a steric effect. Furthermore, since alkaline hydrolysis does not show this effect, the steric effect must be associated with the influence of the cycloamylose. When the concentration of cycloamylose (which is in excess of the ester) is varied, it is found that the rate constant of disappearance of ester (measured by phenol release) is not a linear function of the cycloamylose concentration. Rather it appears to approach a maximum value asymptotically. This behavior is an indication of complex formation. The stereochemical effect on the rate constants must then be related to the stereochemistry of complex formation. That complex formation does indeed take place is seen in the adherence of every set of kinetic experiments to eq. 11.16 or its counterpart when $[C]_0 \gg [S]_0$, which has the same form as eq. 11.16 except that $[S]_0$ is replaced by $[C]_0$. Reciprocal (Lineweaver-Burk) plots of typical data showing adherence to eq. 11.16 or its counterpart are given (Fig. 11.7).[68]

From Lineweaver-Burk (reciprocal) plots of the kinetic data, the maximal rate constant and the binding (dissociation) constant may be obtained. A series of these constants is shown in Table 11.2. The binding constants vary from 10^{-2} to 10^{-3} M, values not far different from some enzymic binding

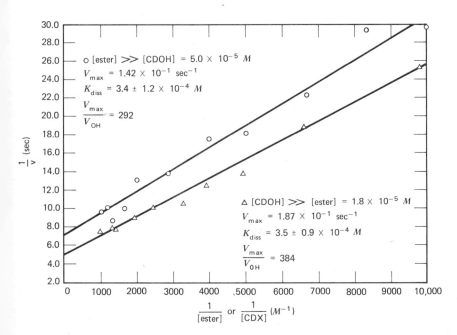

Fig. 11.7 Plot of the hydrolysis of *m-t-*butylphenyl acetate catalyzed by cycloheptaamylose: ○, (ester) ≫ (catalyst); △ (catalyst) ≫ (ester); 10% acetonitrile-water, pH 10.6.

Table 11.2
Catalytic Rate Constants and Accelerations in the Cyclohexaamylose-Catalyzed Hydrolysis of Phenyl Acetates[a,d]

Acetate	$k_c \times 10^2 (\text{sec}^{-1})$[b]	k_c/k_u[b,c]	$K_d \times 10^3 (M)$
p-t-Butylphenyl	0.075	1.2	7.7
p-Tolyl	0.26	4	14
o-Tolyl	1.4	37	48
Phenyl	5.5	69	70
m-Chlorophenyl	26	140	6(4.8)
m-Tolyl	11	160	34
m-t-Butylphenyl	13	260	2.0
3,5-Dimethylphenyl	14	250	20(13)

[a] 25°, 1.5%, (v/v) acetonitrile-water, pH 10.6.
[b] k_c catalyzed.
[c] k_u uncatalyzed.
[d] From M. L. Bender, *Trans. N. Y. Acad. Sci.*, **29**, 301 (1967).

constants. The rate constants are normalized in the middle column of Table 11.2 by the uncatalyzed rate constants, thus showing the maximal accelerations imposed by the cycloamylose. These accelerations vary from 20% for p-t-butylphenyl acetate to 260-fold for m-t-butylphenyl acetate, again showing the clear specificity for *meta*-substituted compounds.[68]

The explanation for this specificity must reside in complex formation. But we must first determine the mode of complex formation. Several pieces of experimental data indicate that phenyl esters bind within the hole of these doughnut-like molecules. For example, the spectrum of the p-t-butylphenol-cyclohexaamylose complex differs from the spectrum of p-t-butylphenol in water or cyclohexane, but is practically superimposable on its spectrum in dioxane solution. This circumstantial evidence may be explained by noting that the inside of the cycloamylose torus is composed of tetrahydropyran units containing five carbons and one ethereal oxygen atom, very similar to the dioxane structure. Competitive inhibition occurs in ester hydrolysis and pyrophosphate reactions catalyzed by cycloamyloses. Competitive inhibition requires a discrete site for which the substrate and the inhibitor compete. The only discrete site associated with the cycloamyloses is the cavity. A series of binding constants of 25 compounds shows a rough correlation with the molar refraction of the compound, indicating the importance of dispersion forces in the binding. These data are consistent only with binding within the cavity of the cycloamylose.

On the basis of binding of the guest molecule within the cavity of the host, and more particularly, the binding of the apolar (hydrocarbon) portion of the guest within the cavity of the host, one may attempt to explain the specificity between *meta-* and *para-*substituted phenyl acetates. Several facts should be kept in mind: (1) *meta/para* specificity is greatest for substituents of largest bulk; (2) the reactivity of an unsubstituted ester is midway between the *meta* and *para* substituted compounds; thus the *para-*substituted compound can be considered to have a negative specificity while the *meta-*substituted compound may be considered to have a positive specificity. From these considerations, the models of Fig. 11.8[71] were assembled. The difference seen in the models is that the ester linkage (silvered atoms) of the *para-*compound (small acceleration) is a considerable distance from the cycloamylose moiety whereas the ester linkage of the *meta-*compound (large acceleration) is directly contiguous to the ring of secondary hydroxyl groups surrounding the cavity of the cycloamylose. Hence the acceleration must stem from some interaction of the hydroxyls of the cycloamylose with the ester linkage.

The models shown in Fig. 11.8 are arbitrary in the sense that the primary hydroxyl groups are placed at the bottom and the secondary hydroxyl groups at the top in this photograph. Complexing could conceivably occur in the opposite sense. However, blocking of all primary hydroxyl groups has no effect on the efficacy of cycloheptaamylose as a catalyst, and thus the models shown are probably correct.

The hydroxyl groups can accelerate reactions by general base, general acid, or nucleophilic catalysis. Nucleophilic catalysis is indicated here on the basis of experiments on the hydrolysis of *m*-nitrophenyl benzoate with cyclohexaamylose. Observations of this reaction at 390 nm show the rapid

Fig. 11.8 Pauling-Corey-Koltun models of the α-cyclodextrin-*p-t-*butylphenyl acetate complex (left) and the α-cyclodextrin-*m-t-*butylphenyl acetate complex (right). From M. L. Bender, *Trans. N.Y. Acad. Sci, Series II,* **29,** No. 3, 301 (1967). © The New York Academy of Sciences 1967; Reprinted by permission.

liberation of *m*-nitrophenol (approximately 20 sec) whereas observations at 245 nm show the slow formation of benzoate ion (thousands of seconds). Thus, the reaction must proceed in two steps, with the formation of *m*-nitrophenol preceding the formation of benzoate ion. This conclusion is reinforced by kinetic experiments which indicate that the rate constant for the liberation of benzoic acid from the three benzoate esters, *m-t*-butylphenyl, *m*-chlorophenyl, and *m*-nitrophenyl, is identical, although electronic considerations would predict that the rate constants should be quite different. This anomaly can be explained by postulating the formation in the reaction of a common intermediate whose decomposition is rate-determining. The common intermediate must contain the common parts of the system, the benzoic acid and cyclohexaamylose. Thus it may be identified as benzoyl-cyclohexaamylose. It has in fact been isolated.

Thus the mechanism of the cyclodextrin reaction may be given as[69]

$$\text{C} + \text{R}\overset{\text{O}}{\overset{\|}{\text{C}}}\text{OAr} \overset{K}{\rightleftarrows} \text{C} \cdot \text{R}\overset{\text{O}}{\overset{\|}{\text{C}}}\text{OAr} \longrightarrow \text{C}\!-\!\overset{\text{O}}{\overset{\|}{\text{C}}}\text{R} \overset{\text{OH}^\ominus}{\longrightarrow} \text{C} + \text{RCO}_2^\ominus + \text{phenol}$$
(11.21)

This pathway is formally analogous to that of the chymotrypsin and trypsin reactions we shall discuss later.

In Chapter 9, the proposition was made that intramolecular catalysis is superior to the corresponding intermolecular catalysis. An implication was that this superiority could be carried over to an intracomplex catalysis of an enzyme-substrate reaction. This prediction can be tested by comparing the intracomplex reaction of cyclohexaamylose-*m-t*-butylphenyl acetate with the corresponding intermolecular reaction, using the lower limit of the interconversion between corresponding intermolecular and intramolecular reactions developed in Chapter 9 (10 *M*). Table 11.3 presents this comparison by attempting to calculate the intracomplex rate constant from the corresponding hydrolytic rate constant. This comparison indicates that in at least

Table 11.3

Kinetic Factors Responsible for the Difference in Rate of Liberation of *m-t*-Butylphenol from *m-t*-Butylphenyl Acetate by Hydroxide Ion and Cyclohexaamylose

Rate Constant of Hydroxide Ion Catalysis	$1.2\ M^{-1}\ \text{sec}^{-1}$
Conversion to rate constant of alkoxide ion reaction of pK 12.1 (fourfold)	$4.8\ M^{-1}\ \text{sec}^{-1}$
Conversion to an intramolecular reaction from an intermolecular reaction (10 M)	$48\ \text{sec}^{-1}$
Experimental cyclohexaamylose, $k_{\text{cat}}(\text{lim})$	$13\ \text{sec}^{-1}$

one instance, the intramolecular-intermolecular comparison in Chapter 9 is mirrored by a quite similar intracomplex-intermolecular comparison.

Since the cycloamylose and chymotrypsin pathways are formally identical, it is of interest to investigate the relationship of each of their catalytic rate constants to the rate constants for the alkaline hydrolysis of the same substrates. Table 11.4 shows such calculations. In each set the second-order hydroxide ion rate constant is compared to the second-order catalytic complexing process, which may be defined as k_{cat}/K (the rate constant when $[S]_0 \ll K$ or $[E]_0 \ll K$). The cycloamylose reactions are seen to be $5 \times 10^3 - 10^5$ times faster than hydroxide ion reactions while the chymotrypsin reactions are $3 \times 10^4 - 10^6$ times faster than hydroxide ion reactions. The conclusion that may be drawn from these comparisons is that the cycloamylose reactions show roughly the same rate enhancement with respect to hydroxide ion reactions as do the chymotrypsin reactions. There is one proviso, however. The rate constant for the chymotrypsin reaction was determined at pH 8, its

Table 11.4
A Comparison of Second-Order Rate Constants[f]

1. Cyclodextrin + *m-t*-butylphenyl acetate		
a. α-CD[d] + MTBPA[e]	$6.5 \times 10^3\ M^{-1}\sec^{-1}$	
OH$^\ominus$ + MTBPA	$1.2\ M^{-1}\sec^{-1}$	
α-CD/OH$^\ominus$		5.1×10^3
b. β-CD + MTBPA	$1.3 \times 10^5\ M^{-1}\sec^{-1}$	
OH$^\ominus$ + MTBPA	$1.2\ M^{-1}\sec^{-1}$	
β-CD/OH$^\ominus$		10^5
2. Chymotrypsin + Substrates		
a. Chymotrypsin + ATrA[a]	$12.6\ M^{-1}\sec^{-1}$	
OH$^\ominus$ + ATrA	$3 \times 10^{-4}\ M^{-1}\sec^{-1}$	
Chymo/OH$^\ominus$		4×10^4
b. Chymotrypsin + ATrEE[b]	$4 \times 10^5\ M^{-1}\sec^{-1}$	
OH$^\ominus$ + ATrEE	$0.61\ M^{-1}\sec^{-1}$	
Chymo/OH$^\ominus$		$\sim 10^6$
c. Chymotrypsin + ATyEE[c]	$1.2 \times 10^4\ M^{-1}\sec^{-1}$	
OH$^\ominus$ + ATyEE	$0.45\ M^{-1}\sec^{-1}$	
Chymo/OH$^\ominus$		3×10^4

[a] Acetyl-L-tryptophanamide.
[b] Acetyl-L-tryptophan ethyl ester.
[c] Acetyl-L-tyrosine ethyl ester.
[d] α-Cyclodextrin.
[e] *m-t*-Butylphenyl acetate.
[f] From M. L. Bender, *Trans. N. Y. Acad. Sci.*, **29**, 301 (1967). © The New York Academy of Sciences 1967. Reprinted by permission.

maximum, whereas the rate constant for the cycloamylose reaction was determined at pH 13, its maximum. Therefore the two sets are not strictly comparable. But the comparison indicates that if one can reduce the pH at which cycloamylose operates, for example by introduction of an imidazole group into the molecule, its rate enhancement might truly parallel the chymotrypsin reaction. The first such attempt at introducing an imidazole group shows only a three to fourfold rate enhancement over a combination of cycloamylose and imidazole,[70] but larger effects will probably be found in the future when stereochemistry and binding are considered.[72]

11.2.3 Polymeric Complexes

Since enzymic catalysts are polymeric systems and show multiple interactions toward their substrates, the use of polymeric catalytic systems that can also show multiple interactions toward substrates has been of considerable interest.

Thermally prepared poly-α-amino acids have catalytic activity in the hydrolysis of p-nitrophenyl acetate at pH 6.8. The activity of the thermal polymers, although weak in comparison to that of contemporary enzymes, is greater than that of the equivalent amount of unpolymerized amino acids. Aging for five to ten years does not lead to deterioration of catalytic activity.[95]

Polymeric acids have been used to catalyze the hydrolysis of esters and amides. The rate constants of hydrolysis using these polymeric catalysts are sometimes but not always larger than the rate constants of hydrolysis of a corresponding concentration of hydrochloric or sulfuric acid catalyst. Dowex-50, a cation exchange resin of the phenolsulfonic acid type, for example, hydrolyzes simple dipeptides with rate constants approximately one hundredfold larger than the corresponding concentration of hydrochloric acid.[73] Somewhat lesser rate enhancements are seen in the hydrolysis of peptides by a soluble polyvinylsulfonic acid polymer.[74] The reason for the superiority of polymeric over monomeric acid catalysts can be attributed primarily to the local concentration of oxonium ions that must be larger in the neighborhood of the polymer than in the bulk of the solution. Therefore, if the substrate is bound to the polymer, it should be hydrolyzed at a rate larger than by an equivalent, homogeneously dispersed concentration of oxonium ions.

Several factors lead to binding of substrates to a polymeric acid catalyst. Apolar binding is a factor of considerable importance. For example, butyl acetate is hydrolyzed tenfold faster by a comparable concentration of homogeneous polymeric sulfonic acid than by hydrochloric acid but ethyl acetate shows no rate enhancement.[75] Likewise, the rate enhancement in the hydrolysis of ethyl acetate by a lightly cross-linked ion exchange resin is greater than in the hydrolysis of methyl acetate.[76] Finally, a water soluble polymeric acid has the greatest effect on the hydrolysis of those esters of greatest chain length.[77] Thus a combination of apolar binding plus the excess local concentration of oxonium ions in the vicinity of the polymer leads to enhanced rates of reaction.

Electrostatic interactions between substrate and polymer also lead to binding. The ion exchange resin, Dowex-50, requires free ammonium ion for its catalytic activity. It catalyzes the hydrolysis of glycylglycine 112 times *faster* than the equivalent concentration of hydrochloric acid, but catalyzes the hydrolysis of *N*-acetylglycine two times *slower* than the equivalent concentration of hydrochloric acid.[78] Likewise, a water soluble polystyrene-sulfonic acid hydrolyzes 2-amino-2-deoxy-β-D-glucopyranoside hydrochloride thirtyfold and the diethylaminoethyl ether of starch hydrochloride twentyfold *faster* than equivalent concentration of hydrochloric acid, but hydrolyzes starch threefold *slower* than an equivalent concentration of hydrochloric acid.[79] These specific catalyses can only be due to one factor: the selective binding of the cationic substrates to the ion exchange resin followed by catalysis by the high local concentration of oxonium ions.

In special circumstances, pi molecular complexes can lead to binding of substrate to an ion exchange resin. The hydrolysis of allyl acetate by a sulfonic acid resin is enhanced by partial exchange of the oxonium ions with silver ions.[80] Although the rate of hydrolysis of propyl acetate is monotonically decreased by increasing concentration of silver ion on the resin, the rate of hydrolysis of allyl acetate goes through a maximum (at about 50%) as the concentration of silver ion on the resin is increased, producing a twofold specificity for the hydrolysis of the olefinic ester because of its increased concentration near the polymer surface.

The polyanions, poly(ethenesulfonate)

$$(-CH_2-CH-)_n$$
$$\quad\quad\quad |$$
$$\quad\quad\quad SO_3^{\ominus}$$

and poly(β-sulfoethyl methacrylate,

$$\quad\quad\quad CH_3$$
$$\quad\quad\quad |$$
$$(-CH_2-C-)_n$$
$$\quad\quad\quad |$$
$$\quad\quad\quad C=O$$
$$\quad\quad\quad |$$
$$\quad\quad\quad OC_2H_4SO_3^{\ominus}$$

accelerate the $Hg^{2\oplus}$-catalyzed aquation of $Co(NH_3)_5Cl^{2\oplus}$ presumably by an increase in electrostatic potential in the polymer domain.[81] The reaction involves a compact inorganic system.

$$Co(NH_3)_5Cl^{2\oplus} + Hg^{2\oplus} + H_2O \longrightarrow Co(NH_3)_5H_2O^{3\oplus} + HgCl^{\oplus} \tag{11.22}$$

If apolar binding is important for catalysis by polymeric acids, the solvent can have an effect on catalytic activity. For example, in acetone–water mixtures, a polymeric acid is often a poorer catalyst per oxonium ion than hydrochloric acid.[77] But in water, the same polymeric acid becomes a much better catalyst than hydrochloric acid. This change can be readily explained on the basis that apolar binding of the substrate to the polymer is better in water solution than in a solution of an organic solvent.

Poly(triethyl(vinylbenzyl)ammonium hydroxide) is a twentyfold more effective catalyst than equivalent amounts of either sodium hydroxide or benzyltrimethylammonium ion for the transformation of glyoxal to glycolic acid.[82] The efficiency of this polymeric base may be explained in similar electrostatic terms to those given above.

Thus, these rudimentary polymeric catalysts do show some of the attributes of binding, enhanced catalysis, and specificity exhibited by enzymes. Now let us look at some more sophisticated systems.

The rates of displacement of bromide ion from bromoacetate ion and bromoacetamide by poly(4-vinylpyridine), poly(methacrylic acid), and poly(vinylpyridine betaine) are sometimes faster than reactions with the corresponding monomeric compounds.[83] The reaction of 4-methylpyridine with either alkylating agent shows only a small positive salt effect, as does the reaction of bromoacetamide with poly(4-vinylpyridine). But the reaction of partially protonated poly(4-vinylpyridine) with bromoacetate ion increases sharply with the degree of protonation of the polymer in the range of α (the degree of protonation) $= 0.1 - 0.5$ and with decreasing ionic strength. These results indicate an electrostatic interaction between bromoacetate ion and the positively-charged polymer that facilitates reaction. The carboxylate anions in partially ionized poly(methacrylic acid) are 4–10 times more reactive toward bromoacetamide than those of simple dicarboxylic acids, their reactivity decreasing sharply with the increasing degree of ionization of the polymer. This rate enhancement can be attributed to hydrogen bonding of the substrate to the unionized carboxylic acid groups of the polymer while the carboxylate ions function as nucleophiles to displace bromide ion. Both bromoacetate ion and bromoacetamide react with poly(4-vinylpyridine zwitter ion) under conditions in which there is no reaction with the corresponding monomer. Since both the neutral and negatively-charged substrates show the same effect, the electrostatics are presumably not involved. But in these reactions, as well as in the other described above, binding of the monomeric substrate to the polymeric reactant must be invoked in order to explain the rate enhancements.

The concept of substrate binding to a polymeric catalyst by electrostatic interaction was applied by Letsinger.[84] A partially protonated poly(4-vinylpyridine) in ethanol-water solution was shown to serve as a particularly effective catalyst, relative to 4-picoline, nonprotonated polymer, or highly-protonated polymer, for the solvolysis of a nitrophenyl ester substrate bearing a negative charge. Figures 11.9[84] and 11.10[84] show the dependence of the rate constant of the solvolysis of neutral and anionic acetate esters on the degree of protonation of monomeric 4-picoline or poly(4-vinylpyridine). The monomer is a better catalyst than the polymer toward the neutral substate (Fig. 11.9), perhaps because of a steric effect in the latter. However, the polymer is a better catalyst than the monomer toward the anionic ester, at essentially all degrees of ionization (Fig. 11.10). With the anionic ester, 3-nitro-4-acetoxybenzenesulfonate ion, the catalytic activity approaches a maximum when the polymer is partially protonated (very close to a bell-shaped curve), implying that both protonated pyridinium ions and neutral pyridine groups are necessary for this catalysis. The pyridine groups serve as

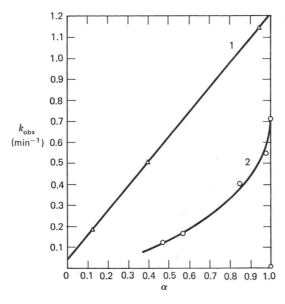

Fig. 11.9 Solvolysis of 2,4-dinitrophenyl acetate catalyzed by 0.0157 M 4-picoline, △ (curve 1), and by 0.010 M pyridine units in poly(4-vinylpyridine), ○ (curve 2). From R. L. Letsinger and T. J. Savereide, *J. Amer. Chem. Soc.*, **84**, 114 (1962). © 1962 by the American Chemical Society. Reprinted by permission of the copyright owner.

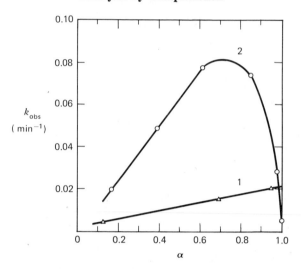

Fig. 11.10 Solvolysis of potassium 3-nitro-4-acetoxybenzenesulfonate catalyzed by 0.0157 M 4-picoline, △ (curve 1) and by 0.010 base molar poly(4-vinylpyridine), ○ (curve 2). From R. L. Letsinger and T. J. Savereide, *J. Amer. Chem. Soc.*, **84**, 114 (1962). © 1962 by the American Chemical Society. Reprinted by permission of the copyright owner.

nucleophilic catalysts. The pyridinium ions serve as electrostatic binding agents increasing the local concentration of the anionic substrate in the region of the polymeric coil. Similar catalytic phenomena are seen in the solvolysis of 5-nitro-4-acetoxysalicylic acid, and 3-nitro-4-acetoxybenzenearsonic acid with poly(4-vinylpyridine). At its maximum, the polymer is a ninefold better catalyst than the monomer toward 3-nitro-4-acetoxybenzenesulfonate ion. The asymmetry of the rate constant-α profile is attributed to an unfavorable conformational change when the acidity of the medium increases.

The behavior of poly(4(5)-vinylimidazole) with the anionic ester, 3-nitro-4-acetoxybenzenesulfonate, is similar.[85] Catalysis by the polymer exceeds that by the monomer by sixfold at maximal activity of the polymer. A copolymer of 4(5)-vinylimidazole and acrylic acid also exhibits a selective catalysis. The copolymer is a better catalyst than imidazole toward a positively-charged *p*-nitrophenyl ester substrate in a region in which the imidazole groups in both the monomer and polymer are largely unprotonated. On the other hand, imidazole is a better catalyst than the copolymer toward a neutral ester. Finally, the copolymer is only 1/20th to 1/30th as effective as the monomer toward negatively-charged ester substrates.[86]

Poly(4(5)-vinylimidazole) is a better catalyst than the corresponding monomer even toward the neutral ester, p-nitrophenyl acetate, when the reaction is carried out at high pH. Likewise, poly(5(6)-vinylbenzimidazole) is a particularly effective catalyst toward neutral esters at high pH. These enhanced reactivities at high pH are attributed to multiple catalysis by a combination of anionic and neutral imidazole groups on the polymers.[83] In addition poly-(5(6)-vinylbenzimidazole) is a better catalyst than the corresponding monomer toward sodium 4-acetoxy-3-nitrobenzenesulfonate at all pH's. At low pH's, electrostatic interactions of the kind described above can operate. At high pH's the multiple catalysis suggested before can operate. At neutral pH, multiple catalysis by two benzimidazole groups can account for the enhanced catalysis. The description of these reactions as multiple catalyses (Chapter 10) is consistent with the significantly lower activation enthalpies in the reactions of p-nitrophenyl acetate with the polymeric catalysts, poly(5(6)-vinylbenzimidazole) and poly(4(5)-vinylimidazole), and in the reaction of 4-acetoxy-3-nitrobenzoic acid with poly(5(6)-vinylbenzimidazole), compared to the reactions of these esters with the corresponding monomers.[87] The lower activation enthalpies are partially compensated by more unfavorable activation entropies. Nevertheless, the sizable difference in activation enthalpies does imply a multiple catalysis, either by two functionalities or by a binding and a functionality.

Selective polymeric catalysis at its best is seen in the reaction of the cationic polymeric catalyst, poly(N-vinylimidazole) with the anionic polymeric substrate, copoly(acrylic acid-2,4-dinitrophenyl p-vinylbenzoate).[88] Solvolysis of neutral and anionic monomeric esters, 2,4-dinitrophenyl p-isopropylbenzoate and p-nitrophenyl hydrogen terephthalate, using the cationic polymeric catalyst, shows the usual selectivity toward the anionic substrate with dependence on the degree of ionization of the polymer, due to electrostatic interactions between monomeric substrate and polymeric catalyst. Solvolysis of the polymeric substrate by the polymeric catalyst is twice as effective as by the corresponding monomeric catalyst, although solvolysis of the monomeric substrate, 2,4-dinitrophenyl p-isopropylbenzoate, is twentyfold less effective with the polymeric than the monomeric catalyst. Thus, some specific interaction of the polymeric substrate and polymeric catalyst is indicated. The dependence of the catalytic rate constant for this reaction on the catalyst concentration substantiates this conclusion (Fig. 11.11).[88] Although the catalytic rate constant of the reaction of the polymeric substrate with the monomeric catalyst, N-methylimidazole, shows a linear (and low) dependence on the catalyst concentration, the catalytic rate constant of the reaction of the polymeric substrate with the polymeric catalyst, poly(N-vinylimidazole), shows a typical saturation phenomenon, corresponding to either eq. 11.9 or 11.10. Since the catalytic rate constant of the latter is greater than the former, the complex formed between catalyst and substrate must be a productive

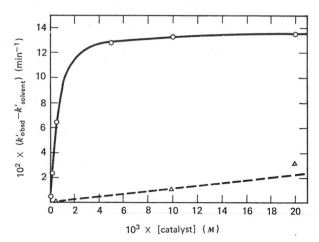

Fig. 11.11 Reaction of a copolymer substrate (6.14 × 10^{-5} M in ester) catalyzed by poly(N-vinylimidazole) (○) and by N-methylimidazole (△) at 25°. The solid line is calculated. From R. L. Letsinger and I. S. Klaus, *J. Amer. Chem. Soc.*, **87**, 3380 (1965). © 1965 by the American Chemical Society. Reprinted by permission of the copyright owner.

one, corresponding to eq. 11.9. The inverse phenomenon in which the polymeric catalyst is saturated by the polymeric substrate is also seen. Competitive inhibition of this reaction by polyacrylic acid, a polymer containing the same binding groups as the substrate, also occurs. All these phenomena—specificity, saturation, and competitive inhibition—can be attributed to formation of a catalyst-substrate complex that increases the probability of encounter between the nucleophilic sites on the catalyst and the ester groups of the substrate. This system thus exhibits many of the characteristics of an enzymic process, although its efficiency is still low. A synthetic macromolecule with high esterolytic activity has recently been reported by Klotz.[89] It consists of a highly branched polyethylenimine containing pendant lauroyl groups that readily hydrolyzes *p*-nitrophenyl laurate or *p*-nitrophenyl acetate (neutral substrates), especially the former. This is an example of an apolar interaction. A synthetic ethylenimine polymer has also been prepared which contains dodecyl groups (to bind small substrate molecules) and imidazolylmethyl side chains (as nucleophilic catalytic sites). This macromolecule, with a high local concentration of binding and catalytic groups, catalyzes the hydrolysis of uncharged nitrophenyl esters in water at pH 7 with rates markedly greater than previously observed with any other synthetic substances under similar conditions. These polymers have been called synzymes.[89]

Electrostatic interactions are also seen in the poly(N-vinylimidazole)-poly(*p*-nitrophenyl polyuridylic acid succinate) system. The hydrolysis of the polyester is catalyzed one hundredfold more efficiently by the polymeric than by monomeric imidazole at the same concentration.[90] In addition, the polymeric catalysis can be almost completely suppressed by the addition of 1.3×10^{-5} M hexadecyltrimethylammonium bromide. Finally, the suppression can be eliminated by 1.2×10^{-5} M sodium dodecyl sulfate, although the latter does not significantly influence the hydrolysis in the absence of the cationic detergent.

Polynucleotides can lead to very specific complexing. For example, although the treatment of nucleoside 5'-phosphates in aqueous solution with water soluble carbodiimides gives no synthesis of phosphodiester bonds, treatment of a mixture of thymidine hexanucleotide and polyadenylic acid, which are known to complex with one another, with water soluble carbodiimides gives a 5% yield of thymidine dodecanucleotide.[91] This reaction was predicted on the assumption of a head-to-tail "Watson-Crick" arrangement of the thymidine oligonucleotides along the polyadenylic acid strand,

$$
\begin{array}{c}
\text{---ApApApApApApA---} \\
\text{---TpTpTpT} \quad \text{TpTpT---} \\
| \quad \quad | \\
\text{HO} \quad \text{O} \\
| \\
^\ominus\text{O---P=O} \\
| \\
^\ominus\text{O}
\end{array}
\qquad (11.23)
$$

the 3'-hydroxyl group of one thymidine oligomer lying in a favorable steric relationship to the terminal 5'-phosphate group of the adjacent oligomer. Although the yield is low, the potential is high.

The fundamental assumption of this chapter is that the proper matrix for bringing together catalyst and substrate is one of the critical factors in devising an efficient catalyst. This assumption implies that the matrix does not take an active part in the catalysis other than holding the substrate and catalyst rigidly in proximity to one another, and orienting the substrate and catalyst correctly with respect one another. But a matrix, especially a macromolecular one, such as one of the polymers we have been discussing or an enzyme, can in the binding reaction raise the ground state energy of the substrate by some process other than rigidification, for example, by distortion of the substrate. A suggestion of this kind has been made to explain the catalysis of the *cis-trans* isomerization of the azo dye, chrysophenine, by the polymer polyvinylpyrrolidone which does not occur with the corresponding monomer.[92]

Certainly proximity and orientation effects are very important in catalysis. Random collision between molecules is too haphazard a process to bring about specific and efficient catalysis. Direction must be given. This direction can be brought about chemically by stereospecific complexing. At the same time that this direction is being imposed on the substrate, the unfavorable translational entropy of bringing two molecules together may be overcome by any of the forces described above. The complexing of cyclohexaamylose with m-chlorophenyl acetate, 3,4,5-trimethylphenyl acetate, and m-ethylphenyl acetate, for example, have favorable free energies of complexing of 3.5–3.6 kcal/mole, due to varying contributions from entropy and enthalpy.[68] This chemical potential for complexing means that an appreciable fraction of the substrate and catalyst molecules will be complexed at quite small concentrations. If the complex possesses, in addition to strength, correct stereochemistry between catalyst and substrate, then the favorable entropy of activation associated with intramolecular reactions and the favorable enthalpy associated with catalyses can combine, leading to efficiency. In addition, the stereochemistry will result in specificity. These are the two goals of catalysis.

The stereochemical aspects of the above statement were enunciated many years ago by Emil Fischer, in terms of the "lock and key" or complementarity theory,[93] specifically referring to enzymic reactions. But some of the nonenzymic catalyses described here, especially those involving polymers and inclusion compounds, show the rudiments of the application of the principles of binding and particularly complementarity of binding to the design of synthetic catalysts of high efficiency and specificity. Perusal of Chapters 1–10 indicates that catalytic concepts are relatively circumscribed. But the application of these catalytic concepts to the proper complementary matrices is a field in its infancy.

REFERENCES

1. J. R. Platt, *Annu. Rev. Phys. Chem.*, **16**, 517 (1965).
2. M. S. Newman and S. Hishida, *J. Amer. Chem. Soc.*, **84**, 3582 (1962).
3. M. L. Bender and M. S. Silver, *J. Amer. Chem. Soc.*, **84**, 4589 (1962).
4. C. Djerassi and A. E. Lippman, *J. Amer. Chem. Soc.*, **77**, 1825 (1955).
5. U. R. Ghatak and J. Chakravarty, *Chem. Commun.*, 184 (1966).
6. H. G. O. Becker, J. Schneider, and H.-D. Steinleitner, *Tetrahedron Lett.*, 3761 (1965).
7. Y. Shalitin and S. A. Bernhard, *J. Amer. Chem. Soc.*, **86**, 2292 (1964).
8. L. Holleck, G. A. Melkonian, and S. B. Rao, *Naturwissenschaften.*, **45**, 438 (1958); see also G. Vavon and J. Scandel, *C. R. Acad Sci.*, **223**, 1144 (1946).
9. M. L. Bender, J. A. Reinstein, M. S. Silver, and R. Mikulak, *J. Amer. Chem. Soc.*, **87**, 4545 (1965).
10. F. Ramirez, B. Hansen, and N. B. Desai, *J. Amer. Chem. Soc.*, **84**, 4588 (1962).
11. C. N. Lieske, E. G. Miller, Jr., J. J. Zeger, and G. M. Steinberg, *ibid.*, **88**, 188 (1966).
12. V. Franzen, *Chem. Ber.*, **88**, 1361 (1955).
13. B. Capon and R. Capon, *Chem. Commun.*, 502 (1965).
14. E. J. Fuller, *J. Amer. Chem. Soc.*, **85**, 1777 (1963).
15. V. I. Rozengart, *Dokl. Akad. Nauk SSSR*, **133**, 1223 (1960); V. I. Rozengart and L. V. Shepshelevich, *Biokhimiya*, **27**, 689 (1962).

References

16. G. R. Schonbaum and M. L. Bender, *J. Amer. Chem. Soc.*, **82**, 1900 (1960).
17. R. L. Letsinger, S. Dandegaonker, W. J. Vullo, and J. D. Morrison, *J. Amer. Chem. Soc.*, **85**, 2223 (1963).
18. R. L. Letsinger and J. D. Morrison, *J. Amer. Chem. Soc.*, **85**, 2227 (1963).
19. See also R. L. Letsinger and D. B. MacLean, *J. Amer. Chem. Soc.*, **85**, 2230 (1963).
20. B. Capon and B. Ch. Ghosh, *J. Chem. Soc. B*, 472 (1966).
21. L. Michaelis and M. L. Menten, *Biochem. Z.*, **49**, 333 (1913).
22. C. A. Kraus, *J. Phys. Chem.*, **60**, 129 (1956).
23. R. M. Fuoss and F. Accascina, *Electrolytic Conductance*, Interscience Publishers, Inc., New York, 1959.
24. I. M. Klotz and H. A. DePhillips, Jr., *J. Phys. Chem.*, **69**, 2801 (1965).
25. I. M. Klotz and J. S. Franzen, *J. Amer. Chem. Soc.*, **84**, 3461 (1962).
26. L. Larsson, *Arkiv. Kemi*, **13**, 259 (1958).
27. B. A. Cunningham and G. L. Schmir, *J. Amer. Chem. Soc.*, **88**, 551 (1966).
28. See G. Briegleb, *Elektronen-Donator-Acceptor-Komplexe*, Springer Verlag, Berlin, 1961; L. J. Andrews and R. M. Keefer, *Molecular Complexes in Organic Chemistry*, Holden-Day, Inc., San Francisco, 1964.
29. M. J. S. Dewar and C. C. Thompson, Jr., *Tetrahedron, Suppl.*, **7**, 97 (1966).
30. T. Higuchi and L. Lachman, *J. Amer. Pharm. Ass., Sci. Ed.*, **44**, 521 (1955).
31. K. A. Connors and J. A. Mollica, Jr., *J. Amer. Chem. Soc.*, **87**, 123 (1965); J. A. Mollica, Jr., and K. A. Connors, *ibid.*, **89**, 308 (1967); see also K. A. Connors and J. A. Mollica, Jr., *J. Pharm. Sci.*, **55**, 772 (1966).
32. F. M. Menger and M. L. Bender, *J. Amer. Chem. Soc.*, **88**, 131 (1966).
33. S. D. Ross and I. Kuntz, *J. Amer. Chem. Soc.*, **76**, 3000 (1954).
34. A. Bruylants and J. B. Nagy, *Bull. Soc. Chim. Belg.*, **75**, 246 (1966).
35. A. K. Colter and S. S. Wang, *J. Amer. Chem. Soc.*, **85**, 114 (1963); A. K. Colter, S. S. Wang, G. H. Megerle, and P. S. Ossip, *ibid.*, **86**, 3106 (1964); A. K. Colter, F. F. Guzik, and S. H. Hui, *J. Amer. Chem. Soc.*, **88**, 5754 (1966).
36. C. G. Swain and L. J. Taylor, *J. Amer. Chem. Soc.*, **84**, 2456 (1962).
37. K. L. Kirk and L. A. Cohen, *J. Org. Chem.*, **34**, 390 (1968).
38. T. C. Bruice and R. M. Topping, *J. Amer. Chem. Soc.*, **85**, 1480 (1963).
39. See also J. W. Eastman, G. Engelsma, and M. Calvin, *ibid.*, **84**, 1339 (1962).
40. T. C. Bruice and R. M. Topping, *J. Amer. Chem. Soc.*, **85**, 1488 (1963).
41. J. A. V. Butler, *Trans. Faraday Soc.*, **33**, 229 (1937).
42. W. A. Kauzmann, *Advan. Protein Chem.*, **14**, 1 (1959).
43. A. Packter and M. Donbrow, *Proc. Chem. Soc., London*, 220 (1962).
44. R. P. Bell, E. Gelles, and E. Möller, *Proc. Roy. Soc., Ser. A*, **198**, 308 (1949).
45. J. Steinhardt and C. H. Fugitt, *J. Res. Nat. Bur. Stand., Sect. A*, **29**, 315 (1942).
46. J. R. Knowles and C. A. Parsons, *Chem. Commun.*, 755 (1967).
47. R. L. Letsinger and T. E. Wagner, *J. Amer. Chem. Soc.*, **88**, 2062 (1966); see also T. E. Wagner, C.-j. Hsu, and C. S. Pratt, *J. Amer. Chem. Soc.*, **89**, 6366 (1967).
48. H. N. Wolkoff, *Diss. Abstr.*, **21**, 2904 (1961).
49. E. F. J. Duynstee and E. Grunwald, *Tetrahedron*, **21**, 2401 (1965).
50. D. G. Herries, W. Bishop, and F. M. Richards, *J. Phys. Chem.*, **68**, 1842 (1964).
51. E. F. J. Duynstee and E. Grunwald, *J. Amer. Chem. Soc.*, **81**, 4540 (1959).
52. M. T. A. Behme, J. G. Fullington, R. Noel, and E. H. Cordes, *ibid.*, **87**, 266 (1965).
53. J. L. Kurz, *J. Phys. Chem.*, **66**, 2239 (1962).
54. F. Cramer, *Rev. Pure Appl. Chem.*, **5**, 143 (1955).
55. F. Cramer, *Einschlussverbindungen*, Springer-Verlag, Berlin-Heidelberg, 1954.
56. F. Cramer, *Angew. Chem.*, **68**, 115 (1956).

57. *Chem. Eng. News*, **45**(20), 69, May 8, 1967.
58. D. French, *Advan. Carbohyd. Chem.*, **12**, 189 (1957); J. A. Hamilton, L. K. Steinrauf, and R. L. Van Etten, *Acta Crystallogr.* submitted.
59. A. Hybl, R. E. Rundle, and D. E. Williams, *J. Amer. Chem. Soc.*, **87**, 2779 (1965); a detailed description of the structure of a cyclohexaamylose-potassium acetate complex is given.
60. J. A. Thoma and L. Stewart in *Starch: Chemistry and Technology*, Vol. I, R. L. Whistler and E. F. Paschall, Eds., Academic Press, Inc., New York, 1965, p. 209.
61. F. R. Senti and S. R. Erlander in *Non-Stoichiometric Compounds*, L. Mandelcorn, Ed., Academic Press, Inc., New York, 1964, p. 588.
62. F. Cramer, *Chem. Ber.*, **84**, 851 (1951).
63. F. Cramer, W. Saenger, and H.-Ch. Spatz, *J. Amer. Chem. Soc.*, **89**, 14 (1967).
64. F. Cramer and W. Dietsche, *Chem. Ind. (London)*, 892 (1958); idem, *Chem. Ber.*, **92**, 378, 1739 (1959).
65. F. Cramer and W. Kampe, *Tetrahedron Lett.*, 353 (1962); idem, *J. Amer. Chem. Soc.*, **87**, 1115 (1965).
66. J. L. Lach and T.-F. Chin, *J. Pharm. Sci.*, **53**, 924 (1964).
67. F. Cramer, *Angew. Chem.*, **73**, 49 (1961); N. Hennrich and F. Cramer, *Chem. Ind. (London)*, 1224 (1961); idem, *J. Amer. Chem. Soc.*, **87**, 1121 (1965).
68. M. L. Bender, R. L. Van Etten, G. A. Clowes, and J. F. Sebastian, *J. Amer. Chem. Soc.*, **88**, 2318 (1966); R. L. Van Etten, J. F. Sebastian, G. A. Clowes, and M. L. Bender, *J. Amer. Chem. Soc.*, **89**, 3242 (1967).
69. M. L. Bender, R. L. Van Etten, and G. A. Clowes, *ibid.*, **88**, 2319 (1966); R. L. Van Etten, G. A. Clowes, J. F. Sebastian, and M. L. Bender, *ibid.*, **89**, 3253 (1967).
70. F. Cramer and G. Mackensen, *Angew. Chem., Int. Ed. Eng.*, **5**, 601 (1966); idem, *Chem. Ber.*, **103**, 2138 (1970).
71. M. L. Bender, *Trans. N.Y. Acad. Sci.*, **29**, 301 (1967).
72. C. van Hooidonk and J. C. A. E. Breebaart-Hansen, *Recl. Trav. Chim. Pays-Bas*, **89**, 289; H. P. Benschop and G. R. Van den Berg, *Chem. Commun.*, 1431 (1970).
73. J. R. Whitaker and F. E. Deatherage, *J. Amer. Chem. Soc.*, **77**, 3360 (1955).
74. W. Kern, W. Herold, and B. Scherhag, *Makromol. Chem.*, **17**, 231 (1956); see also S. Yoshikawa and O.-K. Kim, *Bull. Chem. Soc. Jap.*, **39**, 1729 (1966).
75. I. Sakurada, Y. Sakaguchi, T. Ono, and T. Ueda, *Makromol. Chem.*, **91**, 243 (1966).
76. S. A. Bernhard and L. P. Hammett, *J. Amer. Chem. Soc.*, **75**, 5834 (1953).
77. S. Yoshikawa and O.-K. Kim, *Bull. Chem. Soc. Jap.*, **39**, 1515 (1966); S. A. Bernhard and L. P. Hammett, *J. Amer. Chem. Soc.*, **75**, 1798 (1953).
78. J. R. Whitaker and F. E. Deatherage, *J. Amer. Chem. Soc.*, **77**, 5298 (1955).
79. T. J. Painter and W. T. J. Morgan, *Chem. Ind. (London)*, 437 (1961).
80. S. Affrossman and J. P. Murray, *J. Chem. Soc. B*, 1015 (1966).
81. H. Morawetz, *Accounts Chem. Res.*, **3**, 354 (1970); see also H. Morawetz et al., *J. Amer. Chem. Soc.*, **91**, 563 (1969) and **92**, 7532 (1970).
82. C. L. Arcus and B. A. Jackson, *Chem. Ind. (London)*, 2022 (1964).
83. H. Ladenheim, E. M. Loebl, and H. Morawetz, *J. Amer. Chem. Soc.*, **81**, 20 (1959); H. Ladenheim and H. Morawetz, *ibid.*, **81**, 4860 (1959); see also H. Morawetz and W. R. Song, *J. Amer. Chem. Soc.*, **88**, 5714 (1966).
84. R. L. Letsinger and T. J. Savereide, *J. Amer. Chem. Soc.*, **84**, 114, 3122 (1962).
85. C. G. Overberger, T. St. Pierre, N. Vorchheimer, and S. Yaroslavsky, *ibid.*, **85**, 3513 (1963); C. G. Overberger, T. St. Pierre, N. Vorchheimer, J. Lee, and S. Yaroslavsky, *J. Amer. Chem. Soc.*, **87**, 296 (1965).

References

86. C. G. Overberger, R. Sitaramaiah, T. St. Pierre, and S. Yaroslavsky, *J. Amer. Chem. Soc.*, **87**, 3270 (1965).
87. C. G. Overberger, T. St. Pierre, C. Yaroslavsky, and S. Yaroslavsky, *J. Amer. Chem. Soc.*, **88**, 1184 (1966).
88. R. L. Letsinger and I. Klaus, *J. Amer. Chem. Soc.*, **86**, 3884 (1964); **87**, 3380 (1965).
89. I. M. Klotz and V. H. Stryker, *ibid.*, **90**, 2717 (1968); G. P. Royer and I. M. Klotz, *ibid.*, **91**, 5885 (1969); I. M. Klotz et al., *Biochemistry*, **8**, 4752 (1969); *Proc. Natl. Acad. Sci., U.S.*, **68**, 263 (1971).
90. R. L. Letsinger and T. E. Wagner, *J. Am. Chem. Soc.*, **88**, 2062 (1966).
91. R. Naylor and P. T. Gilham, *Biochemistry*, **5**, 2722 (1966).
92. R. Lovrien, *Abst. 148th Meeting, Amer. Chem. Soc.*, 44C (1964). More recent work by R. Lovrien and T. Linn, *Biochemistry*, **6**, 2281 (1967) indicates that this may be a solvent effect. However, a similar and striking effect is observed with several proteins.
93. E. Fischer, *Chem. Ber.*, **27**, 2985 (1894); *Z. Physiol. Chem.*, **26**, 60, (1898).
94. D. E. Tutt and M. A. Schwartz, *J. Amer. Chem. Soc.*, **93**, 767 (1971).
95. S. W. Fox, *Naturwissenschaften*, **56**, 1 (1969); D. L. Rohlfing and S. W. Fox, *Arch. Biochem. Biophys.*, **118**, 122, 127 (1967); D. L. Rohlfing, *Science*, **169**, 998 (1970).

Part Four

Enzymic Catalysis

Chapter 12

ENZYMES: CLASSIFICATION AND DETERMINATION

12.1	Enzyme Classification	401
12.2	Enzyme Determination	401
12.3	The Stoichiometric Basis of the Titrations of Hydrolytic Enzymes	412
12.4	Titration Theory	414
12.5	Chymotrypsin	418
	12.5.1 Comparison of the Slopes of the Lineweaver–Burk Plots of the Steady-State and Presteady-State Portions of the Reaction	419
	12.5.2 Second-Order Kinetics	419
	12.5.3 Measurement of the Burst of p-Nitrophenol in the Presteady State	419
12.6	Trypsin	424
	12.6.1 The Burst of p-Nitrophenol	424
	12.6.2 Second-order Conditions	427
12.7	Papain	429
	12.7.1 Burst Titration	429
	12.7.2 Relationship Between the Titration and Rate Assay	432
12.8	Subtilisin	433
	12.8.1 Burst Titration	433
	12.8.2 The Use of Rate Assay as Secondary Standard	436
12.9	Elastase	436
	12.9.1 Burst Titration	436
	12.9.2 The Relationship of the Reaction of Elastase with Diethyl p-Nitrophenyl Phosphate to Other Reactions of Elastase	439
	12.9.3 The Use of a Rate Assay as a Secondary Standard	441

12.10 Acetylcholinesterase 442
    12.10.1 Titrations 443
    12.10.2 Rate Assay Using Phenyl Acetate 446
12.11 Summary 447
    12.11.1 Impurities 449

Enzymes are proteins that catalyze the vast majority of chemical reactions in all living organisms. As catalysts, they are characterized by remarkable speed and selectivity. In every living organism there are a host of enzymes that catalyze complicated processes often involving a series of multiple, coupled reactions. In such processes each enzyme catalyzes its own particular reaction, thereby controlling the ordered sequence of physiological transformations by selective pathways.

The efficiency of enzymes as catalysts is often expressed in terms of the turnover number, that is, the number of molecules transformed by a molecule of enzyme per unit time. Because of the complexity of the phenomena and the multiplicity of enzymes, it is nearly hopeless to attempt a definition of an enzyme in terms of necessary and sufficient conditions. As we did with other catalysts, however, we try to give a working definition of an enzyme embodying its most characteristic properties and refine these notions as we progress through the succeeding chapters.

Thus enzymes can be defined as naturally occurring catalysts of protein nature, possessing a high degree of specificity and obeying Michaelis–Menten kinetics. By specificity, especially when referring to enzymes, we understand the ability of parts of the reacting molecule far removed from the reacting center to strongly influence the rate of reaction. As to Michaelis–Menten kinetics, the chief feature is the phenomenon of saturation: that is to say, that at high concentrations of substrate, the rate becomes independent of this concentration. In general, the rate of such a reaction can be expressed mathematically as

$$\text{rate} = \frac{a(\text{catalyst})(\text{substrate})}{b + (\text{substrate})}$$

where a and b are constants.

Since the midnineteenth century, chemists have been able to demonstrate the existence of powerful catalysts extracted from living organisms, although there has been considerable controversy as to the chemical nature and mode of action of these catalysts. Modern enzyme chemistry may be considered to have started with the crystallization of the enzyme urease in 1926 by Sumner,[1] who showed this material to be a crystalline protein. This advance was greeted with considerable skepticism by those who suggested that the enzyme was present as a mere impurity, adsorbed on crystals of an inert protein. A series of quantitative experiments by Northrop and co-workers[2] and Sumner and co-workers[3] conclusively proved that enzymes are indeed proteins. But, in fact, as in many controversies of this kind, both points of view eventually proved to contain elements of the truth. The backbone of all enzymes is, in all known cases, a protein, but in many cases small nonprotein molecules (coenzymes) necessary for the catalysis *are* adsorbed upon the protein. Co-enzymes in themselves do catalyze chemical reactions, but in conjunction with a protein molecule, both the catalytic efficiency and specificity are tremendously increased. Therefore we conclude that the most important common feature of all enzymes is the presence of a protein.

With the advent of purified and often crystallized proteins and enzymes, the biochemist attained a position from which he might probe the structure of these natural polymers and the details of how they catalyze biochemical reactions. The organic chemist asks the same questions, for an understanding of enzyme structure and mechanism should contribute to the fundamental concepts of structure and reactivity. For these reasons the structure and mechanism of individual enzymes have been the objects of much investigation in recent years. The remainder of this book will attempt to assess some of the progress made in this active and growing field.

The synthesis of enzymes is dependent on genetic control. The "one gene to one enzyme" theory, which postulates the existence of a separate gene for every enzyme, is the general principle of the genetic control of enzyme production.[6] Genetic control of enzyme synthesis can be strongly influenced by the presence of metabolites, usually either substrates or products of the enzyme-catalyzed reaction. The presence of an "inducer," most frequently the substrate of the enzyme, or a related low-molecular-weight substance, in many cases may increase the amount of the corresponding enzyme many hundreds or even thousands of times, without much effect on unrelated enzymes. This phenomenon of enzyme induction is especially marked in microorganisms.[6] The converse of the induction effect is enzyme "repression" which is a specific inhibition of the formation of a particular enzyme caused by an accumulation of a product of the enzymic reaction.[7] Since the effect of increasing the amount of a given enzyme will be to decrease the concentration of its substrate and increase that of its product, it is clear that the effects of induction and repression will result in increasing the formation of deficient enzymes and decreasing the formation of those that are present in excess.

Determination of the constitution of a protein requires the analysis of its primary, secondary, tertiary, and quaternary structure. By primary structure, we understand the specification of all the covalent bonds between the constituent amino acid residues—most commonly, peptide and disulfide bonds. The secondary structure comprises the noncovalent interactions of each amino acid with its immediate spatial neighbors—α-helical arrangements, pleated sheets, and other hydrogen-bonded arrays are included in this category. Tertiary structure encompasses the position in space of all the atoms comprising the protein. Quaternary structure represents the interaction of large protein subunits with each other.

From the point of view of the enzymologist, the most revealing information comes from the primary and secondary structures. The factors governing both are well understood and have been clearly described. Much more controversial are the physico-chemical principles determining the tertiary structure. Proteins possess well-defined tertiary structures; the forces maintaining these discrete, folded conformations are probably primarily nonpolar forces. The transfer of alkanes from water to a hydrocarbon solvent results in a small positive entropy change. This model suggests that the association of the nonpolar side chains of amino acids in a folded protein results in an entropy gain.[8] This suggestion also emphasizes the role of water as the stablizer of globular protein structure. Another interpretation focuses on the negative enthalpy gain accompanying the interaction of hydrocarbon molecules with water. In an enzyme, such an interaction (with its attendant favorable enthalpy change) may result in the formation of "icelike" structures surrounding the hydrocarbon residues, resembling crystalline hydrates.[9] At the moment, the model involving hydrocarbon-hydrocarbon interaction appears to be favored.

Assuming this model, the inside of an enzyme may be likened to a hydrocarbon solvent, with the exception that free movement of individual nonpolar groups is not possible. See, however, Chapter 13. The close-packing requirement of the non-polar groups presumably will restrict the conformation to a limited number of thermodynamically stable possibilities. Polar bond formation, most importantly hydrogen bond formation, does not contribute much to the free energy of the conformation, but does have directional requirements. The actual conformation will then result from a compromise between the packing requirements of nonpolar groups and directional requirements of polar bonds.

The three-dimensional structure of a protein may not be static. The denaturation of an enzyme is an important, although ordinarily unwanted, reaction. Although no concrete chemical description of this process may be made at this time, denaturation is usually ascribed to an unfolding of the polymer chain from the distinctive folded conformation to a random conformation. In the case of an enzyme, this large structural change may result in the loss of catalytic activity. For example, the denaturation of trypsin by urea, as measured by the change in intrinsic viscosity of the solution, exactly parallels the decrease in enzyme activity.[9] Denaturation of an enzyme may be brought about by heat, detergents, extremes of pH, or the addition of various organic compounds such as urea or guanidine.[10]

12.1 ENZYME CLASSIFICATION

The large and rapidly increasing number of known enzymes has made the introduction of systematic nomenclature imperative. In the past, enzyme nomenclature has been far from systematic. A comprehensive scheme of classification and nomenclature was adopted by the International Union of Biochemistry in 1961 and was revised in 1964.[11,12] This system of classification is shown in Table 12.1 together with examples from each subclassification and the approximate number of individual enzymes that have been identified in a particular classification.

This classification attempts to group enzymes by the type of chemical reaction they catalyze. Thus, enzymes are arranged into groups catalyzing similar processes, with subgroups specifying more precisely the actual reaction. The six main classes of enzymes are oxidoreductases, transferases, hydrolases, lyases, isomerases, and ligases. The identity of oxidoreductases is obvious from their description. The transferases transfer groups, whether they be alkyl, acyl, glucosyl, or other groups, from one acceptor to another. Hydrolases catalyze hydrolytic reactions. Lyases remove groups from their substrates (otherwise than by hydrolysis), leaving double bonds. Isomerases catalyze various isomerization reactions. Ligases catalyze the joining together of two molecules in a reaction coupled with the breakdown of a pyrophosphate bond in adenosine triphosphate (ATP). From the point of view of organic reaction mechanisms, some of the categories are unfortunate. A particular example is group 2.1, which comprises enzymes transferring one-carbon groups. These reactions might be $S_N 2$ in character or they might be reactions transferring a hydroxymethyl group (and therefore related to aldehyde reactions), or they might be reactions involving the transfer of formyl or other acyl groups. Thus in one category of enzymes we find a multiplicity of mechanistic types. However, for the most part, the classification is a sound one, both the point of view of the type of organic reactions and their mechanisms. The classification of perhaps 1700 enzymes into a reasonably small number of categories gives confidence that these reactions can be concisely discussed in a chemical manner.

12.2 ENZYME DETERMINATION†

Historically the assay of an enzyme solution has been tied to an operational phenomenon: an enzyme is present if catalysis of a particular chemical reaction can be detected, and the concentration of the enzyme is related to the rate of

(continued on p. 410)

† A goodly part of this and subsequent sections in this chapter has been taken from M. L. Bender et al., *J. Amer. Chem. Soc.*, **88**, 5890 (1966).

Table 12.1

Classification and Nomenclature of Enzymes[a,e]

Classification Number	Classification	Number and Trivial Name of Example	

1. Oxidoreductases

1.1	Alcohol donor	1.1.1.1	Alcohol dehydrogenase
1.2	Carbonyl donor	1.2.1.9	Glyceraldehyde phosphate dehydrogenase
1.3	CH—CH donor	1.3.1.3	Cortisone reductase
1.4	CH—NH_2 donor	1.4.3.4	Monoamine oxidase
1.5	C—NH donor	1.5.1.2	Pyrroline-5-carboxylate reductase
1.6	NADH donor	1.6.2.2.	Cytochrome b_5 reductase
1.7	Other nitrogenous compounds as donors	1.7.99.2	Nitric oxide reductase
1.8	Sulfur groups as donors	1.8.1.2	Sulfite reductase
1.9	Heme groups as hydrogen donors	1.9.3.1	Cytochrome oxidase (formerly called cytochrome a_3)
1.10	Diphenols as hydrogen donors	1.10.3.1	o-Diphenol oxidase
1.11	H_2O_2 as hydrogen acceptor	1.11.1.6	Catalase
		1.11.1.7	Peroxidase
1.12	H_2 as reductant	1.12.1.1	Hydrogenase
1.13	O_2 as oxidant of single substrate	1.13.1.1	Catechol 1,2-oxygenase
1.14	O_2 as oxidant of paired substrate	1.14.1.6	Steroid 11-β-hydroxylase

Systematic Name[d]	Active groups[c] (Cofactors)	Chemical Reaction Catalyzed[b,d]	Number of Related Enzymes of Different Specificity
Alcohol: NAD oxidoreductase	$Zn^{2\oplus}$	$CH_3CH_2OH + NAD = CH_3CH=O + NADH + H^\oplus$	72
D-Glyceraldehyde 3-phosphate: NADP oxidoreductase		D-Glyceraldehyde 3-phosphate + $NADP + H_2O$ = 3-phospho-D-glycerate + NADPH	35
4,5-β-Dihydrocortisone: NADP Δ^4-oxidoreductase		4,5-β-Dihydrocortisone + NADP = cortisone + NADPH + $H^\oplus$	12
Monoamine: oxygen oxidoreductase (deaminating)		Monoamine + $H_2O + O_2$ = an aldehyde + $NH_3 + H_2O_2$	12
L-Proline: NADP 5-oxidoreductase		L-Proline + NADP = Δ^1-pyrroline 5-carboxylate + NADPH + $H^\oplus$	8
Reduced-NAD: ferricytochrome b_5 oxidoreductase	F	$NADH + 2$ ferricytochrome b_5 = $NAD + 2$ ferrocytochrome b_5	18
Nitrogen: (acceptor) oxidoreductase	F,Fe	N_2 + acceptor = $2NO$ + reduced acceptor	6
Hydrogen sulfide: NADP oxidoreductase		$H_2S + 3$ NADP + $3 H_2O$ = sulfite + 3 NADPH + $H^\oplus$	7
Cytochrome c: oxygen oxidoreductase	H,Cu	4 Ferrocytochrome $c + O_2$ = 4 ferricytochrome $c + 2H_2O$	3
o-Diphenol: oxygen oxidoreductase	Cu	2 1,2-Dihydroxybenzene + O_2 = 2 o-quinone + $2H_2O$	3
Hydrogen peroxide: hydrogen peroxide oxidoreductase	H	$H_2O_2 + H_2O_2 = O_2 + 2H_2O$	9
Donor: hydrogen peroxide oxidoreductase	H	Donor + H_2O_2 = oxidized donor + H_2O	
	Fe	$H_2 + 2$ ferredoxin = 2 reduced ferredoxin	1
	Fe	Catechol + O_2 = cis,cis-muconate	13
		$NADP + O_2$ + steroid = 11-β-hydroxysteroid + NADP + H_2O	14

(continued)

Table 12.1 (*continued*)

Classification Number	Classification	Number and Trivial Name of Example	
2. *Transferase*			
2.1	Transfer of one-carbon groups	2.1.1.1	Nicotinamide methyltransferase
2.2	Transfer of aldehydic or ketonic groups	2.2.1.1	Transketolase, 1-glycoaldehyde-transferase
		2.2.1.2	Transaldolase, dihydroxyacetone-transferase
2.3	Transfer of acyl groups	2.3.1.16	Aspartate acetyltransferase
2.4	Transfer of glycosyl groups	2.4.1.1	α-Glucan phosphorylase
2.5	Transfer of alkyl or related groups	2.5.1.1	Dimethylallyltransferase, prenyltransferase (originally known as farnesylpyrophosphate synthetase)
2.6	Transfer of nitrogen-containing groups	2.6.1.1	Aspartate aminotransferase (formerly known as glutamic-oxalacetic transaminase)
2.7	Transfer of phosphorus-containing groups	2.7.7.16	Ribonuclease
2.8	Transfer of sulfur-containing groups	2.8.3.3	Malonate CoA-transferase
3. *Hydrolases*			
3.1	Cleavage of esters	3.1.1.7	Acetylcholinesterase
3.2	Cleavage of glycosyl compounds	3.2.1.17	Lysozyme
3.3	Cleavage of ether bonds	3.3.1.1	Adenosylhomocysteinase

Systematic name[d]	Active groups[c] (Cofactors)	Chemical Reaction Catalyzed[b,d]	Number of Related Enzymes of Different Specificity
S-Adenosylmethionine: nicotinamide N-methyltransferase		S-adenosylmethionine + nicotinamide = S-adenosylhomocysteine + N-methylnicotinamide	
D-Sedoheptulose 7-phosphate: D-glyceraldehyde 3-phosphate glycoaldehydetransferase	$Mg^{2\oplus}$, T	D-Sedoheptulose 7-phosphate + D-glyceraldehyde 3-phosphate = D-ribose 5-phosphate + D-xylulose 5-phosphate	2
D-Sedoheptulose 7-phosphate: D-glyceraldehyde 3-phosphate dihydroxyacetonetransferase		D-Sedoheptulose 7-phosphate + D-glyceraldehyde 3-phosphate = D-erythrose 4-phosphate + D-fructose 6-phosphate	
Acyl-CoA: acetyl-CoA C-acyltransferase	A	Acetyl-CoA + L-aspartate = CoA + N-acetyl-L-aspartate	21
α-1,4-Glucan: orthophosphate glucosyltransferase	Py	(α-1,4-glycosyl)$_n$ + orthophosphate = (α-1,4-glucosyl)$_{n-1}$ + α-D-glucose 1-phosphate	43
Dimethylallyl pyrophosphate: isopentenyl pyrophosphate dimethylallyltransferase		Dimethylallyl pyrophosphate + isopentenyl pyrophosphate = pyrophosphate + geranyl pyrophosphate	6
L-Aspartate: 2-oxoglutarate aminotransferase	Py	L-Aspartate + 2-oxoglutarate = oxaloacetate + L-glutamate	21
Polyribonucleotide 2-oligonucleotidotransferase (cyclizing)	$Mg^{2\oplus}$	Cleaves ribonucleic acid	107
Acetyl-CoA: malonate CoA-transferase		Acetyl-CoA + malonate = acetate + malonyl-CoA	12
Acetylcholine hydrolase		Acetylcholine + H_2O = choline + acetic acid	62
N-Acetylmuramide glycanohydrolase		Hydrolyzes β-1,4 links between N-acetylmuramic acid and 2-acetyl-amino-2-deoxy-D-glucose	45
S-Adenosyl-L-homocysteine hydrolase		S-adenosyl-L-homocysteine + H_2O = adenosine + L-homocysteine	1

(continued)

Table 12.1 (*Continued*)

Classification Number	Classification	Number and Trivial Name of Example	
3.4	Cleavage of peptides	3.4.4.5	Chymotrypsin A
		3.4.4.4	Trypsin
		3.4.4.10	Papain
3.5	Cleavage of carbon–nitrogen bonds other than peptide bonds	3.5.2.6	Penicillinase
3.6	Cleavage of acid anhydride bonds	3.6.1.3	ATPase (myosin)
3.7	Cleavage of carbon–carbon bonds	3.7.1.1	Oxaloacetase
3.8	Cleavage of carbon– or phosphorus–halide bonds	3.8.2.1	DFPase
3.9	Cleavage of phosphorus–nitrogen bonds	3.9.1.1	Phosphoamidase
4. *Lyases*			
4.1	Cleavage of carbon–carbon bonds	4.1.1.1	Pyruvate decarboxylase
		4.1.1.4	Acetoacetate decarboxylase
		4.1.2.7	Aldolase
		4.1.2.9	Phosphoketolase
4.2	Cleavage of carbon–oxygen bonds	4.2.1.2	Fumarate hydratase
4.3	Cleavage of carbon–nitrogen bonds	4.3.1.1	Aspartate ammonia lyase (originally known as aspartase)
4.4	Cleavage of carbon–sulfur bonds	4.4.1.1	Cysteine desulfhydrase
4.5	Cleavage of carbon–halogen bonds	4.5.1.1	DDT-Dehydrochlorinase

Systematic Name[d]	Active groups[c] (Cofactors)	Chemical Reaction Catalyzed[b,d]	Number of Related Enzymes of Different Specificity
		Hydrolyzes peptides, amides, esters, and so on	39
		Hydrolyzes peptides, amides, esters, and so on	
		Hydrolyzes peptides, amides, esters, and so on	
Penicillin amidohydrolase		Penicillin + H_2O = penicilloic acid	46
ATP phosphohydrolase	$Ca^{2\oplus}$	ATP + H_2O = ADP + orthophosphate	12
Oxaloacetate acetylhydrolase	$Mn^{2\oplus}$	Oxaloacetate + H_2O = oxalate + acetate	3
Diisopropylphosphorofluoridate hydrolase	$Mn^{2\oplus}$	Diisopropyl phosphorofluoridate + H_2O = diisopropyl phosphate + HF	21
Phosphoamide hydrolase		Phosphocreatine + H_2O = creatine + orthophosphate	
2-Oxo acid carboxy lyase	T	2-Oxo acid = aldehyde + CO_2	66
Acetoacetate carboxy lyase		Acetoacetate = acetone + CO_2	
Ketose 1-phosphate aldehyde lyase		Ketose 1-phosphate = dihydroxyacetone phosphate + an aldehyde	
D-Xylulose 5-phosphate D-glyceraldehyde 3-phosphate lyase (phosphate acetylating)		D-xylulose 5-phosphate + orthophosphate = acetylphosphate + D-Glyceraldehyde 3-phosphate	
L-Malate hydro lyase		L-Malate = fumarate + H_2O	34
L-Aspartate ammonia lyase	$M^{2\oplus}$	L-Aspartate = fumarate + NH_3	8
Cysteine hydrogen sulfide lyase (deaminating)	Py	L-Cysteine + H_2O = pyruvate + NH_3 + H_2S	6
1,1-Trichloro-2,2-bis(p-chlorophenyl)ethane hydrogen chloride lyase	S	1,1,1-Trichloro-2,2-bis(p-chlorophenyl)ethane = 1,1-dichloro-2,2-bis(p-chlorophenyl)ethylene + HCl	1

(continued)

Table 12.1 (*continued*)

Classification Number	Classification	Number and Trivial Name of Example	
5. Isomerases			
5.1	Racemization and epimerization	5.1.1.1	Alanine racemase
5.2	Cis-trans isomerization	5.2.1.1	Maleate isomerase
5.3	Intramolecular oxidation–reduction reactions	5.3.1.1	Triosephosphate isomerase
5.4	Intramolecular transfer	5.4.2.1	Phosphoglycerate phosphomutase
5.5	Intramolecular lyase reactions	5.5.1.1	Muconate cycloisomerase
6. Ligases			
6.1	Formation of carbon–oxygen bonds	6.1.1.1	Tyrosyl-sRNA synthetase
6.2	Formation of carbon–sulfur bonds	6.2.1.1	Acetyl-CoA synthetase
6.3	Formation of carbon–nitrogen bonds	6.3.2.2	γ-Glutamyl-cysteine synthetase
6.4	Formation of carbon–carbon bonds	6.4.1.2	Acetyl-CoA carboxylase

[a] This classification is based on the scheme proposed by the Commission on Enzymes of the International Union of Biochemistry.

[b] The reactions are written in the direction corresponding with the systematic names; therefore the direction written may not always be that which has been demonstrated. For this reason, an equal sign rather than $\leftrightarrows$ or $\rightarrow$ has been used throughout.

[c] Key: A = anions
      Bt = biotin
      B_{12} = vitamin B_{12}
      F = flavoprotein
      H = heme protein
      $M^{2\oplus}$ = bivalent metal ions
      Py = pyridoxal phosphate
      S = sulfhydryl compound
      T = thiamine pyrophosphate

Systematic Name[d]	Active groups[c] (Cofactors)	Chemical Reaction Catalyzed[b,d]	Number of related Enzymes of Different Specificity
Alanine racemase	Py	L-Alanine = D-alanine	19
Maleate cis,trans-isomerase		Maleate = fumarate	4
D-Glyceraldehyde 3-phosphate ketol isomerase		D-Glyceraldehyde 3-phosphate = dihydroxyacetone phosphate	18
D-Phosphoglycerate 2,3-phosphomutase	B_{12}	2-phospho-D-glycerate = 3-phospho-D-glycerate	4
4-Carboxymethyl-4-hydroxyisocrotonolactone lyase (decyclizing)	$Mn^{2\oplus}$	(+)-4-Carboxymethyl-4-hydroxyisocrotonolactone = cis,cis-muconate	1
L-Tyrosine:sRNA ligase (AMP)	$Mg^{2\oplus}$, $K^{\oplus}$	ATP + L-tyrosine + sRNA = AMP + pyrophosphate + L-tyrosyl-sRNA	12
Acetate:CoA ligase (AMP)	$Mg^{2\oplus}$, $K^{\oplus}$	ATP + acetate + CoA = AMP + pyrophosphate + acetyl-CoA	7
L-Glutamate:L-cysteine γ-ligase (ADP)	$Mg^{2\oplus}$	ATP + L-glutamate + L-cysteine = ADP + orthophosphate + γ-L-glutamyl-L-cysteine	20
Acetyl-CoA:carbon dioxide ligase (ADP)	Bt	ATP + acetyl-CoA + CO_2 + H_2O = ADP + orthophosphate + malonyl-CoA	4

[a] Key: NAD = nicotinamide-adenine dinucleotide
NADH = reduced nicotinamide-adenine dinucleotide
NADP = NAD phosphate
NADPH = reduced NAD phosphate
ATP = adenosine triphosphate
ADP = adenosine diphosphate
AMP = adenosine monophosphate
CoA = coenzyme A

From M. L. Bender, in *Encyclopedia of Polymer Science and Technology*, John Wiley & Sons, New York, 1967, Vol. 6, p. 1.

catalysis. Enzymes, however, may now be characterized by more than their catalytic behavior; namely, in terms of chemical constitution and in terms of stoichiometric chemical reactions. Thus the usage of a "rate assay" to determine the concentration of a catalyst such as an enzyme may be questioned, a usage that is tantamount to using a rate assay for determining the concentration of a catalyst such as hydrochloric acid. This commentary is particularly true of the hydrolytic enzymes whose chemical constitution and stoichiometric chemical reactions have become reasonably well known in the past decade. It is the present thesis that the accurate, simple, convenient procedures now available for the direct titration of the concentration of active sites of the hydrolytic enzymes chymotrypsin, trypsin, elastase, subtilisin, papain, acetylcholinesterase, and others offer a better approach to the determination of enzyme concentration than the rate assays of the past. Thus the definition of an enzyme used here is an intact active site that can undergo a specific stoichiometric reaction.

Methods are presented for the determination of the concentration of active sites of certain hydrolytic enzymes. Because of the possible ambiguity with respect to the number of active sites per enzyme molecule, it is convenient to define these determinations in terms of normality of active sites.

There are many other approaches to defining enzyme solutions. One may speak of enzyme units, enzyme activity, enzyme purity, or enzyme concentration.[13] One unit of enzyme is defined as that amount of enzyme that will catalyze the transformation of 1 μmole of substrate per minute under defined conditions.[13] The specific activity is expressed as units of enzyme per milligram of protein. The concentration of an enzyme solution is expressed as units of enzyme per milliliter. The purity of an enzyme solution is expressed as the specific activity of the preparation divided by the specific activity of the pure enzyme. These definitions were developed on the basis of the premise stated earlier that enzymes may be observed to catalyze reactions, but are not sufficiently characterized chemically for direct analysis. Of the above, we will not use units of enzyme nor specific activity of an enzyme. We will, however, use the concentration of an enzyme solution, which we will attempt to express in equivalents of enzyme per liter (normality). Knowing the equivalent weight of the enzymes from independent investigations and the weight of enzyme making up a given solution, the purity of an enzyme solution can then be expressed in terms of the actual normality divided by the normality calculated on a weight basis.

Rate assays for the determination of the concentration of enzyme solutions in terms of units of enzyme per milliliter have been described for many of the enzymes considered here utilizing natural or synthetic substrates.[14,15] In certain instances these assays have been converted from units of enzyme per milliliter to equivalents of enzyme per milliliter, a true concentration, by making the assumption that a reference standard of 100% pure (active) enzyme was available. There is one fundamental problem and a number of subsidiary problems associated with such a procedure.

The fundamental problem is associated with our present inability to define a 100% pure (active) enzyme. Certainly enzymes have been crystallized and shown to be homogeneous by chromatography, by electrophoresis, and by ultracentrifugation. But it is difficult to accept the use of a pure enzyme as an absolute analytical standard when enzymes can adsorb impurities, when crystalline enzymes, even pure ones, contain very appreciable amounts of water that cannot be removed without denaturation of the enzyme, and when crystalline α-chymotrypsin, one of the purest and best characterized enzymes extant, is often found to contain 20–30% of impurities (including water). Analytical chemistry demands that absolute reference standards be extremely well characterized. At the present time no enzyme meets the criteria that analytical chemistry demands for such a characterization.

In addition to this fundamental problem of the absolute reference standard, rate assays are plagued with many uncertainties that limit the accuracy of the determination of an enzyme concentration. These include uncertainties due to the large number of variables in the rate assay that must be controlled such as temperature, ionic strength, cofactors, inhibitors, activators, pH, and so on. The recommended procedure[11,12,13] for carrying out a rate assay is to use initial rates "in order to avoid complications due, for instance, to reversibility of reactions or to formation of inhibitory products." Furthermore, conditions are recommended in which the enzyme is saturated with substrate so that the rate approaches zero-order kinetics and a maximal velocity may be observed, or alternatively that a maximal velocity be determined by a Michaelis–Menten treatment.[13] Many of the rate assays for the enzymes under discussion violate one or more of the recommendations given above. These recommendations, even if followed, beg the questions: of the use of initial rates, a risky, last-resort procedure because initial rates are, many times, hard to determine and may hide a multitude of kinetic sins; of impurities in the enzyme test solution that may completely invalidate the kinetics when compared to a pure enzyme solution; and of the problems of rate assay that in some instances will not allow agreement from two laboratories to better than a factor of 2. Thus both theoretical and practical problems plague the use of rate assays for the determination of enzyme concentration.

A number of the above problems may be obviated by the use of the "optical factor." In essence, this procedure involves a spectrophotometric determination of the concentration of protein in a given solution. Again the 100% pure (active) enzyme is used as absolute reference standard and again the problems inherent in such a procedure arise. Furthermore, the assumption must be made that this solution is free from ultraviolet-absorbing contaminants, particularly of the peptide and protein variety. Since this assumption is ordinarily the point of issue in a determination of enzyme concentration, the use of the "optical factor" is a most unsatisfactory method for the determination of absolute enzyme concentration. The use of an "optical factor" involves not the determination of an enzyme concentration but rather that of a protein concentration. Since the two concentrations are not necessarily identical because of the usual presence of denatured enzyme, this procedure is is unsatisfactory. Thus the use of an "optical factor" begs the question of the reactivity of the enzyme since the activity of the enzyme may easily be lost without any change in gross chemical structure. Thus analytical methods must be developed for the determination of the active sites of enzymes.

One of the knotty problems in dealing with enzymes is the possible presence of an enzymatically active impurity that can react with a substrate or titrant. Such extraneous reaction would be expected to occur at a rate different from that of the principal reaction, and thus should show up in the kinetics. However, it is advisable with any enzyme preparation to determine kinetics with at least two different degrees of purification against several substrates in order to determine if the observed activity is independent of purification. A further check may be made by comparing the rate of reaction of an inhibitor with the enzyme preparation to the rate of inhibition of its activity as determined with several different substrates.

12.3 THE STOICHIOMETRIC BASIS OF THE TITRATIONS OF HYDROLYTIC ENZYMES

The stoichiometric reactions of hydrolytic enzymes with organophosphates provide the ultimate basis for the determination of the concentration of such enzyme solutions.[16] Following the discovery of these stoichiometric reactions, many other stoichiometric reactions have been observed with hydrolytic enzymes; some of these reactions are readily adaptable for the present purposes. In general, what is required for the determination of enzyme concentration is a stoichiometric reaction that is readily and accurately observable. But since an enzyme is a complex organic molecule, care must be taken to ensure that the stoichiometry observed and used is a meaningful stoichiometry. The primary requisite of the stoichiometry is that it must involve the active site of the enzyme. This requirement follows from the objective of these determinations, namely to titrate active sites as a true measure of active enzyme molecules. The stoichiometry may involve either a substrate,[17] an inhibitor,[18] or a coenzyme.[19] However, the latter two reagents do not test the primary chemical reaction of the enzyme and thus are less desirable than the former.

Thus the optimal stoichiometric reaction is a reaction between the active site and something closely resembling a normal substrate of the enzyme.

The stoichiometric basis of the present determinations of hydrolytic enzyme concentrations may be represented by eq. 12.1, which represents the overall pathway for α-chymotrypsin,[20] trypsin,[21] elastase,[22] subtilisin,[23] papain,[24] and acetylcholinesterase[25] reactions.

$$E + S \underset{}{\overset{K_s}{\rightleftarrows}} ES \xrightarrow{k_2} \underset{+ P_1}{ES'} \xrightarrow{k_3} E + P_2 \qquad (12.1)$$

The conversion of enzyme, E, into acyl-enzyme, ES', and product one, P_1, can be considered a stoichiometric reaction with respect to an individual active site. If this process can be observed before the turnover (regeneration) of the enzyme occurs, then a direct measure of the enzyme concentration may be made.

A limiting example of the conversion of an enzyme into acyl-enzyme and P_1 without the regeneration of any enzyme is seen in the reaction of organophosphate and organosulfonate compounds with many hydrolytic enzymes, mentioned above. Hydrolytic enzymes have been chosen as examples because there is more known about their stoichiometry. They have therefore been gone into in detail. Observations of the stoichiometry of the process were made by the amount of reactant (organophosphate) used up, by the liberation of P_1, in terms of hydrogen ion or p-nitrophenol, or by the attachment of phosphorus, isopropyl groups (from diisopropyl phosphorofluoridate), or radioactively-labeled organophosphate to the protein.[14] Any of these observations of stoichiometry could be used as the basis of a procedure for the routine determination of enzyme concentration. Several disadvantages of using this approach may be noted. (1) The organophosphates are, in general, nonspecific compounds; for example, a stoichiometric reaction of bromelain and diisopropyl phosphorofluoridate has been found but it does not involve the active site of this enzyme;[26] (2) except for the observation of p-nitrophenol (P_1), the experimental methods are both tedious and of limited accuracy.

The use of a specific substrate stems not only from the requirement that the stoichiometric reaction take place at the active site, but also from the desirability of being able to titrate the active sites of one enzyme in the presence of a number of other enzymes. Thus the titrant for any given enzyme should optimally be modeled as closely as possible on the substrate of that enzyme to ensure reaction only at the active site and to ensure maximum selectivity between different enzymes.

The stoichiometry used for the titrations presented here depends not on a pure enzyme as absolute standard, but rather on (a change in a physical property of) the substrate as absolute standard. The stoichiometric transformation involves the reaction, $E \cdot S \rightarrow ES' + P_1$. The ultimate test of the stoichiometry, as mentioned above, is that ES' is inactive whereas E is active enzymatically. With this test, the stoichiometry that is pertinent to an enzymatic process is demonstrated. In other words, this stoichiometry is an operational measure of the active site of the enzyme.

Probably the first direct observation of stoichiometric production of P_1 using a substrate was in the α-chymotrypsin-catalyzed hydrolysis of p-nitrophenyl acetate.[27] Observation of the liberation of p-nitrophenol from this substrate led to the conclusion that the reaction proceeds in at least two distinct kinetic steps, specifically a fast initial liberation of approximately 1 mole of p-nitrophenol per mole of enzyme, followed by a slow turnover reaction. This observation, in itself, indicated the possibility of determining the concentration of kinetically active sites on the enzyme in an easy and accurate manner. However, the exact working conditions of a reliable experimental titration method were not established.

The kinetic equations rigorously describing the hydrolysis of p-nitrophenyl acetate by α-chymotrypsin proved to be complex,[28, 29] and furthermore the amount of phenol liberated in the initial step was shown to be a complex function of the equilibrium and rate constants involved in the reaction.[30, 31] Experimental confirmation of these equations was not demonstrated. Subsequently methods were developed to titrate α-chymotrypsin with the substrate N-trans-cinnamoylimidazole[32] and the inhibitor 3-nitro-4-carboxyphenyl N,N-diphenylcarbamate.[33] These methods did not, however, test the validity of the equations referred to above describing the stoichiometry during the enzymatic hydrolysis of a specific substrate.

A number of titrations of hydrolytic enzymes will be described. These conform to the stoichiometry expressed in eq. 12.1 and provide practical methods for the absolute determination of the normality of an enzyme solution. They involve titrations of the concentrations of active sites of α-chymotrypsin, trypsin, elastase, subtilisin, papain, and acetylcholinesterase. A number of these titrations involve substrates specific for the particular enzyme, and thus the titrations are accurate, experimentally facile, and specific.

12.4 TITRATION THEORY

The kinetic equations describing eq. 12.1 have been solved.[17, 28, 34, 35] These equations offer several methods for the determination of the concentration of the (kinetically) active enzyme. Let us consider some of these possibilities. The first of these gives either a relative or absolute enzyme concentration, but the others give absolute enzyme concentration.

1. The initial turnover rate of formation of p-nitrophenol in the enzymatic hydrolysis of a p-nitrophenyl ester following eq. 12.1 may be expressed by eq. 12.2.[28]

$$V_0 = \frac{\left(\frac{k_2 k_3}{k_2 + k_3}\right)[E]_0[S]_0}{[S]_0 + \frac{K_s k_3}{k_2 + k_3}} = \frac{k_{cat}[E]_0[S]_0}{[S]_0 + K_m(app)} \quad (12.2)$$

where $K_m(app)$ is the apparent Michaelis constant and k_{cat} is the catalytic rate constant, both operational parameters; K_s, k_2, and k_3 are defined by eq. 12.1. and may be operational in certain instances.

When $[S]_0 \gg K_m(app)$, the rate reduces to

$$V_0 = \frac{k_2 k_3 [E]_0}{k_2 + k_3} \quad (12.3)$$

Furthermore, when $k_2 \gg k_3$

$$V_0 = k_3 [E]_0 \quad (12.4)$$

In all cases V_0 is proportional to $[E]_0$ at a given $[S]_0$ and given experimental conditions. If the proportionality constant between V_0 and $[E]_0$ is known, the latter may be easily calculated from the former; if the proportionality constant is not known, a relative rate may still be calculated. Many rate assays of hydrolytic enzymes are based on this proportionality. However, as discussed above, such rate assays have severe limitations.

One approach for determining the proportionality constant between V_0 and $[E]_0$ is to determine one V_0 and one $[E]_0$. V_0 may be measured straightforwardly but a 100% pure (active) enzyme must be used in order to determine $[E]_0$. Thus this method depends ultimately on using a pure enzyme as absolute reference standard, with the limitations inherent in this procedure, as outlined above.

Another approach for determining the proportionality constant between V_0 and $[E]_0$ may be utilized when eq. 12.4 is operative. The proportionality constant of eq. 12.4, k_3, may be determined independently if the acyl-enzyme, ES', of either a good or poor substrate is prepared. ES' may be prepared by using an excess of substrate to effect acylation, followed by dilution of the solution so that the substrate concentration drops below $K_m(app)$ and buildup of the acyl-enzyme stops. Alternatively, acylation may be stopped by shifting the pH and/or removing the excess substrate by means of Sephadex filtration or other methods. The decay of ES' to E may be followed in many ways to determine the first-order constant, k_3, the most rigorous being the regeneration of E as measured with a "true" substrate of the enzyme; this reaction, is, of course, independent of the original enzyme concentration. The constant,

k_3, together with the maximal velocity, $V_0 = k_3[E]_0$, then gives the absolute enzyme concentration directly. Essentially, this procedure was used by Wilson and co-workers[38,39] for acetylcholinesterase, although the transformation from k_3 to $[E]_0$ was more complicated.

2. The relationship between the slopes of the Lineweaver–Burk plots of the steady-state and presteady-state portions of reaction 12.1 leads to the determination of the absolute concentration of enzyme. Equation 12.2, describing the initial steady-state (turnover) reaction, may be transformed into

$$\frac{1}{V_0} = \frac{k_2 + k_3}{k_2 k_3} \frac{1}{[E]_0} + \frac{K_s}{k_2[E]_0} \frac{1}{[S]_0} \qquad (12.5)$$

which in terms of Michaelis parameters is identical with

$$\frac{1}{V_0} = \frac{1}{k_{\text{cat}}} \frac{1}{[E]_0} + \frac{K_m(\text{app})}{k_{\text{cat}}[E]_0} \frac{1}{[S]_0}. \qquad (12.6)$$

The comparable equation of the presteady state is

$$\frac{1}{b} = \frac{1}{k_2} + \frac{K_s}{k_2} \frac{1}{[S]_0} \qquad (12.7)$$

if $(k_2 + k_3) > k_3 K_m(\text{app})$[28] ($b$ is the first-order rate constant of the presteady state; see eq. 12.10). Thus the slope of eq. 12.5 (plotted as $1/V_0$ versus $1/[S]_0$) is $K_s/k_2[E]_0$ while the slope of eq. 12.7 is K_s/k_2; by dividing the former slope by the latter, we may obtain $[E]_0$, the absolute concentration of enzyme.[40]

3. Under certain very stringent conditions, the kinetics of acylation ($d[P_1]/dt$) may be observed under second-order conditions.[34,35,41] Under these conditions, which require that $k_2 \gg k_3$ and that $K_m(\text{app}) > [S]_0 \cong [E]_0 < K_s$, the rate of appearance of p-nitrophenol may be represented by

$$V = \frac{d[P_1]}{dt} = (k_2/K_s)([E]_0 - [P_1])([S]_0 - [P_1]) \qquad (12.8)$$

Equation 12.8 may be transformed into

$$\frac{V}{[S]_0 - [P_1]} = \frac{k_2}{K_s}[E]_0 - \frac{k_2}{K_s}[P_1] \qquad (12.9)$$

A plot of $V/([S]_0 - [P_1])$ versus $[P_1]$ gives a straight line whose intercept divided by the slope is $[E]_0$, the absolute concentration of enzyme.

4. The most common set of conditions for observation of enzymatic processes, and the most important set for the determination of enzyme concentration, utilizes $[S]_0 \gg [E]_0$. Under these conditions a presteady state (acylation) and a steady state (deacylation) may be observed, the former often being too fast to measure, and the concentration, π, of P_1 liberated in the presteady state is related to the absolute enzyme concentration. For reaction

12.1, when $[S]_0 \gg [E]_0$, $[P_1]$ (conveniently p-nitrophenol) produced in time t may be described by[28]

$$[P_1] = \frac{k_{cat}[E]_0[S]_0 t}{[S]_0 + K_m(app)} + [E]_0 \left[\frac{\frac{k_2}{k_2 + k_3}}{\left(1 + \frac{K_m(app)}{[S]_0}\right)} \right]$$
$$\times \left[1 - \exp\left(-\frac{(k_2 + k_3)[S]_0 + k_3 K_s}{K_s + [S]_0} t \right) \right] \quad (12.10)$$

At high values of t, the exponential term approaches zero and the production of P_1 can be described as a linear function of t

$$[P_1] = \pi + At \quad (12.11)$$

where $A = k_{cat}[E]_0[S]_0/[[S]_0 + K_m(app)]$ and π may be expressed by

$$\pi = [E]_0 \frac{\left(\frac{k_2}{k_2 + k_3}\right)^2}{\left(1 + \frac{K_m(app)}{[S]_0}\right)^2} \quad (12.12)$$

Since it is the intercept of a plot of $[P_1]$ versus t at $t = 0$, π may be easily determined experimentally. Equation 12.12 may be transformed to eq. 12.13,

$$\frac{1}{\sqrt{\pi}} = \frac{k_2 + k_3}{k_2 \sqrt{[E]_0}} + \frac{(k_2 + k_3) K_m(app)}{k_2 \sqrt{[E]_0}} \frac{1}{[S]_0} \quad (12.13)$$

which indicates that $\pi = [E]_0$ only if $k_2 \gg k_3$ and $[S]_0 \gg K_m(app)$. If the latter condition does not hold, a series of experiments at constant $[E]_0$ but varying $[S]_0$ enables one to circumvent this condition. From eq. 12.13 a plot of $1/\sqrt{\pi}$ versus $1/[S]_0$ gives a straight line whose intercept is $[(k_2 + k_3)/k_3]$ $(1/\sqrt{[E]_0})$. If the condition is met that $k_2 \gg k_3$, the intercept of such a plot will give $1/\sqrt{[E]_0}$ and hence $[E]_0$ directly.

The reactions of diethyl p-nitrophenyl phosphate and related compounds with various hydrolytic enzymes are usually carried out under conditions in which $[S]_0 > [E]_0$. However, these reactions must be treated in a special manner since no evidence for an adsorptive enzyme-substrate complex has as yet been found. Hartley and Kilby,[42] who first investigated the reaction of this reagent with chymotrypsin, considered that the reaction followed eq. 12.14 (where S is the phosphate ester, ES' is the diethylphosphoryl-enzyme, P_1 is nitrophenol, and k_{II} is the second-order rate constant of the reaction of enzyme and phosphate ester) rather than eq. 12.1.

$$E + S \xrightarrow{k_{II}} ES' + P_1 \quad (12.14)$$

However, it is known that diethylphosphoryl-trypsin and diethylphosphoryl-chymotrypsin spontaneously dephosphorylate.[43,44] Therefore, the reaction should be written

$$E + S \xrightarrow{k_{II}} ES' \xrightarrow{k_3} E + P_2 \quad (12.15)$$
$$+ P_1$$

where the symbolism is the same as before with the addition of P_2, diethyl phosphate. If one assumes that the concentration of phosphate titrant, S, is much greater than that of the enzyme and that the reaction between enzyme and phosphate occurs only at the active site of the enzyme, the relationship between π, the amount of p-nitrophenol liberated in the pseudo-first-order reaction, and the true enzyme concentration may be derived. The derivation utilizes two differential equations describing eq. 12.15, $d[P_1]/dt = k_{II}[S]_0[E]$, and $d[ES']/dt = k_{II}[S]_0[E] - k_3[ES']$, and the conservation equation $[E]_0 = [E] + [ES']$, and leads to eqs. 12.16 and 12.17

$$\sqrt{\pi/[E]_0} = \frac{1}{1 + k_3/k_{II}[S]_0} \quad (12.16)$$

or

$$\sqrt{\pi/[E]_0} = \frac{1}{1 + k_3/(k_{obsd} - k_3)} \quad (12.17)$$

where k_{obsd} (or b) is the pseudo-first-order rate constant when the substrate is in great excess over the enzyme, corresponding to k_{II}. From eq. 12.17, it may be seen that $\pi = [E]_0$ when k_{obsd}, the pseudo-first-order rate constant of the inhibition reaction, is much greater than the dephosphorylation rate constant, k_3. Thus there are three requirements for a successful titration corresponding to eq. 12.15: $[S]_0 \gg [E]_0$; $k_{obsd} \gg k_3$; and the titration reaction must occur at the active site and only at the active site of the enzyme.

12.5 CHYMOTRYPSIN

α-Chymotrypsin (chymotrypsin A, EC 3.4.4.5)[13] was titrated using a number of substrates, both specific and nonspecific. Some titrations involve the substrates, p-nitrophenyl acetate and 2,4-dinitrophenyl acetate, using three methods of determining concentration: (1) a comparison of the slopes of the Lineweaver–Burk plots of the steady-state and presteady-state portions of the reaction; (2) second-order kinetics; (3) measurements of the burst of p-nitrophenol in the presteady-state. The three methods give results which are in essential agreement with one another.[36]

12.5.1 Comparison of the Slopes of the Lineweaver–Burk Plots of the Steady-State and Presteady-State Portions of the Reaction

Results were obtained concerning the turnover rates of the α-chymotrypsin-catalyzed hydrolysis of p-nitrophenyl acetate at substrate concentrations sufficiently low that complications arising from extraneous reactions were avoided.[35] From the slopes of (turnover) plots of $1/V_0$ versus $1/[S]_0$, values of $[E]_0 k_2/K_s$ were determined, following eq. 12.5. In addition, the kinetics of the presteady-state were determined, giving from eq. 12.7 an independent value of k_2/K_s. Using these two sets of data, $[E]_0$ was determined as shown in Table 12.2.[35]

Table 12.2
Determination of the Concentration of an α-Chymotrypsin Solution using the Steady-State and Presteady-State Portions of the Hydrolysis of p-Nitrophenyl Acetate[a]

Concentration of Enzyme From wt(g/liter)	pH	$k_2[E]_0/K_s$ (sec^{-1})	k_2/K_s, (M^{-1} sec^{-1})	$[E]_0(M)$	$[E]_0(M)$
1.61×10^{-2}	7.8	0.0017	3530	4.8×10^{-7}	5.0×10^{-7}
1.03×10^{-2}	7.7	0.00093	3410	2.7×10^{-7}	3.1×10^{-7}

[a] From M. L. Bender et al., *J. Amer. Chem. Soc.*, **88**, 5890 (1966). © 1966 by the American Chemical Society. Reprinted by permission of the copyright owner.

12.5.2 Second-Order Kinetics

The α-chymotrypsin-catalyzed hydrolysis of p-nitrophenyl acetate was carried out under second-order conditions ($K_m(\text{app}) > [S]_0 \cong [E]_0 < K_s$). The data were treated according to eq. 12.9. Figure 12.1 shows such a plot. Values of V were obtained by calculating $V \cong \Delta[P_1]/\Delta t$ from experimental results with Δt approximately $0.25 t_{1/2}$. From eq. 12.9 it is seen that the slope of this plot gives $k_2/K_s = 3380\ M^{-1}$ sec^{-1} (a conventional second-order plot gives $k_2/K_s = 3630\ M^{-1}$ sec^{-1} for the same reaction), while the intercept of this plot gives $k_2[E]_0/K_s$. The quotient of the intercept/slope = $[E]_0 = 8.17 \times 10^{-6}\ M$ while an N-*trans*-cinnamoylimidazole titration of the same solution gave $[E]_0 = 7.85 \times 10^{-6}\ M$.

12.5.3 Measurement of the Burst of p-Nitrophenol in the Presteady State

The α-chymotrypsin-catalyzed hydrolysis of p-nitrophenyl acetate was carried out under turnover conditions ($[S]_0 \gg [E]_0$). Plots of eq. 12.3 ($1/\sqrt{\pi}$

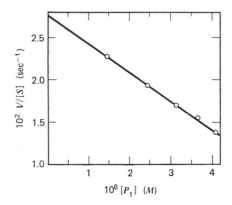

Fig. 12.1 The titration of α-chymotrypsin by p-nitrophenyl acetate using second-order conditions: 25.0°, pH 7.8, 0.067 M phosphate buffer, $(S)_0 = 6.72 \times 10^{-6}$ M, $(E)_0 = 8.17 \times 10^{-6}$ M. From M. L. Bender et al., *J. Amer. Chem. Soc.*, **88**, 5890 (1966). © 1966 by the American Chemical Society. Reprinted by permission of the copyright owner.

versus $1/[S]_0$) are shown in Fig. 12.2 for two different enzyme concentrations. The experimental points lie on fairly good straight lines, whose slopes are in good agreement with the calculated slopes based on values of the various constants of eq.12.13 given previously.[35] The extrapolated π value at $[S]_0 = \infty$ ($1/[S]_0 = 0$) is in good agreement with N-*trans*-cinnamoylimidazole titration values, as shown in Table 12.3. In addition the α-chymotrypsin-catalyzed hydrolysis of 2,4-dinitrophenyl acetate was investigated under turnover conditions.

Solutions of α-chymotrypsin have been titrated in many ways. Most of the titrants are listed in Table 12.4 together with the relationship of the titration value resulting from each titrant to that found with N-*trans*-cinnamoylimidazole, not because the latter is inherently the best titrant, but because it has been available for comparative purposes as an accessible standard for a longer period of time than any other titrant. All of the titrations in Table 12.4 meet the requirements set forth above for measurement of the stoichiometric conversion of enzyme to acyl-enzyme, namely that $[S]_0 \gg K_m(\mathrm{app})$ and that $k_2 \gg k_3$. Furthermore, $[S] > [E]$ in all titrations, although in some titrations, for example, with N-*trans*-cinnamoylimidazole, the [S]/[E] ratio is low (see later). Finally, all the titrations, including those with the nonspecific substrates such as p-nitrophenyl trimethylacetate, those with specific substrates such as p-nitrophenyl benzyloxycarbonyl-L-tyrosinate, and those with inhibitors such as 3-nitro-4-carboxyphenyl N,N-diphenylcarbamate and diisopropyl phosphorofluoridate, give titrations that are essentially equivalent

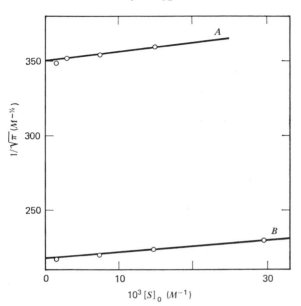

Fig. 12.2 The titration of α-chymotrypsin by *p*-nitrophenyl acetate at 25.0°, pH 7.8, 0.067 M phosphate buffer, 1.6% acetonitrile-water (v/v); A, $(E)_0 = 8.2 \times 10^{-6}$ M; B, $(E)_0 = 21.1 \times 10^{-6}$ M. From M. L. Bender et al., *J. Amer. Chem. Soc.*, **88**, 5890 (1966). © 1966 by the American Chemical Society. Reprinted by permission of the copyright owner.

to one another (some of the comparisons are indirect ones, such as a calculation of molecular weight of the enzyme on the basis of the stoichiometry). Since the titrations are independent of substrate, a strong argument may be made that in all titrations, one and the same active site is being titrated stoichiometrically, and further that no extraneous reaction is being observed. Titrations of the active site of α-chymotrypsin are also independent of pH, various titrations being carried out from pH 9.5 (*p*-nitrophenyl acetate) down to pH 2.5 (*p*-nitrophenyl *N*-benzyloxycarbonyl-L-tyrosinate).

The most specific titrant for α-chymotrypsin is *p*-nitrophenyl *N*-benzyloxycarbonyl-L-tyrosinate; the least specific titrant is diethyl *p*-nitrophenyl phosphate. As an additional check to determine whether these two compounds, and by inference the other titrants of Table 12.4, reflect the same site of the enzyme, the rate constant for the liberation of *p*-nitrophenol from the reaction of diethyl *p*-nitrophenyl phosphate with α-chymotrypsin was compared to the rate constant for the loss of enzymatic activity toward *p*-nitrophenyl *N*-benzyloxycarbonyl-L-tyrosinate. The two rate constants were

Table 12.3
Titration of Some α-Chymotrypsin Solutions Calculated from the Burst of p-Nitrophenol from a p-Nitrophenyl Ester Substrate[c]

	$[S]_0 \times 10^5$ (M)	$[E]_0$ (g/liter)	$[E]_0 \times 10^{6,a}$ (M)	$[E]_0 \times 10^{6,b}$ (M)	$[E]_0 \times 10^{6,c}$ (M)	% purity
p-Nitrophenyl acetate	3.4–67.7	0.694	27.98	20.42	21.08	75.3
p-Nitrophenyl acetate	6.8–67.7	0.277	11.17	7.82	8.18	73.2
2,4-Dinitrophenyl acetate	13.94			2.82	2.86	
2,4-Dinitrophenyl acetate	6.97			2.82	2.82	
2,4-Dinitrophenyl acetate	14.12			1.14	1.15	

[a] Theoretical using weight of enzyme and a molecular weight of 24,800.
[b] Titration with N-*trans*-cinnamoylimidazole.
[c] From M. L. Bender et al., *J. Amer. Chem. Soc.*, **88**, 5890 (1966). © 1966 by the American Chemical Society. Reprinted by permission of the copyright owner.

Table 12.4
Titrants for α-Chymotrypsin Solutions[a]

Titrant	Species Measured	$[S]_0/K_m$ (app)	k_2/k_3	Titration value × 100% Using N-Trans-cinnamoyl-imidazole
p-Nitrophenyl N-benzyloxycarbonyl-L-tyrosinate	p-Nitrophenol	100	1000 (?)	97–101
p-Nitrophenyl N-acetyl-DL-tryptophanate	p-Nitrophenol	10–100	1000 (?)	91–99
p-Nitrophenyl acetate	p-Nitrophenol	1–10[b]	600	88–104
p-Nitrophenyl isobutyrate	p-Nitrophenol			~100 ± 10
p-Nitrophenyl trimethylacetate	p-Nitrophenol	10	2800	~100 ± 10
2,4-Dinitrophenyl acetate	p-Nitrophenol	30	600	101
N-trans-Cinnamoylimidazole	Titrant	100	4000	100
Phenylmethanesulfonyl fluoride	Protons	∞	∞	103
	^{14}C labeling of enzyme	∞	∞	103
3-Nitro-4-carboxyphenyl N,N-diphenyl-carbamate	3-Nitro-4-hydroxy-benzoic acid	∞	∞	100 ± 10
Diisopropyl phosphorofluoridate	Isopropyl groups of enzyme	∞	∞	100 ± 10
	P of enzyme	∞	∞	100 ± 10
	Protons	∞	∞	100 ± 10
Diethyl p-nitrophenyl phosphate	p-Nitrophenol			100 ± 10

[a] From M. L. Bender et al., *J. Amer. Chem. Soc.*, **88**, 5890 (1966). © 1966 by the American Chemical Society. Reprinted by permission of the copyright owner.

found to be identical,† within experimental error, confirming the coherence of the titrants of Table 12.4. Chymotrypsin can be stoichiometrically titrated by p-nitrophenyl N^2-acetyl-N^1-benzylcarbazate (a nitrogen analog of phenylalanine) at pH 7.04 and 25°. However, this reagent also titrates trypsin. More titrants can be expected although many exist at present. No more are needed but they will keep coming.

12.6 TRYPSIN

The stoichiometric titration procedure for the determination of the active sites of trypsin (trypsin, EC 3.4.4.4)[13] in terms of eq. 12.1 is based upon kinetic studies and isolation experiments.[21] However, unlike α-chymotrypsin, few stoichiometric procedures, other than the phosphorylation experiments, have been developed for trypsin. On the other hand, as with many enzymes, a host of rate assay procedures have been developed, involving denatured hemoglobin, denatured casein, α-N-benzoyl-L-arginine ethyl ester, α-N-benzoyl-L-argininamide, and α-N-benzoyl-L-lysinamide as substrates.[4,6] These rate assays suffer from the inadequacies of rate assays as discussed previously. Several titrants for the active sites of trypsin, based on specific substrates of the enzyme, have been developed.

Solutions of bovine and porcine trypsins have been titrated using six titrants and two methods of titration. The titrants include diethyl p-nitrophenyl phosphate, p-nitrophenyl acetate, p-nitrophenyl N-acetyl-L-leucinate, p-nitrophenyl N-acetyl-DL-tryptophanate, p-nitrophenyl N^2-benzyloxycarbonyl-L-lysinate hydrochloride,[36] and p-nitrophenyl p-guanidinobenzoate.[37] The methods of titration involve measurement of the "burst" of p-nitrophenol in the presteady state of the reaction and the use of second-order kinetics.

12.6.1 The Burst of p-Nitrophenol

The most extensive experiments determined the initial "burst" of p-nitrophenol liberated in the presteady state of the trypsin-catalyzed hydrolysis of the p-nitrophenyl esters. Representative results of the "burst" titrations of bovine trypsin with p-nitrophenyl N^2-benzyloxycarbonyl-L-lysinate are given in Fig. 12.3,[36] showing the dependence of the burst on the substrate concentration. The "bursts" in the reaction of trypsin with the other titrants (except diethyl p-nitrophenyl phosphate) are more heavily dependent upon

† This result is similar to the result of Hartley and Kilby[42] who demonstrated the identity of the inhibition of amino acid esterase, amidase, and proteolytic activities with that of p-nitrophenol release in the reaction of α-chymotrypsin and diethyl p-nitrophenyl phosphate.

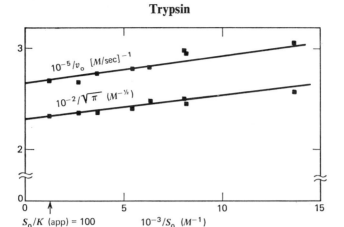

Fig. 12.3 The effect of substrate concentration on the burst titration ($1/\sqrt{\pi}$) and the turnover ($1/v$) of bovine trypsin with p-nitrophenyl N^2-benzyloxycarbonyl-L-lysinate hydrochloride, pH 3.71, 0.04 M formate. From M. L. Bender et al., *J. Amer. Chem. Soc.*, **88**, 5890 (1966). © 1966 by the American Chemical Society. Reprinted by permission of the copyright owner.

substrate concentration than the lysine "burst." A summary of the titration results for both bovine and porcine trypsin is given in Table 12.5. Using accessible concentrations of the titrants, p-nitrophenyl acetate, N-acetyl-L-leucinate, and N-acetyl-DL-tryptophanate, the burst of p-nitrophenol is far from the limiting burst because the ratios $[S]_0/K_m(\text{app})$ and/or k_2/k_3 are not sufficiently high.

The burst produced in the reaction of porcine trypsin and diethyl p-nitrophenyl phosphate can be shown to be stoichiometric. The kinetics show that $b = k_2[S]_0/K_s = 1.65 \times 10^{-5}$ sec^{-1} for reaction with 3.5×10^{-3} M phosphate and $k_3 = 10^{-4}$ sec^{-1}.[43] Thus from eq. 12.17 the enzyme concentration does not differ from the burst by more than 5%.

A near-optimal titrant of trypsin on the basis of its near-stoichiometric burst and its specificity is p-nitrophenyl N^2-benzyloxycarbonyl-L-lysinate hydrochloride. The titration of trypsin with this substance was investigated in detail. Figure 12.4[36] shows a typical titration of trypsin with this substance. Figure 12.3[36] shows that the burst of p-nitrophenol in the presteady-state of the titration reaction is not greatly affected by substrate concentration in the region of substrate concentration ordinarily used. Thus using a substrate concentration of 1.05×10^{-3} M, the usual concentration for the titration, and the data of Table 12.5, eq. 12.12 indicates that the ratio $\pi/[E]_0 = 0.909$. That is, under these conditions, the burst corresponds to 91% of the true enzyme concentration.[45]

Since solutions of bovine trypsin at pH 3 undergo an initial loss of activity,

Table 12.5
Titration of Trypsin with Some p-Nitrophenyl Esters[b]

p-Nitrophenyl Ester Of	$[S]_0 \times 10^5$ (M)	$K_m(\text{app}) \times 10^5$ (M)	k_2/k_3	pH	I	CH_3CN [%(v/v)]	Titrated $[E]_0 \times 10^6$ (M)	Enzyme (% Purity)[a]
Porcine trypsin								
Acetic acid	81–271	15.6	6	6.24		15	15.8	83
N-Acetyl-L-leucine	28–97	32		4.46	0.2	2.4	32.4	80
	16–42	25		4.55	0.2	1.6	32.4	84
N-Acetyl-DL-tryptophan	5.3–11.2	32		3.46	0.05	1.6	3.20	81
	5.3–11.2	7.1		3.32	0.5	1.6	5.50	80
N^2-Benzyloxycarbonyl-L-lysine	105			3.0	0.05	1.3	438	75
Diethylphosphoric acid	305			7.8	0.05	6.1	381	65
Bovine trypsin								
Acetic acid	10–124	7.7	15.2	7.16	0.5	1.6	22.4	69
	15–124	7.7	15.2	7.59	0.5	1.6	20.6	63
	15–124	7.7	15.2	8.00	0.5	1.6	16.9	64
	15–124	7.7	15.2	8.61	0.5	1.6	15.4	59
	28–73	23		4.49	0.02	2.4		51
N-Acetyl-L-leucine		131					181	50
N-Acetyl-DL-tryptophan		104					229	64
N^2-Benzyloxycarbonyl-L-lysine	5.9–29.3	7.9		3.46	0.05	1.6	1.95	49
	7–80	1.0		3.71	0.02	2.0	18.9	57
Lysine	105	1.27	27.6[1]	3.0	0.05	1.3	48	57.5

[a] Calculated as the percentage of the titrated $[E]_0$ of the $[E]_0$ determined on a weight basis assuming a molecular weight of 24,000.
[b] From M. L. Bender et al., *J. Amer. Chem. Soc.*, **88**, 5890 (1966). © 1966 by the American Chemical Society. Reprinted by

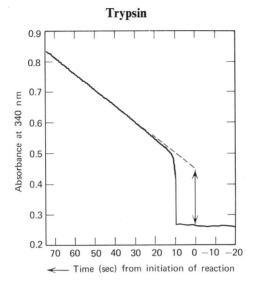

Fig. 12.4 The reaction of p-nitrophenyl N^2-benzyloxycarbonyl-L-lysinate hydrochloride with bovine trypsin: $(E)_0 = 3 \times 10^{-5}$ M, $(S)_0 = 1.05 \times 10^{-3}$ M, pH 3.0, citrate buffer (0.05 M), 1.3% (v/v) acetonitrile-water, 25°. From M. L. Bender et al., *J. Amer. Chem. Soc.*, **88**, 5890 (1966). © 1966 by the American Chemical Society. Reprinted by permission of the copyright owner.

as measured by the rate assay with α-N-benzoyl-L-arginine ethyl ester at pH 8, titrations of trypsin solutions after variable amounts of time were carried out at pH 3, using p-nitrophenyl N^2-benzyloxycarbonyl-L-lysinate as titrant. From 4-min incubation to 3-hr incubation at room temperature, no change in the burst was observed.

The rate of the trypsin-catalyzed hydrolysis of α-N-benzoyl-L-arginine ethyl ester is considerably increased in the presence of calcium ion, 10^{-2} M calcium ion giving nearly 100% increase in rate (partially an ionic strength effect). Therefore, the dependence of the titration on calcium ion was investigated. Bovine trypsin was prepared both in pH 2.92 citrate buffer and in pH 3.0–0.06 M calcium ion. The percentages of active trypsin determined in two titrations of the same trypsin stock using these two solutions were 57 and 60, respectively. Thus the effect of calcium ion on the burst is at best to increase the value 5%. The optimal titrant for trypsin appears to be p-nitrophenyl p-guanidinobenzoate[37] since there is a burst, but negligible turnover.

12.6.2 Second-order Conditions

Several experiments were performed to determine trypsin concentration in this manner, using N-acetyl-DL-tryptophan p-nitrophenyl ester and p-nitrophenyl N^2-benzyloxycarbonyl-L-lysinate hydrochloride as substrates.

These determinations were limited to the bovine enzyme since a large amount of enzyme is consumed in each experiment. Figure 12.5 shows a plot of $V/([S] - [P_1])$ for the tryptophan substrate. Although these experiments are subject to error from both spontaneous hydrolysis and deacylation of the acyl-enzyme, the data lead to results that are reasonably consistent with those of the burst titrations (see Table 12.5).

All titrations of Table 12.5 indicate that the bovine trypsin preparation used is 49–60% active enzyme. Those titrations of Table 12.5, where both the k_2/k_3 and $[S]_0/K_m$(app) (or k_b/k_3) ratios are known indicate that the bovine trypsin preparation contains 59–69% active enzyme and that the porcine trypsin preparation contains 65–83% active enzyme. The purity of porcine trypsin may be appreciably improved by gel filtration using Sephadex G-25 (see Section 12.11).

Although the precision in Table 12.5 between different titrants is not high, it is certainly reasonable considering the uncertainties of the experiments and different conditions used. The question may still be asked as to whether the various titrants of Table 12.5 reflect the same active site. Two approaches have been used to investigate this problem. The rate constant for the phosphorylation of porcine trypsin with diethyl *p*-nitrophenyl phosphate, as reflected by the release of *p*-nitrophenol, was compared to the rate constant for the inactivation of the enzyme, as reflected by the turnover of *p*-nitrophenyl N^2-benzyloxycarbonyl-L-lysinate at pH 6.55. The first-order rate constants of these two reactions were 1.77×10^{-4} and 1.65×10^{-4} sec^{-1},

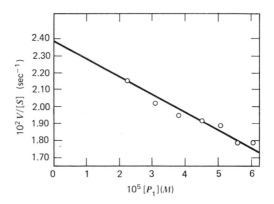

Fig. 12.5 The trypsin-catalyzed hydrolysis of *N*-acetyl-DL-tryptophan *p*-nitrophenyl ester under second-order conditions: $(S)_0 = 2.71$–7.43×10^{-5} M, $(E)_0 = 1.81 \times 10^{-4}$ M, pH 3.11, 25°, 0.05 M citrate buffer, 1.6% acetonitrile-water (v/v). From M. L. Bender et al., *J. Amer. Chem. Soc.*, **88**, 5890 (1966). © 1966 by the American Chemical Society. Reprinted by permission of the copyright owner.

respectively, indicating that the most specific titrant, the lysine ester, and the least specific titrant, the phosphate, react at the same site on the enzyme. Presumably the other intermediate titrants do also. At pH 6.28, the turnover of p-nitrophenyl N^2-benzyloxycarbonyl-L-lysinate with porcine trypsin is 92% inhibited by 3.26×10^{-6} M soybean trypsin inhibitor. This experiment indicates that at least 92% of the burst of p-nitrophenol using the lysine substrate as titrant must arise from the active site as defined by soybean trypsin inhibitor.[46] Thus the titrations of Table 12.5 appear to be independent of both pH and substrate.

For trypsin solutions, a convenient secondary standard is the rate assay based on α-N-benzoyl-L-arginine ethyl ester as substrate. This reaction has been investigated intensively, and found to obey Michaelis–Menten kinetics.[47] The K_m(app) for this reaction at pH 8 is 2×10^{-5} M so that at a concentration of 10^{-3} M, saturation conditions hold and the velocity may be expressed by $V_{max} = k_3[E]_0$. Using a trypsin titrant it is possible to determine $[E]_0$ and using the rate assay it is possible to determine V_{max}. Thus, the coefficient k_3 may be determined. The known k_3 may then be used in conjunction with V_{max} to determine the $[E]_0$ of an unknown solution. Using a trypsin solution incubated in 0.05 M citrate buffer, pH 3, for at least 20 min, and assay conditions of pH 8.0, 25.0°, and 10^{-3} M α-N-benzoyl-L-arginine ethyl ester, k_3 is found to be 1.32×10^4.

p-Nitrophenyl p-guanidinobenzoate is the most specific titrant of trypsin known today,[37] using as criterion that it has a high rate constant for p-nitrophenol production and negligible turnover.

12.7 PAPAIN

Papain (papain, EC 3.4.4.10)[13] is distinguished from the enzymes discussed so far by the fact that its active site contains a sulfhydryl group that acts as a nucleophile during the catalytic process, converting the substrate first into an intermediate acyl-enzyme, a thiol ester; the acyl-enzyme then reacts with water giving product carboxylic acid and regenerating the enzyme.[24] Although the important nucleophile of the active site is different from that of α-chymotrypsin and trypsin (SH vs. OH), the overall pathway of the papain reaction appears to follow eq. 12.1, which serves as the stoichiometric basis of the titration of the enzyme.

12.7.1 Burst Titration

The papain-catalyzed hydrolysis of p-nitrophenyl N-benzyloxycarbonyl-L-tyrosinate shows an initial "burst" of p-nitrophenol when the zero-order portion of the plot of absorbance of *p*-nitrophenol versus time is extrapolated to time zero. This observation is excellent confirmatory evidence for the mechanism of eq. 12.1.

A typical set of data for "bursts" at various substrate concentration is shown in Fig. 12.6 plotted according to eq. 12.13. Such plots were analyzed by a weighted least-squares method using [S]$_0$ as the weighting factor. The results of three sets of experiments are listed in Table 12.6.

Several semiquantitative arguments are presented that k_2/k_3 is much greater than unity in this reaction. (1) A survey of reported $K_m(K_s)$ data for papain substrates shows values ranging from 10^{-3} M and higher. (2) In the presence of 30% methanol, the turnover rate increases twofold, possibly because methanol is acting as a nucleophile in deacylation as it is known to do in the deacylation of *trans*-cinnamoyl-papain. (3) The substrate *p*-nitrophenyl *N*-acetyl-DL-tryptophanate shows a burst with papain, only the L species reacting. (4) A lower limiting value of the k_2/k_3 ratio can be calculated from the experimental observation that the presteady state is complete in the time that elapses between the addition of enzyme and the beginning of observation of the appearance of *p*-nitrophenol (see Table 12.6).

The lowest titrant concentrations in Table 12.6 are only some 60% greater than the enzyme normality in the cell, in violation of the assumption

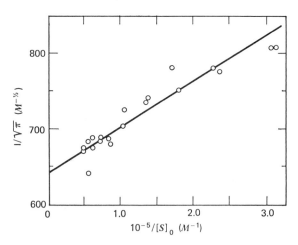

Fig. 12.6 The dependence of the *p*-nitrophenol "burst" on the initial substrate concentration in the titration of papain by *p*-nitrophenyl α-*N*-benzyloxycarbonyl-L-tyrosinate: 1.6% (v/v) acetonitrile-water, pH 3.18, 0.05 M potassium hydrogen phthalate, 25.0°, (E)$_0$ = 2.45 × 10^{-6} M. From M. L. Bender et al., *J. Amer. Chem. Soc.*, **88**, 5890 (1966). © 1966 by the American Chemical Society. Reprinted by permission of the copyright owner.

Table 12.6
The Titration of Papain Using p-Nitrophenyl N-Benzyloxycarbonyl-L-tyrosinate[a]

$[S]_0 \times 10^6$ (M)	$K_m \times 10^6$ (M)	k_2/k_3	$[E]_0^{expt} \times 10^6$ (M)	$[E]_0^{calcd} \times 10^6$ (M)	Enzyme % purity	$([E]_0/k_{BAEE}) \times 10^4$ (M sec)
4.0–21.5	0.63	100 ± 60	2.45 ± 0.03	6.22	39.4	8.78
3.4–21.8	0.63		1.67 ± 0.02	3.98	42.0	8.83
4.2–19.8	0.39	190 ± 110	2.49 ± 0.03	6.22	39.9	8.92
						8.84 av

[a] From M. L. Bender et al., *J. Amer. Chem. Soc.*, **88**, 5890 (1966). © 1966 by the American Chemical Society. Reprinted by permission of the copyright owner.

implicit in the derivation of eq. 12.13 that $[S]_0 \gg [E]_0$. Actually to be more correct, the necessary assumption is that the rate of change of the acyl-enzyme ($d[ES']/dt$) is much less than the rate of change of the substrate ($d[S]/dt$).† In the present experiments, since the presteady-state is essentially instantaneous with respect to the steady state ($b \gg k_3$), the substrate concentration rapidly diminishes by an amount equal to the concentration of acyl-enzyme formed, and then decreases in the steady-state reaction while the acyl-enzyme concentration remains sensibly constant.

12.7.2 Relationship Between the Titration and Rate Assay

The rate of the hydrolysis of α-benzoyl-L-arginine ethyl ester, a specific substrate for papain, was measured simultaneously with the titration experiments. Thus since $K_m(\text{app}) \gg [S]_0$, a pseudo-first-order reaction is observed whose rate constant is $k_{\text{BAEE}} = (k_{\text{cat}}/K_m(\text{app}))[E]_0$. This pseudo-first-order rate constant was shown to be proportional to $[E]_0$ and independent of the substrate concentration provided that 10^{-5} M ethylenediaminetetraacetic acid was present. In the absence of the chelating agent, there is a small decrease in k_{BAEE} as the initial substrate concentration increases, possibly due to a small metal ion impurity in the substrate solution. Using $[E]_0$ from the titrations, the ratio $(K_m(\text{app})/k_{\text{cat}})_{\text{BAEE}}$ (equal to $[E]_0/k_{\text{BAEE}}$) was determined as 8.84×10^{-4} M sec under the conditions denoted in Table 12.6. By this means the normality of any papain solution can be readily determined.

The observation that papain is only 58–75% pure reconciles the data of Smith and Parker[50] with those of Whitaker and Bender.[24] The former workers found a maximal k_{cat} (or k_0) for α-N-benzoyl-L-arginine ethyl ester of 9.0 sec^{-1} assuming a 100% pure enzyme preparation, whereas the latter, who were able to titrate their papain solutions, obtained a value of 16.0 sec^{-1}. These two values are in agreement if Smith's enzyme solutions were, in fact, only 56% pure, a reasonable value.

Presumably any stoichiometric reaction with the single sulfhydryl group of the active site may provide a titration for papain. In early studies of papain, Finkle and Smith[51] noted that papain reacted with less than 1 equivalent of p-chloromercuribenzoate, a sulfhydryl reagent, when the latter was in twofold excess. The fraction of sulfhydryl groups so titrated was observed to be approximately proportional to the activity of the enzyme preparation; thus this procedure appears to afford a titration of the active site of papain. However, this titration has the disadvantage that a given enzyme preparation

† The titration of α-chymotrypsin with *trans*-cinnamoylimidazole[32] illustrates this point; it is successful as long as $[S]_0$ is only slightly greater than $[E]_0$. The deacylation rate is relatively so small that this condition is met.

may contain protein molecules, or fragments thereof, which, though enzymatically inactive, may still contain free sulfhydryl groups capable of reacting with p-chloromercuribenzoate. Liener[53] found that his best preparations of ficin, an enzyme similar to papain, reacted with 0.90–0.95 equivalent of N-ethylmaleimide, based on an enzyme concentration calculated from a nitrogen analysis and a molecular weight of 26,000.[54] The release of iodide ion in the reaction of iodoacetamide with ficin[55] indicated that the preparation used was 77% pure as determined on a dry weight basis and a molecular weight of 26,000. However, these titrations suffer from the same disadvantages as those mentioned above for papain. Furthermore, the reactions of p-chloromercuribenzoate, N-ethylmaleimide, and iodoacetamide are undesirable for titrations because these substances are not related to papain substrates.

12.8 SUBTILISIN

The basis of a titration procedure for the determination of the active sites of subtilisin (subtilopeptidase A, EC 3.4.4.16)[13] in terms of eq. 12.1 is the stoichiometric reaction of the enzyme with diisopropyl phosphorofluoridate[56] and the isolation of the acyl-enzyme, $trans$-cinnamoyl-subtilisin.[23] The latter work[23] mentioned a titrimetric procedure for the active site of subtilisin using N-$trans$-cinnamoylimidazole, although no details were given.

12.8.1 Burst Titration

The subtilisin-catalyzed hydrolysis of N-$trans$-cinnamoylimidazole shows an initial (negative) "burst" in the disappearance of the substrate when the zero-order portion of the plot of absorbance versus time is extrapolated to time zero.[36] This observation is excellent confirmatory evidence for the mechanism of eq. 12.1. A typical set of data for the "burst" at various substrate concentrations and enzyme concentrations is given in Table 12.7.

No presteady-state reaction was observed in a titration when the substrate concentration greatly exceeded that of the enzyme. One can thus conclude on a qualitiative basis that $k_2 \gg k_3$. This same conclusion was also shown quantitatively.[36]

The value of K_m(app) determined from a Lineweaver–Burk plot of the steady-state reaction under the conditions of the titration experiment was 2.87×10^{-5} M. The substrate concentrations ordinarily used in titration experiments are 100-fold greater than this value. The calculated values of the enzyme concentration reported in Table 12.7 utilized eq. 12.12 with the

Table 12.7

The Titration of Subtilisin Using N-$trans$-Cinnamoylimidazole[a]

Titration	Enzyme scln	$[S]_0 \times 10^4$ (M)	π, abs unit	$[E]_{calc} \times 10^5$ (M)[b]	$[E]_{stock} \times 10^3$ (M)	% Purity[c]	$k_3 \times 10^2$ (sec^{-1})
I	1	1.89	0.507	6.02	1.93	57.1	1.37
II	1	1.98	0.263	3.10	1.95	57.7	1.51
III	2	2.02	0.371	4.38	1.40	57.6	1.44
IV	3	2.02	0.285	3.36	1.08	55.7	1.50
V	4	5.81	0.131	4.58	1.47 ± 0.07	—	1.47
VI	4	2.87	0.131 ± 0.0006	4.64	1.48 ± 0.07	—	1.45
VII	4	2.09	0.383 ± 0.0006	4.51	1.44 ± 0.02	—	1.53

[a] From M. L. Bender et al., *J. Amer. Chem. Soc.*, **88**, 5890 (1966). © 1966 by the American Chemical Society. Reprinted by permission of the copyright owner.
[b] Calculated.
[c] Based on the weight of enzyme used and a molecular weight of 27,400.

known $K_m(\text{app})/[S]_0$ ratios for each experiment. The use of this ratio usually amounted to a factor of about 3.2% and in no case more than 4.2% in the overall calculation.

The derivation of eq. 12.12 requires that $[S]_0 \gg [E]_0$. Under the titration conditions $[S]_0/[E]_0$ was usually 5–10, which does not meet this requirement completely. Since the substrate concentration does not change appreciably during the period of the burst, the $[S]_0$ listed in Table 12.7 and used in eq. 12.12 is the concentration of substrate after the burst. Since the k_2/k_3 ratio is so high in this titration, there is no theoretical ambiguity connected with the fact that the $[S]_0/[E]_0$ ratio is not very high.†

N-*trans*-Cinnamoylimidazole is certainly not a "natural" substrate for subtilisin. In addition, it is a fairly reactive acylating agent and is used in large excess. Therefore it was necessary to determine whether N-*trans*-cinnamoylimidazole reacts with the active site which is operative for other (natural) substrates of subtilisin and with no other groups.

Subtilisin has a broad specificity toward acylamino acid derivatives. When subtilisin is titrated with N-*trans*-cinnamoylimidazole at pH 7, a *trans*-cinnamoyl intermediate is formed.[23] Such a reaction was carried out, the pH of the solution was immediately lowered to pH 4.0 (from 7.0), and the solution was passed through a Sephadex G-25 column. This treatment gave a solution of *trans*-cinnamoyl-subtilisin separated from excess substrate and any hydrolysis products. When an aliquot of this acyl-enzyme was added to a solution containing p-nitrophenyl N-benzyloxycarbonyl-L-valinate and the release of p-nitrophenol observed, a sigmoid curve reflecting a first-order approach (deacylation of the cinnamoyl-enzyme) to the first-order ester hydrolysis reaction should be observed, if the enzyme is indeed cinnamoylated at the same active site as that responsible for the ester hydrolysis. The rate constant for the approach to the ester hydrolysis turnover should equal k_3, the deacylation rate constant for *trans*-cinnamoyl-subtilisin, according to eq. 12.18. This phenomenon was observed and the rate constant so obtained was $1.52 \pm 0.15 \times 10^{-2}$ sec^{-1}, which agrees within experimental error with the value of k_3 determined directly (1.45×10^{-2} sec^{-1}). Thus the observed "burst" in the reaction of subtilisin with N-*trans*-cinnamoylimidazole represents acylation of the same active site as that involved in the hydrolysis of a substrate containing a naturally occurring amino acid.

In the above discussion, the applicability of eq. 12.1 has been assumed. There is, however, one piece of information that may be interpreted as being inconsistent with eq. 12.1; namely only a questionable saturation was observed in the presteady-state reaction. Thus, the experimental observations on this titration reaction may require that eqs. 12.15–12.17 be utilized. However,

† See the discussion in Section 12.7.1.

the above data also satisfy the requirements of eq. 12.17 for a rigorous titration.

12.8.2 The Use of Rate Assay as Secondary Standard

Only stock solutions of rather high enzyme concentration ($\sim 1 \times 10^{-3} M$) can be directly titrated by the procedure given above. Since more dilute stock solutions are frequently desired, it is convenient to have a secondary standard that can establish the concentrations in more dilute solutions. The reaction of *N-trans*-cinnamoylimidazole with subtilisin is, in fact, both a titration and rate assay. The turnover constant k_{cat} ($=k_3$ in this instance since $k_2 \gg k_3$) has been determined under the conditions of the titration by dividing the velocity of the steady-state reaction by the enzyme concentration determined from the burst, as shown in Table 12.6. Using $k_3 = 1.45 \times 10^{-2} \sec^{-1}$, the enzyme concentration can then be determined from the equation $V_{max} = k_3[E]_0$. If the activity of the enzyme is not affected by dilution, the enzyme concentration calculated from the rate assay should equal the enzyme concentration calculated by dilution of the titrated stock enzyme solution. Such an identity does, in fact, hold over at least a fortyfold range in enzyme concentration from 1.7×10^{-6} to $4.8 \times 10^{-5} M$.

The titration indicates the presence of approximately 57% active enzyme and 43% (w/w) inert material in a typical sample of crystalline subtilisin. The possibility that other enzymatically active components are present in this material has not been excluded. The pronounced tendency of subtilisin to autolyze during purification has been noted by several workers[57] and it is possible that the inert material consists largely of autodegradation products.

12.9 ELASTASE

The basis of a titration procedure for the determination of the active sites of elastase (pancreatopeptidase E, EC 3.4.4.7)[13] is inhibition of elastase by diisopropyl phosphorofluoridate[58] and the observation of an initial burst of *p*-nitrophenol in the elastase-catalyzed hydrolysis of *p*-nitrophenyl isobutyrate and trimethylacetate (see Section 12.9.2).

12.9.1 Burst Titration

With the other enzymes reported in this series, stoichiometric titrations, corresponding to step k_2 of eq. 12.1, have been observed with specific and/or nonspecific substrates of the enzyme. With elastase, a carboxylic ester has not as yet been found that is satisfactory as a titrant. However, *p*-nitrophenyl diethyl phosphate (Paraoxon), which was shown to react with chymotrypsin and trypsin resulting in a completely inhibited enzyme and the liberation of approximately 1 mole of p-nitrophenol per mole of enzyme,[42, 59] has been used as a titrant of the active sites of elastase. Elastase reacts with excess diethyl p-nitrophenyl phosphate at pH 7.8 in a reaction that is characterized by a first-order liberation of p-nitrophenoxide ion (Fig. 12.2). Before the reaction takes place and after the reaction has reached five half-lives, a slow spontaneous zero-order hydrolysis of diethyl p-nitrophenyl phosphate is evident; the rates of the spontaneous hydrolysis before and after reaction are identical within the experimental error of 4%. That the reaction is essentially irreversible was shown by inhibition of elastase with excess diethyl p-nitrophenyl phosphate, separation of the inhibited enzyme from the excess reactant by gel filtration using a Sephadex G-25 column, and observation of the activity of the enzyme toward p-nitrophenyl trimethylacetate. At the conclusion of the gel filtration, the enzyme had less than 1% of the original activity. After 65 hr at pH 7.8, the inhibited enzyme had gained no activity (less than 5%) while a blank, uninhibited enzyme still retained 35% of its activity

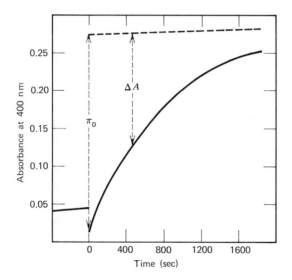

Fig. 12.7 Titration of elastase with diethyl p-nitrophenyl phosphate: pH 7.74 tris-HCl buffer, 7.6% (v/v) acetonitrile-water, $25.0 \pm 0.2°$, $(E)_0 = 1.7 \times 10^{-5}\ M$, $(S)_0 = 7.48 \times 10^{-3}\ M$. The dashed line is extrapolated from the spontaneous hydrolysis of the phosphate after more than eight half-lives of the reaction. Time zero is the time of addition of the enzyme. At negative times the spontaneous hydrolysis of the phosphate alone is seen. The noise level is 0.002 absorbancy unit. From M. L. Bender et al., *J. Amer. Chem. Soc.*, **88**, 5890 (1966). © 1966 by the American Chemical Society. Reprinted by permission of the copyright owner.

toward *p*-nitrophenyl trimethylacetate. These results indicate that there is probably no "turnover" of diethyl *p*-nitrophenyl phosphate during the course of a titration experiment.

By plotting the logarithm of ΔA, the difference between the absorbance at any time and the absorbance extrapolated from the spontaneous hydrolysis of the phosphate after the reaction is complete (see Fig. 12.7), versus time, the pseudo-first-order rate constant, b, of the reaction may be obtained (Fig. 12.8). In addition, by extrapolating this plot to zero time, π, a measure of the *p*-nitrophenol liberated in this reaction, and thus a measure of the enzyme consumed in this reaction, may be obtained.

The *p*-nitrophenol produced from the reaction of diethyl *p*-nitrophenyl phosphate with elastase is proportional to the concentration of enzyme protein present over a tenfold range, while the pseudo-first-order rate constant

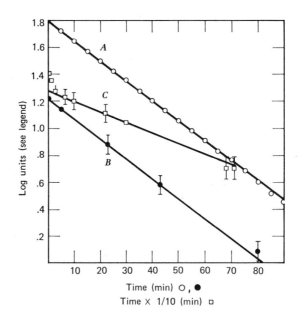

Fig. 12.8 The reaction of diethyl *p*-nitrophenyl phosphate with elastase ESFF 6501 (fraction 29) chromatographed on carboxymethylcellulose and dialyzed against pH 7.8 ± 0.03 phosphate buffer (I = 0.1), $(E)_0 = 4 \times 10^{-5}$ M, $(S)_0 = 2.2 \times 10^{-3}$ M, 3.85% (v/v) acetonitrile-water: A, release of p-nitrophenol, ordinate, log ($10^4 \triangle A$); B, loss of activity of enzyme toward *p*-nitrophenyl isobutyrate, ordinate log ($10^2 V_{net}$) when 100 uliters of phosphate solution was assayed; C, loss of activity of enzyme toward *p*-nitrophenyl *N*-benzyloxycarbonyl-L-tyrosinate, ordinate log ($10^3 V_{net}$) when 200 μliters of phosphate solution was assayed. From M. L. Bender et al., *J. Amer. Chem. Soc.*, **88**, 5890 (1966). © 1966 by the American Chemical Society. Reprinted by permission of the copyright owner.

of the reaction is independent of the enzyme protein concentration over this range. On the other hand, the pseudo-first-order rate constant of the reaction is directly proportional to the phosphate concentration. Both of these results are presented in Table 12.8.

Since the diethyl *p*-nitrophenyl phosphate-elastase reaction is stoichiometric within experimental error, π is a direct measure of the normality of active sites of elastase in the solution. Table 12.8 records the results of a number of such reactions. These experiments give results that are quite consistent with one another. The purity of the three enzyme preparations ($[E]_0$(expt)/$[E]_0$(calcd)) (listed in Table 12.9) is the same within experimental error ($\pm 10\%$ maximum deviation). The pseudo-first-order rate constants for the reactions are reasonably consistent considering the unknown effects of different acetonitrile concentrations and phosphate concentrations.

The titration of elastase with diethyl *p*-nitrophenyl phosphate follows eqs. 12.15–12.17. The condition that $[S]_0 \gg [E]_0$ is easily met by the design of the experiments.

12.9.2 The Relationship of the Reaction of Elastase with Diethyl *p*-Nitrophenyl Phosphate to other Reactions of Elastase

The question may be asked as to whether the titration occurred at—and only at—the active site of the enzyme. A determination was made of the rate of abolition of enzymatic activity toward different substrates during the phosphate inhibition. Three substrates were used: *p*-nitrophenyl trimethylacetate, isobutyrate, and *N*-benzyloxycarbonyl-L-tyrosinate. The results of

Table 12.8
Effect of Enzyme and Substrate Concentrations on the Reaction of Elastase with Diethyl *p*-Nitrophenyl Phosphate[a]

Enzyme Stock Soln (μliters)	pH	Phosphate × 10^3 (M)	π, abs unit	Enzyme stock[b] × 10^4 (M)	b × 10^3 (sec^{-1})
50	7.80	7.48	0.0554	2.11	1.19
250	7.74	7.48	0.250	2.02	1.21
200	7.79	7.48	0.215	2.06	1.24
25	7.87	7.48	0.0253	1.88	1.14
50	7.92	2.99	0.0522	1.99	0.501

[a] From M. L. Bender et al., *J. Amer. Chem. Soc.*, **88**, 5890 (1966). © 1966 by the American Chemical Society. Reprinted by permission of the copyright owner.
[b] Calculated from observed burst of *p*-nitrophenol.

Table 12.9

The Reaction of Elastase with Diethyl p-Nitrophenyl Phosphate[a]

Enzyme	CH_3CN [% (v/v)]	(Phosphate) $\times 10^3$ (M)	pH	π (abs unit)	$[E] \times 10^5$ (M)	$\dfrac{[E]_0 \text{(exptl)}}{[E]_0 \text{(calcd)}}$	$b \times 10^4$ (sec^{-1})
ESFF 6501	3.85	2.2	~7.8	0.65	5.2	0.79	5.75
Crystalline	3.85	2.2	7.77	0.72	5.6	0.83	5.56
ESFF 5691	7.6	7.48	7.80±0.1	0.025	20.2	0.78	12.0

[a] From M. L. Bender et al., *J. Amer. Chem. Soc.*, **88**, 5890 (1966). © 1966 by the American Chemical Society. Reprinted by permission of the copyright owner.

Table 12.10
Kinetics of the Loss of Activity of Elastase during Reaction with Diethyl p-Nitrophenyl Phosphate[a]

Enzyme	ESFF 5691	ESFF 6501
k(release of p-nitrophenol from diethyl p-nitrophenyl phosphate), sec^{-1}	1.22×10^{-3}	5.7×10^{-4}
k(loss of activity toward p-nitrophenyl trimethylacetate), sec^{-1}	1.3×10^{-3}	
k(loss of activity toward p-nitrophenyl isobutyrate), sec^{-1}		5.5×10^{-4}
k(loss of activity toward N-benzyloxycarbonyl-L-tyrosine p-nitrophenyl ester), sec^{-1}	5.6×10^{-5}	$6 \pm 2 \times 10^{-4}$ 3.67×10^{-5}

[a] From M. L. Bender et al., *J. Amer. Chem. Soc.*, **88**, 5890 (1966). © 1966 by the American Chemical Society. Reprinted by permission of the copyright owner.

these experiments are shown in Table 12.10 and partially illustrated in Fig. 12.8.

The rate constant of the liberation of p-nitrophenol in the reaction of elastase with diethyl p-nitrophenyl phosphate is equivalent to the rate constant of the loss of activity of elastase toward p-nitrophenyl trimethylacetate and isobutyrate (Table 12.10). The rate constant of the loss of activity of α-chymotrypsin and trypsin toward p-nitrophenyl N-benzyloxycarbonyl-L-tyrosinate also equals the rate constant of the reaction of these enzymes with diethyl p-nitrophenyl phosphate.

The second-order rate constant for the elastase-catalyzed hydrolysis of a considerable number of substrates of elastase was also determined, using two enzyme preparations. Values of $k_{cat}[E]_0/K_m$(app) were determined either as observed pseudo-first-order rate constants or from Lineweaver–Burk plots. Since $[E]_0$ is known by titration, k_{cat}/K_m(app) is readily evaluated. As seen in Table 12.11, the k_{cat}/K_m(app) determined for a given substrate with the two enzyme preparations were quite similar except for p-nitrophenyl N-benzyloxycarbonyl-L-tyrosinate. This result indicates that the titration of the active sites of elastase with diethyl p-nitrophenyl phosphate gives a consistent measure of the activity of the enzyme toward seven substrates of elastase, including both acylamino acid and carboxylic acid derivatives.

12.9.3 The Use of a Rate Assay as a Secondary Standard

Any of the reactions shown in Table 12.11, which have been shown to give concordant results of k_{cat} or k_{cat}/K_m(app) with two different enzyme preparations (with the exception of p-nitrophenyl N-benzyloxycarbonyl-L-tyrosinate), can be used as a rate assay to determine enzyme concentration.

Table 12.11

The Kinetics of Some Elastase-Catalyzed Reactions Using Two Elastase Preparations[a]

p-Nitrophenyl ester	pH	$[E]_0 \times 10^6$ (M)	$[S]_0 \times 10^5$ (M)	Kinetics	k_{cat}/K_m (app) $\times 10^{-8}$ (M^{-1} sec^{-1}) Enzyme 1[b]	k_{cat}/K_m (app) $\times 10^{-8}$ (M^{-1} sec^{-1}) Enzyme 2[c]
Trimethylacetate	7.33	5	5.5	MM[d]	0.099	0.123
Acetate	7.43	7.0	16	MM[d]	0.47	0.41
Isobutyrate	7.68	0.7–2.6	0.7–15	MM[d]	1.21	1.62
Furoate	7.69	0.5–2.6	0.1–2.0	MM[d]	2.35	1.69
N-Benzyloxycarbonyl-L-isoleucinate	7.70	0.5–2.9	0.53	First-order	0.296	0.250[b]
N-Benzyloxycarbonyl-L-leucinate	7.79	0.1–0.5	0.5	First-order	28.7	30.4
N-Benzyloxycarbonyl-glycinate	7.85	9.6	1.0	First-order	15.2	13.3
N-Benzyloxycarbonyl-L-tyrosinate	7.79	0.53–5.6	0.95	First-order	19.3	0.41

[a] From M. L. Bender et al., *J. Amer. Chem. Soc.*, **88**, 5890 (1966). © 1966 by the American Chemical Society. Reprinted by permission of the copyright owner.
[b] Enzyme 1 was Worthington electrophoretically purified elastase.
[c] Enzyme 2 was Worthington crystalline elastase purified by chromatography on carboxymethyl cellulose at pH 8.0.
[d] Michaelis Menten.

Using the equations $V_{max} = k_{cat}[E]_0$ or $k_{obsd} = (k_{cat}/K_m)[E]_0$ and the values of k_{cat} or k_{cat}/K_m(app) for a particular substrate under the conditions of reaction shown in Table 12.11, $[E]_0$ may be calculated directly from V_{max} or k_{obsd}.

Elastase appears to contain one active site per molecule since the molecular weight determined by sedimentation (25,000)[60] agrees roughly with the equivalent weight determined by diisopropyl phosphorofluoridate. While other serine enzymes would presumably react with diethyl p-nitrophenyl phosphate, the titration of elastase is still valid when the other enzymes are present in trace amounts.

12.10 ACETYLCHOLINESTERASE

The enzyme acetylcholinesterase (acetylcholine acetylhydrolase, EC 3.1.1.7)[13] serves physiologically as a catalyst for the hydrolysis of acetylcholine. However, acetylcholinesterase is similar to the proteolytic enzymes in this series in that its kinetic pathway also follows eq. 12.1.[65]

The following evidence supports this pathway. The enzyme is inhibited by certain organophosphorus compounds (e.g., tetraethyl pyrophosphate and diisopropyl phosphorofluoridate)[66] by forming a phosphoryl–enzyme.[67] As with the other enzymes of this series, the phosphoryl moiety is isolated on a serine hydroxyl group after the degradation of the protein.[68] The phosphoryl enzyme can be reactivated by nucleophilic compounds.[69] Dimethylcarbamylcholine and dimethylcarbamyl chloride also inhibit the enzyme.[70] Both carbamylated enzymes regain their activity in water at the same rate and are also reactivated by hydroxylamine at the same rate.[70] These results indicate the formation of acylenzyme in the enzyme-catalyzed hydrolysis of these substrates. Finally, noncompetitive inhibition[71] and pH dependency[74] studies using acetate esters are consistent with the three-step kinetic equation involving the formation of an acyl-enzyme intermediate in the reaction pathway.

Three titrations of acetylcholinesterase have been made on the basis of the stoichiometry of eq. 12.1. The enzyme has been phosphorylated with ^{32}P-labeled inhibitor from which the concentration of active sites could be calculated.[67] Similar titrations were carried out using diethoxyphosphorylthiolcholine and tetraethyl pyrophosphate.[73] In both these instances possible nonspecific phosphorylation was taken into account. A different approach utilized dimethylcarbamyl fluoride as substrate. The ratio of the activity of the carbamylated enzyme solution compared to the original activity of the enzyme was followed nearly to completion of the carbamylation and decarbamylation reactions; from this determination and the decarbamylation rate constant, the normality of the enzyme solution could be evaluated.[74]

Here is described the direct spectrophotometric observation of a stoichiometric reaction in the eel acetylcholinesterase-catalyzed hydrolysis of a carboxylic ester and its use in the development of a procedure for the determination of a concentration (normality) of the active site of this enzyme o-Nitrophenyl dimethylcarbamate was used as the substrate.[75]

12.10.1 Titrations

The titrations of eel cholinesterase by o-nitrophenyl and p-nitrophenyl dimethylcarbamates are shown in Table 12.12. All the titrations except no. 5 are characterized by an initial rapid release of o-nitrophenol followed by a slow steady-state release of the phenol. The initial release of nitrophenol was usually of the order of 20×10^{-4} absorbance units. Titration 6 is shown in Fig. 12.9. For all titrations using o-nitrophenyl dimethylcarbamate, $k_2/k_3 \geq 4.00$.

The $K_m(\text{app})$ for the acetylcholinesterase-catalyzed hydrolysis of o-nitrophenyl dimethylcarbamate, determined under conditions similar to those of

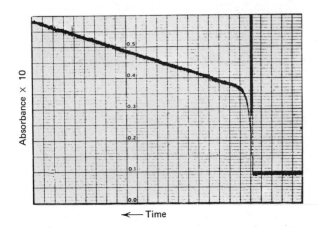

Fig. 12.9 Titration of acetylcholinesterase with o-nitrophenyl dimethylcarbamate at pH 7.70. The extrapolation of the absorbance to time zero, which is represented by the initial rise in absorbance, corresponds to an enzyme concentration of 7.45×10^{-6} M. The initial presteady-state reaction of approximately 1 min is followed by a steady-state (zero-order) reaction. The abscissa is 60 sec/division. From M. L. Bender et al, *J. Amer. Chem. Soc.*, **88**, 5890 (1966). © 1966 by the American Chemical Society. Reprinted by permission of the copyright owner.

the titrations, are listed in Table 12.13. By comparing the titrant concentrations in Table 12.12 and the $K_m(\text{app})$'s of Table 12.13 the $[S]_0/K_m(\text{app})$ ratio may be shown to be greater than 100 for all titrations.

As seen in Table 12.12, the ratio $[S]_0/[E]_0 > 100$ for all the titrations and therefore the requirement that $[S]_0 \gg [E]_0$ is satisfied.

In the steady-state turnover of the titration experiments, $k_2 \gg k_3$ and $[S]_0 \gg K_m(\text{app})$ (except for titration 5). Thus $V = k_3[E]_0$. The values of k_3 so determined are shown in Tables 12.12 and 12.13.

The enzyme preparations used in these titrations consisted of approximately 7% active enzyme based on a comparison of the activity per milligram of protein determined by Kremzner and Wilson[77] on the purest preparation to date and the activity per milligram of protein of the present lot 23B-7587.†

The titration with o-nitrophenyl dimethylcarbamate can be carried out from approximately pH 5 to 10, spanning the entire range where the enzyme is stable.‡ The lower limit of enzyme concentration usable in this titration is

---

† The activity reported by Kremzner and Wilson was corrected for substrate inhibition and pH.

‡ The k_{OH^-} of o-nitrophenyl dimethylcarbamate is quite low (8.7×10^{-5} $M\,\text{sec}^{-1}$ at 25.0°), leading to negligible spontaneous hydrolysis in these experiments.

Table 12.12

The Titration of Acetylcholinesterase Solutions with o-Nitrophenyl Dimethylcarbamate (Numbers 1–4 and 6–8) and p-Nitrophenyl Dimethylcarbamate (Number 5)[a]

Number	[Titrant] $\times 10^4$ (M)	pH	Dilution Factor	$\pi = [E]_0 \times 10^6$	$V \times 10^{10}$ (M sec^{-1})	$k_3 \times 10^4$ (sec^{-1})
1	6.64	8.20	3.750	2.0 ± 0.1	9.2 ± 0.28	4.8 ± 0.4
2	6.87	8.20	4.833	1.5 ± 0.1	6.8 ± 0.14	4.6 ± 0.4
3	10.03	7.73	3.750	3.6 ± 0.15	19.9 ± 0.6	5.5 ± 0.4
4	5.02	7.72	3.750	3.6 ± 0.15	22.5 ± 0.3	6.2 ± 0.3
5	10.33	7.78	7.500	1.8 ± 0.08	5.15 ± 0.13	
6	5.94	7.70	1.100	7.45 ± 0.13	48.9 ± 0.4	6.57 ± 0.17
7	6.01	9.00	1.375	3.46 ± 0.09	21.0 ± 0.2	6.08 ± 0.2
8	5.89	7.01	1.1000	1.4 ± 0.4	6.77 ± 0.07	3.9 ± 0.2

[a] From M. L. Bender et al., *J. Amer. Chem. Soc.*, **88**, 5890 (1966). © 1966 by the American Chemical Society. Reprinted by permission of the copyright owner.

Table 12.13

The Values of K_m[a] for the Acetycholinesterase-Catalyzed Hydrolysis of o-Nitrophenyl Dimethylcarbamate and a Comparison of the Values of k_3 for the Deacylation of the Dimethylcarbamyl-Enzyme Determined by the Rate Assay Method[b] with the Values Determined by the Titrations[a]

From the Rate Assay Method			From the titrations in Table 12.12	
pH	$k_3 \times 10^4$ (sec^{-1})	K_m(app) $\times 10^6$ (M)	$k_3 \times 10^4$ (sec^{-1})	pH
7.85	6.7 ± 0.3		6.2 ± 0.3	7.72
7.98	7.3–7.6		4.8 ± 0.4	8.20
9.24	6.0 ± 0.1		6.1 ± 0.2	9.00
7.03	3.5 ± 0.0		3.9 ± 0.2	7.01
7.00	4.3, 3.2			
7.04		2.9 ± 0.1		
7.82		5.4 ± 0.2		
7.82		3.0 ± 0.1		
9.16		4.5 ± 0.1		

[a] From M. L. Bender et al., *J. Amer. Chem. Soc.*, **88**, 5890 (1966). © 1966 by the American Chemical Society. Reprinted by permission of the copyright owner.
[b] o-Nitroacetanilide was used to assay the enzyme-catalyzed hydrolysis of o-nitrophenyl dimethylcarbamate.[76]

446 Enzymes: Classification and Determination

about 10^{-6} M, where the uncertainty would be about 20% if the titration was carried out above pH 8 in order to obtain the maximal spectral change.

The hydrolyses of o-nitrophenyl and p-nitrophenyl dimethylcarbamates are not catalyzed by chymotrypsin, subtilisin, trypsin, elastase, or papain, using 10^{-5} M enzyme solution at pH 7–8 for 15–30 min. On the other hand, pig liver[78] and ox liver[79] carboxylesterases catalyze the hydrolyses of o- and p-nitrophenyl dimethylcarbamates. In fact, titrations of these enzymes using these substrates have been developed.[78,79]

12.10.2 Rate Assay Using Phenyl Acetate

A spectrophotometric rate assay has been developed for determining enzyme concentration down to at least 5.5×10^{-11} M, using a value of k_{cat} based on the titration described here. Eadie plots were used to determine V_m, the maximal velocity. An example is given in Fig. 12.10. The rate assay was carried out for every titrated enzyme sample. The acetylcholinesterase-catalyzed hydrolysis of phenyl acetate follows Briggs-Haldane kinetics up to 10^{-2} M where substrate inhibition becomes apparent. Thus $k_{cat} = V_m[E]_0$. Using this relationship and the values for $[E]_0$ and V_m, k_{cat} can be calculated, as shown in Table 12.14. The values of k_{cat}(lim) calculated from the various

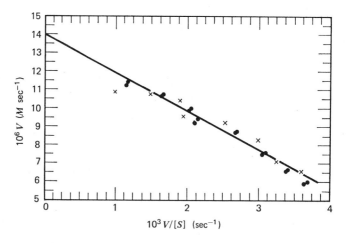

Fig. 12.10 Eadie plots for the acetylcholinesterase-catalyzed hydrolysis of phenyl acetate for rate assay 6: ×, first determination of the initial rates; ●, second determinations carried out 1 week later. From M. L. Bender et al., *J. Amer. Chem. Soc.*, **88**, 5890 (1966). © by the American Chemical Society. Reprinted by permission of the copyright owner.

Table 12.14

Kinetics of the Acetylcholinesterase-Catalyzed Hydrolysis of Phenyl Acetate[a]

Rate Assay	$V_m \times 10^7$ (M sec^{-1})	$[E]_0 \times 10^6$ Before Dilution	Dilution Factor $\times 10^{-3}$	$[E]_0 \times 10^{11}$ After Dilution	$k_{cat} \times 10^{-4}$ (sec^{-1})	K_m(app) $\times 10^3$ M	(pH)	k_{cat}(lim) $\times 10^{-4}$ (sec^{-1})
1–2	10.0 ± 0.3	7.4	122	6.07 ± 0.3	1.65 ± 0.14	1.9	7.62	1.77 ± 0.15
3–5	9.0 ± 0.2	13.5	244	5.53 ± 0.23	1.63 ± 0.10	1.8	7.82	1.71 ± 0.11
6	140 ± 2	8.21	10.32	79.4 ± 1.4		2.1	8.57	1.76 ± 0.06
7	168 ± 2	4.76	5.16	92.2 ± 2.4	1.76 ± 0.06	2.2	8.52	1.82 ± 0.07
8	65.1 ± 1	1.91	5.16	37.0 ± 0.18	1.82 ± 0.07	1.9	8.52	

[a] From M. L. Bender et al., *J. Amer. Chem. Soc.*, **88**, 5890 (1966). © 1966 by the American Chemical Society. Reprinted by permission of the copyright owner.

titrations agree well with those calculated from their corresponding rate assays, the greatest difference being about 6%. The rate assays were determined on enzyme solutions from 5.5×10^{-11} N to 9.2×10^{-10} N, and on enzyme solutions which had been diluted from 5160 to 244,000 times from solutions used in the titrations. The agreement among the k_{cat}(lim) values indicates that the enzyme can be diluted quantitatively to this low concentration range, and indicates that in this range any possible association-dissociation phenomena of the enzyme do not affect k_{cat}(lim).

Since k_{cat}(lim) for the acetylcholinesterase-catalyzed hydrolysis of phenyl acetate is found to be 1.76×10^4 sec^{-1}, the concentration of an enzyme solution can be determined by this rate assay, using the relationship, $[E]_0 = V_m/k_{cat}$(lim). The rate assay may be carried out over the pH range 6.4–9.0 where the pH dependency of k_{cat} is given by the equation

$$k_{cat} = k_{cat}(\text{lim}) \times (1 + H^{\oplus}/K)$$

Rate assays have been reported for the acetylcholinesterase-catalyzed hydrolysis of acetylcholine, leading to a k_{cat} of 1.4×10^4 sec^{-1} at pH 7, 0.1 M NaCl, 0.01 M MgCl$_2$ and 0.02 M sodium phosphate at 25.0°, and a k_{cat} of 1.2×10^4 sec^{-1} at pH 7.4, 38°, 3.4×10^{-2} M MgCl$_2$, 2.5×10^{-2} M NaHCO$_3$, and 1.5×10^{-2} M substrate.[67] Since the acetylcholinesterase-catalyzed hydrolyses of acetylcholine and of phenyl acetate are reported to give k_{cat}'s within 10% of one another,[72] they may be compared. The first of the acetylcholine rate constants gives a k_{cat}(lim) of 1.7×10^4 sec^{-1}, which is in good agreement with the rate constant of phenyl acetate reported here.

12.11 SUMMARY

The titrations described above are based on the mechanistic pathways of eq. 12.1 or 12.15. If these equations do not describe the titration reactions completely (e.g. if another intermediate exists in finite amount) then, of course, the titration is no longer quantitatively valid. However, all the burst titrations described here may be quantitatively described by eq. 12.1 or 12.15, and are considered to be valid. Furthermore, the titrations meet the criteria for a rigorous determination of the concentration of active sites in terms of eqs. 12.12 and 12.17. These criteria require the determination of k_2 or k_{obsd}, k_3 and K_m(app), as well as the knowledge of $[S]_0$ and $[E]_0$. The former constants have been determined in a variety of ways. The constant k_2 was determined by: (1) measurement of the rate of loss of enzyme activity by an external assay (acetylcholinesterase); (2) the determination of b and from it an estimate of k_2 according to eq. 12.22 (acetylcholinesterase and subtilisin); (3) an analysis of the presteady-state at different substrate concentrations (chymotrypsin);[29, 35] (4) an analysis of both the presteady- and steady-state at different substrate concentrations (trypsin).[21] The constant k_3 was ordinarily determined from the steady-state reaction (k_{cat}), knowing that $k_2 \gg k_3$ and thus that $k_{cat} = k_3$ (acetylcholinesterase, elastase, subtilisin, and chymotrypsin).[29,35] For trypsin, a complete analysis of both the presteady-state and steady-state was used.[21] The constant b was determined directly. For papain, the k_2/k_3 ratio was considered directly by: (1) the use of eq. 12.18, (2) the absolute magnitude of K_m(app), (3) the effect of methanol on the kinetics, and (4) the agreement of the bursts from two titrants of different k_2/k_3 ratio.

The titrations of α-chymotrypsin, trypsin, and papain utilized specific substrates of these enzymes, while the titrations of elastase, subtilisin, and acetylcholinesterase did not. However, the titrations of each of the latter enzymes have been shown to involve the active site of the enzyme, as defined by a specific substrate. The titrations of α-chymotrypsin and trypsin appear to be independent of titrant, over a wide range of specificity, and of pH of the titration. The titration of papain gives identical results with two substrates. The titration of elastase gives the same results with p-nitrophenyl trimethylacetate[80] as with the diethyl p-nitrophenyl phosphate. All of these identities indicate the determination of a fundamental quantity, namely the intact active site. Thrombin has also been titrated, as have many other enzymes, by the use of p-nitrophenyl N-benzyloxycarbonyl-L-tyrosinate and N-benzyloxycarbonyl-L-lysinate.[87]

The optimal titrant should (1) be a specific substrate but have little turnover, (2) give a titration in a few minutes in order to obviate denaturation, and for convenience, (3) give a titration around neutrality in order to reproduce physiological conditions, (4) be a stable, available, and soluble reagent whose stoichiometric reaction is easily detectable, and (5) give a titration over a wide range of enzyme concentrations. Not one of the titrants described fulfills all of the requirements listed above. However, each fits these requirements most closely for a given enzyme.†

12.11.1 Impurities

It is the thesis of this chapter that the concentration of active enzyme may be determined by a titration procedure in the presence of impurities and furthermore that once this concentration is known it may be used for subsequent kinetic studies where its value is needed. The presence of impurities in an enzyme preparation could affect the titration procedure by interfering

† The percentages of purity can be expressed in a different manner. The titrations can be used to determine the number of sites per milligram of protein, which also expresses the purity. The chemist (and the author) may prefer the former, the biochemist the latter. The two ways of expressing purity are in fact identical.

with the active site and/or the substrate. Impurities that reversibly block the active site would not be expected to influence a titration since it is independent of such problems. A competitive inhibitor such as a peptide impurity may raise the K_m(app) to some apparent higher value, K_m(app)', but it is likely under such a perturbation that $[S]_0$ will still be much greater than K_m(app)' for many titrants, and thus the titration will remain stoichiometric. Since the presence of a considerable amount of competitive inhibitor would not be expected to affect a titration, the possibility exists that it may lie undetected in the titration and later may affect the kinetics. However, if an enzyme is used in the determination of the kinetics of a specific substrate, the concentration of enzyme, and thus the concentration of any competitive inhibitor associated with it, will be at such a low level that the latter will probably be in too low concentration to affect the former. Irreversible inhibitors will, of course, affect both the titration and the kinetics in the same way. Finally, an impurity which is finitely but slowly reversible may complicate both the titration and the kinetics.

In the titrations reported above, the enzyme purity has varied from a low of 7% with acetylcholinesterase to a high of 83% with porcine trypsin. Even Worthington three-times-crystallized α-chymotrypsin titrates at 90% purity only in certain instances (Table 12.15). These numbers raise substantial doubts about either the purity of the various enzyme preparations or the titrations themselves. The validity of each titration has been discussed above. Therefore, let us look at the impurities.

The purities mentioned above have been calculated from the relationship of the titrated value to that calculated on the basis of the weight and molecular weight of the enzyme. A fraction of the impurities calculated on this basis can be accounted for by the moisture content of these preparations, perhaps 5%. The hygroscopic nature of the lyophilized proteins can possibly increase this percentage. Another fraction of the impurities in α-chymotrypsin[81] and trypsin[82] (and probably other proteolytic enzymes) consists of ninhydrin-sensitive materials, which may be identified most easily as degradation products of the enzyme such as amino acids and peptides. These materials can be separated from α-chymotrypsin or trypsin using Sephadex G-25 or G-50. An additional impurity has been separated from α-chymotrypsin using Sephadex G-25 and must be a small molecule since it is eluted from the column considerably after the enzyme and peptide fractions.[81] This material has not yet been identified.

Recently different batches of α-chymotrypsin have been found to give different results in experiments involving proton transfer from an indicator to the imidazole group of the active site of α-chymotrypsin.[81] As a consequence of these findings the gel filtration purification using Sephadex G-25 mentioned above was developed.[81] In order to test the effect of batch difference

Table 12.15
α-Chymotrypsin- and Trypsin-Catalyzed Hydrolysis of Specific Substrates[a]

Enzyme	Batch Number	Substrate	Purity wt. %	$[E]_0$ Soln ($M \times 10^7$)	$[S]_0$ Soln ($M \times 10^4$)	k_{cat} (sec^{-1})	K_m(app) ($M \times 10^5$)
Chymotrypsin	CDI-6114-5 untreated	Methyl N-acetyl-L-tryptophanate	87.1	1.411	2.041	40.1	13.0
	CDI-6114-5 through G-25 (fine)	Methyl N-acetyl-L-tryptophanate	86.6	2.117	2.041	39.0	9.3
	CDI-6110-1 untreated	Methyl N-acetyl-L-tryptophanate	66.1	0.9869	2.041	40.5	13.8
Trypsin	TRL-6261 untreated	Ethyl α-N-benzoyl-L-argininate	57	2.504	1.623	16.2	1.2
Trypsin	TRL-6261 Sephadexed using G-50	Ethyl α-N-benzoyl-L-argininate	77	1.65	1.584	13.7	1.06

[a] From M. L. Bender et al., *J. Amer. Chem. Soc.*, **88**, 5890 (1966). © 1966 by the American Chemical Society. Reprinted by permission of the copyright owner.

and Sephadex treatment on both the titration of active sites and the kinetics of enzymatic catalysis, the experiments in Table 12.15 were carried out. For one batch of α-chymotrypsin which gave disparate results in temperature-jump experiments,[81] Sephadex filtration changed neither the purity of the enzyme preparation, as determined by *N-trans*-cinnamoylimidazole titration, nor the kinetic constants in the α-chymotrypsin-catalyzed hydrolysis of *N*-acetyl-L-tryptophan methyl ester. Furthermore, two batches of α-chymotrypsin, different in purity, gave identical kinetic constants when their differences in purity were taken into account by determination of the concentration of active enzyme *via* titration. Finally, Sephadex filtration improved the purity of a bovine trypsin preparation, but its kinetic constants remained unchanged. This has been the history of the titration usage so far. For example, the k_{cat}(lim) for the α-chymotrypsin-catalyzed hydrolysis of *N*-acetyl-L-tryptophan methyl ester in aqueous solution carried out by four different workers over a period of 4 years using four different preparations of α-chymotrypsin were: (1) 57 sec^{-1},[83] (2) 55 sec^{-1},[84] (3) 53 sec^{-1},[85] and (4) 50 sec^{-1}.[86]

REFERENCES

1. J. B. Sumner, *J. Biol. Chem.*, **69**, 435 (1926).
2. J. H. Northrop, M. Kunitz, and R. M. Herriott, *Crystalline Enzymes*, Columbia University Press, New York, 1948.
3. J. B. Sumner and G. F. Somers, *Chemistry and Methods of Enzymes*, Academic Press, Inc., New York, 1947.
4. P. A. Srere, *Science*, **158**, 936 (1967).
5. J. C. Gerhart and A. B. Pardee, *J. Biol. Chem.*, **237**, 891 (1962).
6. H. O. Halvorson, *Advan. Enzymol.*, **22**, 99 (1960).
7. F. Jacob and J. Monod, *J. Mol. Biol.*, **3**, 318 (1961).
8. W. Kauzmann, *Advan. Protein Chem.*, **14**, 1 (1959).
9. I. M. Klotz, *Brookhaven Symp. Biol.*, **13**, 25 (1960).
10. C. J. Martin and G. M. Bhatnager, *Biochemistry*, **6**, 1638 (1967).
11. *Report of the Commission on Enzymes of the International Union of Biochemistry*, Pergamon Press, London, 1961.
12. *Enzyme Nomenclature: Recommendations* (1964) *of the International Union of Biochemistry on the Nomenclature and Classification of Enzymes*, Elsevier Publishing Co., New York, 1965.
13. *Science*, **150**, 719 (1965).
14. M. Laskowski, *Methods Enzymol.*, **2**, 8 (1955).
15. H.-U. Bergmeyer (Ed.), *Methods of Enzymatic Analysis*, Academic Press, New York, 1963.
16. A. K. Balls and E. F. Jansen, *Advan. Enzymol.*, **13**, 321 (1952).

17. B. Chance in *Technique of Organic Chemistry*, vol. 8, S. I. Friess and A. Weissberger, Eds., Interscience Publishers, Inc., New York, 1953, p. 627, and references therein.
18. T. Inagami and J. M. Sturtevant, *J. Biol. Chem.*, **235**, 1019 (1960); D. G. Doherty and F. Vaslow, *J. Amer. Chem. Soc.*, **74**, 931 (1952).
19. D. D. Ulmer, T.-K. Li, and B. L. Vallee, *Proc. Nat. Acad. Sci. U.S.*, **47**, 1155 (1961).
20. M. L. Bender and F. J. Kézdy, *J. Amer. Chem. Soc.*, **86**, 3704 (1964) and references therein.
21. M. L. Bender and E. T. Kaiser, *ibid.*, **84**, 2556 (1962), and references therein; M. L. Bender, F. J. Kézdy, and J. Feder, *ibid.*, **87**, 4953 (1965); M. L. Bender, J. V. Killheffer, Jr., and F. J. Kézdy, *ibid.*, **86**, 5331 (1964); C. R. Gunter, Ph.D. Thesis, Northwestern University, 1966.
22. T. H. Marshall, J. R. Whitaker, and M. L. Bender, *Biochemistry*, **8**, 4665, 4671 (1969).
23. S. A. Bernhard, S. J. Lau, and H. Noller, *ibid.*, **4**, 1108 (1965); M. L. Begue and C. G. Miller, unpublished experiments.
24. A. Stockwell and E. L. Smith, *J. Biol. Chem.*, **227**, 1 (1957); G. Lowe and A. Williams, *Proc. Chem. Soc.*, 140, (1964); M. L. Bender and L. J. Brubacher, *J. Amer. Chem. Soc.*, **86**, 5333 (1964); M. L. Bender and J. R. Whitaker, *ibid.*, **87**, 2728 (1965); L. J. Brubacher, unpublished experiments.
25. I. B. Wilson, *Enzymes*, **4**, 501 (1959); R. M. Krupka and K. J. Laidler, *J. Amer. Chem. Soc.*, **83**, 1458 (1961); M. L. Bender and J. K. Stoops, *ibid.*, **87**, 1622 (1965); J. K. Stoops, unpublished experiments.
26. T. Murachi, *Biochim. Biophys. Acta*, **71**, 239 (1963); T. Murachi, T. Inagami, and M. Yasui, *Biochemistry*, **4**, 2815 (1965); T. Murachi and M. Yasui, *ibid.*, **4**, 2275 (1965).
27. B. S. Hartley and B. A. Kilby, *Biochem. J.*, **56**, 288, (1954).
28. H. Gutfreund and J. M. Sturtevant, *Biochem. J.*, **63**, 656 (1956).
29. H. Gutfreund and J. M. Sturtevant, *Proc. Natl. Acad. Sci. U.S.*, **2**, 719 (1956).
30. L. Ouellet and J. A. Stewart, *Can. J. Chem.*, **37**, 737 (1959).
31. J. M. Sturtevant, *Brookhaven Symp. Biol.*, **13**, 151 (1960).
32. G. R. Schonbaum, B. Zerner, and M. L. Bender, *J. Biol. Chem.*, **236**, 2930 (1961).
33. B. F. Erlanger and F. Edel, *Biochemistry*, **3**, 346 (1964).
34. T. Spencer and J. M. Sturtevant, *J. Amer. Chem. Soc.*, **81**, 1874 (1959).
35. F. J. Kézdy and M. L. Bender, *Biochemistry*, **1**, 1097 (1962).
36. M. L. Bender et al., *J. Amer. Chem. Soc.*, **88**, 5890 (1966).
37. E. T. Chase, Jr., and E. Shaw, *Biochem. Biophys. Res. Comm.*, **29**, 508 (1967).
38. I. B. Wilson, M. A. Hatch, and S. Ginsberg, *J. Biol. Chem.*, **235**, 2312 (1960).
39. I. B. Wilson and M. A. Harrison, *J. Biol. Chem.*, **236**, 2292 (1961).
40. J. M. Reiner, *Behavior of Enzyme Systems*, Burgess Publishing Co., Minneapolis, Minn., 1959, p. 58, proposes a method for the absolute determination of enzyme concentration that is essentially a combination of eqs. 12.5 and 12.7. This method is limited, however, because it applies only to initial rates and only when $[E] \cong [S] \cong K_s$.
41. M. L. Bender, G. R. Schonbaum, and B. Zerner, *J. Amer. Chem. Soc.*, **84**, 2562 (1962).
42. B. S. Hartley and B. A. Kilby, *Biochem. J.*, **50**, 672 (1952).
43. W. Cohen, M. Lache, and B. F. Erlanger, *Biochemistry* **1**, 686 (1962).
44. A. L. Green and J. D. Nicholls, *Biochem. J.*, **72**, 70 (1959).
45. This correction was not made in a preliminary communication. M. L. Bender, J. V. Killheffer, Jr., and R. W. Roeske, *Biochem. Biophys. Res. Commun.*, **19**, 161 (1965).

46. Since the binding of soybean trypsin inhibitor diminishes greatly at lower pH, it is not possible to carry out this experiment at the pH of the titration: M. Kunitz, *J. Gen. Physiol.*, **30**, 291 (1947); R. F. Steiner, *Arch. Biochem. Biophys.*, **49**, 71 (1954).
47. H. Gutfreund, *Trans. Faraday Soc.*, **51**, 441 (1955); G. W. Schwert and M. A. Eisenberg, *J. Biol. Chem.*, **179**, 665 (1949); T. Inagami and J. M. Sturtevant, *Biochim. Biophys. Acta*, **38**, 64 (1960).
48. A. N. Glazer and E. L. Smith, *J. Biol. Chem.*, **236**, 2948 (1961).
49. A. Light, R. Frater, J. R. Kimmel, and E. L. Smith, *Proc. Nat. Acad. Sci. U. S.*, **52**, 1276 (1964).
50. E. L. Smith and M. J. Parker, *J. Biol. Chem.*, **233**, 1387 (1958).
51. B. J. Finkle and E. L. Smith, *J. Biol. Chem.*, **230**, 669 (1958).
52. Unpublished experiments of Dr. J. R. Whitaker.
53. I. E. Liener, *Biochim. Biophys. Acta*, **53**, 332 (1961).
54. S. A. Bernhard and H. Gutfreund, *Biochem. J.*, **63**, 61 (1956).
55. M. R. Hollaway, A. P. Mathias, and B. R. Rabin, *Biochim. Biophys. Acta*, **92**, 111 (1964).
56. M. Ottesen and C. G. Schellman, *Compt. Rend. Trav. Lab. Carlsberg*, **30**, 157 (1957).
57. G. Johansen and M. Ottesen, *Compt. Rend. Trav. Lab., Carlsberg*, **34**, 199 (1964).
58. M. A. Naughton and F. Sanger, *Biochem. J.*, **70**, 4P (1958).
59. B. A. Kilby and G. Youatt, *Biochem. J.*, **57**, 303 (1954).
60. U. J. Lewis, D. E. Williams, and N. G. Brink, *J. Biol. Chem.*, **222**, 705 (1956).
61. Personal communication from Dr. W. A. Loeven, The Netherlands Institute for Preventive Medicine, Leiden, The Netherlands. The activity corresponded to 3.4 u/mg of protein when tested against alkali-treated elastin of pH 8.7. This enzyme is free of the α_2 globulin fraction which contains elastase inhibitor.
62. W. A. Loeven, *Acta Physiol. Pharmacol. Neerl.*, **12**, 57 (1963).
63. Dr. W. A. Loeven, personal communication.
64. S. Avrameas and J. Uriel, *Biochemistry*, **4**, 1750 (1965).
65. I. B. Wilson, F. Bergmann, and D. Nachmansohn, *J. Biol. Chem.*, **186**, 781 (1950).
66. K.-B. Augustinsson and D. Nachmansohn, *J. Biol. Chem.*, **179**, 543 (1949).
67. H. O. Michel and S. Krop, *J. Biol. Chem.*, **190**, 119 (1951).
68. F. Sanger, *Proc. Chem. Soc.*, 76 (1963).
69. I. B. Wilson, *J. Biol. Chem.*, **190**, 111 (1951); I. B. Wilson, S. Ginsburg, and C. Quan, *Arch. Biochem. Biophys.*, **77**, 286 (1958), and references therein.
70. I. B. Wilson, M. A. Hatch, and S. Ginsburg, *J. Biol. Chem.*, **235**, 2312 (1960).
71. I. B. Wilson and J. Alexander, *ibid.*, **237**, 1323 (1962); R. M. Krupka, *Biochemistry*, **3**, 1749 (1964).
72. R. M. Krupka, *Biochem. Biophys. Res. Commun.*, **19**, 531 (1965).
73. H. C. Lawler, *J. Biol. Chem.*, **236**, 2296 (1961).
74. I. B. Wilson and M. A. Harrison, *J. Biol. Chem.*, **236**, 2292 (1961).
75. M. L. Bender and J. K. Stoops, *J. Amer. Chem. Soc.*, **87**, 1622 (1965).
76. J. K. Stoops and M. L. Bender, to be published.
77. L. T. Kremzner and I. B. Wilson, *J. Biol. Chem.*, **238**, 1714 (1963).
78. D. J. Horgan, E. C. Webb, and B. Zerner, *Biochem. Biophys. Res. Commun.*, **23**, 18 (1966).
79. M. Runnegar and B. Zerner, personal communication.
80. T. H. Marshall, Ph.D. Thesis, Northwestern University, 1966.
81. A. Yapel, M. Han, R. Lumry, A. Rosenberg, and D. F. Shiao, *J. Amer. Chem. Soc.*, **88**, 2573 (1966).
82. S. P. Jindal, unpublished experiments in this laboratory.

References

83. G. E. Clement and M. L. Bender, *Biochemistry*, **2**, 836 (1963).
84. M. L. Bender and G. A. Hamilton, *J. Amer. Chem. Soc.*, **84**, 2570 (1962).
85. Extrapolated from pH 7.9: J. R. Knowles, *Biochem. J.*, **95**, 180 (1965).
86. M. J. Gibian, unpublished observations in this laboratory using Worthington chymotrypsin CDI 6084–5.
87. F. J. Kézdy, L. Lorand, and K. D. Miller, *Biochemistry*, **4**, 2302 (1965).

Chapter 13

STRUCTURE OF ENZYMES

13.1	Molecular Weight	458
13.2	X-Ray Diffraction	460
13.3	Determination of Enzymes	473

In common with all proteins, enzymes are composed of approximately 20 amino acids. When the amino acid compositions of enzymes are compared to those of structural proteins, no differences are apparent. Dayhoff lists the amino acid composition of approximately 100 enzymes: there are no special amino acids that are found in enzymes; there appear to be no special percentages, nor any unique discernible features in the amino acid compositions.[1] This analysis is not universally accepted. There is the argument that a special amino acid or amino acid interaction may be the unique feature in enzymatic catalysis,[2] but no definitive case has been made for such a proposition.

The methods developed by Sanger[3] for the determination of the amino acid sequence in the protein insulin have been employed to work out the amino acid sequences in a number of enzymes. The first such enzyme sequence so determined was that of ribonuclease (Fig. 13.1).[4] Subsequently the amino acid sequences of a number of other enzymes have been elucidated; these include chymotrypsin,[5] trypsin,[6] lysozyme,[7] papain,[8] subtilisin,[10] and carboxypeptidase A.[11] The sequence of ribonuclease is representative of amino acid sequences of those enzymes that have been determined so far. It contains a single polypeptide chain of 124 amino acids involving 20 different amino acids. No special sequence of amino acids is apparent, nor is there a repeating

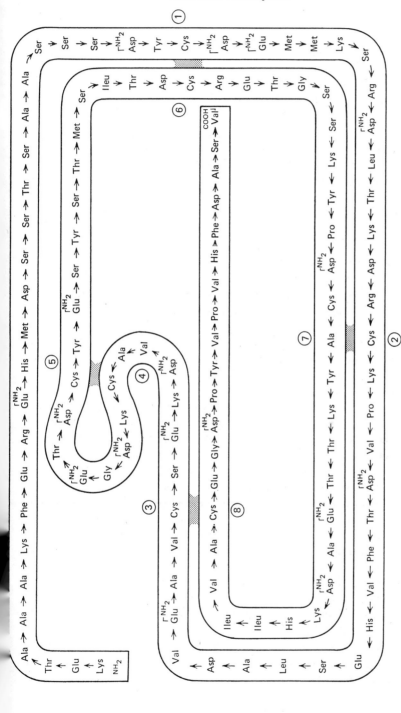

Fig. 13.1 A two-dimensional schematic diagram of the structure of bovine ribonuclease showing the arrangement of the disulfide bonds (indicated by numerals at the crosslinked cysteine residues) and the sequence of the amino acid residues. The arrows indicate the direction of the peptide chain starting from the amino end. From D. H. Spackman, W. H. Stein, and S. Moore, *J. Biol. Chem.*, **235**, 648 (1960).

pattern of any kind; no amino acid is present that is not found in other proteins. The two-dimensional sequence is cross-linked at four places by disulfide bonds between cysteine residues. Unfortunately, it must be concluded that the linear sequences shown here, and all others so far determined, do not give any direct clue to the special chemical characteristics of the enzymes.[3] As noted later, however, the linear sequence of amino acids does, in fact, in itself determine the three-dimensional conformation of the enzyme (tertiary structure).

As intimated above, enzymes are not linear proteins nor are they pleated sheets, but rather they are globular proteins whose overall shape in aqueous solution is usually described as ellipsoidal on the basis of viscosity, light scattering, and sedimentation phenomena.[12]

13.1 MOLECULAR WEIGHT

The molecular weights of many enzymes have been determined by physical methods. Ribonuclease (mol wt 12,700) and lysozyme (mol wt 14,600) are among the smallest enzymes known. Molecular weights of enzymes range from these low figures up to greater than 1,000,000 for the enzyme pyruvate decarboxylase. Some enzymes of high molecular weight reversibly dissociate into several subunits: for example, the enzyme aldolase (mol wt 159,000) consists of four subunits of approximately equal size. Fatty acid synthetase, a crystalline multienzyme complex, consists of 21 subunits of average molecular weight 100,000 (Fig. 13.2).[59]

The relationship between the two-dimensional (secondary) amino acid sequence of an enzyme and the three-dimensional (tertiary), presumably unique, folded structure is an intriguing one. Important experiments with ribonuclease indicate that the two-dimensional sequence completely determines its three-dimensional folding. In these experiments, ribonuclease was reduced so that all the disulfide crosslinks were broken and a completely unfolded structure was obtained. When this inactive, unfolded enzyme was allowed to reoxidize, the native structure was essentially completely reformed.[14] Similar experiments have been carried out with trypsin.[15] Thus the native conformation is probably the one most thermodynamically stable at neutral pH. Genetic information in an organism can thus be transferred from a linear DNA molecule to a linear messenger RNA molecule to a sequence of amino acids in an enzyme, which then assumes a particular three-dimensional shape suitable for efficient and specific catalysis in the direct control of physiological processes.

The synthesis of enzymes is a continuous process in every living organism. In the case of certain animal enzymes, there is direct isotopic evidence of a fairly rapid turnover of enzymes.

Fig. 13.2 The crystallized multi-enzyme complex fatty-acid synthetase magnified 11 times. From *Scientific Research*, **3**, 17, April 1, 1968.

In recent years, increasing investigation into enzyme structure has indicated a trend that foretells that many enzymes and other proteins will consist of basic units with a molecular weight of about 20 to 40 thousand. The equivalent weight of many enzymes can be determined by chemical reaction with the active sites of the molecules. This approach was discussed in Chapter 12.

Isozymes, as the name implies, are multiple molecular forms of an enzyme from the same species.

In the chicken, lactic dehydrogenase[16] occurs in two principal forms derived from the breast muscle and from the heart. These forms are distinguishable on the basis of amino acid composition and physical, immunologic, and catalytic properties. The number of lactic dehydrogenases in the chicken (and in several other species) is five: three exhibit properties that are intermediate between those of muscle and heart types. For example, electrophoresis yields five distinct and equally spaced bands with the muscle and heart types at the extremes of the pattern. Furthermore, the amino acid compositions of the five isozymes show a regular variation from that characteristic of the muscle type to that of the heart type. Moreover, the enzymatic and immunologic properties exhibit a similar gradation from one of the extreme types to the other.

Treatment with solutions of guanidinium chloride in the presence of mercaptoethanol converts lactic dehydrogenase into four subunits. If lactic dehydrogenase is a tetramer, the assumption of two gene products (two types of subunits) accounts for the observed findings. Thus, designating one type of subunit as *H* (heart type) and the other as *M* (muscle type), five structurally distinct lactic dehydrogenases may be formed: *HHHH, HHHM, HHMM, HMMM,* and *MMMM*. This formulation assumes that the subunit sites in the tetramer are equivalent; otherwise one might have, for example, four molecules corresponding to the structure *HHHM*. Different amounts of each of two subunit types obviously accounts for the stepwise variation in the properties of the isozymes between two extremes. All cases of isozymes probably cannot be accounted for on such a basis. In cases such as that of lactic dehydrogenase, the term *hybrid enzymes* has been suggested as a reasonable alternative to isozymes.

Pyruvate dehydrogenase of *E. coli*, of molecular weight 4,000,000, is composed of three types of enzymes, at least one of which is in turn composed of subunits. The complete particle contains a minimum of 88 separate polypeptides. This complex system spontaneously forms, in high yields, upon mixing the components.

Isozymes may or may not be conformers (i.e., molecules that exist in different conformations). The enzyme malate dehydrogenase is considered to consist of isozymes that are not solely conformational in origin.[17]

13.2 X-RAY DIFFRACTION

Many enzymes have been crystallized (Fig. 13.2 and 13.3).[18,59] The most direct determinations of three-dimensional structures have been made by X-ray diffraction analysis of the crystalline substances. The relationship of the structure determined by X-ray diffraction to that of an enzyme in solution has not been established as yet, but the X-ray results for solids have no counterpart for solutes and at the present time they are the only basis for discussion of three-dimensional enzyme structure. In discussing X-ray structure one must keep in mind that many enzymes show polymorphic behavior and have close relatives, called isozymes[15] above, which differ often times in minute respects from one another. Therefore, the single species which is described here may be in the general case only one of a family of related species.

Of the half dozen enzymes whose three-dimensional structures have been attacked by X-ray methods in recent years, the first whose structure was analyzed to atomic dimensions was that of lysozyme.[19] Hen egg white lysozyme crystals were prepared in one molar sodium chloride at pH 4.7. These tetragonal crystals, which contain 33.5 wt% sodium chloride, consist of a unit cell containing eight lysozyme molecules. Analyses at 6 Å resolution, using the heavy atom isomorphous replacement technique, gave essentially the complete structure of the lysozyme molecule. The molecule is an irregular one

X-Ray Diffraction 461

Fig. 13.3 Crystals of ribonuclease in alcohol (3). From M. Kunitz, *J. Gen. Physiol.*, **24**, 15 (1940).

whose axes are roughly 30 × 30 × 45 Å. The α-helix can be recognized in the three-dimensional structure, which is shown in Fig. 13.4.[19] The helical segments total about 55 of the 124 residues in the enzyme, indicating a maximal helix constant of about 42%, in good agreement with the prediction of 45% from optical rotatory dispersion measurements. Four disulfide residues form cross-links of the protein chain. The molecule has a hydrophobic spine consisting of six tryptophan residues together with other hydrophobic residues. It will be noted that the amino acids making up this hydrophobic spine run vertically through the middle of the diagram of Fig. 13.4. Competitive inhibitors (and presumably substrates) of lysozyme bind in the cleft defined by this hydrophobic spine. The location of attachment of N,N'-diacetylchitobiose, a competitive inhibitor of lysozyme, was determined by X-ray diffraction analysis at 6 Å resolution, as shown in Fig. 13.5. This inhibitor attaches to the enzyme, at least in the crystal structure, in a cleft of the enzyme (see also Fig. 13.6). This direct evidence for the binding of a small molecule to a complementary position on the enzyme surface is an important piece of direct evidence supporting the "lock and key" theory originally

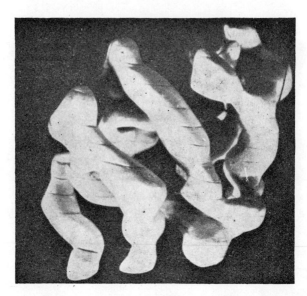

Fig. 13.4 Solid model of lysozyme electron density greater than about 0.5 electrons/Å³ at 6 Å resolution. From C. C. F. Blake, D. F. Koenig, G. A. Mair, A. C. T. North, D. C. Phillips, and V. R. Sarma, *Nature*, **206**, 757 (1965).

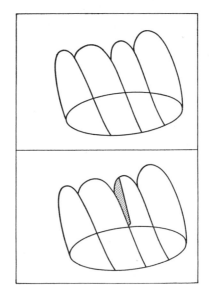

Fig. 13.5 The binding of a competitive inhibitor N, N'-diacetylchitobiose (shaded) to the enzyme lysozyme at 6-Å resolution. From L. N. Johnson and D. C. Phillips, *Nature*, **206**, 761 (1965).

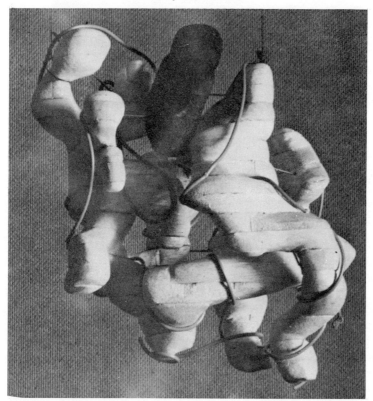

Fig. 13.6 Model of a lysozyme molecule at 6-Å resolution obtained by X-ray analysis showing the general course of the polypeptide chain (wire) and the site at which the tri-*N*-acetylchitotriose is bound to the enzyme in the crystalline state. From C. C. F. Blake, L. N. Johnson, G. A. Mair, A. C. T. North, D. C. Phillips, and V. R. Sarma, *Proc. Roy. Soc.*, **167**, B378 (1967).

suggested by Emil Fischer.[20] In two regions of the cleft, the competitive inhibitor comes very close to the enzyme molecule. These regions include the three tryptophans 62, 63, and 108. In contrast to the above results, molecules that are not competitive inhibitors do not bind specifically to the cleft in the lysozyme but when introduced into the crystal are dispersed throughout the crystal structure in a random fashion.

The structure of chymotrypsinogen has also been investigated.[21] A diagram of the 5 Å resolution of chymotrypsinogen carried out using isomorphous replacement techniques is shown in Fig. 13.7. It is seen that the molecule is roughly ellipsoidal, of dimensions $50 \times 40 \times 40$ Å. There is a small concavity on the underside of the model shown in Fig. 13.7, which has been suggested several times to be related to the active site of this zymogen. There is

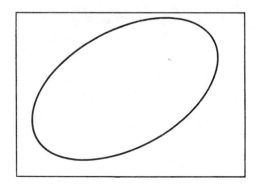

Fig. 13.7 Model of chymotrypsinogen A at 5-Å resolution. From J. Kraut et al., *Proc. Natl. Acad. Sci., U.S.*, **48**, 1417 (1962).

very little helix in this molecule; it has been estimated from optical rotatory dispersion measurements and from hydrogen-deuterium exchange measurements that there is 15% helix in chymotrypsinogen.[22] An analysis of the total length of all links in Fig. 13.7 indicates 650 Å compared to a calculated value for the total length of the 240 residues of chymotrypsinogen in α-helix conformation of 365 Å, or a total length for 240 residues in a fully extended conformation of 850 Å. Thus the chymotrypsinogen molecule may be described as a loosely coiled chain and it is not surprising that a 5-Å X-ray diffraction pattern does not show a contiguous polypeptide chain. Unfortunately, the X-ray pattern of this molecule fades out at 2.5 Å and thus the determination of the atomic structure of this molecule does not appear to be promising.

In addition to those on the zymogen, X-ray diffraction studies of α-chymotrypsin itself have been carried out.[23] By isomorphous replacement techniques and phase transition studies it has been shown that α-chymotrypsin crystallizes in a dimeric form having P_21 symmetry which can be described by Fig. 13.8. The plane of asymmetry can be seen in the packing of these crystals which have been otherwise described as "stacked baby shoes."

Some other enzymes whose X-ray structures are known include carboxypeptidase A,[11] ribonuclease,[13] papain,[24] elastase,[9] and subtilisin.[25]

In future years the three dimensional structure of other enzymes will undoubtedly be elucidated with increasing frequency. Reference 1 lists the status of globular protein crystal structure projects in January 1969, indicating that 10 enzymes were under investigation, including ribonuclease, lysozyme, chymotrypsinogen, chymotrypsin, elastase, papain, carbonic anhydrase, carboxypeptidase, lactate dehydrogenase, and alcohol dehydrogenase. Also, cytochrome C is being studied.[25] Of these, most have met with success.

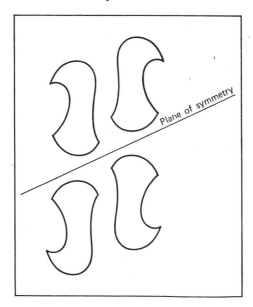

Fig. 13.8 α-Chymotrypsin according to D. M. Blow, M. G. Rossmann, and B. A. Jeffery, *J. Mol. Biol.*, **8**, 65 (1964).

The others are going forward and one can anticipate results in the near future with many of them. When a number of enzymes of differing inherent structure have been elucidated, one will have a better feeling for what features are common to three dimensional enzyme structure. At the present time it appears that one can say only that an enzyme in three dimensions is simply a collection of amino acids with no special features other than a special spatial arrangement. As pointed out before, when viewing the amino acid composition no unique feature is evident (Table 13.1). Furthermore, when looking at the sequence in two dimensions, no special interaction of amino acids is apparent. Finally when viewing three-dimensional enzyme structure there appears to be no special interaction of amino acids (this is also true for the hemoglobin molecule, which is not an enzyme, but is a functional protein). Thus the unique aspect of an enzyme must be discussed as a unique juxtaposition of amino acid side chains that carry out the catalytic function. It appears that the scanning electron microscope holds the promise of 1-Å resolution.[26]

It is important to consider the forces that hold the enzyme protein chains together in the unique three-dimensional structure which is necessary for catalytic activity. Part of these forces consist of strong covalent interactions through the cystine disulfide crosslinks. However the preponderance of the

Table 13.1
Analysis of Some Proteins and Enzymes[61]

Amino Acid	Wool		Erythrocuprein		Dog Plasma		Carboxy-peptidase		Papain	
	Grams of Amino Acid Residue/ 100 g Protein	Average Deviation	Grams of Amino Acid Residue/ 100 g Protein	Average Deviation	Grams of Amino Acid Residue/ 100 g Protein	Average Deviation	Grams of Amino Acid Residue/ 100 g Protein	Average Deviation	Grams of Amino Acid Residue/ 100 g Protein	Average Deviation
Glycine	5.80	0.09	7.85	0.17	1.99	0.32	3.85	0.12	6.39	0.10
Alanine	3.51	0.08	4.75	0.12	6.55	0.66	4.12	0.17	4.49	0.19
Serine	7.25	0.19	5.18	0.10	3.36	0.36	8.36	0.09	4.90	0.03
Threonine	4.61	0.13	5.54	0.14	3.55	0.25	7.82	0.10	3.30	0.05
Proline	5.33	0.05	3.25	0.11	4.51	0.03	3.09	0.04	4.31	0.12
Valine	3.57	0.10	6.87	0.15	6.11	0.17	4.72	0.06	7.13	0.21
Isoleucine	1.97	0.05	3.74	0.16	0.99	0.00	6.60	0.04	5.22	0.14
Leucine	4.90	0.12	6.07	0.14	10.99	0.39	8.12	0.11	5.26	0.02
Phenylalanine	1.75	0.09	4.37	0.18	6.66	0.25	6.38	0.17	2.28	—
Tyrosine	2.97	0.08	0.95	0.02	5.26	0.16	9.32	0.41	13.25	—
Tryptophan	1.73	0.12	—	—	—	—	3.30	0.10	—	—
Cystine/2	7.93	0.05	3.51	0.02	5.12	0.04	0.69	—	—	—
Methionine	0.39	0.02	—	—	0.78	0.04	0.39	0.00	—	—
Aspartic acid	4.24	0.18	14.32	0.29	8.95	0.45	10.12	0.22	9.79	0.16
Glutamic acid	8.58	0.13	10.97	0.39	16.18	0.73	9.36	0.26	10.91	0.22
Amide N	7.46	0.57	1.18	0.00	0.87	0.16	1.05	0.08	1.51	0.06
Arginine	20.32	0.04	4.27	0.01	5.35	0.18	5.54	0.09	6.95	0.11
Histidine	1.46	0.10	5.97	0.09	2.50	0.05	3.07	0.09	0.75	0.06
Lysine	3.25	0.15	7.86	0.08	11.30	0.15	6.85	0.10	4.97	0.08

forces that hold enzymes together consist of many individually weak noncovalent interactions.[26] It is now quite generally agreed that apolar bonds are the most important factor determining protein structure. The interaction between two apolar groups may be described as a Van der Waals interaction, but in the aqueous media in which enzymes exist, the interaction of water with apolar groups must also be considered. In a study of the transfer of alkanes from water to a hydrocarbon solvent it was found that this process is characterized by a small positive enthalpy change and a large entropy change.[41] Kauzmann has expanded on this idea, considering the transfer of hydrocarbons between aqueous and nonaqueous liquid phases. He has concentrated on the fact that, as mentioned above, the entropy of this process is large and is the force that drives alkanes from water into a hydrocarbon solvent. He postulates that when the hydrocarbon lies within an aqueous medium, it affects the stucture of the solvent water in its immediate vicinity.[28] From the consideration of

$$2(\text{hydrocarbon})_{aq} \rightleftharpoons 2 \text{ hydrocarbon} + H_2O\cdots H_2O \qquad (13.1)$$

he points out that there is a gain in entropy entailed in removing hydrocarbon from aqueous solution because of the enhanced interaction of the water molecules that were formerly frozen by the hydrocarbon in aqueous environment. Therefore, hydrocarbon molcules tend to associate and thus one can explain micelle formation as well as the interaction of hydrocarbon parts in the "oily" insides of proteins and enzymes. A substantial stabilizing force thus should result from the association of apolar residudes, the effect being primarily due to entropy changes in the solvent. This is a popular belief, but recent data indicate that many apolar groups are on the surface; the complementary statement—that all polar groups are on the surface—may be made.[29] Némethy and Scheraga[30] have made a theoretical calculation on the basis of statistical mechanical treatment of water clusters indicating that when a hydrocarbon is inserted into the water an increase in the degree of hydrogen bonding occurs. On the basis of these calculations the thermodynamic properties of the hydrophobic bonds from the interaction of side-chains of many amino acids have been calculated giving standard free energy changes of these interactions from -0.2 to -1.5 kcal/mole, with enthalpy changes that are small and positive, and entropy changes of 2–11 entropy units with an average of eight. A typical large calculated free energy change would involve the two phenyl rings of phenylalanine or the two alkyl side-chains of the amino acid isoleucine, as shown in Fig. 13.9.

Klotz, starting with the same basic thermodynamics between hydrocarbons and water as Kauzmann, has taken a different viewpoint. He has emphasized the fact that the enthalpy increases when a hydrocarbon is inserted into aqueous solution and has coupled this with the fact that crystalline hydrates

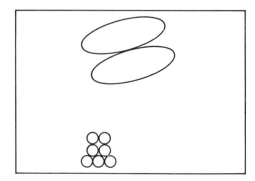

Fig. 13.9 Apolar interactions of 2 phenylalanine residues and 2 isoleucine residues (after G. Némethy and H. A. Scheraga, *J. Phys. Chem.*, **66**, 1773 (1962)). © 1962 by the American Chemical Society. Reprinted by permission of the copyright owner.

are known to be formed from a number of apolar substances including hydrocarbons.[31] This point of view suggests that an isolated hydrocarbon molecule distorts the structure immediately surrounding the protein molecules with the creation of "ice-like" structures that resemble crystalline hydrates. Thus the protein molecule would be stabilized by an iceberg sheath in which the apolar residues of the amino acids were inserted into the surrounding aqueous medium.

The controversy between these two descriptions of apolar bonds has not yet been resolved.[32] Certainly water surrounding a hydrocarbon molecule in aqueous solution has some statistical similarity to clathrates but not the same stability. In a clathrate, stability is provided by the crystal lattice in its three-dimensional extended order, whereas on a protein surface the water layer can have only two-dimensional order. Probably both effects are operative in that "icebergs" may exist as stablizing structures in holes or depressions in the surface of an enzyme, while the apolar interactions between hydrocarbon moieties (the hydrophobic bond, a misnomer because it is the water interaction and not the bonding that is important) are involved in the interior of the protein. One could resolve this controversy by analysis of the three dimensional structures of proteins to atomic dimensions. At the moment, available cases suggest different effects. One of these (hemoglobin) is a rather closed, compact molecule in which the hydrocarbon groups interact with one another in the interior, while another (lysozyme) is a rather open molecule in which many of the hydrocarbon portions of the molecule interact with the aqueous medium to form a stabilizing iceberg sheath.

Of course, polar bonds as well as apolar bonds exist in enzymes. In particular, the hydrogen bond has been discussed for a long time as a factor that

X-Ray Diffraction 469

influences structure in enzymes. Hydrogen bonding can stabilize an α-helix, which has been discussed above as contributing partially to the structure of lysozyme and chymotrypsin, or other helices or nonhelical atomic arrangements. In fibrous proteins the pleated sheet involving hydrogen bonds is an important factor. It must be kept in mind that an intrapeptide hydrogen bond probably cannot provide any driving force for folding an extended peptide chain into a globular protein. This conclusion is based on results with models such as N-methylacetamide in aqueous solution, where the free energy of intrapeptide hydrogen bond formation is not favorable. However, in a hydrocarbon solvent the free energy picture may change and the hydrogen bond may become important. The most important point about a hydrogen bond is that as primary folding occurs by apolar bond formation, this folding will break solvent-polar group hydrogen bonds. These hydrogen bonds must eventually be satisfied and donor-acceptor interactions involving hydrogen bonds will occur. Since hydrogen bonds are directional, there will be a restriction on the permitted conformations in the folded enzyme. The actual conformation will represent a compromise between the apolar group packing requirements and the polar hydrogen bond requirements. Thus nonhelical regions of a protein do not resemble a random coil in solution. That is, the protein has a unique three-dimensional structure.

The primary sequence, as shown in Fig. 13.1, appears to be the complete determiner of the three-dimensional structure of the enzyme. This important result, which indicates that all the information for three-dimensional protein synthesis resides in the two-dimensional sequence, was first arrived at by experiments with the enzyme ribonuclease. The enzyme was fully reduced so that the disulfide linkages were broken and then allowed to reform by reoxidation. The native structure of the enzyme was reformed as determined by a return of 80% of the activity, indicating that the native conformation at neutral pH is probably a thermodynamically more stable one.[33, 34] Comparable experiments have been carried out with trypsin. The reformation of the native structure of ribonuclease is interfered with by urea, guanidine, and phenol, indicating the importance of apolar bonds. Furthermore, polyalanyl derivatives of ribonuclease refold successfully (eight of the 11 amino groups were derivatized) indicating the inherent specificity of the folding process.

In contrast to the spontaneous folding process stands the unfolding process ordinarily described as denaturation. This process may be reversible or irreversible. The irreversible process probably has associated with it a secondary step beyond mere unfolding, such as an autolysis or a precipitation. We discuss only the reversible changes brought about by urea, heat, or extremes of pH. This is a difficult process to describe but the denatured state is not characterized at all except that in the case of an enzyme it is inactive.

Urea inactivates many enzymes. This effect is said to be associated with a change in conformation. For example, Fig. 13.10 indicates a direct relationship between the inactivation of trypsin and the intrinsic viscosity of a trypsin solution.[35] The effect of urea in disrupting enzyme conformation was originally ascribed to a breakage of hydrogen bonds.[36] However, it was shown recently that eight molar urea had no effect on several known hydrogen bonds.[37] Furthermore it was shown that urea influences apolar bonds in several different ways. For example, crystalline inclusion compounds are formed between urea and straight chain hydrocarbons. Furthermore, urea increases the solubility of amino acids with large aromatic or aliphatic side chains.[38] Assuming that the side-chains are completely hidden in the folded enzyme and are completely exposed in the unfolded state, the free energy of the unfolding by urea would amount to 20 kcal/mole, certainly enough to account for the unfolding process.

Different denaturating agents lead to different conformational changes. The same reagent may affect different globular proteins in different ways. Finally, a given denaturing agent may lead to two or more successive reactions.

Heat denaturation of enzymes results in a peculiar effect of temperature: elevation of temperature accelerates the enzymatic reaction and at the same time can destroy the enzyme thus leading to a temperature at which the maximal amount of substrate will react in a given period of time. The inactivation of an enzyme by heat has been shown to be equivalent to the denaturation of the protein as defined by precipitation for the enzymes pepsin

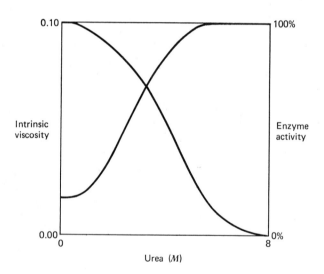

Fig. 13.10 The effect of urea on enzymatic activity and viscosity.

and trypsin.[39] In the process of heat denaturation very high temperature coefficients are found. However, the free energy of the process is a normal one. Thus the enthalpy of the reaction is very high and the entropy of the reaction is very large and positive, consistent with a picture of unfolding involving multiple interaction.[40]

These various pieces of information require that an enzyme structure is not completely static. In fact, an enzyme can unfold more or less completely into a random coil and refold into its unique active structure as shown in the ribonuclease experiments. Furthermore, it can be reversibly denatured by a number of external agents, many of these processes being reversible. Thus an enzyme need not be a static, rigid molecule. Certainly it need not be static when a substrate approaches it to bind for the process of catalytic action. It has been suggested that conformational changes occur on binding that account for a portion of the specificity of enzymes and also perhaps for a portion of the catalysis.[41] This concept of the "induced fit" between enzyme and substrate can be tested by observing what happens to the X-ray structure on binding. This poses a difficulty in the sense that if apolar interactions are involved in binding they would disappear with a mobile enzyme. However, the fact that small substrates are poorer binders than large substrates has led to this concept as an extension of the lock and key theory. This point will be further discussed in the context of binding (Chapter 18).

One of the current topics of investigation is whether an enzyme reacts the same in solution as in the crystalline state. The answer seems to be—sometimes. This question is crucial now that a number of X-ray structures have been determined. Crystalline indoleacryloyl-chymotrypsin appears to deacylate with the same rate and pH-rate profile as in solution.[42] Sluyterman states that papain crystals in 70% methanol-20% sodium sulfate have 100% activity.[43] However, the reactivity of ferrimyoglobin seems to be slightly lower in the crystalline state than in solution.[44] Urease[45] and trypsin[46] appear to catalyze in highly nonaqueous environments. Since protein crystals contain many molecules of water, there may not be much difference between the solid phase and solution. This subject is being actively investigated by Rupley,[60] Richards,[46] and others.

The structural people, important as they are, assume that a knowledge of the structure gives knowledge of the mechanism.[47] Recourse to the controversy surrounding neighboring group participation by saturated carbon in simple molecules (MW less than 1000) immediately shows that a knowledge of structure does not necessarily lead to a knowledge of mechanism. Of course, structure is a prerequisite to mechanism, but the two do not necessarily go together. In fact from reading this book, it would be presumptuous to equate the two.

Many enzymes are composed of several subunits that are held together by polar (metal ions, disoulfide links, etc.) or apolar links. Ocasionally, there is an equilibrium between an inactive form and an active form of the enzyme that differs only in the degree of aggregation. Still more occasionally the substrate affects this equilibrium.[48]

An enzyme can be considered to be a colloid, since it scatters X-rays, can be sedimented in an ultracentrifuge, and has an active site on the surface of a large molecule. In fact, one definition of a colloid is a molecule whose molecular weight is over 1000.[49] The point of all of this is that most enzyme chemists know very little about heterogeneous catalysis where multiple active sites exist and vice versa although the two must be related to one another.

Several probes for the study of proteins have been developed recently. These involve electron spin resonance,[50] nuclear magnetic resonance,[51] and "reporter groups."[52] The first and third specifically involve the active site whereas the second does not. The most sensitive spectrophotometric assay involves fluorescence measurements. It is because of sensitivity that we have favored p-nitrophenyl esters, among other reasons.

Protein conformation has been approached in other ways. For example, hydrogen–deuterium exchange and optical rotatory dispersion have been used. In each case, an attempt was made to assess the percentage of α-helix. Recent work indicates that neither of these methods is infallible.[53, 54]

When the structure of these molecules is known, it becomes possible to postulate the mechanism of action. Of course, there may be many different mechanisms just as there are for organic molecules. In this discussion it has been assumed that enzymes exist as discrete molecules. There is now much evidence to that effect.

The three-dimensional structure of enzymes, as of other proteins, is still largely unknown. Therefore, one of the present focuses of attention is the identification of the active sites of enzymes.

The hypothesis of an active site (Chapter 14), which assumes that only a small part of the protein surface catalyzes the reactions, is a reasonable one on many grounds. Many of the reactions catalyzed by enzymes involve molecules that are quite small, so that their contact with the protein can be only very limited. Also, not only kinetic arguments but direct observations point to the binding of substrates with enzyme in a stoichiometric fashion. For example, the equilibrium dialysis method of Klotz and co-workers[55] gives both the binding constant and the number of binding sites in the enzyme molecule. The binding constant obtained in this manner has been shown to be identical to the kinetically determined binding constant, K_m, in several instances. Spectrophotometric methods have made possible the determination of the existence of a combination of NAD (nicotinamide-adenine dinucleotide) with alcohol dehydrogenase, as well as that of the ternary complex of NAD, substrate, and enzyme.[56] Ultracentrifugation has been used to show the binding of NADH by yeast alcohol dehydrogenase. Finally, the X-ray diffraction study of the complex of N,N'-diacetylchitobiose with lysozyme (Fig. 13.5) gives convincing evidence both of the binding of substrate to enzyme and of the function of a small portion of the protein surface as an active site to which the substrate is bound. Figure 13.5 depicts the location of attachment of the molecule N,N'-diacetylchitobiose, and Fig. 13.6 shows the attachment of the inhibitor N,N',N''-triacetylchitotriose to the enzyme.

These molecules become attached, at least in the crystal structure, in a cleft in the surface of the enzyme that runs nearly vertically down the middle in Fig. 13.4.

13.3 DETERMINATION OF ENZYMES

The International Union of Biochemistry has adopted a system of units[57, 58] which can be applied to all enzymes, which expresses their activity in absolute terms, and which makes possible the comparison between the activities of different enzymes. The definition recommended is as follows: one unit (U) of an enzyme is that amount that will catalyze the transformation of one micromole of the substrate per minute under standard conditions. It is recommended that wherever possible enzyme assays be based on measurements of initial rates of reaction and not on amounts of substrates changed by the end of a period of time, unless it is known that the velocity remains constant throughout this period. To facilitate measurement of the initial velocity, the substrate concentration should whenever possible be high enough to saturate the enzyme, so that the kinetics in the standard assay will approach zero-order.

The basis of the above assay method is, of course, the operational definition of an enzyme as a catalyst for a given reaction. The particular assay for a given enzyme will thus vary with the reaction it catalyzes. The optimal assay is one that is experimentally convenient and one that uses a specific substrate of the enzyme. For chymotrypsin, the hydrolysis of N-acetyl-L-tyrosine ethyl ester is often used; for ribonuclease, the hydrolysis of cytidine cyclic phosphate is often used.

As more enzymes are purified and as more stoichiometric reactions of enzymes are discovered, more titrations (all-or-none assays) can be used to determine the absolute concentration of active enzyme in a solution. Spectrophotometric titrations of chymotrypsin, trypsin, subtilisin, papain, elastase, acetylcholinesterase, and others are now possible; these utilize stoichiometric reactions of substrates or inhibitors with the respective enzymes. They have been discussed earlier (Chapter 12).

REFERENCES

1. M. O. Dayhoff, *Atlas of Protein Sequence and Structure, 1969*, National Biomedical Research Foundation, 1969.
2. C. Viswanatha, *Proc. Nat. Acad. Sci.*, **51**, 1117 (1964).
3. F. Sanger, *Adv. Protein Chem.*, **7**, 1 (1952).
4. C. H. W. Hirs, S. Moore, and W. H. Stein, *J. Biol. Chem.*, **235**, 633 (1960); D. H. Spackman, W. H. Stein, and S. Moore, *J. Biol. Chem.*, **235**, 648 (1960); and D. G. Smyth, W. H. Stein, and S. Moore, *J. Biol. Chem.*, **238**, 227 (1963).
5. B. S. Hartley, "The Structure and Active Groups of Chymotrypsin," 6th Int. Cong. of Biochem., N.Y., July–August, 1964.
6. R. A. Bradshaw, H. Neurath, and K. A. Walsh, *Proc. Natl. Acad. Sci., U.S.*, **63**, 406 (1969); *Sci. Res.*, **3**, April 1, 1968, p. 17.
7. J. Jolles and P. Jolles, *C. R. Acad. Sci.*, **253**, 2773 (1961).
8. J. Drenth et al., *Nature*, **218**, 929 (1968).
9. D. M. Shotton et al., *J. Mol. Biol.*, **32**, 155 (1968); D. M. Shotton and H. C. Watson, *Nature*, **225**, 811 (1970).
10. E. L. Smith, F. S. Markland, C. B. Kasper, R. J. DeLange, M. Landon, and W. H. Evans, *J. Biol. Chem.*, **241**, 5974 (1966).
11. W. N. Lipscomb, J. A. Hartsuck, G. N. Reeke, Jr., F. A. Quiocho, P. H. Bethge, M. L. Ludwig, T. A. Steitz, H. Muirhead, and J. Coppolla, *Brookhaven Symp. Biology*, 24 (1968); G. N. Reeke, Jr., J. A. Hartsuck, M. L. Ludwig, F. A. Quiocho, T. A. Steitz, and W. N. Lipscomb, *Proc. Nat. Acad. Sci., U.S.*, **58**, 2220 (1967).
12. J. T. Edsall and J. Wyman, *Biophysical Chemistry*, Academic Press, New York, 1958; C. Tanford, *Physical Chemistry of Macromolecules*, J. Wiley and Sons, New York, 1961.
13. C. L. Sai and B. L. Horecker, *Arch. Biochem. Biophys.*, **123**, 186 (1968).
14. E. Stellwagen and H. K. Schachman, *Biochemistry*, **1**, 1056 (1962); W. C. Deal, W. J. Rutter, and K. E. van Holde, *Biochemistry*, **2**, 246 (1963); H. K. Schachman and S. J. Edelstein, *Biochemistry*, **5**, 2681 (1966).
15. E. W. Holmes, Jr., J. I. Malone, A. I. Winegrad and F. A. Oski, *Science*, **159**, 650 (1968).
16. R. D. Cahn, N. O. Kaplan, L. Levine, and E. Zwilling, *Science*, **136**, 962 (1962).
17. A. N. Schechter and C. J. Epstein, *Science*, **159**, 997 (1968); N. O. Kaplan, *7th Int. Cong. of Biochem.*, **3**, 179 (1967).
18. M. Kunitz, *J. Gen. Physiol.*, **24**, 15 (1940).
19. C. C. F. Blake, D. F. Koenig, G. A. Mair, A. C. T. North, D. C. Phillips, and V. R. Sarma, *Nature*, **206**, 757 (1965); L. N. Johnson and D. C. Phillips, *Nature*, **206**, 761 (1965); see also R. A. Harte and J. A. Rupley, *J. Biol. Chem.*, **443**, 1663 (1968).
20. E. Fischer, *Ber.*, **27**, 2985 (1894); *Z. Physiol. Chem.*, **26**, 60 (1898).
21. J. Kraut et al., *Proc. Nat. Acad. Sci., U.S.*, **48**, 1417 (1962).
22. P. Urnes and P. Doty, *Adv. Protein Chem.*, **16**, 401 (1961).
23. D. M. Blow, M. G. Rossmann, and B. A. Jeffery, *J. Mol. Biol.*, **8**, 65 (1964); B. W. Mathews, P. B. Sigler, R. Henderson, and D. M. Blow, *Nature*, **214**, 652 (1967); P. B. Sigler, D. M. Blow, B. W. Mathews, and R. Henderson, *J. Mol. Biol.*, **35**, 143 (1968).
24. S. Wright, R. A. Alden, and J. Kraut, *Nature*, **221**, 235 (1969).
25. R. E. Dickerson et al., *J. Biol. Chem.*, **242**, 3015 (1967).
26. *Chem. Eng. News*, 18 (June 17, 1968).
27. J. A. Schellman and C. Schellman, in *The Proteins*, H. Neurath (Ed.), Vol. 2, Academic Press, New York, 1964, p. 1.

References

28. W. Kauzmann, *Adv. Protein Chem.*, **14**, 1 (1959).
29. See, however, I. M. Klotz, *Arch. Biochem. Biophys.*, **138**, 704 (1970).
30. G. Nemethy and H. A. Scheraga, *J. Phys. Chem.*, **66**, 1773 (1962).
31. H. S. Frank and M. W. Evans, *J. Chem. Phys.*, **13**, 507 (1945); H. S. Frank and W.-Y. Wen, *Disc. Faraday Soc.*, **24**, 133 (1957).
32. I. M. Klotz, *Ciba Foundation Symposium on the Frozen Cell*, 1970; G. E. W. Wolstenholme and M. O'Connor (Eds.), J. and A. Churchill Publishing Co., London.
33. H. White, Jr., *J. Biol. Chem.*, **235**, 383 (1960); **236**, 1353 (1961).
34. C. B. Anfinsen et al., *J. Biol. Chem.*, **236**, 422, 1361 (1961); **237**, 1839 (1962).
35. J. I. Harris, *Nature*, **177**, 471 (1956).
36. A. E. Mirsky and L. Pauling, *Proc. Nat. Acad. Sci., U.S.*, **22**, 439 (1936).
37. M. Levy and J. P. Magoulas, *J. Amer. Chem. Soc.*, **84**, 1345 (1962).
38. P. L. Whitney and C. Tanford, *J. Biol. Chem.*, **237**, PC1735 (1962).
39. J. H. Northrop, *J. Gen. Physiol.*, **15**, 29 (1931); J. H. Northrop and M. Kunitz, *Erg. Enzymforsch*, **2**, 104 (1933).
40. A. E. Stearn, *Adv. Enzymol.*, **9**, 25 (1949).
41. D. E. Koshland, Jr., J. A. Yankeelov, Jr., and J. A. Thoma, *Fed. Proc.*, **21**, 1031 (1962).
42. G. L. Rossi and S. A. Bernhard, *J. Mol. Biochem.*, **49**, 85 (1970).
43. A. L. Sluyterman, personal communication.
44. B. Chance, A. Ravilly, and N. Rumen, *J. Mol. Biol.*, **17**, 525 (1966).
45. J. J. Skujins and A. D. McLaren, *Science*, **158**, 1569 (1967).
46. F. M. Richards, personal communication.
47. D. C. Phillips, *Sci. Amer.*, **215**, 78 (1968).
48. G. Hathway and R. S. Criddle, *Proc. Nat. Acad. Sci., U.S.*, **56**, 680 (1966).
49. R. L. Burwell, Jr., personal communication.
50. L. J. Berliner and H. M. McConnell, *Proc. Nat. Acad. Sci., U.S.*, **55**, 708 (1966).
51. T. R. Stengle and J. D. Baldeschwieler, *J. Amer. Chem. Soc.*, **89**, 3045 (1967); J. R. Markley, I. Putter and O. Jardetsky, *Science*, **161**, 1249 (1968).
52. M. B. Hille and D. E. Koshland, Jr., *J. Amer. Chem. Soc.*, **89**, 5945 (1967).
53. B. H. Leichtling and I. M. Klotz, *Biochemistry*, **5**, 4026 (1966).
54. L. I. Katzin and E. Gulyas, *J. Amer. Chem. Soc.*, **86**, 1655 (1964).
55. I. M. Klotz, F. M. Walker, and R. B. Pivan, *J. Amer. Chem. Soc.*, **68**, 1486 (1946).
56. *Report of the Commission of Enzymes of the International Union of Biochemistry*, Vol. 20, Pergamon Press, New York, 1961.
57. *Enzyme Nomenclature, IUB Recommendations*, Elsevier Publishing Co., Amsterdam, London, New York, 1964.
58. M. L. Bender et al., *J. Amer. Chem. Soc.*, **88**, 5890 (1966).
59. *Scientific Research*, **3**, 17, April 1, 1968.
60. J. A. Rupley, personal communication.
61. G. R. Tristram and R. H. Smith, *Adv. Protein Chem.*, C. B. Anfinsen, Jr., M. L. Anson, and J. T. Edsall, Eds., **18**, Academic Press, New York, © (1963); see also D. Kirschenbaum, *Anal. Biochem.*, in press (1971).

Chapter 14

THE CONCEPT OF THE ACTIVE SITE

14.1 Constituents of the Active Site 478
14.2 Binding and Specificity 484

It is a general characteristic of enzyme-catalyzed reactions that the formation of a complex between the enzyme and its substrate precedes any covalent bond-making or bond-breaking steps. This binding occurs on, or in, a limited part of the extensive enzyme surface. The situation could hardly be otherwise in the numerous cases in which the substrate is a small molecule—it is difficult to envision significant interactions between the 17 atoms of p-nitrophenyl acetate with more than, say, 50 or 100 atoms of bovine trypsin, which contains about 3200 atoms. (Similar disparities obviously exist between relative sizes or numbers of functional groups, etc.) Yet the hydroysis of p-nitrophenyl acetate is catalyzed by trypsin by the same mechanism as those of macromolecular substrates, the mechanism including initial complex formations with an apparent Michaelis constant indicating very effective binding.

The same phenomenon can be inferred from the action of enzymes upon substrates that are of high molecular weight. Regardless of the number of points at which the two molecules could be in contact, covalent change in the substrate is highly localized: a proteolytic enzyme cleaves only one bond at a time in its substrate. Moreover, the number of structural features common to the specific substrates of a particular enzyme is small—often only three or four—suggesting that only a few strategic groups are bound, presumably to a similar number of complementary loci on the enzyme. These loci are, so to speak, clustered in a relatively small region—the *active site*— of the three-dimensional structure.

The binding of proteolytic enzymes is largely hydrophobic binding[1] (in an aqueous medium). Suggestions of ionic binding, particularly in acetylcholinesterase[2] and related enzymes, are probably exaggerated unless this ionic binding takes place in a medium of low dielectric. For example, the apolar analog of the usual acetylcholinesterase substrate, which has a cationic charge, binds with approximately the same binding constant and reacts with approximtely the same rate constant.[3] The binding in lysozyme is shown dramatically in Fig. 14.1[22] (see later). A similar phenomenon is seen in the binding to the cyclodextrins (cycloamyloses).[4]

An integral part of the concept of the active site is that there are a small number of groups on the enzyme that produce its high catalytic activity. In chymotrypsin the number is two or possibly three, excluding the binding. In other enzymes (a group on) the substrate may be involved or a small number of cofactors may be involved.

Fig. 14.1 Photograph of the model of a lysozyme molecule obtained by X-ray analysis at 6 Å resolution together with the increase in electron density observed in the presence of di-*N*-acetylchitobiose (hatched). The increase in electron density due to *N*-acetylglucosamine is shown as the darker part of the chitobiose. From L. N. Johnson and D. C. Phillips, *Nature*, **206**, 761 (1965).

The Concept of the Active Site

All in all, one may say that the enormous complexity of enzyme-substrate interaction is markedly reduced by this concept. Without this concept, it would be impossible to deal with enzymes as catalysts. With it, it is difficult but possible.

The simple template "lock and key" theory, which assumes immobility, is only a first approximation for the binding of substrate to enzyme.

14.1 CONSTITUENTS OF THE ACTIVE SITE

The constituents of the active site can be grossly subdivided into those entities responsible for catalytic action and those responsible for the enzyme's specificity. The catalytic entities may consist of prosthetic groups (coenzymes) or constituents of the protein itself. That is, in those enzymes requiring coenzymes, the active site includes the prosthetic group in the active enzyme. Table 14.1 gives a list of the catalytic constituents of the active sites of enzymes. In a given enzyme the catalytic entities of the active site are probably relatively few in number whereas the specificity residues may be manifold. It is very difficult to pinpoint the specificity residues in an enzyme. Our best information is seen in the cleft of the enzyme lysozyme where the competitive inhibitor N,N'-diacetylchitobiose appears to contact the side chains of many amino acids, notably of several tryptophan residues.

It is difficult in some cases to distinguish between a prosthetic group and a reactant. For example, NAD (nicotinamide-adenine dinucleotide) participates in a stoichiometric reaction with the substrate alcohol in a reaction catalyzed by the enzyme alcohol dehydrogenase to give NADH (reduced nicotinamide-adenine dinucleotide) and acetaldehyde. However, the nucleotide is subsequently reoxidized in a separate step involving a flavin, and thus is regenerated. A number of the substances listed in Table 14.1 undergo similar reactions.

Methods used for identifying the catalytic entities of the active site have been varied and ingenious. Careful purification and analysis have indicated the presence of many of the cofactors listed in Table 14.1. In many instances these cofactors may be removed from the protein component by dialysis, chromatography, or other methods, with subsequent loss of enzyme activity; the enzyme may then be reactivated by replacing the cofactors.

As mentioned earlier, reversible inhibition of an enzyme by a substrate-like molecule may block the active site from its usual function and reduce the enzymatic catalysis to a very low level. One example of such a reversible inhibition is the use of the drug iproniazid (1-isonicotinyl-2-isopropylhydrazine) which was shown to inhibit the enzyme monoamine oxidase.[5] In the brain, an enzyme of this type converts 5-hydroxytryptophan to 5-hydroxytryptamine

Table 14.1
Catalytic Constituents of Enzymatic Active Sites[a]

Coenzymic Catalysts or Reactants

Oxidation-Reduction Systems
    Nicotinamide-adenine dinucleotide (NAD)
    Nicotinamide adenine-dinucleotide phosphate (NADP)
    Flavin nucleotides such as flavin mononucleotide (FMN) and flavin adenine dinucleotide (FAD)
    Metal porphyrin complexes such as those found in the cytochromes, cobamide, peroxidase, and catalase
    Ascorbic acid
    Lipoic acid (thioctic acid)
    Coenzyme Q (ubiquinone)
    Metal ions such as $Cu^{2\oplus}$

Nonoxidation-Reduction Systems
    Thiamine pyrophosphate
    Pyridoxal phosphate
    Folic acid (pteroyl-L-glutamic acid)
    Biotin
    Glutathione
    S-Adenosylmethionine
    Coenzyme A
    Adenosine monophosphate, diphosphate, and triphosphate (AMP, ADP, ATP)
    Uridine phosphate
    Various metal ions, mainly $Zn^{2\oplus}$, $Mn^{2\oplus}$, $Mg^{2\oplus}$, $Cu^{2\oplus}$, $Co^{2\oplus}$
    4′-Phosphopantetheine

Constituents of the Protein

Carboxylate ion
Alcoholic hydroxyl group
Phenolic hydroxyl group
Ammonium ion (amine)
Imidazolium ion (imidazole)
Guanidinium ion
Indole ring
SH group
SCH_3 group
Peptide or amide group

[a] From M. L. Bender, in *Encyclopedia of Polymer Science and Technology*, John Wiley & Sons, Inc., New York, 1967, vol. 6, p. 1.

(serotonin), a substance that has pressor activity. The inhibition of this reaction by iproniazid presumably accounts for its use as a "psychic energizer" for depressed mental states. Using inhibitors with different structures as probes, the topography of the enzymatic active site can be deduced.

Inhibition of an enzyme may also occur by an irreversible (covalent) interaction. This kind of inhibition, which often obeys simple stoichiometry, is of great importance to enzyme chemistry and mechanism, for by use of it the active site of the enzyme can be investigated in a direct chemical manner. A classical example of such inhibitions is the demonstration by Balls and co-workers that chymotrypsin can be stoichiometrically inhibited by the reaction of one mole of this enzyme with one mole of O,O'-diisopropyl phosphorofluoridate.[6] This reaction was further shown to involve attachment of an O,O'-diisopropylphosphoryl residue to the hydroxyl group of serine 195 of the enzyme. A large number of enzymes, including acetylcholinesterase, are subject to inhibition by compounds containing the O,O'-diisopropylphosphoryl residue. The inhibition of acetylcholinesterase by organophosphorus compounds has been exploited in the development of the "nerve gases" since acetylcholinesterase controls transmission of nerve impulses in the body through its control of the hydrolysis of acetylcholine.

Many other stoichiometric reactions utilizing specific reagents or substrates have been used to identify constituents of the active sites of enzymes. Some of these reactions are shown in Table 14.2. For example, the imidazole group of histidine 57 of α-chymotrypsin reacts stoichiometrically with L-1-tosylamido-2-phenylethyl chloromethyl ketone, a reactive alkylating agent with a backbone corresponding to that of phenylalanine, a substrate of chymotrypsin.[7] The product of this reaction is a completely inactive enzyme. This result implies either that the imidazole group is a necessary part of the active site of chymotrypsin or that the bulk of the alkylating agent prevents access of substrate to the site. Methylation of histidine 57 by methyl p-nitrobenzenesulfonate also produces an inactive enzyme.[8] In the enzyme ribonuclease, iodoacetate ion carboxymethylates either histidine 119 or histidine 12, the former in higher yield.[9] Neither of these carboxymethylated proteins reacts further with iodoacetate and neither is enzymatically active. However, if the two inactive ribonuclease derivatives are mixed with one another, enzymatic activity reappears.[10] The stoichiometric inhibition of papain and glyceraldehyde-3-phosphate dehydrogenase by reagents such as p-mercuribenzoate has been interpreted to mean that sulfhydryl groups are constituents of the active site of these enzymes.

Several criteria must be met for the identification of constituents of the active site by inhibition reactions. First, a stoichiometric reaction must be observed. Second, the reaction must be specific for a particular group. Third, the product must be identified and the inactivity of the enzyme must be

Constituents of the Active Site

Table 14.2[a]
Some Reactions Used in Identification of Active Sites of Enzymes

---

(α-chymotrypsin)—OH + ((CH$_3$)$_2$CHO)$_2$P(=O)F ⟶ ((CH$_3$)$_2$CHO)$_2$P(=O)—O—(α-chymotrypsin) + HF

(α-chymotrypsin)—OH + CH$_3$CO—C$_6$H$_4$—NO$_2$ ⟶ (α-chymotrypsin)—O—C(=O)CH$_3$ + HO—C$_6$H$_4$—NO$_2$

(α-chymotrypsin)—(imidazole)NH + C$_6$H$_5$—CH$_2$—CH(NHTs)—C(=O)—CH$_2$Cl ⟶ (α-chymotrypsin)—(imidazole)N—CH$_2$—C(=O)—CH(NHTs)—CH$_2$—C$_6$H$_5$ + HCl

transaminase—NH$_2$ + $^{2\ominus}$O$_3$POCH$_2$—(pyridine ring with CHO, O$^\ominus$, CH$_3$, $\overset{\oplus}{N}$H) ⟶ transaminase—N=CH—(pyridine ring with O$^\ominus$, CH$_3$, $\overset{\oplus}{N}$H, CH$_2$OPO$_3^{2\ominus}$) + H$_2$O

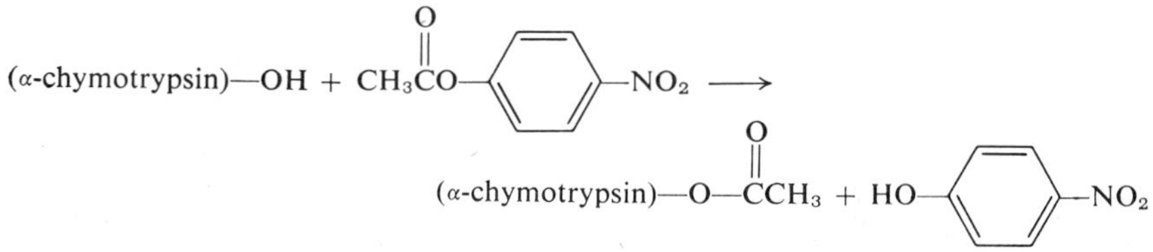

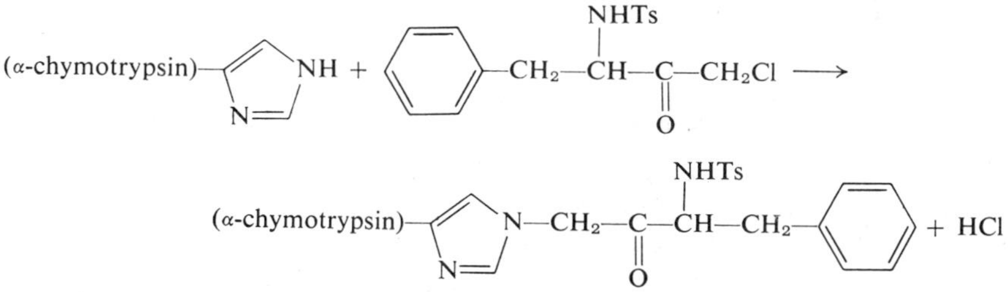

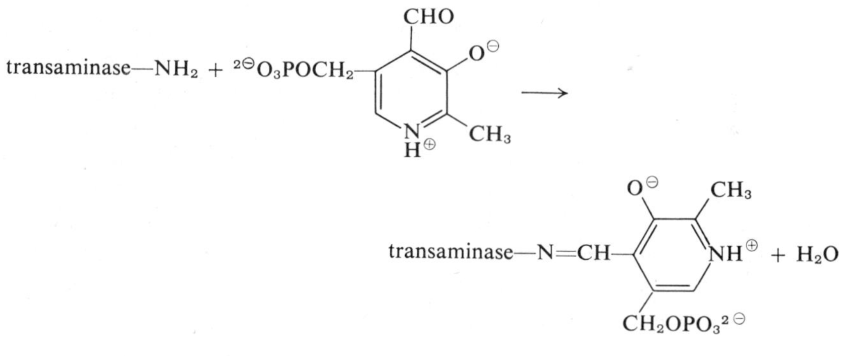

transaldolase—(CH$_2$)$_4$NH$_2$ + fructose 6-phosphate ⟶
CH$_2$OH—C(=N(CH$_2$)$_4$—transaldolase)—CH$_2$OH + glyceraldehyde 3-phosphate

---

(continued)

Table 14.2 (*continued*)

---

aldolase—$(CH_2)_4NH_2$ + dihydroxyacetone phosphate $\longrightarrow$

$$\begin{array}{c} CH_2OPO_3^{2\ominus} \\ | \\ C=N(CH_2)_4\text{—aldolase-} + H_2O \\ | \\ CH_2OH \end{array}$$

acetoacetate decarboxylase—$(CH_2)_4NH_2$ + acetoacetate $\longrightarrow$

$$\begin{array}{c} H_3C \\ \diagdown \\ C=N(CH_2)_4 \text{ acetoacetate decarboxylase} + CO_2 \\ \diagup \\ H_3C \end{array}$$

---

[a] From M. L. Bender, in *Encyclopedia of Polymer Science and Technology*, John Wiley & Sons, Inc., New York, 1967, vol. 6, p. 1.

demonstrated. Finally, the distinction must be made between chemical modification by the reagent and blocking of the site through steric hindrance of the reagent.

The pH dependence of enzyme reactions indicates in many instances the participation of one or more ionizable groups in enzymatic catalysis, either an acid, a base, or both. Of course, one does not know a priori whether the ionizable group is necessary for the binding of the substrate to the enzyme, for the catalytic reaction, or for the maintenance of the proper conformation of the protein. However, the pH dependence of enzyme reactions, in conjunction with other evidence, has been a useful guide to the identification of groups involved in catalytic action of enzymes. The identity of the group can be deduced from the pK of the ionizable group. For example, hydration of fumarate to malate by the enzyme fumarase is characterized by a bell-shaped pH-rate profile. From this pH dependence, Alberty showed that two groups, one acidic and one basic, are involved in the enzymatic catalysis. He tentatively identified the groups, which had pK_a values of 5.3 and 7.3, as imidazole groups of histidine moieties of the enzyme.[11]

Since protein conformation is pH-sensitive, the range of ionizable groups that may be identified by the pH dependency of an enzymatic reaction is quite limited, usually to between pH 3 and 10. The vast majority of enzymatic reactions depend on groups with pK values from 5 to 9 although a few exceptions exist: e.g., pepsin catalyzes reactions at pH values below 5. Correlations of these pK values with the known pK values of the various protein constituents listed in Table 14.1 indicate that only a few groups such as carboxylate ions, imidazole groups, ammonium ions, and phenol groups are

Constituents of the Active Site

implicated by the pH dependency of enzymatic reactions. Unfortunately no information can be gained about protein constituents whose pK values fall below 3 or above 10.

Nonenzymatic systems that reproduce in some measure the constituents of the active site of an enzyme can confirm designation of the active site. For example, the monoanion of fumaric acid undergoes a hydration to malic acid that is characterized by a bell-shaped pH-rate profile just as the fumarase reaction.[12] The acidic and basic groups responsible for this intramolecular reaction are the carboxylic acid and the carboxylate ion of the fumaric acid monoanion, and not the imidazole groups presumably operative in the enzymatic reaction (Chapter 10). However, the demonstration that an acid and a base can catalyze the hydration of an olefin lends credence to the view that an acid and a base on the enzyme fumarase operate in a like fashion.

Even more important nonenzymatic models of enzymatic reactions include reactions catalyzed by various coenzymes. Pyridoxal phosphate, thiamine pyrophosphate, nicotinamide-adenine dinucleotide, and various metal ions, without their corresponding enzymes, have been shown to catalyze the same general chemical transformation that is carried out by the cofactors in association with the protein, albeit at a much slower rate. These results, coupled with the fact that the enzymatic processes are dependent on these cofactors, indicate that the active site of the enzyme is centered around the cofactor.

The binding of detergent anions to bovine serum albumin provides rudimentary knowledge of the concept of the active site. However, in such bindings, an *individual* site is not seen and binding may be different from enzymatic binding.[13] Further rudimentary knowledge on active sites may be obtained by a comparison of enzymic active sites with those of Ziegler-Natta (heterogeneous) active sites.[14] Nuclear magnetic resonance has been used to identify sites on proteins that interact with drugs in solution.[15]

Enzyme-substrate intermediates can be observed or inferred in a number of ways. The formation of an adsorptive complex between enzyme and substrate before the catalytic process takes place was postulated, as mentioned above, on kinetic and equilibrium evidence. By skillful combination of spectrophotometry and fast reaction kinetic techniques, Chance was able to demonstrate the presence of several enzyme-substrate intermediates in enzymatic redox reactions involving catalase and peroxidase.[16] This observation was possible because the visible absorption of the iron-porphyrin complex in the active site of these enzymes is changed on formation of the enzyme-substrate intermediate(s).

Intermediates with substrate covalently bonded to the enzyme protein have been trapped in several instances. Four examples of such trapping reactions include the stoichiometric reactions of the enzymes transaminase,[17] transaldolase,[18] aldolase,[17] and acetoacetate decarboxylase[19] shown in Table 14.2.

484   The Concept of the Active Site

In each case the intermediate compound involves a Schiff base bond between the enzyme and the substrate (or cofactor). Although Schiff base bonds are unstable in aqueous solution, treatment of the unstable Schiff base with sodium borohydride reduces it to a stable amine derivative of the enzyme, which can be degraded and identified.

Evidence for formation of enzyme-substrate intermediates also comes from sources such as tracer experiments involving carbon, hydrogen, and oxygen isotopes, whose outcome can be explained only by the intermediacy of a particular compound. Furthermore, isotopic exchange experiments also indicate intermediate formation. An example is the pair of isotopic exchange reactions between dihydroxyacetone phosphate and tritiated water catalyzed by the enzymes aldolase and triose phosphate isomerase. In each case a different α-hydrogen of the ketone is exchanged with water, leading to two discrete monotritiated derivatives, **1** and **2**:

$$\begin{array}{cc}
\text{H} & \text{H} \\
| & | \\
\text{T—C—OH} & \text{HO—C—T} \\
| & | \\
\text{C=O} & \text{C=O} \\
| & | \\
\text{CH}_2\text{OPO}_3\text{H}^\ominus & \text{CH}_2\text{OPO}_3\text{H}^\ominus \\
\text{Labeled by isomerase} & \text{Labeled by aldolase} \\
\mathbf{1} & \mathbf{2}
\end{array}$$

These two exchange reactions indicate stereospecific formation of the corresponding enolate ion intermediates on the surfaces of the two enzymes.[20, 21]

The complex of N,N'-diacetylchitobiose with lysozyme (Fig. 14.1) gives convincing evidence both of the binding of substrate to enzyme and of the function of a small portion of the protein surfaces as an active site to which the substrate is bound. Fig. 14.1 shows the location of attachment of the molecule, N,N'-diacetylchitobiose, which is a competitive inhibitor, to the enzyme. This molecule becomes attached, at least in the crystal structure, in a cleft in the surface of the enzyme which runs nearly vertically down the middle of Fig. 14.1.

Both the kinetic and equilibrium arguments for binding of the substrate to an active site of the enzyme were anticipated by the "lock and key" theory. This theory, due to the insight of Emil Fischer, explained the specificity of enzymes on the basis of a stereochemical (template) relationship between the substrate and enzyme. The orderly sequence of reactions in a cell is made possible by the fact that enzymes are specific.

14.2 BINDING AND SPECIFICITY

Enzymes exhibit specificity with regard to the gross reaction, the position of cleavage, the geometry of the molecule, and the optical activity of the molecule. Some enzymes have a very broad specificity; for example, chymotrypsin will catalyze the hydrolysis of essentially every amino acid derivative to some extent and will hydrolyze many carboxylic acid derivatives that bear little relationship to the natural protein substrates of this enzyme. On the other hand, some enzymes have very narrow specificity; for example, it was not until 1965 that a substrate other than urea was found for the enzyme urease. The specificity, whether broad or narrow, can most easily be explained by a reasonably close template correspondence; the enzyme can present an asymmetric template to the substrate molecule and thus distinguish between enantiomorphs. The simplest description for the operation of this template specificity involves the binding of the substrate to the enzyme. Interestingly enough, specificity is not completely defined by differences in binding constants (K_m or K_s). For example, chymotrypsin binds a D substrate slightly better than it binds an L substrate. But the L substrate is subsequently hydrolyzed approximately 10^4 times faster than the D substrate, and thus the enzyme manifests its optical specificity kinetically. This result may be very easily rationalized on the basis that the correct stereochemical fit for reactivity is not a prerequisite for binding.

In order to explain the unreactivity of smaller analogs of substrates, both those that do not bind and those that bind but do not react, the "induced fit" theory of specificity has been proposed.[23] This theory involves three postulates: (1) that a change in geometry of the protein is caused by the substrate as it fits into the active site; (2) that a delicate orientation of catalytic groups is required for enzyme action; and (3) that the substrate "induces" this proper orientation by the change it causes in geometry of the enzyme.

Another hypothesis that attempts to explain the unreactivity of substrate analogs is embodied in the concept of nonproductive binding, which proposes that molecules that differ in structure from a specific substrate may bind to the active site of an enzyme in more than one mode. Many of the modes of binding will be nonproductive (i.e., will not lead to reaction).[24] An example of this is in the unreactivity of the D substrate mentioned above.

The "induced fit" theory requires a flexible enzyme and presumably a rigid substrate. The hypothesis of nonproductive binding requires a rigid enzyme and a rigid substrate. Other combinations are obviously possible. But the important point is that the simple template "lock and key" theory is only a first approximation for the binding of the substrate to enzyme. We will take this point up again in Chapter 19.

REFERENCES

1. B. H. J. Hofstee, *Nature*, **213**, 42 (1967).
2. F. Bergmann and A. Shimoni, *Biochim. Biophys. Acta*, **8**, 347 (1952).
3. R. M. Krupka, *Biochemistry*, **5**, 1983 (1966).

4. M. L. Bender, *Trans. N.Y. Acad. Sci.*, **29**, 301 (1967).
5. E. A. Zeller and J. Barsky, *Proc. Soc. Exptl. Biol. Med.*, **81**, 459 (1952).
6. E. F. Jansen, M.-D. F. Nutting, and A. K. Balls, *J. Biol. Chem.*, **179**, 201 (1949); E. F. Jansen, M.-D. F. Nutting, R. Jang, and A. K. Balls, *ibid.*, **179**, 189 (1949); **185**, 209 (1950).
7. G. Schoellmann and E. Shaw, *Biochemistry*, **2**, 252 (1963), see also B. L. Vallee and J. F. Riordan, *Ann. Rev. Biochem.*, **38**, 733 (1969).
8. Y. Nakagawa and M. L. Bender, *J. Amer. Chem. Soc.*, **91**, 1566 (1969); *Biochemistry*, **9**, 259 (1970).
9. A. M. Crestfield, W. H. Stein, and S. Moore, *J. Biol. Chem.*, **238**, 2413 (1963).
10. A. M. Crestfield, W. H. Stein, and S. Moore, *J. Biol. Chem.*, **238**, 2421 (1963).
11. R. A. Alberty, *J. Cell. Comp. Physiol.*, **47**, Suppl. 1, 245 (1956).
12. M. L. Bender and K. A. Connors, *J. Amer. Chem. Soc.*, **84**, 1980 (1962).
13. J. A. Reynolds, S. Herbert, H. Polet, and J. Steinhardt, *Biochemistry*, **6**, 937 (1967).
14. L. Rodriguez and H. M. Van Looy, *J. Polymer Sci.*, **4**, A-1, 1951 (1966).
15. O. Jardetsky, J. S. Cohen, D. H. Meadows, and J. L. Markley, *Proc. Nat. Acad. Sci. U.S.*, **58**, 1307 (1967).
16. B. Chance in *Technique of Organic Chemistry*, Vol. VIII, Part 2, S. L. Friess, E. S. Lewis, and A. Weissburger, Eds., Interscience Publishers Inc., New York, 1962, Chapter 22.
17. R. C. Hughes, W. T. Jenkins, and E. H. Fischer, *Proc. Nat. Acad. Sci. U.S.*, **48**, 1615 (1962).
18. J. C. Speck, Jr., P. T. Rowley, and B. L. Horecker, *J. Amer. Chem. Soc.*, **85**, 1012 (1963).
19. F. H. Westheimer, *Proc. Chem. Soc.*, 253 (1963).
20. S. V. Reider and I. A. Tose, *J. Biol. Chem.*, **234**, 1007 (1959).
21. See also T. McL. Spotswood, J. M. Evans, and J. H. Richard, *J. Amer. Chem. Soc.*, **89**, 5052 (1967).
22. L. N. Johnson and D. C. Phillips, *Nature*, **206**, 761 (1965).
23. D. E. Koshland, Jr., *J. Cell. Comp. Physiol.*, **54**, Suppl. 1, 245 (1959).
24. G. E. Hein and C. Niemann, *J. Amer. Chem. Soc.*, **84**, 4495 (1962); S. A. Bernhard and H. Gutfreund, *Proc. Int. Symp. Enzyme Chem., Tokyo and Kyoto, 1957* Maruzen, Tokyo, 1958, p. 124; E. A. Zeller et al., *Biochem. J.*, **95**, 262 (1965).

Chapter 15

ENZYME KINETICS

15.1	Steady-State Kinetics	487
15.2	Allosteric Kinetics	491
15.3	Stepwise Kinetics	494
15.4	Transient Kinetics	497

15.1 STEADY-STATE KINETICS

The characteristic property and function of enzymes is the catalysis of chemical reactions. Any fundamental study of this catalytic function must be based on quantitative measurements of the rate of catalyzed reactions. From the effect of varying conditions on the rate, inferences may be made about the mechanism of enzyme action. The effect of substrate concentration on the rate of enzymatic reactions is the most instructive.†

In nearly all cases when initial velocity of an enzyme-catalyzed reaction is plotted against substrate concentration a section of a rectangular hyperbola is obtained, as shown in Fig. 15.1. Such a result is obtained whenever a process depends upon a simple equilibrium between two reacting entities. A theory involving such an equilibrium was put forward by Michaelis and Menten[1] based on an earlier suggestion of Henri.[2] It postulates that the enzyme first forms a complex with its substrate and this subsequently breaks

† The following treatment is not intended to be comprehensive. See also Ref. 11 and Chapter 12.

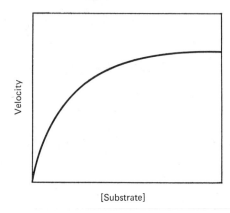

Fig. 15.1 The effect of substrate concentration on the rate of an enzyme-catalyzed reaction.

down yielding the free enzyme and products of the reaction. The process may be formulated as

$$E + S \rightleftharpoons ES \longrightarrow E + \text{products} \tag{15.1}$$

When $[S]_0 \gg [E]_0$, the velocity of this reaction may be expressed as

$$v = \frac{V}{1 + (K_m/[S])} \tag{15.2}$$

where v is the velocity of the reaction, V the maximum velocity (the velocity at infinitely high substrate concentration), $[S]$ the concentration of the substrate, and K_m the substrate concentration at half the maximum velocity. In simple cases K_m is a function of the degree of binding of the substrate to the enzyme. Thus the two parameters V and K_m of the Michaelis–Menten equation denote the velocity at which the enzyme-substrate complex decomposes and the tendency of the enzyme to bind to the substrate, respectively.

Fig. 15.1 and the Michaelis–Menten mechanism indicate that the velocity of an enzymatic reaction is first-order[5,6] in substrate at very low substrate concentration, that is, when the enzyme is not saturated with substrate, and zero-order in substrate at very high concentration, that is, when the enzyme is saturated with substrate.[3,4] This description, of course, could be reversed: one could saturate the substrate with the enzyme rather than vice versa, but, of course, a catalyst is ordinarily present in small concentration and thus the condition in which the concentration of substrate is greater than that of enzyme is usually the one of interest.

This idealized treatment is beset with many complications. One of the interesting ones involves the presence of a "competitive" inhibitor in the system where the inhibitor competes with the substrate for the binding site of the enzyme. The binding of the substrate to the enzyme implies that a molecule similar in size and shape to the substrate, but unable to react with the enzyme, would bind to the same place, that is, the active site, on the enzyme surface. Binding of the competitive inhibitor in place of the substrate leads to a decrease in the velocity of the enzymatic reaction. Such a decrease is observed and can be described by eqs. 15.3a and 15.3b, where E is the enzyme, S the substrate, I the inhibitor, and P the reaction products.

$$E + S \rightleftharpoons ES \longrightarrow E + P \quad (15.3a)$$

$$E + I \rightleftharpoons EI \quad (15.3b)$$

Such inhibitions are important physiologically in the control of enzyme reactions by addition of specific inhibitors. The phenomenon of competitive inhibition provides strong evidence that ES is on the reaction pathway.

Many other more complex kinetic behaviors have been observed such as hydrolysis in nonaqueous medium,[7,8] measurement of individual steps,[9,10] relaxation techniques,[11] "product inhibition," "substrate inhibition," and oscillation.[12]

Michaelis–Menten kinetics describes the simplest possible enzymatic process involving one substrate and a unidirectional system. In the analysis of a more complicated system, the difficulties mount rapidly. For example, one could discuss equilibrium systems, systems involving two reactants and two products, and systems involving intermediates.[6,13,14]

Inhibition may be either competitive, noncompetitive, or uncompetitive. These concepts are fully treated elsewhere,[15-17] and are beyond the scope of this book. An enzyme-catalyzed reaction may be activated (accelerated) by excess substrate[18] or by a foreign material.[19-22] A striking example of the acceleration of hydrolysis of a poor substrate of the enzyme trypsin is shown in Fig. 15.2.[21]

A further consideration when two or more substrates bind to the enzyme is whether the binding is ordered ("ping-pong kinetics") or not (random kinetics).[23] One example of an enzymatic process that goes through an ordered mechanism involves liver alcohol dehydrogenase.[24] This was discovered mainly through the use of competitive inhibitors. Another example involves the enzyme pyruvate carboxylase.[25] In the former case, intermediates involving the enzyme and $NAD^{\oplus}$; the enzyme, $NAD^{\oplus}$, and the alcohol; the enzyme, NADH, and the aldehyde; and the enzyme and NADH were postulated on the basis of the kinetics. In the latter case, a carboxybiotin-enzyme intermediate was postulated to explain the kinetics. Examples of enzymatic

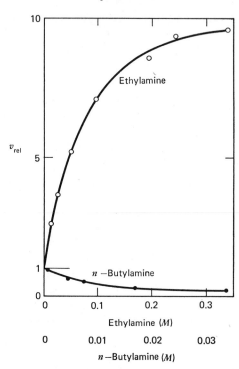

Fig. 15.2 Action of alkylamines on the trypsin-catalyzed hydrolysis of AGEE (*N*-acetylglycine ethyl ester). Relative rates are plotted taking unaccelerated rate as unity. The concentration of AGEE is 0.376 M in the case of ethylamine and 0.350 M for *n*-butylamine. From T. Inagami and T. Murachi, *J. Biol. Chem.*, **238**, PC1905 (1963).

processes that proceed through a random mechanism are few and as time goes on become fewer. Perhaps one example is the enzymatic conversion of a sulfonium salt, *S*-adenosylmethionine, to a thioether.[26] This reaction apparently involves no intermediate nor would one be predicted on chemical grounds. This method in general allows one to use steady-state kinetics and yet to say something about individual steps in the reaction. What "ordered" means is that there is a (covalent) intermediate in the reaction. What "random" means is that there is not.

Kinetics, especially steady-state, can not *prove* mechanism. Kinetics can only *disprove* mechanism.

The simple Michaelis–Menten treatment described above assumes an equilibrium between the enzyme and the substrate. Briggs and Haldane[27] showed that the same rate equation is obtained when the concentration of ES does not change with time, compared to the rates of change of other reactants in eq. 15.1. This is the steady-state treatment mentioned above and was predicated on some earlier work of Bodenstein.[28] The Briggs-Haldane treatment gives the equivalent of eq. 12.2, but the meaning of the experimentally observed parameters is changed. K_m is replaced by the ratio of rate constants $(k_{-1} + k_2)/k_1$.

When an added nucleophile is present, the concept of quasi-equilibrium must be considered.[29,30] This concept, which is between a steady-state and a Michaelis–Menten treatment, demands that the binding of the nucleophile to the enzyme (in the presence of the substrate) be identical with the binding of water to the enzyme (in the presence of the substrate). Sometimes this condition is met and at other times it is not.[31] When quasi-equilibrium occurs, the concentration of the equilibrium complex involving enzyme substrate, and nucleophile may be perturbed from its equilibrium value if binding of the substrate is not perturbed by binding of the reactant. Since most enzymatic processes proceed through such a triple complex, where the third component is usually water, this concept can be of great importance.

15.2 ALLOSTERIC KINETICS

A number of enzymes, involved mostly in biosynthetic pathways of intermediary metabolism, display unique kinetic behavior both in terms of reactivity with their substrates and with some other specific compounds, called effectors. These enzymes are called allosteric enzymes. This concept was first enunciated in 1963.[32]

A sigmoidal v versus [S] relationship is observed with allosteric enzymes (Fig. 15.3).

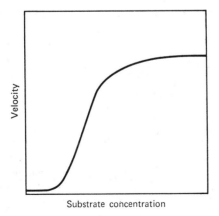

Fig. 15.3 The effect of substrate concentration on the velocity of a typical allosteric enzymatic reaction.

The rate, unlike the case of enzymes that display Michaelis–Menten behavior, shows a higher than first-order dependency on substrate concentration. The upward curvature (reminiscent of the oxygen saturation curve of hemoglobin) suggests cooperativity for binding the substrate molecules. Inasmuch as these enzymes seem to be composed of multisubunit assemblies, the individual binding sites located on different subunits must interact cooperatively.

The catalytic behavior of these enzymes is regulated (or modulated) by effectors, either positively or negatively (moves the sigmoid to the left or right, respectively). It appears that the effectors are intermediary metabolites on the same or parallel pathway that the enzyme serves. An earlier metabolite is usually a positive effector and a later one takes on the role of a negative effector. Often the product of a pathway inhibits the first critical enzyme of the same pathway without affecting any of the in-between steps on the pathway ("feedback inhibitors"), as in eq. 15.4.

$$A_1 \xrightarrow{E_1} A_2 \xrightarrow{E_2} A_3 \xrightarrow{E_3} A_4 \xrightarrow{E_4} P \qquad (15.4)$$

Thus one is dealing with an important concept in the control of metabolic reactions, because physiologically occurring concentrations of the particular effectors will suffice to show the activation or inhibition described.

With allosteric enzymes the added modifier or effector has little, if any, relation to the actual substrate. This implies that they must be bound at another site (allosteric site) on the enzyme and not at the catalytic site. In fact (e.g., with aspartate transcarbamylase) the effector is often bound to a different (so-called regulatory) subunit that together with the catalytic subunits forms the multisubunit native enzyme.[33] Allosteric effectors do not directly participate in the enzymatic reaction. Their action may be interpreted either in terms of increasing or decreasing the velocity of the actual reaction or in terms of altering the affinity of the enzyme for the substrate.

A concerted model for changes of a tetrameric enzyme complex was proposed as shown in Fig. 15.3; a stepwise process was suggested later[34] (Fig. 15.4). In this figure a tetrameric enzyme was assumed to exist although it is known that many other oligomeric enzyme aggregations exist.[35] An attempt to differentiate between the concerted and the stepwise models of allostericism (multiple conformational change) by relaxation techniques has been described recently.[36]

Antagonistic allosteric effectors[37] and homosterism in which the substrate

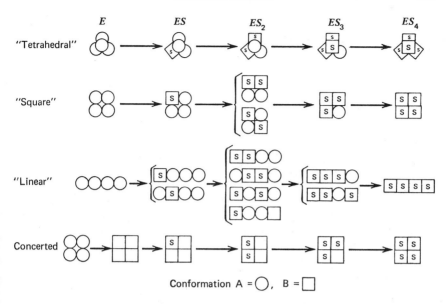

Fig. 15.4 Schematic illustration of the various modes of binding the ligand S to the tetrameric enzyme for four major cases. Conformation A of the subunit is denoted by circles. Conformation B, the one capable of binding S, is denoted by squares. In the models involving a progressive change it is assumed that a subunit in conformation B is present only when S is bound to it. In the concerted model all conformations change to B together. From D. E. Koshland, Jr., G. Némethy, and D. Filmer, *Biochemistry*, **5**, 365 (1966). © 1966 by the American Chemical Society. Reprinted by permission of the copyright owner.

modifies its own reactivity[38] have been reported. The behavior of the enzyme yeast glycogen synthetase has been interpreted in terms of allostericism.[39,40] Models for allosteric effects in a dehydrogenase enzyme[41] and in the binding of small molecules to a polymerizing protein system[42] have been suggested.

A number of different modifications of the native enzyme lead (often reversibly) to the loss of its allosteric properties (desensitization). Such modifications range from natural enzyme mutations to treatment of the isolated wild-type enzyme with mild heat; exposures to urea or other protein reagents, for example, mercurials. Desensitization often involves a marked drop in molecular weight, that is, dissociation into subunits. These subunits are catalytically active per se but will now display a classical Michaelis–Menten behavior (Fig. 15.1). Thus cooperativity is lost in a manner reminiscent of a comparison between the oxygenation of hemoglobin and myoglobin. The $V_{\max}$ of the desensitized enzyme is essentially the same as that of the native one. For example, aspartate transcarbamylase (MW ~310,000) on treatment with p-chloromercuribenzoate dissociates into two types of subunits that can be separated from each other. Only one of these (63% by weight) is catalytically active. The other (37%) is responsible however, for binding the allosteric effectors (e.g., cytidine triphosphate). The isolated catalytic units show typical Michaelis–Menten behavior and are insensitive to the presence of the triphosphate in the reaction medium. Addition of the regulatory protein, however, changes the v versus [S] profile to a sigmoidal curve.[43] The native (or reconstituted) enzyme assembly consists of a ring of three pairs of regulatory subunits (i.e., 6 × 17,000 MW); above and below this ring, on each side there is a set of three catalytic subunits (3 × 33,000 MW). Allostery means interaction at a distance. The fact that allosteric enzymes consist of subunits does not prove this concept, for the "allosteric site" and the "catalytic site" may be adjacent to one another or far apart (in different subunits).

15.3 STEPWISE KINETICS

The kinetics and mechanism associated with the enzyme chymotrypsin can serve as an illustration of an enzymatic process involving one covalent intermediate. It is therefore taken up in some detail below.

α-Chymotrypsin reacts with diisopropyl phosphorofluoridate in a 1:1 stoichiometric reaction to give a crystalline product that contains one gram atom of phosphorus per mole and which is completely inactive enzymatically.[44] Chymotrypsin can also be inactivated by diphenylcarbodiimide,[45] diphenyldiazomethane,[46] and cupric ion.[47] Phenyl-(p-phenylazophenyl)-carbamyl chloride also inactivates chymotrypsin. Since this compound exists in cis and $trans$ isomers and since the cis isomer is more reactive, exposure of the reaction mixture to light (which converts the $trans$ isomer to the cis) produces more inactivation.[48] The reaction of α-chymotrypsin with p-nitrophenyl acetate,[50,51] N-acetyl-L-tryptophan,[52] N-furylacryloyl-L-tryptophan,[53] (the latter two are specific) lead to acyl-enzymes (isolable at low pH). In addition to these derivatives, a diphenylcarbamyl-enzyme,[54] several sulfonyl-enzymes,[55,56] and a host of other acyl-enzymes,[50,57–62] have been prepared from chymotrypsin and in several cases crystallized.[47,58] These compounds are usually stoichiometric, inactive compounds. The phosphoryl, carbamyl, or acyl group may be removed from the enzyme by a nucleophile with varying degrees of ease. These enzymatic processes provide a solid chemical background for the applicability of eq. 15.5 to chymotrypsin reactions.

$$E + S \underset{}{\overset{K_s}{\rightleftharpoons}} E \cdot S \xrightarrow{k_2} \underset{+ P_1}{ES'} \xrightarrow{k_3} E + P_2 \qquad (15.5)$$

where E is enzyme, S substrate, E·S complex, ES' acyl-enzyme, P_1 and P_2 the first and second products, k_2 and k_3 the rate constants of acylation and deacylation, and K_s the dissociation constant (as are all equilibrium constants in this chapter) of E·S.

The formation and decomposition of acyl-enzyme intermediates may be detected spectrophotometrically during enzymatic processes. Such observations may be made at pH 7.8, as well as at the lower pH's usually used to isolate an acyl-enzyme. An acyl-enzyme has been detected spectrophotometrically in α-chymotrypsin-catalyzed hydrolysis of labile substrates[57] such as o-nitrophenyl cinnamate and N-trans-cinnamoylimidazole, and in the α-chymotrypsin-catalyzed hydrolysis of a nonlabile ester.[52,53,63] The direct observation of step k_3 from an isolated specific (fast) acyl-enzyme indicates that it is an intermediate in the reaction pathway.[50]

In the hydrolysis of p-nitrophenyl acetate, an initial rapid liberation ("burst") of one mole of p-nitrophenol per mole of enzyme was observed, followed by a slow (zero-order) reaction of the remaining substrate.[64] The kinetics of the initial rapid reaction (the presteady-state, measuring acylation) and the subsequent slow reaction (the steady-state, measuring deacylation) in the α-chymotrypsin-catalyzed hydrolysis of p-nitrophenyl acetate is consistent with eq. 15.5.

Direct evidence for the applicability of eq. 15.5 to specific substrates has been reported (see earlier). An acyl-enzyme intermediate was seen spectrophotometrically in the α-chymotrypsin-catalyzed hydrolysis of N-acetyl-L-tryptophan methyl ester at low pH.[65] A burst of p-nitrophenol was found at low pH for the specific chymotrypsin substrates, N-acetyl-L-tryptophan and N-benzyloxycarbonyl-L-tyrosine p-nitrophenyl esters.[65]

N-Acetyl-L-tryptophan reacts with α-chymotrypsin at low pH to form an equilibrium concentration of acyl-enzyme, as demonstrated spectrophotometrically by the decrease in the normality of active sites,[65] and by isotopic labeling.[66] The rate of deacylation of N-furylacryloyltryptophanyl-chymotrypsin equals the rate of turnover of the corresponding ester at pH 8 (33 sec^{-1}).[53]

Indirect evidence for the application of eq. 15.5 to the reactions of specific ester substrates follows. The catalytic rate constants of the α-chymotrypsin-catalyzed hydrolysis of the ethyl, methyl, and p-nitrophenyl esters of N-acetyl-L-tryptophan are essentially identical, while the K_m(app)'s of these derivatives of common backbone differ markedly. The nucleophilic character of deacylations of chymotrypsin and trypsin predicts substantially different rates, but a common rate-determining deacylation explains the results.[67] The different K_m(app) in the former reaction may be explained by differing k_2/k_3 ratios affecting K_m(app).[67] The kinetics of the α-chymotrypsin-catalyzed hydrolysis of specific substrates in the presence of methanol may be successfully analyzed in terms of the competitive partitioning of the acyl-enzyme intermediate by the nucleophiles, water and methanol.[68] The individual acylation and deacylation rate constants of the α-chymotrypsin-catalyzed hydrolysis of N-acetyl-L-tryptophan methyl and ethyl esters determined from relative rate data, the effect of pH on the kinetics, and alcoholysis data are consistent with one another.[69] These rate constants, together with those of the alcoholysis of the acyl-enzyme and the acylation of the enzyme by the free acid (determined from isotopic oxygen exchange), give an equilibrium constant for the overall reaction consistent with independently determined equilibrium constants.[69,70]

Indirect evidence for the application of eq. 15.5 to the reactions of specific amide substrates follows. The rates of the α-chymotrypsin-catalyzed hydrolysis of N-acetyl-L-tyrosinamide and of N-acetyl-L-phenylalaninamide are not affected by 20% methanol.[68,71] Plots of k_{cat}/K_m(app) versus pH for both N-acetyl-L-tryptophanamide and ethyl ester are identical in shape to one another, and also are identical to the pH-k_2 profile of a discrete acylation step.[72] The synthesis of N-benzoyl-L-tyrosylglycinanilide from N-benzoyl-L-tyrosine and glycinanilide[73] must proceed through an acyl-enzyme intermediate, since α-chymotrypsin plus N-acetyl-L-tryptophan and N-furylacryloyltryptophan have been shown to produce the corresponding acyl-chymotrypsins,[65,53] and since several acyl-enzymes have been shown to react with ammonia and amines such as hydroxylamine.[68] The principle of microscopic reversibility indicates that the reverse process, the hydrolysis of an amide, may proceed via an acyl-enzyme intermediate. The pH dependence of the reaction of N-benzoyl-L-tyrosine and hydrazine is consistent with this view.[74]

In opposition to the above recital of consistency, much criticism has been leveled at eq. 15.5.[76-79] This criticism states: (1) it may be impossible to describe the mechanism of an enzyme reaction in a discrete way[80]; (2) the acyl-enzyme mechanism may apply to restricted (nonspecific) substrates, but not to all substrates, particularly the natural peptide substrates; and (3) certain kinetic and trapping experiments involving hydroxylamine are inconsistent with the acyl-enzyme hypothesis.

Answers to these criticisms follow. If a compound has a discrete structure, its reactions must have a discrete (not necessarily unique) mechanism. Since an acyl-enzyme intermediate can be demonstrated in the reactions of the number of representative families of substrates, such as p-nitrophenyl esters, ethyl esters, acids, and amides, particularly of specific substrates, the intermediate may be concluded to be a general one. Alkylation of chymotrypsin by α-bromo-4-nitroacetophenone results in a new band at 350 nm.[75] So it is not surprising that the cinnamoyl derivatives have a slightly perturbed spectrum. The criticism of experimental inconsistency (all involving hydroxylamine[76-79]) is the most serious. However, the experimental basis of two important criticisms have themselves subsequently been criticized.[78,80-83] The remaining criticisms include the 10% difference between the partitioning of N-acetyl-L-tyrosine ethyl ester and N-acetyl-L-tyrosinehydroxamic acid by hydroxylamine and water;[78] and the apparent linear dependency of the rate of the α-chymotrypsin-catalyzed reaction of N-acetyl-L-tyrosine with hydroxylamine on hydroxylamine concentration.[78] The former experiments are not far from theory. In the latter case, the theoretical curve is derived from an enzyme-saturated system, whereas the experimental data were obtained using an enzyme-unsaturated system; however, this does not explain the linearity of the rate with respect to hydroxylamine concentration. Furthermore, hydroxylamine had no effect on the rate of decomposition of N-acetyl-L-tyrosine p-nitroanilide, but much hydroxamic acid was produced, in agreement with the acyl-enzyme hypothesis.[84] Thus since hydroxylamine reactions have many complications, since neither of the alleged inconsistencies is large, since the introduction of 1-3 M hydroxylamine may perturb the normal enzymatic mechanism, and since the reactions of the nucleophile, methanol, show complete consistency,[72] the above criticisms cannot be given much weight. Furthermore, the presteady-state parameters of the chymotrypsin-catalyzed hydrolyses of specific ester substrates such as N-acetyl-L-phenylalanine methyl ester, N-acetyl-L-tryptophan ethyl ester, and N-furylacryloyl-L-tyrosine ethyl ester were found to be in agreement with eq. 15.5 by a proflavin displacement method.[85]

There have been a number of questions raised with regard to hippurate substrates. For example, indole is a noncompetitive inhibitor of methyl hippurate although it is a competitive inhibitor of hippuramide. p-Nitrophenyl hippurate shows the expected burst,[86] but on the other hand, this substrate has been shown to hydrolyze nonenzymatically via an oxazolone intermediate.[87] In opposition, the effect of hydroxylamine on the chymotrypsin-catalyzed hydrolysis of the methyl ester at pH 8.0 is consistent with the acyl-enzyme theory.[88]

15.4 TRANSIENT KINETICS

There are several approaches to transient kinetics. In all cases, one wishes to determine the kinetics of individual steps. Sometimes these are slow and fairly conventional methods (steady-state or simple spectrophotometry), (see earlier) can be used. Sometimes the steps are fast and relaxation techniques alluded to earlier have to be used. The dividing line between the two is not absolute but a good rule of thumb is that the former applies to steps with half-lives of the order of seconds whereas the latter applies to steps of millisecond (stopped-flow) or microsecond duration (relaxation). Complex formation and acylation (in the case of hydrolytic enzymes) often belong to the latter class while deacylation often belongs to the former. This description implies the presence of a reactive intermediate. In reactions of chymotrypsin and other proteolytic enzymes, one reactive (covalent) intermediate is found but in the general case any number of covalent intermediates may be found. As time progresses more covalent intermediates are being found in enzymatic reactions.

Enzyme kinetics in the transient state have been amply treated by Chance.[89] He described their application in the millisecond region to such oxidative enzymes as catalase and peroxidase, because reactions catalyzed by these enzymes had favorable spectral properties. Subsequently, the same fast techniques were applied to proteolytic enzymes when it was discovered how to introduce favorable spectral changes into reactions catalyzed by these enzymes. For example, it was found that the dye proflavin binds to chymotrypsin and trypsin in the absence of substrate or acyl group. The displacement of the dye from the chymotrypsin and trypsin complexes has been studied using a stopped-flow device.[90,91] This, however, introduces a foreign substance into an enzyme-catalyzed reaction. A major advance was made when it was found that arylacryloyl-substrates (where the acryloyl group is on nitrogen) have favorable spectral changes in reactions catalyzed by proteolytic enzymes.[92]

A stopped-flow trace of the change in absorbance observed on the displacement of trypsin-bound proflavin by benzoylarginine ethyl ester is shown by Gutfreund.[93] Although not the first, a stopped-flow trace of the deacylation of N-furylacyloyl-L-tryptophanyl-α-chymotrypsin is shown in Fig. 15.5.[94] This experiment is important because it indicates that eq. 15.5 applies to specific (fast) substrates at the pH of optimal reaction. As indicated before, the rate of deacylation shown in Fig. 15.5 is equivalent to the turnover of the corresponding ethyl ester by the enzyme.

When a reaction is close to equilibrium, the approach to equilibrium may be characterized by a spectrum of relaxation times that are related to the rate constants in the mechanism and that allow one to look at the microsecond region. This applies to nonenzymatic processes in previous chapters and also to enzymatic mechanisms.[95] This approach is being exploited by Eigen and co-workers.[95-97]

Reaction rates are usually measured on systems far from equilibrium. However, the study of reactions very close to equilibrium may be used for determining the rate constants of reaction in solution that are too rapid to permit the use of mixing methods. In the relaxation method a reaction mixture is displaced slightly from equilibrium by changing an independent variable such as electric field strength, pressure, or temperature. If the reaction

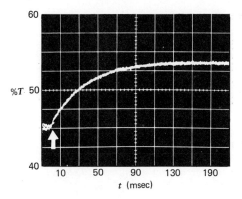

Fig. 15.5 Deacylation of N-2-furylacryloyl-L-tryptophanyl-α-chymotrypsin. Storage oscilloscope trace: percentage transmittance at 340 nm versus time. Arrow indicates where the flow stops. Three consecutive runs are superimposed. After mixing: $[E]_0 = 5.25 \times 10^{-5}\ M$, $[S]_0 = 5.00 \times 10^{-5}\ M$, 0.05 M Tris, pH 7.90, I = 0.15, acetonitrile 0.5% (v/v); 25.0°. From C. G. Miller and M. L. Bender, *J. Amer. Chem. Soc.*, **90**, 6850 (1968). © 1968 by the American Chemical Society. Reprinted by permission of the copyright owner.

is not at equilibrium, it is possible to use a mixing technique (stopped-flow) to approach the equilibrium situation. A simple enzymatic reaction such as S ⇌ P is characterized by a simple relaxation time, which may be composed of several components. If an enzymatic reaction goes through one intermediate, a spectrum of relaxation times may be observed. Since perturbation from equilibrium is small, concentrations are taken to be equilibrium concentrations.

Let us consider a system in which only a single step (of arbitrary order) takes place.[96] The rate equation can be expressed as

$$\frac{-dx_e}{dt} = kx_e - k\,\Delta x_e \tag{15.6}$$

where x_e is the equilibrium concentration of a substance and Δx_e its deviation from equilibrium. The solution of the differential eq. 15.6 for small perturbations from equilibrium leads to†

$$\frac{x}{x_e} = e^{-kt} \tag{15.7}$$

which is a usual first-order equation. The reason that only small deviations from equilibrium are acceptable is that x_e and squared terms in x_e must

† The rate constant k is sometimes defined as $1/\tau$ where τ is the relaxation time or $1/\lambda$ where λ is the relaxation time.

disappear. For systems involving more steps the mathematics is more complicated, but the results are comparable in that the rate constants are a summation of first-order constants (in various combinations), and so the rate constants of individual steps may be determined (Chapters 16 and 17).

REFERENCES

1. L. Michaelis and M. Menten, *Biochem. Z.*, **49**, 333 (1913).
2. V. Henri, *Compt. Rend. Acad. Sci., Paris*, **135**, 916 (1902).
3. E. R. Stadtman, G. D. Novelli, and F. Lipmann, *J. Biol. Chem.*, **191**, 365 (1951).
4. F. J. Kézdy and M. L. Bender, *Biochemistry*, **1**, 1097 (1962) and references therein.
5. E. S. Swinbourne, *J. Chem. Soc.*, 2371 (1960).
6. F. J. Kézdy, J. Jaz, and A. Bruylants, *Bull. Soc. Chim. Belges*, **67**, 687 (1958).
7. J. J. Skujins and A. D. MacLaren, *Science*, **148**, 1569 (1967).
8. F. R. Bettelheim, A. Smith, and R. Lukton, *Chem. Eng. News*, 41 (May 14, 1962).
9. R. Bernhard, *Sci. Res.*, 29 (November 2, 1968).
10. G. G. Hammes, personal communication.
11. M. L. Bender, J. R. Whitaker, and F. Menger, *Proc. Nat. Acad. Sci., U.S.*, **53**, 711 (1965).
12. A. K. Ghosh, R. W. Estabrook, and B. Chance, *Sci. Amer.*, **217**, no. 4. 50 (1967); G. D. Rose and J. A. Quinn, *Science*, **159**, 636 (1968).
13. V. Bloomfield, L. Peller, and R. A. Alberty, *J. Amer. Chem. Soc.*, **84**, 4367 (1962).
14. M. Krongelb, T. A. Smith, and B. H. Abeles, *Biochim. Biophys. Acta*, **167**, 473 (1968).
15. M. Dixon and E. C. Webb, *Enzymes*, 2nd ed. Academic Press, New York, 1964, p. 966.
16. See also M. L. Kremer, *Israel J. Chem.*, **5**, 137 (1967).
17. See also C. Walter, *Steady State Applications in Enzyme Kinetics*, Ronald Press Co., New York, 1965.
18. M. L. Bender and F. J. Kézdy, *Ann. Rev. Biochem.*, **34**, 49 (1965).
19. J. Botts and M. Morales, *Trans. Faraday Soc.*, **49**, 696 (1953).
20. J. P. Wolf, II, and C. Niemann, *J. Amer. Chem. Soc.*, **81**, 1012 (1959).
21. T. Inagami and T. Murachi, *J. Biol. Chem.*, **238**, PC 1905 (1963).
22. C. R. Gunter, Ph.D. Thesis, Northwestern University, 1966.
23. W. W. Cleland, *Biochim. Biophys. Acta*, **67**, 104, 173, 188 (1963).
24. C. C. Wratten and W. W. Cleland, *Biochemistry*, **4**, 2442 (1965).
25. H. A. Feir and I. Suzuki, *Can. J. Biochem.*, **47**, 697 (1969).
26. W. P. Jencks, in *Current Aspects of Biochemical Energetics*, N. O. Kaplan and E. P. Kennedy, Eds., Academic Press, New York, 1966, p. 273.
27. G. E. Briggs and J. B. S. Haldane, *Biochem. J.*, **19**, 338 (1925).
28. M. Bodenstein, *Z. Physik. Chem.*, **85**, 329 (1913).
29. S. A. Bernhard and H. Gutfreund, in *Proc. Int. Symp. Enzyme Chem., Tokyo and Kyoto, 1957*, Maruzen Co., Tokyo, p. 124.
30. J. Z. Hearon, S. A. Bernhard, S. L. Friess, D. J. Botts, and M. F. Morales in *The Enzymes*, 2nd ed., Vol. 1, P. D. Boyer, H. Lardy, and R. Myrbäck, Eds., Academic Press, New York, 1959, p. 130.
31. S. A. Bernhard, W. C. Coles, and J. F. Nowell, *J. Amer. Chem. Soc.*, **82**, 3043 (1960).
32. J. Monod, J. P. Changeux, and F. Jacob, *J. Mol. Biol.*, **6**, 306 (1963).
33. J. C. Gerhart and H. K. Schachmann, *Biochemistry*, **4**, 1054 (1965).

References

34. D. E. Koshland, Jr., G. Némethy, and D. Filmer, *Biochemistry*, **5**, 365 (1966).
35. *Proc. 7th Meeting F.E.B.S.*, E. Kvamme and E. Pihl, Eds., Academic Press, New York, 1968.
36. H. d'A. Heck, personal communication.
37. A. Piérard, *Science*, **154**, 1572 (1966).
38. W. D. McElroy, personal communication.
39. L. B. Rothman and E. Cabib, *Biochemistry*, **6**, 2098 (1967).
40. L. B. Rothman and E. Cabib, *Biochemistry*, **6**, 2107 (1967).
41. B. D. Sanwal, C. S. Stachow, and R. A. Cook, *Biochemistry*, **4**, 410 (1965).
42. L. W. Nichol, W. J. H. Jackson, and D. J. Winzor, *Biochemistry*, **6**, 2449 (1967).
43. D. S. Wiley and W. N. Lipscomb, *Nature*, **218**, 1119 (1968).
44. A. K. Balls and E. F. Jansen, *Adv. Enzymol.*, **13**, 321 (1952).
45. B. K. Blossey and J. A. Shafer, *Fed. Proc.*, 447 (1967).
46. A. A. Aboderin and J. S. Fruton, *Proc. Nat. Acad. Sci. U.S.*, **56**, 1252 (1967).
47. I. V. Berezin et al., *Mol. Biol.*, **1**, 719, 843 (1967); R. E. Benesch and R. Benesch, *J. Amer. Chem. Soc.*, **77**, 5877 (1955).
48. H. Kaufman, S. M. Vratsanos, and B. F. Erlanger, *Science*, **162**, 1487 (1968).
49. A. K. Balls and H. N. Wood, *J. Biol. Chem.*, **219**, 245 (1956).
50. C. R. Gunter, Doctoral Dissertation, Northwestern University, 1966.
51. C. E. McDonald and A. K. Balls, *J. Biol. Chem.*, **221**, 993 (1956).
52. F. J. Kézdy and M. L. Bender, *J. Amer. Chem. Soc.*, **86**, 938 (1964).
53. C. G. Miller and M. L. Bender, *J. Amer. Chem. Soc.*, **89**, 6850 (1968).
54. B. F. Erlanger and W. Cohen, *J. Amer. Chem. Soc.*, **85**, 348 (1963).
55. A. M. Gold and D. Fahrney, *Biochemistry*, **3**, 783 (1964); J. Kallos and D. Rizok, *J. Mol. Biol.*, **7**, 599 (1963); **9**, 255 (1964).
56. D. H. Strumeyer, W. N. White, and D. E. Koshland, Jr., *Proc. Nat. Acad. Sci. U.S.*, **50**, 931 (1963).
57. M. L. Bender, G. R. Schonbaum, and B. Zerner, *J. Amer. Chem. Soc.*, **84**, 2540 (1962).
58. A. K. Balls and F. L. Aldrich, *Proc. Nat. Acad. Sci. U.S.*, **41**, 190 (1955); A. K. Balls, C. E. McDonald, and A. S. Brecher, *Proc. Int. Symp. Enzyme Chem.*, Tokyo and Kyoto, 1957, Maruzen Co., Tokyo, 1958, p. 392.
59. A. Singh, E. R. Thornton, and F. H. Westheimer, *J. Biol. Chem.*, **237**, PC3006 (1962).
60. M. Caplow and W. P. Jencks, *Biochemistry*, **1**, 883 (1962).
61. J. Berliner and H. M. McConnell, *Proc. Nat. Acad. Sci. U.S.*, **55**, 708 (1966).
62. A. G. Marshall, *Biochemistry*, **7**, 2450 (1968).
63. M. L. Bender and B. Zerner, *J. Amer. Chem. Soc.*, **84**, 2550 (1962).
64. B. S. Hartley and B. A. Kilby, *Biochem. J.*, **50**, 672 (1952); **56**, 288 (1954).
65. F. J. Kézdy, G. E. Clement, and M. L. Bender, *J. Amer. Chem. Soc.*, **86**, 3690 (1964).
66. F. Kawahara, unpublished results.
67. G. W. Schwert and M. A. Eisenberg, *J. Biol. Chem.*, **179**, 665 (1949).
68. M. L. Bender, G. E. Clement, C. R. Gunter, and F. J. Kézdy, *J. Amer. Chem. Soc.*, **86**, 3697 (1964).
69. M. L. Bender and F. J. Kézdy, *J. Amer. Chem. Soc.*, **86**, 3704 (1964).
70. A. Himoe, K. Brandt, G. Czerlinski, and G. P. Hess, *Fed. Proc.*, **26**, 447 (1967); see, however, G. P. Hess et al., *Phil. Trans. Roy. Soc., Lond.*, B **257**, 89 (1970).
71. S. Kaufman and H. Neurath, *J. Biol. Chem.*, **180**, 181 (1949).
72. M. L. Bender, G. E. Clement, F. J. Kézdy, and H. d'A. Heck, *J. Amer. Chem. Soc.*, **86**, 3680 (1964).
73. O. Gawron, A. J. Glaid III, R. E. Boyle, and G. Odstrechel, *Arch. Biochem. Biophys.*, **95**, 293 (1961).
74. W. H. Schuller and C. Niemann, *J. Amer. Chem. Soc.*, **74**, 4630 (1952).

75. D. S. Sigman and E. R. Blout, *J. Amer. Chem. Soc.*, **89**, 1747 (1967).
76. S. A. Bernhard and H. Gutfreund, *Proc. Int. Symp. Enzyme Chem.*, *Tokyo and Kyoto, 1957*, Maruzen Co., Tokyo, 1958, p. 124.
77. S. A. Bernhard and H. Gutfreund, *Prog. Biochem. Biophys. Chem.*, **10**, 115 (1960).
78. M. Caplow and W. P. Jencks, *J. Biol. Chem.*, **238**, PC1907 (1963).
79. J. Z. Hearon, S. A. Bernhard, S. L. Friess, D. J. Botts, and M. F. Morales in *The Enzymes*, 2nd ed., Vol. 1, P. D. Boyer, H. Lardy, and K. Myrbäck, Eds., Academic Press, New York, 1959, p. 130.
80. C. Niemann, *Science*, **143**, 1287 (1964).
81. F. J. Kézdy, G. E. Clement, and M. L. Bender, *J. Biol. Chem.*, **238**, PC3141 (1963).
82. R. Epand and I. B. Wilson, *J. Biol. Chem.*, **238**, 3148 (1963).
83. M. Caplow and W. P. Jencks, *J. Biol. Chem.*, **238**, PC3140 (1963).
84. T. Inagami and J. M. Sturtevant, *Biochem. Biophys. Res. Commun.*, **14**, 69 (1964).
85. A. Himoe, K. G. Brandt, R. J. DeSa, and G. P. Hess, *J. Biol. Chem.*, **244**, 3483 (1969); see also R. M. Epand and I. B. Wilson, *ibid.*, **240**, 1104 (1965).
86. F. J. Kézdy and M. L. Bender, unpublished results.
87. J. de Jersey, A. A. Kortt, and B. Zerner, *Biochem. Biophys. Res. Comm.*, **23**, 745 (1966).
88. R. M. Epand and I. B. Wilson, *J. Biol. Chem.*, **239**, 4138 (1964).
89. B. Chance, in *Technique of Organic Chemistry*, S. L. Friess, E. S. Lewis, and A. Weissberger, Eds., 2nd ed. Vol. VIII, Part II, 1963, p. 1314. Interscience, New York.
90. S. A. Bernhard and H. Gutfrend, *Proc. Nat. Acad. Sci., U.S.*, **53**, 1238 (1965).
91. T. E. Barman and H. Gutfreund, *ibid.*, **53**, 1243 (1965).
92. E. Charney and S. A. Bernhard, *J. Amer. Chem. Soc.*, **89**, 2726 (1967).
93. H. Gutfreund, *An Introduction to the Study of Enzymes*, J. Wiley and Sons, New York, 1965, p. 259.
94. C. G. Miller and M. L. Bender, *J. Amer. Chem. Soc.*, **90**, 6850 (1968).
95. G. G. Hammes and R. A. Alberty, *J. Amer. Chem. Soc.*, **82**, 1564 (1960).
96. M. Eigen and L. DeMaeyer, *Technique of Organic Chemistry*, A. Weissberger, Ed., Vol. 8, Part 2, Chap. 18, Interscience Div., Wiley & Sons, New York, 1963; G. Czerlinski, *7th Int. Cong. Biochem. Tokyo*, 1967, Symp. III, p. 183.
97. G. H. Czerlinski, *Chemical Relaxation*, Marcel Dekker, New York, 1966.

Chapter 16

ENZYME MECHANISMS: PROTEINS

16.1	Functional Groups of the Active Site	505
16.2	Chymotrypsin	505
16.3	Mechanism of Action of α-Chymotrypsin	510
16.4	Other Serine Proteinases	514
	16.4.1 Trypsin	515
	16.4.2 Elastase	517
	16.4.3 Subtilisin	517
	16.4.4 Thrombin	517
	16.4.5 Acetylcholinesterase	518
	16.4.6 Carboxylesterases	519
	16.4.7 Thiolsubtilisin (An Artificial Enzyme)	519
16.5	Other Protein Enzymes	520
	16.5.1 Sulfhydryl Enzymes: Papain	520
	16.5.2 Sulfhydryl Enzymes: Ficin	522
	16.5.3 Sulfhydryl Enzymes: Bromelain	522
	16.5.4 Pepsin	522
	16.5.5 Ribonuclease	523
	16.5.6 Fumarase	524
	16.5.7 Sucrose Phosphorylase	525
	16.5.8 Amylases	525
	16.5.9 Translocase	525
	16.5.10 Lysozyme	527
16.6	Active Transport Proteins	530
16.7	Conclusion	530

Although no enzyme mechanism has been elucidated in complete detail, a knowledge of the mechanism of a number of enzyme reactions has progressed to the point where discussion is possible. In particular, the mechanisms of several enzymes containing coenzymes, and of several enzymes containing no coenzymes, have been elucidated.

When the active site of the enzyme consists solely of protein components, it is more difficult to investigate the mechanism of the process than when the ready-made handle of a coenzyme exists. Sometimes model organic systems can be used to investigate the enzymatic process, but the choice of a model is not always as obvious as when a coenzyme is present.

When a coenzyme is present it is reasonable to assume that the coenzyme will participate directly in the enzymatic catalysis. On the other hand, a purely protein enzyme may act in two different mechanistic ways, as suggested by Koshland.[1] The enzyme may combine with the substrate(s) in an adsorptive complex, but may not participate in the subsequent catalytic reaction by a covalent interaction. On the other hand, formation of the adsorptive complex may be followed by *covalent* reaction of the enzyme (Chapter 15) with one of the substrates, leading to an enzyme-substrate compound that then reacts further to form the products of the reaction and to regenerate the enzyme. The second of these mechanistic pathways includes a *reactive intermediate*; much effort has been made to identify such intermediates, as has been done in organic reactions.

The mechanism of enzyme action, one of the central problems in biochemistry, may be met in many forms, but nowhere is it met as directly as in hydrolytic reactions catalyzed by proteolytic enzymes. These enzymes include the serine proteinases such as chymotrypsin, trypsin, elastase, cocoonase, plasmin, subtilisin, the esterases, and thrombin; the cysteine proteinases such as papain, ficin, and bromelain; the metal-containing peptidases such as the aminopeptidases and carboxypeptidase; and peptidases active at acid pH's such as pepsin and rennin. These enzymes catalyze the hydrolysis of peptides, amides, and esters, reactions whose nonenzymatic mechanisms have been investigated in great detail. Except for the metal-containing peptidases, these enzymes are strictly protein in nature, and thus one can ask the ultimate question associated with enzyme mechanism: how does a protein catalyze an organic reaction? We shall also consider such enzymes as lysozyme, ribonuclease, fumarase, and sucrose phosphorylase, all of which are also strictly protein in nature.

To answer this question, many laboratories have put forth a prodigious effort using a variety of approaches. Although no final answers to this question have come forth, much progress has been made and many controversies have arisen. It is the purpose of this chapter to examine critically and selectively some of the progress and controversies.

The viewpoint taken here is that a true mechanistic description requires

Chymotrypsin

the identification of the individual steps of the reactions and the description of the transition state of each step both stereochemically and electronically.

Covalent intermediates are often found in reactions catalyzed by protein enzymes. The many approaches to the identification of covalent enzyme-substrate intermediates in reactions catalyzed by protein enzymes include: (1) isolation of stoichiometric, inactive enzyme derivatives; (2) observation of transient intermediates; (3) various kinetic arguments including relative rates and Michaelis constants of different carboxylic acid derivatives, effect of added nucleophiles on the kinetics, and pH dependence of the various steps of the reaction; and (4) isotopic exchange reactions.

16.1 FUNCTIONAL GROUPS OF THE ACTIVE SITE

Many approaches have been taken in mapping the active site of protein enzymes. The major trends include: (1) X-ray diffraction analysis of the entire enzyme molecule[2]; (2) determination of the amino acid sequence of the entire enzyme or of the regions surrounding specific groups of the active site; (3) comparison of amino acid sequences of different enzymes; (4) derivatization of groups near the active site after specific binding (either covalently or noncovalently) to the site[3]; (5) complete derivatization of certain functionalities of the enzyme; (6) introduction of chromophores into the active site; (7) kinetic and thermodynamic studies, including the effect of pH and the structure of the substrate on rate; (8) the heat of ionization of groups together with their pK's; and (9) studies of model systems. Specific inhibitions can be effected at the active site. The primary method for determining whether a reagent reacts at the active site is to see whether it is a (competitive) inhibitor of a specific substrate of the enzyme. pH-rate-constant profiles of the enzymatic reaction have often been used to give a clue as to the reactive entity.

If both the pK value of a certain group in a protein and its heat of ionization are known, the nature of the group can often be specified. It has been suggested, however, that this technique may not apply to groups in the active sites of enzymes, where local electrostatic fields may exist. So far there is no evidence of this kind and it is necessary to wait until more data have been accumulated. The heats of ionization of the groups may be determined by observing the effect of alteration of temperature on the pK's. Table 16.1 shows representative pK's and heats of ionization.[4]

16.2 CHYMOTRYPSIN

Many of the approaches enumerated above have been used to identify the functional groups in the active site of the enzyme chymotrypsin.

Table 16.1
pK Values and Heats of Ionization of Some Groups Present in Proteins[a]

Group	pK (25°)	ΔH_i (cal/mol)
Carboxyl (α)	3.0–3.2	($\sim$1,500)
Carboxyl (aspartyl)	3.0–4.7	($\sim$1,500)
Carboxyl (glutamyl)	ca. 4.4	($\sim$1,500)
Phenolic hydroxyl (tyrosine)	9.8–10.4	6,000
Sulfhydryl	8.3–8.6	—
Imidazolium (histidine)	5.6–7.0	6,900–7,500
Ammonium (α)	7.6–8.4	10,000–13,000
Ammonium (α, cystine)	6.5–8.5	—
Ammonium (ϵ, lysine)	9.4–10.6	10,000–12,000
Guanidinium (arginine)	11.6–12.6	12,000–13,000

[a] From M. Dixon and E. C. Webb, *Enzymes*, 2nd Ed., Academic Press, New York, 1964, p. 144.

The three-dimensional structures of tosyl-α-chymotrypsin and indoleacryloyl-α-chymotrypsin are known and are similar.[52]

Much evidence suggests that the hydroxyl group of a serine moiety of α-chymotrypsin is the group that is acylated. (1) A dialkylphosphoryl,[7] an acetyl,[8] or a nitrotyrosyl group[8] resides on the hydroxyl group of a serine moiety after hydrolysis, and also after partial enzymatic hydrolysis to peptides containing the dialkylphosphoryl or acetyl or nitrotyrosyl group.[8–10] The spectrum of the cinnamoyl group of the denatured acyl-enzyme is identical with that of the model *trans*-cinnamoyl-*N*-acetylserinamide.[11,12] This result indicates a perturbation of the π–π transition of the cinnamoyl group in the native enzyme. This same perturbation, observed in four different cinnamoyl-enzymes (Table 16.2)[13] indicates that some common phenomenon is occurring; the simplest explanation is a common physical perturbation[14] rather than a common chemical change,[12] since the chemistry of these enzymes is different. (3) The rates of the nonenzymatic (alkaline) hydrolyses of *trans*-cinnamoyl-α-chymotrypsin and acetyl-α-chymotrypsin in 8 *M* urea are almost identical with the rates of alkaline hydrolyses of *O*-cinnamoyl-*N*-acetylserinamide and *N*,*O*-diacetylserinamide,[15] respectively. (4) The equilibrium constant of the reaction *N*-acetyl-L-tryptophanyl-α-chymotrypsin $\rightleftharpoons$ *N*-acetyl-L-tryptophan + α-chymotrypsin is that of a normal lactone-hydroxyacid equilibrium; that is, the acyl-enzyme of a specific substrate may be thermodynamically identified as an ester (and not as an acyl-imidazole).[16–18] (5) The rate constant for the ethanolysis of *N*-acetyl-L-tryptophanyl-α-chymotrypsin is 54.5 M^{-1} sec^{-1} while the rate constant of

Table 16.2
Difference Spectra of Some Cinnamoyl Enzymes and Cinnamoyl Esters[a,c]

trans-Cinnamoyl-Enzyme or Ester	λ_{max} (nm)	ϵ_{max}	λ_{max} (enzyme) $-\lambda_{max}$ (model) (nm)	kcal
α-Chymotrypsin	292	17,700	10.5	3.65
Trypsin	296	19,400	14.5	4.98
Subtilisin	289	21,000	7.5	2.65
Papain	326	26,500	20	5.73
N-Acetylserinamide[a]	281.5	24,300	—	—
Cysteine[b]	306	22,600	—	—

[a] O-cinnamoyl.
[b] S-cinnamoyl.
[c] From M. L. Bender and L. J. Brubacher, *J. Amer. Chem. Soc.*, **86**, 5333 (1964). © 1964 by the American Chemical Society. Reprinted by permission of the copyright owner.

the reaction of N-acetyl-L-tryptophan ethyl ester with the presumed serine hydroxyl group of chymotrypsin is ca. 1000 sec^{-1},[96] a relationship found frequently in corresponding intermolecular and intramolecular reactions of esters.[19] (6) An inactive monotosylated enzyme was converted via an elimination reaction to an inactive "anhydro" enzyme, indicating that the hydroxyl group of serine plays a positive role of acyl acceptor, and not a negative role of steric hindrance.[20]

Several kinetically important ionizable groups of α-chymotrypsin are known. A group with apparent pK_a of 7 is operative in both acylation and deacylation reactions. A second group of pK_a 9 is seen in acylation while a group of variable pK_a is sometimes seen in deacylation.

The ionizable group of pK_a 7 is usually identified as the imidazole group of a histidine moiety of the enzyme,[21] the only group on the enzyme with a pK_a around neutrality, with the exception of the terminal α-ammonium ions. Of course, the kinetically determined pK_a is only an apparent constant that may be perturbed by pre-equilibria. However, photooxidation studies[22] and studies using both rate and "all or none" assays[21] indicate that when a histidine residue is photooxidized, the enzymatic activity disappears. Furthermore, L-1-tosylamido-2-phenylethyl chloromethyl ketone, which probably binds at the active site since it has the backbone of a specific substrate, inactivates chymotrypsin completely;[23] it has been found that alkylation by this reagent occurs at histidine 57 of the enzyme.[24,25] Finally, model studies indicate that imidazole is indeed an efficient catalyst of ester hydrolysis around neutrality.[26-28]

The two histidine residues in chymotrypsin are very close in space to one another. The relationship found is –His–Phe–Cys–S–S–Cys–His.[29,30] The identical sequence is found in trypsin.[31] This coincidence may be significant in view of the mechanistic similarities between these two enzymes.[29,31,32]

The acidic group of $pK_a \sim 9$ seen solely in acylation may be the α-ammonium group of the N-terminal isoleucine residue.[33]* Activation of fully acetylated chymotrypsinogen gave an acetylated δ-chymotrypsin containing one free amino group, the α-ammonium group of N-terminal isoleucine; when this group is acetylated the enzyme is inactive. This prototropic group may be titrated in acetylated δ-chymotrypsin (and thus should be seen in acylation) but not in DIP-acetylated δ-chymotrypsin (and thus should not be seen in deacylation). The reason for its disappearance in the acyl-(phosphoryl-)enzyme must be determined. The group of pK_a 9 cannot be the group acylated, the serine hydroxyl group, since it has a pK_a of 13.5.[34]

The involvement of other amino acids in chymotrypsin catalysis has been reviewed.[21] One or two methionine residues near the active site have been modified by alkylation with reagents that bind adsorptively or covalently to the active site,[35,36] or by oxidation.[21,37,38] These modified enzymes have partial activity, due in one instance to a change in K_m (app) (binding) rather than to a change in V_{max} (catalysis).[21] The involvement of tyrosine residues in chymotrypsin action has been probed by the effect of iodination of the protein.[39-41] Probably none of the four tyrosines of chymotrypsin is essential for catalytic action.[41] When 6.3 atoms of iodine per molecule have been incorporated, V_{max} remains unchanged while K_m (app) increases by a factor of 6.

Modification of a tryptophan residue of chymotrypsin by treatment with DFP or N-bromosuccinimide indicates that the tryptophan environment changes during catalytic action.[42-44] Two alkylating agents containing chromophoric groups, 2-bromomethyl-4-nitrophenol and 2-acetoxy-5-nitrobenzyl chloride, alkylate only a tryptophan of chymotrypsin[45,53]. One to two tryptophan residues of chymotrypsin have been oxidized by a mixture of hydrogen peroxide and acetyl-L-tyrosine ethyl ester.[38] This mixture probably forms N-acetyl-L-tyrosine peroxyacid in the active site and may react with groups nearby. One cystine is oxidized to cysteic acid[38] in this reaction.

The intriguing possibility that an arginine derivative may be involved in the active site of α-chymotrypsin[46,47] should be investigated further.

In trypsin the acyl group of the acyl-enzyme probably resides on the serine hydroxyl group since *trans*-cinnamoyl-trypsin resembles *trans*-cinnamoyl-α-chymotrypsin spectrally, in rate of deacylation, and in rate of hydrolysis of the denatured acyl-enzyme.[48] Histidine has been implicated in trypsin action

* An N-terminal valine also seems to be involved. (P. Valenzuela and M. L. Bender, *J. Amer. Chem. Soc.*, **93**, in press (1971).)

Chymotrypsin

by dependence of the reaction rate on a basic group of pK_a 6–7[32,48] and specific inhibition of trypsin by L-5-amino-1-tosylamidopentyl chloromethyl ketone.[49] As mentioned above, two histidines of trypsin have been found in close proximity to one another.[31]

The rates of elastase-[50] and subtilisin-catalyzed[51] reactions depend on a basic group of $pK_a \sim 7$, indicating the possible involvement of a histidine group in their catalytic action.

The sequence (see Table 16.3)[62] of chymotrypsinogen (x),[30] trypsinogen (y),[63] and elastase (z),[64] show very great similarities, especially in the region of the active histidines:

(x) His.Phe.Cys.Gly.Gly.Ser.Leu.Ile.Asn.Gly.Asn.
    40
                    Trp.Val.Val.Thr.Ala.Ala.His.Cys.
                                                58

(y) His.Phe.Cys.Gly.Gly.Ser.Leu.Ile.Asn.Ser.Gln.
    29
                    Trp.Val.Val.Ser.Ala.Ala.His.Cys.
                                                47

(z) His.Thr.Cys.Gly.Gly.Thr.Leu.Ile.Arg.Gln.Asn.
    40
                    Trp.Val.Met.Thr.Ala.Ala.His.Cys.
                                                58

and of the active serine of these enzymes:

(x) Cys.Met.Gly.Asp.Ser.Gly.Gly.Pro.Leu.Val.Cys.
    191   201

(y) Cys.Gln.Gly.Asp.Ser.Gly.Gly.Pro.Val.Val.Cys.
    179   189

(z) Cys.Gln.Gly.Asp.Ser.Gly.Gly.Pro.Leu...

Table 16.3

Sequences near the Active Serine of Chymotrypsin and some other Serine Proteinases[a]

DFP-Inhibited Enzymes	Sequence
Trypsin	Gly–Asp–Ser–Gly
Chymotrypsin	Gly–Asp–Ser–Gly
Elastase	Gly–Asp–Ser–Gly
Thrombin	Asp–Ser–Gly
Subtilisin	Thr–Ser–Met–Ala
Mold Protease	Thr–Ser–Met–Ala

[a] From F. Sanger, *Proc. Chem. Soc.*, 76 (1963).

16.3 MECHANISM OF ACTION OF α-CHYMOTRYPSIN

The concept of the acyl-enzyme intermediate and its effect on the kinetics of chymotrypsin-catalyzed reactions have been covered in Chapter 15. Of the two catalytic steps in α-chymotrypsin-catalyzed reactions, deacylation is the more amenable to mechanistic description. However, since deacylation can be (for selected cases) simply the microscopic reverse of acylation, a description of these deacylations should suffice for the entire catalytic reaction. Reversible systems with a forward reaction of acylation and a reverse reaction of deacylation include the equilibrium acylation by N-acetyl-L-tryptophan, the acylation of α-chymotrypsin by N-acetyl-L-tryptophan methyl ester[96] plus the methanolysis of N-acetyl-L-tryptophanyl-α-chymotrypsin,[96] and the radioactive exchange of methanol-^{14}C with N-acetyl-L-phenylalanine methyl ester.[65] The hypothesis that acylation is the microscopic reverse of deacylation rests on the assumption, thus far valid, that an ester and an (undissociated) carboxylic acid or methanol and water are mechanistically equivalent. For reactions that involve a strong acylating agent, this principle cannot be used.

Since the group of pK_a 9, identified earlier as N-terminal isoleucine, is seen kinetically in acylation but not in deacylation, its identification as a catalytic moiety would violate microscopic reversibility, unless a secondary equilibrium perturbs the prototropic equilibrium differently in ES and ES'. Furthermore, it cannot be the group acylated (see above). It has been suggested that it is a group responsible for conformational stabilization,[32] a suggestion supported by the pH dependence of specific rotation of the enzyme[66-68] and by other data concerning possible conformational changes during α-chymotrypsin action.[69]

The mechanistic information most pertinent to a description of the deacylation step includes the following. (1) The acyl-enzyme is an ester in which the acyl group of the substrate is attached to the hydroxyl group of a serine residue. (2) A base of pK_a 7 is kinetically involved in the deacylation. (3) This base is an imidazole group as discussed above. (4) The deacylation reaction is a nucleophilic reaction as characterized by a Hammett ρ-constant of of +2.1.[70] Polycationic and polyanionic derivatives lead to the suggestion of a Brønsted plot.[54] (5) The reaction is first-order in the nucleophile water, as determined kinetically and by analogy with the kinetics of the methanolysis reaction. (6) The nucleophile reacts in its protonated form,[71] based on the deacylations of acetyl-,[72] methyl isopropoxyphosphonyl-, and diethylphosphoryl-chymotrypsins with the nucleophiles isonitrosoacetone, glycinehydroxamic acid, and phenylacetohydroxamic acid.[73,74] These reactions exhibit bell-shaped curves, depending on two groups, one with a pK_a of ~7 and the other with a pK_a of the nucleophile. (7) The reactivity of amine and alcohol nucleophiles in deacylation is only slightly dependent on their basicities.[18]

(8) No detectable intermediate is observed in deacylation.[75] (9) The hydrogen atom on imidazole appears to be important (see later).

Although imidazole has been shown to catalyze ester hydrolysis by both nucleophilic[26,27] and general basic[28] mechanisms, two arguments rule out nucleophilic catalysis. The dependence of the deacylation rate on the water concentration requires that the formation of an acyl-imidazole intermediate (of nucleophilic catalysis) be fast and its decomposition be slow, whereas the observation of no intermediate in the reaction (and the pH dependency of the reaction) requires that the formation of an acyl-imidazole intermediate be slow and its decomposition be fast; these data are obviously contradictory. The deuterium oxide isotope effects are consistent with the rate-determining proton transfer of general basic catalysis, but are inconsistent with nucleophilic catalysis.[76,77] A deacylation involving the pre-equilibrium formation of a (small concentration of) acylimidazole followed by rate-determining attack of water may be ruled out on the basis of its pH dependency.

The components of the transition state of the deacylation reaction must then include the acyl-serine ester, an imidazole group, and a molecule of water. The rate-determining proton transfer, which must shift a proton from a water molecule (or other nucleophile) to the imidazole base, is probably concerted with attack by the nucleophile.

Deacylation has a Hammett ρ-constant similar to that found in the base-catalyzed hydrolysis and alcoholysis of esters, which are known to proceed through tetrahedral addition intermediates,[19] suggesting but not proving that deacylation also does. The greater reactivity of a carbamyl fluoride than that of a carbamyl chloride supports this hypothesis.[78] A symmetrical, general base-catalyzed system, such as the alcoholysis of the acyl-enzyme (ester) to produce an ester product and enzyme (alcohol), requires such a tetrahedral intermediate.[71,78,79]

Two kinetically indistinguishable mechanisms based on the above information can be proposed.[71,80–82] Equation 16.1 describes the catalysis as a general base-catalyzed formation and a general acid-catalyzed decomposition of the tetrahedral intermediate (in both directions).

$$\begin{bmatrix} -B & H & H \\ & \diagdown & \diagup \\ & O & \\ & | & \\ & O-C-R \\ & \| & \\ & O \end{bmatrix} \rightleftharpoons \begin{bmatrix} -BH^{\oplus} & OH \\ & | \\ & O-C-R \\ & | \\ & O^{\ominus} \end{bmatrix} \rightleftharpoons \begin{bmatrix} -B & OH \\ & H & | \\ & O & C-R \\ & & \| \\ & & O \end{bmatrix} \quad (16.1)$$

Equation 16.2 describes the mechanism in exactly the opposite order.

$$\left[\begin{array}{c}-BH^{\oplus}\\O\end{array}\overset{O}{\underset{OH}{\overset{\ominus}{\underset{}{C-R}}}}\right] \rightleftharpoons \left[\begin{array}{c}-B\\O\end{array}\overset{H-O}{\underset{OH}{\underset{}{C-R}}}\right] \rightleftharpoons \left[\begin{array}{c}-BH^{\oplus}\\O^{\ominus}\end{array}\overset{O}{\underset{OH}{\underset{}{C-R}}}\right] \quad (16.2)$$

It may be ruled out on the basis that when applied to the deacylation of N-acetyl-L-tyrosyl-α-chymotrypsin, its rate constant would be over 10^9 M^{-1} sec^{-1},[71] approximating diffusion-control and thus much faster than known reactions involving such covalent changes.

Equation 16.2 satisfies all data pertinent to the deacylation reaction except that it does not imply a bell-shaped pH-k_3 profile, but rather a double sigmoid, the first depending on the ionization of the imidazolium ion,[71] and the other on the ionization of the nucleophile.[72] The experimental data thus demand that the nucleophile be present in its protonated form, implying that a catalytic function be given to the proton of the nucleophile.[71] Figure 16.1, which satisfies this requirement, agrees with all known experimental data pertinent to the α-chymotrypsin mechanism.[71] This mechanism utilizes the unique ability of imidazole to serve simultaneously as a general base and general acid. The reaction has the attributes of a concerted reaction, which should enhance its kinetic efficacy. All transition states should be neutral, predicting no effect of ionic strength or dielectric constant on the rates, as found experimentally. Due to the contribution from general acidic catalysis, the enzymatic deacylation should be faster than a corresponding intramolecular general basic catalysis, as found experimentally.[71] The stereochemistry of Fig. 16.1 does not permit collinear proton transfer from oxygen to nitrogen or vice versa. However, if one or more water molecules intervene,[83] collinear proton transfers can be achieved.

Many other mechanisms of α-chymotrypsin catalysis using imidazole and hydroxyl groups may be ruled out.[71] Nucleophilic catalysis by imidazole is not consistent with the deuterium oxide and methanolysis data. General basic catalysis by imidazole implies that the anion of the nucleophile should be more reactive than the protonated nucleophile; this is not found. General acidic catalysis by an unknown acid and general basic or nucleophilic catalysis by imidazole is inconsistent with microscopic reversibility and symmetry arguments that demand that if such a general acidic catalysis is seen in acylation, it must also be seen in deacylation; this is not found. General basic catalysis by imidazole in acylation and nucleophilic catalysis by imidazole in deacylation is incompatible with the microscopic reversibility and symmetry arguments.

No isotopic oxygen exchange was found in the deacylations of cinnamoyl-chymotrypsin-carbonyl-O^{18} and p-nitrobenzoyl-chymotrypsin-carbonyl-O^{18}.[84] Therefore a tetrahedral intermediate has no direct support but

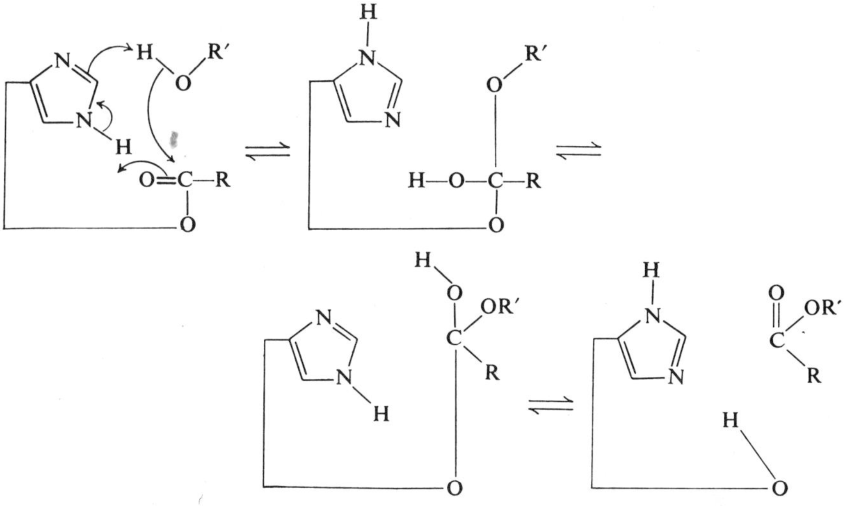

Fig. 16.1 Mechanism of Chymotrypsin Catalysis

circumstantial evidence indicates that it exists; this problem is being worked on in this laboratory.[85]

Recent results on chymotrypsin are the following: there appears to be a profound difference in the binding of α- and δ-chymotrypsin at high pH, the latter leveling off to a value that is perhaps 30% lower at pH 11 than at pH 7 (at 1.0 M) whereas the former shows a profound decrease in binding (with *no* leveling off) at pH 11 compared to pH 7.[86] Modification of the amino group of isoleucine 16 has been claimed not to deactivate the enzyme,[55] but this has been disputed.[56] An intramolecular reaction of a substrate acylamido group has been suggested as a step in chymotrypsin catalysis.[57]

Modification of α-chymotrypsin by methyl *p*-nitrobenzenesulfonate specifically attacks histidine 57 and completely inhibits enzyme activity, as determined by rate and titration. (The *N*-methylchymotrypsin product may have a minute amount of activity.) This reagent does not modify trypsin or subtilisin, two related enzymes. Methylation occurs at the active site and indicates the importance of the hydrogen atom on the imidazole group of the active site.[87]

The binding of several inhibitors results in proton uptake with δ- and α-chymotrypsin, but not with the zymogen. A single ionization of apparent pK 8.8 determines binding. Both pH-recorder and spectrophotometric pH-indicator dye techniques agree with one another. At a given pH, proton absorption exhibits a saturation phenomenon, in parallel to substrate binding. These experiments provide direct proof of the intimate relationship at high pH between the enzyme's state of protonation and its ability to bind small molecules.[88]

Whereas proton uptake by the enzyme appears to be minimal at pH 7.6, Hess and co-workers have reported that it never exceeds 0.5 protons per molecule and is less at pH 10.2 than at 9.6. Proton uptake does indeed go through a maximum at high pH with δ-chymotrypsin but not with α-chymotrypsin. The proton uptake of the two enzymes correlates well with the change in binding with pH.[90]

The binding of the competitive inhibitor, N-acetyl-D-tryptophanamide to α-chymotrypsin was studied up to pH 10.6 by equilibrium dialysis. The binding depends on a group of apparent pK 9.3 at 5°, tentatively identified as that group responsible for an enzyme conformational change.[91] The situation with respect to α-chymotrypsin at high pH is undoubtedly very complicated and is not resolved yet.[92]

The enhanced nucleophilicity of the serine in the active site of chymotrypsin has been suggested to be due to an interior aspartic acid which is hydrogen bonded to the histidine of the active site which in turn is hydrogen bonded to the serine of the active site. It is suggested that polarization of the whole system makes the serine oxygen strongly nucleophilic and explains its reactivity toward esters and amides. This proposal came from a re-examination of the amino acid sequence near the active site serine,[93] but may not be correct,[94] as this is only circumstantial evidence, as hydrogen bonds (especially involving carboxylate) are not very strong, and as hydrogen bonding may not lead to the required reactivity. Alternatively the rate enhancement associated with chymotrypsin-catalyzed hydrolyses is proposed to be partly the result of an enzyme-induced conformational alternation about the peptide (acylamino) bond of the substrates. This theory is consistent with data from enzyme and enzyme model studies. The enzyme studies involved substrates of free and restricted rotation. This idea may explain the anomalous absorptions of Table 16.2.[94]

The specificity of α-chymotrypsin has been treated several times. The constant k_{cat}/K_m (app) which obviates nonproductive binding has been treated[96] as has the deacylation constant k_3.[97]

16.4 OTHER SERINE PROTEINASES

In addition to chymotrypsin, a number of other enzymes can be classified as serine proteinases. They include trypsin,[98] elastase,[99] subtilisin,[100] cocoonase,[101] plasmin,[102] thrombin,[103] and various esterases. The identification is based on the formation of an inactive phosphorylated enzyme with the phosphorus atom residing on a serine hydroxyl group.

16.4.1 Trypsin

The pathways of reactions catalyzed by α-chymotrypsin and trypsin are similar, as are the functionalities of the active site (see preceding). Two trypsins have been separated by chromotography on an SE Sephadex column.[89] In addition, the mechanistic characteristics of the individual steps of these enzymes are similar. The thousandfold faster hydrolysis of a specific ester than that of an amide substrate by α-chymotrypsin[5] is paralleled by a threehundredfold difference in the trypsin-catalyzed reactions.[104,105] In addition to the basic group of apparent pK_a 6–7 seen in the acylation and deacylation steps of both enzymes, a second (acidic) group is seen in the acylation reactions of these two enzymes, but not in the deacylation. The pK_a of this group is ~9 in chymotrypsin and ~10 in trypsin.[106]

In addition to its primary function of splitting carboxylic acid derivatives, trypsin will also split carbon–carbon bonds,[107] that is, β-keto esters, as will chymotrypsin, when the stereochemistry is correct and when the carbon–carbon bond is activated. The trypsin-catalyzed hydrolysis of p-nitrophenyl acetate, using "stopped flow," where the presteady-state and the steady-state can be seen,[108] and the reaction in an aprotic solvent (dimethyl sulfoxide) have been investigated.[109]

Trypsin is inhibited by many substances. Calcium ion presumably stabilizes the enzyme against autolysis. On the other hand, the enzyme is inhibited by diethyl p-nitrophenyl phosphate, O,S-diethyl p-nitrophenyl phosphorothioate,[110] L-5-amino-1-tosylamidopentyl chloromethyl ketone,[111] which has been shown to act on a histidine residue,[112] and heavy metal ions.[113]

The catalytic activity of trypsin may be enhanced; in the specific case of ethyl N-acetylglycinate, a nonspecific substrate, 0.91 M ethylamine produced a ninefold increase in rate, whereas aliphatic amines in general are competitive inhibitors of specific substrates, as mentioned in Chapter 15.[114]

There has been considerable controversy over a trypsin enzyme of molecular weight 6000 which allegedly contains no histidine but is still active.[115] This has been disputed with the claim that the activity is due to the presence of some unchanged trypsin.[116]

Treatment of trypsin with sodium nitrite leads to inactivation and the stoichiometric loss of one lysine residue.[117] This indicates the additional involvement of an ammonium ion in the active site of trypsin. The ammonium ion may be involved in the electrostatic interaction between substrate and enzyme that must occur in trypsin.

Trypsin has many high molecular weight, naturally occurring inhibitors, among them the basic inhibitor of the bovine pancreas. The complete sequence of this inhibitor is known.[118] Another is soybean inhibitor. In the latter case it is known that inhibition involves cleavage of a peptide bond.[119]

This implies the formation of an acyl-enzyme in addition to noncovalent interaction.

The kinetics of trypsin-catalyzed reactions have been well studied but perhaps not as much as those of chymotrypsin. The specific synthetic substrates are derivatives of arginine[120] and lysine.[121] The α-N-tosyl-derivatives have a k_2 comparable to k_3 unlike the α-N-benzoyl derivatives.[120] Trypsin shows an apparent substrate activation[122] that may be related to several forms of the enzyme mentioned earlier.

The pH dependence of trypsin has been investigated. Early work shows a bell-shaped curve that probably reflects k_{cat}/K_m.[123] A very significant profile was obtained by Gutfreund who investigated the trypsin-catalyzed hydrolysis of ethyl α-N-benzoylargininate, found a pK of 6.25 for the catalytic step, and concluded that a histidine (imidazole) is part of the active site.[124] The trypsin-catalyzed hydrolysis of α-N-benzoylargininamide shows a bell-shaped curve for the rate step with a maximum at about pH 8.[125] Replacement of the side chain of arginine by the isosteric hydrocarbon radical (charged versus uncharged) resulted in a decrease of 10^2–10^6 in k_{cat}/K_m.[126]

The deacylation of diethylphosphoryl-trypsin with formohydroxamic acid[127] exhibits a bell-shaped deacylation curve, as do similar chymotrypsin reactions. The methanolysis of cinnamoyl-trypsin is dependent on the first power of the methanol concentration, as is the corresponding chymotrypsin reaction. The partitionings of cinnamoyl-α-chymotrypsin and cinnamoyl-trypsin are similar: $k_{methanolysis}/k_{hydrolysis}$ are 79 and 82, respectively.[6] No observable buildup of an unstable intermediate occurs in the deacylation of either cinnamoyl-trypsin or of cinnamoyl-α-chymotrypsin.[48] The hydrolysis of the specific substrate, N-benzoyl-L-arginine ethyl ester, by trypsin shows a deuterium oxide kinetic isotope effect of 2.5,[106] as do reactions catalyzed by α-chymotrypsin. The reactions of nonspecific substrates of both α-chymotrypsin and trypsin may be accelerated by the addition of a molecule of suitable structure.[18,128,129] The extensive nature of the mechanistic parallelism between α-chymotrypsin and trypsin permits one to extend the mechanism of Fig. 16.1 to trypsin as well as α-chymotrypsin.

In addition to the mechanistic similarity between α-chymotrypsin and trypsin, there exists a profound similarity in rate constants of deacylation (although not in binding constants). Over a 10^5-fold range in the rate of constants of deacylation of various acyl-α-chymotrypsins and acyl-trypsins, the ratio, $k_{acyl-α-chymotrypsin}/k_{acyl-trypsin}$, is 1 ± 0.5.[130] Of the 10^5-fold range of rate constants, electronic differences between the various acyl groups could account for perhaps 10-fold of this range. But some interaction of the various acyl groups with the two enzymes must explain the other 10^4-fold range. This interaction is obviously the same for both α-chymotrypsin and trypsin, implying at least partial common specificity sites of both enzymes. Common portions of the sequences of α-chymotrypsin and trypsin, including the reactive imidazole(s) and serine of the active sites, have been shown above. Of the sequence of 19 amino acids including histidine 57 of α-chymotrypsin, 16 correspond to the sequence in trypsin. Of the sequence of 11 amino acids including serine 195 of α-chymotrypsin, nine correspond to the sequence in trypsin. Three of the five differences are merely differences of a CH_2 group. Since the sequences shown above contain both the mechanistically important hydroxyl and imidazole groups, these sequences may be postulated to make up the common portion of the active sites of these enzymes that produce identical deacylation rate constants for nonionic substrates.[32,130]

16.4.2 Elastase

The elastase-catalyzed hydrolysis of p-nitrophenyl trimethylacetate shows both a presteady-state (first-order) and a steady-state (zero-order) release of p-nitrophenol.[50] Crystalline elastase has only a minor component of true elastase. Electrophoretically purified elastase is better but even it has a small impurity which is specific for tyrosine bonds.[50] This impurity can be removed by chromatography.[131] There is kinetic evidence for an extended active center in elastase.[59]

16.4.3 Subtilisin

Several acyl derivatives of subtilisin have been prepared: *trans*-cinnamoyl-subtilisin, *trans*-furylacryloyl-subtilisin, *trans*-indoleacryloyl-subtilisin, and furoyl-subtilisin.[9,12,61]

The histidine of the active site of subtilisin has been identified as residue 64.[132] Many aspects of subtilisin reactions resemble those of chymotrypsin.[133] Several strains of subtilisin can be reversibly inactivated by the dye 4-(4'-aminophenylazo)phenylarsonic acid.[34] This inactivation is apparently tied to the acidic form of the imidazole (of histidine) from the pH dependence. On the other hand, one subtilisin is inactivated by phenylboronic acid and the pH dependence is on the basic form of imidazole and the acidic form of a group of pK 9.[135] Some other inactivators of subtilisin include the bromomethyl ketone corresponding to Z†-phenylalanine[136] and the choromethyl ketone corresponding to Z-alanyl-glycyl-phenylalanine,[137] which is even faster.

16.4.4 Thrombin

Another serine proteinase having a similar sequence to chymotrypsin[138] is the enzyme thrombin that is involved in blood clotting. It was found early

† Z = benzyloxycarbonyl.

that the enzyme could be inhibited by diisopropyl phosphorofluoridate.[139] Preliminary evidence indicates that the thrombin-catalyzed hydrolysis of N-acetyl-L-tryptophan p-nitrophenyl ester proceeds with an initial burst of p-nitrophenol.[131] The thrombin-catalyzed hydrolyses of a number of p-nitrophenyl esters including the acetate, the Z-glycinate, the Z-tyrosinate, the Z-phenylalaninate, the acetyl-leucinate, and the iodoacetate were investigated. Also investigated was the hydrolysis of tosylarginine methyl ester.[140] Since the thrombin-catalyzed hydrolyses of three esters of benzoylarginine have the same k_{cat}, it was concluded that k_3 is the rate-determining step; on the other hand, since the hydrolyses of four esters of tosylarginine have different k_{cat}'s, it was concluded that k_2 is partially rate-determining in those cases.[141]

16.4.5 Acetylcholinesterase

Other serine proteinases are the esterases. There are many esterases including acetylcholinesterase, carboxylesterases, atropine esterase, and so on.

Acetylcholinesterase, which is an important enzyme connected with nerve transmission, is a serine proteinase that is irreversibly inhibited by low concentrations of certain organophosphorus compounds such as diisopropyl phosphorofluoridate, tetraethyl pyrophosphate, and isopropyl methylphosphonofluoridate. Some of these compounds are chemical warfare agents (nerve gases) and others are used as insecticides. Their toxicity derives from the remarkable facility with which they inhibit acetylcholinesterase[142,143] by a phosphorylation of the enzyme.[144-146] The inhibited enzyme can be dephosphorylated and thus reactivated by nucleophilic reagents.[147,148] The best reactivator, which was designed on the basis of molecular complementarity, is pyridine-2-aldoxime methiodide.[149-151] A method for determining the turnover number (and normality) of the enzyme is based on the fact that certain compounds that are generally regarded as inhibitors are, in reality, extremely poor substrates. For example, the reaction of dimethylcarbamyl fluoride with acetylcholinesterase yields the rate constants for the individual steps. On this basis, the turnover number for the acetylcholinesterase-catalyzed hydrolysis of acetylcholine at 25° and pH 7.0 was calculated to be 7.4×10^5 min^{-1} (substrate = 2.5×10^{-3} M).[152]

In addition to the phosphate and carbamate inhibitors, esters of methanesulfonic acid inhibit the enzyme. In the latter case, a methanesulfonyl-enzyme is formed.[153] In addition, substrate inhibition can be inferred.[154] The methanesulfonyl-enzymes can be reactivated by the proper nucleophile.[155] As with trypsin, the rate of formation of the sulfonyl-enzyme is increased by a number of substituted ammonium ions.[156]

Recently, the molecular weight of acetylcholinesterase was determined by sedimentation equilibrium and found to be 260,000. The enzyme was split in the presence of guanidine and mercaptoethanol into four subunits each having one fourth the molecular weight of the native enzyme. Two different examinations of the C terminal residues indicated two kinds of polypeptide chains in acetylcholinesterase. This suggests that acetylcholinesterase has a dimeric hybrid structure, with two α and two β chains, like hemoglobin.[151]

16.4.6 Carboxylesterases

Amino acid sequences of carboxylesterases from a variety of species have been determined, including horse, sheep, pig, ox, and chicken. A great similarity can be seen around the active serine as shown in Table 16.4.[158] These sequences are reasonably similar to those in chymotrypsin. The mechanisms may also be similar.

Table 16.4
Amino Acid Sequences of Active-site Peptides of Carboxylesterases [b]

Species	Amino Acid Sequence [a]
Horse	Gly–Glu–Ser*–Ala–Gly–Gly–Glu–Ser
Pig, sheep	Gly–Glu–Ser*–Ala–Gly–Gly–Glu–Ser
Ox	Gly–Glu–Ser*–Ala–Gly–Ala–Glu–Ser
Chicken	Gly–Glu–Ser*–Ala–Gly–Gly–Ile–Ser

[a] The asterisk indicates the serine of the active site.
[b] From R. C. Augusteyn, J. de Jersey, E. C. Webb, and B. Zerner, *Biochim. Biophys. Acta*, **171**, 128 (1969).

16.4.7 Thiolsubtilisin (An Artificial Enzyme)

Between the serine proteinases (above) and the cysteine proteinases (below) lie artificial enzymes. In the simplest, an oxygen atom of the active site is replaced by sulfur. This is not easy with the enzymes containing disulfide bonds for the sulfur reagent necessary for the transformation may ruin the tertiary structure of such enzymes, by nucleophilic attack on the disulfide bonds which are responsible in part for the tertiary structure. But the transformation has been accomplished with a number of bacterial enzymes which contain no disulfide links.[159-162] There has been some controversy over whether the modified enzyme is a true enzyme, for it hydrolyzes nitrophenyl ester substrates but not amides. In one sense, toward p-nitrophenyl ester substrates it is a better enzyme than the original at pH 6 (k_{cat}/K_m)[161] but considering another criterion (k_{cat}) it is a poorer enzyme. Whatever criterion is used, the cysteine enzyme is real since it hydrolyzes p-nitrophenyl acetate much faster than cysteine itself. Other serine proteinases, which are not discussed, include urokinase, cathepsin, and chymopapain.

16.5 OTHER PROTEIN ENZYMES

16.5.1 Sulfhydryl Enzymes: Papain

Reactions catalyzed by papain follow eq. 16.1; acylation is dependent on groups with pK_a's of 4.3 and 8.4 (as an acid-base pair); and deacylation is dependent on a single group with a pK_a of 3.9. Furthermore, the acyl-papain is a thiol ester. Thus as with α-chymotrypsin, papain-catalyzed reactions show a pK_a of 8.4 in acylation, but not in deacylation. However, in contrast to α-chymotrypsin, this group may be directly identified as a sulfhydryl group whose pK_a disappears on acylation. By process of elimination, the catalytic entity (seen in both acylation and deacylation), which is an imidazole group of histidine in α-chymotrypsin, may be a perturbed histidine group in papain. Papain reactions show a deuterium oxide isotope effect like that in chymotrypsin reactions,[13] implying a rate-determining proton transfer.

Two prototropic groups of pK_a 4.3 and 8.4 control the rate of acylation of papain by α-N-benzoyl-L-arginine ethyl ester and amide, while one prototropic group of pK_a 3.9 controls the deacylation of α-N-benzoyl-L-arginyl-papain.[163] The group of pK_a 3.9 to 4.3 has been identified as a carboxylic acid group[164,165] on the basis of pH dependencies and heats of ionization. A sulfhydryl group is implicated in papain action on the basis of (1) chemical inhibition studies[164,165]; (2) the involvement of a group of pK_a 8.4; (3) its appearance in acylation but not in deacylation[163]; (4) the spectrophotometric assignment of the acyl-enzymes thionohippuryl-papain[166] and $trans$-cinnamoyl-papain[13] as thiol esters; and (5) the greater ability of papain than chymotrypsin to catalyze transpeptidation, understandable on the basis that thiol esters react more readily than oxygen esters with amines.

The location of the active center cysteine residue has been elucidated by means of a radioactive iodoacetate[167] and a chloroketone analog of a specific substrate.

A histidine residue is also at the active site of papain on the basis of model studies,[169] X-ray crystallographic studies, and irreversible inhibition by 1,3-dibromoacetone.[170] The latter reagent is a bifunctional one and further

Other Protein Enzymes

indicates that the sulfhydryl group and the imidazole (histidine) group of the active site are approximately 5 Å from one another.

The amino acid sequences around the cysteine residue of papain[171] and ficin[174] are known and are reasonably similar.

The amino acid sequences around the active site serine in animal proteases resembles that around the active site cysteine in plant proteases.[171] Likewise the amino acid sequences about the active site histidine are similar.[171] The papain site corresponds to seven amino acid residues[172] and one of the subsites specifically interacts with phenylalanine residues.[173]

N-Benzoyl-L-arginine derivatives and N-benzoyl-L-lysine derivatives are known to be specific substrates for the enzymes papain and ficin (one coming from the papaya tree and the other coming from the fig tree). Esters, amides, and other derivatives have been investigated. Transesterification is known as well as hydrolysis.[175]

Papain can be inhibited by L-5-amino-1-tosylamidopentyl chloromethyl ketone[176] and by the chloromethyl ketone of tosyl-L-phenylalanine; they react with the sulfhydryl group of the active site.[177]

The reaction of papain with ethyl hippurate (*carbonyl*-^{18}O) shows no oxygen exchange[178] so there exists no direct evidence for a tetrahedral intermediate. The unprotonated imidazole group is usually assumed to be involved in the catalysis, but the imidazolium ion has also been proposed to participate, on the basis of a Hammett plot of acylation.[179]

The effect of alcohol on deacylation, which is a criterion of general basic or nucleophilic catalysis has taken some interesting turns. Cinnamoyl-papain shows kinetic dependence on the presence of added nucleophiles.[13] On the other hand, hippuryl-papain has been reported to show none.[180] On the basis of the latter result a 4-step mechanism was proposed in which one acyl-papain is converted to another. The experimental work on which this is based was contradicted by subsequent work.[181] The 4-step scheme was made more firm by identical ratios of acetonitrile to alcohol in the rate of deacylation of cinnamoyl- and hippuryl-papain even though the absolute rates differ by 10^3.[182] Papain catalysis of reactions of the specific substrate, p-nitrophenyl N-acetyl-L-tryptophanate with the alcohols, methanol, ethanol, n-butanol, and n-pentanol shows a wide variety of behavior, the lower alcohols showing a decrease in the rate of deacylation with a slight curvature to the plot.[183] The best fit to the data is encompassed in the following:

$$\begin{array}{c} \text{P}_1 \\ + \\ \text{E} + \text{S} \rightleftharpoons \text{E·S} \longrightarrow \text{ES}' \longrightarrow \text{E} + \text{P}_2 \\ \updownarrow \quad\quad \updownarrow \quad\quad\quad\quad \updownarrow \\ \quad\quad\quad \text{E·S·N} \\ \text{E·N} + \text{S} \quad\quad\quad \text{E—S}'\text{·N} \longrightarrow \text{EN} + \text{P}_2 \end{array} \quad (16.3)$$

which indicates the possibility of the binding of the nucleophile to the enzyme substrate complex and/or to the acyl-enzyme. Obviously, the final word has not been said.

The mechanism of eq. 15.5 involving an acyl-enzyme intermediate has been suggested for the action of the sulfhydryl-containing proteolytic enzymes, papain, ficin, and bromelain.[184,185] The evidence for this mechanism includes: (1) the equivalence of k_{cat} for the papain- and ficin-catalyzed hydrolyses of α-N-benzoyl-L-arginine ethyl ester and amide;[184,186] (2) the equivalence of k_{cat} for the papain-catalyzed hydrolysis of ten esters of hippuric acid[187]; (3) the essentiality of a single sulfhydryl group of the enzymes on the basis of chemical inhibition studies[184,188]; (4) the spectrophotometric observation of the acyl-enzymes, thionohippuryl-papain[166] and *trans*-cinnamoyl-papain,[13] and (5) identical pH dependencies of k_{cat}/K_m(app) but different pH dependencies of k_{cat} for the papain-catalyzed hydrolysis of α-N-benzoyl-L-arginine ethyl ester and amide, which are consistent only with the two-step acyl-enzyme mechanism of eq. 15.5.[163] Kinetic work further supports this.[189]

16.5.2 Sulfhydryl Enzymes: Ficin

As mentioned earlier, this enzyme from fig latex resembles papain to a great degree. Therefore it is not discussed in detail, but it should be noted that ordinary ficin contains nine components which have been separated.[190]

16.5.3 Sulfhydryl Enzymes: Bromelain

Another sulfhydryl enzyme is stem bromelain which is obtained from the juice of the pineapple fruit. Its existence has been known for almost a century but it was not until recently that it was shown to consist of five components,[191] and furthermore, four of its components were shown to have the specificity of the serine proteinases, trypsin and elastase.[192] In other words, there may be a parallelism emerging between the serine (animal) proteinases and the cysteine (plant) proteinases with regard to specificity.

16.5.4 Pepsin

The formation of L-tyrosyl-L-tyrosine, as well as the expected products of hydrolysis from N-acetyl-L-tyrosyl-L-tyrosine using pepsin as catalyst, can be explained only in terms of a transpeptidation such as

$$2\text{AcTyrTyr} \longrightarrow \text{AcTyr} + \text{AcTyrTyrTyr}$$

followed by a subsequent splitting of the tripeptide.[193] This transpeptidation reaction requires an amino-enzyme (in contrast to an acyl-enzyme) intermediate,[193] if nucleophilic reactions by the amino group of tyrosine, an unfavored reaction in acidic solution, are excluded. The radioactive exchange reaction between ZTyrTyr and ZTyr–C[14] catalyzed by pepsin led to Tyr*–Tyr, whereas the alternate exchange reaction of ZTyrTyr with Tyr-C[14] did not occur under the same conditions.[194] This experiment indicates that exchange via an amino-enzyme may occur but that exchange via an acyl-enzyme is not possible or is much slower. Without any enzyme intermediate, the exchange would require an unlikely four-center reaction. Thus, this exchange indicates an amino-enzyme intermediate. However, an acyl-enzyme intermediate cannot be excluded since pepsin catalyzes the oxygen-18 exchange between water and Z-L-phenylalanine, but not the D-compound.[195]

16.5.5 Ribonuclease

Ribonuclease is an enzyme which splits nucleic acids; its sequence[196] and X-ray structure have been completely worked out, and it has been the subject of much investigation. It consists of 124 amino acids and is specifically split into fragments of about 100 and of about 20 amino acids by the enzyme subtilisin. Neither of the fragments is enzymatically active, but a 1:1 mixture of the two is.[197] Recently a group at Merck has synthesized the larger fragment, combined it with the naturally derived smaller fragment and found activity.[198] Furthermore the entire ribonuclease molecule has been synthesized by the "solid-state" method and shown to have activity.[199]

There seems to be no doubt that the cleavage of a nucleic acid by ribonuclease proceeds in two covalent steps, the formation of a cyclic phosphate and its subsequent hydrolysis.[201] Kinetic studies have been carried out on each step.[202,203]

The magnitude of the catalysis by ribonuclease has been estimated to be 10^9–10^{10}.[204] The phosphatases are presumably similar to ribonuclease.[205] In ribonuclease, alkylation of histidines 12 and 119 by iodoacetate ion but not by iodoacetamide has implicated these two residues. Histidine 119 is alkylated on the one position and histidine 12 on the three position.[206] Lysine 41 was also shown to be in the active site by the specific inactivation of the enzyme by dinitrofluorobenzene.[207] By contrast eight or nine of the lysines, but not lysine 41, can be guanylated without diminishing activity. It therefore appears that the amino group on lysine 41 is necessary. One of the important experiments that have been carried out with this enzyme has been the synthesis of an active hybrid enzyme by mixing two inactive enzymes, one of which was alkylated on histidine 119 and the other on histidine 12.[208–210]

The two kinetic pK's for the ribonuclease-catalyzed hydrolysis of cytidine 2′,3′-phosphate are close to normal, being 6.78 and 5.22 for the free enzyme and 8.10 and 6.30 for the enzyme-substrate complex.[211] Two general mechanisms have been proposed for ribonuclease action; one involves conjugate (multiple) catalysis of the general acid-general base variety.[212] An imidazolium ion and an imidazole would serve these functions. On the other hand, a hydrogen atom of the substrate has been proposed to be involved.[213] Of these suggestions, the former seems preferable since it is analogous to what has been said earlier and since it seems less artificial. A model of ribonuclease has been constructed on the basis of the above evidence which, however, still leaves considerable freedom of movement.[204,214] Imidazole and morpholine act as general bases and acids (Chapter 5) to catalyze the hydrolysis of a nucleic acid analog.[215] It has been suggested on the basis of these results that a general acid-general base mechanism occurs in ribonuclease.[61]

Ribonuclease (and probably some other protein enzymes) are in reality glycoproteins; that is, there is a sugar moiety attached to the protein.[216]

16.5.6 Fumarase

Fumarase is an enzyme that hydrates a double bond according to

$$HO_2CCH=CHCO_2H + H_2O \rightleftharpoons HO_2CCH_2CHOHCO_2H$$

That is, it catalyzes the interconversion of fumarate and malate. Related enzymes are aconitase, which catalyzes the interconversion of aconitic acid and isocitric acid, and aspartase which adds the elements of ammonia instead of water to fumaric acid. Until recently, fumarase was thought to be a completely specific enzyme; this attitude was changed by the finding that fumarase catalyzes the conversion of tartrate to oxaloacetate.[213] Early studies of fumarase-catalyzed reactions showed that the initial rate of the forward reaction (hydration of fumarate) bears no obvious relation to the initial rate of the reverse reaction.[218] The pH dependence of four competitive inhibitors of fumarase-catalyzed reactions was investigated over an extensive pH range and the binding constants to both malate and fumarate were calculated.[219] An estimate of k_1 from experiment and from diffusion theory puts it at 10^{10} M^{-1} sec^{-1} or greater.[220] The temperature dependence of fumarase-catalyzed reactions indicates that the steady-state constants can be interpreted in terms of many parameters used earlier.[221]

The view of the stereochemistry of the addition has taken an about-face. On the basis of deuterium labeling experiments, the stereochemistry of addition was originally thought to be *cis*.[222] Then it was unequivocally shown in two laboratories to be a *trans* addition, again using deuterium labeling and nuclear magnetic resonance.[223,224] This conclusion was generalized to the enzyme-catalyzed hydration of aconitic acid[225] and to the whole Krebs cycle.[224,226]

The mechanism of this enzymatic process is a perplexing one. Incorporation of deuterium from deuterium oxide into L-malate during dehydration to fumarate occurs at a lower rate than the production of fumarate, and furthermore, the rate of deuterium incorporation is in quantitative agreement with that expected for the reversal of the overall reaction. This indicates no deuterium exchange; furthermore, there is no deuterium isotope effect. Therefore a carbonium ion mechanism has been proposed.[226,227] Results on competitive inhibitors mentioned earlier are consistent with a *trans* stereochemistry.[219] Several mechanisms based on the erroneous stereochemistry have been presented.[229] The deuterium isotope effect is also disputed, the initial rates being reported to have k_H/k_D of 1.5 to 2.5.[228]

16.5.7 Sucrose Phosphorylase

The kinetics of sucrose phosphorylase (disaccharide transferase) from *Pseudomonas saccharophilia* were examined using a purified sample (molecular weight 80,000–100,000) and found to be consistent with a double displacement mechanism involving formation of a glucosyl-enzyme intermediate and subsequent reaction of this intermediate with an acceptor to form the reaction products. Further evidence of the formation of the intermediate was obtained by isolating a denatured form of the intermediate. It contained one mole of glucose per mole of enzyme.[230] A glucosyl-enzyme intermediate was originally proposed on the basis of exchange experiments.[231] This general problem has also been discussed by Koshland.[232]

16.5.8 Amylases

The amylases are enzymes that hydrolyze various components of starch. α-Amylase hydrolyzes the major component (80%) of starch, amylose, which is composed of α-1,4-linked glucose molecules. β-Amylase hydrolyzes the minor component (20%) of starch, amylopectin, which has in addition to the α-1,4 links, some 1,6 cross links. These two enzymes are compared in Table 16.5.

The difference in stereochemistry observed in the amylases implies that there may be a difference in stereochemistry with other pairs of enzymes. This is clearly seen in the enzymes that split sugars, as shown in Table 16.6.

16.5.9 Translocase

Another protein enzyme is "PMp" translocase which gives a "burst" analogous to that seen in chymotrypsin reactions when one of the substrates is in deficient quantity. It is essentially a transferase enzyme, as is

Table 16.5
Comparison of α- and β-Amylase

Property	α	β
Molecular weight	45,000–59,000	152,000
Groups determining pH dependence of V_{max}	4.22 and 7.45[a]	3.7 and 7.0[b]
Known essential groups		SH
Mode of attack	Indiscriminate	Penultimate unit
Cations	1 gram atom of $Ca^{2\oplus}$/mole	None
Anions	$Cl^{\ominus}$?	
Cleavage	At anomeric carbon atom	At anomeric carbon atom
Stereochemistry	Retention	Inversion[b]
Intermediate	Glucosyl-enzyme	Carbonium ion

[a] S. Ono et al., *Bull. Chem. Soc. Japan*, **31**, 957 (1958).
[b] J. A. Thoma and D. E. Koshland, Jr., *J. Biol. Chem.*, **235**, 2511 (1960).

chymotrypsin under certain conditions, and is involved in the first stage of cell wall metabolism.[234]

Another enzyme involves the transfer of acetate from acetoacetate to succinyl coenzyme *A*. It has been shown that the reaction is insensitive to pH from 6 to 8 and is inhibited by acetoacetyl coenzyme *A* and by *p*-nitrophenyl acetate. The kinetics can be explained only by formation of an intermediate.[235,236]

Table 16.6
Pairs of Enzymes Catalyzing Reactions at Anomeric Carbon Atoms

Family	Inversion	Retention	Substrate
Amylase	β-Amylase	α-Amylase	Amylose Amylopectin
Phosphorylase	Maltose phosphorylase[a]	Sucrose phosphorylase[a]	Sucrose
		Muscle phosphorylase[b]	Glycogen
Glucoside	Takamaltase	Pancreatic maltase α- and β-Glucosidase	Maltose Glycosyl-X
Nucleosides	Nucleoside phosphorylase[a]	Nucleoside glycosidase	Nucleoside

[a] Exchange.
[b] No exchange.

16.5.10 Lysozyme

Lysozyme is an enzyme that cleaves certain cell walls. As noted earlier its complete three-dimensional structure is known. It is a globular protein, roughly in the shape of a butterfly, with one half containing a good deal of helix, and the other, β pleated sheet. The active site is between these two halves and has been described as a cleft or groove. The amino acid sequence and three-dimensional structure of lysozyme are given in Figs. 16.2 and 16.3.[237] This cleft is reasonable since the natural substrate of this enzyme is a long-chain polymer. More specifically lysozyme cleaves two glucose units between an *N*-acetylglucosamine and an *N*-acetylmuramic acid moiety. In addition to the natural substrate of the enzyme, chitin, the polysaccharide of *N*-acetyl-β-D-glucosamine is a substrate as are some degradation products of chitin[238] and a number of β-aryl di-*N*-acetylchitobiosides.[239] As does the natural substrate, the smaller substrates obey Michaelis–Menten kinetics.

The catalytically active functional groups have been suggested to be the carboxyl groups of glutamic acid 35 and aspartic acid 52, presumably one in the ionized form and one in the nonionized form,[240] and possibly the imidazole group of histidine 15.[241] The latter suggestion comes from a study of the lysozyme-catalyzed hydrolysis of six carboxylic esters. These compounds are not similar to the natural glucose substrates of the enzyme, and furthermore, their hydrolyses are not inhibited by the usual inhibitors of lysozyme. It may therefore be concluded that the ester hydrolyses, while interesting, are not occurring at the same site as the usual lysozyme cleavage.

The usual lysozyme is derived from egg white. However, lysozyme can also be obtained from papaya (and fig) latex. Papaya lysozyme has a molecular weight of 25,000 in opposition to egg white lysozyme, which has a molecular weight of about 13,000.[242] This suggests dimer formation in the former case. There appear to be some differences in kinetic properties of the two enzymes. This is to be contrasted with the serine proteinases (of animal and plant origin). Iodine oxidation of a single tryptophan inactivates the egg white enzyme[243] but not the papaya enzyme.[242]

The binding of the inhibitor 2-acetamido-2-deoxy-D-glucose to lysozyme was studied using an nmr technique. A chemical shift of the acetamido methyl proton resonance resulted from association with the enzyme.[244]

The mechanism of lysozyme action, the enzyme whose structure is known best, may be complicated. The determination of the structure by Phillips has led to his suggestion for its mode of action.[245] It is claimed that in lysozyme there is a slight movement on binding the substrate; this point will be amplified in Chapter 19. Three possible mechanisms have been proposed for lysozyme action: (1) the neighboring *N*-acetylamino group acts as a

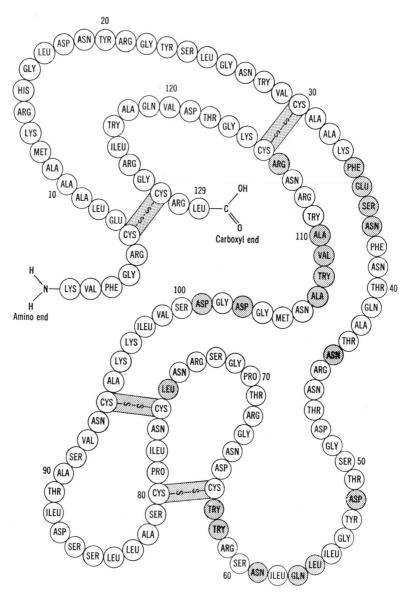

Fig. 16.2 The sequence of lysozyme. The dark circles adjoin the active site. The light circles do not. From D. C. Phillips, *Sci. Am.*, **215**, no. 5, 78 (1966).

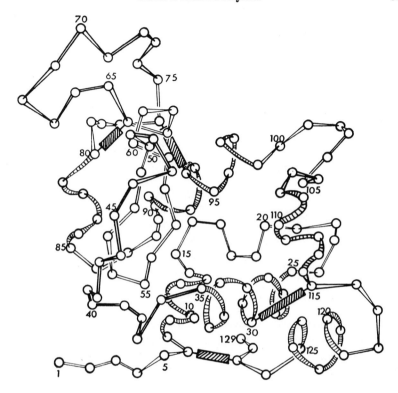

Fig. 16.3 The three-dimensional structure of lysozyme. ▨—disulfide links. From C. C. F. Blake, D. F. Koenig, G. A. Mair, A. C. T. North, D. C. Phillips, and V. R. Sarma, *Nature*, **206**, 757 (1965).

neighboring group and aspartic acid as a general acid; (2) the carboxylate anion of glutamic acid 35 further assists the neighboring *N*-acetylamino group by acting as a general base; and (3) the carboxylate anion of aspartic acid 52 acts as a nucleophile and glutamic acid 35 as a general acid (see Chapter 9).[246] The last is Phillips' suggestion. Recently it was shown that lysozyme catalyzes the hydrolysis of *p*-nitrophenyl 2-acetamido-4-*O*-(2-acetamido-2-deoxy-β-D-glucopyranosyl)-2-deoxyglucopyranoside with a value of k_{cat} 20 times greater than *p*-nitrophenyl 4-*O*-(2-acetamido-2-deoxy-β-D-glucopyranosyl)-β-D-glucopyranoside.[247] These results suggest acetamido participation.[248]

A concerted acid-base or acid-nucleophilic catalysis is favored on the basis of kinetic results with small substrates.[249] Transglycosylation was demonstrated in lysozyme-catalyzed reactions. This reaction has been used to synthesize *p*-nitrophenyl β-D-glucoside and a corresponding oligomer. Since lysozyme hydrolyzes these glycosidic bonds and since they contain no acetylamino group, anchimeric assistance by the *N*-acetylamino group can be ruled out as a necessary pathway for the enzyme-catalyzed cleavage of glycosidic bonds.[251] The lysozyme active site, while not completely characterized, must certainly be large, for the enzyme can interact with a hexasaccharide and there is strong kinetic dependence on the length of the saccharide. This implies the importance of nonproductive complexes (complexes of enzyme and substrate that do not lead to reaction).[251] Evidence for carbonium ion formation and participation of glutamic acid 35 is particularly convincing.[58] Retention at the anomeric center means that if a carbonium ion is formed, it must be constrained by the enzyme.[60]

There are many other protein enzymes such as cellulase, clarase, penicillinase, the lipases, urease, hyaluronidase, and invertase that have not been discussed. Some have been mentioned in other places. Others do not fit into the present scheme. Still others are discussed elsewhere.

16.6 ACTIVE TRANSPORT PROTEINS

Active transport proteins are strictly not enzymes since their function is to transport small molecules through a membrane barrier, but since they are protein in nature and bind small molecules as do enzymes, one is mentioned here. For example, sulfate ion is bound firmly to *Salmonella typhimurium*. The binding site is located near the cell surface and has been suggested as playing a role in active transport of sulfate ion. A simple procedure involving chromatography on DEAE-cellulose leads to the sulfate binding material in homogeneous form. The protein has a molecular weight of 32,000 and a typical amino acid composition, except that it lacks sulfur-containing amino acids. One sulfate ion is bound per protein molecule at saturation. The dissociation constant is low and is highly dependent on ionic strength.[233]

There are many other proteins that are involved in active transport but will not be considered here.

16.7 CONCLUSION

There appear to be several families of protein enzymes. The families appear to catalyze by a mechanism involving a nucleophile (on the protein) assisted by a general base or two general bases or a general acid and a general base (on the protein). The main families of protein enzymes are called serine enzymes and cysteine enzymes. In the former, the nucleophile (Chapter 6) is an hydroxyl group; in the latter, a sulfhydryl group. The general base in

Conclusion

both serine and cysteine enzymes is usually the imidazole moiety of a histidine residue of the protein, for reasons given earlier (Chapter 4). In addition to the nucleophile and general base mentioned above, these enzymes contain a binding site that imparts specificity to the enzymatic catalysis.

In general the serine enzymes are more efficient catalysts than the corresponding cysteine enzymes, presumably because an unionized hydroxyl group is a better nucleophile than an unionized sulfhydryl group. Of course neither of these groups may be acting per se but possibly as the corresponding anion produced by a nearby general base. In general the Brønsted relation discussed earlier (Chapter 4) suggests that the nucleophile of greatest basicity (hydroxyl) will be the most reactive.

Specificity and the binding site were alluded to above. In essentially all protein enzymes, there are three kinds of specificity, which may involve positive, negative, or neutral substrates. The positive substrates may be alkylammonium ions; the negative substrates usually contain a carboxylate group. The neutral substrates bind by hydrophobic binding, as discussed earlier. The charged substrates have a component of an electrostatic interaction in addition to the hydrophobic interaction. The relative importance of these kinds of binding may vary in individual cases. The electrostatic interaction is larger when the dielectric constant is lower. The hydrophobic interaction is larger when the apolar part of the molecule is larger. Whether the electrostatic interaction is larger than the hydrophobic interaction depends on the particular situation. There has been a tendency to say one or the other is predominant, but when both are possible, both need to be considered. In both the serine enzymes and cysteine enzymes, the overall mechanism appears to be the same and the specificities appear to be parallel: a cationic group, a large neutral group, or a small neutral group. Anionic specificity does not seem to exist.

One may ask why no coenzymes (Chapter 17) are needed here. The answer appears to be that the enzyme protein has many nucleophiles (Chapter 6) and general bases (Chapter 4), and that these are sufficient to catalyze the hydrolytic reactions described here (either by stabilization of the intermediate or by nucleophilic attack). This question will be discussed in more detail in the next chapter.

One may also probe the difference between an ordinary protein and an enzyme protein. For example, the protein insulin does catalyze the hydrolysis of *p*-nitrophenyl acetate but it does so at a rate that can be accounted for fully by the corresponding concentration of imidazole nucleophiles on the protein (Chapter 4). Ordinary proteins usually contain many binding sites whereas enzymes have single sites. Proteins seem, however, to lack the catalytic components (in the proper stereochemistry) to effect both binding and catalysis (which is essential for reaction). It would be of interest to see if one could transform an ordinary protein into an enzyme protein by the introduction of the proper catalytic components in the proper organization. This description implies that an enzyme protein catalyzes because it has both binding and catalytic components (side chains) in the correct juxtaposition.

REFERENCES

1. D. E. Koshland, Jr., in *The Mechanism of Enzyme Action*, W. D. McElroy and B. Glass, Eds., The Johns Hopkins Press, Baltimore, 1954, p. 357.
2. J. Kraut, D. F. High, and L. C. Sieker, *Proc. Natl. Acad. Sci., U.S.*, **51**, 839 (1964); D. M. Blow, M. G. Rossmann, and B. A. Jeffery, *J. Mol. Biol.*, **8**, 65 (1964).
3. B. R. Baker, *J. Pharm. Sci.*, **53**, 347 (1964).
4. M. Dixon and E. C. Webb, *Enzymes*, 2nd ed., Academic Press, New York, 1964, p. 144.
5. B. Zerner, R. P. M. Bond, and M. L. Bender, *J. Amer. Chem. Soc.*, **86**, 3674 (1964).
6. C. R. Gunter, Doctoral Dissertation, Northwestern University, 1965.
7. N. K. Schaffer, S. C. May, Jr., and W. H. Summerson, *J. Biol. Chem.*, **206**, 201 (1954).
8. R. A. Oosterbaan and M. E. van Adrichem, *Biochim. Biophys. Acta*, **27**, 423 (1958); Y. Shalitin and J. R. Brown, *Biochem. Biophys. Res. Commun.*, **24**, 817 (1966).
9. N. K. Schaffer, S. C. May, Jr., and W. H. Summerson, *J. Biol. Chem.*, **202**, 67 (1953); N. K. Schaffer, L. Simet, S. Harshman, R. R. Engle, and R. W. Drisko, *J. Biol. Chem.*, **225**, 197 (1957).
10. F. Turba and G. Gundlach, *Biochem. Z.*, **327**, 186 (1955).
11. J. Mercouroff and G. P. Hess, *Biochem. Biophys. Res. Commun.*, **11**, 283 (1963).
12. S. A. Bernhard, S. J. Lau, and H. Noller, *Biochemistry*, **4**, 1108 (1965).
13. M. L. Bender and L. J. Brubacher, *J. Amer. Chem. Soc.*, **86**, 5333 (1964).
14. J. R. Platt, *Science*, **129**, 372 (1959); *J. Chem. Phys.*, **34**, 862 (1961).
15. B. M. Anderson, E. H. Cordes, and W. P. Jencks, *J. Biol. Chem.*, **236**, 455 (1961).
16. M. L. Bender, F. J. Kézdy, and C. R. Gunter, *J. Amer. Chem. Soc.*, **86**, 3714 (1964).
17. R. M. Epand and I. B. Wilson, *J. Biol. Chem.*, **239**, 4138 (1964).
18. P. W. Inward and W. P. Jencks, *J. Biol. Chem.*, **240**, 1986 (1965).
19. M. L. Bender, *Chem. Rev.*, **60**, 53 (1960).
20. D. H. Strumeyer, W. N. White, and D. E. Koshland, Jr., *Proc. Natl. Acad. Sci., U.S.*, **50**, 931 (1963).
21. D. E. Koshland, Jr., D. H. Strumeyer, and W. J. Ray, Jr., *Brookhaven Symp. Biol.*, **15**, 101 (1962).
22. L. Weil and A. R. Buchert, *Federation Proc.*, **11**, 307 (1952).
23. G. Schoellman and E. Shaw, *Biochemistry*, **2**, 252 (1963).
24. L. B. Smillie and B. S. Hartley, *Abstr. Fed. Eur. Biol. Soc., London*, 26 (March 1964).
25. D. Pospisilova, B. Meloun, and F. Sorm, *Abstr. Fed. Eur. Biol. Soc., London*, 27 (March 1964).
26. M. L. Bender and B. W. Turnquest, *J. Amer. Chem. Soc.*, **79**, 1652, 1656 (1957).
27. T. C. Bruice and G. L. Schmir, *J. Amer. Chem. Soc.*, **79**, 1663 (1957).

References

28. W. P. Jencks and J. Carriuolo, *J. Amer. Chem. Soc.*, **83**, 1743 (1961).
29. J. R. Brown and B. S. Hartley, *Biochem. J.*, **89**, 59P (1963); B. Keil, Z. Prusik, and F. Sorm, *Biochim. Biophys. Acta*, **78**, 559 (1963).
30. B. S. Hartley, *Nature*, **201**, 1284 (1964).
31. K. A. Walsh, D. L. Kauffman, K. S. V. S. Kumar, and H. Neurath, *Proc. Natl. Acad. Sci., U.S.*, **51**, 301 (1964).
32. M. L. Bender, J. V. Killheffer, Jr., and F. J. Kézdy, *J. Amer. Chem. Soc.*, **86**, 5331 (1964).
33. B. Labouesse, H. L. Oppenheimer, and G. P. Hess, *Biochem. Biophys. Res. Commun.*, **14**, 318 (1964); H. L. Oppenheimer, B. Labouesse, K. Carlsson, and G. P. Hess, *Federation Proc.*, **23**, 315 (1964).
34. T. C. Bruice, T. H. Fife, J. J. Bruno, and N. E. Brandon, *Biochemistry*, **1**, 7 (1962).
35. G. Gundlach and F. Turba, *Biochem. Z.*, **335**, 573 (1962).
36. W. B. Lawson and H. J. Schramm, *J. Amer. Chem. Soc.*, **84**, 2017 (1962); H. J. Schramm and W. B. Lawson, *Z. Physiol. Chem.*, **332**, 97 (1963).
37. H. Schachter and G. H. Dixon, *J. Biol. Chem.*, **239**, 813 (1964).
38. G. H. Dixon and H. Shachter, *Can. J. Biochem.*, **42**, 695 (1964).
39. A. N. Glazer and F. Sanger, *Biochem. J.*, **90**, 92 (1964).
40. S. K. Dube, O. A. Roholt, and D. Pressman, *J. Biol. Chem.*, **239**, 1809 (1964).
41. D. L. Filmer and D. E. Koshland, Jr., *Biochem. Biophys. Res. Commun.*, **17**, 189 (1961).
42. J. F. Wooton and G. P. Hess, *Nature*, **188**, 726 (1960).
43. T. Viswanatha and W. B. Lawson, *Arch. Biochem. Biophys.*, **93**, 128 (1961).
44. K. Simon, *Experientia*, **18**, 150 (1962).
45. D. E. Koshland, Jr., Y. D. Karkhanis, and H. G. Latham, *J. Amer. Chem. Soc.*, **86**, 1448 (1964).
46. B. F. Erlanger, *Proc. Natl. Acad. Sci., U.S.*, **46**, 1430 (1960); W. Cohen and B. F. Erlanger, *Biochim. Biophys. Acta*, **52**, 604 (1961).
47. T. Viswanatha, *Proc. Natl. Acad. Sci., U.S.*, **51**, 1117 (1964).
48. M. L. Bender and E. T. Kaiser, *J. Amer. Chem. Soc.*, **84**, 2556 (1962).
49. M. Mares-Guia and E. Shaw, *Federation Proc.*, **22**, 528 (1963).
50. T. H. Marshall, J. R. Whitaker, and M. L. Bender, *Biochemistry*, **8**, 4665, 4671 (1969).
51. M. L. Begue (unpublished results).
52. R. Henderson, *J. Mol. Biol.*, **54**, 341 (1970).
53. H. R. Horton and G. Young, *Biochim. Biophys. Acta*, **194**, 272, (1969).
54. L. Goldstein and E. Katchalski, *Fresenius' Z. Anal. Chem.*, **243**, 375 (1968).
55. S. P. Agarwal et al., *Biochem. Biophys. Res. Comm.*, **43**, 510 (1971).
56. P. Valenzuela, Doctoral dissertation, Northwestern Univ., 1971.
57. M.-A. Coletti-Previero et al., *Fed. Eur. Biochem. Letters*, **11**, 213 (1970).
58. J. A. Rupley, V. Gates, and R. Bilbrey, *J. Amer. Chem. Soc.*, **90**, 5633 (1968).
59. R. C. Thompson and E. R. Blout, *Proc. Natl. Acad. Sci., U.S.*, **67**, 1734 (1970).
60. M. A. Raftery, personal communication.
61. D. A. Usher, *Nature*, **228**, 4716 (1970).
62. F. Sanger, *Proc. Chem. Soc.*, 76 (1963).
63. K. A. Walsh and H. Neurath, *Proc. Natl. Acad. Sci., U.S.*, **51**, 884 (1964).
64. B. S. Hartley, *6th Intern. Congr. Biochem.*, New York, 1964, p. 253; D. M. Shotton and H. C. Watson, *Nature*, **225**, 811 (1970); see also H. Neurath and R. A. Bradshaw, *Accounts Chem. Res.*, **3**, 249 (1970).
65. M. L. Bender and W. A. Glasson, *J. Amer. Chem. Soc.*, **82**, 3336 (1960).
66. H. Neurath, J. A. Rupley, and W. J. Dreyer, *Arch. Biochem. Biophys.*, **65**, 243 (1956).

67. H. Parker and R. Lumry, *J. Amer. Chem. Soc.*, **85**, 483 (1963).
68. B. H. Havsteen and G. P. Hess, *J. Amer. Chem. Soc.*, **85**, 791 (1963).
69. B. Labouesse, B. H. Havsteen, and G. P. Hess, *Proc. Nat. Acad. Sci. U.S.*, **48**, 2137 (1962).
70. M. Caplow and W. P. Jencks, *Biochemistry*, **1**, 883 (1962).
71. M. L. Bender and F. J. Kézdy, *J. Amer. Chem. Soc.*, **86**, 3704 (1964).
72. F. C. Wedler, F. L. Killian, and M. L. Bender, *Proc. Natl. Acad. Sci., U.S.*, **65**, 1120 (1970).
73. A. L. Green and J. D. Nicholls, *Biochem. J.*, **72**, 70 (1959).
74. W. Cohen and B. F. Erlanger, *J. Amer. Chem. Soc.*, **82**, 3928 (1960).
75. M. L. Bender, *J. Amer. Chem. Soc.*, **84**, 2582 (1962).
76. M. L. Bender and G. A. Hamilton, *J. Amer. Chem. Soc.*, **84**, 2570 (1962).
77. M. L. Bender, E. J. Pollock, and M. C. Neveu, *J. Amer. Chem. Soc.*, **84**, 595 (1962).
78. H. P. Metzger and I. B. Wilson, *Federation Proc.*, **23**, 316 (1964).
79. S. L. Johnson, *J. Amer. Chem. Soc.*, **86**, 3819 (1964).
80. M. A. Marini and G. P. Hess, *J. Amer. Chem. Soc.*, **82**, 5160 (1960).
81. B. R. Rabin, *Mechanismen Enzymatischer Reactionen*, Springer Verlag, Berlin, 1964, p. 74.
82. F. Bergmann, I. B. Wilson, and D. Nachmansohn, *J. Biol. Chem.*, **186**, 693 (1950).
83. A. Lowenstein and S. Meiboom, *J. Chem. Phys.*, **27**, 1067 (1957); Z. Luz and S. Meiboom, *J. Amer. Chem. Soc.*, **86**, 4764 (1964).
84. M. L. Bender and H. d'A. Heck, *J. Amer. Chem. Soc.*, **89**, 1211 (1967).
85. M. L. Bender and J. W. Amshey, unpublished results.
86. P. Valenzuela and M. L. Bender, *Proc. Natl. Acad. Sci., U.S.*, **63**, 1214 (1969); A. Himoe, P. C. Parks, and G. P. Hess, *J. Biol. Chem.*, **242**, 919 (1967).
87. Y. Nakagawa and M. L. Bender, *J. Amer. Chem. Soc.*, **91**, 1566 (1969); *Biochemistry*, **9**, 259 (1970).
88. M. L. Bender and F. C. Wedler, Jr., *J. Amer. Chem. Soc.*, **89**, 3052 (1967); F. C. Wedler and M. L. Bender, *J. Amer. Chem. Soc.*, **91**, 3894 (1969); see also J. Keizer and S. A. Bernhard, *Biochemistry*, **5**, 4127 (1966); A. Y. Moon, J. M. Sturtevant, and G. P. Hess, *Fed. Proc.*, **21**, 229 (1962); B. F. Erlanger, H. Castleman, and A. G. Cooper, *J. Amer. Chem. Soc.*, **85**, 1872 (1963); G. P. Hess, *7th Int. Cong. Biochem.*, p. 173 (1967); K. G. Brandt, A. Himoe, and G. P. Hess, *J. Biol. Chem.*, **242**, 3973 (1967); B. H. Havsteen, *J. Biol. Chem.*, **242**, 767 (1967).
89. D. D. Schroeder and E. Shaw, *J. Biol. Chem.*, **243**, 2943 (1968); J. F. Hruska, J. H. Law, and F. J. Kézdy, *Biochem. Biophys. Res. Comm.*, **36**, 272 (1969).
90. P. Valenzuela and M. L. Bender, *Biochemistry*, **9**, 2440 (1970).
91. C. H. Johnson and J. R. Knowles, *Biochem. J.*, **101**, 56 (1966).
92. See also J. H. Wang and L. Parker, *Proc. Natl. Acad. Sci., U.S.*, **58**, 2451 (1967).
93. D. M. Blow, J. J. Birktoft, and B. S. Hartley, *Nature*, **221**, 337 (1969).
94. F. M. Menger, *Yale Sci.*, **45**, 8 (1970).
95. L. Polgar, personal communication.
96. M. L. Bender and F. J. Kézdy, *Ann. Rev. Biochem.*, **34**, 49 (1965).
97. A. Williams, *Chem. Commun.*, **715** (1967); M. L. Bender, J. V. Killheffer, Jr., and F. J. Kézdy, *J. Amer. Chem. Soc.*, **86**, 5330, 5331 (1964); D. W. Ingles, J. R. Knowles, and J. A. Tomlinson, *Biochem. Biophys. Res. Comm.*, **23**, 619 (1966); J. R. Knowles, *J. Theoret. Biol.*, **9**, 213 (1965).
98. T. Inagami and J. M. Sturtevant, *Biochem. Biophys. Res. Comm.*, **14**, 69 (1964); trypsin is more difficult to saturate and less stable than chymotrypsin.
99. M. L. Bender and T. H. Marshall, *J. Amer. Chem. Soc.*, **90**, 201 (1968).

100. C. S. Wright, R. A. Alden, and J. Kraut, *Nature*, **221**, 235 (1969) and references therein.
101. F. C. Kafatos, A. M. Tartakoff, and J. H. Law, *J. Biol. Chem.*, **242**, 1477 (1967); F. C. Kafatos, J. H. Law, and A. M. Tartakoff, *ibid.*, **242**, 1488 (1967).
102. R. C. Augusteyn, J. de Jersey, E. C. Webb, and B. Zerner, *Biochim. Biophys. Acta*, **171**, 128 (1969) and earlier papers; K. Myrback, *Acta Chem. Scand.*, **1**, 142 (1947); M. C. Wall and K. J. Laidler, *Arch. Biochem. Biophys.*, **43**, 299 (1953); E. Heilbron, Ed., *Structure and Reaction of DFP Sensitive Enzymes*, Research Institute of National Defense, p. 69; H. P. Benschop, G. R. Van Den Berg, and H. L. Boter, *Rec. Trav. Chim. Pays-Bas*, **87**, 387 (1968).
103. F. J. Kézdy, L. Lorand, K. D. Miller, *Biochemistry*, **4**, 2302 (1965); L. Lorand, *Fed. Proc.*, **24**, 784 (1965); S. Magnusson, *Proc. Chem. Soc.*, 25 (September 21, 1968).
104. S. A. Bernhard, *Biochem. J.*, **59**, 506 (1955).
105. T. Inagami and J. M. Sturtevant, *J. Biol. Chem.*, **235**, 1019 (1960).
106. J. V. Killheffer, Jr., unpublished results.
107. J. Roget and F. Calvet, *Ann. Real. Soc. Espan. Fis. Quim. Ser. B.*, **58**, 357 (1962).
108. J. A. Stewart and L. Ouellet, *Can. J. Chem.*, **37**, 751 (1959).
109. F. A. Bettelheim and A. Lukton, *Nature*, **198**, 357 (1963).
110. B. A. Kilby and G. Youatt, *Biochem. J.*, **57**, 303 (1954).
111. M. Mares-Guia and E. Shaw, *Fed. Proc.*, **22**, 528 (1963).
112. P. H. Petra, W. Cohen, and E. N. Shaw, *Biochem. Biophys. Res. Comm.*, **21**, 612 (1965).
113. N. M. Green and H. Neurath, *J. Biol. Chem.*, **204**, 379 (1953).
114. T. Inagami and T. Murachi, *J. Biol. Chem.*, **238**, PC1905 (1963).
115. I. E. Liener and T. Viswanatha, *Biochim. Biophys. Acta*, **36**, 250 (1959).
116. J. Chevallier, Y. Jacquot-Armand, and J. Yon, *Biochim. Biophys. Acta*, **92**, 521 (1964).
117. T. Hofmann and S. G. Scrimger, *Fed. Proc.*, **25**, 589 (1966).
118. B. Kassell and M. Laskowski, Sr., *Biochem. Biophys. Res. Comm.*, **20**, 463 (1965).
119. R. Finkenstadt and M. Laskowski, Jr., *J. Biol. Chem.*, **242**, 771 (1967).
120. D. T. Elmore and N. J. Baines, *Biochem. J.*, 25P, **76** (1960).
121. H. Werbin and A. Palm, *J. Amer. Chem. Soc.*, **73**, 1382 (1951).
122. C. G. Trowbridge, A. Krehbiel, and M. Laskowski, Jr., *Biochemistry*, **2**, 843 (1963).
123. E. F. Jansen, A. L. Curl, and A. K. Balls, *J. Biol. Chem.*, **189**, 671 (1951).
124. H. Gutfreund, *Trans. Faraday Soc.*, **51**, 441 (1955).
125. J. Chevallier and J. Yon, *Biochim. Biophys. Acta*, **122**, 116 (1966).
126. B. M. Sanborn and G. E. Hein, *Biochemistry*, **7**, 3616 (1968).
127. W. Cohen, M. Lache, and B. F. Erlanger, *Biochemistry*, **1**, 686 (1962).
128. R. J. Foster, *J. Biol. Chem.*, **236**, 2461 (1961).
129. T. Inagami and T. Murachi, *J. Biol. Chem.*, **238**, PC1905 (1963); **239**, 1394 (1964).
130. M. L. Bender, J. V. Killheffer, Jr., and F. J. Kézdy, *J. Amer. Chem. Soc.*, **86**, 5330 (1964).
131. L. Lorand, K. D. Miller, and F. J. Kézdy, unpublished results.
132. F. S. Markland, E. Shaw, and E. L. Smith, *Proc. Natl. Acad. Sci., U.S.*, **61**, 1440 (1968).
133. A. N. Glazer, *J. Biol. Chem.*, **242**, 433 (1967).
134. A. N. Glazer, *Proc. Natl. Acad. Sci., U.S.*, **59**, 996 (1968).
135. M. Philipp and M. L. Bender, *Proc. Natl. Acad. Sci., U.S.*, **68**, 478 (1971).
136. E. Shaw and J. Ruscica, *J. Biol. Chem.*, **243**, 6312 (1968).
137. K. Morihara and T. Oka, *Arch. Biochem. Biophys.*, **138**, 526 (1970).
138. B. S. Hartley, personal communication.

139. J. A. Gladner and K. Laki, *J. Amer. Chem. Soc.*, **80**, 1263 (1958).
140. L. Lorand, W. T. Brannen, Jr., and N. G. Rule, *Arch. Biochem. Biophys.*, **96**, 147 (1962).
141. E. F. Curragh and D. T. Elmore, *Biochem. J.*, **93**, 163 (1964).
142. D. Nachmansohn and E. A. Feld, *J. Biol. Chem.*, **171**, 715 (1947).
143. D. Nachmansohn, *Ergeb. Physiol. Biol. Chem. Exptl. Pharmakol.*, **48**, 575 (1955).
144. E. F. Jansen, M.-D. F. Nutting, and A. K. Balls, *J. Biol. Chem.*, **179**, 201 (1949).
145. E. F. Jansen, M.-D. F. Nutting, R. Jang, and A. K. Balls, *J. Biol. Chem.*, **185**, 209 (1950).
146. I. B. Wilson and F. Bergmann, *J. Biol. Chem.*, **185**, 479 (1950).
147. I. B. Wilson, *J. Biol. Chem.*, **190**, 111 (1951).
148. I. B. Wilson, *J. Biol. Chem.*, **199**, 113 (1952).
149. I. B. Wilson and S. Ginsburg, *Biochim. Biophys. Acta*, **18**, 168 (1955).
150. A. F. Childs, S. R. Davies, A. L. Green, and J. P. Rutland, *Brit. J. Pharmacol.*, **10**, 462 (1955).
151. See also I. B. Wilson, S. Ginsburg, and C. Quan, *Arch. Biochem. Biophys.*, **77**, 286 (1958).
152. I. B. Wilson and M. A. Harrison, *J. Biol. Chem.*, **236**, 2292 (1961). See also I. B. Wilson, M. A. Hatch, and S. Ginsburg, *J. Biol. Chem.*, **235**, 2312 (1960), I. B. Wilson, M. A. Harrison, and S. Ginsburg, *J. Biol. Chem.*, **236**, 1498 (1961).
153. R. Kitz and I. B. Wilson, *J. Biol. Chem.*, **237**, 3245 (1962).
154. I. B. Wilson and J. Alexander, *J. Biol. Chem.*, **237**, 1323 (1962).
155. J. Alexander, I. B. Wilson, and R. Kitz, *J. Biol. Chem.*, **238**, 741 (1963).
156. R. Kitz and I. B. Wilson, *J. Biol. Chem.*, **238**, 745 (1963).
157. W. Leuzinger, M. Goldberg, and E. Cauvin, *J. Mol. Biol.*, **40**, 217 (1969).
158. R. C. Augusteyn, J. de Jersey, E. C. Webb, and B. Zerner, *Biochim. Biophys. Acta*, **171**, 128 (1969).
159. L. Polgar and M. L. Bender, *J. Amer. Chem. Soc.*, **88**, 3153 (1966); K. E. Neet and D. E. Koshland, Jr., *Proc. Natl. Acad. Sci., U.S.*, **56**, 1606 (1966).
160. L. Polgar and M. L. Bender, *Biochemistry*, **6**, 610 (1967); **8**, 136 (1969).
161. L. Polgar and M. L. Bender, *Biochim. Biophys. Acta*, submitted.
162. L. Polgar and M. L. Bender, *Adv. Enzymology*, **33**, 381 (1970); M. Philipp, L. Polgar, and M. L. Bender, *Methods in Enzymology*, **19**, 215 (1970).
163. J. R. Whitaker and M. L. Bender, *J. Amer. Chem. Soc.*, **87**, 2728 (1965).
164. A. Stockell and E. L. Smith, *J. Biol. Chem.*, **227**, 1 (1957).
165. E. L. Smith, V. J. Chauvre, and M. J. Parker, *J. Biol. Chem.*, **230**, 283 (1958).
166. G. Lowe and A. Williams, *Proc. Chem. Soc.*, 140 (1964).
167. A. Light, R. Frater, J. R. Kimmel, and E. L. Smith, *Proc. Natl. Acad. Sci., U.S.*, **52**, 1276 (1964).
168. S. S. Husain and G. Lowe, *Chem. Commun.*, 345 (1965).
169. G. Lowe and A. Williams, *Biochem. J.*, **96**, 194 (1965).
170. S. S. Husain and G. Lowe, *Chem. Commun.*, 310 (1968).
171. G. Lowe, *Nature*, **212**, 1263 (1966).
172. I. Schechter and A. Berger, *Biochem. Biophys. Res. Comm.*, **27**, 157 (1967).
173. I. Schechter and A. Berger, *Biochem. Biophys. Res. Comm.*, **32**, 898 (1968).
174. R. C. Wong and I. E. Liener, *Biochem. Biophys. Res. Comm.*, **17**, 470 (1964).
175. A. N. Glazer, *J. Biol. Chem.*, **241**, 3811 (1966).
176. E. Shaw, personal communication.
177. J. R. Whitaker and J. Perez-Villasenor, *Arch. Biochem. Biophys.*, **124**, 70 (1968).
178. J. F. Kirsch and E. Katchalski, *Biochemistry*, **4**, 884 (1965).
179. G. Lowe and A. Williams, *Biochem. J.*, **96**, 199 (1965).

180. A. W. Lake and G. Lowe, *Biochem. J.*, **101**, 402 (1966).
181. A. C. Henry and J. F. Kirsch, *Biochemistry*, **6**, 3536 (1967).
182. G. Lowe, unpublished symposium, Oxford, 1968.
183. A. L. Fink and M. L. Bender, *Biochemistry*, **8**, 5109 (1969).
184. B. R. Hammond and H. Gutfreund, *Biochem. J.*, **72**, 349 (1959).
185. H. Gutfreund, *Discuss. Faraday Soc.*, **20**, 167 (1955).
186. E. L. Smith, *J. Biol. Chem.*, **233**, 1392 (1958).
187. A. Williams, "The Mechanism of Action of Some Hydrolytic Enzymes," Doctoral thesis, Oxford University, 1964.
188. A. Stockell and E. L. Smith, *J. Biol. Chem.*, **227**, 1 (1957).
189. J. F. Kirsch and M. Igelstrom, *Biochemistry*, **5**, 783 (1966).
190. V. C. Sgarbieri, S. M. Gupte, D. E. Kramer, and J. R. Whitaker, *J. Biol. Chem.*, **239**, 2170 (1964).
191. M. El-Gharbawi and J. R. Whitaker, *Biochemistry*, **2**, 476 (1963).
192. R. M. Silverstein and F. J. Kézdy, *Fed. Proc.*, **29**, 929 (1970).
193. H. Neumann, Y. Levin, A. Berger, and E. Katchalski, *Biochem. J.*, **73**, 33 (1959).
194. J. S. Fruton, S. Fujii, and M. H. Knappenberger, *Proc. Natl. Acad. Sci., U.S.*, **47**, 759 (1961).
195. N. Sharon, V. Grisaro, and H. Neumann, *Arch. Biochem. Biophys.*, **97**, 219 (1962).
196. C. H. W. Hirs, S. Moore, and W. H. Stein, *J. Biol. Chem.*, **235**, 633 (1960); D. H. Spackman, W. H. Stein, and S. Moore, *J. Biol. Chem.*, **235**, 648 (1960).
197. F. M. Richards and P. J. Vithayathil, *J. Biol. Chem.*, **234**, 1459 (1959).
198. R. G. Denkwalter, D. F. Veber, F. W. Holly, and R. Hirschmann, *J. Amer. Chem. Soc.* **91**, 501 (1969).
199. B. Gutte and R. B. Merrifield, *J. Amer. Chem. Soc.*, **91**, 501 (1969).
200. B. Witkop, *Science*, **162**, 318 (1968).
201. E. M. Crook, A. P. Mathias, and B. R. Rabin, *Biochem. J.*, **74**, 230 (1960), and references cited therein.
202. H. Witzel and E. A. Barnard, *Biochem. Biophys. Res. Comm.*, **7**, 289, 295 (1962).
203. A. M. Myers, Doctoral Dissertation, Northwestern University, 1965.
204. F. H. Westheimer, *Adv. Enzymol.*, **24**, 441 (1962).
205. F. J. Kézdy, personal communication.
206. A. M. Crestfield, W. H. Stein, and S. Moore, *J. Biol. Chem.*, **238**, 2413 (1963).
207. C. H. W. Hirs, M. Halmann, and J. H. Kycid, *Proceedings of the First IUB Symposium on Macromolecular Structure and Biological Function, Stockholm*, Academic Press, London, 1961.
208. A. M. Crestfield, W. H. Stein, and S. Moore, *J. Biol. Chem.*, **238**, 2421 (1963).
209. For earlier work on the histidine residues in ribonuclease, see E. A. Barnard and W. D. Stein, *J. Mol. Biol.*, **1**, 339 (1959); W. D. Stein and E. A. Barnard, *ibid.*, **1**, 350 (1959); H. G. Gundlach, W. H. Stein, and S. Moore, *J. Biol. Chem.*, **234**, 1754 (1959).
210. The spectrophotometric titration of the imidazole groups of ribonuclease shows normal pK's for these groups; J. W. Donovan, *Biochemistry*, **4**, 823 (1965).
211. D. G. Herries, A. P. Mathias, and B. R. Rabin, *Biochem. J.*, **85**, 127 (1962).
212. D. Findlay, D. G. Herries, A. P. Mathias, B. R. Rabin, and C. A. Ross, *Nature*, **190**, 781 (1961).
213. H. Witzel, *Ann.*, **635**, 182, 191 (1960).
214. Scheraga and J. A. Rupley, *Adv. Enzymol.*, **24**, 161 (1962).
215. D. G. Oakenfull, D. I. Richardson, Jr., and D. A. Usher, *J. Amer. Chem. Soc.*, **89**, 5491 (1967).
216. W. A. Hirs, personal communication.

217. S. Nakamura and H. Ogata, *Biochem. J.*, **103**, 77 P (1967).
218. E. M. Scott and R. Powell, *J. Amer. Chem. Soc.*, **70**, 1104 (1948).
219. P. W. Wigler and R. A. Alberty, *J. Amer. Chem. Soc.*, **82**, 5482 (1960).
220. R. A. Alberty, Seminar, Northwestern University, 1961.
221. D. A. Brant, L. B. Barnett, and R. A. Alberty, *J. Amer. Chem. Soc.*, **85**, 2204 (1963).
222. H. F. Fisher, C. Frieden, J. S. M. McKee, and R. A. Alberty, *J. Amer. Chem. Soc.*, **77**, 4436 (1955); T. C. Farrar, H. S. Gutowsky, R. A. Alberty, and W. G. Miller, *J. Amer. Chem. Soc.*, **79**, 3978 (1957).
223. O. Gawron and T. P. Fondy, *J. Amer. Chem. Soc.*, **81**, 6333 (1959).
224. F. A. L. Anet, *J. Amer. Chem. Soc.*, **82**, 994 (1960).
225. S. Englard, *J. Biol. Chem.*, **235**, 1510 (1960).
226. O. Gawron, A. J. Glaid, III, and T. P. Fondy, *J. Amer. Chem. Soc.*, **83**, 3634 (1961).
227. R. A. Alberty, W. G. Miller, and H. F. Fisher, *J. Amer. Chem. Soc.*, **79**, 3973 (1957).
228. J. F. Thompson, *Arch. Biochem. Biophys.*, **90**, 1 (1960).
229. S. Englard, *J. Biol. Chem.*, **233**, 1003 (1958); A. I. Krasna, *J. Biol. Chem.*, **233**, 1010 (1958).
230. R. Silverstein, J. Voet, D. Reed, and R. H. Abeles, *J. Biol. Chem.*, **242**, 1338 (1967).
231. M. Doudoroff, H. A. Barker, and W. Z. Hassid, *J. Biol. Chem.*, **168**, 725 (1947).
232. D. E. Koshland, Jr., *Proc. Natl. Acad. Sci., U.S.*, **44**, 98 (1958).
233. A. B. Pardee, *J. Biol. Chem.*, **241**, 5886 (1966); see also A. B. Pardee, L. S. Prestidge, M. B. Whipple, and J. Dreyfuss, *J. Biol. Chem.*, **241**, 3962 (1966); J. Dreyfuss and A. B. Pardee, *Biochim. Biophys. Acta*, **104**, 308 (1965); A. B. Pardee and L. S. Prestidge, *Proc. Natl. Acad. Sci., U.S.*, **55**, 189 (1966); J. Dreyfuss, *J. Biol. Chem.*, **239**, 2292 (1964); J. Dreyfuss and A. B. Pardee, *J. Bacteriol.*, **91**, 2275 (1966).
234. M. Heydanek, W. G. Struve, and F. C. Neuhaus, *Biochemistry*, **8**, 1214 (1969).
235. L. B. Hersh and W. P. Jencks, *J. Biol. Chem.*, **242**, 3468 (1967).
236. L. B. Hersh and W. P. Jencks, *J. Biol. Chem.*, **242**, 3481 (1967).
237. D. C. Phillips, *Sci. Amer.*, **215**, No. 5, 78 (1966).
238. J. A. Rupley, *Biochim. Biophys. Acta*, **83**, 245 (1964).
239. G. Lowe, G. Sheppard, M. L. Sinnott, and A. Williams, *Biochem. J.*, **104**, 893 (1967); T. Osawa and Y. Nakazawa, *Biochim. Biophys. Acta*, **130**, 56 (1966).
240. D. C. Phillips, *Proc. Natl. Acad. Sci., U.S.*, **57**, 484 (1967); S. M. Parsons et al., *Biochemistry*, **8**, 700 (1969); S. M. Parsons and M. A. Raftery, *Biochemistry*, **8**, 4199 (1969).
241. D. Piszkiewicz and T. C. Bruice, *Biochemistry*, **7**, 3037 (1968).
242. J. B. Howard and A. N. Glazer, *J. Biol. Chem.*, **242**, 5715 (1967).
243. F. J. Hartdegen and J. A. Rupley, *Biochim. Biophys. Acta*, **92**, 625 (1964).
244. F. W. Dahlquist and M. A. Raftery, *Biochemistry*, **7**, 3269 (1968).
245. D. C. Phillips, *Sci. Amer.*, **215**, No. 5, 78 (1966).
246. G. Lowe, *Proc. Roy. Soc., B*, **167**, 431 (1967).
247. G. Lowe and G. Sheppard, *Chem. Commun.*, 529, (1968).
248. D. Piszkiewicz and T. C. Bruice, *Biochemistry*, **7**, 3037 (1968).
249. G. Lowe, G. Sheppard, M. L. Sinnott, and A. Williams, *Biochem. J.*, **104**, 893 (1967).
250. M. A. Raftery and T. Rand-Meir, *Biochemistry*, **7**, 3281 (1968).
251. J. A. Rupley, personal communication.

Chapter 17

ENZYME MECHANISM: COENZYMES

17.1	Redox Coenzymes	540
	17.1.1 NAD Coenzymes	540
	17.1.2 Flavin Coenzymes	548
	17.1.3 Other Redox Enzymes Containing Coenzymes	551
	17.1.4 Cobalamin and Cobamide Coenzymes (Vitamin B_{12})	573
	17.1.5 Lipoic Acid	577
	17.1.6 Coenzyme Q (Ubiquinone)	580
	17.1.7 Metal Ions That Can Change Valence	580
	17.1.8 Ferridoxin	581
17.2	Nonredox Coenzymes	581
	17.2.1 Thiamine Pyrophosphate (Vitamin B_1)	582
	17.2.2 Pyridoxal Phosphate (Vitamin B_6)	586
	17.2.3 Tetrahydrofolic Acid	594
	17.2.4 Biotin	596
	17.2.5 S-Adenosylmethionine	600
	17.2.6 Coenzyme A	602
	17.2.7 Purine Phosphates	604
	17.2.8 Pyrimidine Phosphates	607
	17.2.9 Metal Ion-Containing Enzymes	607
	17.2.10 Glutathione	611
	17.2.11 Vitamins A, D, E, and K	612
17.3	Summary	614

There are a multitude of enzymatic reactions that proceed via coenzymes. Coenzymes may be viewed as: (1) other substrates[1]; (2) vitamins needed for proper nutrition; or (3) vital parts (organic or inorganic) of the active site of certain enzymes. The occurrence of a coenzyme implies a knowledge of the active site of the enzyme, since the coenzyme is presumably at the active site.

Such enzymes can be grossly subdivided into enzymes containing nonredox coenzymes and those containing redox coenzymes. Some enzymes that require templates are considered later. Some enzymes that involve Schiff base intermediates are considered here although they strictly do not have coenzymes. As do most of the chapters, this one emphasizes mechanism.

The mechanisms of a number of coenzyme reactions have been brilliantly elucidated, for example, for thiamine-containing enzymes by Breslow,[2] for pyridoxal-containing enzymes by Snell, Braunstein, and co-workers,[3,4] and for NAD-containing enzymes by Westheimer, Vennesland, and co-workers.[5]

17.1 REDOX COENZYMES

There are many enzymes containing redox coenzymes, including nicotinamide coenzymes such as NAD (nicotinamide adenine dinucleotide) and NADP (nicotinamide adenine dinucleotide phosphate); flavin nucleotides such as FMN (flavin mononucleotide) and FAD (flavin adenine dinucleotide); metal porphyrins such as peroxidase, catalase, the cytochromes, heme proteins, and others; metal porphyrins containing an organometallic bond such as the cobamide (vitamin B_{12}) coenzymes; ascorbic acid; lipoic acid (thioctic acid); coenzyme Q (ubiquinone); various metal ions that can change valence; and ferridoxin. Most of these coenzymes will be covered here.[6]

17.1.1 NAD Coenzymes

There are many enzymes containing NAD or NAD phosphate (NADP) coenzymes. Among these are alcohol dehydrogenase, acetaldehyde dehydrogenase, L-lactate dehydrogenase, L-malate dehydrogenase, D-glycerate dehydrogenase, dihydroorotate dehydrogenase, α-glycerophosphate dehydrogenase, 3-phosphoglyceraldehyde dehydrogenase, L-glutamate dehydrogenase, D-glucose dehydrogenase, β-hydroxysteroid dehydrogenase, NADH cytochrome reductase, NADPH (transhydrogenase), NADH diaphorase, L-β-hydroxybutyrylCoA dehydrogenase, and others.

NAD coenzymes may be named as such (nicotinamide adenine dinucleotide), or earlier as DPN (diphosphopyridine nucleotide), or still earlier as CoI (coenzyme I) in honor of Warburg and Christian who first put this coenzyme on a chemical basis, although they were not the first persons to recognize the existence of a dialyzable cofactor.[7]

Nicotinamide is not to be confused with nicotine, which may have very harmful effects on the body. Nicotinamide is so often confused with nicotine that its name was shortened to niacin, which is often found as an additive in breakfast cereals. Nicotinamide is a preventive of pellagra and thus is beneficial nutritionally.

The structure of $NAD^\oplus$, the oxidized form, consists of a nicotinamide moiety, an adenine moiety, two ribose groups, and two phosphate groups, as shown in **1**.

$NAD^\oplus$
1

As mentioned earlier, there are many dehydrogenases that participate in redox reactions. Perhaps the most studied enzyme is alcohol dehydrogenase (ADH). The equation for its reaction is

$$CH_3CH_2OH + NAD^\oplus \xrightleftharpoons{ADH} CH_3CHO + NADH + H^\oplus \qquad (17.1)$$

Notice the direct involvement of the $NAD^\oplus$ in the reaction. The implication is that it is used up and that there is a *direct* hydrogen transfer. This is proved later. The enormous effort that has been put into these enzymes derives from their importance and from the fact that there is a large change in absorbancy at 340 nm, which makes these reactions easy to follow. This change in absorbancy results from the reduction of the pyridine ring to a dihydropyridine ring.

At one time it was believed that redox reactions in biochemical systems, in which metabolites are oxidized indirectly although the eventual oxidant is oxygen, involved one-electron transfers, but the above equation cannot be reconciled with such a scheme and was considered revolutionary in its time. Thus kinetics rather than equilibrium is important. The advent of deuterium tracers upset the one-electron ideas and further indicated that with organic metabolites, the first step of a long sequence involves $NAD^\oplus$ (followed by flavin nucleotides, cytochromes, cytochrome oxidase, and eventually oxygen).

A more detailed reaction of alcohol dehydrogenase is that shown in eq. 17.2 with the R group representing ribose phosphate phosphate ribose adenosine.

$$\text{(nicotinamide-R)} + CH_3CH_2OH \underset{}{\overset{ADH}{\rightleftharpoons}} \text{(dihydronicotinamide-R)} + CH_3CHO + H^{\oplus} \quad (17.2)$$

In 1936 Karrer showed that reduction takes place in the pyridine ring. Furthermore, the reduced form of $NAD^{\oplus}$ has its added hydrogen on the 4 position only, as indicated in eq. 17.2 and as demonstrated by deuterium tracer experiments; the α-pyridones of the product contain the same amount of deuterium as the original reactant.[8] This reaction also demonstrated that a 4-deuterium atom is enzymatically transferred to pyruvate. The deuterium atom cannot be in a 3-position for this is an aromatic nucleophilic substitution reaction like that shown in eq. 17.3.

$$\text{(2,4-dinitrofluorobenzene)} + RNH_2 \xrightarrow{\text{aprotic solvent}} \text{(Meisenheimer intermediate)} \quad (17.3)$$

Equation 17.3 is now a well-known reaction in organic chemistry; it was elucidated as part of the work for which Sanger received the Nobel prize. The product of eq. 17.3 (actually an intermediate) was isolated by him.

Hydrogen transfer directly from the substrate to the coenzyme is most convincingly shown in Table 17.1 for alcohol dehydrogenase. (There is a possible confusion concerning D since it can stand for dinucleotide or deuterium.) The table indicates that no hydrogen atoms come from the medium; rather the hydrogen is directly transferred from the substrate to the $NAD^{\oplus}$. Furthermore, "turnover" with excess CH_3CD_2OH introduces no more than one deuterium atom into NADD. This result indicates that the enzymatic reaction is stereospecific for $NAD^{\oplus}$.

The stereochemistry of NADD can be visualized in eq. 17.4.

$$\text{(D,H isomer)} \quad \text{or} \quad \text{(H,D isomer)} \quad (17.4)$$

Table 17.1
Deuterium Analysis in Alcohol Dehydrogenase Reactions[a]

	Analyzed Substance	Atoms D/Molecule
$CH_3CH_2OH + NAD^\oplus \xrightarrow{D_2O}$ $CH_3CHO + NADH + H^\oplus$	NADH	0.02
$CH_3CD_2OH + NAD^\oplus \xrightarrow{H_2O}$ $CH_3CDO + NADD + H^\oplus$	NADD	1.01; 0.99
	CH_3CDO	1.00

[a] From B. Vennesland and F. H. Westheimer, in *The Mechanism of Enzyme Action*, W. D. McElroy and B. Glass, Eds., Johns Hopkins Press, Baltimore, © 1954, p. 357.

Since all the deuterium is transferred as shown above, what is being observed is not an isotopic effect. Therefore one can be sure NADD is one stereoisomer.

In opposition, chemically produced NADD that is enzymatically oxidised with acetaldehyde results in about 50% of the deuterium being transferred, not 100%. Enzymatically produced NADD with pyruvate in the presence of lactic dehydrogenase gives a lactate in which *one* atom of deuterium per molecule has been transferred.[5] Mahler and Cordes list extensive stereospecificity for many substrates; this stereospecificity falls into two classifications. The first, denoted *A*, refers to the first compound in eq. 17.4 and the second, denoted *B*, refers to the second compound.[10] Some typical dehydrogenases having the *A* specificity are alcohol, L-lactate, and L-malate dehydrogenases; those showing the *B* specificity include β-hydroxysteroid dehydrogenase, pyridine transhydrogenase, and the enzymes that convert testosterone to 4-androstene-3,17-dione and androstane to androstane-3,7-dione. It has been suggested that metabolic coupling of reactions of different stereospecificity may occur with association of different stereoisomers of NAD. This is a hypothesis that requires more facts.

Analogous to the stereochemistry of $NAD^\oplus$ is the stereochemistry of the substrate. This has been explored in detail; for example, the enzymatic reduction of pyruvate by lactic dehydrogenase leads to L-lactate and vice versa. This implies an anomaly in alcohol dehydrogenase but in fact it is not an anomaly at all. Two enantiomorphic materials are produced by the action of alcohol dehydrogenase (containing reduced NAD) as shown in eqs. 17.5 and 17.6.

The two isotopic ethanol products were shown to be enantiomorphic since the first product lost only deuterium on enzymatic reoxidation whereas the second lost only hydrogen on enzymatic reoxidation. Furthermore, the

$$CH_3CHO + NADD \xrightarrow{ADH} CH_3-\underset{D}{\overset{H}{\underset{|}{\overset{|}{C}}}}-OH + NAD^\oplus \qquad (17.5)$$

$$CH_3CDO + NADH \xrightarrow{ADH} CH_3-\underset{H}{\overset{D}{\underset{|}{\overset{|}{C}}}}-OH + NAD^\oplus \qquad (17.6)$$

alcohol product of eq. 17.6 was converted to the tosylate (retention) and then treated with sodium hydroxide (inversion) leading to a product that lost only deuterium on enzymatic reoxidation. Finally, enough of the alcohol product of eq. 17.5 was isolated to show that it was optically active.[5]

Many model systems of these reactions have been constructed. The first real model system is shown in eq. 17.7,[11] although earlier malachite green was used in the oxidation of quinones, but there the proton can equilibrate with the medium and thus it is not possible to talk unequivocally about direct hydrogen transfer.

$$\underset{R}{\underset{|}{\overset{D\quad D}{\text{pyridine-CONH}_2}}} + C_6H_5-\underset{\underset{C_6H_4NMe_2}{|}}{C}=C_6H_4=N^\oplus Me_2 \xrightarrow[R.T.]{pH\ 7}$$

$$\underset{R}{\underset{|}{\overset{D}{\text{pyridinium-CONH}_2}}} + C_6H_5-\underset{\underset{C_6H_4NMe_2}{|}}{\overset{D}{\underset{|}{C}}}-C_6H_4NMe_2 \qquad (17.7)$$

Another model system involves thioketones (which are more reactive toward nucleophiles than ordinary ketones) as shown in eq. 17.8.

$$\underset{R}{\underset{|}{\overset{H\quad D}{\text{pyridine-CONH}_2}}} + \phi_2C=S + H^\oplus \longrightarrow$$

$$\underset{R}{\underset{|}{\overset{D}{\text{pyridinium-CONH}_2}}} + \phi_2\underset{|}{\overset{H}{\underset{|}{C}}}-SH \qquad (17.8)$$

The rate is first-order in each component and the reaction proceeds more rapidly in polar than nonpolar solvents. As is seen in eq. 17.8, a hydrogen atom is transferred rather than a deuterium atom. The reaction is unaffected by typical free-radical inhibitors. A mechanism that is consistent with these facts is the direct transfer of a hydride ion. Such a mechanism is consistent with the kinetics, the solvent effect, and the isotope effect. A neighboring hydroxyl group accelerates the reduction of a thioketone presumably by polarization of the thiocarbonyl group.[12] Furthermore, electron-withdrawing substituents accelerate the reaction.

Still another model system involves pyruvic acid or another α-keto acid[13] as shown in eq. 17.9.

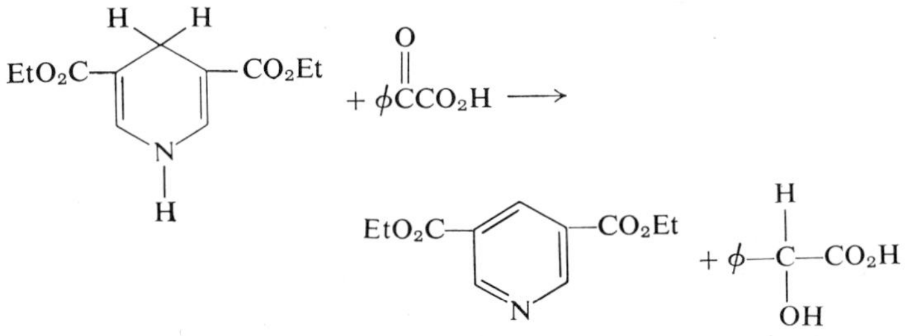

$$\text{EtO}_2\text{C}\underset{\underset{H}{N}}{\overset{H\ \ H}{\bigcirc}}\text{CO}_2\text{Et} + \phi\overset{O}{\underset{\|}{C}}\text{CO}_2\text{H} \longrightarrow$$

$$\text{EtO}_2\text{C}\underset{N}{\bigcirc}\text{CO}_2\text{Et} + \phi\overset{H}{\underset{OH}{-C-}}\text{CO}_2\text{H} \qquad (17.9)$$

In the reaction a direct hydrogen transfer was demonstrated since there was no deuterium in the product when the reaction was run in deuterium oxide. The implication is that the enzymatic reaction can proceed with direct hydrogen transfer as does the nonenzymatic reaction. This is partially verified by eq. 17.10,

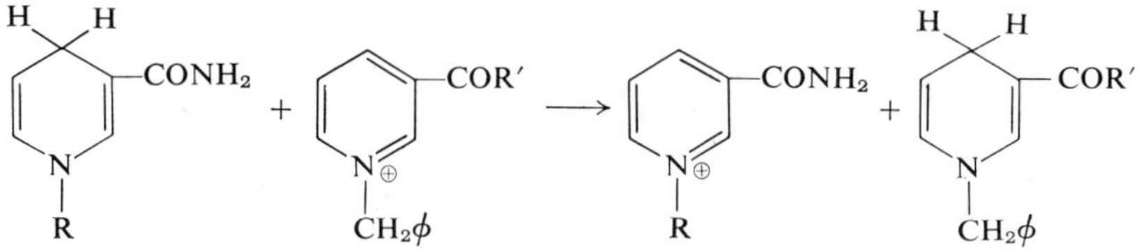

$$\underset{R}{\overset{H\ H}{\bigcirc}}\text{-CONH}_2 + \underset{\underset{CH_2\phi}{N^\oplus}}{\bigcirc}\text{-COR}' \longrightarrow \underset{R}{\overset{}{\bigcirc}}\text{-CONH}_2 + \underset{\underset{CH_2\phi}{N}}{\overset{H\ H}{\bigcirc}}\text{-COR}' \qquad (17.10)$$

which indicates the direct transfer of a hydrogen in a model transhydrogenase reaction (the molecule containing the R group corresponds to NAD⊕ or NADP⊕ or its reduced equivalent).[14] Kinetically the above reaction is found to be second-order, which is consistent with a direct hydrogen transfer.

The mechanism of the dehydrogenases has been extensively discussed. In particular, the mechanism of the alcohol dehydrogenase shows an addition of hydroxylamine and other nucleophiles to the 4-position of the pyridine ring.† Possibly an ethoxide ion adds to the 4-position of the pyridine ring also. One can ask whether there is an addition followed by an internal oxidation-reduction reaction but there is no analogy for this possibility. It is conceivable that an addition compound in a second mole of NAD⊕ occurs since there is an even number of NAD molecules in all these enzymes. Something of this possibility is shown in eq. 17.11.

$$(17.11)$$

Complexes involving charge transfer interaction were indicated before. It has been proposed by Kosower that ethanol can form a charge complex with the pyridine ring and eventually lead to acetaldehyde as shown in eq. 17.12, which shows a charge transfer interaction between the pyridinium ion of NAD⊕ and the substrate. This has some analogy.[16]

$$(17.12)$$

Zinc ion binding, which has not been mentioned previously, has been incorporated into a mechanism for dehydrogenase action that is due to Theorell as shown in eq. 17.13,[15]

$$(17.13)$$

where the lines emanating from the zinc show its various coordinations.

† Adducts of NAD and carbonyl compounds have been reported by G. Di Sabato, *Biochemistry*, **9**, 4594 (1970).

A kinetic isotope effect is found when the reaction of reduced alcohol dehydrogenase is carried out in deuterium oxide. The binding of NADH to the enzyme is 1.3-fold less and the rate constant of hydrogen transfer is twofold to threefold less.

What is really known is that a complex is formed of zinc ion, NAD, substrate, and enzyme. Furthermore, there is stereospecificity with respect to both substrate and NAD. Finally, it is known that there is a direct transfer of a hydride ion, from labeling experiments and isotope effects as well as model experiments. From the above, the best mechanism for dehydrogenase enzymes involving pyridine coenzymes seems to be eq. 17.13. It should be pointed out that some intramolecular analogs of eq. 17.13 have been attempted and that complexing of the zinc cation with the oxyanion of ethoxide will make it easier for a hydride ion to move.

The existence of both binary and ternary complexes of the enzyme, the substrate, and $NAD^{\oplus}$ follows from fluorometric observations that yield equilibrium constants, from kinetic observations, and from changes in optical rotatory dispersion, specifically Cotton effects. Essentially, rotatory dispersion titrations can be carried out. With NADH and liver alcohol dehydrogenase, Cotton effects suggest asymmetry about the zinc site that has not been mentioned heretofore. Two atoms of zinc ion are found in one molecule of liver alcohol dehydrogenase. Furthermore, complexing with 1,10-phenanthroline inhibits the enzyme.

The kinetics of dehydrogenase systems can be expressed by either eqs. 17.14 or 17.15. In the first, due to Theorell and Chance,[15] and the second, due to Mahler, Baker, and Shiner, [17] E stands for enzyme, R for reduced NAD, S' for oxidized substrate, and O for oxidized NAD. The essential difference between the two mechanisms is whether the substrate can independently bind or not. The first is general; the second applies to a particular dehydrogenase.[18]

$$E \begin{smallmatrix} k_1 \\ \rightleftarrows \\ k_2 \end{smallmatrix} ER \begin{smallmatrix} k_3 \\ \rightleftarrows \\ k_4 \end{smallmatrix} ERS \underset{k'}{\overset{k}{\rightleftarrows}} EOS \begin{smallmatrix} k_3' \\ \rightleftarrows \\ k_4' \end{smallmatrix} EO \begin{smallmatrix} k_1' \\ \rightleftarrows \\ k_2' \end{smallmatrix} E \quad (17.14)$$

with ES and ES' branches via k_7, k_8, k_5, k_6 and k_7', k_8', k_5', k_6'

$$E \rightleftarrows ER \rightarrow ERS \underset{k'}{\overset{k}{\rightleftarrows}} EOS' \rightarrow EO \rightleftarrows E \quad (17.15)$$

Pyridine coenzymes can oxidize secondary alcohols including cyclohexanol[18,19] when liver alcohol dehydrogenase is present, and a model for the reduction of (an activated) double bond has been found.[20] Many inhibitors of the dehydrogenases are known. These are particularly important in the metabolism of alcohol that can lead to drunkenness. When the inhibitor happens to be a barbiturate, death has been known to occur, presumably from a powerful inhibition of alcohol dehydrogenase.

When NADH is coupled with liver alcohol dehydrogenase, the fluorescence intensity of NADH is enhanced.[21]

17.1.2 Flavin Coenzymes

There are two principal flavin coenzymes (called flavin or riboflavin), mononucleotide (FMN) and flavin adenine dinucleotide (FAD). The structures of these materials are shown in **2** and **3**. The early history of these coenzymes has been amply told by Mahler and Cordes.[22] Flavoproteins utilizing either FAD or FMN as coenzyme are numerous. Substrates dehydrogenated by flavoproteins include pyridine nucleotides, as in the last section, α-amino acids, α-hydroxy acids, and substances containing activated saturated carbon–carbon bonds that can be converted to olefins. When oxidations occur, the isoalloxazine ring of FMN or FAD becomes reduced. The reduced flavoproteins, in turn, become substrates for reactions involving other electron acceptors, regenerating the oxidized form of the coenzyme.

Both enzymatic and nonenzymatic reductions of flavoproteins are characterized by the appearance of transient electron paramagnetic resonance signals and absorption bands in the 550–700 nm range. These data are consistent with the reduction occurring via two consecutive one-electron transfers, involving the flavin semiquinone as intermediate as shown in eq. 17.16.[23-28]

Riboflavin is usually called Vitamin B_2, and FMN is very closely related to riboflavin except that riboflavin does not contain a phosphate group. This phosphate group may be supplied by the body; therefore riboflavin is a vitamin. The general differentiation between a coenzyme and a vitamin is the phosphate group.

Of the many reactions catalyzed by flavoproteins, perhaps the most studied has been the reaction of succinic dehydrogenase which catalyzes the interconversion of succinate and fumarate. A clever double deuterium labeling experiment has demonstrated *trans*-stereochemistry for this reaction.[29] L-Chlorosuccinic acid and L-methylsuccinic acid were shown to be substrates of this enzyme.[30,31] An electron spin resonance analysis of this reaction (the electron spin signals arise from free radicals) indicates that the concentration of free radicals is proportional to the first power of the succinate dehydrogenase concentration.[32] The view of this reaction has been complicated

Redox Coenzymes

[Structure of FMN, labeled **2**]

[Structure of FAD, labeled **3**]

[Reaction scheme showing oxidized flavin → semiquinone → reduced flavin] (17.16)

by the report that some (nonheme) iron is found in the enzyme and is reduced as well as the flavin.[33] D-Methylsuccinate is known to be an inhibitor of this system.[30]

Molecular complexes of flavins are well known. Presumably these complexes are charge transfer complexes. Some stability constants of these complexes in aqueous solution are shown in Table 17.2.[34]

There is evidence that in some flavoenzymes, tyrosine residues are involved in flavin binding.[35,36] Both oxidized and reduced forms of flavins can form

Table 17.2
Stability Constants of Some Flavin Molecular Complexes in Aqueous Solution[a]

Components	Temperature (°C)	$K\,(M^{-1})$
Riboflavin-adenosine	17	120
Riboflavin-caffeine	17	91
FAD-caffeine	17	83
Riboflavin-chlorpromazine	—	1000
FMN-tryptophan	—	59
FMN-tryptophan	25	92
FMN-catechol	25	10.4
FMN-2,3-naphthalenediol	25	242
FMN(protonated)-catechol	25	0.68
FMN(protonated)-2,3-naphthalenediol	25	68
Riboflavin-hydroquinone	25	2.9
Riboflavin-catechol	25	3.9
Riboflavin-1,3,5-trihydroxybenzene	25	7.4
Riboflavin-3,5-dihydroxytoluene	25	9.2
Riboflavin-trimethylhydroquinone	25	10
Riboflavin-1,4-naphthalenediol	25	55
Riboflavin-1,7-naphthalenediol	25	98
Riboflavin-2,7-naphthalenediol	25	102
Riboflavin-1,5-naphthalenediol	25	111
Riboflavin-2,3-naphthalenediol	25	162

[a] From G. Tollin, in *Molecular Associations in Biology*, Academic Press, New York, 1968, p. 400.

complexes with tryptophan as well as tyrosine residues. The low fluorescence yield of most FMN enzymes[37] could be due to complexing, as could the modification of flavin chemistry. A number of flavoenzymes have been shown to exhibit charge transfer spectra.[38–40]

Rapid reaction studies have been carried out with flavoprotein dehydrogenases. At least three enzymes belong to the category of flavoprotein dehydrogenases containing a redox active disulfide link.[41]

A number of model systems of flavins have been investigated under anaerobic conditions. Simple flavins are readily changed in the presence of a number of phenylacetate anions such as phenylacetate, phenylpropionate, phenylglycinate, indole-3-acetate, and naphthylacetate concomitant with the stoichiometric production of carbon dioxide. The product from phenylacetate contains a benzyl radical covalently attached to the flavin nucleus.[42] The mode of production of flavin free radical was investigated in a model system. A comparison of the various biological analogs of the coenzymes was also made. FAD had an enhanced rate of flavin free radical formation as compared with FMN.[43] Compounds that complex with flavins decrease the rate of reaction in this model system.[44]†

Riboflavin has been shown to participate in many photochemical reactions. Many but not all involve the cleavage of water.[45-48]

Magnetic circular dichroism and circular dichroism spectra of various riboflavin analogs have also been measured. These measurements permit one to speculate about the nature of the molecules and their transitions.[49]

17.1.3 Other Redox Enzymes Containing Coenzymes

These will include ascorbate (vitamin C), various oxidases, various hydroxylases, catalase, peroxidase, the cytochromes, and certain heme proteins.[50] According to the classifical definition, an oxidation reaction is one in which electrons are removed from an atom or molecule and a reduction reaction is one in which electrons are added to an atom or molecule. For most inorganic reactions this is a satisfactory definition; usually one can easily recognize when such reactions have occurred. For example, ferrous ion is oxidized to ferric ion by removing one electron and ferric ion is reduced to ferrous ion by adding an electron. In such oxidation-reduction reactions we say that the atom changes its valence when it is oxidized or reduced. However, the main elements in organic compounds, namely, carbon, hydrogen, oxygen, and even nitrogen usually do not change their valence in the same sense and generally maintain the same number of bonds to neighboring elements in the compound even though the compound may be oxidized or reduced. Nevertheless, the most common organic oxidation-reduction reactions can be classified in two general categories, as shown in Fig. 17.1.

1. Removal of 2 hydrogens (dehydrogenation).

$$RCH_2OH \longrightarrow RCH=O \xrightarrow{\cdot H_2O} RCH(OH)_2 \longrightarrow RCOOH$$
alcohol + 2(H) + 2(H)
aldehyde acid

2. Replacement of hydrogen by a more electronegative element.

$$R_3CH \xrightarrow{Br_2} R_3CBr + HBr$$

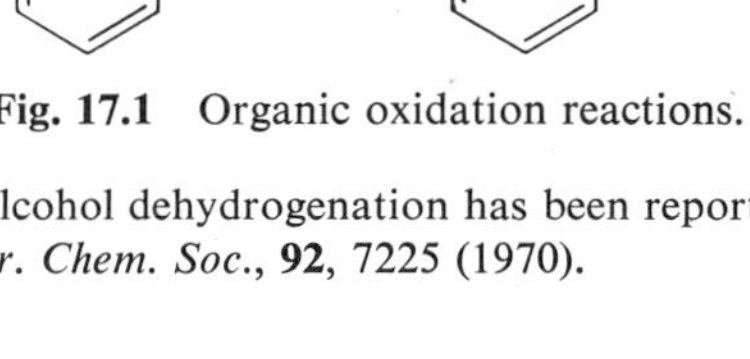

⌬ + (O) ⟶ ⌬–OH

Fig. 17.1 Organic oxidation reactions.

† A flavin model for alcohol dehydrogenation has been reported by L. E. Brown and G. A. Hamilton, *J. Amer. Chem. Soc.*, **92**, 7225 (1970).

Probably the most important category of organic redox reactions is the removal or addition of two hydrogens. An example is the oxidation of a primary alcohol to an acid that can be thought of as involving two dehydrogenation steps—the first dehydrogenation giving the aldehyde that can add a molecule of water and the second giving the acid. Another category of organic oxidation reactions is the replacement of H by a more electronegative element—examples shown in Fig. 17.1 are the replacement by bromine to give an alkyl bromide or replacement of a hydrogen on an aromatic ring to give a phenol. During the first part of this section we discuss possible mechanisms for dehydrogenation-hydrogenation reactions and then lead into reactions similar to those of the second category, especially hydroxylation reactions. When one considers possible mechanisms for transferring hydrogen, one must bear in mind that each hydrogen can be transferred in three ways—as a hydrogen radical, as a hydride ion, or as a proton. By these three methods the hydrogen is transferred, respectively, with one electron, two electrons, or none. Each of these species has been detected in either solution or the gas phase and they have all been suggested as being involved in many organic reactions. Most organic oxidation-reduction reactions involve an overall 2-electron (hydrogen) change—a one electron change would give a free radical and although organic free radicals are known they are not usually the final product in a reaction. Consequently for the overall reaction where two H's are removed or added there are six possible methods by which this can be done, as shown in Table 17.3.[50] (1) Both hydrogens can be transferred as hydrogen radicals. (2) One can be transferred as a hydride ion and the other as a proton. (3) Both can be transferred as protons but then two electrons must be transferred also by some pathway. The other mechanisms are self-explanatory. For all of the last four cases electrons must be either added or removed in order to maintain electrical neutrality and consequently if these are the methods by which the oxidation or reduction takes place then there must be a mechanism for the electron transfer as well as for the

Table 17.3

Possible Mechanisms for Addition or Removal of 2 Hydrogens[a]

$$
\begin{aligned}
&(1)\ H\cdot + H\cdot \\
&(2)\ H^{\ominus} + H^{\oplus} \\
&(3)\ H^{\oplus} + H^{\oplus} + 2e \\
&(4)\ H\cdot + H^{\oplus} + 1e \\
&(5)\ H^{\ominus} + H\cdot - 1e \\
&(6)\ H^{\ominus} + H^{\ominus} - 2e
\end{aligned}
$$

[a] From G. A. Hamilton, personal communication.

hydrogen transfer. Mechanism 1 is a free radical chain reaction, such as the thermal decomposition of alkanes to give lower alkenes and hydrogen. A nonenzymatic example of mechanism 2 is the reduction of various carbonyl compounds with metal hydrides. A hydride ion is transferred from the metal hydride and a proton is picked up from water. An enzymatic example is the reaction of NADH (DPNH) with aldehydes. Both the nonenzymatic and enzymatic reactions are believed to proceed by mechanism 2, as shown in eq. 17.16.[50] Mechanism 3 is particularly interesting because both hydrogens are removed or added as protons. Thus these reactions can be related to many acid- and base-catalyzed reactions that take place in aqueous solutions where a proton is either removed by a base or donated by an acid. One example of this is a simple tautomerization reaction, typified by the enolization of ketones, as shown in Fig. 17.2.[50] This may seem like an unlikely redox reaction but it will be clearer with the tautomerization shown in this figure. If one had a molecule like phenylazocatechol, then by two proton shifts one could obtain phenylhydrazino-*o*-quinone, which is really only a tautomer of the original. However, the catechol part of the molecule has been oxidized and the azo part reduced.

For this type of mechanism the following is necessary. (1) Some pathway must be available for transferring the protons from one position to another without the positions coming in contact. This requirement is met by aqueous or other protonic solvents. (2) A pathway must be available for moving the electrons from the part that is being oxidized to the part that is being reduced. In the present case this is accomplished by means of a conjugated pi electron system.

The preceding is an example of an internal oxidation-reduction reaction; however, in most cases at least two molecules are involved; one is oxidized

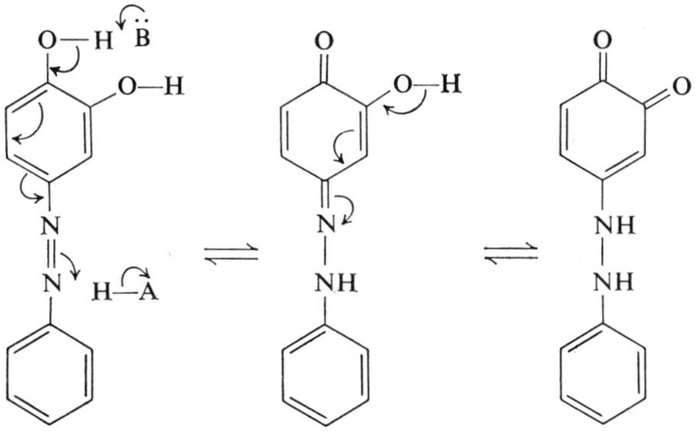

Fig. 17.2 Internal oxidation-reduction reactions.

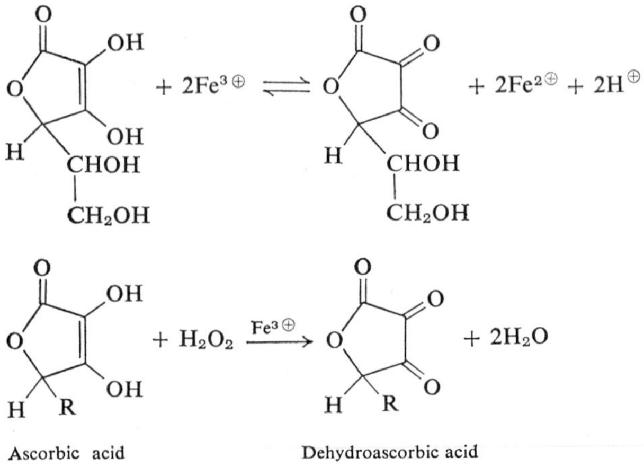

Fig. 17.3 Oxidation of ascorbic acid.

and one is reduced. In aqueous solution a pathway is available for transporting protons from one molecule to the other but there is no conjugated pi electron system for transferring electrons between two molecules. Metal ions may serve as a conjugating link between two molecules, allowing facile electron transfer as illustrated in Figs. 17.3 and 17.4.[50]

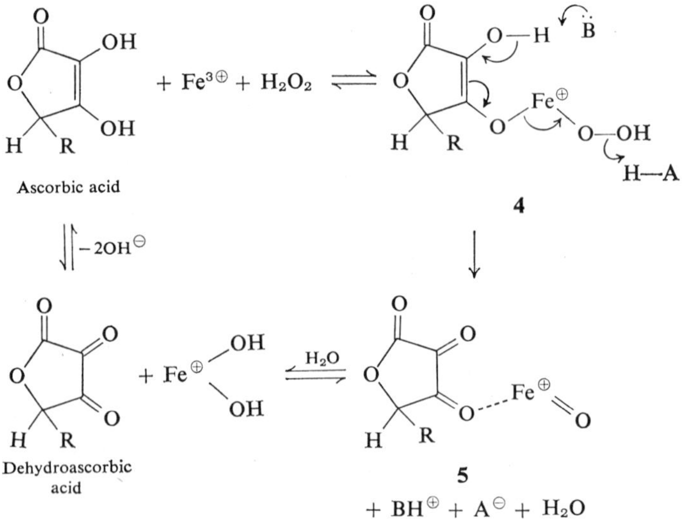

Fig. 17.4 Ferric ion-catalyzed oxidation of ascorbic acid by hydrogen peroxide.

Ascorbic Acid. Grinstead[51] has studied the oxidation of ascorbic acid with ferric ion to give ferrous ion and dehydroascorbic acid (the first reaction) and also the ferric ion-catalyzed oxidation with hydrogen peroxide (the second reaction). He found that the rate of the second reaction was approximately 40 times the rate of the first reaction. However, he proposed that the first step for both was the same; namely, the cleavage of a ferric ion ascorbate complex to give ferrous ion and the semiquinone-like ascorbate free radical. In order to explain the increased rate of the second reaction it was necessary to invoke a chain mechanism. However there is no good evidence for such a mechanism and another possibility for the hydrogen peroxide reaction is shown in Fig. 17.4.[50]

The first step shown here is assumed to be an equilibrium formation of ferric-ascorbate-peroxide complex giving this first intermediate (see Chapter 8). For convenience in Fig. 17.4 and for the remainder of this section a bond between a metal ion and a negatively charged species is illustrated as a solid line and that between a metal ion and a neutral species as a dotted line. The charge which is marked on the metal ion is the charge of that part of the complex, in other words, the valence of the metal ion adjusted by the number of negative species with which it is coordinated—therefore the iron is in the ferric form in complex I(**6**) of catalase or peroxidase (see p. 563), but the charge on the complex is only plus one because the iron is complexed with two negative species. The next step shown here is the step in which the ascorbate is oxidized. This is accomplished by the elimination of a molecule of water. A proton is removed from the ascorbate and the electrons migrate through the iron and onto the peroxide causing the cleavage of the oxygen–oxygen bond. This intermediate has an iron to oxygen double bond, which in aqueous solution one might expect to add water, and the complex in the final step can dissociate to dehydroascorbic acid and dihydroxyferric ion. In this mechanism the ascorbic acid is oxidized and the hydrogen peroxide reduced by the shift of protons through the solvent and electrons through the metal complex. Another reaction[54] for which one can write a similar mechanism is the copper ion-catalyzed oxidation of ascorbic acid by oxygen. The oxidation of ascorbic acid by oxygen is a very slow reaction in the absence of metal ions—copper is one of the most effective ions in catalyzing this reaction. A possible mechanism is shown in Fig. 17.5.[50] Again the first step is assumed to be an association of the three reactants to give this first complex. Then by a proton shift through the solvent and an electron shift through the metal complex the enediol is oxidized and the oxygen reduced. This second complex can then dissociate to the products. The reaction is probably not as simple as shown since the product hydrogen peroxide can react to oxidize the ascorbic acid. A copper-containing enzyme called ascorbic acid oxidase also catalyzes this reaction. The ascorbate free radical

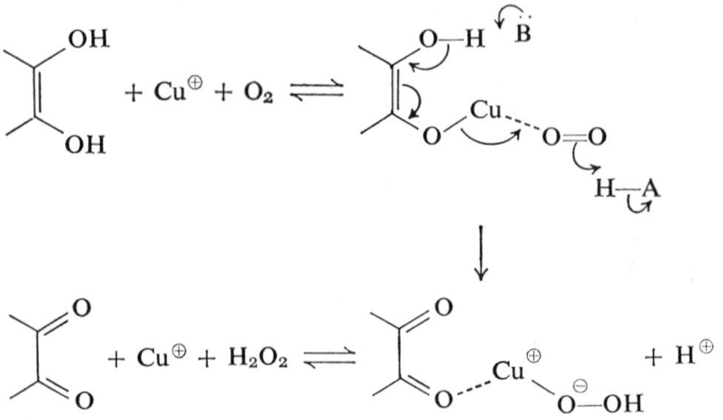

Fig. 17.5 Possible mechanism for the copper ion-catalyzed autoxidation of enediols.

is involved in the enzymatic reaction so presumably the enzymatic mechanism is not similar to this.[52] However another copper-containing enzyme, tyrosinase, catalyzes the autoxidation of catechol to o-benzoquinone. The enzymatic oxidation occurs by a two electron change.[52] Consequently, it is reasonable that the enzymatic mechanism is similar to the one shown where the enediol is catechol instead of ascorbic acid. Probably the rate-determining step is step two of this mechanism. The enzyme may increase the rate of the reaction by a general acid-general base mechanism (Chapter 10) with the acid and base being correctly positioned on the surface of the enzyme for maximum efficiency. This explanation has been put forward to explain the efficiency of many non-oxidation-reduction enzyme reactions and one can now apply this mechanism to the following cases. The catalytic decomposition of hydrogen peroxide to oxygen and water in the presence of manganous or ferric chelates of triethylenetetramine is probably another example of this mechanism. These reactions have been studied by Wang as possible models for the enzyme catalase (see Fig. 17.6).[50] Wang observed that for the manganous reaction the rate of the reaction depended on the hydrogen peroxide concentration to greater than the first power but less than second power. This indicates that an equilibrium between two molecules of peroxide and the manganous complex is involved. Then by a proton shift, a molecule of water is lost and the result is that one molecule of peroxide is oxidized to oxygen and the other reduced to water. This may be the mechanism for the ferric complex catalysis too, although the ferric ion catalysis shows a more complicated behavior. That these chelates are much better catalysts than the free ions can be ascribed completely to a pH effect. The rapid catalysis occurs at high pH where the free ions are hydrolyzed or insoluble. No evidence eliminates these mechanisms as possibilities.

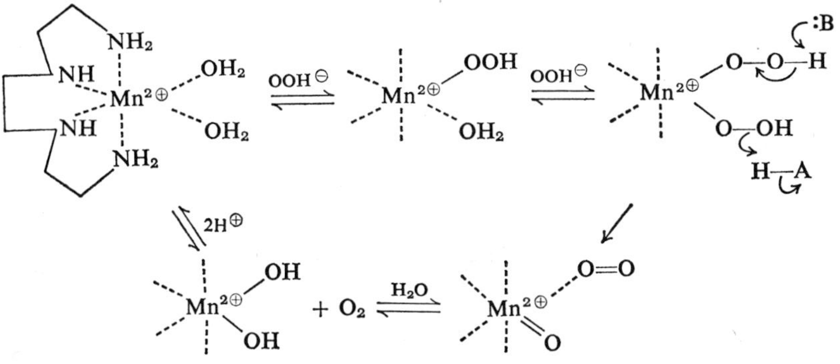

Fig. 17.6 Manganous ion-catalyzed decomposition of hydrogen peroxide.

Udenfriend and coworkers[56] observed that a system composed of ascorbic acid, ferric ion, EDTA, and either oxygen or hydrogen peroxide hydroxylated various aromatic compounds to give phenols (Fig. 17.7).[50] The system would work without EDTA; its function is to keep the iron in solution at higher pH's. The original workers concluded from the orientation of the entering hydroxyl group on the aromatic ring that the reaction was an electrophilic hydroxylation; in other words, that the attacking species was positively charged.

However, the orientation of the groups is not too different from what would be expected for a free-radical reaction and for this reason it is widely felt that the reaction is free radical in character. It is easy to write a free-radical mechanism because it is known that ferrous ions react readily with hydrogen peroxide to give hydroxyl radicals that can attack aromatic rings. This is called the Fenton reaction and is shown in the first two lines of Fig. 17.7.[50] The reaction has been studied extensively and the products with various aromatic compounds are those one would expect for free-radical attack. Consequently, various authors[57] proposed that in the Udenfriend system the ferric ion reacted with ascorbic acid to give ferrous ion and dehydroascorbic acid, the third reaction in Fig. 17.7. This reaction is known to occur in systems without H_2O_2. Then the ferrous ion can react with peroxide as in the Fenton reaction. Either oxygen or hydrogen peroxide can react in this system and early investigations indicated that the products were similar in the two reactions. To explain the oxygen reaction it was assumed that oxygen reacted with ascorbic acid to give H_2O_2 and dehydroascorbic acid, a reaction that is known to occur readily, especially in the presence of metal ions. This is shown on the bottom line of Fig. 17.7. Even though this free-radical mechanism is fairly reasonable, the yields of products

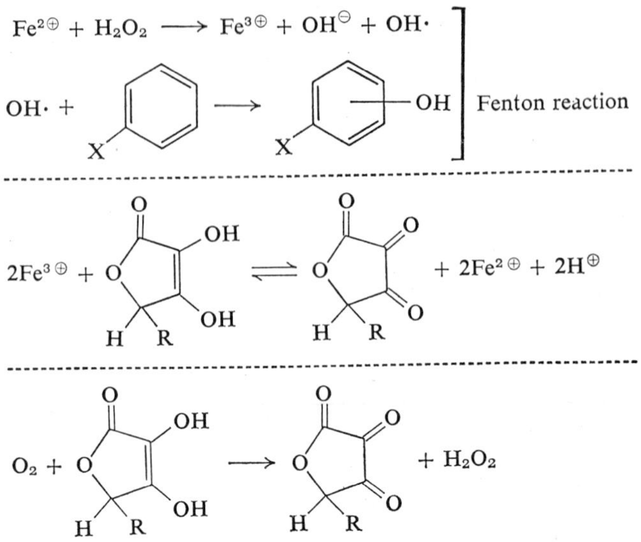

Fenton reaction:

$$Fe^{2\oplus} + H_2O_2 \longrightarrow Fe^{3\oplus} + OH^{\ominus} + OH\cdot$$

OH· + [benzene-X] ⟶ [benzene-X-OH]

$$2Fe^{3\oplus} + \text{(ascorbate enediol)} \rightleftharpoons \text{(dehydroascorbate)} + 2Fe^{2\oplus} + 2H^{\oplus}$$

$$O_2 + \text{(ascorbate enediol)} \longrightarrow \text{(dehydroascorbate)} + H_2O_2$$

Fig. 17.7 Hydroxyl radical hydroxylation.

were very low and the products in some cases did not appear to be the same as those one would expect for attack by hydroxyl radical.

In the ferric ion-catalyzed oxidation of ascorbic acid with hydrogen peroxide, (Fig. 17.4) **4** was suggested to react to give **5** by a proton shift with elimination of water and then **5** added water across the iron–oxygen double bond to give this ferric oxidized enediol complex which then dissociated to give products (see Fig. 17.13).[50] However, several resonance contributing structures can be written for intermediate **5**, the most important of which are considered in Fig. 17.8.

If one breaks the iron–oxygen double bond homolytically then two more fairly good structures can be drawn, one with the iron in the ferric state and the ascorbate as a semiquinone. Consequently one has almost the universal intermediate and one could explain the oxygen acting as either $O^{\ominus}$, $O^{\oplus}$, or O·. The structure of interest for an electrophilic substitution mechanism is C. This does not look like a particularly stable structure with a positive charge on oxygen but the important point is that such an intermediate, that is, **5**, can act as if the oxygen has a positive charge because one can be developed on the oxygen merely by a shift of electrons through the iron and onto the ascorbic acid. Consequently, it is fairly easy to write an electrophilic mechanism with this type of intermediate.

A similar reaction strongly suggests that an intermediate such as **5** is involved.[58] In the ascorbic acid hydroxylation system it is necessary to have

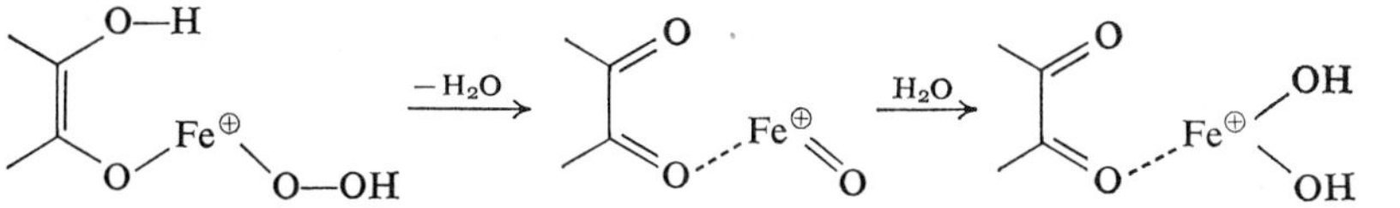

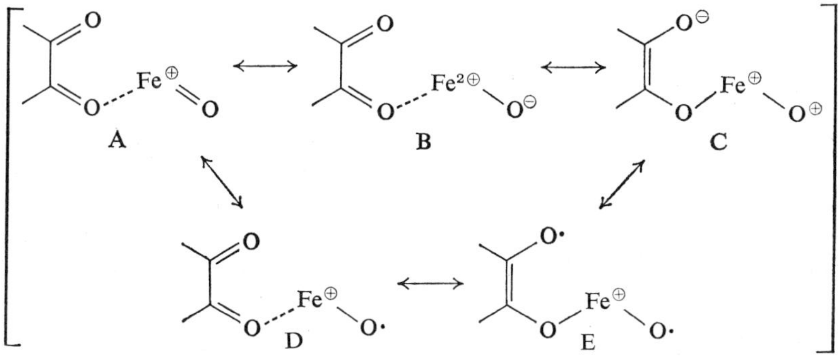

Fig. 17.8 Possible hydroxylating species. A, B, C, D, and E are five canonical structures of the resonance hybrid.

a large excess of ascorbic acid over both H_2O_2 and the aromatic compound being hydroxylated in order to observe any hydroxylation. It appears that the reaction that occurs to the greatest extent is merely the oxidation of ascorbic acid to dehydroascorbic acid. This compound can further oxidize to give other products with the result that the system is complicated even more.

Since, presumably, the part of the ascorbic acid that is necessary for the hydroxylation reaction is the enediol part, other enediols were tried to see if a cleaner reaction could be obtained. The reaction studied the most is one using catechol instead of ascorbic acid (see Chapter 8). Typical reaction conditions are shown in Table 17.4.[50] In this system the disappearance of H_2O_2 was followed by a polarographic method and thus it is necessary to have sodium perchlorate as a supporting electrolyte. The acetate buffer does participate in the reaction. With this system a very smooth first-order disappearance of hydrogen peroxide was observed; the reaction appears to be first-order for at least 60 or 70% reaction and the H_2O_2 reacts until it is all used up. If Fe^{3+} is omitted the reaction does not go. If everything else is present except the catechol the reaction does not go. If anisole is omitted a slightly different reaction occurs but it proceeds only until about 20 to 30% of the hydrogen peroxide is used up. When the catechol concentration is in excess of the ferric ion (at least twofold to threefold excess) the rate of the reaction is linearly dependent on the concentration of ferric ion. The effect of catechol concentration on the rate is interpreted to mean that the ferric ion forms a mono complex with the catechol which is the active species in the reaction. If the concentration of catechol is very large, then two and three to one complexes of catechol with $Fe^{3\oplus}$ can be formed and these apparently are inactive in the hydroxylation reaction. The dependence of the rate of the reaction on the anisole concentration was not studied thoroughly but preliminary indications are that the rate does not depend on the anisole concentration as long as enough is present to react with all the hydrogen peroxide. The products were analyzed by a gas chromatographic procedure and summarized in Table 17.5.[50] The experimental hydroxylation conditions in the first column and the yield of hydroxylated products (based on the amount of hydrogen peroxide present which is the limiting factor in all cases except the final reaction that used oxygen instead of hydrogen peroxide) are shown in the second column. A communication on some of these reactions has been published, but the yields of their products or the exact conditions of the experiments are not given. The first reaction here is the catechol-catalyzed reaction and the yield of hydroxylated products is between 50 and 60%. In this reaction it is found that nearly all hydroxylation occurs in the *ortho-* and *para-*positions and the ratio of *ortho* to *para* is almost exactly 2:1. A very small amount of *meta* product is seen.

In addition to catechol other compounds will act as catalysts: for example, hydroquinone gives a very similar reaction, and 3,4-dihydroxybenzoic acid also is a catalyst. The Fenton reaction, with ferrous ion and hydrogen

Table 17.4

Ferric-Catechol-H_2O_2 Hydroxylation of Anisole[a]

Typical Conditions	
$Fe^{3\oplus}$	$4 \times 10^{-5}\ M$
Catechol	$15 \times 10^{-5}\ M$
Anisole	$0.01\ M$
H_2O_2	$1.8 \times 10^{-3}\ M$
Acetate buffer	$0.005\ M$, pH 4.3
$Na^{\oplus}ClO_4^{\ominus}$	$0.15\ M$
Characteristics	
Disappearance of H_2O_2 follows first-order kinetics. First order in $Fe^{3\oplus}$ with catechol in excess. With above concentrations half-time ca. 15 min	

[a] From G. A. Hamilton, personal communication.

Table 17.5
Products from the Hydroxylation of Anisole[a]

Conditions		% Yield of Phenols (based on H_2O_2)	Phenol Isomer Distribution %		
			Ortho	Meta	Para
Catechol	15×10^{-5} M				
Fe^{3+}	7.9×10^{-5} M	50 to 55	64	3	33
H_2O_2	1.75×10^{-3} M				
Same as above but with hydroquinone in place of catechol		57	65	<5	35
Ascorbate	1.0×10^{-2} M				
Fe^{3+}	2.2×10^{-3} M	5	88	0	12
EDTA	7.8×10^{-3} M				
H_2O_2	1.8×10^{-2} M				
Norman and Radda (Ref. 59)			84	0	16
Fe^{2+}	2.0×10^{-3} M				
H_2O_2	1.75×10^{-3} M	20	86	0	14
Norman and Radda (Ref. 59)			84	0	16
Ascorbate	2.0×10^{-2} M				
Fe^{3+}	4.4×10^{-3} M		45	26	29
Oxygen atmosphere					
Norman and Radda (Ref. 59)			61	0	30

[a] From G. A. Hamilton, personal communication.

peroxide, involves the hydroxyl radical and thus presumably the distribution of isomers found here is characteristic for attack upon anisole by the hydroxyl radical. The ascorbic acid and hydrogen peroxide system also involves the hydroxyl radical as the hydroxylating species because the isomer distribution is very similar to that for the Fenton reaction. Consequently the free-radical reaction discussed earlier for this reaction and proposed by various authors is probably correct. However the ascorbic acid system with oxygen appears to be different from the hydrogen peroxide reaction.

It is very evident that the mechanism of the catechol and hydroquinone reactions must also be different first because it is a catalytic reaction and second because of the different isomer distribution. The very least one can say about this reaction is that it does not involve a free hydroxyl radical as does the Fenton reaction. The isomer distribution indicates it is an electrophilic reaction. The kinetics indicate that iron, H_2O_2, and catechol are involved in the rate-determining step. Any scheme requires that the kinetics depend on at least one of the reagents to half-order or zero-order. A mechanism that is consistent with all these data is shown in Fig. 17.9.[50] The first step involves the formation of a catechol-ferric-hydrogen peroxide complex which in the second step eliminates a molecule of water to give the next intermediate, which is proposed as the active hydroxylating species. According to this mechanism the overall reaction is between anisole and H_2O_2 giving the methoxyphenols, the catechol and Fe^{2+} ion being merely catalysts. This is the stoichiometry that is observed. Another reaction that one might expect the second intermediate to undergo is the addition of water across the iron–oxygen double bond to give catechol-quinone and ferric hydroxide. The water reaction essentially removes the catalyst catechol from the reaction and thus if it occurs to any great extent it should stop the catalytic reaction.

If anisole is replaced by chlorobenzene with everything else the same in the system only about 30% of the hydrogen peroxide is used up and then the reaction stops. This is consistent with this mechanism because chlorobenzene is much less reactive than anisole toward electrophilic reagents.

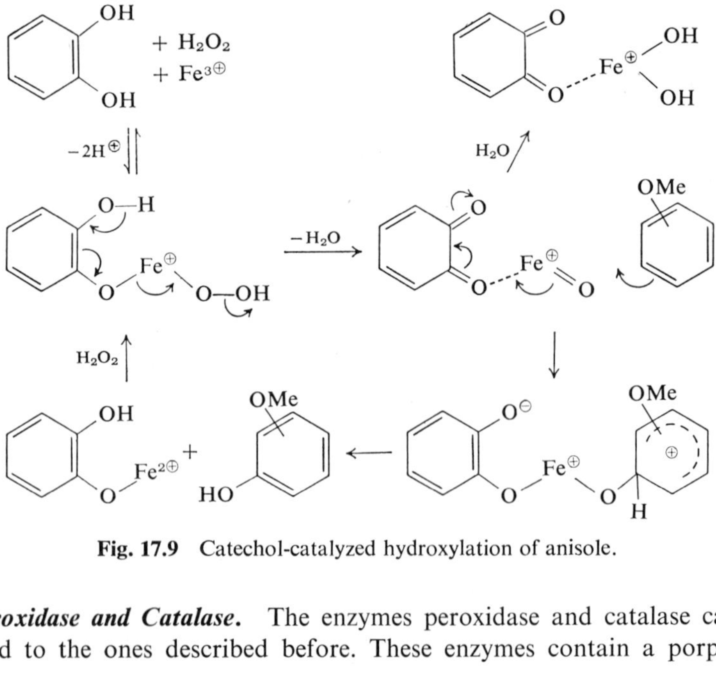

Fig. 17.9 Catechol-catalyzed hydroxylation of anisole.

Peroxidase and Catalase. The enzymes peroxidase and catalase can be related to the ones described before. These enzymes contain a porphyrin ring and iron in the form of ferric ion. Catalase catalyzes the disproportionation of hydrogen peroxide to oxygen and water while peroxidase[50] catalyzes the oxidation of a large number of organic compounds by hydrogen peroxide. By spectroscopic studies various workers, in particular Chance and co-workers, have shown that there are two main intermediates in the reactions of these enzymes with hydrogen peroxide (Chapter 15). The intermediates called complexes I (6) and II (7) are shown in eq. 17.17.[61]

The following shows catalase and peroxidase intermediates,

$$E + H_2O_2 \longrightarrow \underset{\text{(green)}}{\text{Complex I (6)}} \xrightarrow{AH_2} \underset{\text{(red)}}{\text{Complex II (7)}} + AH\cdot$$

$$\downarrow{\scriptsize H_2O_2 \atop \text{(catalase)}} \qquad\qquad \downarrow{\scriptsize AH_2 \atop \text{(peroxidase)}}$$

$$E + O_2 \qquad\qquad\qquad E + AH\cdot$$

(17.17)

where complex I (6) has $FeO^{3\oplus}$ and complex II (7) has $FeO^{2\oplus}$. Both of these enzymes react with hydrogen peroxide or alkyl hydrogen peroxides to give the green complex I (6).[62] With catalase this complex usually reacts with hydrogen peroxide to regenerate the enzyme and give oxygen.

With catalase, the hydrogen peroxide is oxidized to oxygen by the disulfide bridges in the rate-determining step, the peroxide being combined with the iron of the heme group. The principal piece of evidence is a plot of log absorbance versus pH which shows a pK at 9.2.[65] This mechanism is obviously quite different from what has been described previously.

Catalase has a histidine residue at its active site, as determined by amino acid analysis after irreversible inhibition. The histidine residue is number 74, counting in the usual way from the amino terminus.[64]

Peroxidase on the other hand reacts very readily with electron donors such as phenols, aromatic amines, and so on, to give the red complex II (7). This complex then reacts again with an electron donor to give the free enzyme. Catalase can be made to give complex II (7) but not as easily as peroxidase; catalase complex II (7) is not as reactive as peroxidase complex II (7). It is known from titrimetric studies that complex I (6) has the oxidizing equivalents of the original enzyme and one peroxide and that complex II (7) has one electron less oxidizing equivalent; in other words in going from complex I (6) to II (7) some other substance must be oxidized by a one electron change as shown here. George has proposed that complexes I (6) and II (7) have iron in higher oxidation states—he proposes that complex I (6) has a $FeO^{3\oplus}$ species that essentially has the iron in the $+5$ state and that complex II (7) has a $FeO^{2\oplus}$ species that has iron formally in the $+4$ state. In the catechol hydroxylation system catechol is a catalyst for hydroxylations by hydrogen peroxide because it can chelate with ferric

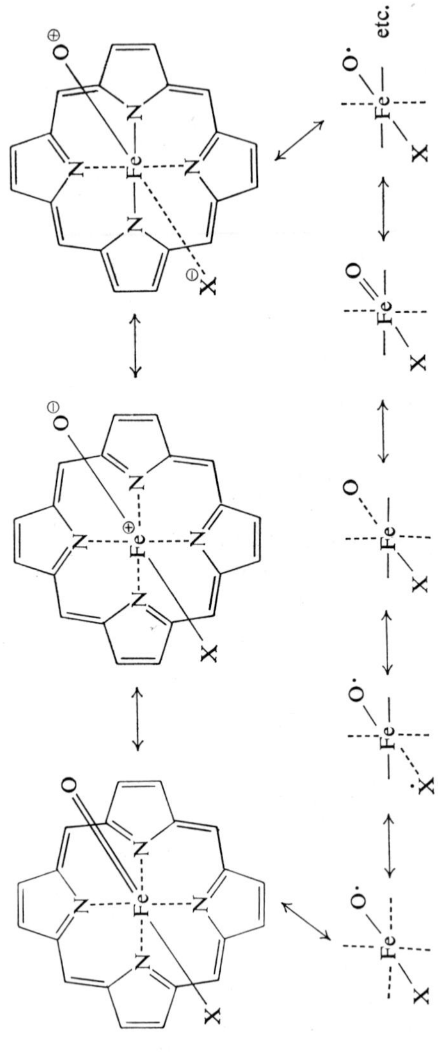

Fig. 17.10 Complex I (6).

Redox Coenzymes

ion and be reversibly oxidized and reduced by two electrons. For the enzymes, peroxidase and catalase, the ferric ion is chelated to the porphyrin ring. This system should be capable also of reversible oxidation and reduction by two electrons. Consequently for these complexes one can write structures similar to those proposed earlier and a mechanism for obtaining these structures.[65] In the first step a hydrogen peroxide molecule displaces the water and loses a proton. In the same type of mechanism as proposed before for the catechol catalysis this complex could lose a molecule of water to give the next intermediate. In this step the porphyrin ring has become oxidized by two electrons; there is one extra double bond in the final structure. The structure looks somewhat peculiar with the iron–oxygen double bond and the oxidized porphyrin ring, but this is only one structure of many different resonance forms. Some of these are shown in Fig. 17.10.[50] These are only a few of the possibilities. In complexes like these, where a metal is chelated to a reversibly oxidized and reduced system, one cannot specify the valence state of the metal because if one can write resonance forms such as these the valence of the metal loses its meaning. The factor that would determine whether complex I (**6**) reacts in one electron steps as in the case of peroxidase or in a two electron step is probably the nature of the group X. Much controversy as to the nature of X in these enzymes exists and the importance of the various resonance forms would be affected by X. In peroxidase the complex appears to act as if the oxygen was a radical and a mechanism for the formation of complex II (**7**) and regeneration of the peroxidase is shown in Fig. 17.11.[50] Only one resonance form of complex I (**6**) has been shown that could react with a donor AH_2 to give the species with the hydrogen on oxygen or X—these should be interconvertible by a proton shift. One of these is probably complex II (**7**) and the second one is favored because resonance forms can be written for this in the same way as for complex I (**6**); for the first structure fewer canonical forms exist. Either

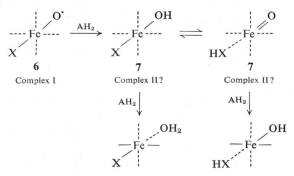

Fig. 17.11 Possible peroxidase reactions.

of these by another one electron change would give back essentially the free peroxidase. The structures shown for complexes I (**6**) and II (**7**) are not too different from those proposed by George except that one cannot speak of the iron being in the $+4$ or $+5$ states because one cannot write resonance structures with the iron in discrete valence states. Consequently one would expect far more stability for such oxygen–porphyrin iron complexes than for the simple oxygen–iron complexes proposed by George.[61] These processes may be related to the many redox enzymes that are discussed by Hayaishi.[66]

In organic compounds one can write resonance forms for molecules which have a conjugated pi system but this restricts one essentially to planar regions of molecules. However for metal complexes one can get overlap not only between the p electron orbitals but also between the d electron orbitals of the metal ion and the p orbitals on the ligands, and consequently it is possible to get conjugation in three dimensions, as shown in formula **8**.

The molecular oxygen hydroxylation reaction with ascorbic acid appears to be different from the hydrogen peroxide reaction—the isomer distributions are not the same. The perhydroxyl radical has been suggested to be involved.[59] With no more evidence than these product studies, a speculative but very intriguing mechanism is proposed for this reaction; it is shown in Fig. 17.12.[50] The first step shown here is an association of ferrous ion, ascorbic acid, and oxygen to give a complex. This mechanism is written assuming the iron acts as ferrous ion, but it might be ferric ion. This complex is assumed to react in the next step with the aromatic compound as shown. Electrons are drawn out of the aromatic ring as in a typical electrophilic substitution, then drawn toward the iron. The ascorbic acid system is oxidized and then eventually the negative charge is placed on the oxygen atom which becomes attached to the aromatic ring. Thus eventually one is left with the complex of iron, dehydroascorbic acid, and an oxygen atom that can add water and dissociate to give dehydroascorbic acid and ferrous ion once again. The other product is a zwitterion that by a proton shift gives the product phenol. This mechanism uses molecular oxygen and by merely shifting electrons accomplishes an overall four electron oxidation—the ascorbate part is oxidized by two electrons and the aromatic ring is oxidized by two electrons. This step is a bimolecular reaction and the aromatic intermediate formed should be one of the most highly stabilized of all electrophilic substitution intermediates because of the two opposite charges. This figure indicates that the reaction with anisole is an electrophilic substitution. This is because the products with anisole indicate an electrophilic reaction and also nitrobenzene gives mainly *meta*-substitution with this system.[67] However other possibilities are available. If one forgets about the aromatic compound for a moment one can see that this type of mechanism gives essentially atomic oxygen, in other words, an oxygen atom with

Redox Coenzymes

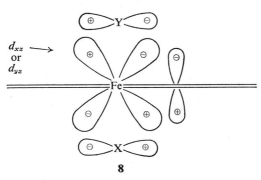

8

six electrons around it. This may be shown more clearly in Fig. 17.13.[50] By a shift of electrons as shown the enediol structure is oxidized and essentially atomic oxygen is produced. This species with only six electrons reminds one very much of a carbene such as CH_2 which also has only six electrons. Typical reactions of carbenes include addition to double bonds giving cyclopropanes and insertion into carbon–hydrogen bonds giving methyl groups. Consequently one would expect a species such as this oxygen species to

Fig. 17.12 Iron-ascorbic acid-oxygen hydroxylation.

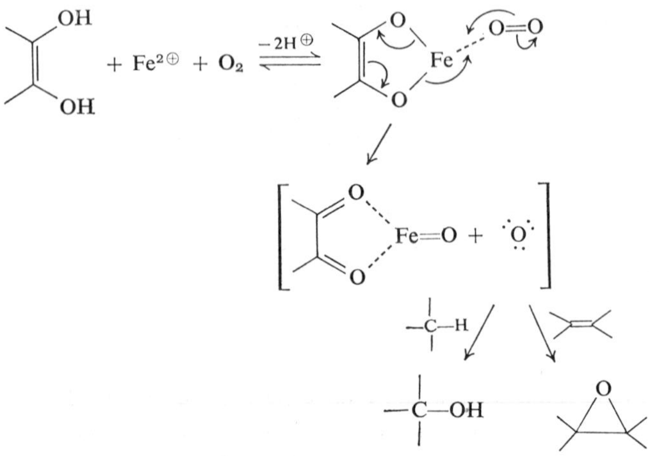

Fig. 17.13 Oxene mechanism.

undergo similar reactions giving epoxides with double bonds and alcohols with saturated hydrocarbons or phenols with aromatic compounds. A species such as this free oxygen atom may not actually exist, the important point being that an intermediate such as the iron–enediol–oxygen complex can by merely a shift of electrons give a species which one might expect to be able to give epoxides with double bonds, and insert into carbon–hydrogen bonds. Thus the hydroxylation of aromatic compounds with the ascorbic acid–iron–oxygen system can be viewed as merely a carbon–hydrogen insertion reaction by a species which is electronically similar to a carbene. Because of the similarity of this species to a carbene this is called the oxene mechanism. Many enzymes present in biological systems catalyze the hydroxylation of various compounds by molecular oxygen.

Mixed Function Oxidases. One large group of such enzymes are what Mason[67] calls "mixed function oxidases." In these reactions two of the oxidizing equivalents of the oxygen are utilized to hydroxylate an organic compound and the other two equivalents are used usually to dehydrogenate some other compound. In many cases the compound that is dehydrogenated is NADH or NADPH. An example is the enzyme phenylalanine hydroxylase. This enzyme catalyzes the hydroxylation of phenylalanine to tyrosine but this reaction goes only if some reducing substance, usually NADPH, is present to take up the other two oxidizing equivalents of oxygen. Kaufman has characterized an enzyme that carries out the hydroxylation of phenylalanine but can use other reduced compounds such as tetrahydrofolic acid derivatives instead of NADPH. Two of the oxidizing equivalents of the oxygen are used to hydroxylate the phenylalanine while the other two are used to oxidize some other reduced substance and this other reduced substance must be present in order for the hydroxylation to occur. This can be readily rationalized if the mechanism of the hydroxylation is similar to one of those shown in Fig. 17.13.[50] According to this mechanism the ascorbic acid must be oxidized during the hydroxylation and thus it could not serve as a catalyst because it is being used up. Thus in the enzyme reaction, if this mechanism holds, some group on the enzyme would be oxidized during the hydroxylation and this group would have to be reduced, in order to give the active enzyme again. This presumably is the function of the reducing agents that must be present in stoichiometric amounts during the enzymatic reaction. A cofactor for phenylalanine hydroxylase from rat liver is a dihydropteridine derivative,[68] which is believed to have the structure shown in formula **9**.[50] The evidence suggests that it is either this or the oxidized

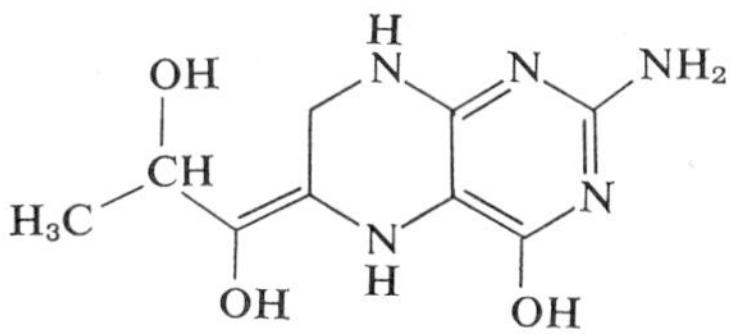

form where the alpha enol amine is oxidized to the alpha keto imine by removal of the two hydrogens. This part of the molecule is very similar to the enediol structure of catechol and thus one can write a mechanism similar to that in Fig. 17.13 using this as catalyst. In order for the enzymatic hydroxylation to proceed, the enzyme folate reductase must be present. Other enzymes with very similar characteristics to these aromatic compound-hydroxylases are the enzymes that hydroxylate saturated hydrocarbons, in particular, the enzymes that hydroxylate steroids. These, also, use oxygen and some reduced compound, usually NADPH, in stoichiometric amounts. The oxene mechanism shown in Fig. 17.13 probably applies to these reactions. Therefore similar types of cofactors for these enzymes should be looked for. Crude preparations from guinea pigs that had been fed on a diet deficient in ascorbic acid could not hydroxylate proline to hydroxyproline nearly as efficiently as preparations from normal animals, indicating that ascorbic acid may be the cofactor.[69]

Cytochromes. The cytochromes are part of the electron transport system in the body.[6] Again the early history of the cytochromes has been well told.[7] There are many cytochromes and some of these are shown in Table 17.6.[7] An enormous amount of work has been done on the primary structure of cytochrome from many different species and from this has come the

Table 17.6
Properties of the Cytochromes*

Class	Name	Source	Spectra λ_{max} [ϵ_{mM}] Reduced			Oxidized		E'_0	Molecular Weight ($M \times 10^{-3}$)	Physiological Reductant	Oxidant	Reaction with O_2	CO	$CN^\ominus$, $Fe^\ominus$, $N_3^\ominus$
			α	β	γ	γ								
A	a	Mitochondria[b]	600 [23]	—[a]	439 [94]	425 [76]	+0.28		c	a_3	?	?	+
	a_3	Mitochondria[b,c]	(603.5) [19.4]	—	443 [79] 430 (CO)		(+0.30)	240	$c + a$	O_2	+	+	+
A_2	a_1	Bacteria	590										
	a_2	Pseudomonas[d]	652 549	629 523	460 418	412		90	C _C551	O_2 $O_2, NO_2^\ominus$	+ +	+ +	+ +
B	b	Mitochondria[b,p]	563 [21]	532	429 [114]		0.05[e]	30	CoQ, NHI	c_1	–	–	–
	b_1	E. coli[f]	560	530	426	415	0.250		fp?	Q_1, Q_2	–	–	–
		Other bacteria	559	528	426	415			fp	$NO_3^\ominus$			
	b_2	Yeast	557 [33]	528 [17]	424 [198]	413 [117]		170[g]	fp	c_1 or c	–	–	–
	b_3	Plant microsomes	559	529	425			40	fp	?	+	–	–
	b_5	Microsomes, liver mitochondria	556 [26]	526 [13]	423 [171]	413 [115]	0.02	14	fp	c, P450	–	–	–
	b_6	Chloroplasts	563				–0.06		fp?	f?	+		
	b_7	Spadix of Arum plant	560	529			–0.03		b	O_2?			
	B562	E.coli	562	532	427		0.254	7	fp?	C?	+		
	B559	Streptomyces fradiae	559	530	408	410		13	?	?			
	B420	Microsomes[h]	—[i]	—	420 450 (CO)				b_5, hydroxylase	O_2		+	
C	o	Acetobacter, E. coli	568 (CO)						C	O_2	+	–	–
	c	Mitochondria[b]	550 [27.7]	521 [15.9]	415 [125]	407	0.254	13	c_1	a	–	–	–
	c_1	Mitochondria[b]	554 [24.1]	524 [11.6]	418 [116]	410–12	0.220	37	b	c	–	–	–
	c_2	Denitrifying bacteria	550–52	522–25	416–18	409	0.25–	32	fp?	$NO_3^\ominus$?	–	–	±†
		Rhodospirillum rubrum	550	521	416		0.32	14	fp?	a, o?	–	–	
	c_3	Desulfovibrio desulfuricans	552	522	418	410	–0.205	11.3[k,o]	fp?	$SO_4^{2\ominus}$	+	–	
	c_4	Azotobacter, others[h]	551	522	414	414	0.30		B	c_5	–	–	–
	c_5	Azotobacter, others[i]	555	524	418	414	0.32		c_4	a_1, a_2	–	–	–

C551	Pseudomonas	551		416	409	0.286	7.6	B fp?	a_2 ?
C552	Chromatium	552	523	416	410	+0.010	97[l]	?	?	+	–
C554	Halotolerant bacteria[j]	554	521	415				?	?		
C555 or f	Chloroplasts	555	525	423	413	0.365	110[m]	b_6?	Chl a	–	–
C556 or h	Hepatopancreas of gastropods	556	527	422	408		18.5	?	?		
Cytochromoid C	Rhodospirillum rubrum	568 550		424 416 (CO)	390	–0.008	28[n,o]	?	?	+	+
Cytochromoid C	Chromatium	565 547		426	400	–0.005	36[n,o]	C552?	?	+	–

* From H. R. Mahler and E. H. Cordes, *Biological Chemistry*, Harper and Row, New York, 1966, p. 594.
† Slow; high K_{diss}

[a] The dash indicates component or reaction absent; an empty space indicates information not available.
[b] Component has been identified in (and in some cases isolated from) the mitochondrial electron transport chain in animals, plants, yeasts, and molds (*Neurospora*).
[c] Defined as the entity in cytochrome oxidase that reacts with CO in the reduced and with HCN, HNO_3, and probably O_2 in the oxidized state. Monomeric cytochrome oxidase has an $M \simeq 240{,}000$ and contains 2 hemes per molecule [$a \cdot Cu : a_3 \cdot Cu = 1$]. The minimal functional unit is three times larger ($M \simeq 700{,}000$). In this complex, the ratio may be $(4a)Cu_2 : (2a_3)Cu_4$.
[d] *Pseudomonas* cytochrome oxidase (9) and the terminal oxidase of certain bacteria do not possess cyt a_3. The isolated protein contains one mole of heme plus one mole of a_2. The systematic name is cytochrome CD.
[e] Estimated value for particulate form. The isolated solubilized forms show considerably lower values. Similarly, partially denatured particulate enzymes show different E_0' values and become autoxidazable.
[f] The b-type cytochrome of many bacteria.
[g] Part of the L-lactate dehydrogenase. M shown is for dimeric form containing 2 hemes and 2FMN.
[h] Perhaps the terminal oxidase for microsomal electron transport with either cyt b_5 or the microsomal hydroxylase functioning as an electron donor. Only the γ peak is evident in the spectrum; this is also referred to as P450 because of λ_{max} of CO compound of particulate form.
[i] Cyts c_4 and c_5 are the counterparts in aerobic bacteria of cyts c_1 and c.
[j] Formerly called cyt b_4.
[k] Two hemes per molecule.
[l] Three hemes per molecule.
[m] Two hemes per molecule.
[n] Two hemes per molecule.
[o] Also contains 1 NHI per heme.
[p] Pure preparations of these cytochromes with enzymatic activity have not yet been obtained from mitochondrial sources.

postulate of phylogenetic trees.[70,71] Chance evaluates the 1967 status of the experimental work and the ideas about the nature and control of electron transport mechanisms in multienzyme sequences of cytochromes.[72] The immunological properties of the cytochromes have also been discussed,[73] as well as the electron density map of a particular cytochrome.[74] There exists a cytochrome chain that transfers electrons finally to oxygen (cytochrome oxidase). This chain has been the object of study.[75]

Rate constants were measured for the reversible redox reaction between ferricytochrome c and ferrocyanide.

$$\text{Cyt–}c^{3\oplus} + [\text{Fe}^{2\oplus}(\text{CN})_6]^{4\ominus} \rightleftharpoons \text{Cyt–}c^{2\oplus} + [\text{Fe}^{3\oplus}(\text{CN})_6]^{3\ominus} \quad (17.18)$$

Kinetic measurements around neutrality were made by the temperature jump technique, and comparative measurements were determined by a spectrophotometric titration based on an absorbance difference at 550 nm between ferricytochrome and ferrocytochrome c.[76]

Electron spin resonance studies of cytochrome oxidase show that a copper ion associated with this enzyme goes from the cupric to the cuprous state, when the enzyme is reduced by substrate.[77] There has been a dispute as to the occurrence of histidine in the heme peptide of cytochrome c; it has been settled in the affirmative.[258]

Amino Acid Oxidases. The nonenzymatic oxidation of tryptophan to kynurenine shown in eq. 17.19 probably goes through a formylkynurenine intermediate[78] as does the enzymatic reaction.

$$\text{[tryptophan]} \xrightarrow[\text{2 hematin}]{\text{alkali}} \text{[kynurenine]} \quad (17.19)$$

In the enzyme tryptophan oxygenase, three things have been noted. First, an enzyme-substrate complex can be seen spectrally. Second, there is a cyclic valence change during catalysis. Third, p-mercuribenzoate inhibits the enzyme.[80] In D-amino acid oxidase systems, ketimine intermediates have been postulated on the basis of isotopic nitrogen compounds.[81,82] A flavin seems to be involved in amino acid oxidases. For example, an enzyme-substrate intermediate has been noted. Furthermore the substrate could be affixed covalently to the enzyme by the use of sodium borohydride. The enzyme contained at least one bound flavin. In particular, radioactively labeled D-alanine was affixed to the epsilon-amino group of a lysine residue. It was therefore postulated that the flavin on the enzyme was attacked by the amino acid as the lysine amino group on the enzyme attacked the amino acid as shown in eq. 17.20.[83,84] The flavin semiquinoid form of L-amino acid oxidase was found not to be reduced further by L-amino acids.[85]

$$\text{(structure 17.20)}$$

(17.20)

17.1.4 Cobalamin and Cobamide Coenzymes (Vitamin B_{12})

Vitamin B_{12} is a complicated cobalt complex. The structure of B_{12} is given in formula **10**.[86] If B_{12} is treated as a metal complex, it may be considered to be built of four regions. The cobalt atom is in the center. Working outward, one encounters a corrin ring that provides four coordinating nitrogen atoms to the cobalt. Finally one encounters substituents on the ring. The cobalt also has bound to it two (or one) axial ligands, X and Y. If one of the axial ligands Y is the benzimidazole of the naturally occurring nucleotide, the series of complexes is called cobalamins whereas if the base is removed by hydrolysis and Y = H_2O the series is referred to as the cobamides. There is special interest in the series of molecules in which X is a carbon ligand for a metal–carbon bond was shown to be present in one of the B_{12} coenzymes[87] and this appears to be mechanistically important.

One of the important reactions of cobamide coenzyme (with enzyme) reactions is the isomerization of methylmalonylCoA to succinylCoA. This is a reversible process and corresponds on the surface to the movement of a methyl group under mild conditions. By tracer experiments, it was shown that 20% of the reaction occurs by movement of the carboxyl group and 80% of the reaction occurs by movement of the –SCoA group.[88] Radioactive CoA of the medium is not incorporated into the products during the enzymatic reaction.[89] Further there is no exchange with added acrylic acid.[90] A mechanism for this important reaction was predicated on the chemistry of metal carbonyls (Chapter 8).[90] It embodies the intramolecular 1,2 shift of a CoA bound carboxyl group. In particular, the enzymatic process was postulated to consist of: (1) acylation of reduced B_{12} to form methylmalonyl B_{12}; (2) migration to form succinyl B_{12}; and (3) regeneration of the reduced B_{12}.

Another important reaction involving cobamide coenzymes is the conversion of 1,2-propanediol to propionaldehyde. This is a classical organic

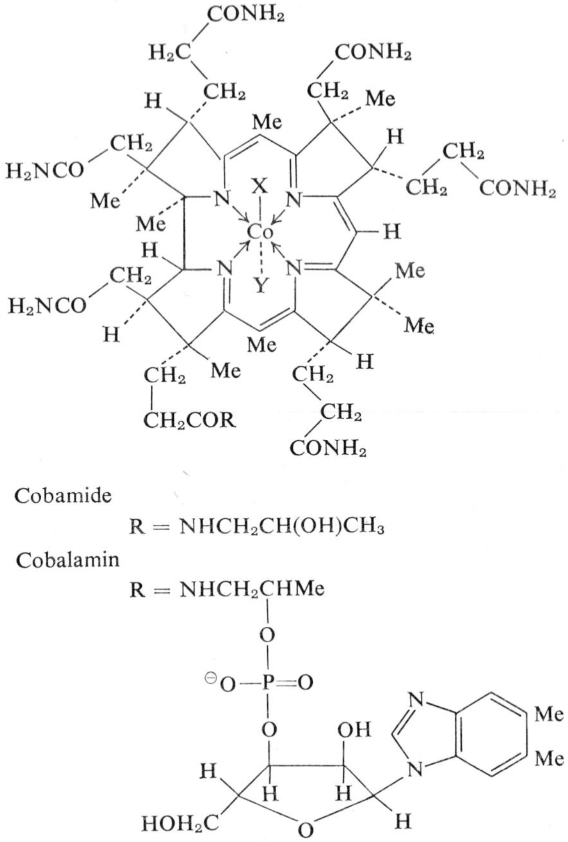

Cobamide   R = NHCH$_2$CH(OH)CH$_3$

Cobalamin   R = NHCH$_2$CHMe

The structural formulae of cobalamins and cobamides.

10

reaction that is usually carried out with sulfuric acid, but with a B$_{12}$ enzyme it can be carried out under mild conditions. The reaction is essentially an intramolecular redox reaction and it is known that B$_{12}$ is essential since when it is removed by charcoal, the enzyme is inactive, but when B$_{12}$ is readded the enzyme is once again active.[91] When dl-1,2-propanediol-1-t is converted to propionaldehyde in the presence of the enzyme dioldehydrase and cobamide coenzyme, tritium is transferred to the coenzyme. The tritiated coenzyme so obtained transfers tritium to the reaction product when reacted with dl-1,2-propanediol and apoenzyme. The coenzyme is tritiated exclusively at the C-5′ position of the adenosyl moiety. The conversion of dl-1, 2-propanediol-1-t to propionaldehyde proceeds with intermolecular and intramolecular tritium transfer. The results led to the following postulate.

Hydrogen is abstracted from C-1 of dl-1,2-propanediol and transferred to the coenzyme, where it becomes equivalent with at least one, but probably both, hydrogens of the C-5' position. This results in a reduced form of the coenzyme and a molecule derived through the oxidation of the substrate. In a subsequent step the hydrated form of propionaldehyde is formed by a transfer of hydrogen from the reduced coenzyme to the intermediate derived from the substrate.[92,93]

Labeling data have clearly established the mechanism of the dioldehydrase reaction to be the transfer of hydride ion to the 5' carbon of the coenzyme to the substrate.[93,94]

Another suggestion has recently been made for the mechanism of action of cobamide-containing enzymes. The requirements for alkylation and binding of nitrogenous ligands to cobalamins and cobamides suggest that the central cobalt atom can exist either approximately in the plane of the corrin ring or axially displaced above this plane. Distortion of the normal octahedral complex by axial displacement may result when the dimethylbenzimidazole moiety of the coenzyme is prevented from coordinating with the cobalt. A transition from $Co^{III}-R^{\ominus}$ to $Co^{I}-R^{\oplus}$ is proposed.[95]

The stereochemistry of B_{12}-containing enzymes has been much worked on. Dioldehydrase, an enzyme that requires a cobamide coenzyme, catalyzes the conversion of 1,2-propanediol to propionaldehyde as mentioned above. Whereas propanediol-1-d_1, produced by the reduction of R-lactaldehyde with NADD and ADH, is converted to propionaldehyde-1-d, S-lactaldehyde[96] is converted to propionaldehyde-2-d. This was established by nmr spectra.

When d-1,2-propanediol is converted to propionaldehyde in the presence of dioldehydrase and dimethylbenzimidazolylcobamide coenzyme, one of two C-1 hydrogens of the substrate is stereospecifically transferred to position C-2.[97]

The stereochemistry of the reaction of methylmalonyl coenzyme A has also been investigated. Propionyl coenzyme A was carboxylated with epimerase-free carboxylase and the resulting methylmalonyl coenzyme A was reduced with Raney nickel. Isolation of (R)-β-hydroxyisobutyric acid N-phenylcarbamate from the reaction mixture showed that methylmalonyl coenzyme A has the (S) configuration. Propionyl coenzyme A was also converted to deuteriosuccinate with a D_2O extract of a beef liver mitochondrial acetone powder. The deuteriosuccinate had a plain positive optical rotatory dispersion curve corresponding to the (S) configuration. It was concluded that (S)-methylmalonyl coenzyme A, formed by carboxylation of propionyl coenzyme A, was epimerized in D_2O to (RS)-methylmalonyl (coenzyme A)-2-d and that methylmalonyl(coenzyme A)-2-d was isomerized to succinyl coenzyme A with net retention of configuration of C-2.[98]

The stereochemistry of the corresponding glutamate reaction was also investigated. 4-Deuterioglutamic acid was prepared by incubation in D_2O of ammonium mesaconate with an extract of *Clostridium tetranomorphum*. In these extracts methyl aspartase catalyzed the formation of *threo*-3-methyl-L-aspartate-3-*d* which was then rearranged to glutamate by the cobamide coenzyme-dependent glutamate mutase. The monodeuteriosuccinate obtained by chloramine-*T* oxidation of the glutamate showed a plain negative optical rotatory disperson curve, and was therefore (*R*)-succinate-2-*d*. It was concluded that the intramolecular rearrangement of *threo*-3-methyl-L-aspartate to L-glutamate proceeded with methyl inversion of configuration of carbon atom 3. (*R*)-Succinate-2-*d* was also prepared from (2*S*, 3*R*)-aspartic acid-3*d*.[99] Since the propionate and glutamate mutase reactions proceed differently, the former with retention of configuration and the latter with inversion, they may involve at least two steps, one or more of which are analogous, but at least one taking place with different stereochemistry. Perhaps the cobamide coenzyme releases an electrophilic 5'-adenosyl hydride group (perhaps irreversibly to a methinyl residue on the enzyme), and the resulting one-carbon complex participates in the rearrangement by removal and transfer of a proton or cationic group.

Models of B_{12} have been prepared by synthesizing the "cobaloximes" which are identical to the natural cobalamins in that the cobalt atom is surrounded by a planar ligand ring, although a much simpler one. In the models the ligand is made up of two dimethylglyoximate groups. In both cases the ligands are held to the cobalt through four nitrogen atoms. An alkyl group and a basic group are also coordinated to the cobalt above and below the plane of the ligand (formula **11**).[100,101] A number of cobaloximes

A cobaloxime

B = basic group (e.g. pyridine, trimethylamine)

11

containing substituted alkyl, alkenyl, and hydroxylalkyl groups were prepared. The β-hydroxylalkylcobaloximes are more reactive than the simple alkylcobaloximes. The Co–C bonds are readily cleaved. Perhaps the most dramatic demonstration of how much the cobaloximes resemble cobalamins is the work with extracts of the methane-producing bacterium *Methanobacillus omelanskii*,[102] although recent tracer evidence indicates that cobaloximes are not good models for the coenzyme in enzyme systems.[103] One of the most unusual features of vitamin B_{12} is that it can be reduced to B_{12S}, the highly reactive species which is a Co(I) complex. The Co(I) derivatives react with alkylating agents to yield stable organocobalt compounds. Vitamin B_{12S} and the Co(I) cobaloximes are the most powerful nucleophiles known at present, being 10^{14} times better than methanol solvent (see Chapter 6).[101] The synthesis and reactions of 5′-deoxyadenosylcobaloximes have been described. These are better models for B_{12} than the cobaloximes.[259]

17.1.5 Lipoic Acid

Lipoic acid was discovered independently in several laboratories in the late 1940s as a growth factor and a requirement for pyruvate oxidation for certain microorganisms.[104] When the substance was isolated in crystalline form and its structure determined, the name 6-thioctic acid was suggested. The name lipoic acid, however, was adopted as the trivial designation of 1,2-dithiolane-3-valeric acid (formula **12**).[104] Lipoic acid is widely distrib-

$$H_2C \underset{S-S}{\overset{CH_2}{\diagup \diagdown}} CH(CH_2)_4CO_2H$$

Lipoic Acid

12

uted among microorganisms, plants, and animals. Most nutritional investigations with higher animals have failed to show a growth response to added lipoic acid. However, there is no doubt that this substance plays a vital role in animal metabolism. Lipoic acid can be reduced easily to dihydrolipoic acid by disulfide interchange. It has been suggested that because of the unique reactivity of the lipoic acid ring that it participates in the primary quantum conversion reaction of photosynthesis.[105,106] Lipoic acid of pyruvate and α-ketoglutarate dehydrogenation complexes is bound covalently to protein, specifically to the ε-amino groups of lysine residues.[107,108]

The effects of lipoic acid and its derivatives have been noted with a variety of biological systems, but it appears that a majority of these effects are due to a thiol-disulfide interchange rather than a vitamin-like function of lipoic acid. At present the only established role of lipoic acid is in the CoA- and NAD-linked oxidative decarboxylation of α-ketoacids. The available evidence indicates that the oxidative decarboxylation of pyruvate and α-ketoglutarate proceeds via Fig. 17.14.[109] In the decarboxylation itself, thiamine pyrophosphate is involved to produce an "active aldehyde" (see p. 584), as shown in eq. 17.21.[110]

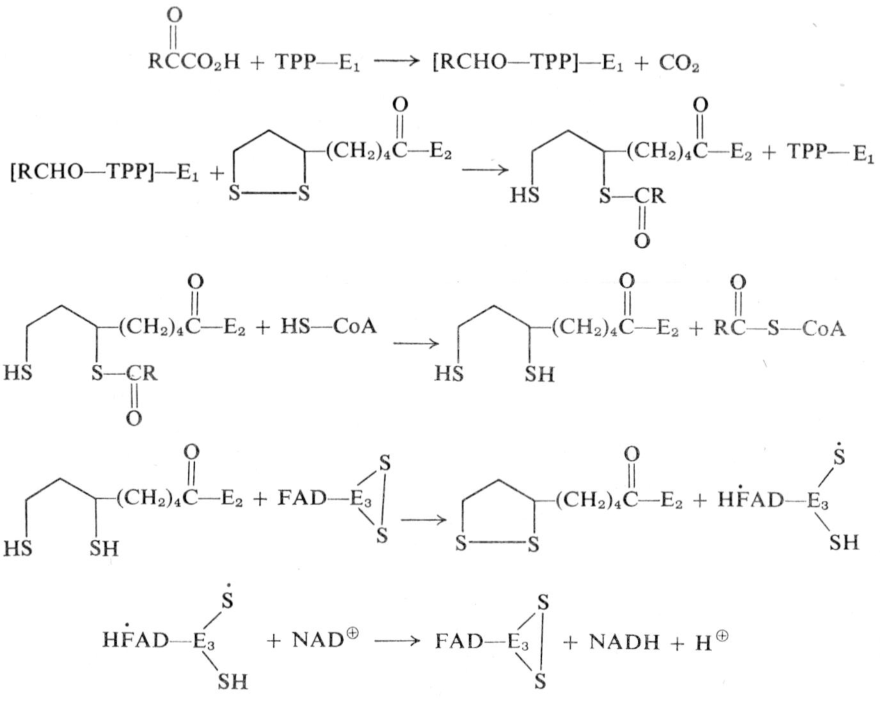

$$RCCO_2H + TPP\text{---}E_1 \longrightarrow [RCHO\text{---}TPP]\text{---}E_1 + CO_2$$

[RCHO—TPP]—E$_1$ + [lipoyl-(CH$_2$)$_4$C(O)—E$_2$ with S—S] $\longrightarrow$ [lipoyl-(CH$_2$)$_4$C(O)—E$_2$ with HS and S—CR(O)] + TPP—E$_1$

[lipoyl-(CH$_2$)$_4$C(O)—E$_2$ with HS and S—CR(O)] + HS—CoA $\longrightarrow$ [lipoyl-(CH$_2$)$_4$C(O)—E$_2$ with HS and SH] + RC(O)—S—CoA

[lipoyl-(CH$_2$)$_4$C(O)—E$_2$ with HS and SH] + FAD—E$_3$ $\longrightarrow$ [lipoyl-(CH$_2$)$_4$C(O)—E$_2$ with S—S] + HFAD—E$_3$ (with S, SH)

HFAD—E$_3$ + NAD$^{\oplus}$ $\longrightarrow$ FAD—E$_3$ + NADH + H$^{\oplus}$

R = CH$_3$, HOOC(CH$_2$)$_2$

Fig. 17.14 Oxidative decarboxylation of α-keto acids. From L. J. Reed, in *Comprehensive Biochemistry*, M. Florkin and E. H. Stotz, Eds., Elsevier, Amsterdam, 1966, vol. 14, p. 105.

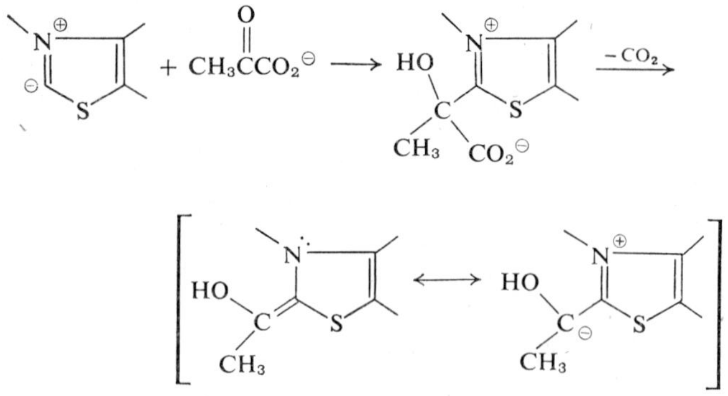

$$\left[\text{thiazolium ylid} + CH_3CCO_2^{\ominus} \longrightarrow HO\text{-}C(CH_3)(CO_2^{\ominus})\text{-thiazolium} \xrightarrow{-CO_2} \right.$$

$$\left. HO\text{-}C(CH_3)=\text{thiazoline} \longleftrightarrow HO\text{-}C^{\ominus}(CH_3)\text{-thiazolium} \right] \quad (17.21)$$

Active acetaldehyde

The acyl-generation reaction is visualized as a reductive acylation of protein bound lipoic acid. Specifically, "active acetaldehyde" is believed to attack the disulfide linkage of bound lipoic acid in a nucleophilic displacement reaction, followed by a reverse condensation as shown in eq. 17.22.[111]

$$\tag{17.22}$$

Lipoyl dehydrogenase, which is involved in the last step of the sequence shown in Fig. 17.14, was first discovered in bacteria and shown to be involved with α-keto acids.[112,113] In the α-oxoglutarate dehydrogenase complex, it was shown that this enzyme is a flavoprotein.[114,115] This is implied in the last line of Fig. 17.14. The intermediate in this final step is presumed to be the semiquinone or free radical state of the flavin (FADH) as in other reactions of flavoenzymes.[116]

Pig heart lipoyl dehydrogenase is relatively stable in 6.5 M urea. On the anaerobic addition of reducing substrates (such as reduced NAD) in 6.5 M urea, irreversible denaturation occurs. The molecular weight of lipoyl dehydrogenase was found to be 100,000 with 2 molecules of flavin adenine dinucleotide, and two active centers per molecule of protein. Amino acid analysis, end group analysis, and peptide mapping after digestion with tryp-

$$\tag{17.23}$$

sin suggest that the enzyme consists of two identical peptide chains, which are covalently linked by the two active center disulfide bonds.[117] A model for this enzyme exists in the oxidation of dihydrolipoic acid by flavin mononucleotide. The rate is first-order in each substituent and increases with increasing pH. The effect of varying R paralleled its effect on other two electron transfer reactions.[118]

17.1.6 Coenzyme Q (Ubiquinone)

Coenzyme Q is synthesized from acetate in the usual way and is a substituted benzoquinone as ubiquinone implies. It is involved in photosynthesis and in redox (electron transport) systems.[22] The structure of ubiquinone is shown in formula **13**, where $n = 6 - 10$.

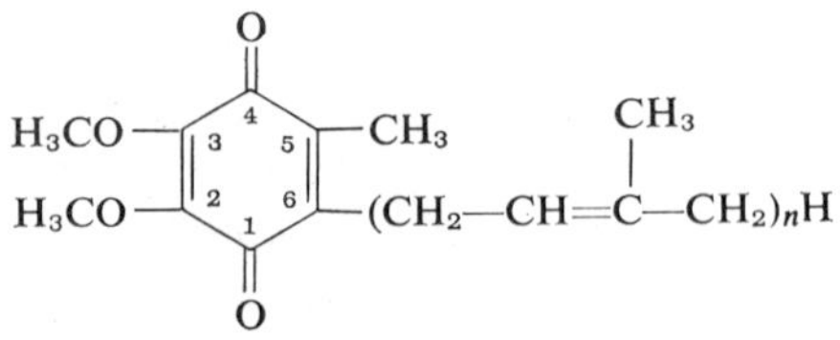

Ubiquinone
13

17.1.7 Metal Ions That Can Change Valence

Metal ions such as cupric ion and ferric ion that can undergo one- or two-electron changes are potentially operative in enzymatic redox systems. Some heme iron complexes and nonheme iron complexes have been considered already. Copper in some instances can replace iron, but it is not as common.[18] The probable structure of the copper complex with tetraglycine is shown in formula **14** ($R = CH_2CO_2H$).

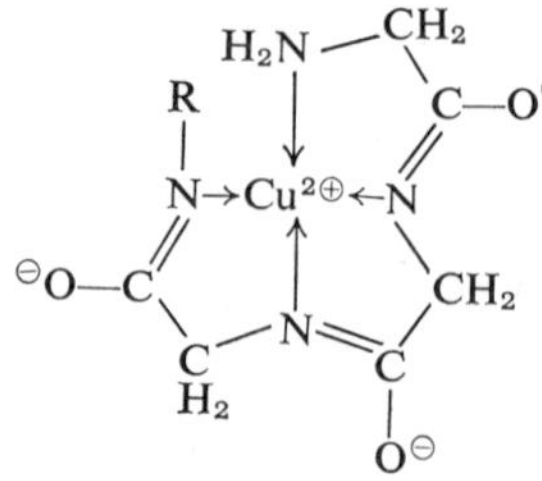

14

17.1.8 Ferridoxin

A nonheme iron-containing coenzyme is ferridoxin, which is involved in pyruvic acid metabolism and nitrogen fixation. Ferridoxins are low molecular weight proteins having strongly negative redox potentials, functioning as electron carriers on the hydrogen side of the pyridine nucleotides.[120-122] They have been obtained from many green plants as well as from some photosynthetic and nonphotosynthetic bacteria. Ferridoxin is reported to contain four to seven iron atoms per molecule, depending on the species (seven in *Clostridia*) probably linked to an equal number of cysteine residues in the protein. It also contains an equal number of acid-labile sulfur atoms (either inorganic sulfide or very labile organic residues giving rise to H_2S on acidification). A model of the active site[18] is shown in formula **15**. Amino

$$
\begin{array}{c}
(aa)_8 \quad (aa)_3 \quad (aa)_2 \quad (aa)_2 \\
\text{Cys} \quad \text{Cys} \quad \text{Cys} \quad \text{Cys} \\
X-S \quad S-S \quad S-S \quad S-S \\
\text{Fe} \quad \text{Fe} \quad \text{Fe} \quad \text{Fe} \quad \text{Fe} \quad \text{Fe} \quad \text{Fe} \quad (aa)_{18} \\
S \quad S \quad S \quad S \quad S \quad S \quad S-X \\
\text{Cys} \quad \text{Cys} \quad \text{Cys} \quad \text{Cys} \\
(aa)_7 \quad (aa)_2 \quad (aa)_2 \quad (aa)_3
\end{array}
$$

Ferridoxin
15

acid sequences have been determined for many ferridoxin proteins, including alfalfa ferridoxin.[123] A rigorous statistical comparison of alfalfa and clostridial ferridoxins demonstrates a degree of amino acid sequence similarity clearly greater than would be expected on the basis of random occurrences. Although results support the possibility of a common evolutionary origin for all ferridoxins, without knowledge of the minimal number of residues required for function, such analyses cannot conclusively prove ancestral function.

17.2 NONREDOX COENZYMES

There are a multitude of enzymes containing nonredox coenzymes, in fact, more than the redox coenzymes, including thiamine pyrophosphate (vitamin B_1), pyridoxal phosphate (vitamin B_6), folic acid (pteroylglutamic acid), biotin, *S*-adenosylmethionine, vitamins A, D, E, and K, Coenzyme A, metal ions such as $Zn^{2\oplus}$, $Mg^{2\oplus}$, $Cu^{2\oplus}$, and $Co^{2\oplus}$ (most of which do not undergo easy electron transfer), the "acyl-carrier protein," adenosine monophosphate, diphosphate, and triphosphate, glutathione, and others. Most of these coenzymes will be covered here.[124] Corresponding nonenzymatic reactions have been discussed in Chapter 6.

17.2.1 Thiamine Pyrophosphate (Vitamin B_1)

Thiamine pyrophosphate, a coenzyme of universal occurrence in living systems, was originally detected as a nutritional factor required for the prevention of polyneuritis in birds and beriberi in man. The substance has been known for about 30 years and its structure is given in formula **16**. The

Thiamine pyrophosphate

16

crystalline vitamin (B_1) was first obtained in 1925 and, as can be seen from the structure, incorporates two heterocyclic rings, a pyrimidine ring and a thiazolium ring. This vitamin participates in three kinds of enzymatic reactions: (1) nonoxidative decarboxylation of α-keto acids; (2) oxidative decarboxylation of α-keto acids (which has been discussed before); and (3) the formation of α-ketols (acyloins).[125] Reaction 1 requires a thermostable organic cofactor in addition to an enzyme. Ugai et al. demonstrated a thiamine-dependent nonenzymatic reaction mechanistically analogous to (2) and (3) above.[126] Later, thiamine itself was shown to catalyze these reactions.[127]

A clue to the mechanism of these reactions is provided by the relatively specific but nonenzymatic catalyst, cyanide ion. This substance catalyzes the benzoin condensation reaction (eq. 17.24), which is similar to the enzymatic acetoin condensation described below; it also catalyzes the decarboxylation of α-keto acids such as pyruvic (eq. 17.25, where B is a base). In both eqs. 17.24 and 17.25 the negative charge in the intermediate is stabilized by the cyanide ion.

The thiazolium system contains two possible ylids (compounds containing a $C^\ominus$–$N^\oplus$ group) which are analogs of cyanide ion. Of these two possible compounds, Breslow showed by deuterium exchange studies that the ylid of thiazolium ion is formed in neutral solution (see later).[130] The ylid and cyanide ion are analogs of one another in that they are both resonance

$$\text{Ph-CHO} \underset{}{\overset{CN^{\ominus}}{\rightleftarrows}} \text{Ph-C(O}^{\ominus}\text{)(H)(CN)} \underset{B}{\overset{BH^{\oplus}}{\rightleftarrows}} \text{Ph-C(OH)(H)(CN)} \underset{BH^{\oplus}}{\overset{B}{\rightleftarrows}}$$

$$\left[\text{Ph-C}^{\ominus}(\text{OH})(\text{CN}) \longleftrightarrow \text{Ph-C(OH)=C=N}^{\ominus} \right] \overset{PhCHO}{\rightleftarrows} \text{Ph-C(HO)(CN)-C(O}^{\ominus}\text{)(H)(Ph)} \rightleftarrows$$

$$\text{Ph-C(O}^{\ominus}\text{)(CN)-C(OH)(H)(Ph)} \underset{}{\overset{CN^{\ominus}}{\rightleftarrows}} \text{Ph-C(=O)-C(OH)(H)(Ph)} \qquad (17.24)$$

hybrids in which one of the canonical forms has a neutral six-electron carbon atom (a carbene). As in cyanide ion, the thiazolium ion ylid would be expected to be a good nucleophile and also to stabilize a negative charge, giving the catalyst a dual functionality.

$$\text{CH}_3\text{-C(=O)-CO}_2^{\ominus} \overset{CN^{\ominus}}{\rightleftarrows} \text{CH}_2\text{-C(O}^{\ominus}\text{)(CN)-C(=O)(O}^{\ominus}\text{)} \underset{B}{\overset{BH^{\oplus}}{\rightleftarrows}} \text{CH}_2\text{-C(OH)(C≡N)-C(=O)(O}^{\ominus}\text{)} \rightleftarrows$$

$$\left[\text{CH}_3\text{-C(OH)=C=N}^{\ominus} \longleftrightarrow \text{CH}_3\text{-C}^{\ominus}(\text{OH})(\text{C≡N}) \right] \underset{B}{\overset{BH^{\oplus}}{\rightleftarrows}} \text{CH}_3\text{-CH(OH)(CN)} \underset{BH^{\oplus}}{\overset{B}{\rightleftarrows}}$$

$$\text{CH}_3\text{-CH(O}^{\ominus}\text{)(CN)} \underset{CN^{\ominus}}{\rightleftarrows} \text{CH}_3\text{-CH(=O)} \qquad (17.25)$$

The question arises as to which heterocyclic ring (or both or none) is involved in thiamine reactions. Westheimer thought the pyrimidine ring was involved,[129] but Breslow conclusively showed that the thiazolium ring is involved.[130,131] He showed by nmr measurements that the 2-position of the thiazolium ring exchanges deuterium with the solvent. In other words, there is a carbanion formed (at pH 5) which can catalyze aldehyde reactions by stabilizing the carbanion formed. This is analogous to catalysis by cyanide ion. In fact the thiazolium ring or thiamine can be called "biological cyanide," since, in the decarboxylation at least, the reactions are strictly analogous.

The question may be legitimately asked as to whether the same thing

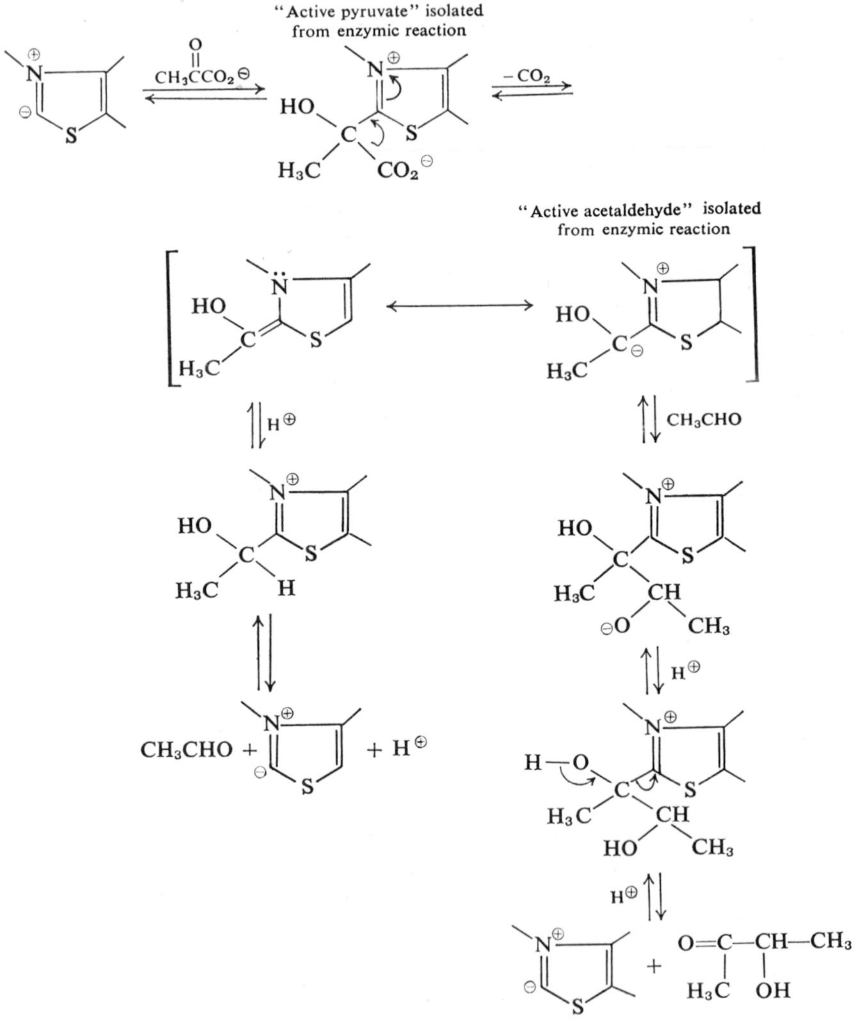

Fig. 17.15 Decarboxylation and acetoin condensation of pyruvate with the enzyme carboxylase.

happens in enzyme systems. The answer to this question is a decided yes;[132] "active pyruvate" and "active acetaldehyde" were detected in the decarboxylation and oxidation of pyruvate.[133] These are indicated in Fig. 17.15. In fact, "active acetaldehyde" (2-α-hydroxyethylthiamine) has been isolated from an incubation of pyruvic acid with pyruvic oxidase showing that it is an intermediate in this enzymatic process,[134] has been synthesized,[135] and has been isolated from *Escherichia coli*.[136]

In addition to the decarboxylation reaction, thiamine also participates in transketolase and phosphoketolase reactions, as shown in Figs. 17.16 and 17.17, respectively. In the decarboxylation reaction, an α-keto acid is transformed to a β-keto acid acid, which decarboxylates more readily. Furthermore, a positive nitrogen electron sink is introduced, which speeds the decarboxylation. In the transketolase reaction, a reverse aldol condensation is followed by an aldol condensation. In the phosphoketolase reaction, the enzyme must participate in the removal of an hydroxyl group and there must be stereospecific proton removal.

Acylthiazolium salts are labile. For example, 2-benzoylthiazolium cation in methanol solvent gives methyl benzoate immediately. Also, 2-acetylthiamine and phosphate ion give acetyl phosphate.[137] Furthermore, the hydrolysis of 2-benzoyl-3,4-dimethylthiazolium iodide is very exergonic,[138,139]

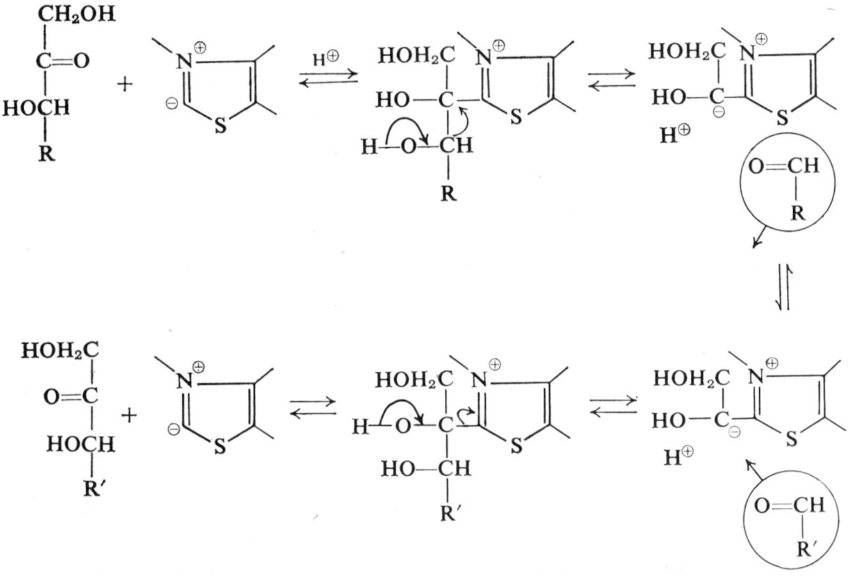

Fig. 17.16 Transketolase: fructose-6-phosphate + glyceraldehyde-3-phosphate → tetrose phosphate + xylulose-5-phosphate.

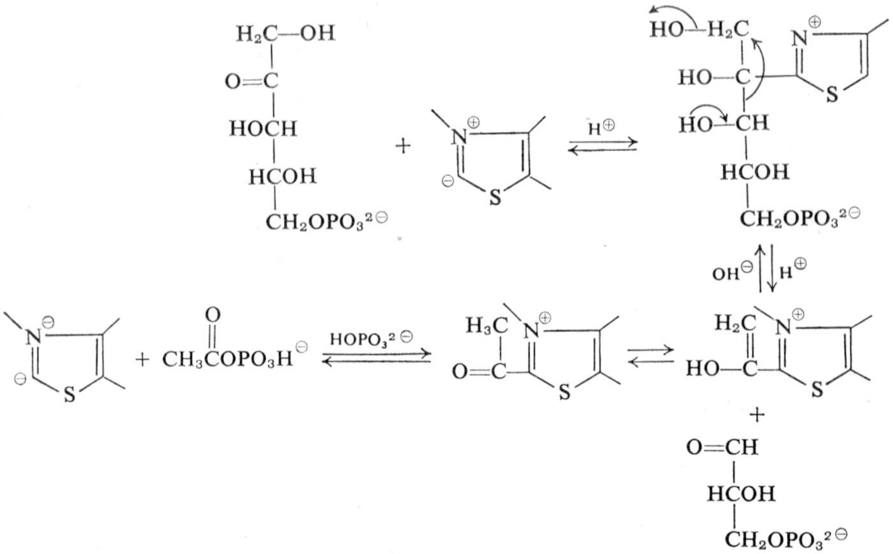

Fig. 17.17 Phosphoketolase: xylulose-5-phosphate + $HOPO_3^{2\ominus}$
→ glyceraldehyde-3-phosphate + acetyl phosphate.

and the hydrolysis of 2-acetyl-3,4-dimethylthiazolium ion is rapid; under certain conditions it is general base-catalyzed.[140]

A model for the formation of the acetyl group in the phosphoketolase reaction was discovered, involving a derivative of *ortho*-phenylenediamine.[141] In the transketolase reaction, a thiamine pyrophosphate-glycolaldehyde intermediate has been detected, using radioactive measurements.[142]

Thiamine has been implicated in other reactions, specifically the decarboxylation of oxalyl coenzyme A[143] and pyruvate[144] and the carboligase reaction (glyoxylate and carbon dioxide to hydroxymalonate).[145]

Many analogs of thiazolium ion and thiamine have been investigated[146-148] and the chemistry of thiamine has been investigated.[149] There is an enzyme thiaminase that decomposes it.[150]

17.2.2 Pyridoxal Phosphate (Vitamin B_6)

Pyridoxal phosphate (Vitamin B_6) represents the epitome of coenzyme action. It is involved in many transformations of amino acids, specifically racemization, decarboxylation, transamination, β-substitution, elimination, and condensation, among others. The structure of pyridoxal phosphate is shown in formula **17**. Other forms of the vitamin exist, namely the alcohol (pyridoxine or pyridoxol), and the corresponding amine (pyridoxamine).

The question may be asked as to the function of the various groups in pyridoxal phosphate. The phosphate group is probably involved in electrostatic binding to the enzyme. On the other hand, methylpyridoxal, in which the phenolic hydroxyl group is replaced by a methyl group, does not catalyze transamination or dehydration.[151] Thus the phenolic hydroxyl group appears to be essential. In addition it appears that the aldehyde group of pyridoxal phosphate forms a Schiff base with an amino group of the enzyme.

$$\text{H}_2\text{O}_3\text{POCH}_2 - \text{pyridine ring with CHO, OH, CH}_3$$

Pyridoxal phosphate

17

Probably the best piece of evidence for this conclusion comes from the inhibition of many pyridoxal phosphate enzymes by carbonyl reagents.[151] On the other hand, neither the hydroxymethyl group nor the methyl group of pyridoxal appears to be essential. The nitrogen atom in the ring serves as an electron sink and source and can be replaced by an *ortho*-nitrogen atom[151] which will have the same conjugation, or by an *ortho* or *para*-nitro group, but not by a *meta*-nitro group. This is reinforcement both for the conjugation and the role of electron sink and source.

The implication is that these reactions proceed via Schiff bases. There many enzymes that are known to react via Schiff bases. Among them are dehydratase,[152] aldolase,[153] acetoacetate decarboxylase,[154] and pyrrole synthetase.[155] In a number of cases, the Schiff base intermediate has been trapped by borohydride reduction.[156] The first indication that a Schiff base occurred in pyridoxal reactions was a very old observation that amino acids and carbonyl compounds led to incorporation of deuterium from a deuterated solvent.[157] This is a rather indirect argument, but there appears to be no good way to explain the incorporation of deuterium other than by formation of a Schiff base and tautomerization. Since a coenzyme is a substrate that is regenerated, and pyridoxal phosphate conforms to this generalization, the extension of this idea to the pyridoxal case would involve the following, where PyCHO is the aldehyde form and PyCH$_2$NH$_2$ is the amine form of the coenzyme.

Equation 17.26 of course applies to the transamination reaction; this is the crux of the pyridoxal phosphate coenzyme reactions. A mechanism for this reaction is depicted in Fig. 17.18 where the wavy lines stand for the surface of the enzyme. The question may be asked as to what the enzyme actually does. For one thing it probably binds the coenzyme through

$$\text{PyCHO} + \text{R}-\underset{\underset{\text{NH}_2}{|}}{\overset{\overset{\text{H}}{|}}{\text{C}}}-\text{CO}_2\text{H} \rightleftharpoons \left[\text{PyCH}=\text{N}-\underset{\underset{\text{CO}_2\text{H}}{\diagdown}}{\overset{\overset{\text{R}\quad\text{H}}{\diagup}}{\text{C}}} \right] \rightleftharpoons$$

$$\left[\text{PyCH}_2-\text{N}=\underset{\underset{\text{CO}_2\text{H}}{|}}{\overset{\overset{\text{R}}{|}}{\text{C}}}-\text{CO}_2\text{H} \right] \rightleftharpoons \underset{\underset{\text{O}}{\|}}{\overset{\overset{\text{R}}{|}}{\text{C}}}-\text{CO}_2\text{H} + \text{PyCH}_2\text{NH}_2$$

$$\text{PyCH}_2\text{NH}_2 + \text{R}'-\overset{\overset{\text{O}}{\|}}{\text{C}}-\text{CO}_2\text{H} \rightleftharpoons \text{PyCH}_2\text{N}=\underset{\underset{\text{R}'}{|}}{\overset{\overset{\text{R}'}{|}}{\text{C}}}-\text{CO}_2\text{H} \rightleftharpoons$$

$$\text{PyCH}=\text{N}-\underset{\underset{\text{H}}{|}}{\overset{\overset{\text{R}'}{|}}{\text{C}}}-\text{CO}_2\text{H} \rightleftharpoons \text{PyCHO} + \text{R}'-\underset{\underset{\text{NH}_2}{|}}{\overset{\overset{\text{H}}{|}}{\text{C}}}-\text{CO}_2\text{H}$$

$$\text{R}-\underset{\underset{\text{NH}_2}{|}}{\overset{\overset{\text{H}}{|}}{\text{C}}}-\text{CO}_2\text{H} + \text{R}'-\overset{\overset{\text{O}}{\|}}{\text{C}}-\text{CO}_2\text{H} \rightleftharpoons \text{R}'-\underset{\underset{\text{NH}_2}{|}}{\overset{\overset{\text{H}}{|}}{\text{C}}}-\text{CO}_2\text{H} + \text{R}\overset{\overset{\text{O}}{\|}}{\text{C}}\text{CO}_2\text{H} \quad \text{Total}$$

(17.26)

hydrophobic and electrostatic interactions, as mentioned before. In model systems a metal ion may be required and the initial formation of a metal chelate, as shown in formula **18**, was postulated, the purpose of the metal

Metal chelate

18

ion presumably being to labilize the pi electron system. Subsequently it was found that an acid and base acting together can more efficiently cause the tautomerization.[159-162] The enzyme could perform these functions.

If indeed a Schiff base is formed, and all indications point to this occurrence, the Schiff base between an amino acid and pyridoxal phosphate could also lead to decarboxylation, as shown in Fig. 17.19, or racemization, as shown in Fig. 17.20. Thus the cleavage of different bonds in the Schiff base

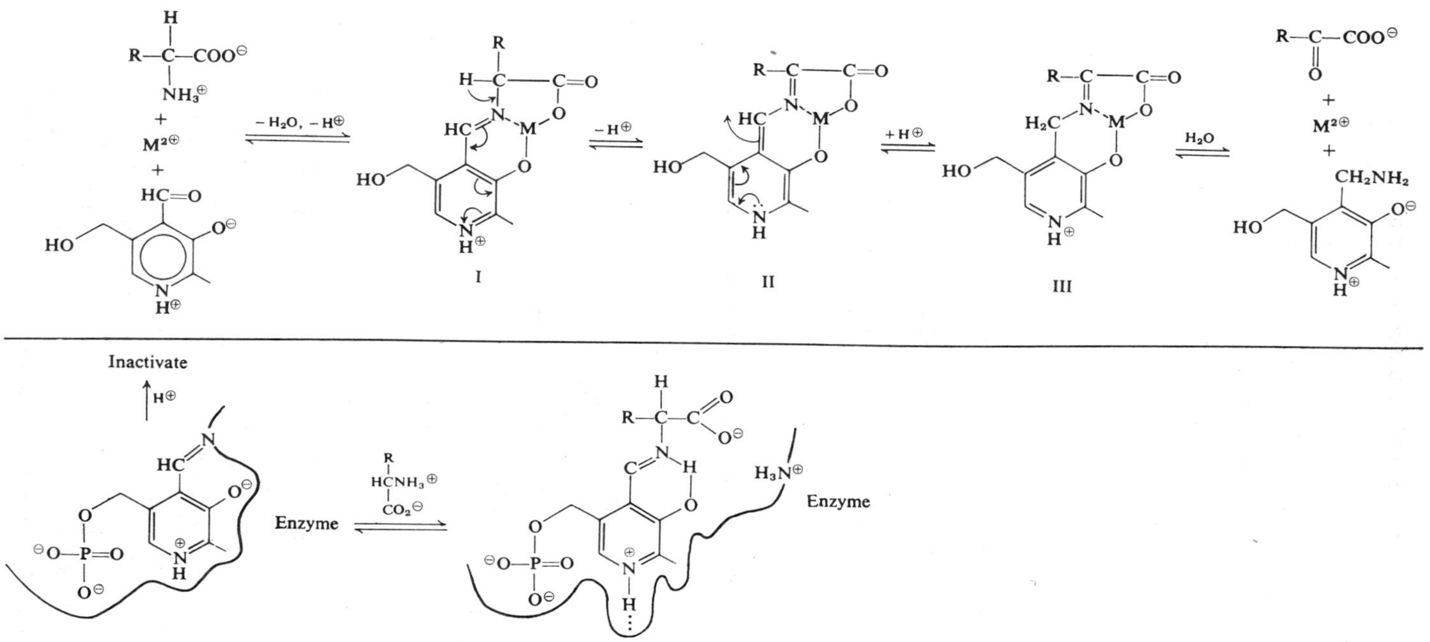

Fig. 17.18 Coenzyme reactions of pyridoxal: transamination.

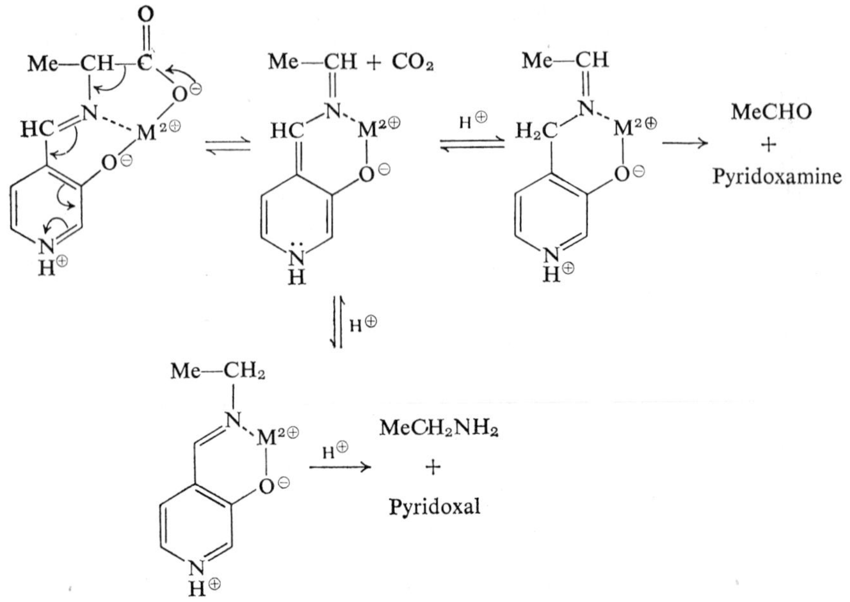

Fig. 17.19 Coenzyme reactions of pyridoxal: decarboxylation.

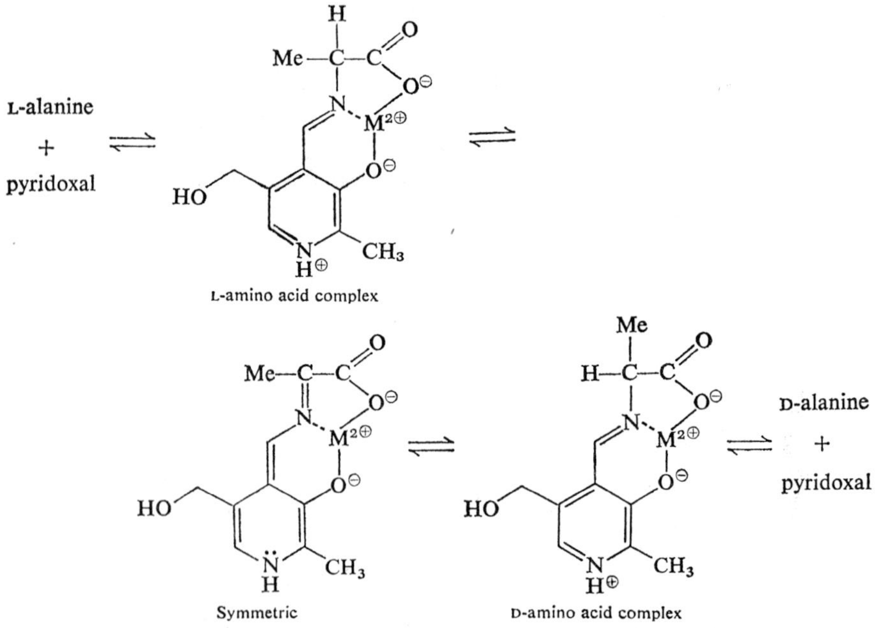

Fig. 17.20 Coenzyme reactions of pyridoxal: racemization.

can lead to transamination, a in **18**, decarboxylation, b in **18**, or condensation, c in **18**. An elimination reaction is shown in Fig. 17.21; it has been shown that a carbanion is formed in the further conversion of the product to α-ketobutyrate.[163] A typical β-substitution that leads to the formation of an important amino acid is shown in Fig. 17.22, and a condensation reaction is shown in Fig. 17.23. In these mechanisms, a metal ion is again shown as chelator and labilizer of the pi electron system, but presumably this could be better done by acid-base catalysis. The racemization, transamination, and dehydration of serine (which leads eventually to a new amino acid, tryptophan) are suggested to involve the ionization of the proton attached to the α-carbon of the amino acid moiety in the Schiff base chelate. Readdition of the proton to the same site in this optically inactive intermediate, followed by hydrolysis, completes the racemization reaction. Readdition of the proton to the original carbonyl carbon atom yield transamination.

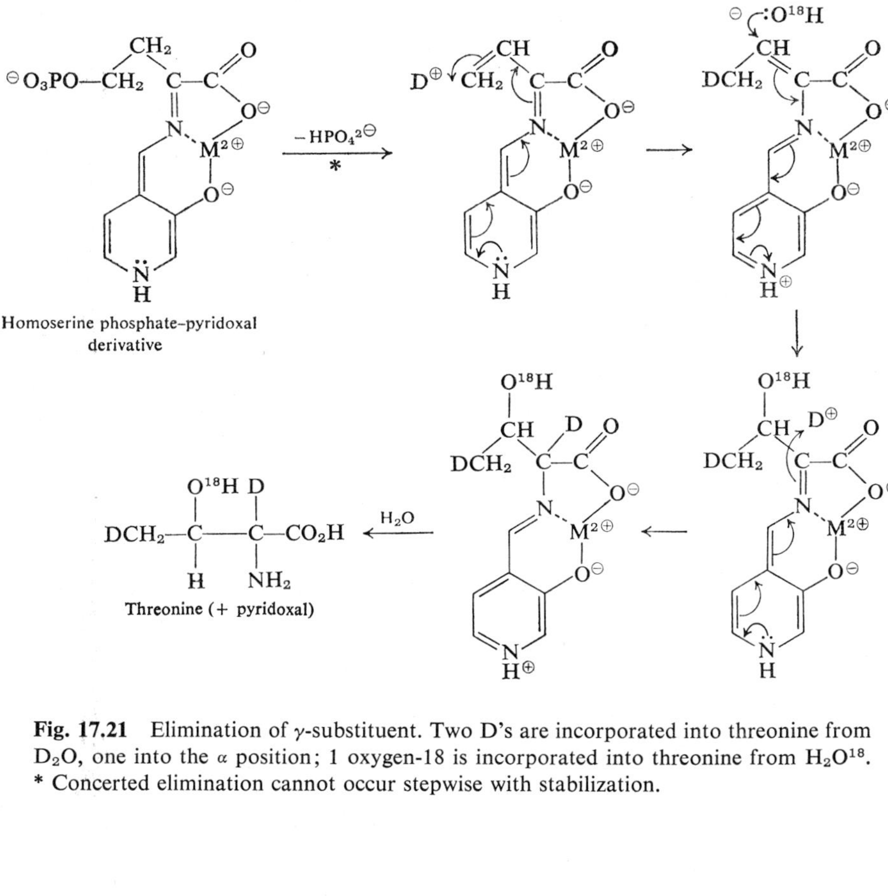

Fig. 17.21 Elimination of γ-substituent. Two D's are incorporated into threonine from D_2O, one into the α position; 1 oxygen-18 is incorporated into threonine from H_2O^{18}.
* Concerted elimination cannot occur stepwise with stabilization.

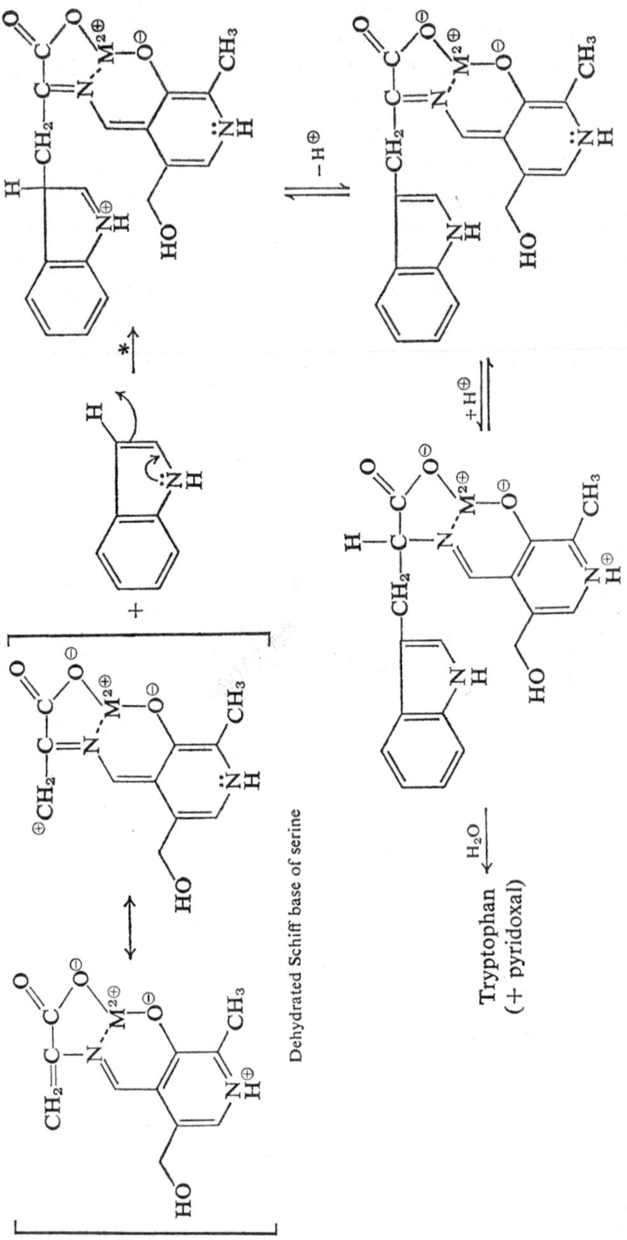

Fig. 17.22 Coenzyme reactions of pyridoxal: serine + indole → tryptophan (β-substitution).
* Enamine adding in a Michael reaction to an α,β-unsaturated system.

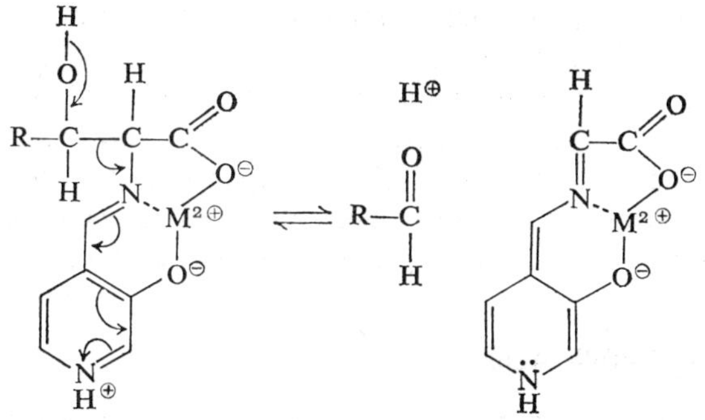

Fig. 17.23 Coenzyme reactions of pyridoxal: condensation.

The overall transamination reaction may be completed by reversal of this process. The desulfurization of homocysteine, a β,γ-elimination reaction, may be formulated in the same general fashion, although the reaction is somewhat more complicated. The first two steps, identical to those for transamination, are followed by a tautomerization, elimination, another tautomerization, and finally hydrolysis. The condensation of serine and indole to yield tryptophan involves addition of the nucleophilic reagent indole to the unsaturated intermediate in serine dehydration, followed by tautomerization and hydrolysis. Although most of the reactions have not been the subject of rigorous kinetic investigation, the internal consistency of the mechanisms augments the plausibility of the whole scheme. In pyridoxal-dependent nonenzymatic transamination, a kinetic investigation has fully corroborated the suggested mechanism and has revealed, in addition, that this reaction involves rapid pre-equilibrium Schiff base formation followed by rate-determining tautomeric rearrangement (Chapter 6).[164]

Pyridoxal-dependent enzymatic and nonenzymatic reactions differ in important respects. Most of the enzymatic reactions do not require metal ions, but the nonenzymatic reactions may depend on them. The enzymatic reactions are specific with regard to both substrate and type of reaction catalyzed, but the nonenzymatic reactions are nonspecific. The enzymatic reactions occur more rapidly than do their nonenzymatic counterparts. Finally, the enzymatic reactions involve a transamination reaction, rather than Schiff base formation, as the initial step.[165] This speeds the enzymatic reaction, since transamination proceeds faster than Schiff base formation.[166]

An elegant kinetic study of the glutamic-aspartic transaminase reactions, utilizing temperature jump techniques, has confirmed that the pyridoxal phosphate-dependent enzymatic reactions proceed via a reaction pathway similar to that for the nonenzymatic reactions.[167] The two tautomeric Schiff bases were identified as reaction intermediates and their rates of formation, interconversion, and decomposition were measured. The rate-determining step was found to be the interconversion of the two Schiff bases. A study of the turnover of this reaction gave rate constants approximately one-half that of the temperature jump study.[168] Perhaps a difference in concentration can explain the difference in rate constants. The stereochemistry of transaminating systems is being investigated by deuterium tracer methods.[169]†

17.2.3 Tetrahydrofolic Acid

Tetrahydrofolic acid (FH_4) is composed of a reduced pteridine, p-aminobenzoic acid, and L-glutamic acid. Its structure is shown in formula **19**.

Tetrahydrofolic acid (FH_4)

19

FH_4 coenzymes are involved in the transfer of one-carbon fragments at the oxidation level of formate, formaldehyde, or methanol. At the oxidation level of formate, the adducts may exist in four structural forms: N^{10}-formyl FH_4, N^5-formyl FH_4, N^5-formimino FH_4, and N^5,N^{10}-methenyl FH_4. The syntheses and interconversions of the formyl derivatives of FH_4 are summarized in Fig. 17.24. N^{10}-Formyl-FH_4 can be synthesized from FH_4 and formate in the presence of FH_4 formylase and adenosine triphosphate. The reaction probably proceeds through a phosphorylated FH_4 enzyme intermediate.[170-173]

In contrast to transfer reactions of one-carbon fragments at the level of oxidation of formate, in which several FH_4 coenzymes are involved, such reactions at the formaldehyde level of oxidation apparently involve only one coenzyme, N^5,N^{10}-methylene FH_4.[174] The structure of this material is shown in formula **20**. The principal sources of this coenzyme are the reduction of the corresponding methenyl compound and an amino acid source to be mentioned later. A chemical synthesis from formaldehyde and FH_4 is also possible. The reduction of the methylene coenzyme by NADH or

† Aspartate-glutamate transaminase is reported to consist of two enzymes (M. Martinez-Carrion, *Biomed. News*, November 1970, p. 12.

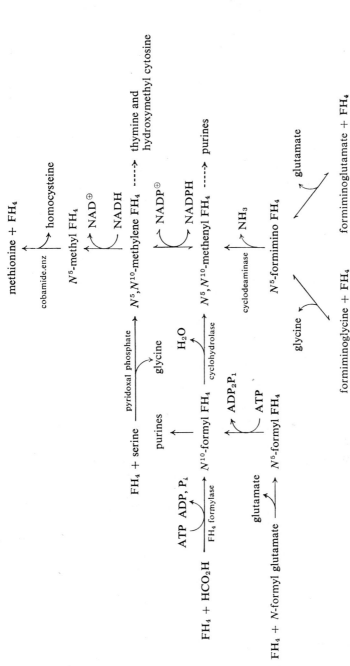

Fig. 17.24 Metabolic interconversions of tetrahydrofolate coenzymes. From H. R. Mahler and E. H. Cordes, *Biological Chemistry*, Harper and Row, New York, 1966, p. 353.

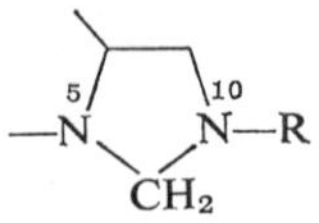

$$\underset{\substack{\text{N^5,N^{10}-Methylene FH}_4\\\mathbf{20}}}{-\text{N}\underset{\text{CH}_2}{\overset{5\diagup\diagdown 10}{}}\text{N}-\text{R}}$$

NADPH through the intermediacy of a flavoprotein yields N^5-methyl FH$_4$ (formula **21**).[175,176] Thus there are many interconversions possible in the oxidation and reduction of activated one-carbon units as shown in Fig. 17.25.[177]

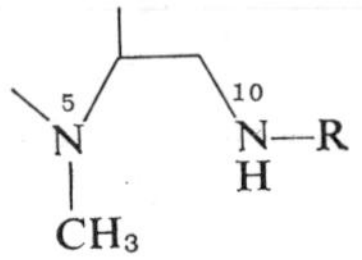

$$\underset{\substack{N^5\text{-Methyl FH}_4\\\mathbf{21}}}{\underset{\text{CH}_3}{\overset{5\diagup\diagdown 10}{\text{N}}}\overset{}{\underset{\text{H}}{\text{N}}}-\text{R}}$$

Probably the most significant model for the methylene derivative consists of the reaction of an N,N'-diarylethylenediamine with formaldehyde. The resulting compound can act as a transhydroxymethylating agent or can be reversibly oxidized to the corresponding unsaturated compound which can serve as a formylating agent.[178]

N^{10}-Formyl FH$_4$ transfers its one-carbon fragment in the synthesis of purines. N^5,N^{10}-Methenyl FH$_4$ is involved in the synthesis of the amino acid serine from the amino acid glycine. This has been alluded to before. In this process, pyridoxal phosphate activates the amino acid (by Schiff base formation) and FH$_4$ activates the formaldehyde. Such a carbon–carbon condensation in an organic system would necessitate the use of strong conditions, but in the enzymatic system, only mild conditions are required. N^5-Methyl FH$_4$ serves as the methyl group donor for the biosynthesis of methionine from homocysteine.

17.2.4 Biotin

Biotin, a growth factor for both yeast and humans (Vitamin H) was first isolated by Kögl and the structure shown in formula **22** was established by du Vigneaud. Pioneering studies by Lardy and others strongly suggested that biotin functions as a coenzyme for carboxylation reactions.[179] These

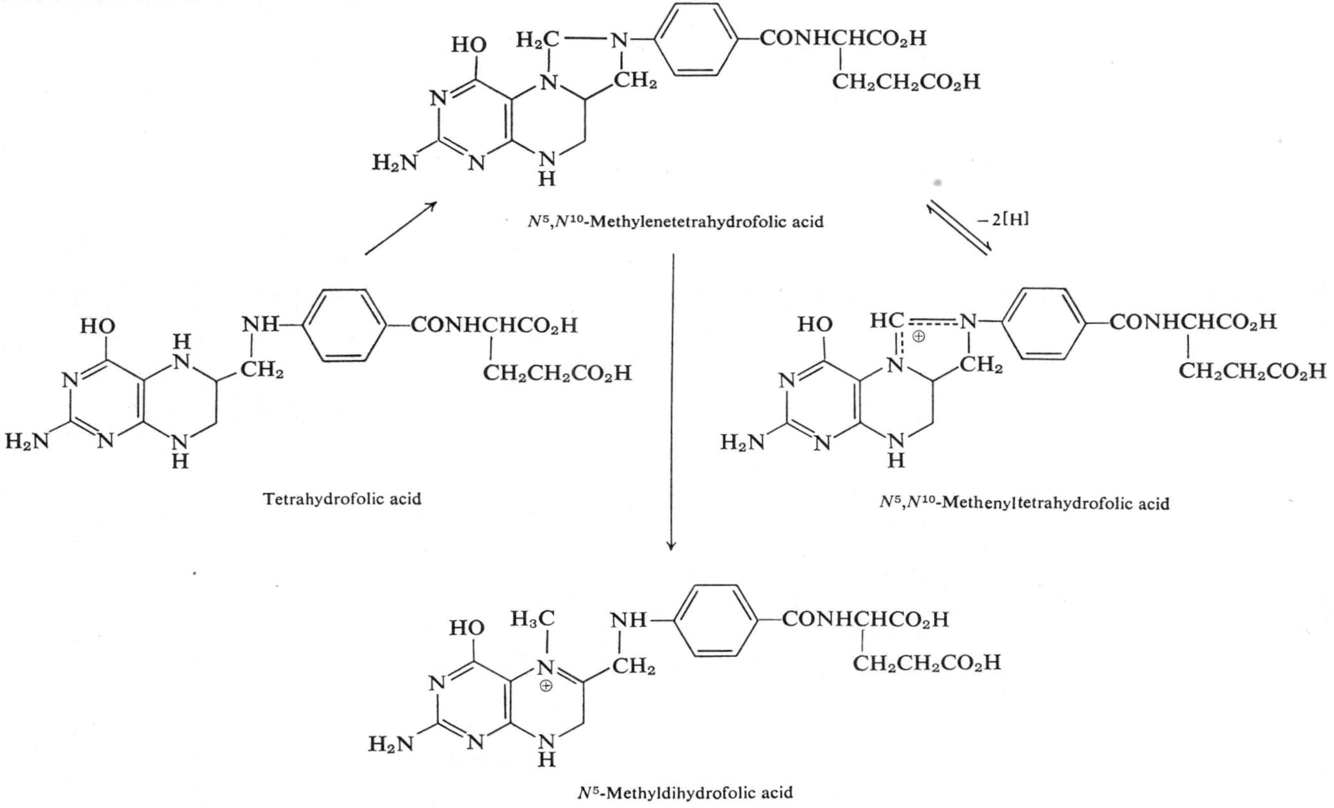

Fig. 17.25 Oxidation and reduction of activated one-carbon units. From L. Jaenicke, in *The Mechanism of Action of Water-soluble Vitamins*, A. V. S. de Reuck and M. O'Connor, Eds., Little, Brown and Co., Boston, 1961, p. 44.

experimental studies were preceded by the suggestion that biotin was involved in carbon dioxide fixation.[180]

$$\text{Biotin structure: imidazolidinone fused with tetrahydrothiophene, bearing } CH_2CH_2CH_2CH_2CO_2H$$

Biotin
22

Biotin serves as coenzyme for several types of enzymatic reactions. One class includes ATP-dependent carboxylations, with the cleavage of ATP to ADP and inorganic phosphate.[181–184]

Another type of biotin-dependent enzymatic reaction is exemplified by the methylmalonate-oxaloacetate transcarboxylase reaction. That is, a carboxyl group is exchanged between two substrates.[185]

Still another role for biotin is the synthesis of carbamyl phosphate.[186]

Biotin is bound to the ϵ-amino group of lysine of the enzyme through the carboxyl group on the side chain of the coenzyme.

The key to the understanding of biotin came about when free (+)-biotin was incubated with β-methylcrotonyl carboxylase in the absence of substrates other than carbon dioxide.[187] The result of this incubation was the production of carboxybiotin, a labile compound, whereas the corresponding dimethyl ester is not labile. These results can be explained only by the sequence shown in eq. 17.27.

$$CO_2 + ATP + \text{biotinylenzyme} \longrightarrow ADP + P_i + \text{carboxybiotinylenzyme}$$
$$\text{carboxybiotinylenzyme} + \text{substrate} \longrightarrow$$
$$\text{biotinylenzyme} + \text{carboxysubstrate} \quad (17.27)$$

In the initial step of this sequence, biotin on the enzyme reacts with ATP-activated carbon dioxide (or bicarbonate) to form carboxybiotin. This product, often thought to be N-carboxylated, is in fact O-carboxybiotin; this intermediate subsequently acts as a carboxylating agent, usually for an acyl coenzyme A.[188] Biotin probably functions as a nucleophilic catalyst in these reactions, and the utilization of this cofactor may provide the enzyme with certain catalytic advantages as compared with a path involving the direct reaction of an acyl coenzyme A with ATP-activated carbon dioxide or bicarbonate. In carbamyl phosphate synthesis, a carbamate-biotin-enzyme complex was postulated in addition to the intermediate shown earlier.

Nonredox Coenzymes

In all reactions dependent on biotin, the protein avidin (from egg white) inhibits the enzyme. This inhibition can often be overcome by adding biotin. Furthermore, biotin when added to the enzyme gives enhanced reactivity. In certain instances analysis can reveal the presence of biotin per se.

In particular the demonstration of the intermediate carboxylated biotinylenzyme is based on experiments with $(1,3,5\text{-}^{14}\text{C})\beta$-methylglutaconyl coenzyme A which is formed from $(1,3,5\text{-}^{14}\text{C})\beta$-hydroxy-$\beta$-methylglutaryl coenzyme A in the presence of methylglutaconase. When this compound is incubated with unlabeled β-methylcrotonyl coenzyme A and biotinylenzyme, it gives rise to labeled β-methylcrotonyl coenzyme A. These two reaction sequences are depicted in Fig. 17.26.[187]

2-Imidazolidone was used as an analog of biotin in an attempt to elucidate the enzymatic mechanism by which this cofactor is carboxylated and subsequently transfers a carboxyl group to a nucleophilic agent. Studies of the carboxylation reaction were carried out by investigating the reactions of 2-imidazolidone with activated acyl compounds that serve as models for ATP-activated bicarbonate. Model studies of the transcarboxylation from an N-carboxybiotin were carried out. Infrared and rate studies indicate that N-carboxybiotin is not particularly nucleophilic. In addition to the usual metal ion catalysis of biotin reactions which involves magnesium ion,[187] calcium ion was also found to serve in this model system.[189] On the other hand, Cu(II) and Mn(II) prevent the decarboxylation of N-carboxy-2-imidazolidone, and the hydrolysis of the imidazolidone anion is insensitive to general acid or general base catalysis.[190] These results imply that biotin must be carboxylated on oxygen.

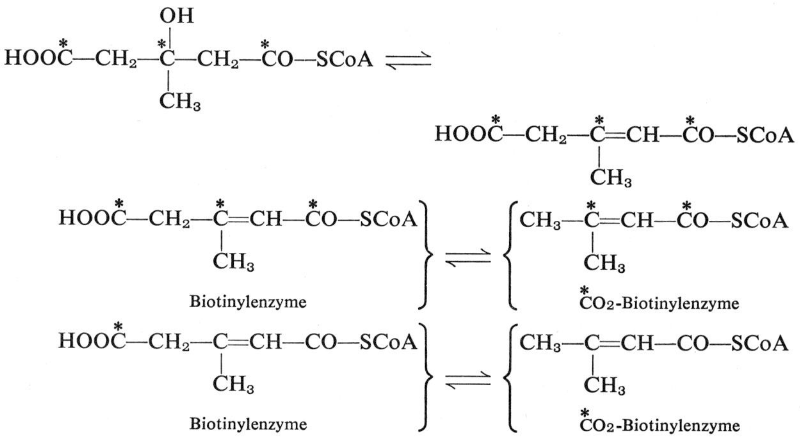

$$\overset{*}{\text{HOOC}}-\text{CH}_2-\overset{*}{\underset{\underset{\text{CH}_3}{|}}{\overset{\overset{\text{OH}}{|}}{\text{C}}}}-\text{CH}_2-\overset{*}{\text{CO}}-\text{SCoA} \rightleftharpoons$$

$$\overset{*}{\text{HOOC}}-\text{CH}_2-\overset{*}{\underset{\underset{\text{CH}_3}{|}}{\text{C}}}=\text{CH}-\overset{*}{\text{CO}}-\text{SCoA}$$

$$\left\{ \overset{*}{\text{HOOC}}-\text{CH}_2-\overset{*}{\underset{\underset{\text{CH}_3}{|}}{\text{C}}}=\text{CH}-\overset{*}{\text{CO}}-\text{SCoA} \atop \text{Biotinylenzyme} \right\} \rightleftharpoons \left\{ \text{CH}_3-\overset{*}{\underset{\underset{\text{CH}_3}{|}}{\text{C}}}=\text{CH}-\overset{*}{\text{CO}}-\text{SCoA} \atop \overset{*}{\text{CO}}_2\text{-Biotinylenzyme} \right\}$$

$$\left\{ \overset{*}{\text{HOOC}}-\text{CH}_2-\underset{\underset{\text{CH}_3}{|}}{\text{C}}=\text{CH}-\text{CO}-\text{SCoA} \atop \text{Biotinylenzyme} \right\} \rightleftharpoons \left\{ \text{CH}_3-\underset{\underset{\text{CH}_3}{|}}{\text{C}}=\text{CH}-\text{CO}-\text{SCoA} \atop \overset{*}{\text{CO}}_2\text{-Biotinylenzyme} \right\}$$

Fig. 17.26 From F. Lynen, J. Knappe, and E. Lorch, in *The Mechanism of Action of Water-soluble Vitamins*, A. V. S. de Reuck and M. O'Connor, Eds., Little, Brown, and Co., Boston, 1961, p. 80.

17.2.5 S-Adenosylmethionine

Methionine has been known to be a biochemical "methyl donor" for a number of years and the sulfonium salt S-adenosylmethionine has been demonstrated to be the compound corresponding to "active" methionine which performs the transfer of a methyl group to various nucleophiles. A typical reaction is given (eq. 17.28), which illustrates the synthesis of creatine from guanidylacetic acid:

$$\text{Adenosine}-\overset{\oplus}{\underset{CH_3}{S}}-CH_2CH_2\underset{NH_2}{C}HCOO^{\ominus} + NH_2-\overset{\overset{\oplus}{NH_2}}{\underset{}{C}}-NH-CH_2COO^{\ominus} \longrightarrow$$

$$\text{Adenosine}-S-CH_2CH_2\underset{NH_3^{\oplus}}{C}HCOO^{\ominus} + NH_2-\overset{\overset{\oplus}{NH_2}}{\underset{}{C}}-\underset{CH_3}{N}-CH_2COO^{\ominus}$$

(17.28)

This reaction in organic chemical terms is the reaction of a nucleophile (amine) with a sulfonium salt yielding a sulfide and a substituted amine. There are many similar reactions of S-adenosylmethionine with other nucleophiles such as nicotinamide, imidazole, histamine, and catechols of the epinephrine type.[191] Since the sulfonium salt, S-adenosylmethionine, has three different alkyl groups attached to the sulfur atom, it is conceivable that nucleophiles will react with groups other than the methyl group. An enzymatic reaction of this kind has been observed in the conversion of S-adenosylmethionine to α-amino-γ-butyrolactone (eq. 17.29); this reaction presumably proceeds by intramolecular nucleophilic attack of carboxylate ion on the four-carbon chain attached to the sulfur atom (eq. 17.29).[192] Other reactions of sulfonium salts include the processes of dimethylthetin, homocysteine, and spermidine formation.

$$\text{Adenosine}-\overset{\oplus}{\underset{CH_3}{S}}-CH_2 \quad \overset{O^{\ominus}}{\underset{CH_2-CHNH_3^{\oplus}}{\diagdown}}C=O \longrightarrow$$

$$\text{Adenosine}-S-CH_3 + \underset{H_2C-CHNH_3^{\oplus}}{\overset{H_2C-O-C=O}{|\quad\quad\quad|}} \quad (17.29)$$

Quite analogous to the reactions of sulfonium compounds are the reactions of ammonium compounds with nucleophiles, such as the reaction of thiamine and amines such as *m*-nitroaniline, *m*-aminobenzoic acid, or pyridine catalyzed by the enzyme carp thiaminase (eq. 17.30).[193]

$$\text{thiamine} + \text{amine} \longrightarrow \text{pyrimidine-CH}_2\text{-amine} + \text{thiazole} + H^{\oplus} \quad (17.30)$$

Methionine can be formed from homocysteine in two different ways. One has been mentioned before (from methyl-FH$_4$ and an enzyme). The other way is through the use of betaine as a methyl donor:

$$\underset{\underset{NH_3^{\oplus}}{|}}{HSCH_2CH_2CHCO_2^{\ominus}} + \underset{\text{Betaine}}{(CH_3)_3\overset{\oplus}{N}CH_2CO_2^{\ominus}} \xrightarrow{\text{enzyme}}$$

$$\text{methionine} + \text{dimethylglycine} \quad (17.31)$$

S-Adenosylmethionine can be synthesized from methionine and ATP according to the following:

$$\text{adenosyl-O-CH}_2\text{O-P(O)(OH)-O-P(O)(OH)-O-P(O)(OH)-OH} + CH_3SCH_2CH_2\underset{\underset{NH_3^{\oplus}}{|}}{CHCO_2^{\ominus}} \longrightarrow$$

$$\text{adenosyl-}\overset{\oplus}{\underset{\underset{CH_3}{|}}{S}}\text{-CH}_2CH_2\underset{\underset{NH_3^{\oplus}}{|}}{CHCO_2^{\ominus}} + \text{phosphate} + \text{pyrophosphate} \quad (17.32)$$

The enzyme mechanism might be a single or double displacement, but there is no evidence for methyl transfer to the enzyme so that we will assume a direct reaction of ATP with the nucleophile on the enzyme surface.

One of the reactions of *S*-adenosylmethionine which must be viewed with suspicion is shown in eq. 17.33, postulated since it is known that this compound undergoes a spontaneous as well as an enzymatic reaction.

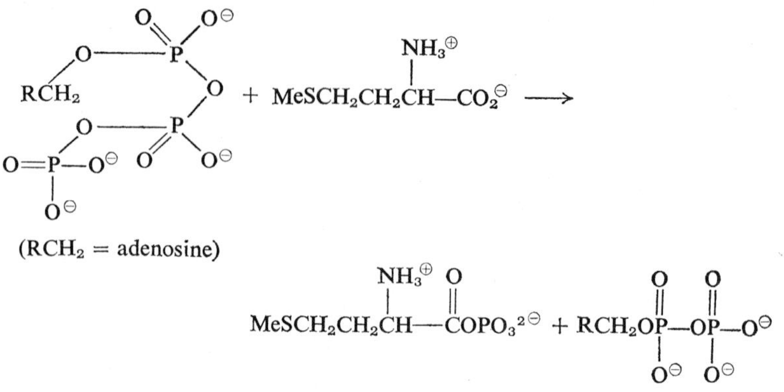

$$\begin{array}{c} \text{RCH}_2\text{-O-P(=O)(O}^{\ominus}\text{)-O-P(=O)(O}^{\ominus}\text{)-O-P(=O)(O}^{\ominus}\text{)O}^{\ominus} + \text{MeSCH}_2\text{CH}_2\overset{\overset{\displaystyle\text{NH}_3^{\oplus}}{|}}{\text{CH}}-\text{CO}_2^{\ominus} \longrightarrow \end{array}$$

(RCH$_2$ = adenosine)

$$\text{MeSCH}_2\text{CH}_2\overset{\overset{\displaystyle\text{NH}_3^{\oplus}}{|}}{\text{CH}}-\overset{\overset{\displaystyle\text{O}}{\|}}{\text{C}}\text{OPO}_3^{2\ominus} + \text{RCH}_2\text{OP(=O)(O}^{\ominus}\text{)-OP(=O)(O}^{\ominus}\text{)-O}^{\ominus}$$

(17.33)

17.2.6 Coenzyme A

Coenzyme A is involved in acyl group transfer. It was identified as a heat-stable cofactor required for certain acetylations,[194,195] and was implicated as a cofactor for the incorporation of acetate into acetoacetate and citrate.[188,189] The structure shown in formula **23** was elucidated by at least three people.[195] This molecule contains a multiplicity of functional groups, but the most significant group from the point of view of mechanism is the sulfhydryl group. It is this sulfhydryl group that is involved in the formation of acetyl coenzyme A.[200,201] In other words, acyl derivatives of coenzyme A are thiol esters. In fact, coenzyme A is often abbreviated as CoASH. One may ask the question as to what is special about thiol esters (as opposed to oxygen esters). In alkaline hydrolysis the two show about the same rate constants. In acid hydrolysis, thiol esters react faster than corresponding oxygen esters. Thiol esters have been shown to be more reactive in condensation, presumably involving a carbanion. Acyl derivatives of CoASH are usually formed in ATP-dependent reactions and undergo a number of reactions that involve either attack of a nucleophile on the acyl carbon atom, with transfer of the acyl moiety to the nucleophile with release of CoASH, additions to species having a double bond in the acyl moiety, condensation at the α-carbon atom of the thiol ester, or reactions involving acyl group interchange. The preference in biological systems for thiol esters is not clear in reactions involving nucleophilic attack, but seems to correlate with nonenzymatic condensation reactions. The free energy of a thiol ester has been stated to be larger than that of the corresponding oxygen ester, but this may be a function of the different pK's of alcohols and mercaptans. Of course, if the above is correct a thermodynamic reason for the preference of thiol esters would exist. The alkaline hydrolysis of corresponding oxygen and thiol esters has been exhaustively treated.[202]

Nonredox Coenzymes

Coenzyme A
23

The dehydrogenation of butyryl coenzyme A to crotonyl coenzyme A has been studied and presumably proceeds via a flavin intermediate involving two one-electron changes.[203] This is really a redox reaction and belongs in that section. An enzyme-coenzyme A intermediate was postulated on the basis of isolation and kinetic experiments in the succinyl coenzyme A-acetoacetate coenzyme A transferase system.[204]

17.2.7 Purine Phosphates

There are two purine phosphates of interest, guanosine and adenine phosphates. We concentrate on the latter. The structure of adenosine triphosphate is shown in formula **24**. There are of course the corresponding monophosphate and diphosphate. The triphosphate (ATP) is involved in many biological oxidations and therefore there is a question of whether it should be

Adenosine triphosphate (ATP)
24

considered here or in the section on redox systems. The fundamental role of the hydrolysis of ATP is as the driving force for biochemical processes which is presumably derived from the ultimate hydrolysis of the pyrophosphate linkages of the molecule (see Chapter 8).[205] The number of individual enzymes known to require ATP as coenzyme, substrate, or energy source is very large.[198,206] ATP-dependent reactions may be placed in two broad classifications: the cleavage of ATP (1) to ADP and inorganic phosphate or (2) its cleavage to AMP and pyrophosphate, in either case providing the driving force for otherwise unfavorable reactions. A simple analog of ATP, γ-phenylpropyl triphosphate, is readily hydrolyzed by potato apyrase, muscle myosin, or inorganic pyrophosphatase in the presence of zinc ion.[207] The standard free energy changes for a number of ATP, ADP, and pyrophosphate reactions were calculated as a function of acid or metal ion (super acid) concentration. The metal ion involved is the magnesium ion, which presumably chelates to two phosphate oxygen atoms (either terminal or internal). Entropies were also calculated. The relative contribution of the two depends on the reaction.[208] Many bivalent metal ions catalyze the nonenzymatic hydrolysis of ATP. Cupric ion is unusually effective.[209,210] Some possible mechanisms in the hydrolysis of ATP and ADP in the presence of metal ions serving as super acids are shown in Fig. 17.27.[210] A Schardinger (cyclo)dextrin has been reported to catalyze the hydrolysis of ADP to AMP and phosphate ion. It is suggested that this catalysis occurs by prior complexing of the substrate with the dextrin since Schardinger dextrins can form inclusion complexes.[211]

The question may be asked as to the position of the metal ion chelate. The structure of the complexes of ADP and ATP with divalent metal ions was investigated by an nmr study of the hydrogen and phosphorus nuclei. Chemical shifts indicate that calcium and zinc form chelates with the β- and γ-phosphate groups of ATP and with the α- and β-phosphate groups of ADP. On the other hand, cupric ion interacts solely with the α- or β-phosphate groups of ATP, but cupric as well as manganous ion interacts with the α- and β-phosphate groups of ADP.[212]

All ATP-dependent biosynthetic processes involve the formation of a covalent bond between two substrate molecules coupled with the cleavage of one of the pyrophosphate links of ATP. A number of ATP-dependent reactions involve carboxyl group activation. ATP can be cleaved either to ADP and phosphate ion or to AMP and pyrophosphate, as indicated before.

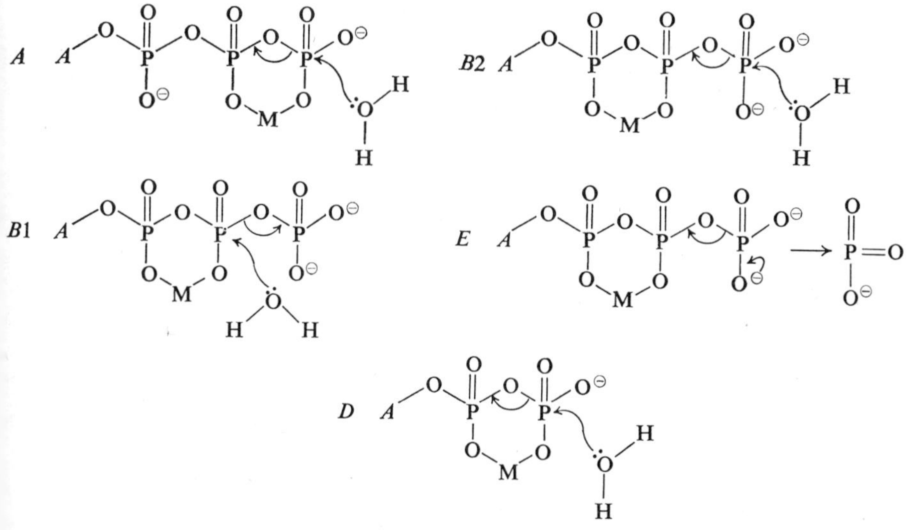

Fig. 17.27 Hydrolysis of ATP. From M. Tetas and J. M. Lowenstein, *Biochemistry*, 2, 350 (1962).

Acetyl coenzyme *A* is often involved. An enzyme-bound acyl adenylate (formula **25**) is formed in the latter reactions. Acetyl CoA is presumably synthesized as shown in eq. 17.34.[198]

$$R-\overset{O}{\underset{\|}{C}}-O-\overset{O^\ominus}{\underset{\|}{\underset{O}{P}}}-O-CH_2-\text{(ribose with OH, OH)}-\text{adenine}$$

Acyl adenylate
25

An enzyme that splits pyrophosphate has been purified[213] and kinetic studies indicate that the true substrate is a complex of magnesium ion and pyrophosphate.[214] The metal ion forms a six-membered ring with the pyrophosphate as shown in formula **26**[214] which can speed the hydrolysis either by

$$\text{enz} + CH_3-CO_2{}^{18}H + ATP \xrightleftharpoons{Mg^{2\oplus}} \left[CH_3-\overset{O}{\underset{\|}{C}}-O^{18}-AMP\right]\text{enz} + PP_i$$

$$\left[CH_3-\overset{O}{\underset{\|}{C}}-O^{18}-AMP\right]\text{enz} + HSCoA \rightleftharpoons \quad (17.34)$$

$$CH_3-\overset{O}{\underset{\|}{C}}-SCoA + \text{enz} + \text{adenosine}-5'-O-\overset{O^\ominus}{\underset{\|}{\underset{O}{P}}}-O^{18}H$$

polarizing the system, neutralizing the negative charge, or changing the position of the active functionalities on the enzyme.

$$\underset{\textbf{26}}{\text{(Mg}^{2+}\text{ chelate with pyrophosphate, 5 Å span)}}$$

17.2.8 Pyrimidine Phosphates

There are several pyrimidine phosphates: uridine, cytosine, and inosine phosphates. We concentrate on the first. Uridine nucleotides have been shown to be important in carbohydrate metabolism.[215] A recent Nobel prize was given to Leloir for this work. The general structure for UDP-sugars is shown in formula **27**. The role of UDP-sugar coenzymes is twofold. Such coenzymes are involved in a variety of glycosyl transfer reactions. Also the sugar itself can be transformed either by oxidation or by oxidation and decarboxylation.

General structure for UDP sugars
27

Both RNA and DNA synthesis have been investigated in recent years. These molecules are large, and involved in genetic transfer. Templates are usually needed for these reactions which are essentially polymerizations. The complexity can be seen in Fig. 17.28 where A and T stand for adenosine and thymidine, respectively.[216] In RNA polymerase, a protein is said to stimulate the initiation of RNA chains.[217]

17.2.9 Metal Ion-Containing Enzymes

Metal ions that can change valence were important in our discussion of redox systems. What of those metal ions such as zinc, manganous, magnesium, and cobalt ions, to name a few, which do not participate easily in valence changes? They are still important as "super acid" catalysts in enzymatic reactions of nucleophiles with acyl compounds. Notice that the ions mentioned are all polyvalent so that it is possible that they can form chelates with the substrate. The concept of super acid catalysis has been mentioned before (Chapter 8) and presumably the same mechanism applies to these enzymatic reactions as the nonenzymatic reactions.

Depending on the equilibrium constant between the metal ion and the

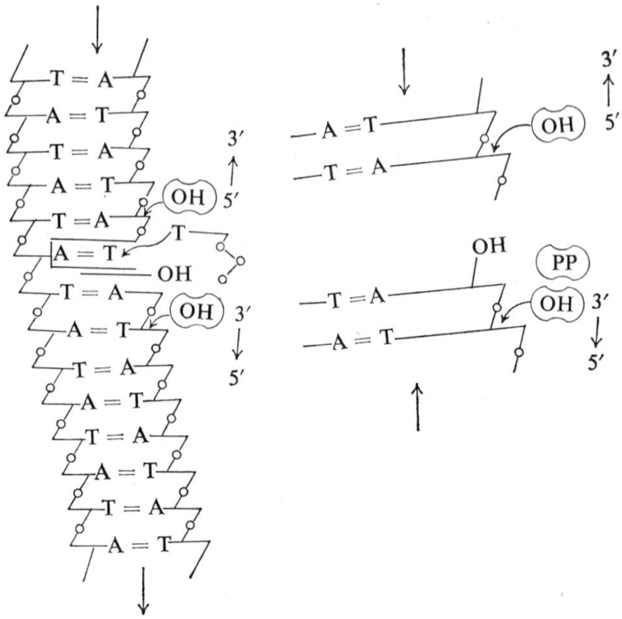

Fig. 17.28 Synthesis of DNA. From A. Kornberg, *Science*, **163**, 1416 (1969). © 1963 by the American Association for the Advancement of Science.

enzyme, one may talk of a metal ion complex with the enzyme (which is dissociable) or a metalloenzyme (which is not dissociable in the usual sense). The term metalloenzyme has been used to include the former as well as the latter, and furthermore in doing so, metal ions involved in redox systems and those involved in nonredox systems have been lumped together, as opposed to the treatment in this book.[237] Two properties of metal ions make them unique: (1) metal ions can serve as electronic markers, for example, for various kinds of spectroscopic studies; and (2) different metal ions can be substituted for one another in a graded series, and information can be gained from these substitutions.[237] The hydrolysis of substrates bound to enzymes is not solely attributable to the metal ion. This implies that the enzyme provides specificity directly to the substrate, or complexes with the metal ion, thus changing its reactivity. Many active-site reagents attack nearby amino acids because they may complex with the metal ion first.[238]

Carboxypeptidase. Carboxypeptidase contains 1 atom of zinc per molecular weight of 34,300.[218] A variety of complexing agents inhibit carboxypeptidase by removing zinc.[219] Crystalline metal-free apocarboxypeptidase is enzymatically inactive; activity is restored by addition of zinc to the apoenzyme. Reactivation is proportional to the amount of zinc bound, up to 1 g atom per mole of protein.[220] Silver ion, p-mercuribenzoate, or ferricyanide titrate only one SH group in the apocarboxypeptidase and none in the metallocarboxypeptidase.[221] Reactions of the apo-enzyme and metalloenzyme with 1-fluoro-2,4-dinitrobenzene or phenyl isothiocyanate indicate a binding of the metal to an α-amino group of N-terminal asparagine. Chelation of the zinc ion occurs to two imidazole groups and a carboxylate ion.

Acetylation of carboxypeptidase A by N-acetylimidazole increases (first-order) esterase activity toward hippuryl-DL-β-phenyllactate, whereas it decreases (second-order) peptidase activity toward N-benzyloxycarbonyl-glycyl-L-phenylalanine virtually to zero in certain instances.[222,223] The acetylation of two tyrosine residues is involved. The major effect of acetylation appears in a change in K_m(app) in both esterase and peptidase reactions.

Although the zinc ion of carboxypeptidase is probably involved in its catalysis, no details of the reaction can be given. Metal ion catalysis of ester and amide hydrolysis is known.[224,225] The most definitive work[225] suggests that in cobalt ion-promoted glycine ester hydrolysis, a complex of the form shown below occurs.

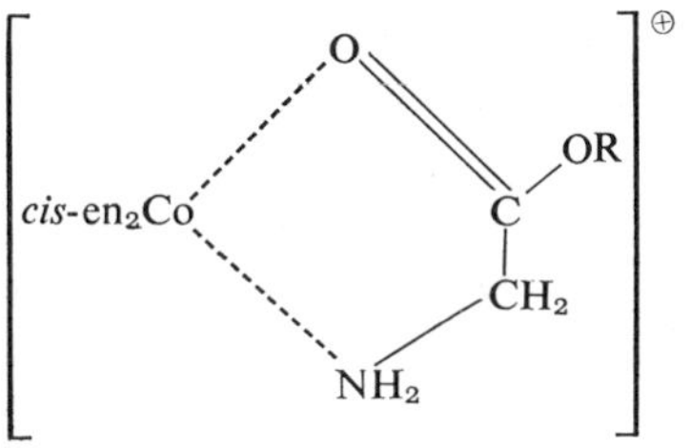

$$\left[cis\text{-en}_2\text{Co} \overset{O}{\underset{NH_2}{\diagdown}} \overset{}{\underset{}{C}} \overset{OR}{\underset{CH_2}{\diagup}} \right]^{\oplus} \qquad (17.35)$$

The zinc ion of the enzyme may function in a similar manner to this cobalt ion; the cobalt ion-catalyzed ester hydrolysis is subject to further catalysis by acetate ion[225]; a basic group on the enzyme may thus serve to enhance the activity of the zinc ion of the enzyme. It has been suggested that the metal ion serves as a general acid catalyst in dehydration.[260]

Zinc and cobaltous ions hydrate 2-pyridinealdehyde at a rate which is 10^7 times faster than reaction with water alone.[226]

Aminopeptidases have also been studied.[227,228] The substrate specificity and kinetics of lenticular β-leucine aminopeptidase (3.4.1.1.) may be concerted.

Carbonic Anhydrase. Carbonic anhydrase is a metal-containing enzyme that ordinarily has one zinc atom per molecule of enzyme protein. The enzyme is ordinarily associated with the reversible hydration of carbon dioxide but this enzyme catalyzes the reversible hydration of acetaldehyde[230,231] and pyridine aldehydes[232] as well as the hydrolysis of p-nitrophenyl acetate.[233] Infrared studies with azide, nitrate, and bicarbonate ions lead to a picture of the hydration.[234,235]

Neutral Protease. Microorganisms produce a group of enzymes called neutral proteases which contain one gram atom of zinc per molecule of protein. These enzymes which include thermolysin, enzymes from *B. subtilis*, *B. megaterium, aeromonas proteolyticus*, and *B. cereus* are endopeptidases and chelator sensitive. Again the metal ion appears to be involved in the catalysis.[236]

Oxaloacetate Decarboxylase. The mechanisms of enzymatic decarboxylation of β-keto acids can be distinguished by their dependence upon or independence of metal ions. For example, the oxaloacetate decarboxylases from *Micrococcus lysodeikticus*,[239] from cod[240] and from other sources[241] require metal ions for activation. By way of contrast, the decarboxylase from *Clostridium acetobutylicum* catalyzes the decarboxylation of acetoacetic acid by way of a Schiff base between enzyme and substrate as intermediate.[242,243] The latter enzyme utilizes a specific lysine residue[244] but does not require metal ions. A similar situation obtains with respect to the aldol condensation, where enzymes of class II require metal ions for activity, whereas enzymes of class I react by way of Schiff bases as intermediates.[245]

Sodium borohydride has no effect upon the activity of the oxaloacetate decarboxylase from cod, regardless of whether substrate is present or absent, or whether pyruvate (the reaction product) is present or absent. However, the reduction of pyruvate by borohydride in the presence of manganous ions and enzyme leads to the formation of excess D-lactate, as shown by the dispersion curve for optical activity and by specific enzyme assays with D- and L-lactate dehydrogenases. The enzymic catalysis of borohydride reduction depends upon the presence of metal ions and does not occur in the presence of EDTA. These facts are consistent with the hypothesis that reduction and decarboxylation occur by way of a metal ion-enzyme-substrate complex analogous to that previously suggested for relevant model systems as shown in Fig. 17.29.[246]

Acyl-Carrier Protein. Strictly speaking, acyl-carrier protein is not an enzyme (Chapter 16), but it is involved in enzymatic fatty acid synthesis, has a molecular weight of 9600, and the prosthetic group (coenzyme) which carries the acyl moiety has been identified as 4-phosphopantetheine.[250] The acyl-carrier protein is like an acyl-enzyme intermediate. Many different acyl

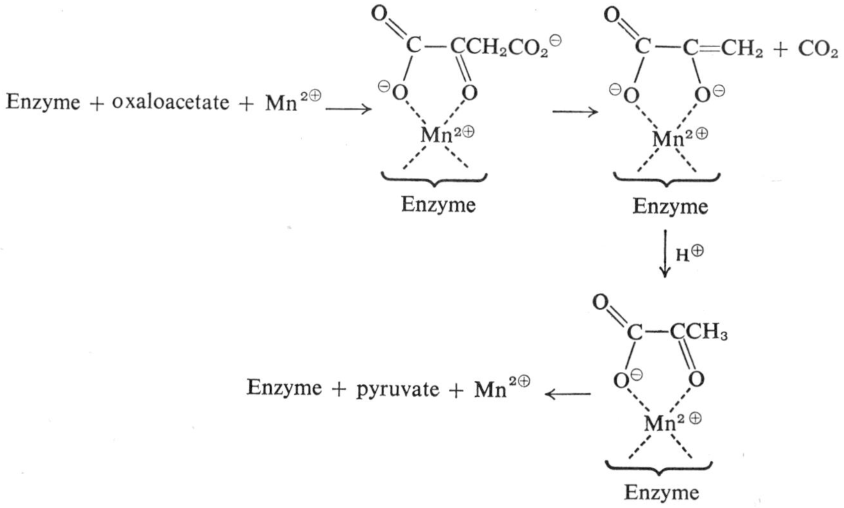

Fig. 17.29 Decarboxylation by oxaloacetate decarboxylase. From G. W. Kosicki and F. H. Westheimer, *Biochemistry*, **7**, 4303 (1968). © 1968 by the American Chemical Society. Reprinted by permission of the copyright owner.

groups can become attached to the acyl-carrier protein, but apparently even-chain acyl groups are preferred and 16 carbon atoms is the limit.[251] Acetylation of the four lysine residues and of the amino group of the terminal serine residue in *E. coli* acyl-carrier protein has no effect on the ability of this protein to function in fatty acid synthesis, which is the usual analytical tool.[252]

17.2.10 Glutathione

The structure of glutathione was established as the tripeptide L-glutamyl-L-cysteylglycine as shown in formula **28**. Although a large number of

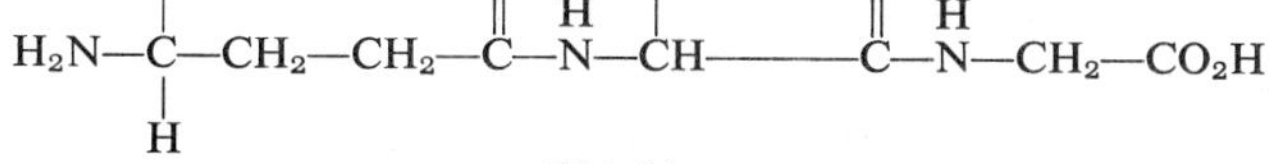

$$H_2N-\underset{H}{\overset{CO_2H}{\underset{|}{C}}}-CH_2-CH_2-\overset{O}{\overset{\|}{C}}-\underset{}{\overset{H}{\underset{|}{N}}}-\underset{}{\overset{CH_2-SH}{\underset{|}{CH}}}-\overset{O}{\overset{\|}{C}}-\underset{}{\overset{H}{\underset{|}{N}}}-CH_2-CO_2H$$

Glutathione
28

biochemical functions have been ascribed to glutathione, the number of instances in which glutathione is known to function as a coenzyme is limited. Of these probably the most widely studied is the glyoxalase reaction:

$$CH_3-\overset{O}{\underset{\|}{C}}-\overset{O}{\underset{H}{C}} \xrightarrow{\text{glutathione}} CH_3-\overset{OH}{\underset{H}{\underset{|}{C}}}-CO_2H \qquad (17.36)$$

An elegant model system has been developed for the glyoxalase reaction.[253,254]

$$\phi-\overset{O}{\underset{\|}{C}}-\overset{O}{\underset{\|}{C}}-H + HS-CH_2-CH_2-N\overset{CH_3}{\underset{CH_3}{\diagdown}} \rightleftharpoons \phi-\overset{}{C}-\overset{}{\underset{H}{C}} \rightleftharpoons$$

$$\phi-\overset{O^{\ominus}}{\underset{H}{\underset{|}{C}}}-\overset{O}{\underset{\|}{C}}-S-\overset{CH_3}{\underset{|}{\underset{CH_2}{N^{\oplus}-CH_3}}}-CH_2 \rightleftharpoons \phi-\overset{OH}{\underset{H}{\underset{|}{C}}}-CO_2H + HS-CH_2-CH_2-N(CH_3)_2 \qquad (17.36)$$

Phenylglyoxal is smoothly converted to mandelic acid in aqueous solution at room temperature in the presence of N,N-dimethyl-β-mercaptoethylamine. Neither thiols alone nor amines alone are catalysis for this reaction, although a mixture of the two is somewhat effective. The free thiol group is required for catalysis, since the S-methyl derivative of the catalyst above is inactive. Likewise, the trimethylammonium derivative of the catalyst above is devoid of catalytic activity, indicating the necessity of a basic site on the catalyst molecule. The model reaction, as is the case with glyoxalase I, proceeds with internal hydrogen transfer. On the basis of this information, the mechanism in eq. 17.36 appears reasonable for the model system.

17.2.11 Vitamins A, D, E, and K

In addition to the coenzymes discussed above, vitamins A, D, E, and K should be considered. Dihydrobiopterin, a vitamin A derivative, is a coenzyme for the hydroxylation of phenylalanine.[255]. The structures of vitamins E and K are shown in formula 29. These coenzymes are involved in electron transport (redox) systems and could have been discussed in the previous section. The term vitamin D originally designated the antirachitic principle in preparations of the "fat-soluble A" factor. However, before the actual isolation (now called vitamin D_3 or cholecalciferol) from fish

$$\text{Vitamin E } (\alpha\text{-tocopherol})$$

$K_1: R = CH=\overset{CH_3}{\overset{|}{C}}(CH_2)_3\overset{CH_3}{\overset{|}{C}H}(CH_2)_3\overset{CH_3}{\overset{|}{C}H}(CH_2)_3\overset{CH_3}{\overset{|}{C}H}CH_3$

$K_2: R = CH=\overset{CH_3}{\overset{|}{C}}CH_2(CH_2CH=\overset{CH_3}{\overset{|}{C}})_4CH_2CH=C\overset{\diagup CH_3}{\diagdown CH_3}$

$K_3: R = H$ (menadione)

Vitamins K

29

liver oils, it was shown that an antirachitic compound (vitamin D_2, calciferol, or ergocalciferol) could be produced in the laboratory by the irradiation of the plant sterol ergosterol. The term vitamin D_1 has been discarded since the material to which it was first applied has been found to be a mixture of calciferol and other sterols. Observations that vitamin D activity[256] could be produced by the irradiation (preferably by ultraviolet light) of plants or of certain sterols had, in fact, been foreshadowed by the demonstration that sunlight had a pronounced curative action on rachitic children or laboratory animals. This effect is due to the formation in vivo of vitamins from provitamins. The latter term, as applied to vitamin D, implies a sterol which can be converted by irradiation into a D vitamin. Thus 7-dehydrocholesterol yields vitamin D_3, and 22-dihydroergosterol yields vitamin D_4. Several other provitamins have been reported to be present in natural materials, and all the provitamins appear to be sterols in which carbon atom 3 bears an hydroxyl group and ring B contains the $\Delta^{5,7}$-dienic group.

In higher animals, vitamin D deficiency causes abnormalities in calcium and phosphate metabolism and results in structural changes in bones and teeth; the syndrome characteristic of a severe deficiency in children is called rickets; in adults, osteomalacia. The ingestion of excessive amounts of vitamin D also produces toxic symptoms; initially there is a rise in the blood calcium level followed by metastatic calcification of various internal organs and, ultimately, by decalcification of skeletal structures.

The D vitamins stimulate the absorption of $Ca^{2\oplus}$ from the intestinal tract, but do not appear to exert a direct effect on the absorption of phosphate; the lowered accumulation of bone salts on avitaminotic animals is chiefly a result of an impaired ability to absorb calcium. In addition, vitamin D appears to function in the internal tissues. For example, the amount of citrate present in the bones and internal organs (kidney, heart) of vitamin D-deficient rats rises rapidly when the vitamin is given. Although it is generally agreed that the D vitamins play an important role in the process of growth and especially in the formation and maintenance of bones, the biochemical functions of this group of vitamins remain obscure.[257]

17.3 SUMMARY

One may legitimately ask what is a coenzyme (prosthetic group) and why are they needed, since strictly protein enzymes exist (Chapter 16). One may ask what is the underlying feature of a coenzyme. The answer may be found in the multitude of mechanisms that enzymes containing cofactors catalyze. Protein enzymes (Chapter 16) catalyze hydrolytic reactions of various kinds by nucleophiles or general bases present in the protein. With enzymes containing cofactors (coenzymes, prosthetic groups) (either small inorganic or organic molecules) many reactions such as redox, decarboxylation, elimination, condensation, and others, which a protein alone cannot catalyze, are found to occur. Of course, many coenzymes are vitamins when they are phosphorylated by the body. In addition all coenzymes must be bound to the enzyme protein. The binding between coenzyme and enzyme may be either covalent or noncovalent. In many instances the coenzyme can function without the protein at a diminished rate very much akin to the catalyses described in Chapters 6 and 8. Therefore the crucial factor is that the coenzyme possesses the ability to catalyze many chemical reactions that the protein alone cannot. So in the old argument of whether an enzyme is a protein or whether the protein is simply a carrier for something much smaller in molecular weight (Chapter 12), both sides as usual are correct. In some instances, an enzyme is solely a protein. In others, it is a protein associated with a small moiety.

REFERENCES

1. F. J. Kézdy, personal communication.
2. R. Breslow, *J. Amer. Chem. Soc.*, **80**, 3719 (1958).
3. A. E. Braunstein, in *The Enzymes*, P. D. Boyer, H. Lardy, and K. Myrbäck, Eds., vol. 2, 2nd ed., Academic Press, New York, 1960, p. 113.
4. E. E. Snell, A. E. Braunstein, E. S. Severin, and Y. M. Torchinsky, *Pyridoxal Catalysis: Enzymes and Model Systems*, Interscience Publishers, New York, 1968.
5. B. Vennesland and F. H. Westheimer, in *The Mechanism of Enzyme Action*, W. D. McElroy and B. Glass, Eds., Johns Hopkins Press, Baltimore, 1954, p. 357.

6. M. Florkin and E. H. Stotz, Eds., *Comprehensive Biochemistry*, Vol. 14, Elsevier, Amsterdam, 1966.
7. H. R. Mahler and E. H. Cordes, *Biological Chemistry*, Harper and Row, New York, 1966, p. 354.
8. M. E. Pullman, A. San Pietro, and S. P. Colowick, *J. Biol. Chem.*, **206**, 129 (1954).
9. J. Jarabak and P. Talalay, *J. Biol. Chem.*, **235**, 2147 (1960).
10. H. R. Mahler and E. H. Cordes, *Biological Chemistry*, Harper and Row, New York, 1966, p. 357.
11. D. Mauzerall and F. H. Westheimer, *J. Amer. Chem. Soc.*, **77**, 2261 (1955).
12. R. H. Abeles, R. F. Hutton, and F. H. Westheimer, *J. Amer. Chem. Soc.*, **79**, 712 (1957).
13. R. Abeles and F. H. Westheimer, *J. Amer. Chem. Soc.*, **80**, 5459 (1958); see also D. C. Dittmer and R. A. Fouty, *J. Amer. Chem. Soc.*, **86**, 91 (1964); D. C. Dittmer et al., *Tetrahedron Lett.*, 827 (1961).
14. G. Cilento, *Arch. Biochem. Biophys.*, **88**, 352 (1960).
15. H. Theorell, in *The Mechanism of Action of Water Soluble Vitamins*, A.V.S. de Reuck and M. O'Connor, Eds., Little, Brown and Co., Boston, 1961, p. 5; H. Theorell and B. Chance, *Acta Chem. Scand.*, **5**, 1127 (1951).
16. E. M. Kosower, *J. Amer. Chem. Soc.*, **78**, 3497 (1956).
17. H. Mahler, B. R. Baker, Jr., and V. J. Shiner, *Biochemistry*, **1**, 47 (1962).
18. K. Dalziel and F. M. Dickinson, *Biochem. J.*, **100**, 34 (1966).
19. K. Dalziel and F. M. Dickinson, *Biochem. J.*, **100**, 491 (1966).
20. B. E. Norcross, P. E. Klinedinst, Jr., and F. H. Westheimer, *J. Amer. Chem. Soc.*, **84**, 797 (1962).
21. H. Theorell, *7th Int. Cong. of Biochem.*, Tokyo, 1967, Symp. III, p. 177.
22. H. R. Mahler and E. H. Cordes, *Biological Chemistry*, Harper and Row, New York, 1966, p. 359; see also M. Florkin and E. H. Stotz, Eds., *Comprehensive Biochemistry*, Vol. 14, Elsevier, Amsterdam, 1966, p. 127.
23. H. Beinert, *J. Amer. Chem. Soc.*, **78**, 5323 (1956); H. Beinert and R. H. Sands, *Free Radicals in Biological Systems*, M. S. Blois et al., Eds., Academic Press, New York, 1961, p. 17.
24. J. L. Fox and G. Tollin, *Biochemistry*, **5**, 3865 (1966).
25. G. R. Penzer and G. K. Radda, *Quart. Rev.*, **21**, 43 (1967).
26. R. A. White and G. Tollin, *J. Amer. Chem. Soc.*, **89**, 1253 (1967).
27. A. V. Guzzo and F. Tollin, *Arch. Biochem. Biophys.*, **105**, 380 (1964).
28. J. M. Lhoste, A. Haug, and P. Hemmerich, *Biochemistry*, **5**, 3290 (1966).
29. T. T. Tchen and H. van Milligan, *J. Amer. Chem. Soc.*, **82**, 4115 (1960).
30. O. Gawron, A. J. Glaid, T. P. Fondy, and M. M. Bechtold, *Nature*, **189**, 1004 (1961).
31. O. Gawron et al., *J. Amer. Chem. Soc.*, **84**, 3877 (1962).
32. T. C. Hollocher, Jr., and B. Commoner, *Proc. Natl. Acad. Sci., U.S.*, **47**, 1355 (1961).
33. E. C. Slater, *Conference on Homogeneous Catalysis*, Haarlem, Holland, 1966 (unpublished).
34. G. Tollin, *Molecular Associations in Biology*, Academic Press, New York, 1968, p. 400.
35. P. Strittmatter, *J. Biol. Chem.*, **236**, 2329 (1961).
36. K. Yagi and T. Ozawa, *Biochim. Biophys. Acta*, **42**, 381 (1960).
37. J. C. M. Tsibris, D. B. McCormick, and L. D. Wright, *J. Biol. Chem.*, **241**, 1138 (1966).
38. V. Massey and H. Ganther, *Biochemistry*, **4**, 1161 (1965).

39. V. Massey and G. Palmer, *J. Biol. Chem.*, **237**, 2347 (1962).
40. C. Veeger, D. V. DerVertanian, J. F. Kalse, A. DeKok, and J. F. Koster, *Flavins and Flavoproteins*, E. C. Slater, Ed., Elsevier, Amsterdam, 1966, p. 114.
41. V. Massey et al., *7th Int. Cong. of Biochem., Tokyo*, 1967, Symp. III, p. 165.
42. P. Hemmerich, V. Massey, and G. Weber, *Nature*, **213**, 728 (1967).
43. J. L. Fox and G. Tollin, *Biochemistry*, **5**, 3865 (1966).
44. J. L. Fox and G. Tollin, *Biochemistry*, **5**, 3873 (1966).
45. W. J. Nickerson and G. Strauss, *J. Amer. Chem. Soc.*, **82**, 5007 (1960).
46. G. Strauss and W. J. Nickerson, *J. Amer. Chem. Soc.*, **83**, 3187 (1961).
47. B. Holmstrom and G. Oster, *J. Amer. Chem. Soc.*, **83**, 1867 (1961).
48. J. T. Spence and J. Tocatlian, *J. Amer. Chem. Soc.*, **83**, 816 (1961).
49. G. Tollin, *Biochemistry*, **7**, 1720 (1968).
50. G. A. Hamilton, personal communication; *Adv. Enzymol.*, **26**, (1969); G. A. Hamilton, J. W. Hanifin, Jr., and J. P. Friedman, *J. Amer. Chem. Soc.*, **88**, 5269 (1966).
51. R. R. Grinstead, *J. Amer. Chem. Soc.*, **82**, 3464 (1960).
52. I. Yamasaki and L. H. Piette, *Biochim. Biophys. Acta*, **50**, 62 (1961).
53. H. S. Mason, H. Spencer, and I. Yamasaki, *Biochem. Biophys. Res. Comm.*, **4**, 236 (1961).
54. V. S. Butt and M. Hollaway, *Arch. Biochem. Biophys.*, **92**, 24 (1961) and references therein.
55. J. H. Wang, *J. Amer. Chem. Soc.*, **77**, 822, 4715 (1955); R. C. Jarnagin and J. H. Wang, *J. Amer. Chem. Soc.*, **80**, 786, 6477 (1958).
56. S. Udenfriend et al., *J. Biol. Chem.*, **208**, 731, 741 (1954).
57. R. Breslow and L. N. Lukens, *J. Biol. Chem.*, **235**, 292 (1960).
58. G. A. Hamilton, J. W. Hanifin, Jr., and J. P. Friedman, *J. Amer. Chem. Soc.*, **88**, 5269 (1966); G. A. Hamilton, J. P. Friedman, and P. M. Campbell, *J. Amer. Chem. Soc.*, **88**, 5266 (1966).
59. R. O. C. Norman and G. K. Radda, *Proc. Chem. Soc.*, 138 (1962).
60. I. Yamasaki, H. S. Mason, and L. Piette, *J. Biol. Chem.*, **235**, 2444 (1960).
61. P. George, *Currents in Biochemical Research*, D. E. Green, Ed., Interscience Publishers, New York, 1956, p. 338.
62. B. Chance and G. R. Schonbaum, *J. Biol. Chem.*, **237**, 2391 (1962).
63. H. Hermel and R. Havemann, *Biochim. Biophys. Acta*, **128**, 283 (1966).
64. E. Margoliash, personal communication.
65. B. Chance, *Currents in Biochemical Research*, D. E. Green, Ed., Interscience Publishers, New York, 1956, p. 308.
66. O. Hayaishi, *Oxygenases*, Academic Press, New York, 1962.
67. H. S. Mason, *Advan. Enzymology*, **19**, 79 (1957).
68. S. Kaufman, *J. Biol. Chem.*, **237**, PC2712 (1962).
69. N. Stone and A. Meister, Abstracts, *FASEB Meetings*, 1963, p. 19c; *Nature*, **194**, 555 (1962).
70. E. Margoliash, W. M. Fitch, and R. E. Dickerson, *Brookhaven Symp. Biology*, **21**, 259 (1968).
71. W. M. Fitch and E. Margoliash, *Brookhaven Symp. Biology*, **21**, 217 (1968).
72. B. Chance, *Biochem. J.*, **103**, 1 (1967).
73. E. Margoliash, M. Reichlin, and A. Nisonoff, *Proc. Symp. Struct. Chem. Aspects of Cytochromes*, Osaka, Japan, 269 (1967).
74. R. E. Dickerson, M. L. Kopka, J. E. Weinzierl, J. C. Varnum, D. Eisenberg, and E. Margoliash, *Biochemistry*, **6**, 225 (1967).
75. B. Chance, C.-P. Lee, and L. Mela, *Fed. Proc.*, **26**, 1341 (1967).

References

76. K. G. Brandt, P. C. Parks, G. H. Czerlinski, and G. P. Hess, *J. Biol. Chem.*, **241**, 4180 (1966).
77. D. E. Griffiths and M. Beinert, *Biochem. J.*, **81**, 42P (1961).
78. W. E. Knox and A. H. Mehler, *J. Biol. Chem.*, **187**, 419, 431 (1950); W. E. Knox, *Biochim. Biophys. Acta*, **14**, 117 (1954).
79. H. R. Mahler and E. H. Cordes, *Biological Chemistry*, Harper and Row, New York, 1966, p. 693.
80. H. Maeno and P. Feigelson, *J. Biol. Chem.*, **242**, 596 (1967).
81. A. Meister, D. Wellner, and S. J. Scott, *J. Natl. Cancer Inst.*, **24**, 31 (1960).
82. A. Meister and D. Wellner, in *The Enzymes*, P. D. Boyer, H. Lardy, and K. Myrbäck, Eds., Vol. 7, 2nd ed., Academic Press, New York, 1963, p. 609.
83. L. Hellerman and D. S. Coffey, *J. Biol. Chem.*, **242**, 582 (1967).
84. D. S. Coffey, A. H. Niems, and L. Hellerman, *J. Biol. Chem.*, **240**, 4058 (1965); A. H. Niems and L. Hellerman, *Ann. Rev. Biochem.*, **39**, 867 (1970).
85. V. Massey and B. Curti, *J. Biol. Chem.*, **242**, 1259 (1967).
86. H. A. O. Hill, J. M. Pratt, and R. J. P. Williams, *Chem. Brit.*, 156 (1969).
87. P. G. Lenhert and D. C. Hodgkin, *Nature*, **192**, 937 (1961).
88. H. Eggerer, P. Overath, F. Lynen, and E. R. Stadtman, *J. Amer. Chem. Soc.*, **82**, 2643 (1960).
89. J. Retey, U. Coy, and F. Lynen, *Biochem. Biophys. Res. Comm.*, **22**, 274 (1966).
90. H. W. Whitlock, Jr., *J. Amer. Chem. Soc.*, **85**, 2343 (1963).
91. R. H. Abeles and H. A. Lee, Jr., *J. Biol. Chem.*, **236**, PC1 (1961).
92. P. A. Frey and R. H. Abeles, *J. Biol. Chem.*, **241**, 2732 (1966).
93. P. A. Frey, M. K. Essenberg, and R. H. Abeles, *J. Biol. Chem.*, **242**, 5369 (1967).
94. J. A. Retey, A. Umani-Ronchi, J. Seibl, and D. Arigoni, *Experientia*, **22**, 502 (1966).
95. J. D. Brodie, *Proc. Natl. Acad. Sci., U.S.*, **62**, 461 (1969).
96. R. H. Abeles, P. A. Frey, J. S. Fleming, and G. J. Karabatsos, *Abstr. 149th ACS Meeting*, 1965, p. 19P.
97. B. Zagalak, P. A. Frey, G. J. Karabatsos, and R. H. Abeles, *J. Biol. Chem.*, **241**, 3028 (1966).
98. M. Sprecher, M. J. Clark, and D. B. Sprinson, *J. Biol. Chem.*, **241**, 872 (1966).
99. M. Sprecher, R. L. Switzer, and D. B. Sprinson, *J. Biol. Chem.*, **241**, 864 (1966).
100. G. N. Schrauzer and J. Kohyle, *Chem. Ber.*, **97**, 3056 (1964).
101. G. N. Schrauzer et al., *J. Amer. Chem. Soc.*, **89**, 143 (1967); **92**, 7078 (1970).
102. *Chem. Eng. News*. October 21, 1968, p. 42; J. W. Sibert and G. N. Schrauzer, *J. Amer. Chem. Soc.*, **92**, 1421 (1970).
103. P. A. Frey, M. K. Essenberg, R. H. Abeles, and S. S. Kerwar, *J. Amer. Chem. Soc.*, **92**, 448 (1970); T. C. Stadtman, *Science*, **172**, 859 (1971).
104. L. J. Reed, in *Comprehensive Biochemistry*, Vol. 14, M. Florkin and E. H. Stotz, Eds., Elsevier, Amsterdam, 1966, Chapter 2.
105. M. Calvin, *Fed. Proc.*, **13**, 697 (1954).
106. J. A. Barltrop, P. M. Hayes, and M. Calvin, *J. Amer. Chem. Soc.*, **76**, 4348 (1954).
107. L. J. Reed, M. Koike, M. E. Levitch, and F. R. Leach, *J. Biol. Chem.*, **232**, 143 (1958).
108. K. Suzuki and L. J. Reed, *J. Biol. Chem.*, **238**, 4021 (1963).
109. L. J. Reed, in *Comprehensive Biochemistry*, Vol. 14, M. Florkin and E. H. Stotz, Eds., Elsevier, Amsterdam, 1966, p. 105.
110. L. J. Reed, in *Comprehensive Biochemistry*, Vol. 14, M. Florkin and E. H. Stotz, Eds., Elsevier, Amsterdam, 1966, p. 107.
111. L. L. Ingraham, *Biochemical Mechanisms*, John Wiley and Sons, New York, 1962, p. 85.

112. I. C. Gunsalus, *Fed. Proc.*, **13**, 715 (1954).
113. L. J. Reed, *Adv. Enzymol.*, **18**, 319 (1957).
114. D. R. Sanadi, M. Langley, and F. White, *J. Biol. Chem.*, **234**, 183 (1959).
115. V. Massey, *Biochim. Biophys. Acta.*, **30**, 205 (1958); **37**, 314 (1960).
116. V. Massey, Q. H. Gibson, and C. Veeger, *Biochem. J.*, **77**, 341 (1960).
117. V. Massey, T. Hofmann, and G. Palmer, *J. Biol. Chem.*, **237**, 3820 (1962).
118. I. M. Gascoigne and G. K. Radda, *Chem. Commun.*, 211 (1965).
119. H. R. Mahler and E. H. Cordes, *Biological Chemistry*, Harper and Row, New York, 1966, p. 582.
120. D. I. Arnon, *Science*, **149**, 1460 (1965) and references therein.
121. A. San Pietro, Ed., *Non-Heme Iron Proteins*, The Antioch Press, Yellow Springs, Ohio, 1965.
122. R. Malkin and J. C. Rabinowitz, *Ann. Rev. Biochem.*, **36**, 113 (1967).
123. S. Keresztes-Nagy, F. Perini, and E. Margoliash, *J. Biol. Chem.*, **244**, 981 (1969).
124. M. Florkin and E. H. Stotz, Eds., *Comprehensive Biochemistry*, Vol. 2, Elsevier, Amsterdam, 1962, Chapter 1.
125. R. Breslow, in *The Mechanism of Action of Water-Soluble Vitamins*, A. V. S. de Reuck and M. O'Connor, Eds., Little, Brown and Co., Boston, 1961, p. 65.
126. T. Ugai, S. Tanaka, and S. Dokawa, *J. Pharm. Soc. Japan*, **63**, 269 (1943).
127. S. Mizuhara, *J. Japan Biochem. Soc.*, **22**, 102 (1950).
128. F. H. Westheimer and L. L. Ingraham, *Chem. Ind.*, 846 (1956).
129. K. Fry, L. L. Ingraham, and F. H. Westheimer, *J. Amer. Chem. Soc.*, **79**, 5225 (1957).
130. R. Breslow, *J. Amer. Chem. Soc.*, **79**, 1762 (1957).
131. R. Breslow, *J. Amer. Chem. Soc.*, **80**, 3719 (1958); this paper is definitely worth reading.
132. L. O. Krampitz, *Ann. Rev. Biochem.*, **38**, 213 (1969).
133. H. Holzer and K. Beaucamp, *Angew. Chem.*, **71**, 776 (1959).
134. H. Holzer et al., *Biochem. Biophys. Res. Comm.*, **3**, 599 (1960).
135. L. O. Krampitz et al., *J. Amer. Chem. Soc.*, **80**, 5893 (1958).
136. G. L. Carlson and G. M. Brown, *J. Biol. Chem.*, **235**, PC3 (1960).
137. F. G. White and L. L. Ingraham, *J. Amer. Chem. Soc.*, **82**, 4114 (1960).
138. C. P. Nash, C. W. Olsen, F. G. White, and L. L. Ingraham, *J. Amer. Chem. Soc.*, **83**, 4106 (1961).
139. F. G. White and L. L. Ingraham, *J. Amer. Chem. Soc.*, **84**, 3109 (1962).
140. G. E. Lienhard, *J. Amer. Chem. Soc.*, **88**, 5642 (1966).
141. A. B. Turner and H. C. S. Wood, *Proc. Chem. Soc.*, 61 (1964).
142. H. Holzer, R. Katterman, and D. Busch, *Biochem. Biophys. Res. Comm.*, **7**, 167 (1962).
143. W. B. Jakoby, E. Ohmura, and O. Hayaishi, *J. Biol. Chem.*, **222**, 435 (1956).
144. J. Ullrich, J. H. Wittorf, and C. J. Gubler, *Biochim. Biophys. Acta*, **113**, 595 (1966).
145. L. Jaenicke and J. Koch, *Biochem. Zeit.*, **336**, 432 (1962).
146. R. G. Yount and D. E. Metzler, *J. Biol. Chem.*, **234**, 738 (1959).
147. J. Biggs and P. Sykes, *J. Chem. Soc.*, 1849 (1959).
148. W. Hafferl, R. Lundin, and L. L. Ingraham, *Biochemistry*, **2**, 1298 (1963).
149. G. D. Maier and D. E. Metzler, *J. Amer. Chem. Soc.*, **79**, 4386 (1957).
150. J. L. Wittliff and R. L. Airth, *Biochemistry*, **7**, 736 (1968).
151. D. E. Metzler, M. Ikawa, and E. E. Snell, *J. Amer. Chem. Soc.*, **76**, 648 (1954).
152. D. Portsmouth, A. C. Stoolmiller, and R. H. Abeles, *J. Biol. Chem.*, **242**, 2751 (1967).
153. D. E. Morse and B. L. Horecker, *Science*, **161**, 813 (1968).

154. W. Tagaki and F. H. Westheimer, *Biochemistry*, **7**, 891 (1968).
155. D. L. Nandi and D. Shemin, *J. Biol. Chem.*, **243**, 1236 (1968).
156. See the review by E. H. Fischer in *Structure and Activity of Enzymes*, T. W. Goodwin, J. I. Harris, and B. S. Hartley, Eds., Academic Press, New York, p. 111.
157. R. M. Herbst, *Adv. Enzymol.*, **4**, 75 (1944).
158. H. R. Mahler and E. H. Cordes, *Biological Chemistry*, Harper and Row, New York, 1966, p. 343.
159. T. C. Bruice and R. M. Topping, *J. Amer. Chem. Soc.*, **85**, 1480 (1963).
160. T. C. Bruice and R. M. Topping, *J. Amer. Chem. Soc.*, **85**, 1488 (1963).
161. T. C. Bruice and R. M. Topping, *J. Amer. Chem. Soc.*, **85**, 1493 (1963).
162. D. S. Auld and T. C. Bruice, *J. Amer. Chem. Soc.*, **89**, 2090 (1967).
163. M. Krongelb, T. A. Smith, and R. H. Abeles, *Biochim. Biophys. Acta*, **167**, 473 (1968).
164. B. E. C. Banks, A. A. Diamantis, and C. A. Vernon, *J. Chem. Soc.*, 4235 (1961).
165. W. T. Jenkins and I. W. Sizer, *J. Biol. Chem.*, **234**, 1179 (1959).
166. E. E. Snell and W. T. Jenkins, *J. Cell Comp. Physiol.*, **54**, Suppl. 1, 161 (1959).
167. G. G. Hammes and P. Fasella, *J. Amer. Chem. Soc.*, **84**, 4644 (1962).
168. S. F. Velick and J. Vavra, *J. Biol. Chem.*, **237**, 2109 (1962).
169. H. C. Dunathan et al., *Biochemistry*, **7**, 4532 (1968).
170. L. Jaenicke and E. Brode, *Biochem. Z.*, **334**, 342 (1961).
171. H. R. Whiteley and F. M. Huennekens, *J. Biol. Chem.*, **237**, 1290 (1962).
172. R. H. Himes and J. C. Rabinowitz, *J. Biol. Chem.*, **237**, 2915 (1962).
173. L. Jaenicke and E. Brode, *Biochem. Z.*, **334**, 108 (1961).
174. M. J. Osborn, P. T. Talbert, and F. M. Huennekens, *J. Amer. Chem. Soc.*, **82**, 4921 (1960).
175. A. R. Larrabee, S. Rosenthal, R. E. Cathou, and J. M. Buchanan, *J. Amer. Chem. Soc.*, **83**, 4094 (1961).
176. A. R. Larrabee, S. Rosenthal, R. E. Cathou, and J. M. Buchanan, *J. Biol. Chem.*, **238**, 1025 (1963).
177. L. Jaenicke, in *The Mechanism of Action of Water-soluble Vitamins*, A. V. S. de Reuck and M. O'Connor, Eds., Little, Brown and Co., Boston, 1961, p. 44.
178. L. Jaenicke and E. Brode, *Ann.*, **624**, 120 (1959); see also R. G. Kallen and W. P. Jencks, *J. Biol. Chem.*, **241**, 5851 (1966) for the mechanism of this reaction.
179. H. A. Lardy and R. Peansky, *Physiol. Rev.*, **33**, 560 (1953).
180. D. Burk and R. J. Winzler, *Science*, **97**, 57 (1943).
181. S. J. Wakil, E. B. Titchener, and E. B. Gibson, *Biochim. Biophys. Acta*, **29**, 225 (1958).
182. S. J. Wakil and D. M. Gibson, *Biochim. Biophys. Acta*, **41**, 122 (1960).
183. S. Ochoa and Y. Kaziro, *Fed. Proc.*, **20**, 982 (1961).
184. B. K. Bachhawat, W. G. Robinson, and M. J. Coon, *J. Biol. Chem.*, **219**, 539 (1956).
185. H. G. Wood et al., *J. Biol. Chem.*, **238**, 547 (1963).
186. *Chem. Eng. News*, 54 (September 23, 1968).
187. F. Lynen, J. Knappe, and E. Lorch, in *The Mechanism of Action of Water-Soluble Vitamins*, A. V. S. de Reuck and M. O'Connor, Eds., Little, Brown and Co., Boston, 1961, p. 80; F. Lynen et al., *Angew. Chem.*, **71**, 481 (1959).
188. F. Lynen, J. Knappe, and E. Lorch, *Proc. 5th Int. Cong. Biochem.*, *Moscow*, 1961, **4**, 225 (1961); J. Knappe, *Ann. Rev. Biochem.*, **39**, 757 (1970).
189. M. Caplow, *J. Amer. Chem. Soc.*, **87**, 5774 (1965).
190. M. Caplow and M. Yager, *J. Amer. Chem. Soc.*, **89**, 4513 (1967).
191. J. Axelrod and R. Tomchick, *J. Biol. Chem.*, **233**, 702 (1958).

192. S. H. Mudd, *J. Biol. Chem.*, **234**, 1784 (1959).
193. D. W. Woolley, *Nature*, **171**, 323 (1953).
194. F. Lipmann, *J. Biol. Chem.*, **160**, 173 (1945).
195. D. Nachmansohn and M. Berman, *J. Biol. Chem.*, **165**, 551 (1946).
196. M. Soodak and F. Lipmann, *J. Biol. Chem.*, **175**, 999 (1948).
197. J. R. Stern and S. Ochoa, *J. Biol. Chem.*, **179**, 491 (1949).
198. H. R. Mahler and E. H. Cordes, *Biological Chemistry*, Harper and Row, New York, 1966, Chap. 8.
199. J. Baddiley, *Adv. Enzymol.*, **16**, 1 (1955).
200. F. Lynen, E. Reichert, and L. Rueff, *Ann.*, **574**, 1 (1951).
201. F. Lynen and E. Reichert, *Angew. Chem.*, **63**, 47 (1951).
202. K. A. Connors and M. L. Bender, *J. Org. Chem.*, **26**, 2498 (1961).
203. H. Beinert and E. Page, *J. Biol. Chem.*, **225**, 479 (1957).
204. L. B. Hersh and W. P. Jencks, *J. Biol. Chem.*, **242**, 339 (1967).
205. F. Lipmann, *Adv. Enzymol.*, **1**, 99 (1941).
206. M. Dixon and E. C. Webb, *Enzymes*, 2nd ed., Academic Press, New York, 1964.
207. D. L. Miller and F. H. Westheimer, *J. Amer. Chem. Soc.*, **88**, 1511 (1966).
208. R. A. Alberty, *J. Biol. Chem.*, **244**, 3290 (1969).
209. J. M. Lowenstein, *Biochem. J.*, **70**, 222 (1958).
210. M. Tetas and J. M. Lowenstein, *Biochemistry*, **2**, 350 (1963).
211. A. R. Todd, *J. Cell Comp. Physiol.*, **54**, Suppl. 1, p. 1 (1959).
212. M. Cohn and T. R. Hughes, Jr., *J. Biol. Chem.*, **237**, 176 (1962).
213. J. Josee, *J. Biol. Chem.*, **241**, 1938 (1966).
214. J. Josee, *J. Biol. Chem.* **241**, 1948 (1966).
215. R. Caputto, L. F. Leloir, C. E. Cardini, and A. C. Paladini, *J. Biol. Chem.*, **184**, 333 (1950).
216. A. Kornberg, *Science*, **163**, 1416 (1969).
217. *Sci. Res.*, 13 (May 26, 1969); K. Imahori, I. Shimada, and T. Ohta, *7th Int. Cong. of Biochem.*, *Tokyo*, 1967, Symp. III, p. 175.
218. B. L. Vallee and H. Neurath, *J. Amer. Chem. Soc.*, **76**, 5006 (1954).
219. B. L. Vallee, *Fed. Proc.*, **20**, Suppl. 10, 71 (1961).
220. J. E. Coleman and B. L. Vallee, *J. Biol. Chem.*, **235**, 390 (1960).
221. B. L. Vallee, T. L. Coombs, and F. L. Hoch, *J. Biol. Chem.*, **235**, PC45 (1960).
222. R. T. Simpson, J. F. Riordan, and B. L. Vallee, *Biochemistry*, **2**, 616 (1963).
223. B. L. Vallee, *Fed. Proc.*, **23**, 8 (1964).
224. M. L. Bender and B. W. Turnquest, *J. Amer. Chem. Soc.*, **79**, 1889 (1957); H. Kroll, *J. Amer. Chem. Soc.*, **74**, 2036 (1952); M. L. Bender, *Adv. Chem.*, **37**, 19 (1963); J. P. Collman and D. A. Buckingham, *J. Amer. Chem. Soc.*, **85**, 3039 (1963).
225. M. D. Alexander and D. H. Busch, *Abstr. 148th Meeting, Amer. Chem. Soc.*, September 1964, 25-O.
226. Y. Pocker and J. E. Meany, *J. Amer. Chem. Soc.*, **89**, 631 (1967).
227. P. F. Palmberg, Doctoral Dissertation, Northwestern University, 1969.
228. G. F. Bryce and B. R. Rabin, *Biochem. J.*, **90**, 513 (1964); $k_H/k_D = 3.0$ at pH 7.6, 1.9 at pH 9.45.
229. See for example R. P. Davis, *The Enzymes*, P. D. Boyer, H. Lardy and K. Myrbäck, Eds., Academic Press, **5**, 545 1961.
230. Y. Pocker and J. E. Meany, *J. Amer. Chem. Soc.*, **87**, 1809 (1965).
231. Y. Pocker and J. E. Meany, *Biochemistry*, **4**, 2535 (1965).
232. Y. Pocker and J. E. Meany, *Biochemistry*, **6**, 239 (1967).
233. Y. Pocker and J. T. Stone, *J. Amer. Chem. Soc.*, **87**, 5497 (1965); *Biochemistry*, **7**, 4139 (1968).

References

234. M. E. Riepe and J. H. Wang, *J. Amer. Chem. Soc.*, **89**, 4229 (1967).
235. M. E. Riepe and J. H. Wang, *J. Biol. Chem.*, **243**, 2779 (1968).
236. J. Feder, *Biochem. Biophys. Res. Comm.*, **32**, 326 (1968); *Biochemistry*, **6**, 2088 (1967); J. Feder and J. M. Schuck, *Biochemistry*, **9**, 2784 (1970).
237. B. L. Vallee and R. J. P. Williams, *Chem. Brit.*, **4**, 397 (1968).
238. See, for example, J. F. Riordan, M. Sokolovsky and B. L. Vallee, *Biochemistry*, **6**, 3609 (1967).
239. D. Herbert, *Methods Enzymol.*, **1**, 753 (1955).
240. A. Schmitt, I. Bottke, and G. Siebert, *Z. Physiol. Chem.*, **347**, 18 (1966).
241. M. Utter, *The Enzymes*, P. D. Boyer, H. Lardy and K. Myrbäck, Eds., Academic Press, *New York*, 1961 **5**, 319.
242. W. Tagaki and F. H. Westheimer, *Biochemistry*, **7**, 891 (1968) and references therein.
243. S. Warren, B. Zerner, and F. H. Westheimer, *Biochemistry*, **5**, 817 (1966).
244. R. A. Laursen and F. H. Westheimer, *J. Amer. Chem. Soc.*, **88**, 3426 (1966).
245. E. Grazi, T. Cheng, and B. L. Horecker, *Biochem. Biophys. Res. Comm.*, **7**, 250 (1962).
246. G. W. Kosicki and F. H. Westheimer, *Biochemistry*, **7**, 4303 (1968).
247. E. E. van Tamelen et al., *J. Amer. Chem. Soc.*, **88**, 4752 (1966).
248. E. E. van Tamelen and T. J. Curphey, *Tetrahedron Lett.*, **3**, 121 (1962).
249. E. E. Van Tamelen et al., *J. Amer. Chem. Soc.*, **88**, 5937 (1966).
250. P. W. Majerus, A. W. Alberts, and P. R. Vagelos, *Proc. Natl. Acad. Sci., U.S.*, **53**, 410 (1965).
251. P. R. Vagelos, personal communication, 1969.
252. P. W. Majerus, *Science*, **159**, 428 (1968).
253. V. Franzen, *Chem. Ber.*, **89**, 1020 (1956).
254. V. Franzen, *Chem. Ber.*, **90**, 623 (1957).
255. S. Kaufmann, *Proc. Natl. Acad. Sci., U.S.*, **50**, 1085 (1963).
256. C. Niemann and H. J. Klein Obbink, *Vitamins Hormones*, **12**, 69 (1954); R. Nicolaysen and N. Eeg-Larsen, *Vitamins Hormones*, **11**, 29 (1953).
257. G. E. Wolstenholme and C. M. O'Connor, *Bone Structure and Metabolism*, Little, Brown and Co., Boston, 1956; G. H. Bourne, *The Biochemistry and Physiology of Bone*, Academic Press, New York, 1956.
258. M. A. Cusanovich, T. Meyer, S. M. Tedro, and M. D. Kamen, *Proc. Natl. Acad. Sci., U.S.*, **68**, 629 (1971).
259. G. N. Schrauzer and J. W. Sibert, *J. Amer. Chem. Soc.*, **92**, 1022 (1970); G. N. Schrauzer and R. J. Holland, *J. Amer. Chem. Soc.*, **93**, 1503 (1971).
260. M. Caplow, *J. Amer. Soc.*, **93**, 230 (1971).

Chapter 18

ENZYME CATALYSIS: BINDING AND SPECIFICITY

18.1 Binding 622
18.2 Specificity 626
    18.2.1 Specificity of Proteolytic Enzymes 627
    18.2.2 Other Enzymes 638

18.1 BINDING

In this chapter binding and specificity are discussed: in an arbitrary fashion, binding is discussed first and specificity second, although the two really cannot be separated. In discussing binding, two major aspects are treated: binding forces and binding constants, although again the two cannot be separated. In discussing specificity, the specificity of the proteolytic and other enzymes will be treated. In all enzymatic processes, a (presumably stereospecific) binding precedes the catalytic step (Chapter 17) and it is one of the tenets of this book that such complexing assists catalysis and is thus necessary for a speedy reaction. In this chapter, it will be assumed that the enzyme is rigid[1]; in the next chapter a flexible enzyme is considered.

One may consider the possible forces that are involved in binding a substrate to an enzyme. This binding ordinarily occurs in aqueous solution, so the principal driving force must be apolar. Long ago Fischer[2,3] enunciated the "lock and key" theory, and recently a model system for this process, in particular a monomolecular inclusion complex, has been described.[4] The apolar nature of binding to chymotrypsin can be seen in Figs. 18.1[5] and 18.2.[6] In these figures, the free energy of the binding is plotted versus the surface area of various hydrocarbons (and other materials) but the free energy could equally well be plotted against the refractive index, as it has been, or against the molecular weight. All will give similar plots. The important thing is that binding primarily involves apolar forces or, as it is sometimes called, the "extraction model."[5] In addition to apolar forces between enzyme and substrate, there also can be electrostatic forces and charge transfer forces. The former are presumably important in the reactions of trypsin;[7] acetylcholinesterase, which is alleged to have an anionic site as well as an esteratic site;[8] and aldolase, which binds all diphosphates 100 times more tightly than monophosphates.[9]

Charge transfer forces are probably not important in the binding of substrate to enzyme. Analogous aromatic and hydroaromatic substrates of α-chymotrypsin were found to have about the same binding constant.[10,11]

There is general agreement that the contribution of the charge-transferred form to the ground state (the "intermolecular semi-polar bond") is small.[12] The observation of charge-transfer absorption maxima from interaction between coenzymes (e.g., $FADH_2$ and $NAD^\oplus$ in lipoamide dehydrogenase[13]) is not as relevant to the present discussion as that between $NAD^\oplus$ and glyceraldehyde-3-phosphate dehydrogenase,[14] for which there are indications that the cofactor interacts with a tryptophan residue of the protein.

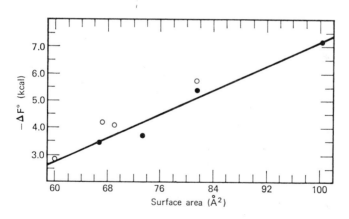

Fig. 18.1 $\Delta F°$ of formation of EI complex plotted against surface area of the inhibitor. The filled circles represent, in ascending order of surface area, benzene, toluene, ethylbenzene, naphthalene, and anthracene. The open circles represent, in the same order, chlorobenzene, nitrobenzene, and azulene. From J. L. Miles, D. A. Robinson, and W. J. Canady, *J. Biol. Chem.*, **238**, 2936 (1963).

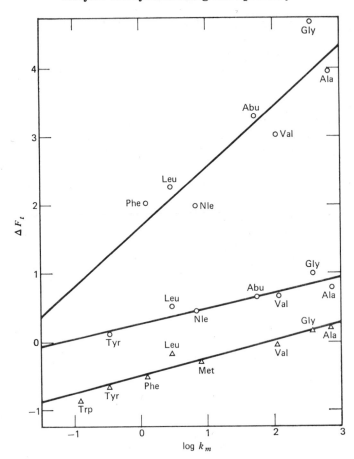

Fig. 18.2 The free energies of transfer of amino acids (ΔF_t) from water to ethanol (●), to 40% aq. ethanol (○) and to 8 M aq. urea (▲), at 25°; versus log K_m values (K_m in millimolar) for the reactions of the *N*-acetyl-L-amino acid methyl esters with α-chymotrypsin. From J. R. Knowles, in *Proc. 9th European Peptide Symp.*, E. Bricas, Ed., 310 (1968) North Holland Publishing Co., Amsterdam. Abu = L-α-aminobutyric acid; Nle = L-norleucine.

Here the electron-deficient nicotinamide ring of $NAD^{\oplus}$ is thought to be π-complexed with the electron-rich indole ring of a tryptophan residue, and this interaction is responsible for the charge-transfer spectrum. In a similar vein, the specificity of Koshland's reagent (2-hydroxy-5-nitrobenzyl bromide) for tryptophan residues in proteins[15] was thought to arise from a prior π-complexation between reagent and tryptophan side chains.[16] It is now known, however, that the absorption maximum thought to be due to the charge transfer spectrum is due to the ready ionization of a hydrogen atom of the ylid.[17] The small contribution of charge-transfer forces to stability of π- (or charge-transfer) complexes points to the fact that the interaction is a cooperative one, that is, intermolecular forces other than charge-transfer forces lead to the possibility of the charge-transfer interaction.

In the previous discussion it is implied that only one force between substrate and enzyme is operative. Probably this is not true, with apolar forces being assisted by some other force. In most of the previous investigations, a small substrate interacts with a large enzyme. But what if the substrate is also large?[12] Many enzymes such as lysozyme and papain have been shown by X-ray diffraction to possess a cleft or crevasse where the active site resides. Perhaps in such a cleft, a further hole exists that explains the apolar binding of chymotrypsin. The phenomenally small dissociation constants of some trypsin-trypsin inhibitor systems[18] may be most easily explained by multiple interactions. As indicated before, papain probably has multiple interactions also and these are specifically being identified.[19] The term apolar has been used deliberately so the mechanism by which the binding occurs, either hydrophobic or hydrotactic (see earlier), is avoided. A combination of an electrostatic interaction and a hydrogen bonding may occur between carboxyl groups of substrates and arginine residues of enzymes.[4] Five such pairs have been found in crystalline papain. It has been postulated that there is a connection between binding and catalysis,[20] and even that "better binding" leads to "better catalysis."[6] The data quoted are with methyl esters for which the rate-determining step is not known with certainty but the suggestion may have some worth.

Binding constants are indicative either of the forces or of the specificity between enzyme and substrate. A large number (136) of binding constants of aromatic inhibitors determined in the α-chymotrypsin-catalyzed hydrolysis of a valine substrate have been tabulated.[21] In this tabulation, it is seen that anionic inhibitors bind less readily than corresponding neutral inhibitors. This is borne out in this laboratory where an anionic compound was not hydrolyzed by the enzyme whereas the corresponding neutral compound was.[22] This is presumably due to a lack of binding by the anionic compound. The dye proflavin binds to the active site of α-chymotrypsin.[23-25] Biebrich Scarlet [4-(2-hydroxy-1-naphthylazo)azobenzene-3,4'-disulfonic acid] also binds specifically to the active site of α-chymotrypsin.[26] The dye thionine (3,6-diaminophenothiazine) binds specifically to the active site of trypsin.[27] The dye trypan blue binds to the protein, bovine serine albumin.[28]

A thorough analysis of the possible modes of binding of substrates to α-chymotrypsin has been made and a set of empirical rules has been laid down that involves many parameters.[29] A more sophisticated approach involves a quantitative analysis of the binding of α-amino amides to α-chymotrypsin.[30] Amides were chosen to stay away from the acyl-enzyme problem (Chapter 14) which one meets with esters.

A further analysis of binding constants is made in the specificity section, but it is pertinent to consider two aspects here. One is a thermodynamic study of chymotrypsin complexes that indicates that sometimes enthalpy is important and sometimes entropy.[31,32] Another is the binding of amidines and guanidines, which possess a positive charge in the normal enzymatic pH range, to trypsin, which has an affinity for substances possessing a positive charge. When the amidine or guanidine possesses a benzene ring, the dissociation constants are of the order of 10^{-5} to 10^{-6}, which means strong binding; when the amidine or guanidine is attached to an aliphatic chain or ring, the binding is not as great,[33] indicating that apolar binding is still important.

18.2 SPECIFICITY

Let us now consider specificity. The specificity of proteolytic and other enzymes is considered. Of course, such specificity cannot be divorced from binding, and as mentioned before, they are probably connected. Several general concepts have emerged from a study of specificity. One is nonproductive binding, sometimes called wrong-way binding or the formation of dystopic and eutopic complexes (Chapter 14). In any event the postulate is that one form has the correct stereochemistry to react while the other does not. This idea has been enunciated by many people.[35–37] Non-productive binding will affect both the rate and equilibrium constant as shown in eqs. 18.1–18.8.[36]

$$E + S \underset{k_{-1}}{\overset{k_1}{\rightleftharpoons}} ES \overset{k_2}{\longrightarrow} E + P \qquad (18.1)$$

$$E + S \underset{k_{-3}}{\overset{k_3}{\rightleftharpoons}} ES' \overset{k_4}{\longrightarrow} E + P \qquad (18.2)$$

where the binding constants are $K_{S.1} = (k_{-1} + k_2)/k_1$ and $K_{S.2} = (k_{-3} + k_4)/k_3$, and the existence of two kinds of binding leads to the relation

$$\frac{1}{K_0} = \frac{1}{K_{S.1}} + \frac{1}{K_{S.2}} \qquad (18.3)$$

or for the more general case of n kinds of binding

$$\frac{1}{K_0} = \sum^n \frac{1}{K_{S.j}} \qquad (18.4)$$

Specificity

In binding, if one has both productive and nonproductive binding, that is, if $k_4 = 0$

$$k_0 = \frac{K_{S.2}k_2}{K_{S.1} + K_{S.2}} \qquad (18.5)$$

and, generally,

$$k_0 = \frac{k_2}{1 + K_{S.1} \sum_{j+1}^{n} \frac{1}{K_{S.j+1}}} \qquad (18.6)$$

If $k_4 \neq 0$

$$k_0 = \frac{K_{S.2}k_2 + K_{S.1}k_4}{K_{S.1} + K_{S.2}} \qquad (18.7)$$

and, generally,

$$k_0 = \frac{\sum_{j}^{n} \frac{k_{2j}}{K_{S.j}}}{\sum_{j}^{n} \frac{1}{K_{S.j}}} \qquad (18.8)$$

The effect of nonproductive binding on the rate constant will be to decrease it so that for any series an unusually low rate constant associated with a low dissociation constant implies the existence of nonproductive binding. The stereochemistry of binding has been utilized in at least two different ways. One way is to invert the usual L-stereospecificity of α-chymotrypsin substrates, particularly with 3-carbomethoxyisocarbostyril,[38] which is a cyclic analog of phenylalanine and a specific (fast) substrate. This observation created quite a furor at the time, since it was thought that enzymes were completely stereospecific. Many explanations were forthcoming and it was subsequently shown that an inversion of stereospecificity need not be associated with a fused ring system, for a simple interchange of the size (bulk) of the two groups on the α-carbon atom would lead to an inversion of stereospecificity.[38] Another way to utilize the stereochemistry of binding is to start with an optically inactive (racemic) material and produce an optically active one from it. This can be done by α-chymotrypsin[39-41] and other proteolytic enzymes, and other kinds of enzymes also.

18.2.1 Specificity of Proteolytic Enzymes

The striking property of enzyme specificity may be observed in at least three forms: medium (solvent, pH, etc.); reaction; and substrate specificity. In this view, specificity is a "relative kinetic specificity," higher specificity being defined as a higher rate of reaction with respect to some reference substrate or reaction.[20] This definition implies that specificity is not unique to enzyme-catalyzed reactions; for example, hydroxide ion catalyzes the hydrolysis of esters but not of orthoesters, and it catalyzes the hydrolysis of methyl trichloroacetate much more readily than that of methyl acetate. Thus to measure the special contribution of the enzyme to the catalysis, we compare the velocity of the enzymatic reaction to the velocity of a non-enzymatic reaction involving the same gross pathway.[42-45] However, except possibly for α-chymotrypsin,[46] mechanism (or even pathway) is beyond our present understanding.

Enzyme specificity can, however, be discussed fruitfully using presently available data by the concept of "relative enzymatic specificity," the ratio of the reactivity of a given substrate to the reactivity of a reference substrate, under identical conditions. In the same way, "relative binding specificity" may be defined by the ratio of the binding constants, K_s, to the substrate and reference substrate.

The purpose of the following considerations is to establish a relationship between substrate structure and specificity, discussing mainly α-chymotrypsin, for which the most extensive data are available.

Hydrolytic reactions catalyzed by α-chymotrypsin may be described kinetically by the three-step mechanism of eqs. 18.9 and 18.10 (see Chapter 16)

$$E + S \underset{}{\overset{K_s}{\rightleftharpoons}} ES \xrightarrow{k_2} ES' \xrightarrow{k_3} E + P_2 \qquad (18.9)$$
$$K_i \downarrow \qquad +P_1$$
$$ES_x$$

$$V = \dfrac{\dfrac{k_2}{1 + \dfrac{K_s}{K_i} + \dfrac{k_2}{k_3}} E_0 S_0}{S_0 + \dfrac{K_s}{1 + \dfrac{K_s}{K_i} + \dfrac{k_2}{k_3}}} \qquad (18.10)$$

Each of these steps may have its own specificity. Although some kinetic data are available on the individual rate constants of eqs. 18.1 or 18.9,[20] most of the data in the literature have been determined in terms of k_{cat} and K_m(app);[47] hence their meaning is not simply related to the individual rate constants. Moreover, if the existence of more than one binding site is postulated, the meaning of these experimental constants may be complicated by possible "nonproductive (unreactive) complex" formation[35,48,49] and incomplete interaction between the enzyme and substrate.[49,50]

What constants are kinetically meaningful? Certainly K_m(app), K_s, and K_i are not meaningful since they may be a combination of individual binding constants of multiple forms of binding (*inter alia*). Certainly k_{cat} is not unequivocal, since it may pertain to different steps of the reaction, or in the general case be a complex constant. Furthermore, k_2 may be affected by multiple forms of binding.

If multiple forms of binding are ignored, $k_{cat}/K_m(app) = k_2/K_s$.[51] If multiple forms of binding are taken into account, and it is reasonably postulated that only one kind of binding leads to reaction, then the ratio $k_{cat}/K_m(app)$ is still equal to k_2 divided by the binding constant of the productive mode, as shown in eqs. 18.9 and 18.10, and hence $k_{cat}/K_m(app) = k_2/K_s$. Thus $k_{cat}/K_m(app)$ is probably always a meaningful kinetic constant. Furthermore, when independent experimental proof exists that $k_2 \gg k_3$, then the value of $k_{cat}(=k_3)$ may be used as a meaningful constant since it is unaffected by multiple binding (assuming K_i is of the same order of magnitude as K_s).

The above considerations indicate that in a search for a correlation between structure and specificity, only k_3 and the complex constant $k_{cat}/K_m(app)$ may be used.

The data of α-chymotrypsin, while the best in the literature, still adhere to poorer standards than nonenzymic kinetic data, mainly because of inherent difficulties in handling very low enzyme concentrations, problems in determining the purity of enzyme preparations, and the extreme sensitivity of enzymatic reactions toward minute changes of the reaction medium. Thus reproducibility within a factor of 2 from different laboratories must be accepted, extensive ranges of rate constants must be sought, and limitations must be applied to any theory based on these data.

Data on chymotrypsin deacylation rate constants (k_3) have been collected.[20,52,53] Several major conclusions can be drawn from these data: specificity in deacylation is a pure entropy effect; and the interaction of the enzyme with the two groups on the alpha-carbon atom of the acyl group, the aralkyl (alkyl) and acylamido groups, leads to an increase of the reaction rate (an increase in specificity). In this context, the two groups are roughly equally effective and, furthermore, act independently of one another, as can be seen from the approximate additivity of the individual effects. A linear free energy relation was found between deacylation rates and the rate constants for the alkaline hydrolysis of the corresponding ethyl esters. In addition to general base catalysis, general acid catalysis and strain were invoked to explain the rates of deacylation.[54,55] Furthermore when one plots $\log k_3$ versus $\log K_i$ there appears to be a saturation phenomenon occurring,[6] as shown in Fig. 18.3.

Let us now consider the more extensive $k_{cat}/K_m(app)$ data (which have all been reduced to 25°, pH 7.9, and water) of the group of L-amino acid substrates, R_1–CONH–CH(R_2)–C(O)X. The specificity of the leaving group X may be investigated by keeping R_1 and R_2 constant and varying X. Variation of the X group from NH_2 to p-nitrophenoxy in the N-acetyl-L-tryptophan and N-acetyl-L-phenylalanine series led to a large difference in

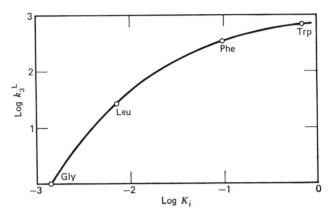

Fig. 18.3 Plot of log k_3^{rel} for L-acyl-chymotrypsins versus $-\log K_i$ (K_i in mM) for the corresponding D-acyl amides. From J. R. Knowles, in *Proc. 9th European Peptide Symp.*, E. Bricas, Ed., 310 (1968). North Holland Pub. Co., Amsterdam.

reactivity, but these relative rates closely parallel those of the hydroxide ion-catalyzed hydrolysis of the corresponding compounds.

An even closer parallelism between $k_{\text{cat}}/K_m(\text{app})$ of an enzymatic reaction and k_{OH^-} of a nonenzymatic reaction is seen in the papain-catalyzed hydrolyses of hippuric acid derivatives, as shown in Fig. 18.4.[56]

The effect of variation of the group X on α-chymotrypsin rate constants, $k_{\text{cat}}/K_m(\text{app})$ (Table 18.1), shows Hammett-type linear free energy relationships. In the hydrolysis of substituted phenyl acetates, $\rho = 1.8$, while a

Table 18.1

Effect of the Variation of the Leaving Group, X, on α-Chymotrypsin-Catalyzed Reactions[b]

Leaving Group, X	$k_{\text{cat}}/K_m(\text{app})$ M^{-1} sec^{-1}		Relative $k_{\text{OH}^\ominus}$ of Acetate Derivatives
	N-Acetyl-L-Tryptophan	N-Acetyl-L-Phenylalanine	
NH$_2$[a]	7.2	1.53	1
OC$_2$H$_5$	5.8–12.4 × 10^5	13–32 × 10^4	2,750
OCH$_3$	7.0–13.1 × 10^5	2.7, 4.2, 15.3 × 10^4	5,500
OC$_6$H$_4$NO$_2$-p	6.8 × 10^7	1.5 × 10^7	315,000

[a] The Taft substituent constants, ρ^* and ρ, are practically identical for aliphatic amides and esters in hydroxide ion-catalyzed hydrolysis.[57,58] Thus it is reasonable to include reactions of amides and esters in the same comparison.
[b] From M. L. Bender and F. J. Kézdy, *Ann. Rev. Biochem.*, **34**, 49 (1965).

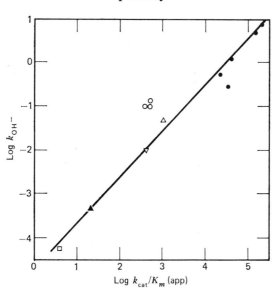

Fig. 18.4 The hydroxide ion- and papain-catalyzed hydrolysis of hippurate derivatives: ● (phenyl esters); ○ (esters of primary alcohols); esters of $(HS(CH_2)_2CO_2-)$; △ (C_2H_5SH); ▲ (isopropyl alcohol); and □ (amide). From M. L. Bender and F. J. Kézdy, *Ann. Rev. Biochem.*, **34**, 49 (1965).

corresponding hydroxide ion-catalyzed reaction gives a $\rho = 1.3$.[59] In the hydrolysis of substituted N-benzoyl-L-tyrosine anilides, $\rho = -1.63$ (25% dimethyl sulfoxide-water), while a corresponding hydroxide ion-catalyzed hydrolysis of anilides gives a $\rho = 0.1$ (water).[60] In the aliphatic series, the k_{cat}/K_m(app) of the α-chymotrypsin-catalyzed hydrolyses of four esters of N-acetyl-L-valine and of three esters of N-acetyl-L-norvaline[61] yield quite good straight lines when plotted versus σ^* of Taft,[58] yielding $\rho^* = 2.5$ and 2.9, respectively, while the ρ^* for esters of glyceric acid is 3.3, and that for lactic or acetic acid is 1.47.[58] No steric factor need be used for the valine and norvaline correlations as is necessary in hydroxide ion-catalyzed reactions, implying that the intramolecular enzymatic reaction reduces the importance of steric hindrances that are important in collisional reactions.

The deacylations of the esters, substituted-benzoyl-α-chymotrypsins, obey the Hammett equation with a ρ of 2.1,[62] while the corresponding hydroxide ion-catalyzed hydrolysis of ethyl benzoates has a ρ of 2.2–2.5. These reactions, where the leaving group is constant, show the same relationship to hydroxide ion-catalyzed substituent effects as is seen in the enzymatic k_{cat}/K_m(app) correlations for esters listed above. It must therefore be concluded that the latter correlations reflect solely electronic effects inherent in the group X and not some binding that is substituent-sensitive. Thus the effect of a variation of the group X on k_{cat}/K_m(app) appears, in general, not to involve productive binding of X to the enzyme, but rather to reflect electronic influences of X. For L-amino acid derivatives, one can therefore use a form of the Taft equation to describe effects of X on the rate:

$$\log \frac{\left(\frac{k_{cat}}{K_{m(app)}}\right)_{R_1,R_2,X}}{\left(\frac{k_{cat}}{K_{m(app)}}\right)_{R_1,R_2,X_0}} = \rho_X^* \sigma_X^* \qquad (18.11)$$

where R_1, R_2, and X refer to the various components of a chymotrypsin substrate.

Data pertinent to the influence of R_1 of R_1–CONH–CH(R_2)–C(O)X on k_{cat}/K_m(app) of α-chymotrypsin reactions are given in Tables 18.2 and 18.3. In Table 18.2, relative specificities of the R_1 groups are calculated, using the acetyl group as standard. In Table 18.3, relative specificities of R_1 groups

Table 18.2

$$\alpha\text{-Chymotrypsin Reactions}^{a,b,e} \quad \frac{\left(\frac{k_{cat}}{K_{m(app)}}\right)_{R_1,R_2,X}}{\left(\frac{k_{cat}}{K_{m(app)}}\right)_{acetyl,R_2,X}}$$

Amino Acid Derivative	R_1CO					
	Nico-tinyl	Pico-linyl	Benzoyl	Isonico-tinyl	Chloro-acetyl	2-Furoyl
L-Alanine	7.6	1.86	11.4	6.7		5.0
L-α-Aminobutyric acid	—	2.1				
L-Norvaline	—	1.41				
L-Valine	6.5	2.17	6.5		1.9	6.15
L-Phenylalanine	4.1[c]					
L-Tryptophan	5.5[c]					
L-Tyrosine	2.8[d]		6.1[d]			
	5.6[e]		21.9[c]	8.6[c]	1.97	

[a] Methyl esters, unless otherwise indicated. The influence of R_1 on the hydroxide ion-catalyzed reactions is in general less than a factor of 2, and thus need not be taken into account.[63]
[b] Refs. 61, 63–69.
[c] Amide.
[d] Hydrazide.
[e] From M. L. Bender and F. J. Kézdy, *Ann. Rev. Biochem.*, **34**, 49 (1965).

Specificity

Table 18.3

α-Chymotrypsin Reactions[a,c] $\dfrac{\left(\dfrac{k_{cat}}{K_{m(app)}}\right)_{YR'_1,Tyr,NH_2}}{\left(\dfrac{k_{cat}}{K_{m(app)}}\right)_{YGly,Tyr,NH_2}}$

Y	R'_1						
	L-Leu	L-Norleu	L-Norval	L-Val	L-Abu[b]	L-Ala	Gly
H	26.9	22.6	23.2	15.0	14.5	3.6	1
Gly	20.8	22.0	—	15.7	—	4.3	1

[a] 30° C; pH 7.8–8.0; 0.1 M phosphate.
[b] L-α-Aminobutyric acid.
[c] From M. L. Bender and F. J. Kézdy, *Ann. Rev. Biochem.*, **34**, 49 (1965).

are calculated in which a part of that group remains constant. The experimental results of both series [over a range of several hundred in $k_{cat}/K_m(app)$] indicate that the relative specificity of a given R_1 is independent of the nature of X or R_2 over a wide range of reactivity. Thus we can formulate the influence of R_1 by the free energy differences of eq. 18.12:

$$\log \dfrac{\left(\dfrac{k_{cat}}{K_{m(app)}}\right)_{R_1,R_2,X}}{\left(\dfrac{k_{cat}}{K_{m(app)}}\right)_{R_{1_0},R_2,X}} = S_{R_1} \qquad (18.12)$$

where the specificity factor, S_{R_1}, is dependent only on the nature of R_1 and is independent of R_2 or X. The variations in steric and electronic properties of R_1 (and probably R_2) as reflected in hydroxide ion reactions do not appear to be important.[63]

If the independence of the specificity of the X and R_1 groups from the other substituents of the substrate is accepted, as postulated above, the independence of the specificity of the R_2 group is already defined. For example, if, from the independence of the X groups

$$\dfrac{N\text{-acetyl-L-alanine amide}}{N\text{-acetyl-L-alanine methyl ester}} = \dfrac{N\text{-acetyl-L-tyrosine amide}}{N\text{-acetyl-L-tyrosine methyl ester}}$$

then

$$\dfrac{N\text{-acetyl-L-alanine amide}}{N\text{-acetyl-L-tyrosine amide}} = \dfrac{N\text{-acetyl-L-alanine methyl ester}}{N\text{-acetyl-L-tyrosine methyl ester}}$$

A corresponding equivalence may be demonstrated from the independence

Table 18.4

α-Chymotrypsin Reactions[a,b] $\dfrac{\left(\dfrac{k_{cat}}{K_{m(app)}}\right)_{R_1,\,tryptophan,\,X}}{\left(\dfrac{k_{cat}}{K_{m(app)}}\right)_{R_1,\,phenylalanine,\,X}}$

Y	Amide	Ethyl Ester	Methyl Ester	p-Nitrophenyl Ester	Amide
R_1	Acetyl	Acetyl	Acetyl	Acetyl	Nicotinyl
Tryptophan Phenylalanine	4.7	4.4, 3.9	8.5, 16.2	4.7	5.7

[a] Data of Table 18.1 used, except for the last entries whose source is Ref. 65.
[b] From M. L. Bender and F. J. Kézdy, *Ann. Rev. Biochem.*, **34**, 49 (1965).

of R_1 groups. Table 18.4 gives an example of the independence of the specificity of R_2 groups. With the exception of the methyl ester, which shows experimental difficulties, these data confirm the independence of the specificity of R_2 groups. Thus we can express the specificity of R_2 in free energy terms by eq. 18.13.

$$\log \dfrac{\left(\dfrac{k_{cat}}{K_{m(app)}}\right)_{R_1,R_2,X}}{\left(\dfrac{k_{cat}}{K_{m(app)}}\right)_{R_1,R_{20},X}} = S_{R_2} \quad (18.13)$$

where the specificity factor S_{R_2} is dependent solely on the nature of R_2 and is independent of R_1 and X.

By combining all equations, we arrive at eq. 18.14, which describes the $k_{cat}/K_m(app)$ of any N-acyl-L-amino acid derivative

$$\log \dfrac{\left(\dfrac{k_{cat}}{K_{m(app)}}\right)_{R_1,R_1,X}}{\left(\dfrac{k_{cat}}{K_{m(app)}}\right)_{R_{10},R_{20},X_0}} = \rho_X^* \sigma_X^* + S_{R_1} + S_{R_2} \quad (18.14)$$

in terms of three independent factors: one nonenzymatic reactivity term and two specificity terms. The total enzymatic rate may arise from two sources: catalysis and binding, which may be manifest in the strength of binding and in the orientation of the bound form. Since all mechanistic characteristics of α-chymotrypsin reactions have been found to be identical for the most and least specific substrates, the same catalysis probably occurs in the reaction of every substrate.[46] Thus the problem of specificity is reduced to the problem of binding, and two independent specificity sites must mean two independent binding sites.[70] This modifies the meaning of $k_{cat}/K_m(app)$. Since the two bindings are independent, they would not be expected to occur simultaneously. The dimerization of acetic acid[71] and the association of ferric ion and bipyridyl,[72] both reactions involving two bindings, appear to occur in two steps. A two-step adsorption scheme is shown below in which a and b are partial complexes formed between enzyme and substrate, and ab is the complete (double) complex, postulated to be the only reactive species.

$$E + S \underset{}{\overset{K_1}{\rightleftharpoons}} a \qquad (18.15)$$

$$K_2 \updownarrow \qquad \updownarrow K_3$$

$$b \underset{}{\overset{K_4}{\rightleftharpoons}} ab \underset{}{\overset{k_2}{\rightleftharpoons}} ES' + P_1 \underset{}{\overset{k_3}{\rightleftharpoons}} E + P_2$$

If enzyme-substrate complex formation is much faster than the kinetic steps, the steady-state assumption for ES' in eq. 18.15 yields the rate expression

$$\frac{dP_1}{dt} = V = \frac{\dfrac{k_2}{K_1 K_3}[S][E]_0}{[S] + 1 \Big/ \left[\dfrac{1}{K_1} + \dfrac{1}{K_2} + \dfrac{1}{K_1 K_3}\left(1 + \dfrac{k_2}{k_3}\right)\right]} \qquad (18.16)$$

If nonproductive binding occurs, a similar, complicated expression is obtained. But with or without nonproductive binding, $k_{cat}/K_m(app) = k_2/K_1K_3 = k_2/K_2K_4$. Since we identify a and b as enzyme-substrate complexes involving binding by R_1 and R_2 of the substrate to the corresponding enzymatic adsorption site, we define the specificity of R_1 and R_2 on $k_{cat}/K_m(app)$ in terms of their influence on both the rate constant, k_2, and the binding constants, K_1 and K_3 (or K_2 and K_4).

The nature of α-chymotrypsin substrates listed above indicates that binding involves only hydrogen bonds and apolar bonds. The presence of a hydrogen bond in the binding of the acylamino group, involving both the NH and C=O groups of the peptide bond, is indicated by the following data: the absence of the NH group leads to considerable loss of reactivity and stereospecificity,[73–75] and $K_m(app)$ is reduced much more in D_2O with an acylamino acid substrate than with a substrate containing no acylamino group.[76] But yet R_1 of the acylamino group also exerts specificity, as seen in Tables 18.2 and 18.3. The apolar nature of R_1 combined with the hydrogen bonding of the NHC(O) group defines the binding of the acylamino group as a mixture of hydrogen and apolar binding. On the other hand, binding by the R_2 group appears to be wholly apolar in nature, as seen in the nature of the R_2 groups.

The two-point attachment of the substrate to the enzyme suggested above is a necessary and sufficient condition for explaining the stereospecificity of α-chymotrypsin, that is, its ability to distinguish between D- and L-isomers, since the relative position of the catalytic and nucleophilic groups of the enzyme with respect to the reactive link of the substrate provides the third point necessary for determination of complete stereospecificity.

An incomplete substrate such as methyl hippurate can be expected to follow eqs. 18.14 or 18.15 only with modifications. This is particularly seen in the noncompetitive inhibition of the hydrolysis of methyl hippurate by indole.[77] To explain this phenomenon, only one additional postulate need be added to eq. 18.15, namely that methyl hippurate, which contains only one binding group, can bind productively in either of two ways to the enzyme—with the phenyl group bound to the R_2 apolar site or with the peptide bond and phenyl group bound to the R_1 site. The inhibitor, indole, would be expected to bind only at the R_2 site because of its lack of polarity. The kinetic scheme accommodating these possibilities,

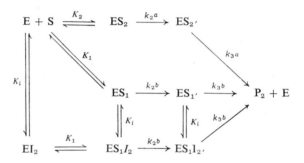

indicates that indole will competitively inhibit substrate binding at site 2 but will have no effect on substrate binding at site 1. A steady-state treatment of this system leads to the prediction that a plot of $[k_{cat}/K_m(\text{app})] \times (1 + [I]/K_i)$ versus $(1 + [I]/K_i)$ should yield a straight line.

This explanation of noncompetitive inhibition considers the chemical nature of the binding as a basis for the kinetics.

Other optical specificity and activation phenomena[63,78-81] may formally be analyzed according to schemes related to that shown above for methyl hippurate, postulating that the substrate and the activator, if present, may bind to either or both of the two independent binding sites and that the different reactivities are associated with different binding sites.

The above analysis provides a working hypothesis for the specificity of α-chymotrypsin on the basis of a catalytic center and two independent binding sites, an apolar site and a site involving both hydrogen bonding and apolar bonding. The analysis assumes a rigid enzyme and a flexible substrate.

Specificity

There have been an enormous number of specificity studies on α-chymotrypsin, mainly by Niemann and co-workers. Some time ago it was shown that fatty acid esters served as substrates for both α-chymotrypsin and trypsin.[82] The rates followed the Hofmeister series, that is the rates (probably second-order) peaked at six or seven carbon atoms. Ordinarily fatty acid esters of *ortho*- or *meta*-hydroxybenzoic acids were used for solubility purposes.[83] The (second-order) rate constants for a similar series of esters with horse liver esterase shows a continual rise up to 14 carbon atoms.[84] The peaking at six or seven carbon atoms may be due to folding. In a further study, homologous acyl-chymotrypsins from two to six carbons were prepared and the thermodynamic parameters of the k_3 step with these compounds were investigated. Nothing unusual was found with the enthalpies although the largest chain seemed to have the highest enthalpy. The entropies, however, did change almost fortyfold and the trend seen in the entropies suggests that either more unreactive configurations relative to the active site are available to the shorter chain substrates, or that a protein conformation change enhancing activity is more efficiently induced by the longer chains.[85] When one or more of the methionines near the active site is oxidized, the specificity of chymotrypsin is changed.[86] Hammett sigma-rho correlations on chymotrypsin reactions have taken many turns. Some years ago, notwithstanding the warning about nonproductive binding given earlier, a two-point line was utilized to investigate the acylation step using nonspecific substrates, and k_2 was found to increase with electron-withdrawing substituents[87] as was k_3, the rate constant of the deacylation step.[88] Of course, a two-point line is insufficient and there were some discrepancies in the k_3 correlation, so both of these reactions are being investigated in this laboratory at present. The discrepancies mentioned above are compounded by a negative rho obtained for five anilides and a positive rho obtained for two substrates.[89,90] The former results were interpreted as indicating the participation of a general acid catalyst, but could be equally well interpreted as evidence for a tetrahedral intermediate with a change in the rate-determining step. The discrepancies in the k_3 step suggest the opposite form of the Hammett plot and again suggest a tetrahedral intermediate. The second-order rate constant (k_{cat}/K_m) has been further confused by a study of a number of anilides,[91] since this is a complex constant.

In line with the lack of absolute specificity by α-chymotrypsin, both D- and L-acyl-chymotrypsins can be deacylated, the D-compounds of course going more slowly than the L-compounds. In fact, the faster the deacylation of the L-acyl-enzyme, the slower the deacylation of the D-acyl-enzyme, as shown in Fig. 18.5.[6] A significant advance was made when it was found that an acylamino group was not even needed. That is, 2-naphthoic acid derivatives would serve as α-chymotrypsin substrates.[92]

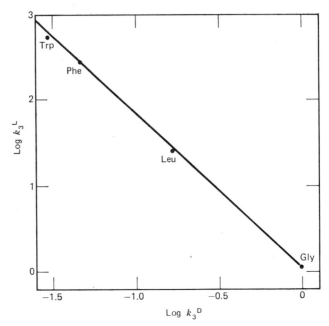

Fig. 18.5 Plot of log k_3^{rel} for D- versus L-acyl-chymotrypsins. From J. R. Knowles, in *Proc. 9th European Peptide Symp.*, E. Bricas, Ed., 310 (1968). North Holland Pub. Co., Amsterdam.

As mentioned before, trypsin requires a positive charge as well as an apolar group for its specificity. On the basis that trypsin hydrolyzes neutral analogs of arginine or lysine (at a lower rate) and that with positively-charged substrates, competitive inhibition was found whereas with neutral substrates either noncompetitive or no inhibition was found; it was postulated that trypsin contains two binding sites, one for neutral substrates and the other for positively-charged substrates, the sites being close in space to one another.[93] Whether this is the only explanation of the results is not clear at this time. A good deal of attention has been paid to apolar (hydrophobic) binding sites. A significant advance was made when a methylene group was inserted into an alanine residue of trypsin at the active site of that enzyme.[94]

The stereospecificity of the acetylcholinesterase-catalyzed hydrolysis of sarin (isopropyl methylphosphonofluoridate) was investigated.[95] Chymotrypsin has been used in a study of enzymic structure-activity relationships.[98]

18.2.2 Other Enzymes

The specificity of enzymes other than proteolytic has not been investigated nearly as much. The reason for this is that they are usually not extracellular and are more difficult to obtain in crystalline form. L-Amino acid oxidase has been alluded to [37] and the specificity of a number of other enzymes can be inferred from the past discussion (Chapters 16, 17 and above). The interactions of ribonuclease with the dinucleoside phosphate, cytidyl-3′:5′-cytidine, and cytidine 2′,3′-cyclic phosphate were investigated using a stopped-flow, temperature jump approach.[96,97] At least two relaxation processes were observed which implies that with this enzyme binding occurs stepwise.

REFERENCES

1. W. P. Jencks, *Catalysis in Chemistry and Enzymology*, McGraw-Hill, New York, 1969.
2. E. Fischer, *Ber.*, **27**, 2985 (1894).
3. E. Fischer, *Z. Physiol. Chem.*, **26**, 60 (1898).
4. R. L. VanEtten et al., *J. Amer. Chem. Soc.*, **89**, 3242 (1967).
5. J. L. Miles, D. A. Robinson, and W. J. Canady, *J. Biol. Chem.*, **238**, 2936 (1963); see also A. J. Hymes, D. A. Robinson, and W. J. Canady, *J. Biol. Chem.*, **240**, 134 1965; L. J. Miles et al., *J. Biol. Chem.*, **237**, 1319 (1962).
6. J. R. Knowles, in *Proc. 9th European Peptide Symp.*, E. Bricas, Ed., k7 (1968).
7. H. R. Mahler and E. H. Cordes, *Biological Chemistry*, Harper and Row, New York, 1966 p. 661.
8. F. Bergmann, *Discuss. Faraday Soc.*, **20**, 126 (1955).
9. F. C. Hartmann and R. Barker, *Biochemistry*, **4**, 1068 (1965).
10. R. R. Jennings and C. Niemann, *J. Amer. Chem. Soc.*, **75**, 4687 (1953).
11. J. B. Jones and C. Niemann, *Biochemistry*, **2**, 498 (1963).
12. E. M. Kosower, in *Flavins and Flavoproteins*, E. C. Slater, Ed., Elsevier Publishing Co., Amsterdam, 1966, p. 1.
13. V. Massey, G. Palmer, C. H. Williams, Jr., B. E. P. Swoboda, and R. H. Sands, in *Flavins and Flavoproteins*, E. C. Slater, Ed., Elsevier Publishing Co., Amsterdam, 1966, p. 133.
14. S. Shifrin, *Biochim. Biophys. Acta*, **81**, 205 (1964).
15. H. R. Horton and D. E. Koshland, Jr., *J. Amer. Chem. Soc.*, **87**, 1126 (1965).
16. D. S. Sigman and E. R. Blout, *J. Amer. Chem. Soc.*, **89**, 1747 (1967).
17. M. Gibian, personal communication.
18. M. Laskowski, Jr., personal communication.
19. N. Abramowitz, I. Schechter, and A. Berger, *Biochem. Biophys. Res. Comm.*, **29**, 869 (1967).
20. M. L. Bender, F. J. Kézdy and C. R. Gunter, *J. Amer. Chem. Soc.*, **86**, 3714 (1964).
21. R. A. Wallace, A. N. Kurtz, and C. Niemann, *Biochemistry*, **2**, 824 (1963).
22. F. J. Kézdy and M. L. Bender, unpublished results.
23. S. A. Bernhard and B. F. Lee, *Abstr. 6th Int. Cong. Biochem.*, **32**, 297 (1964).
24. S. A. Bernhard and H. Gutfreund, *Proc. Natl. Acad. Sci., U.S.*, **53**, 1238 (1965).
25. S. A. Bernhard, B. F. Lee, and Z. H. Tashjian, *J. Mol. Biol.*, **18**, 405 (1966).
26. A. N. Glazer, *J. Biol. Chem.*, **242**, 4528 (1967).
27. A. N. Glazer, *J. Biol. Chem.*, **242**, 3326 (1967).
28. J. H. Lang and E. C. Lasser, *Biochemistry*, **6**, 2403 (1967).
29. G. E. Hein and C. Niemann, *J. Amer. Chem. Soc.*, **84**, 4495 (1962).
30. C. L. Hamilton, C. Niemann, and G. S. Hammond, *Proc. Natl. Acad. Sci., U.S.*, **55**, 664 (1966).
31. D. G. Doherty and F. Vaslow, *J. Amer. Chem. Soc.*, **74**, 931 (1952).

32. F. Vaslow and D. G. Doherty, *J. Amer. Chem. Soc.*, **75**, 928 (1953).
33. M. Mares-Guia and E. Shaw, *J. Biol. Chem.*, **240**, 1579 (1965).
34. D. H. Meadows, J. L. Markley, J. S. Cohen, and O. Jardetzky, *Proc. Natl. Acad. Sci., U.S.*, **58**, 1307 (1967).
35. S. A. Bernhard and H. Gutfreund, *Proc. Int. Symp. Enzyme Chem.*, *Tokyo and Kyoto 1957*, Maruzen Co., Tokyo, 1958, p. 124.
36. G. E. Hein and C. Niemann, *J. Amer. Chem. Soc.*, **84**, 4495 (1962).
37. E. A. Zeller et al., *Biochem. J.*, **95**, 262 (1965).
38. G. E. Hein and C. Niemann, *J. Amer. Chem. Soc.*, **84**, 4487, 4495 (1962).
39. S. G. Cohen, L. H. Klee, and S. Y. Weinstein, *J. Amer. Chem. Soc.*, **88**, 5302 (1966).
40. S. G. Cohen, Z. Neuwirth, and S. Y. Weinstein, *J. Amer. Chem. Soc.*, **88**, 5306 (1966).
41. S. G. Cohen, R. M. Schultz, and S. Y. Weinstein, *J. Amer. Chem. Soc.*, **88**, 5315 (1966); S. G. Cohen and J. Crossley, *J. Amer. Chem. Soc.*, **86**, 4999 (1964); and other papers.
42. D. E. Koshland, Jr., *Adv. Enzymol.*, **22**, 45 (1960).
43. D. E. Koshland, Jr., *J. Theoret. Biol.*, **2**, 75 (1962).
44. D. E. Koshland, Jr., in *Horizons in Biochemistry*, M. Kasha and B. Pullman, Eds., Academic Press, New York, 1962, p. 265.
45. D. E. Koshland, Jr., J. A. Yankeelov, Jr., and J. A. Thoma, *Fed. Proc.*, **21**, 1031 (1962).
46. M. L. Bender and F. J. Kézdy, *J. Amer. Chem. Soc.*, **86**, 3704 (1964).
47. B. Zerner, R. P. M. Bond, and M. L. Bender, *J. Amer. Chem. Soc.*, **86**, 3674 (1964); B. Zerner and M. L. Bender, *J. Amer. Chem. Soc.*, **86**, 3669 (1964).
48. H. T. Huang and C. Niemann, *J. Amer. Chem. Soc.*, **74**, 4634, 5963 (1952).
49. G. E. Hein and C. Niemann, *J. Amer. Chem. Soc.*, **84**, 4495 (1962).
50. G. E. Hain and C. Niemann, *Proc. Natl. Acad. Sci., U.S.*, **47**, 1341 (1961).
51. L. Peller and R. A. Alberty, *J. Amer. Chem. Soc.*, **81**, 5907 (1959).
52. M. L. Bender and F. J. Kézdy, *J. Amer. Chem. Soc.*, **86**, 3704 (1964).
53. M. L. Bender, F. J. Kézdy, and C. R. Gunter, *J. Amer. Chem. Soc.*, **86**, 3714 (1964).
54. A. Williams, *Proc. Chem. Soc.*, 816 (1966).
55. A. Williams, personal communication.
56. A. Williams, Doctoral Thesis, Oxford University, (1964).
57. A. Bruylants and F. J. Kézdy, *Rec. Chem. Progr.*, **21**, 213 (1960).
58. R. W. Taft, Jr., in *Steric Effects in Organic Chemistry*, M. S. Newman, Ed., John Wiley and Sons, New York, 1956, p. 556.
59. M. L. Bender and K. Nakamura, *J. Amer. Chem. Soc.*, **84**, 2577 (1962); E. Tommila and C. N. Hinshelwood, *J. Chem. Soc.*, 1801 (1938).
60. M. L. Bender and R. J. Thomas, *J. Amer. Chem. Soc.*, **83**, 4183 (1961).
61. J. B. Jones and C. Niemann, *Biochemistry*, **1**, 1093 (1962); H. R. Waite and C. Niemann, *Biochemistry*, **1**, 250 (1962).
62. M. Caplow and W. P. Jencks, *Biochemistry*, **1**, 883 (1962).
63. J. P. Wolf, III, R. A. Wallace, R. L. Peterson, and C. Niemann, *Biochemistry*, **3**, 940 (1964).
64. J. P. Wolf, III, and C. Niemann, *Biochemistry*, **2**, 493 (1963).
65. J. Rapp, Doctoral Thesis, California Institute of Technology, Pasadena, California, 1963.
66. R. J. Foster and C. Niemann, *J. Amer. Chem. Soc.*, **77**, 1886 (1955).
67. J. T. Braunholtz, R. J. Kerr, and C. Niemann, *J. Amer. Chem. Soc.*, **81**, 2852 (1959).
68. R. Lutwack, H. F. Mower, and C. Niemann, *J. Amer. Chem. Soc.*, **79**, 5690 (1957).
69. R. J. Kerr and C. Niemann, *J. Amer. Chem. Soc.*, **80**, 1469 (1958).

70. H. Gutfreund, in *The Enzymes*, P. D. Boyer, H. Lardy, and K. Myrbäck, Eds., 2d. ed., Vol. 1, Academic Press, New York, 1959, p. 233; J. M. Reiner, *Behavior of Enzyme Systems*, Burgess Publ. Co., Minneapolis, 1959, p. 189.
71. D. Tabuchi, *Z. Elektrochem.*, **63**, 141 (1960).
72. F. Basolo and R. G. Pearson, *Mechanisms of Inorganic Reactions*, John Wiley and Sons, New York, 1958, p. 153.
73. J. E. Snoke and H. Neurath, *Arch. Biochem.*, **21**, 351 (1949).
74. M. L. Bender and B. W. Turnquest, *J. Amer. Chem. Soc.*, **77**, 4271 (1955); S. G. Cohen and S. Y. Weinstein, *J. Amer. Chem. Soc.*, **86**, 5326 (1964).
75. N. S. Isaacs and C. Niemann, unpublished results.
76. W. Cohen and B. F. Erlanger, *J. Amer. Chem. Soc.*, **82**, 3928 (1960).
77. T. H. Applewhite, R. B. Martin, and C. Niemann, *J. Amer. Chem. Soc.*, **80**, 1457 (1958).
78. R. J. Foster, *J. Biol. Chem.*, **236**, 2461 (1961).
79. G. E. Hein and C. Niemann, *J. Amer. Chem. Soc.*, **84**, 4487 (1962).
80. F. J. Kézdy and M. L. Bender, *Biochemistry*, **1**, 1097 (1962); M. L. Bender and F. J. Kézdy, *Ann. Rev. Biochem.*, **34**, 49 (1965).
81. A. Platt and C. Niemann, *Proc. Natl. Acad. Sci., U.S.*, **50**, 817 (1963).
82. B. H. J. Hofstee, *Biochim. Biophys. Acta*, **24**, 211 (1954).
83. B. H. J. Hofstee, *Biochim. Biophys. Acta*, **32**, 182 (1959).
84. B. H. J. Hofstee, *J. Biol. Chem.*, **207**, 219 (1954).
85. W. P. Cane and D. Wetlaufer, *Abstr. 152nd ACS Mtg.*, 1966, p. C-110.
86. H. Weiner, C. W. Batt, and D. E. Koshland, Jr., *J. Biol. Chem.*, **241**, 2687 (1966).
87. M. L. Bender and K. Nakamura, *J. Amer. Chem. Soc.*, **84**, 2577 (1962).
88. M. Caplow and W. P. Jencks, *Biochemistry*, **1**, 883 (1962).
89. T. Inagami, S. S. York, and A. Patchornik, *J. Amer. Chem. Soc.*, **87**, 126 (1965).
90. B. F. Bundy and C. L. Moore, *Biochemistry*, **5**, 808 (1966).
91. W. F. Sager and P. C. Parks, *J. Amer. Chem. Soc.*, **85**, 2678 (1963) and other papers.
92. M. Silver and T. Sone, *J. Amer. Chem. Soc.*, **89**, 457 (1967).
93. B. M. Sanborn and G. E. Hein, *Biochemistry*, **7**, 3616 (1968).
94. R. J. Vaughan and F. H. Westheimer, *J. Amer. Chem. Soc.*, **91**, 217 (1969).
95. P. J. Christen, Doctoral Dissertation, University of Leiden, The Netherlands, 1967.
96. J. E. Erman and G. G. Hammes, *J. Amer. Chem. Soc.*, **88**, 5614 (1966).
97. J. E. Erman and G. G. Hammes, *J. Amer. Chem. Soc.*, **88**, 5607 (1966).
98. C. Hansch et al., *J. Pharm. Sci.*, **59**, 731 (1970).

Chapter 19

THEORIES OF ENZYME CATALYSIS

19.1	Enzymatic Acceleration	642
19.2	Thermodynamic Parameters	644
19.3	Other Enzymatic Factors	647
19.4	Interaction Between Enzyme and Substrate	648
19.5	Conformational Changes	650
19.6	Specificity of Proteolytic Enzymes	656
19.7	Specificity of Other Enzymes	658

19.1 ENZYMATIC ACCELERATION

Many estimates have been made of the superiority of enzymatic catalysis over acid or base catalysis. For example, this ratio has been quoted as 10^3 or 10^7, among other values. In all cases, enzymatic catalysis is superior to acidic or basic catalysis.[1,2] One may ask the reasons for this superiority. The 10^7 came from a calculation based on eq. 19.1[1]

$$\frac{(V_E)\text{calc}}{V_0} = \frac{(E_T)(55)^4 \theta_A \theta_B \theta_R \theta_S \theta_T}{(A)(B)(R)(S)(T)} \quad (19.1)$$

where V_0 represents the velocity of the nonenzymatic reaction under standard conditions and (V_E)calc is the turnover number of a theoretical enzyme operating by the mechanism giving V_0, but having substrates A and B and catalytic groups R, S, and T in perfect proximity and orientation. The factors $(55)/(B)$, $(55)/(R)$, and the like give the acceleration in rate achieved from proximity (55 is the molarity of pure water). The θ factors give the acceleration achieved because the groups held in proximity are oriented ideally in relation to their neighbors. A more sophisticated treatment is given in Table 19.1,[2] where the only questionable number is the 10^3 for "freezing" of the substrate (the others have a direct, simple analog) (see Chapters 2, 6, 9, 15). Freezing is where the 10^3 figure arises.† Implicit in the above calculation is that all enzyme reactions take place by prior binding of the substrate to the enzyme.[3] In fact, intramolecular reactions have been talked about as analogs of intracomplex reactions (see Chapter 9). One is really not interested in a hydroxide reaction but rather in general base-catalyzed reactions involving imidazole pulling a proton from water. One possible relation between these is that the enthalpy of the hydroxide ion reaction is equal to the enthalpy of the reaction involving imidazole and water. Another possible relation is shown kinetically in the table below.

Table 19.1

Kinetic Factors Responsible for the Difference Between the Hydroxide Ion and α-Chymotrypsin Catalyzed Hydrolyses of N-Acetyl-L-Tryptophan Amide[2]

Rate Constant of Hydroxide Ion Catalysis	3×10^{-4} M^{-1} sec^{-1}
1. Conversion to an intermolecular general base-catalyzed reaction involving imidazole (1.6×10^{-6})	4.8×10^{-10} M^{-1} sec^{-1}
2. Conversion to an intramolecular general base-catalyzed reaction involving imidazole (10 M)	4.8×10^{-9} sec^{-1}
3. Change in rate-determining step (10^2)	4.8×10^{-7} sec^{-1}
4. Freezing of the substrate specificity (10^3)	4.8×10^{-4} sec^{-1}
5. General acidic catalysis by imidazole (10^2)	4.8×10^{-2} sec^{-1}
Total calculated enzymatic rate on the basis of the above five factors	4.8×10^{-2} sec^{-1}
Experimental rate constant	4.4×10^{-2} sec^{-1}

† Probably the best simple model for "freezing" involves the acid-catalyzed lactonization of hydrocoumaric acid, whose rate is increased up to 10^{11}-fold by methyl substitution. (S. Milstein and L. A. Cohen, *Proc. Natl. Acad. Sci., U.S.*, **67**, 1143 (1970).) Evidence from nmr studies of enzyme-substrate and enzyme-inhibitor systems shows that substrates are confined at the active site of the enzyme, have a relatively long residence time, and tumble in solution as an enzyme-substrate complex. This has been called "substrate anchoring," an alternative term to "freezing," and has been stated to account for the high catalytic power of enzymes. (J. Reuben, *Proc. Natl. Acad. Sci., U.S.*, **68**, 563 (1971).)

The functionalities of enzymes have been discussed many times.[4,5] One is dealing with pH's around neutrality and thus the catalytic groups that one needs to consider are the carboxyl, imidazole, thiol, amine, and perhaps hydroxyl groups. Of course, binding by aliphatic or aromatic groups is important also. The pK of a group may be perturbed (changed) by complexing with a metal ion or other agent which has a strong electrostatic effect or by the protein environment. Such perturbations must be taken into account in analyzing an enzyme reaction.

In an enzyme active site, a new catalytic group may be introduced (see Chapter 16) by binding it to something, either covalently or noncovalently. In Chapter 16 the introduction of a new binding group is also described.

The three-dimensional structure of an enzyme and its complexes have been said to lead to a description of its mode of action.[6] But this seems very doubtful since there is still much controversy concerning the reactions of the norbornyl cation[7] which contains only seven carbon atoms and whose three-dimensional structure is known completely. If one extrapolates from the norbornyl cation to an enzyme that contains thousands of large atoms, there is no wonder about the controversy surrounding the mechanism of enzyme action. Nevertheless, knowledge of the three-dimensional structure of an enzyme, as mentioned before, is a prerequisite to determining mechanism. An interesting activity recently has been the comparison of amino acid sequences of a single enzyme from diverse species. This has been carried out with many enzymes, but only one will be mentioned, cytochrome c, which performs the same oxidative function (by presumably the same mechanism) in species from man to microorganisms, and many in between, and whose amino acid sequence has been determined in that same range. From the amino acid sequences of these enzymes, phylogenetic trees have been constructed, which are more basic than gross morphological features.[8,9]

19.2 THERMODYNAMIC PARAMETERS

There has been a continual controversy on whether entropy or enthalpy is the major factor in determining whether enzyme reactions are more efficient than acid or base reactions. Fig. 19.1[2] shows the activation parameters of the α-chymotrypsin-catalyzed hydrolysis of N-acetyltryptophan amide. This reaction exhibits the classical kind of catalysis with a downhill slope from the reactants to the intermediate to the (ionized) product. From Table 19.2,[2] the enzymatic acceleration appears to be due to the entropy, but this has been questioned.[32] The activation parameters of a series of homologous acyl-chymotrypsins mentioned in the previous chapter is also pertinent. The activation parameters for the basic, acidic, and enzymatic isomerization of androst-5-ene-3,17-dione are given in Table 19.3.[10] Here both factors seem to be of importance.

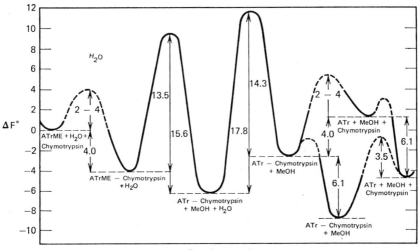

Fig. 19.1 Standard free energy versus reaction coordinate diagram for the α-chymotrypsin-catalyzed hydrolysis of *N*-acetyl-L-tryptophan methyl ester at 25°. From M. L. Bender, F. J. Kézdy and C. R. Gunter, *J. Amer. Chem. Soc.*, **86**, 3714 (1964). © 1964 by the American Chemical Society. Reprinted by permission of the copyright owner.

Bruice has emphasized entropy in model systems, but possibly some of the enormous rate factors he has observed are due to the reduction in Van der Waals interactions as well as entropy of activation. The data, which are really extraordinary, are given in Table 19.4.

A similar phenomenon has been found by Koshland.[12] Acid-catalyzed esterification was compared with γ-lactonization. A variation of the attacking oxygen atom relative to the carbon atom of the carboxylic acid was

Table 19.2

The Activation Parameters of the Deacylation of Some Acyl-α-Chymotrypsins

Acyl-α-chymotrypsin	Rel. k_3 (corrected)	$\Delta H^{\ddagger}$ (kcal mole^{-1})	$\Delta H^{\ddagger}$ (kcal mole^{-1})	$-T\Delta S^{\ddagger}$ (kcal mole^{-1})
N-Acetyl-L-tyrosyl-	3540	14.3	10.3	4.0
N-Acetyl-L-tryptophanyl-	942	17.9	12.0	5.9
trans-Cinnamoyl-	14.7	20.1	11.2	8.9
Acetyl-	1	20.4	9.7	10.7

Table 19.3
Activation parameters for the Isomerization of Androst-5-ene-3,17-dione[a,10]

	HCl pH 0.88[b]	Catalyst Tris–HCl, pH 8.82[b]	Enzyme[c]
Enthalpy of activation, $\Delta H^{\ddagger}$ (kcal mole^{-1})	14.0 ± 0.1	11.4 ± 0.1	5.0 ± 0.1
Entropy of activation, $\Delta S^{\ddagger}$ (cal deg^{-1} mole^{-1})	−19.6 ± 0.4	−15.5 ± 0.4	−16.8 ± 0.4

[a] Data obtained with 1.6% aqueous methanol solutions and a steroid concentration of 0.05 μmole/ml.
[b] Temperature range 15–40°.
[c] Temperature range 15–30°.

Table 19.4
Intramolecular Reactions of Monophenyl Esters of Dicarboxylic Acid Anions[11]

Ester[a]	$k_{\text{hydrol}} / k_{\text{hydrol (glutarate)}}$
–COOR, –COO⁻ (succinate)	1.0
Me₂C(COOR)(COO⁻)	20
–COOR, –COO⁻ (dimethyl)	230
–COOR, –COO⁻ (maleate)	10,000
bicyclic –COOR, –COO⁻	53,000

[a] R = C₆H₅– or p-Br–C₆H₄–.

achieved by using bicyclic systems. The highest rate was achieved with a norbornyl system very similar to that observed by Bruice. A factor of 2×10^4 was observed in this system even after corrections for proximity and torsional strain were made. The orientation factor was attributed to the ability of the reacting atoms to "steer" their orbitals along a path that takes advantage of this strong directional preference. The catalytic efficiency of enzymes is attributed to both juxtaposition and this factor. This laboratory has been working on a synthetic chymotrypsin that contains an imidazole and a hydroxyl group on a norbornane system.[13]

In the binding step itself, entropy can enter in by reduction of the number of translational degrees of freedom, by restriction of rotation, or by a change in the conformation of the enzyme, which Koshland has termed the "induced fit" theory. This will be considered later. The enthalpy of binding could help by a distortion of bond angles or bond lengths. This will also be considered later. This idea gets very close to the well-worn theory of the "rack" in which the enzyme in essence breaks the substrate molecule into pieces. The enthalpy of binding could also help the catalysis by various polarizations. The big question is whether binding can assist catalysis at all and the foregoing are some concrete examples that it can by various means, all of which reduce to the fact that the bound state can be closer to the transition state. One would calculate that the bound state is further from the transition state, for the equilibrium constants of binding are generally favorable. But the transition state may be lower or the free energy of the bound state may in fact be higher (more favorable for reaction because of the various factors mentioned above). Obviously, the last word has not been said about enthalpy and entropy.

19.3 OTHER ENZYMATIC FACTORS

An enzyme can do many things besides bind a substrate. An enzyme can transfer protons to or from it; it can absorb protons on binding; it can utilize various covalent interactions; and finally it can utilize various secondary valence forces, such as London dispersion forces, micellar forces, charge transfer forces, and electrostatic or ion dipole forces such as the reaction of anions with esters containing electron-withdrawing groups. These forces are considered here.

Proton transfer (to or from the enzyme) has been used in many enzymatic mechanisms (see Chapters 16 and 17). A comprehensive treatment of proton transfer is given in Ref. 14. It has also been proposed that a facile proton transfer along a strategically positioned channel can enable the system to reach the transition state faster.[15] The previous postulates about proton transfer suggest a smaller difference between the ground state and the transition state due to the proton transfer. This notion suggests that the speed of proton transfer is also important. A redetermination of the amino acid sequence of α-chymotrypsin indicates an aspartic acid next to the active histidine instead of asparagine. This result is in concert with many other serine proteinases. It is therefore postulated that the active site of chymotrypsin contains an interior aspartic acid hydrogen-bonded to a histidine which in turn is hydrogen-bonded to a serine.[16] Polarization of the system due to the buried negative charge of the aspartic acid residue is said to make the serine oxygen atom strongly nucleophilic and would explain its high reactivity toward amides and esters. There, however, appear to be some inconsistencies in this argument involving hydrogen bonding (see Chapter 16).[17] As mentioned before, the binding of several inhibitors at high pH was shown to cause proton absorption by α-chymotrypsin by a pH recorder and a spectrophotometric pH indicator dye technique.[18] At a given pH, proton binding reached a maximum as saturation with inhibitor or substrate was approached. This phenomenon could be explained by either intramolecular competitive inhibition (by the uncharged base) or the involvement of a salt bridge in the active site of the structure $RCO_2^{\ominus}NH_3^{\oplus}R$.

Secondary valence forces were mentioned before. More precisely, catalysis by these forces is important. They have, however, been looked at only in model systems so far but their applicability to enzymatic catalysis is certainly quite possible. The various kinds are charge transfer forces,[19–21] London dispersion forces,[22] micellar forces,[23] and various electrostatic effects.[24]

Cooperative interactions result in different kinetics and equilibria of binding to four hemoproteins and dehydrogenases.[25] These interactions are linked with allosteric interactions, conformational changes, and multienzyme complexes (Chapters 13–19). Negative cooperativity has been found.[26] Cooperative interactions are in fact not far different from association-dissociation phenomena or isozymes containing more than one unit of enzyme. A particularly well-understood case is that of lactic dehydrogenase.[27] α-Chymotrypsin was claimed to exist as an acyl-enzyme dimer at pH's below neutrality,[29] a form of association.

Formation of multienzyme complexes is closely related to association. Pyruvate dehydrogenase of *E. coli*, of molecular weight 4,000,000, is composed of three types of enzymes, one of which, at least, is in turn composed of subunits. The complete particle contains a minimum of 88 separate polypeptides (Chapter 13).

19.4 INTERACTION BETWEEN ENZYME AND SUBSTRATE

If one talks of the ultimate interaction between enzyme and substrate, one can assume a rigid enzyme, a rigid substrate, a flexible enzyme, and/or

Other Enzymatic Factors 649

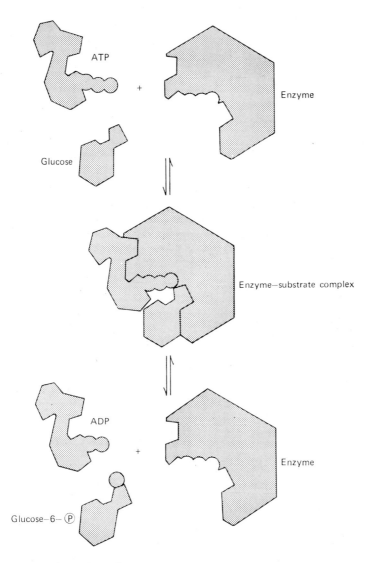

Fig. 19.2 The formation of an enzyme-substrate complex, followed by catalysis. From J. D. Watson, *Molecular Biology of the Gene*, W. A. Benjamin, Inc., New York, 1965, p. 52.

a flexible substrate. This will obviously lead to many possible combinations.[30] Anthropomorphic views of enzymatic catalysis have been extant (see for example, Fig. 19.2).[74] In the extreme form, an enzyme has been proposed to bite the substrate or to break it into pieces. More recently more sophisticated views have prevailed. One thing is certain: enzyme catalysis is not mysterious or awesome; it is complicated perhaps but amenable to experiment.

One of the early postulates about enzyme catalysis was the "rack" theory, which postulated binding of the enzyme to both sides of the reacting bond of the substrate, thus inducing lability by mechanical strain, electron displacement, or both.[32]

Another suggestion has been made on the basis of a rigid enzyme that Emil Fischer assumed long ago. In particular for α-chymotrypsin, catalysis of hydrolysis of specific and nonspecific substrates can be differentiated by a differential freezing[33] of the substrate to the enzyme, as shown in Fig. 19.3.[33]

On the other hand, a flexible enzyme can be assumed. This is certainly reasonable since enzymes are large molecules, and even some small molecules are significantly flexible. Enzyme flexibility has been mentioned several times as a crucial factor in enzyme catalysis[34,35] and conformational differences are crucial to certain small molecule reactions.[36] While this is reasonable, one word of caution should be given: flexibility of enzymes has become a fashion and fashions are not based on rational arguments. The arguments for a rigid enzyme consist of the following: (1) an enzyme could not be formed if a rigid molecule were not being produced; (2) the structure of an enzyme could not be determined if one were dealing with something other than rigid; and (3) the lock and key theory of Emil Fischer is a sufficient theory for most enzymatic processes and some model systems. Arguments for a flexible enzyme follow.

19.5 CONFORMATIONAL CHANGES

Much has been made of conformational changes in enzymes, sometimes called isomerizations, sometimes allosteric interactions, and sometimes other things. In fact, conformational change may be a phrase that is equivalent to ignorance, for it is often difficult to obtain direct evidence about this subject. It may be related to denaturation.[76] The evidence is best summarized for α-chymotrypsin as kinetic, light scattering, spectroscopic, chemical studies, titration experiments, polarimetric studies, and phase transitions.[37] The quantity of arguments does not belie the fact that these are indirect experiments. The only direct evidence would be X-ray crystallographic or electron microscopic studies which show a movement of the active site. One

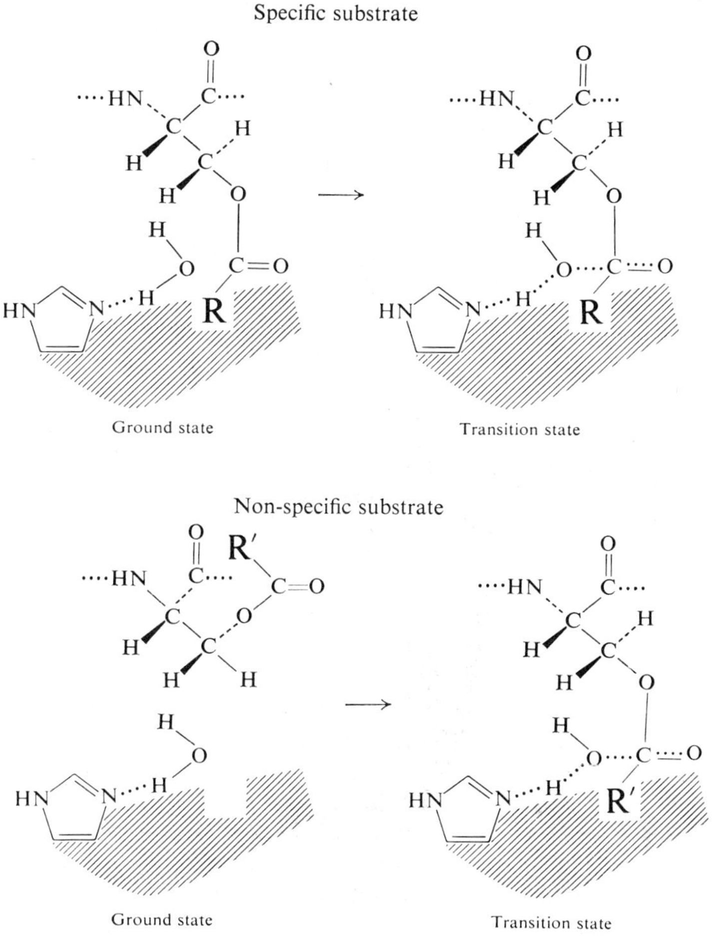

Fig. 19.3 Specificity in the deacylation of acyl-α-chymotrypsins. From M. L. Bender, F. J. Kézdy, and C. R. Gunter, *J. Amer. Chem. Soc.*, **86**, 3718 (1964).

such study has been made on glutamine synthetase from *E. coli* as shown schematically in Fig. 19.4[38] and directly in Fig. 19.5. When association of enzyme molecules occurs as with glutamine synthetase as shown in Fig. 19.5,[37,39] the issue of a *single* active site becomes clouded, since the site may be composed of *more than* one subunit.[75]

Other enzyme-like molecules where association is important and where a conformational change has been postulated are myoglobin and hemoglobin that bind various ions.[40,41] The same critique of these systems applies. The ultimate question is whether a monomeric enzyme undergoes a

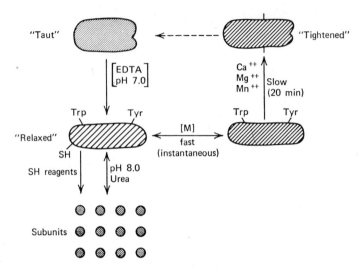

Fig. 19.4 Scheme for the interconversion of taut, relaxed, and tightened enzyme forms. From B. M. Shapiro and A. Ginsburg, *Biochemistry*, **7**, 2153 (1968). © 1968 by the American Chemical Society. Reprinted by permission of the copyright owner.

conformational change. A conformational change, which is intimately tied to strain, is argued for because: (1) there is no obvious way in which an enzyme can catalyze a displacement reaction of a thioether and a sulfonium salt; (2) the best binder sometimes resembles the hypothetical transition state as shown in Fig. 19.6[30] for proline racemase and Fig. 19.7[30] for Δ⁵-3-ketoisomerase; specificity is sometimes manifested in high rates rather than tight binding; and (3) binding forces may be utilized to induce strain in an enzyme-substrate complex as shown in Fig. 19.8.[30]

The enzyme glutamate dehydrogenase has received a lot of attention; it is related to the glutamine synthetase mentioned earlier.[42,43] Another enzyme for which a conformational change has been postulated is pyruvate kinase: it is claimed that potassium ion changes the shape of the enzyme as shown in Fig. 19.9.[44] Strain has also been mentioned as a factor in small molecule systems.[45]

Conformational changes have been postulated in the enzymes fructose diphosphatase on the basis of a change of inhibition,[46] luciferase on the basis of altered reactivity,[47] ribonuclease on the basis of spin labeling,[48] citric acid cycle enzymes on the basis of the same K_m and different k_{cat} for a series of diphosphates,[49] D-amino acid oxidase on the basis of fluorescence measurements,[50,51] and chymotrypsin on the basis of fluorescence measurements,[52] among others. The relation of physical to conformational

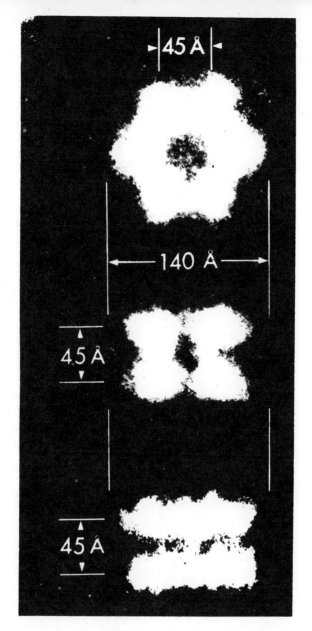

Fig. 19.5 A high-magnification picture of five superimposed images of unfixed glutamine synthetase molecules in the three characteristic orientations. The mean dimensions are indicated. When the molecule rests on a face, the subunits appear as a hexagonal ring (top). Molecules seen on edge show two layers of subunits, either as four spots (center) when viewed exactly down a diameter between subunits, or in general as two lines (bottom). The molecular structure is thus based on twelve identical subunits in two hexagonal layers, the units in one layer being directly above those in the other. Magnification ×3,160,000. From R. C. Valentine, B. M. Shapiro, and E. R. Stadtman, *Biochemistry*, **7**, 2143 (1968). © 1968 by the American Chemical Society. Reprinted by permission of of the copyright owner.

654 Theories of Enzyme Catalysis

Fig. 19.6 Substrate (b) and transition state analog (a) of proline racemase. From W. P. Jencks, in *Current Aspects of Biochemical Energetics*, N. O. Kaplan and E. P. Kennedy, Eds., Academic Press, New York, 1966, p. 273.

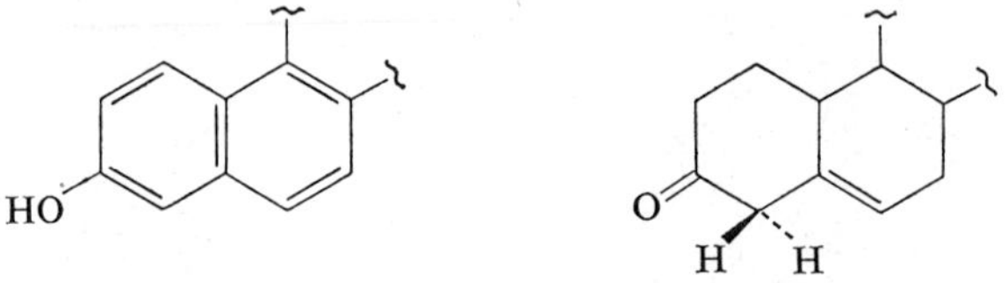

Fig. 19.7 A substrate (b) and transition state analog (a) of Δ^5-3-ketoisomerase. From W. P. Jencks, in *Current Aspects of Biochemical Energetics*, N. O. Kaplan and E. P. Kennedy, Eds., Academic Press, New York, 1966, p. 273.

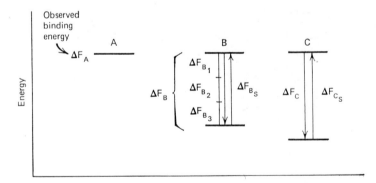

Fig. 19.8 Scheme to show how enzyme-substrate binding forces may be utilized to induce strain in the enzyme-substrate complex. From W. P. Jencks in *Current Aspects of Biochemical Energetics*, N. O. Kaplan and E. P. Kennedy, Eds., Academic Press, New York, 1966, p. 273.

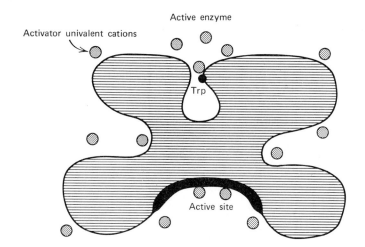
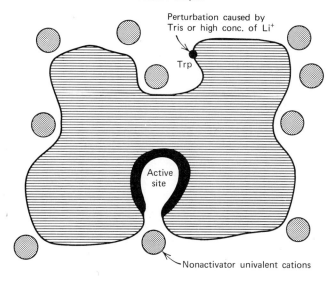

Fig. 19.9 Activator cations such as potassium and rubidium "poise" the pyruvate kinase in the proper conformation for catalysis. Thus they may keep the active site open for combination with the substrate. Inactivator ions such as lithium probably cause perturbations of tryptophan residues and in turn block off the active site. From *Chem. Eng. News*, 40 (July 1, 1968).

change is best argued in an enzyme such as lysozyme for which the three-dimensional structure is known. As mentioned previously, isomerization is a term sometimes used instead of conformational change. In particular, this term has been used with transaminase. This term possesses less mechanistic meaning than conformational change. It implies a change in the enzyme but not necessarily at the active site crucial to catalysis.

A flexible substrate is highly probable, as mentioned earlier. There are many small molecules that exist in more than one conformation. Flexibility of both enzyme and substrate is another possibility, but is a complicated situation with which to work.

Macromolecular catalysis of the *cis* → *trans* isomerization of an azo dye was studied, primarily with proteins. On the basis of these studies, mutual distortion mechanisms were suggested as the cause of macromolecular catalysis.[53] One kind of distortion could be the rack that was mentioned before. The strain mentioned before is also a variation of the rack theory. Strain (in the transition state) has been the interpretation of the reactivity of many model systems,[54-59] and the strained molecule hypothesis of enzyme action was suggested long ago.[60] It has been repeatedly used; a recent study, which does not entertain this idea, concerns the enzyme carbonic anhydrase.[61]

19.6 SPECIFICITY OF PROTEOLYTIC ENZYMES

Enzymes are extremely specific. When this was discovered, it rapidly became apparent that some three-dimensional fit of substrate and enzyme would be required to explain discrimination between chemical compounds of similar structure. Emil Fischer proposed a template model in which (1) the enzyme was a fairly rigid negative of the substrate and (2) a fit between enzyme and substrate was necessary for enzyme action. It follows from this model that substrate analogs which lack necessary binding groups (and hence are not attracted) and substrate analogs which are too large (and hence would not fit) would fail to react. These two requirements are sufficient to explain a large amount of specificity data, but there are a number of anomalies. A typical one is the failure of water to react with ATP in the presence or absence of glucose in the hexokinase reaction. Steric repulsion certainly would not prevent the access of a smaller molecule than the substrate, and the fact that water is the solvent and will fill all "holes" invalidates the argument that it is not attracted to the active site.

To explain anomalies of this sort, an "induced fit model" for specificity was proposed[62] which had the following features: (1) the enzyme exists in a natural conformation which is not necessarily a negative of the substrate;

(2) the substrate induces a change in protein conformation; and (3) the changed conformation produces the correct alignment of catalytic groups so that reaction occurs. In this case, the buttressing action of groups of a certain minimum size may be required to produce correct alignment and hence the unreactivity of smaller substrate analogues, such as water in the hexokinase reaction, is explained. Other evidence for this flexibility at the site during enzyme action has been summarized elsewhere.[63,64] It would seem, therefore, that the active site exists in a thermodynamic valley of the type shown in Fig. 19.10;[63] that substrate interaction produces a new protein geometry, perhaps in a way similar to that in which urea affects the shape of restricted portions of a protein; that the new alignment allows enzyme action; and that the release of products from the enzyme surface allows the protein to revert to its original shape. The peculiar reactivity of the enzyme aldolase mentioned earlier, in which sugar diphosphates containing two phosphate groups show different rates (the six-carbon sugar is faster than the five, which is equal to the eight) but identical binding constants, has been interpreted in terms of induced fit.[65]

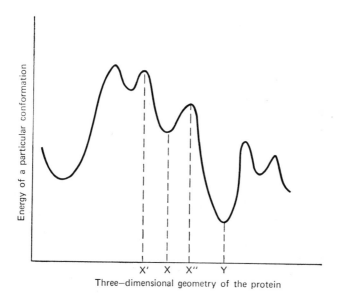

Fig. 19.10 The energy content of various protein conformations. The abscissa represents a theoretical parameter which accurately measures the position in three-dimensional space of each atom in the protein. The ordinate represents the energy of the particular protein conformation. From D. E. Koshland, Jr., *Adv. Enzymol.*, **22**, 45 (1960).

Precise orientation has been emphasized above and was once more called they key to enzyme specificity.[77] Further, the catalytic efficiency of enzymes has been postulated to depend not only on their ability to juxtapose reacting atoms but to "steer" the orbitals of these atoms into a precise alignment that makes rapid reactions possible.[78] This has been disputed on the grounds that it ignores attractive or repulsive forces[79] and this controversy has yet to be resolved.

19.7 SPECIFICITY OF OTHER ENZYMES

In metalloenzymes such as have been discussed previously, it has been suggested that the metal ion functions as a "hotspot,"[66] but the specificity associated with a number of the enzymes that contain metal ions does not lend support to this idea. The use of metals provides a means of probing enzyme mechanism during catalysis.[80]

The specificity of other (non-metal ion-containing) enzymes has been discussed in Chapter 16. Generalizations that can be drawn are few. Sometimes an enzyme will act on a small molecule; sometimes on a large (polymeric) molecule. Both kinds of enzymes can be specific, although the idea of absolute specificity is disappearing with time.

REFERENCES

1. D. E. Koshland, Jr., *Adv. Enzymol.*, **22**, 45 (1960).
2. M. L. Bender, F. J. Kézdy, and C. R. Gunter, *J. Amer. Chem. Soc.*, **86**, 3714 (1964).
3. C. Niemann, *Science*, **143**, 1287 (1964).
4. M. L. Bender, G. R. Schonbaum, and G. A. Hamilton, *J. Polymer Sci.*, **49**, 75 (1961).
5. H. Gutfreund and J. R. Knowles, *Essays Biochem.*, **3**, 25 (1967).
6. D. C. Phillips, *Sci. Amer.*, **215**, No. 5, 78 (1966).
7. B. L. Murr et al., *J. Amer. Chem. Soc.*, **89**, 1730 (1967); G. J. Gleicher and P. von R. Schleyer, *J. Amer. Chem. Soc.*, **89**, 582 (1967).
8. W. M. Fitch and E. Margoliash, *Science*, **155**, 279 (1967).
9. See also W. P. Winter, K. A. Walsh, and H. Neurath, *Science*, **162**, 1433 (1968).
10. J. B. Jones and D. C. Wigfield, *J. Amer. Chem. Soc.*, **89**, 5294 (1967).
11. T. C. Bruice and U. K. Pandit, *Proc. Natl. Acad. Sci., U.S.*, **46**, 402 (1960).
12. D. R. Storm and D. E. Koshland, Jr., *Proc. Natl. Acad. Sci., U.S.*, **66**, 445 (1970).
13. M. Utaka and M. L. Bender, unpublished results.
14. J. H. Wang, *Science*, **161**, 328 (1968).
15. R. Bernhard, *Sci. Res.*, 20 (July 7, 1969).
16. D. M. Blow, J. J. Birktoft, and B. S. Hartley, *Nature*, **221**, 337 (1969).
17. L. Polgar and M. L. Bender, *Proc. Natl. Acad. Sci., U.S.*, **64**, 1335 (1969).
18. F. C. Wedler and M. L. Bender, *J. Amer. Chem. Soc.*, **91**, 3894 (1969).
19. C. G. Swain and L. J. Taylor, *J. Amer. Chem. Soc.*, **84**, 2456 (1962).
20. F. M. Menger and M. L. Bender, *J. Amer. Chem. Soc.*, **88**, 131 (1966).
21. J. A. Mollica, Jr., and K. A. Connors, *J. Amer. Chem. Soc.*, **89**, 308 (1967).

References

22. J. F. Bunnett, *J. Amer. Chem. Soc.*, **79**, 5969 (1957); J. F. Bunnett and L. A. Retallick, *J. Amer. Chem. Soc.*, **89**, 423 (1967).
23. R. B. Dunlap and E. H. Cordes, *J. Amer. Chem. Soc.*, **90**, 4395 (1968).
24. K. Koehler, R. Skora, and E. H. Cordes, *J. Amer. Chem. Soc.*, **88**, 3577 (1966); M. L. Bender and Y.-L. Chow, *J. Amer. Chem. Soc.*, **81**, 3929 (1959); T. C. Bruice and B. Holmquist, *J. Amer. Chem. Soc.*, **89**, 4028 (1967); J. Epstein et al., *J. Amer. Chem. Soc.*, **89**, 2937 (1967); M. A. Schwartz, *J. Pharm. Sci.*, **54**, 1308 (1965).
25. B. Chance and N. Rumen, *Science*, **156**, 536 (1967).
26. A. Levitzki and D. E. Koshland, Jr., *Proc. Natl. Acad. Sci., U.S.*, **62**, 1121 (1969).
27. R. D. Cahn et al., *Science*, **136**, 962 (1962).
28. E. Antonini et al., *J. Biol. Chem.*, **241**, 2358 (1966).
29. M. L. Bender and F. J. Kézdy, *Biochemistry*, **4**, 104 (1965).
30. W. P. Jencks, in *Current Aspects of Biochemical Energetics*, N. O. Kaplan and E. P. Kennedy, Eds., Academic Press, New York, 1966, p. 273.
31. J. D. Watson, *Molecular Biology of the Gene*, W. A. Benjamin, Inc., New York, 1965.
32. H. Eyring, R. Lumry, and J. D. Spikes, in *Mechanism of Enzyme Action*, W. D. McElroy and B. Glass, Eds., The Johns Hopkins Press, Baltimore, 1954, p. 35.
33. M. L. Bender, F. J. Kézdy, and C. R. Gunter, *J. Amer. Chem. Soc.*, **86**, 3714 (1964).
34. G. E. Hein and C. Niemann, *J. Amer. Chem. Soc.*, **84**, 4495 (1962).
35. D. E. Koshland, Jr., *J. Cell. Comp. Physiol.*, **54**, Suppl. 1, 245 (1959).
36. T. C. Bruice and W. C. Bradbury, *J. Amer. Chem. Soc.*, **87**, 4838, 4851, 4846 (1965).
37. B. Labouesse, B. H. Havsteen, and G. P. Hess, *Proc. Natl. Acad. Sci., U.S.*, **48**, 2137 (1962).
38. B. M. Shapiro and A. Ginsburg, *Biochemistry*, **7**, 2153 (1968).
39. R. C. Valentine, B. M. Shapiro, and E. R. Stadtman, *Biochemistry*, **7**, 2143 (1968).
40. B. Chance, A. Ravilly, and N. Rumen, *J. Mol. Biol.*, **17**, 525 (1966).
41. B. Chance and A. Ravilly, *Fed. Proc.*, **25**, 648, abs. no. 2953 (1966); H. C. Watson and B. Chance, in *Hemes and Hemoproteins*, B. Chance, R. W. Estabrook, and T. Yonetani, Eds., Academic Press, New York, 1966, p. 149.
42. A. Bitensky et al., *J. Biol. Chem.*, **240**, 663, 668 (1965).
43. G. M. Tomkins et al., *J. Biol. Chem.*, **240**, 3793 (1965).
44. Chem. Eng. News, 40 (July 1, 1968).
45. H. C. Brown, R. Bernheimer, and K. J. Morgan, *J. Amer. Chem. Soc.*, **87**, 1280 (1965) and other papers by Brown.
46. Symposium, Oxford University, 1968.
47. W. D. McElroy, M. DeLuca, and J. Travis, *Science*, **157**, 150 (1967).
48. I. C. P. Smith, *Biochemistry*, **7**, 745 (1968).
49. A. H. Mehler, personal communication.
50. V. Massey and B. Curti, *J. Biol. Chem.*, **241**, 3417 (1966).
51. V. Massey, B. Curti, and H. Ganther, *J. Biol. Chem.*, **241**, 2347 (1966).
52. Y. D. Kim, Doctoral Dissertation, University of Minnesota, 1968.
53. R. Lovrien and T. Linn, *Biochemistry*, **6**, 2281 (1967).
54. O. R. Zaborsky and E. T. Kaiser, *J. Amer. Chem. Soc.*, **88**, 3084 (1966).
55. E. T. Kaiser et al., *J. Amer. Chem. Soc.*, **87**, 3781 (1965).
56. R. T. LaLonde and L. S. Forney, *J. Amer. Chem. Soc.*, **85**, 3767 (1963).
57. K. B. Wiberg and B. R. Lowry, *J. Amer. Chem. Soc.*, **85**, 3188 (1963).
58. E. A. Dennis, D. H. Usher, and F. H. Westheimer, *J. Amer. Chem. Soc.*, **87**, 2320 (1965).
59. E. A. Dennis and F. H. Westheimer, *J. Amer. Chem. Soc.*, **88**, 3431 (1966).
60. J. H. Quastel, *Biochem. J.*, **20**, 166 (1926).

61. M. E. Riepe and J. H. Wang, *J. Amer. Chem. Soc.*, **89**, 4229 (1967).
62. D. E. Koshland, Jr., *Proc. Natl. Acad. Sci., U.S.*, **44**, 98 (1958).
63. D. E. Koshland, Jr., *Adv. Enzymol.*, **22**, 45 (1960).
64. D. E. Koshland, Jr., *J. Cell. Comp. Physiol.*, **54**, Suppl. 1, 245 (1959).
65. A. H. Mehler, personal communication.
66. R. J. P. Williams, personal communication.
67. J. R. Williams, Jr., and R. H. Steele, *Biochemistry*, **4**, 814 (1965).
68. See also S. W. Weidman et al., *J. Amer. Chem. Soc.*, **89**, 4555 (1967).
69. See also T. Higuchi and K.-H. Gensch, *J. Amer. Chem. Soc.*, **88**, 3874 (1966).
70. See also T. Higuchi, I. H. Pitman, and K.-H. Gensch, *J. Amer. Chem. Soc.*, **88**, 5676 (1966).
71. See also C. C. Schubert and R. N. Pease, *J. Amer. Chem. Soc.*, **78**, 5553 (1956).
72. See also Y. Wolman et al., *J. Amer. Chem. Soc.*, **84** 1889 (1962).
73. See also P. J. Zandstra and S. I. Weissman, *J. Amer. Chem. Soc.*, **84**, 4408 (1962).
74. J. D. Watson, *Molecular Biology of the Gene*, W. A. Benjamin, Inc., New York, 1965, p. 52.
75. See also P. Dimroth et al., *Proc. Natl. Acad. Sci., U.S.*, **67**, 1353 (1970).
76. S. A. Bernhard and G. L. Rossi, in *Structural Chemistry and Molecular Biology*, A. Rich and N. Davidson, Eds., W. H. Freeman and Co., San Francisco, 1968 p. 98.
77. B. H. J. Hofstee, personal communication.
78. D. R. Storm and D. E. Koshland, Jr., *Proc. Natl. Acad. Sci., U.S.*, **66**, 445 (1970).
79. T. C. Bruice, A. Brown, and D. A. Harris, *Proc. Natl. Acad. Sci., U.S.*, **68**, 658 (1971).
80. *Chem. Eng. News*, 43 (December 14, 1970).

POSTSCRIPT

As stated in the Introduction, this book began with a very old and simple kind of homogeneous catalysis, acid-base catalysis, and proceeded through more complicated catalyses to enzyme catalysis which may be the ultimate in catalytic efficiency and specificity.

The important concept in this book is that nonenzymic and enzymic catalysis can be bridged (see Chapters 9–11). There is a continuum of catalysis without a break from the simple to the complex. It is the thesis of this book that there is not a sharp dividing line.

There are many diseases that are caused by enzyme defects. The diseases include Gaucher's, generalized gangliosidosis, metachromatic leukodystrophy, Fabry's, and Tay Sachs', among others. The corresponding enzymes involved are sphingomyelinase, β-galactosidase, β-glucosidase, and sulfitidase. Perhaps the mechanisms discussed in the enzyme chapters have shed some light on these diseases.

The disappearance of vitalism is a corollary of this book. Bohr once said that some biological phenomena could not be fully accounted for in terms of conventional physical concepts, but there is nothing in this book of that nature. Perhaps at some higher level his comment is correct.

The intensive trend toward the discovery of new enzymes and the development of new techniques for their production started about two decades ago and is ever increasing. Four main avenues project promises for further development: the use of microorganisms, the use of tissue cultures, the application of hormones, and the periodic or continuous harvesting of exoenzymes from cells without destroying them.

A section on oxidative catalysis, which is very important in enzymatic reactions, could very well have been included in this book. Scattered mention of redox systems in enzymatic processes have been made in Chapter 17. However, no explicit section on oxidative catalysis has been presented nor have the important concepts of oxidative phosphorylation and coupled redox reactions, these being outside the scope of this book. Possible models for a photochemical aromatic hydroxylation have been described (Refs. 67–73, Chapter 19).

In the microorganism *E. coli*, there are estimated to be between 2000 and 3000 proteins in a single cell, mainly enzymes (Ref. 74, Chapter 19). If this figure can be extrapolated to a mammalian organism, one has to contend with a multitude of enzymes.

Obviously there are many things missing from this book, but it is understandable in a rapidly growing field. One thing is obvious: a greater understanding of enzyme mechanism is needed. While much work has been done in this area, the problems associated with large molecules makes progress slow, although much has been accomplished in the last decade.

Index

A−1 reactions, 51−53
A−2 reactions, 54
Acenaphthene, 365
Acetal, hydrolysis, 296
Acetaldehyde, addition of nucloephiles, 128
   aldol condensation, 66, 92, 220
   decomposition, 6
   hydrate, dehydration, 80, 82, 98, 130
   hydration, 100, 610
   reduction (NADD, NADH), 543
Acetaldehyde dehydrogenase, 540
Acetals, hydrolysis, 48, 52, 53, 123, 127
Acetamide, basicity, 40
2-Acetamido-2-deoxy-D-glucose, 527
Acetate ion, acetylimidazole hydrolysis, 106
   anhydride hydrolysis, 101, 156
   basicity, 103
   cupric complex, 218
   ester hydrolysis, 102, 157, 313, 347
   nucleophilicity, 103, 151
   sulfinyl sulfone hydrolysis, 153
Acetic acid, aromatic chlorination, 159
   basicity, 40
   $CH_2(OH)_2$ dehydration, 129
   dimerization, stepwise, 635
   in enolization, 119
   in glucose mutarotation, 129
   ionization in, 198
   proton exchange in, 30, 32
   reactions, CH_3OH, 98
   H_2O, 21
Acetic anhydride, hydrolysis, 52
   by anionic catalysis, 101, 156, 157, 170
   by neutral catalysis, 106, 170
   steric effects, 105

   as intermediate, trapping, 102
Acetic benzoic anhydride, 157
Acetic formic anhydride, 156
Acetoacetate decarboxylase, 482, 588
Acetoacetic acid, Mg complex, alkylation, 219
   biosynthesis, 602
   bromination, 81
Acetone, basicity, 40
   enolization, 119, 165−167
   halide ion nucleophilicity in, 206
   halogenation, 22, 88
      ammonium ion catalysis, 168
      base catalyzed, 117, 118
      Brønsted relation, 81
      entropy of activation, 76
      GA and GB catalysis, 73, 98
   Lewis basicity, 86
   in olefin polymerization, 185
   oximation, 107
   reaction with $OH^\ominus$, 20, 22
Acetonedicarboxylic acid, decarboxylation, 169, 216, 218
Acetonimine, enolization, 166
Acetonitrile, 214
Acetonylacetone, 81
Acetophenone, 7
   basicity, 40
   oxidation, 237
   oximes, rearrangement, 52
   protonation, 41
Acetoximate ion, 151
Acetoxime acetate, 178, 326
o-Acetoxybenzaldehyde, 354
4-Acetoxy-3-nitrobenzaldehyde, 387
8-Acetoxyquinoline-5-sulfonate ion, 234
Acetyl (radical), 6, 7

Index

Acetylacetone, enol, reaction with $OH^{\ominus}$, 22
  enolate ion, protonation, 22, 98
  hydrolysis, 111
  reaction with, bases, 20, 29, 77
    water, 25
Acetyl chloride, 60
Acetylcholine, hydrolysis, 305
  acetylcholinesterase assay, 447
Acetylcholinesterase, determination, 416
  reactions with organophosphorous compounds, 480, 638
  substrate binding, 477, 623
N-Acetyl-O-cinnamoylserinamide, 506
Acetylcoenzyme A, 606
2-Acetyl-3,4-dimethylthiazolium ion, 586
Acetylene, hydration, 255
  oligomerization, 258
Acetyl fluoride, 60, 99, 101
N-Acetyl-β-D-glucosamine, 527
N-Acetylglycine, 383
N-Acetylhomocysteine thiolactone, 159
Acetyl hypochlorite, 159
N-Acetylimidazole, with carboxypeptidase, 609
  hydrolysis, 101, 106, 157, 161
  intermediate, 102, 161, 177
  nucleophilic reactions, 99
  thiolation, 101
1-Acetyl-3-methylimidazolium ion, 99
Acetyl nitrite, 158
N-Acetyl-L-norvaline esters, hydrolysis, 631
N-Acetylphenylalaninamide, 496
N-Acetylphenylalanine derivatives, hydrolysis, 510, 629
Acetyl phenyl phosphate, 288
Acetyl phosphate, aminolysis, 100
  hydrolysis, 157, 160, 288
  nucleophilic reactions, 230
N-Acetylpyridinium acetate, 160
N-Acetylserinamide phthalate monoester, 302
2-Acetylthiamine, 585
N-Acetyl-L-tryptophan, acylation of chymotrypsin, 494, 495, 506, 510
  derivatives, hydrolysis, 495, 496, 510, 629, 643
N-Acetyl-D-tryptophanamide, 514
N-Acetyl-L-tyrosine derivatives, amide, 496
  ethyl ester, 473
  partitioning, 497
  peroxy acid, 508

N-Acetyl-L-tyrosyl-L-tyrosine, 522
N-Acetyl-L-valine esters, 631
Aconitic acid, 524
Acrylic acid copolymers, 386, 387
Acrylonitrile, 68
"Active acetaldehyde," 578, 585
"Active pyruvate," 585
Activity, specific, definition, 410
  of water, 44
Acyl-carrier protein, 610
Acyl cyanides, 99
Acyl halides, 329
Acylium ions, 58
Acyloin condensation, 178, 582
Acyl transfer (S→N), 100
Adenosine, phosphates, 230, 231, 604
  riboflavin complex, 550
S-Adenosylmethionine, 490
Adipic acid, ester, hydrolysis, 223
Aeromonas proteolyticus, 610
Alanine, deamination, 337
Alberty, R. A., 482
Alcohol dehydrogenase, 478, 540, 543
  NAD binding, 472, 473
  structure, 464
Alcohols, esterification, 306
  reaction with RNCO, 165
  tertiary, dehydration, 52
Aldehydes, basicity, 40
  unsaturated, hydrogenation, 252
Aldolase, 458, 483, 484, 588, 623, 657
Aldol condensation, 60, 99, 170, 220
  mechanisms, 66, 92
Aldonamides, 310
Alginates, 285
Alkaline earth ions in phosphate hydrolysis, 200
Alkenes, hydration, 54
Alkoxymethyl esters, 51
Alkylation, 61
Alkyl halides, transesterification with P esters, 165
Alkyl hydrogen sulfates, 203
Alkyl hydroperoxides, 241
Alkyl hypoiodites, 185
Alkylmercuric halides, 201
All-or-none assay, 473
Allothreonine, 220
Allyl acetate, 383
Allylbenzene, 257
π-Allyl complexes, 260

Index

π-Allylnickel aluminum tetrachloride, 261
Alpha effect, 105, 150, 159
Aluminum, alkoxides, 227
   bromide, 182
   chloride, 7, 183–185
   halides, 182
   ion (decarboxylation), 217
   ion (pyridoxal reactions), 187
   trialkyls, 254, 263
Amide ion, 68
Amides, hydrogen exchange, 100
   hydrolysis, 56, 98, 99, 132, 226
   hydroxylaminolysis, 98, 115
   nucleophilic reactions, 108
   protonation, 41
Amidines, hydrolysis, 100, 132
Amines, addition to benzaldehydes, 128
D-Amino acid oxidase, 652
L-Amino acid oxidase, 638
α-Amino acids, esters, 221
   oxidative deamination, 337
   Strecker degradation, 186
6-Aminobenzimidazole, 162
α-Amino-γ-butyrolactone, 600
2-Amino-2-deoxy-β-D-glucopyranoside
   hydrochloride, 383
α-Aminoisobutyric acid, 188
3-Amino-2-oxindoles, 170
Aminopeptidases, 504, 609
o-Aminophenol, oxidation, 237
α-Aminophenylacetic acid, 335, 366
4-(4′-Aminophenylazophenylarsonic acid), 517
o-Aminophenyl 2,6-dimethylbenzoate, diazonium salt, 312
2-Aminopyridine, 202, 331
L-5-Amino-1-tosylamidopentylchloromethyl ketone, papain inhibition, 521
   trypsin inhibition, 509, 515
Ammonia, nucleophilicity, 151, 152
   reaction with water, 23
Ammonium ion, 31, 231
Amylases, 525
Amylopectin, 525
Amylose, 525
Anchimeric assistance, 281
Androstane, 543
Androstane-3,7-dione, 543
4-Androstene-3,17-dione, 543
5-Androstene-3,17-dione, 644

Anhydrides (carboxylic), formation, 158
   hydrolysis, 52, 99
Anhydro-chymotrypsin, 507
Aniline, anhydride trapping, 102, 157
   basicity, 40
   derivatives, Brønsted relation, 86
   imine (ethyl oxaloacetate), 169
   nucleophilicity, 151–152
   reaction with, chlorobenzaldehyde, Brønsted relation, 128
   chlorodinitrobenzene, 365
Anilinium ions, 167
Anion catalysis, 200
Anisole, basicity, 40
   hydroxylation, 560, 561, 566
p-Anisyl acetate, hydrazinolysis, 346
   hydrolysis, 178, 345
Anserine, 162
Anthracene, 366
Antimony halides, 182
Apocarboxypeptidase, 608
Apolar bonds, 467
Arginine in, chymotrypsin, 508
   enzyme-substrate binding, 625
Aromatic compounds, chlorination, 159
   hydrogen exchange, 98, 116
   protonation, 47, 117
   substitution, 107, 133
Arrhenius intermediate, 13
Arsenate ion, nucleophilicity, 157
Aryl phosphates, 54
Aryl phosphinates, 54
Ascorbic acid, 540, 551
   in aromatic hydroxylation, 240, 557, 566
   oxidation, 238, 554, 555
   in proline hydroxylation, 569
Ascorbic acid oxidase, 555
Asparagine, in carboxypeptidase, 609
Aspartate transcarbamylase, 494
L-Aspartic acid, esters, derivatives, cyclization, 291
   hydrolysis, 222
   in lysozyme site, 527
Aspartic-glutamic transaminase, 593
Aspirin, anion, hydrolysis, 100
   hydrolysis, 337
      intramolecular catalysis, 282
      mechanism, 297
Autoprotolysis (water), 10
Avidin, 599

666 Index

Azide ion, nucleophilicity, 152
Azobenzene-4-sulfonate ion, 369
Azodicarbonate ion, 98, 203
Azulene, 116

Bacillus cereus, 610
Bacillus megaterium, 610
Bacillus subtilis, 610
Bacitracin, 162
Baker, B. R., Jr., 547
Baker, J. W., 148
Balls, A. K., 480
Barbiturates, 548
Barium ion, 223, 229
Bartlett, P. D., 118
Basicity of organic compounds, 39
Basicity functions, 63
Bell, R. P., 91, 119
Bender, M. L., 432
Benkovic, S. J., 306
Benzalacetophenone, 48, 54
Benzaldehyde, acyloin condensation, 172–173
  basicity, 40
  derivatives, nucleophile reactions, Brønsted relations, 128
  semicarbazones, 167
  hydrogenation, 252
  in hydrolysis of leucine ester, 357
Benzamide, basicity, 40
  hydrolysis, 50, 55
Benzamidine, 330
Benzene, 7, 258
Benzenesulfonazide, 237
Benzhydryl chloride, 181
Benzimidazole, 162, 364
Benzoic acid, basicity, 40
  $CH_2(OH)_2$ dehydration, 129
  and derivatives, 33
  protonation, 41
  tetramethylglucose mutarotation, 331
Benzonitrile, 224
o-Benzoquinone, 556
p-Benzoquinone, 248
Benzoylacetone, 369
N-Benzoylanthranilic acid, 339
Benzoylanthranils, 306, 308
α-*N*-Benzoylarginine derivatives with, ficin, 521
  papain, 432, 520, 521
  thrombin, 518

  trypsin, 424, 516
Benzoyl chloride, 183
2-Benzoyl-3,4-dimethylthiazolium iodide, 585
Benzoylformic acid, 545
α-*N*-Benzoyl-L-lysinamide, 424
α-*N*-Benzoyl-L-lysine, 521
N-Benzoylphosphoramidate, 165
2-Benzoylthiazolium ion, 585
N-Benzoyl-L-tyrosine, 496, 631
N-Benzoyl-L-tyrosylglycinanilide, 496
1-Benzyl-2-acetyl-6-oxo-9-carbomethoxy-decahydroisoquinoline, 353
Benzyl adenosine 3'-phosphate, 310
Benzyl cation, 8
Benzyl chloride, 8
1-Benzyl-1,4-dihydronicotinamide, 98
Benzyldimethylsulfonium ion, 366
Benzyl fluoride, 52
N-Benzylideaneanilines, 127, 167
Benzyllithium, 69
1-(Benzyloxycarbamido) 2-phenylethyl bromomethyl ketone, 517
1-(Benzyloxycarbonyl-L-alanyl-glycinamido)-2-phenylethyl chloromethyl ketone, 517
N-Benzyloxycarbonyl-L-asparagine methyl ester, 307
N-Benzyloxycarbonylglycyl-L-phenylalanine, 609
N-Benzyloxycarbonyl-L-phenylalanine, 523
N-Benzyloxycarbonyl-L-tyrosine *p*-nitrophenyl ester, 495
Beriberi, 582
Berzelius, J. J., 1, 2
Betaine, 601
Biacetyl, 9
Bicarbonate ion, 20
Biebrich scarlet, 625
Biotin, 596
Bipyridyl, 635
Bis-π-allylchromium iodide, 261
Bis-π-allylcobalt iodide, 261
Bis-π-allylnickel, 259
Bis(benzonitrile)dichloropalladium, 257
Bis-*p*-chlorophenyl pyrophosphate, 375
Bis(ethylenediamine)cobalt (III) ion, 221
Bisulfate ion, 129
Bisulfite ion, 98, 129
Bjerrum, N., 195
Bodenstein, M., 491
Bohr, N., 661

Index

Borate buffers, 360
Boric acid, 129
Boron fluoride, 182
Braunstein, A. E., 540
Breslow, R., 173, 540, 582, 583
Briggs, G. E., 490
Bromelain, 413, 504
Bromide ion, 152, 154, 159
Bromine, 185
Bromoacetamide, 384
Bromoacetate ion, 215, 384
Bromoacetone, 81, 119
3-Bromoacetylacetone, 81, 88
2-Bromocyclohexyl p-bromobenzenesulfonate, 296
4-Bromoimidazole, 162
2-Bromomethyl-4-nitrophenol, 508, 624
α-Bromo-4-nitroacetophenone, 496
2-Bromooctane, 214
Bromophenol blue, 203
N-Bromosuccinimide, 508
Brønsted, J. N., 75, 78, 195
Brønsted catalysis law, 75
　mechanistic interpretation, 82–84
　in nonaqueous media, 82
　nonlinearity, 78, 81–82
　statistical corrections, 76
Brown, J. F., Jr., 323
Bruice, T. C., 306, 645, 647
Bunnett, J. F., 49
　hypothesis, 47–48
Butadiene, cyclodimerization, 262
　cyclotrimerization, 259
　dimerization, 261
　hydrogenation, 253
　polymerization, 261
1-Butene, 264
n-Butyl acetate, 382
tert-Butyl acetate, 48, 51
tert-Butyl alcohol, 30, 47
n-Butylamine, 330
n-Butylammonium acetate, 171
Butylammonium isobutyrates, 362
n-Butyl p-bromobenzenesulfonate, 205
tert-Butyl chloride, 181, 185, 204
Butyllithium, 175–176
tert-Butyl methacrylate, 263
m-tert-Butylphenyl acetate, 378
p-tert-Butylphenyl acetate, 378
m-tert-Butylphenyl benzoate, 380
n-Butylphosphine, 40

n-Butyl thiolacetate, 133
S-Butyl thiophosphate, 289
γ-Butyrolactone, 196
Butyrylcoenzyme A, 604

Cacodylic acid, 129
Cadmium salts, 220, 221, 230
Caffeine, 364, 550
Calciferol, 613
Calcium ion, catalyst for, acetyl phosphate reactions, 230
　bromination, 220
　hydrolyses, 223, 230
　pyrophosphate ester solvolysis, 229, 336
　complexes with, amino acid esters, 221
　ATP, 231
　effect on trypsin, 427, 515
　metabolism, 613
Carbamyl phosphate, 289, 598
Carbanions, 68, 76
2-Carbethoxycyclopentanone, 220
Carbinolamines, 127, 128, 137
Carbodiimides, 165
Carboligase, 586
3-Carbomethoxyisocarbostyril, 627
Carbon-14, kinetic isotope effects, 125
Carbon acids, 22–25, 76
Carbonate ion, 151
Carbon-carbon bonds, cleavage, 100, 132
Carbon dioxide, hydration, 99
Carbonic acid, 20, 22, 23
Carbonic anhydrase, 464, 610, 656
Carbonium ions, 60, 61
Carbon monoxide, 260, 267
Carbonyl group, addition of water, 54
　Brønsted relations, 128
　catalysis of, 98, 107, 124, 125, 128
Carboxamides, hydrolysis, 48, 54
o-Carboxyacetophenone, 220
3-(2-Carboxybenzylidene) phthalide, 286
Carboxybiotin, 598
N-Carboxy-2-imidazolidone, 599
Carboxylase, 175
Carboxylation of keto and nitro compounds, 219
Carboxylesterase, 519
Carboxylic acids, derivatives, nucleophilic reactions, 107, 132
　catalysis mechanisms, 101
　proton exchange, 32
　unsaturated, hydrogenation, 252

Carboxypeptidase A, 456, 464–466, 504, 608
o-Carboxyphenyl β-D-glucoside, 285, 316
N-(o-Carboxyphenyl)phthalamic acid, and p isomer, 339
o-Carboxyphthalimide, 112, 298
Carnosine, 162
Casein, 424
Catalase, 240, 498, 540, 551, 555
Catechol, catalyst for aromatic hydroxylation, 239, 560
  complexes, 363, 550
  derivatives, reaction with S-adenosylmethionine, 600
  monoanion, 329
  monoesters, 293, 340, 357
  oxidation, 556
Cation catalysis, 200
Cellulose, reaction with epoxides, 306
Cerium (III) hydroxide, 229
  ion, 215, 228
Cerium (IV) ion, 9, 236, 250
Cesium ion, 231
Cetanesulfonic acid, 369
Cetyltrimethylammonium bromide, 371
Cevadine, diacetate, 341
  orthoacetate diacetate, 341
Chance, B., 483, 498, 547, 563, 572
Charge redistribution in proton transfer, 22, 24, 35
  Brønsted relation, 85
Charge transport, 34
Chitin, 527
Chloral hydrate, 156
Chloramphenicol, 157, 369
Chloride ion, catalyst for sulfinyl–O cleavage, 154
  indiazotization, 159
  nucleophilicity, 152
  reaction with iodomethane, 205
Chlorine, 236
Chloroacetic acid, 124, 129
Chloroacetone, 81
p-Chlorobenzaldehyde, 128
Chlorobenzene, 562
p-Chlorobenzenediazonium ion, 115
4-Chlorobutanol, 80, 100
1-Chloro-2,4-dinitrobenzene, 365
2-Chloro-1,2-diphenylethanol, 359
Chloroethanol, 359

Chloroethyl hydrogen phthalate, 301
2-Chloro-1-indanol, 359
p-Chloromercuribenzoate, 432
p-Chloronitrobenzene, 114
2-Chlorooctane, 214
Chloropentamminecobaltate (III) ion, 383
m-Chlorophenyl acetate, 378, 390
p-Chlorophenyl acetate, 102, 178
m-Chlorophenyl benzoate, 380
p-Chlorophenyl ethyl pyrophosphate, 375
3-Chloro-1-propanol, 359
L-Chlorosuccinic acid, 548
Chlorpromazine, 550
Cholecalciferol, 612
Cholestanediol monoacetates, 294
Christian, W., 541
Chrysophenine, 389
α-Chymotrypsin, 456, 494–497, 504
  N-acetyl-L-tryptophanyl-, 510
  N-acetyl-L-tyrosyl, 512
  active site, 477, 509
  acyl-, deacylation, 629, 631, 644
  alkylation, 480
  anhydro-, 507
  assay, 473
  binding, 623, 626, 635
  catalyst of ester hydrolysis, 414
  cinnamoyl-, 506, 507, 512
  compared with cyclodextrin, 381
  conformation changes, 652
  crystal structure, 464
  diethylphosphoryl-, 418
  dimerization, 648
  N-furylacryloyltryptophanyl-, 495, 498
  indoleacryloyl-, 471
  inhibition, 480, 625
  p-nitrobenzoyl-, 512
  purity, 411
  reaction with Paraoxon, 417
  specificity, 425, 628
Chymotrypsinogen, acetylated, activation, 508
  crystal structure, 463, 464
Cinnamic acid, 129
S-Cinnamoylcysteine, 507
N-trans-Cinnamoylimidazole, assay of subtilisin, 436
  hydrolysis by chymotrypsin, 495
  titrant, for chymotrypsin, 414, 423
  for subtilisin, 433

Cinnamoylsalicylaldehyde, 354, 355
Citrate ion, 218, 602
Claisen condensation, 66
Clostridium acetobutylicum, 610
Clostridium tetranomorphum, 576
Cobaloximes, 576
Cobalt hydrotetracarbonyl, 257, 266
Cobalt (I) complexes, 248, 251
Cobalt (II) ion, in autoxidations, 242
    in butadiene polymerization, 264
    catalyst for, acetyl phosphate reactions, 230
      amide hydrolysis, 226
      peroxide decompositions, 241
    complexes, hydrolysis, 99, 221, 609
      hydrogenation catalysts, 248
      olefin polymerization catalysts, 255
Cobalt (III) complexes, in aldol condensation, 221
    ester hydrolysis catalysts, 225
    peptide hydrolysis catalysts, 226
Cobamide enzymes, 540
Cocoonase, 504
Cod, decarboxylase from, 610
Coenzyme Q, 540, 580
2,4,6-Collidine, 105, 139
Common ion effect, 106
Complementarity, 352
Concentration (enzyme), definition, 410
Condensation of amino acids, pyridoxal catalysis, 591
Copper (0), 237
Copper (I) catalyst for, aryl halide reduction, 237
    hydrogenation, 248
    Meerwein and Sandmeyer reactions, 236
Copper (II) catalyst, autoxidation, 242
    decarboxylation, 217
    hydrogenation, 250
    hydrolysis of, amides, 226
      organophosphorous compounds, 228, 230
    oxidation, 237, 555
    peroxide decomposition, 241
    pyridoxal reactions, 187
    V(III)-Fe(III) reaction, 236
    chymotrypsin inactivation by, 494
    complexes, with amino acid esters, 221
      with ATP, 231
      with glycine, aldol condensation, 220

    hydrogenation catalysts, 248
      with salicylidene amino esters, 220
      with tetraglycine, 580
    in keto ester bromination, 220
    reduction by hydrogen, 248
    regeneration of Pd, 266, 270
Coprostane-3β,5β-diol 3-monoacetate, 335
Coprostanol acetate, 335
Cordes, E. H., 543, 548
Corrin ring, 573
Cotton effects, 547
Creatine, synthesis, 600
p-Cresol, esters, hydrolysis, 326
    in mutarotation of tetramethylglucose, 323
Crotonaldehyde, 98
Crotonylcoenzyme A, 604
Cyanide ion, carbonyl addition, Brønsted relation, 128
    catalyst for acyloin condensation and decarboxylation, 172–173, 582
    nucleophilicity, 151–152
Cyanoacetic acid, 32, 374
o-Cyanobenzoic acid, 302
β-Cyanobutyric acid, 302
Cyanohydrins, 99
Cyanophenanthroline, 224
β-Cyanopropionic acid, 302
Cycloamyloses, 374
Cyclodecadiene, 262
Cyclodextrins, 374, 477, 605
Cyclododecatriene, 259
Cycloheptaamylose, 374
Cycloheptatriene, 63
Cyclohexaamylose, 374, 390
    benzoyl-, 380
1,2-Cyclohexanedicarboxamide, 306
1,2-Cyclohexanedicarboxylic acid, 369
1,3-Cyclohexanediol monoacetates, 294
Cyclohexanol, 547
Cyclohexylamine, 35
Cyclohexyl p-toluenesulfonate, 297
Cyclooctaamylose, 374
Cyclooctadiene, 262
Cyclooctatetraene, 258
Cyclopentadiene, 63
Cyclopentanediol monoesters, 294
Cysteine, 188, 189
    esters, hydrolysis, 222
Cytidine-2$'$,3$'$-phosphate, 164, 473, 524

Cytochromes, 464, 540, 551, 644

Dawson, H. M., 120
Dayhoff, M. O., 456
Deacylation of acyl-enzymes, 510
Debye-Huckel theory, 196
cis-Decalin, 370
Decanesulfonic acid, 369
Decarboxylation, amino acids, 588, 591
  α-keto acids, 169–170, 172
  β-keto acids, 165, 169, 374
Decylamine, 369
Decyltrimethylammonium ion, 369
Dehydratase, 588
Dehydroascorbic acid, 555, 557
Dehydrocevadine orthoacetate diacetate, 342
7-Dehydrocholesterol, 613
DeMaeyer, L., 34
Denaturation, 469, 470
4-Deuterioglutamic acid, 576
Deuterium, exchange, with hydrogen, 248
  of ketones, 118
  with water, 253
  isotope effects, 88
  oxide, reaction with $DO^\ominus$, 20
  solvent isotope effects, 106
Deyrup, A. J., 41
DFP, see Diisopropyl phosphorofluoridate
Diacetone alcohol, 66, 67, 165, 171
N,N'-Diacetylchitobiose, 478, 484
N,O-Diacetyl-N-methyl hydroxylamine, 178
Diacyl peroxides, 241
Dialkyl sulfides, 157
N,N'-Diaryformamidinium ions, 112
Diazoacetate ion, 98
Dibenzyldimethylammonium ion, 69
1,3-Dibromoacetone, 520
α-Dicarbonyl compounds, 186
1,3-Dichloroacetone, halogenation, 81
  hydration, 324, 325, 345
4,4'-Dichlorobenzophenone, 325
Dichromate ion, 250
Dicobalt octacarbonyl, 266
N,N-Dicyclohexylbenzamide, 308
Diethoxyphosphoryl thiolcholine, 443
Diethyl acetylethylmalonate, 111
Diethyl acetylmalonate, 111, 115
Diethylaluminum chloride, 185
N-(Diethylaminoethyl)imidazole, 334

Diethylammonium 4-nitrophenoxide, 331
Diethyl n-butylmalonate, 206
Diethyl o-carboxyphenyl phosphate, and p isomer, 290
Diethyl 1,4-dihydropyridine-3,5-dicarboxylate, 545
Diethyl ether, basicity, 40
  enolate alkylation in, 206
  hydrolysis, 48, 54
  ionizations in, 198
Diethyl malonate, 25
  anion, 152
Diethyl p-nitrophenyl phosphate, reaction with, chymotrypsin, 417, 423
  elastase, 436
  trypsin, 424, 515
O,S-Diethyl p-nitrophenyl phosphorothioate, 515
P,P-Diethyl pyrophosphate, 157
Diethyltitanium dichloride, 263
Diffusion control, criterion of mechanism, 102, 123
  in nitramide decomposition, 79
  in reactions, with charge separation, 27
  of hydroxide ion, 26
  of oxonium ion, 22, 26, 76
  of carbanions, 35
Diglyme, 206
Dihydrobiopterin, 612
22-Dihydroergosterol, 613
Dihydrogen arsenate ion, 129
Dihydrogen phosphate ion, 30, 129
Dihydrogen phosphite ion, 129
Dihydrolipoic acid, 580
Dihydoorotate dehydrogenase, 540
Dihydroxyacetone, 98, 99
  phosphate, 484
Dihydroxybenzenes, 52
3,4-Dihydroxybenzoic acid, 560
Dihydroxyfumaric acid, 215
4-(2',4'-Dihydroxyphenyl) imidazole, 163
Dihydroxytartaric acid, 216
3,5-Dihydroxytoluene, 550
Diiodobis(tributylphosphine) palladium (II), 270
Diisopropyl phosphorochloridate, 60
Diisopropyl phosphorofluoridate (DFP), hydrolysis, 52, 60, 100, 228
  reactions with, acetylcholinesterase, 443, 518
  bromelain, 413

Index

chymotrypsin, 480, 494
  elastase, 436
  polyhydric phenols, 329
  thrombin, 518
  titrant for, chymotrypsin, 423
    subtilisin, 433
1,2-Diketones, 173, 252
Dimedone, 173
1,3-Dimethoxybenzene, 116
Dimethoxybenzenes, 52
Dimethylacetoacetic acid, 169
Dimethylaluminum chloride, 263
Dimethylamine, 40, 113
γ-Dimethylaminobutyric acid, 306, 315, 347
2-Dimethylaminoethanethiol, 356, 612
Dimethylaminoethyl acetate, 305
Dimethylamino group, 314
δ-Dimethylaminovaleric acid, 306, 315, 347
Dimethylammonium ion, 31
Dimethylcarbamyl chloride, 443
Dimethylcarbamylcholine, 443
Dimethylcarbamyl fluoride, 443, 518
Dimethyl(1,2-diphenylethyl) amine, 69
Dimethyl ether, 9
Dimethylformamide, 165, 206
β,β-Dimethylglutaric acid, monoester, 300, 646
N,N-Dimethyl-β-mercapto ethylamine, 356, 612
Dimethyl(o-methylbenzhydryl)amine, 69
Dimethyloxaloacetic acid, 216–218
3,5-Dimethylphenyl acetate, 378
Dimethyl phosphate anion, 288
Dimethylphosphine, 40
Dimethyl phosphoacetoin, 355
Dimethyl sulfide, 40
Dimethyl sulfoxide, anion, 69
  basicity, 40
  desolvation of anions, 207
  mixtures, catalysis in, 209
  proton transfers in, 35
  strong base media containing, 64
Dimethylthetin, 600
N,N-Dimethyl-2,4,6-trinitroaniline, 64
β-Dinitriles, 66
2,4-Dinitroaniline, 64
2,4-Dinitroanisole, 115
3,5-Dinitrobenzoate ion, 365
2,4-Dinitrofluorobenzene, see Fluorodinitrobenzene

Dinitromethane, 25
2,4-Dinitrophenyl acetate, hydrolysis, 157, 178
  titrant for chymotrypsin, 418, 423
2,4-Dinitrophenyl benzoate, 157
2,4-Dinitrophenylhydrazones, 125
2,4-Dinitrophenyl p-isopropylbenzoate, 387
2,4-Dinitrophenyl phenyl ether, 115
2,4-Dinitrophenyl p-vinylbenzoate copolymer, 387
Dioldehydrase, 574, 575
Diolefins, conjugated, 252
Diphenyl β-aminoethyl phosphate, 311
Diphenylcarbodiimide, 494
Diphenyldiazomethane, 494
1,1-Diphenylethylene, 40, 69
Discrimination (by substrates among catalysts), 89
Disulfides (organic), 165, 215
β-Disulfones, 66
1,2-Dithiolane-3-valeric acid, see Lipoic acid
Dowex–50, 382
DPN, see Nicotinamide adenine dinucleotide
duVigneaud, V., 596
Dyes, enzyme binding, 625
  reactions with OH$^\ominus$, 371, 372

Edwards, J. O., 148
Effectors, 491–492
Egg albumin, hydrolysis, 369
Eigen, M., 27, 34, 78, 79, 130, 324, 498
Elastase, 464, 504, 509
Electron, solvated, 20
Electron spin resonance, 472
Electrophiles, classification, 180–181
Δ^4 - 3,6-Endoxotetrahydrophthalic acid, monoester, 300, 646
Enolate ions, 25, 206
Enolization of ketones, 98, 119, 128, 283
Enzyme, definition, 398
Epichlorohydrin, 93
Epinephrine, 600
Epoxides, hydrolysis, 52
  reaction with cellulose, 306
Equilibrium, and catalysis, 8, 11
  dialysis, 472
  and proton transfer rates, 22
Ergocalciferol, 613
Ergosterol, 613
Erythrocuprein, 466

Index

Escherichia coli, 611, 661
Esterification, 61, 98
Esters, aminolysis, 99, 100, 132
   hydrolysis, 54, 56, 66, 221
   nucleophilic reactions, 108, 155
   from oxo reaction, 269
Ethanol-1-*d,* 543
Ethers, tertiary, hydrolysis, 52
Ethoxide ion, 68, 152
Ethyl acetate, basicity, 40
   hydrolysis, 57, 177, 383
Ethyl acetoacetate, 76, 98, 220
Ethyl *N*-acetylglycinate, 515
Ethylamine, 169, 369, 515
Ethyl *p*-aminobenzoate, 364, 375
Ethyl benzimidate, 112
Ethyl benzoate, 40, 58
Ethyl α-*N*-benzoyl-L-argininate, 516
Ethyl chloroformate, 153–154
Ethyl cyclohexanone-2-carboxylate, 81
Ethyl dichloracetate, 80, 103, 106, 176
Ethylene, dimerization, 264
   hydrogenation, 254
   oxidation, 270
   polymerization, 263
Ethylene phosphate, 311
Ethylenimine, 48, 54
Ethyl hippurate, 521
Ethyl hydrogen phthalate, 301
N-Ethylmaleimide, 433
Ethyl *m*-nitrobenzimidate, 112
Ethyl orthoformate, 93
Ethyl orthopropionate, 372
Ethyl orthovalerate, 372
m-Ethylphenyl acetate, 390
Ethyl trifluoroacetate, 99, 113
Ethyl trifluorothiolacetate, 99, 109, 113, 133
Ethyl vinyl ether, 52
Exergonic reaction, 26
"Extraction model," 623

Fabry's disease, 660
Fatty acid synthetase, 458
FDNB, *see* Fluorodinitrobenzene
"Feedback inhibitors," 492
Fenton reaction, 238, 557, 560, 561
Ferricyanide ion, 253, 609
Ferricytochrome *c,* 572
Ferridoxin, 540

Ferrimyoglobin, 471
Ferrocyanide ion, 572
Ficin, 433, 504
Finkle, B. J., 432
Fischer, E., 390, 463, 484, 622, 650, 656
Fischer-Tropsch process, 268
Flavin adenine dinucleotide, 540, 548
Flavin mononucleotide, 540, 548
Fluoradene, 63
Fluorene, 35, 63
   anion, 35
Fluorenyllithium, 35
Fluoride ion, catalyst for hydrolysis of acyl fluorides, 101, 153, 155
   effect on Ag^+ catalyst, 250
   nucleophilicity, 151
   reaction with acids, 20, 21
   trap for phosphorylated amines, 160
1-Fluoro-2,4-dinitrobenzene, 371, 372, 523, 609
Folate reductase, 569
Formaldehyde, catalyst for hydrolysis of phosphoramidate, 186
   hydrate, dehydration, 80, 129, 130, 324
   reaction with, glycine-Cu^{2+}, 220
   nucleophiles, 135
   tetrahydrofolate, 99, 108
   thiourea, 128
   urea, 334
Formate ion, 156
Formic acid, 52, 129
Formocholine chloride, 127
Free energy of activation, 4, 8, 9
   and of reaction, 85, 85
Frequency factor, 9
Friedel-Crafts reaction, 182, 183
Fructose diphosphatase, 652
Fructose 6-phosphate, 175
Fumarase, 482, 504
Fumaric acid, addition of NH_3, 524
   hydration, 343, 483, 524
   hydrogenation, 252
   from succinic acid, 548
N-Furylacryloyl-L-tryptophan, 494

β-Galactosidase, 661
Gallium halides, 182
Gaucher's disease, 660
General catalysis, 73
Generalized gangliosidosis, 660

Index

Genetic control of enzyme synthesis, 399
George, P., 563, 566
Germanes, 254
Glucagon, 302
Glucosamines, N-aryl-, 98
Glucose, mutarotation, GA and GB catalysis, 80, 100, 129, 130
 GB catalysis, 73, 76, 139
 6-phosphate, 154, 291
 reaction with water, 23
 tetraacetyl-, 199
 tetramethyl, 199, 322, 330–332, 348
D-Glucose dehydrogenase, 540
Glucosidases, 526, 661
Glucosides, 52
L-Glutamate dehydrogenase, 540, 652
Glutamate mutase, 576
Glutamic acid, 527
 esters, cyclization, 115, 334
 in lysozyme site, 527
 from methylaspartic acid, 576
 transamination, 335
Glutamine, cyclization, 115, 334
 synthetase, 651
L-Glutamyl-L-cysteylglycine(Glutathione), 611
Glutaric acid, monoaryl esters, 298–300, 313, 347, 646
Glyceraldehyde, 99
 3-phosphate, 175
 dehydrogenase, 480, 623
D-Glycerate dehydrogenase, 540
α-Glycerophosphate dehydrogenase, 540
Glyceryl methyl phosphates, 310
α-Glyceryl phosphate, 229
Glycerylphosphorylcholine, 310
Glycinamide, 226
Glycine, conversion to serine, 596
 in decarboxylation, 169
 esters, hydrolysis, 99, 222, 609
 oxidative deamination, 337
 reaction with acetyl phosphate, 230
 in semicarbazone formation, 167
Glycinehydroxamic acid, 510
Glycogen, 526
Glycolic acid, 129, 384
Glycosides, 127
Glycyl-L-asparagine, 302
Glycylglycine, hydrolysis, 226, 383
 reaction with FDNB, 371, 372

Glyoxal, 384
Glyoxalase, 611
Glyoxylic acid, 187, 586
Gold (III) ion, 251
Grignard reaction, 228
Grinstead, R. R., 555
Guanine, 364
Guanosine phosphates, 604
Gutfreund, H., 498

h_0 acidity function, 42
h_R acidity function, 44
$H_\ominus$ acidity function, 63, 64
H_0 acidity function, 42, 43
H_R acidity function, 44, 53
$H_{R'}$ acidity function, 44
Hafnium tetrachloride, 263
Haldane, J. B. S., 490
Halide ions, 153, 206
Halogenation (ketones), 118
Hammett, L. P., 41, 45
Hammond, G. S., 135
Hansen, B., 306
Hard acids, bases, electrophiles, and nucleophiles, 150
Hartley, B. S., 417, 424
Hayaishi, O., 566
Heme proteins, 540
Hemoglobin, 468
 binding and conformation, 651
 peroxide decomposition, 241
 trypsin substrate, 424
Henri, V., 487
n-Heptaldehyde, 270
Hess, G. P., 514
Hexacyanocobaltate(III) ion, 244
1,5-Hexadiene, 259
Hexamethylbenzene, 40
1-Hexene, 270
Hexokinase, 656
Hinshelwood, C. N., 4
Hippuramide, 497
Hippuric acid, 630
Hippuryl-DL-β-phenyl lactate, 609
Histamine, 162, 600
Histidine, and derivatives, catalysts for ester hydrolysis, 161–163
 esters, hydrolysis, 222, 225
 in site of, carboxypeptidase, 609
 catalase, 563

chymotrypsin, 507
elastase, 509
lysozyme, 527
subtilisin, 509
trypsin, 508, 509
Homocysteine, 593, 596, 600
Homogeneous alkylation, 68
Homosterism, 492–493
Horse liver esterase, 637
Hydrazine, *N,N*-diacyl-, 334
 nucleophilicity, 150, 152
 reaction with phenyl esters, 346, 347
Hydrazones, 125
Hydride ion, 248
Hydrindantin, 173
Hydrocarbons, acidity, 35
Hydrofluoric acid, 129
Hydroformylation, 266
Hydrogen, addition to, Ir complexes, 251
 organic compounds, 252
 exchange with H_2O_2, 253
 ortho-para conversion and deuterium exchange, 248
 reaction with $Fe(CN)_6^{3\ominus}$, 253
Hydrogenase, 251
Hydrogen bonds, in ice, 34
 in protein structure, 468
 and proton transfer, 20, 21, 30, 35
Hydrogen bromide, 185
Hydrogen chloride, 184, 185
Hydrogen fluoride, 20, 185
Hydrogen iodide, 6
Hydrogen peroxide, aromatic hydroxylation, 560
 decomposition, 237
 exchange with hydrogen, 253
 oxidation of ascorbate, 238
Hydrogen phosphate ion, 30, 103, 364
Hydrogen sulfide, 20, 23
Hydrogen transfer, modes, 552
Hydroperoxide ion, 105, 150, 151
Hydroperoxyl radical, 237
"Hydrophobic bond," 467
Hydroquinone, 550, 560
Hydrosulfide ion, 21
Hydroxamate ions, 150
Hydroxamic acids, 98
Hydroxide ion, basicity and nucleophilicity, 105, 152
 Brønsted deviations, 79

catalysis by, 1, 37, 62, 63, 72, 73
 intermediates in, 67, 68
 catalysis of ketone halogenation, 81
 reactions with, acids, 19, 20
 carbon acids, 76
 dyes, 371, 372
 glycine ester and derivatives, 222
 methyl bromide, 366
 phenyl esters, 347
 sulfonium ion, 366
 water, 23
Hydroxyacetic acid, 129
γ-Hydroxybutyramide, 310
γ-Hydroxybutyric acid, 51
L-β-Hydroxybutyrylcoenzyme *A* dehydrogenase, 540
β-Hydroxyisobutyric acid, 575
Hydroxylamine, addition to NAD, 546
 in chymotrypsin reactions, 497
 nucleophilicity, 150, 151
 toward ortho esters, 123
 reactions with, acyl-enzymes, 496
 amides, 115
 benzaldehyde, 128
 benzimidates, 112
Hydroxylammonium ion, 129
Hydroxylation, 52, 239
Hydroxyl radical, 237
Hydroxymalonic acid, 586
β-Hydroxy-β-methyl glutarylcoenzyme *A*, 599
4-Hydroxymethylimidazole, 100, 162
2-Hydroxy-4-nitrobenzaldehyde, 188
2-Hydroxy-5-nitrobenzyl bromide, 508, 624
2-Hydroxypyridine, association, 363
 catalyst for, ester hydrolysis, 334
 peptide synthesis, 334
 tetramethylglucose mutarotation, 323, 330, 331
3-Hydroxypyridine, 331
4-Hydroxypyridine, 331
2-Hydroxypyridine-4-carboxaldehyde, 188
3-Hydroxypyridine-4-carboxaldehyde, 335
β-Hydroxysteroid dehydrogenase, 540, 543
5-Hydroxytryptamine (serotonin), 478–480
5-Hydroxytryptophan, 478–480
δ-Hydroxyvaleramide, 310
Hypochlorite ion, catalyst for phosphoramidate hydrolysis, 186
 sarin hydrolysis, 158
 nucleophilicity, 150, 151

Index

Hypochlorous acid, 159

Ice, proton transfer in, 34
   reactions in, 98
"Iceberg," 468
Imidates (esters), 132
Imidazole, basicity and nucleophilicity, 103, 161
   catalysis, of acetylimidazole reactions, 101
   of aryl ester hydrolysis, 326
   characteristics, 160–161
   of ester hydrolysis, 106, 176–177
   of ketone hydration, 325
   of NPA hydrolysis, 102
   of phosphate hydrolysis, 163–164
   solvent isotope effects, 106
   of transamination, 366
   derivatives, in NPA hydrolysis, 161–163
   hydrolysis of, 100
   inhibition of ester hydrolysis, 364
   reactions with, acids, 20, 28
   adenosylmethionine, 600
   phenyl esters, 347
   water, 23
2-Imidazolidone, 599
Imidazolium ion, 20, 23
4-(4'-Imidazolyl)butyramide, 305
4-(4'-Imidazolyl)butyric acid esters, hydrolysis, 304
Imidazolyl group, 314
Imides, 298, 306
Imido esters, 100
Imines, addition reactions, 107, 128
   hydration, 137
   hydrolysis, 108, 227
   intermediates, 165–167
2,2'-Iminodiethanethiol, 234
Indene, 63
Indole, chymotrypsin inhibitor, 497
   conversion to tryptophan, 188, 593
Indoleacryloylimidazole, 364
"Induced fit" theory, 647
Inducer, 399
Ingold, C. K., 51
Initiator, 7
Inorganic pyrophosphatase, 604
Insertion reactions, 247
Insulin, 302, 531
Intermediates, 5, 12, 13
Iodate ion, 250

Iodide ion, 5, 10
   activation of Ni carbonyl, 269
   catalyst for, iodooctane racemization, 155
   MeBr hydrolysis, 153
   nucleophilicity, 152
   in ortho ester hydrolysis, 123
Iodination, 118, 283
Iodine, catalyst for, acetaldehyde decomposition, 6
   aromatic bromination, 185
   tertiary alcohol dehydration, 185
   reaction with dialkyl sulfides, 157
Iodoacetamide, 433
Iodoacetate ion, 480, 523
$trans$-2-Iodocyclohexyl p-toluenesulfonate, 297
2-Iodooctane, 5, 155
Ion association, 206
Ionic radius, 198
Ionic strength, 282
Ionization, heat of, 505
Ionizing power, 204
Ion pairs, in alkyl halide solvolysis, 214
   assistance to ionization, 199
   in glucose mutarotation, 200
   in proton transfer, 30, 33, 35
Ion quadruplets, 199, 200, 215
Iproniazid, 478
Iridium (I) complexes, addition of hydrogen, 251
   olefin hydrogenation, 254
   olefin isomerization, 255
   catalysts for, hydrogen reactions, 248
   containing molecular N, 271
Iridium (III) ion, 264
Iron (0), 251
Iron complexes, in olefin isomerization, 255
Iron (II) ion in, aliphatic hydroxylation, 240
   autoxidation, 242
   decarboxylation, 217
   peroxide decomposition, 241
Iron (III) ion, association with bipyridyl, 635
   catalyst for, ascorbate oxidation, 238
   decarboxylation, 217
   hydroxylation, 239–240, 557, 560
   pyridoxal reactions, 187
   chelate with TETA, 240, 556
   halides, 182
   reduction, by hydrogen, 250
   by vanadium (III), 236
Iron oxide, 239
Iron pentacarbonyl, 268

Isobutene, 47, 185
Isobutyraldehyde, 99, 168
o-Isobutyrylbenzoic acid, 283, 313
Isocitric acid, 524
Isocyanates, 165, 224
Isoleucine, 508, 510
1-Isonicotinyl-2-isopropyl hydrazine, 478
Isonitrosoacetone, 510
Isoprene, 176, 185
Isopropyl methylphosphonofluoridate
    (sarin), catechol complex, 329, 363
  hydrolysis, 228
  racemization, 155
  reactions with, acetylcholinesterase, 518,
    638
  nucleophiles, 151
Isotactic polymers, 264
Isotope effects, 125, 139
Isozymes, 459

J_- basicity function, 64
J_0 basicity function, 44
Jencks, W. P., 177
Jones, P., 119

Karrer, P., 542
Kaufman, S., 568
Kauzmann, W., 467
Ketals, 52, 127
α-Keto acids, 169–170, 172
β-Keto acids, 165, 169, 374
β-Keto esters, enzymatic cleavage, 515
α-Ketoglutaric acid, 577
Δ^5-3-Ketoisomerase, 652
Ketones, bromination, 99
  deuterium exchange, 118
  enolization, 48
    catalysis, 100, 128, 283
    effect of pK, 119
  halogenation, 118, 324
  iodination, 105
  proton transfer, 24
  racemization, 118
Kilby, B. A., 417, 424
Kirsch, J. F., 177
Klotz, I. M., 388, 467, 472
Knoevenagel, E., 170
Knoevenagel condensation, 165, 170
Kögl, F., 596
Koshland, D. E., Jr., 504, 645

Koshland reagent, 508, 624
Kosower, E. M., 546
Kremzner, L. T., 444
Kynurenine, 572

Lactaldehyde, 575
Lactate dehydrogenase, 459, 460, 540, 610
  cooperative interactions, 648
  specificity, 543
  structure, 464
Lactic acid, 543
Lactones, 99, 310
Langenbeck, W., 148, 169
Langmuir isotherm, 14
Lanthanum (III) salts, 215, 220, 229
Lapworth, A., 117, 172
Lardy, H. A., 596
Lauryltrimethylammonium chloride, 370
Lead (II) ion, 227
Lead (IV) ion, 271
Leffler, J. E., 117, 135
Leloir, L. F., 607
Letsinger, R. L., 385
Leuchs' anhydride, 99
L-Leucyl-L-asparagine, 302
Leucyl-RNA, 294
Leveling effect, 42
Levulinic acid, 283, 313
Lewis acids, 86
Liener, I. E., 433
Ligand migration reactions, 247
Lipoamide dehydrogenase, 623
Lipoic acid, 540, 577
Lipoyl dehydrogenase, 578, 579
Lithium aluminum hydride, 228, 254
Lithium ion, 230, 231
Lithium perchlorate, 198, 199
Lithium tetrachloropalladate (II), 266
Liver alcohol dehydrogenase, 489, 547
"Lock and Key" theory, 484
Luciferase, 652
Lutidine, 105, 139
  in solvolysis of tetrabenzyl pyrophosphate,
    163, 229, 336
Lyonium/lyoxide catalysis, 37
L-Lysine, in acyl-carrier protein, 611
  copolymer with L-histidine, 162
  in decarboxylase, 610
Lysozyme, 456, 504
  active site, 625

complexes with glucosamine derivatives, 473, 478, 484
interaction with water, 468
structure, 460, 464
substrate binding, 477

Maclaurin's series, 28, 78
Magnesium ion, binding to ATP, 231
  catalyst for, acetyl phosphate reactions, 230
    biotin reactions, 599
    bromination of keto esters, 220
    hydrolysis of polyphosphates, 230
  complexes with amino acid esters, hydrolysis, 221
Magnesium methyl carbonate, 219
Mahler, H. R., 543, 547, 548
Malachite green, 544
Malate dehydrogenase, 460, 540, 543
Maleamide, 306
Maleic acid, hydrogenation, 252
  monoaryl ester, hydrolysis, 300, 646
Malic acid, 343, 524
Malonic acid, dianion, 218
  monoethyl ester, hydrolysis, 223
Malononitrile, 25
Mandelonitrile, anion, 172–173
Manganese (II) ion, in autoxidation reactions, 242
  binding to ATP, 231
  catalyst for, acetyl phosphate reactions, 230
    bromination of keto esters, 220
    Ce(IV)-Tl(I) reaction, 9, 236
    decarboxylation, 217
    hydrolysis of polyphosphates, 230
    hydrolysis of sarin, 228
  chelate with TETA, 240, 556
  complex with amino acid esters, 221
Manganese (III) ion, 236, 337
Mason, H. S., 568
Mechanism, alteration in catalysis, 4, 5
Meerwein-Ponndorf reduction, 227
Meerwein reaction, 236
Menten, M. L., 487
Mercaptans, 128
Mercaptoacetate ion, 230
o-Mercaptobenzoic acid, 358
p-Mercuribenzoate, 432, 480, 572, 609
Mercuric halides, 201

Mercury (I) complexes, 248
Mercury (II) ion, catalyst for alkyl halide solvolysis, 214
  benzyl chloride solvolysis, 8
  formation of carbonyl compounds, 271
  hydration of acetylene, 255
  hydrogen reduction, 250
  cleavage of thiol esters, 227
  complexes, catalysts for hydrogen reactions, 248
Mesaconic acid, 576
Mesitoic acid, 52
Mesityl oxide, 98
Metachromatic leukodystrophy, 660
Metaphosphate ion, 228, 230, 231
Methane, 6, 39, 63
Methanesulfonic acid, esters, 518
Methanobacillus omelanskii, 577
Methanol, basicity, 40
  co-catalyst for olefin polymerization, 185
  effect on chymotrypsin kinetics, 495–496
  esterification, 98
  nucleophilicity, 152
  proton exchanges in, 33
Methionine, biosynthesis, 596
  in chymotrypsin site, 508, 637
  conversion to adenosyl derivatives, 601
  formation from homocysteine and betaine, 601
  phosphorylation, 601
Methoxide ion, 152
Methoxyacetaldehyde diethyl acetal, 296
ω-Methoxyalkyl p-bromobenzenesulfonates, 295
4-Methoxybenzimidazole, 163
o-Methoxybenzoic acid, 129
Methoxylamine, 112
p-Methoxy-α-methylstyrene, 98
p-Methoxyneophyl p-toluene sulfonate, 198
p-Methoxyphenol, esters, 326
p-Methoxyphenyl acetate, hydrazinolysis, 346
  hydrolysis, 345
o-Methoxyphenyl benzoate, 293
Methyl (radical), 6, 7
N-Methylacetamide, 362, 469
Methyl acetate, 157, 383
N-Methylacetonimine, 166
Methylamine, 40, 167, 168
Methylammonium ion, 29, 31
2-Methylanthracene, 63
Methylaspartase, 576

3-Methyl-L-aspartic acid-3-*d*, 576
2-Methylbenzimidazole, 162
Methyl benzoate, 57, 58
Methyl 2-benzoyl-6-methylbenzoate, 353
Methyl bromide, 153, 366
Methyl chloride, 152
Methyl cinnamate, 364
β-Methylcrotonyl carboxylase, 598
β-Methylcrotonylcoenzyme *A*, 599
Methyl 2,6-dihydroxy benzoate, 340
Methylene-azomethines, 325
Methylene blue, 6
3,4-Methylenedioxy-β-nitrostyrene, 171
4,5-Methylenephenanthrene, 63
Methylenetetrahydrofolate, 108
Methyl ethylene phosphate, 100
Methyl *o*-formylbenzoate, 353, 355
Methylglutaconase, 599
β-Methylglutaconylcoenzyme *A*, 599
N-Methylglycine, 31
3-Methylheptatriene, 261
Methyl hippurate, 497
1-Methylhistidine, 162
Methyl hydrogen phosphate, 287
Methyl hydrogen phthalate, 301
2-Methyl-4-hydroxy-6-aminobenzimidazole, 163
Methyl 2-hydroxycyclohexyl phosphate, 310
N-Methylhydroxylamine, 127
2-Methyl-4-hydroxy-6-nitrobenzimidazole, 163
N-Methylimidazole, 162, 163, 325, 336
2-Methylimidazole, 162
4-Methylimidazole, 162
Methyl 4-(4′-imidazol) butyrate, 304
Methyl iodide, 6, 153, 205, 234
Methylmalonate-oxaloacetate transcarboxylase, 598
Methylmalonylcoenzyme *A*, 573, 575
Methyl mercaptan, 40
Methyl mesitoate, 51, 57
Methyl methacrylate, 68
Methyl *N*-methylphthalamate, 307
Methyl 1-naphthoate, 370
Methyl *p*-nitrobenzenesulfonate, 480, 513
Methyl *o*-nitrobenzoate, 353
Methyl orthobenzoate, 124, 371–372
Methyl 21-oxo-oleanolate, 353
4-Methyl-1,4-pentanediol, 290
p-Methylphenyl acetate, 345, 346

2-Methyl-3-phenylpropionitrile, 64, 205, 207
4-Methylpyridine, 160, 384
Methyl salicylate, 291
D-Methylsuccinic acid, 549
L-Methylsuccinic acid, 548
Methylthioacetaldehyde diethyl acetal, 296
Methyl *o*-toluate, 341
Methyl *p*-toluenesulfinate, 154
Methyl α-*N*-tosylargininate, 518
N-Methyltrifluoroacetanilide, 100
Micelles, 370
Michael condensation, 66, 170
Michaelis, L., 487
Michaelis constant (K_m), 488
Michaelis-Menten equation, 14
Micrococcus lysodeikticus, 610
Microscopic reversibility, 11
Mold protease, 509
Monoamine oxidase, 478
Monohydrogen phosphate ion, 30, 103, 364
Mono-*p*-nitrophenyl glutarate, 299
Morpholine, 168, 355
3-Morpholinophthalide, 355
Muscle myosin, 604
Myoglobin, 651

NAD$^\oplus$ structure, 541
NADH cytochrome reductase, 540
NADH diaphorase, 540
NADPH transhydrogenase, 540
Naphthalene, 370
Naphthalenediols, complexes, 550
2-Naphthoic acid, 637
2-Naphthol-6,8-disulfonic acid, 115
α-Naphthylamine, 158
Natta, G., 263
Neighboring group participation, 281
Némethy, G., 467
Neopentyl alcohol, 32
Neopentyl iodide, 214
Neopentyl phosphate, 52
"Nerve gas," 480
Neutral protease, 610
Niacin, 541
Nickel, "bare," 259
Nickel cyanide, 258
Nickel (II) ion, in acetyl phosphate reactions, 230
  in autoxidations, 242
  catalyst for, amide hydrolysis, 226

keto acid decarboxylation, 217
keto ester bromination, 220
chelates, 220
complexes, 251, 258
halides, complexes with amino acid esters, 221
Nickel tetracarbonyl, 266, 269
Nicotinamide, 541, 600
Nicotinamide adenine dinucleotide (NAD), 6, 540
  binding to alcohol dehydrogenase, 472
  nonenzymic reactivity, 483
  reaction with ethanol, 478
Nicotine, 541
Niemann, C., 637
Ninhydrin, 173
Nitramide decomposition, Brønsted non-linearity, 78, 79, 81
  entropy of activation, 76
  general catalysis, 73, 99
  tautomerization, 79
Nitration (aromatic), 186
Nitriles, 56
Nitrite ion, 151, 152, 158
Nitroacetone, 25, 111
α-Nitroacetophenone, 111
3-Nitro-4-acetoxybenzene sulfonate ion, 385
3-Nitro-4-acetoxyphenyl arsonic acid, 386
5-Nitro-4-acetoxysalicylic acid, 386
p-Nitrobenzaldehyde, 127
p-Nitrobenzamide, 50
Nitrobenzene, basicity, 40
  hydrogenation, 252
  hydroxylation, 566
  nitration, 53
6-Nitrobenzimidazole, 162
p-Nitrobenzoic acid, 129
3-Nitro-4-carboxyphenyl N,N-diphenylcarbamate, 414, 423
Nitroethane, 25
  anion, 98
Nitrogen fixation, 271, 581
4-Nitroimidazole, 162
Nitromethane, acidity, 34, 35, 139
  anion, 98
  basicity, 40
  bromination, 98
  condensation with piperonal, 171
  mutarotation of glucose derivatives in, 199
Nitrones, 98, 127

Nitronium ion, 53
p-Nitrophenyl acetamido diglucosides, 529
m-Nitrophenyl acetate, 178, 346
p-Nitrophenyl acetate, hydrazinolysis, 346
  hydrolysis, acetate ion, 314
    carbonic anhydrase, 610
    chloral hydrate anion, 156
    chymotrypsin, 414, 495
    elastase, 442
    hydroxide ion, 178
    2-hydroxypyridine, 334
    imidazole, 102, 106, 160–161, 178, 326
    imidazole derivative, 334
    insulin, 531
    phosphate ion, 157
    polymers, 382, 387, 388
    thiolsubtilisin, 520
    thrombin, 518
    trimethylamine, 306
    trypsin, 515
  inhibition of enzymic acetate transfer, 526
  reaction with, benzamidine, 330
    butylamine, 330
    chymotrypsin, 419, 494
    o-mercaptobenzoic acid, 358
    nucleophiles, 103, 105, 151
    2-pyridinealdoxime, 234
    titrant for, chymotrypsin, 418
    trypsin, 424
p-Nitrophenyl N^2-acetyl-N^1-benzylcarbazate, 424
p-Nitrophenyl N-acetyl-L-leucinate, 425, 518
p-Nitrophenyl N-acetyl-DL-tryptophanate, 423, 425, 430
p-Nitrophenyl N-acetyl-L-tryptophanate, 518, 521
4-(p-Nitrophenylazo)resorcinol, 20, 21
m-Nitrophenyl benzoate, 379
p-Nitrophenyl benzoylsarcosine, 308
p-Nitrophenyl N-benzyloxy carbonylglycinate, 442, 518
p-Nitrophenyl N-benzyloxy carbonylglycyl-L-phenylalanine, 308
p-Nitrophenyl N-benzyloxy carbonyl-L-isoleucinate, 442
p-Nitrophenyl N-benzyloxy carbonyl-L-leucinate, 442
p-Nitrophenyl N^2-benzyloxy carbonyl-L-lysinate hydrochloride, 424
p-Nitrophenyl benzyloxy carbonylphenyl-alaninate, 518

Index

p-Nitrophenyl *N*-benzyloxy carbonyl-L-
  tyrosinate, hydrolysis, catalysis by,
    elastase, 441, 442
    thrombin, 518
  titrant for, chymotrypsin, 420
    papain, 424
p-Nitrophenyl *N*-benzyloxy carbonyl-L-
  valinate, 435
o-Nitrophenyl cinnamate, 495
p-Nitrophenyl decanoate, 369
p-Nitrophenyl 4-(*N*,*N*-dimethylamino)butyrate, 317
o-Nitrophenyl dimethylcarbamate and *p*-isomer, 443, 446
p-Nitrophenyl furoate, 442
p-Nitrophenyl β-D-glucoside, 529
p-Nitrophenyl glutarate, 314
p-Nitrophenyl glycosides, 311
p-Nitrophenyl *p*-guanidinobenzoate, 424
p-Nitrophenyl hexanoate, 371, 372
p-Nitrophenyl hippurate, 308, 497
p-Nitrophenyl hydrogen terephthalate, 387
p-Nitrophenyl 4-(4'-imidazolyl)butyrate, 304
p-Nitrophenyl *trans*-indoleacrylate, 364
p-Nitrophenyl iodoacetate, 518
p-Nitrophenyl isobutyrate, 423, 436, 442
p-Nitrophenyl laurate, 388
p-Nitrophenyl leucinate, 357
p-Nitrophenyl 2-methoxy-5-nitrobenzoate, 293
p-Nitrophenyl 5-nitrosalicylate, 291, 316
o-Nitrophenyl oxalate anion, 202
p-Nitrophenyl phenacyl methylphosphonate, 355
p-Nitrophenyl phosphate, 113, 288
p-Nitrophenyl polyuridylate, 369
p-Nitrophenyl trimethyacetate, hydrolysis
  by elastase, 436, 442, 517
  titrant for chymotrypsin, 423
4-Nitrophthalimide, 365
Nitrostyrene, 99
Nitrous acid, 159, 186
Northrop, J. H., 399
NAP, *see* *p*-Nitrophenyl acetate
Nucleic acids, 523
Nucleophilicity, correlation with basicity and polarizability, 148
Nucleoside glycosidase, 526
Nucleoside phosphorylase, 526
Nucleosides, 526

n-Octane, 262
Olefines, addition of silanes and germanes, 254
  isomerization, 256
  polymerization, 175–176, 263–264
"One anion" catalysis, 201
"One encounter" reaction, 130
Oppenauer oxidation, 227
Optical factor, 411
Orbital "steering," 647
Organoboron compounds, 98
Organomercury compounds, 98
Organotin compounds, 98
Ornithine, 167
Orthoesters, hydrolysis, 98, 122, 123
Osmium complexes, 251, 254
Osteomalacia, 613
Ostwald, W., 2
Oxalate ion, in $CH_2(OH)_2$ dehydration, 129
  chromium (III) complexes, 215
  oxidation by chlorine, 236
Oxalic acid, monoethyl ester, 223
Oxaloacetate decarboxylase, 610
Oxaloacetic acid, 169, 324–325, 524
  decarboxylation, 215–216
Oxalosuccinic acid, 216
Oxalylcoenzyme *A*, 586
Oxazoline intermediates, 306, 309
Oxazolone intermediates, 306
Oxene, in hydroxylations, 568
Oximes, anions, nucleophilicity, 150
  formation, 98, 127
Oxonium ion, catalysis, 1, 37, 39, 72–73
  classification, 50
  of orthobenzoate hydrolysis, 124
  in ice, 34
  reactions with, bases, 19
    carbanions, 76
    N or O bases, 76
    water, 23
Oxo reaction, 266
Oxygen, in hydroxylations, 240, 566

Palladium (II), chloride, 270
  complexes, hydrogen reactions, 248, 251
  olefin reactions, 256, 266
  ion, in decarboxylation, 217
  diene polymerization, 264
Pancreatic maltase, 526
Pancreatopeptidase E, *see* Elastase
Papain, 456, 504

Index

N-acetyl-L-tryptophanyl-, 521
  active site, 625
  α-N-benzoyl-L-arginyl-, 520
  composition, 466
  crystalline, activity, 471
  hippuryl-, 521
  inhibition, 480
  structure, 464
  thionohippuryl-, 520
Paraffins, alkylation, 61
Paraldehydes, depolymerization, 51
Paraoxon, *see* Diethyl *p*-nitrophenyl phosphate
Parker, M. J., 432
Pearson, R. G., 150, 180
Pedersen, K. J., 75, 78, 120
Pellagra, 541
Penicillins, 375
Pentachlorophenol, 129
Pentacyanocobaltate (II) ion, 244, 251, 252
Pentakis(acetonitrile)cobalt (II) ion, 251
Pepsin, 470, 504
Perchloric acid, 154
Perhydroxyl radical, 566
Perkin condensation, 66
Peroxidase, 490, 540, 551
  ferric complex, 551
Phenanthrene, 366
1,10-Phenanthroline, 218, 547
α-Phenethylamine, 325
Phenol, 30, 34, 40
Phenols, methylated, 184
Phenoxide ion, nucleophilicity, 34, 151, 152, 366
Phenoxyacetic acid, 129
Phenyl acetate, aminolysis, 99, 330
  hydrazinolysis, 346
  hydrolysis (enzymatic), 466
    (nonenzymatic), 178, 326, 378
  substituted, 99, 376
Phenylacetate ion, and analogs, 550
Phenylacetic acid, 129
Phenylacetohydroxamic acid, 510
Phenylalanine, ethyl ester, 222
  hydroxylation, 612
Phenylalanine hydroxylase, 568, 569
Phenylalanylglycinamide, 226
4-(Phenylazo)catechol, 239, 553
Phenylboronic acid, 517
3-Phenyl-1-butene, 325
Phenyl chloroacetate, 329
α-Phenylethyl chloride, 181, 182

Phenylglyoxal, 356, 612
Phenylhydrazine, 128
Phenylhydrazino-*o*-quinone, 553
Phenylhydrazones, 98, 99
Phenyl hydrogen phthalate, 302
N-Phenyliminotetrahydrofuran, 333, 364
D-α-Phenylisocaprophenone, 119
Phenyl isothiocyanate, 609
Phenylmethanesulfonyl fluoride, 423
2-Phenyloxazolin-5-one, 308
Phenyl-(*p*-phenylazophenyl)carbamyl chloride, 494
β-Phenylpropionate ion, 369
γ-Phenylpropyl diphosphate, 231
γ-Phenylpropyl triphosphate, 231, 604
Phenyl radical, 237
Phenyl salicylate, 291, 360
Phenylsodium, 174–175
γ-Phenyltetronic acid, 99
Phillips, D. C., 527
Phloroglucinol, 116, 117, 550
Phosphate esters, 98, 165
Phosphate metabolism, 613
Phosphinate esters, 165
Phosphine oxides, 165
Phosphines, 165, 261
3-Phosphoglyceraldehyde dehydrogenase, 540
O-Phosphohomoserine, 188
Phosphoketolase, 585
Phosphonate esters, 165
4-Phosphopantetheine, 610
Phosphoramidate ion, 186
Phosphoramidates, N-aryl, 289
Phosphoric acid, esters and anhydrides, 200, 229
Phosphoric carbonic anhydride, 99
Phosphorochloridates, 165
Phosphorylases, 526
Phosphorylation, 230
Phosphoryl halides, 329
Photooxidation, 507
Photosynthesis, 577
pH-rate constant profiles, 37, 38, 74
Phthalamic acid, 302, 316
Phthalamide, 306
Phtalanilic acid, 339
Phthalic acid, 301
Phthalonic acid, 173
Phylogenetic trees, 572

2-Picoline, 105
Picric acid, 331
Pinacolone, 111
Pinacol rearrangement, 52
2,5-Piperazinedione, 98
Piperidine, 114, 115, 152
Piperonal, 171
Plasma (dog), 466
Plasmin, 504
Platinum (II) complexes, hydrogen reactions, 248, 251
   olefin reactions, 254, 256
Polarizability, 10
Polyadenylic acid, 389
Poly-1,2-butadiene, 261
Poly-1,4-butadiene, *cis,* 261
   trans, 264
Poly(ethenesulfonic acid), 382, 383
Poly(methacrylic acid), 384
Polyneuritis, 582
Poly(*p*-nitrophenyl polyuridylic acid succinate), 388
Polyphosphates, 230
Polystyrenesulfonic acid, 383
Poly(β-sulfoethyl methacrylate), 383
Poly[triethyl(vinylbenzyl)ammonium hydroxide], 384
Poly[5(6)-vinylbenzimidazole], 387
Poly[4(5)-vinylimidazole], 386
Poly(*N*-vinylimidazole), 387
Poly(4-vinylpyridine), 384
Poly(vinylpyridine betaine), 384
Poly(vinylpyrrolidone), 389
Porphyrin enzymes, 540
Potassium acetate, 374
   ion, 231
Potato apyrase, 604
Potential energy diagrams, 84, 85
Primary structure of proteins, 399
Proflavin, binding to chymotrypsin, 497, 498, 625
Proline, 168, 569
Proline racemase, 652
Promoter, 7
1,2-Propanediol, 573, 574
Propargyl hydrogen phthalate, 302
β-Propiolactone, 51
Propionaldehyde, 573, 574
Propionic anhydride, 156
Propionylcoenzyme *A*, 575

Propylene, 262, 263
Propylene oxide, 228
Propyl 4-(4'-imidazol)thiolbutyrate, 304
Proton, mobility of, 61
   modes of catalysis, 59, 62
Proton jump, 34
Provitamin, 613
Purines, 364, 596
Purity (enzyme), definition, 410
Pyrazole, catalysis of mutarotation of tetramethyl glucose, 331
Pyridine, basicity and nucleophilicity, 103, 151–152
   catalyst for, acylations, 159
      anhydride hydrolysis, 105, 106, 160
      mutarotation of glucose derivatives, 199, 323
      solvolysis of esters, 336
   decarboxylation promoter, 218
   N-oxide, 165
   strong base media containing, 64
Pyridinealdehydes, 609, 610
2-Pyridinealdoxime, 234
   methiodide, 159, 518
Pyridine transhydrogenase, 543
Pyridinium, cresolate, 323
   ion, 129
2-Pyridone, 323–324
Pyridoxal, catalysis, 186–187, 220
   in oxidative deamination, 337
   in transamination, 335, 366
   phosphate, 167, 483
Pyridoxamine, 187–188
Pyrophosphoric acid and derivatives, 54, 289, 606
Pyrrole synthetase, 588
2-Pyrrolidone-5-carboxylic acid, 334
Pyruvate, carboxylase, 489
   decarboxylase, 458
   dehydrogenase, 460, 648
   kinase, 652
   oxidase, 585
Pyruvic acid, acyloin condensation, 174–175
   decarboxylation, 174–175, 586
      (oxidative), 557–558
   enolization, 119
   iodination, 220, 283
   metabolism, 581
   reduction, 542–543

Quasi-equilibrium, 491
Quaternary structure of proteins, 399
8-Quinolineboronic acid, 359

Racemization (amino acids), 588, 591
  (ketones), 118
Rate assay, 410
"Reacting bond rule," 127
Reduced nicotanamideadenine dinucleotide (NADH), 473, 478
Rehybridization, see Charge redistribution
Relaxation technique, 498–500
Rennin, 504
"Reporter groups," 472
Repression of enzyme synthesis, 399
Rhodium (I) complexes, hydrogen reactions, 248, 251, 254
  olefin reactions, 254
Rhodium (III), chloride, C_2H_4 dimerization, 264
  complexes, hydrogen reactions, 248
    olefin reactions, 256
  salts, butadient polymerization, 264
Riboflavine, 548, 550, 551
Ribonuclease, 456, 458, 504
  assay, 473
  cleavage by subtilisin, 523
  conformational changes, 652
  partial hydrolysis, 302
  reaction with iodoacetate, 480
  reduction and re-formation, 469
  structure, 464
  substrate binding, 639
Richards, F. M., 471
Rickets, 613
RNA polymerase, 607
Rothstein, E., 148
Rubidium ion, 231
Rupley, J. A., 471
Ruthenium (0) complexes, 251
Ruthenium (II) complexes, containing molecular N, 271
  hydrogenation catalysts, 248, 252
Ruthenium (III) compounds, catalysts for, acetylene hydration, 255
  butadiene polymerization, 264
  D_2-H_2O exchange, 250
  hydrogen reductions, 248, 250

Salicoyl phosphate, 290

Salicylaldehyde, 220, 227
Salicylaldoximate ion, nucleophilicity, 151
Salicylamide, hydrolysis, 291
Salicylic acid, 129
  anions, 20, 21
Salicyl methyl formal, 285
Salicyl phosphate, 289
Salicyl salicylate, 340
Salmonella typhimurium, 530
Sandmeyer reaction, 236
Sanger, F., 456, 542
Sarin, see Isopropyl methylphosphonofluoridate
Saturation of catalyst by substrate, 14
Saville, B., 152
Schardinger dextrins, 374
Scheraga, H. A., 467
Schiff bases, 98, 99
Schowen, R. L., 139
Seconary structure of proteins, 399
Selectivity, distinction from reactivity, 89, 90
Semicarbazide, addition to benzaldehydes, 127, 128
  in $CH_2(OH)_2$ dehydration, 129
  in ortho ester hydrolysis, 123
  reaction with benzimidates, 112
Semicarbazone formation, catalysis, 108
  by amines, 165, 167
  by anilinium ions, 167
  ^{14}C isotope effect, 125
  by general acids, 98, 115
  by general bases, 99
Sensitizer, 7
Serinamides, 307
Serine, in acyl-carrier protein, 611
  biosynthesis, 596
  in chymotrypsin site, 506
  cleavage to glycine, 327
  conversion to, pyruvate, 188
    tryptophan, 488, 591
  copolymer with histidine, 162
  formation from glycine, 220
  in trypsin site, 508
Shiner, V. J., 547
Silanes, 254
Silicon-oxygen bond, cleavage, 100
Silver ion, catalyst of, alkyl halide solvolysis, 214
  hydrogenation, 250
  complexes, hydrogen reactions, 248, 251

reaction with carboxypeptidase, 609
thiol ester cleavage, 227
Sluyterman, A. L., 471
Smith, E. L., 432
Snell, E. E., 540
Sodium, amyl sulfate, 372
  borohydride, 98, 484, 572
  chloride, 196
  decyl sulfate, 372
  dodecyl sulfate, 372
  ethyl sulfate, 372
  hexadecyl sulfate, 372
  iodide, 196
  ion, 231
  lauryl sulfate, 370–372
  methyl sulfate, 372
  nitrite, 515
  octadecyl sulfate, 372
  oleyl sulfate, 372
  perchlorate, 196
  sulfate, 196
  tetradecyl sulfate, 372
Soft acids, bases, electrophiles and nucleophiles, 150, 176
"Solvation rule," 126
Solvent isotope effects, 106, 119, 139
Solvents, catalysis by, 194
  in proton transfer, 20, 28, 29, 35
Specific catalysis, 37, 43
Spermidine, 600
Sphingomyelinase, 661
Spivey, E., 120
Starch diethylaminoethyl ether hydrochloride, 383
Steady state, 12, 15
Steric hindrance, 105
Stevens, J. B., 49
Strecker degradation, 186
Strophanthidin 3-acetate, 335
Styrene, polymerization, 68
Subtilisin, 456, 504
  active site sequence, 509
  furoyl-, 517
  furylacryloyl-, 517
  indoleacryloyl-, 517
  structure, 464
Succinamic acid, 302
Succinamide, 306
Succinanilic acid, 302, 303
Succinic acid, dehydrogenation, 548

  monoesters, 298, 646
Succinic dehydrogenase, 548
"Succinylaspirin," 337
Succinylcoenzyme A, 573
  -acetoacetylcoenzyme A transferase, 604
Sucrose, hydrolysis, 51
  phosphorylase substrate, 526
Sucrose phosphorylase, 504
Sulfate ion, 530
Sulfides (organic), 165, 215
Sulfinic acids, 165
Sulfinyl sulfones, 153
Sulfite ion, 128, 152, 158
Sulfitidase, 661
Sulfolane, 64
Sulfonyl halides, 329
Sulfuric acid, 8, 23, 51
Sulfurous acid, 23
Sumner, J. B., 398
"Superacid" catalysis, 212–213
Swain, C. G., 93, 120, 135, 323, 328
Synartetic assistance, 281
Synzymes, 388

Takamaltase, 526
Tartaric acid, 524
Tautomeric catalysis, 323, 331
Tay Sachs' disease, 660
Telluric acid, 129
Tertiary amines, catalysts, 159
Tertiary structure of proteins, 399
Testosterone, 543
TETA, see Triethylenetetramine
Tetraacetylglocose, 199
Tetrabenzyl pyrophosphate, 99, 163, 229, 336
Tetra-n-butylammonium, 4-nitrophenoxide, 331
  phenoxide, 323
Tetracarbonylcobaltate anion, 267
Tetraethylammonium nitrate, 214
Tetraethyl pyrophosphate, 157, 443, 518
Tetraglycine, copper complex, 580
Tetrahedral intermediates, in stepwise general-catalyzed reactions, 108–115, 132–133, 223
Tetrahydrofolate formylase, 594
Tetrahydrofolic acid, 99, 108
Tetrahydrofuran, 40
Tetrahydrofuran-3,4-diol monoesters, 294

Index

2-Tetrahydropyranol, 331
Tetramethylglucose, 199, 322, 330–332, 348
Tetramethylphosphorodiamidic chloride, 158
Tetraphenyltin, 263
Thallium (I) ion, 9, 223, 236
Theophylline, 364
Theorell, H., 546, 547
Thermolysin, 610
Thiaminase, 586, 601
Thiamine, 173, 601
　pyrophosphate, 483, 578
Thiazolium zwitterion, 172–174
Thioaspirin, hydrolysis, 358
Thiobenzophenone, 544
Thiocholine esters, 157
6-Thioctic acid, *see* Lipoic acid
Thiocyanate ion, 152
Thiolate ions, 128
Thiol esters, acetylenic, hydration, 98
　aminolysis and ammonolysis, 99, 100, 132
　cleavage by heavy metal ions, 227
Thiols, 98, 128
Thionine, 625
Thioparaconic acid, 159
Thiophenoxide ion, 151, 152
Thiosulfate ion, 10, 151–152, 215
Thiourea, 10, 98, 128, 152
Thorium hydroxide, 229
Thornton, E. R., 135
Threonine, 188, 189, 220
Thrombin, 504, 509
Thymidine hexanucleotide, 389
Tin (II) ion, 230
Tin (IV) halides, 182
Tischenko reaction, 228
Titanium (III) complexes, 264
Titanium tetrachloride, 263
2-Tolancarboxylic acid, 286
2,2′-Tolandicarboxylic acid, 286
2,4′-Tolandicarboxylic acid, 286
Toluene, 63
p-Toluenesulfonic acid, 34
Toluic acids, 129
m-Tolyl acetate, 378
o-Tolyl acetate, 378
p-Tolyl acetate, hydrazinolysis, 346
　hydrolysis, 178, 345, 378
L-1-Tosylamido-2-phenylethyl chloromethyl ketone, 480, 507, 521

N-Tosylarginine, esters, hydrolysis, 518
Transaldolase, 483
Transaminase, 483
Transamination, 99, 588, 591
Transition state theory, 195
Transketolase, 175, 585
Trapping, 102
N,N',N''-Triacetylchitotriose, 473
1,2,4-Triazole, 331, 334
Trichloroacetic acid, 32, 185
Trichloromethyl lithium, 200
Trichlorosilane, 100
Trichlorotris(triphenyl phosphine)ruthenium (III), 270
Triethylamine, 165
Triethylenetetramine (TETA), chelates, 240, 556
2,2,2-Trifluoroacetanilide, 100
　N-methyl-, 76, 110, 115, 139
Trifluoroethanol, 106
Trifluoroethyl acetate, 106, 178, 326
Trifluoroethyl hydrogen phthalate, 302
1,3,5-Trihydroxy benzene, 116, 117, 550
1,3,5-Trimethoxybenzene, 116, 117
Trimethylacetic acid, 129
Trimethylacetic anhydride, 54
Trimethylamine, basicity, 40
　in hydrolysis of, acetyl phosphate, 160
　　esters, 306, 336, 347
　N-oxide, 40
　proton exchange, 30
Trimethylammonium ion, 23, 30
Trimethylhydroquinone, 550
3,4,5-Trimethylphenyl acetate, 390
Trimethylphosphine and oxide, 40
2,4,7-Trinitro-9-fluorenyl tosylate, 365
Triose phosphate isomerase, 484
Trioses, condensation of, 98
Triphenylethyllithiums, 1,1,2- and 2,2,2-, 69
Triphenylmethyl anion, 35
Triphenylphosphine, 269
Triphenylpropyllithiums, 1,1,3- and 2,2,3-, 69
Triphosphates, hydrolysis, 230
Tris-π-allylchromium, 261
Tris-π-allylcobalt, 261
Tropanol, 20
Tropylium ion, 20
Trypan blue, 625
Trypsin, 456, 458, 504
　active site sequence, 509
　activity in nonaqueous media, 471

activity-viscosity relationship, 470
binding of amidines and guanidines, 626
denaturation, 400, 471
diethylphosphoryl-, 418, 516
inhibitors, 515
reduction and re-formation, 469
specificity requirements, 638
substrate binding, 476, 623
Tryptophan, in chymotrypsin, 508
FMN complex, 550
formation, 188, 593
in glyceraldehyde phosphate dehydrogenase, 623
in lysozyme, 527
oxidation, 572
in proteins, with Koshland's reagent, 624
Tryptophan oxygenase, 572
Tunnelling, in proton transfers, 139
Turnover number, 398
"Two anion," catalysis, 201
Tyrosinase, 556
Tyrosine, in carboxypeptidase, 609
in chymotrypsin, 508
in flavin binding, 550
formation, 568
L-Tyrosyl-L-tyrosine, 522

Ubiquinone, see Coenzyme Q
Udenfriend, S., 557
Ugai, T., 582
Unit (enzyme, definition, 410
Uracil, 364
Urea, 128, 334, 400
Urease, 398, 471, 485
Uridine phosphates, 607

Vanadium (III) ion, 236
Vanadium (IV), complexes, polymerization catalysts, 264
salts, in cysteine desulfhydration, 189
Vanadium oxychloride, 263
Vanadium tetrachloride, 263
van't Hoff intermediate, 13
Vennesland, B., 540
Vinylcyclohexene, 262
Vinyl esters, from olefins, 271

4(5)-Vinylimidazoleacrylic acid copolymer, 386
Vitamin B_2, 548
Vitamin B_6, see Pyridoxal
Vitamin B_{12}, 540, 573
Vitamin B_{12s}, 577
Vitamin C, see Ascorbic acid
Vitamin H, see Biotin

Wang, J. H., 556
Warburg, O., 541
Warfarin, 171
Water, in acid catalyzed reactions, 45
acidity in metal complexex, 232
addition to acetaldehyde, 128
autoprotolysis, 10, 20
in $CH_2(OH)_2$ dehydration, 129
exchange with D_2, 253
in glucose mutarotation, 129
nucleophilicity, 151, 152
in olefin polymerization, 185
reactions with, N and O acids, 76
OH^- and H_3O^+, 20
Westheimer, 540, 583
Whitaker, J. R., 432
Wilson, I. B., 416, 444
Winstein, S., 281
Wool, 466

Xylenes, 183–184
Xylulose, 5-phosphate, 175

Yates, K., 49
Yeast alcohol dehydrogenase, 473
Yeast glycogen synthetase, 493

Zero point energy of vibration, 106
Ziegler, K., 263
Ziegler catalysts, 254
Zinc, bromide, Lewis acidity, 86
ion, binding to ATP, 231
catalyst of, acetyl phosphate reactions, 230
keto acid halogenation, 220
in dehydrogenases, 546
Zirconium tetrachloride, 263
Zucker, L., 45
Zucker-Hammett hypothesis, 45